Lecture Notes in Artificial Intelligence 9195

Subseries of Lecture Notes in Computer Science

LNAI Series Editors

Randy Goebel
   University of Alberta, Edmonton, Canada
Yuzuru Tanaka
   Hokkaido University, Sapporo, Japan
Wolfgang Wahlster
   DFKI and Saarland University, Saarbrücken, Germany

LNAI Founding Series Editor

Joerg Siekmann
   DFKI and Saarland University, Saarbrücken, Germany

More information about this series at http://www.springer.com/series/1244

Amy P. Felty · Aart Middeldorp (Eds.)

Automated Deduction – CADE-25

25th International Conference on Automated Deduction
Berlin, Germany, August 1–7, 2015
Proceedings

 Springer

Editors
Amy P. Felty
University of Ottawa
Ottawa
Canada

Aart Middeldorp
University of Innsbruck
Innsbruck
Austria

ISSN 0302-9743   ISSN 1611-3349 (electronic)
Lecture Notes in Artificial Intelligence   '
ISBN 978-3-319-21400-9   ISBN 978-3-319-21401-6 (eBook)
DOI 10.1007/978-3-319-21401-6

Library of Congress Control Number: 2015943367

LNCS Sublibrary: SL7 – Artificial Intelligence

Springer Cham Heidelberg New York Dordrecht London

Printed on acid-free paper

Springer International Publishing AG Switzerland is part of Springer Science+Business Media
(www.springer.com)

Preface

This volume contains the papers presented at the 25th Jubilee Edition of the International Conference on Automated Deduction (CADE-25), held August 1–7, 2015, in Berlin, Germany. CADE is the major forum for the presentation of research in all aspects of automated deduction, including foundations, applications, implementations, and practical experience.

The Program Committee (PC) accepted 36 papers (24 full papers and 12 system descriptions) out of a total of 85 submissions. Each submission was reviewed by at least three PC members or external reviewers appointed by the PC members in charge. The program also included invited lectures given by Ulrich Furbach (University of Koblenz, Germany) and Edward Zalta (Stanford University, USA). In addition, Michael Genesereth (Stanford University, USA) gave an invited lecture in conjunction with the co-located event RuleML (9th International Web Rule Symposium).

To celebrate the 25th jubilee edition of the conference, additional invited speakers were featured at several events. A Special Session on the Past, Present, and Future of Automated Deduction included talks by Ursula Martin (University of Oxford, UK), Frank Pfenning (Carnegie Mellon University, USA), David Plaisted (University of North Carolina at Chapel Hill, USA), and Andrei Voronkov (University of Manchester, UK). Also, the conference reception and the banquet dinner featured speakers Wolfgang Bibel and Jörg Siekmann. In addition, the program was enriched by several affiliated events that took place before the main conference. These events included eight workshops, seven tutorials, three competitions, and one poster event.

During the conference, the Herbrand Award for Distinguished Contributions to Automated Reasoning was presented to Andrei Voronkov in recognition of his numerous theoretical and practical contributions to automated deduction and the development of the award-winning Vampire theorem prover. The Selection Committee for the Herbrand Award consisted of the CADE-25 Program Committee members, the trustees of CADE Inc., and the Herbrand Award winners of the last ten years.

The Best Paper Award was conferred to Vijay D'Silva (Google, Inc., USA) and Caterina Urban (École Normale Supérieure, France) for their paper entitled "Abstract Interpretation as Automated Deduction." In addition, the first Thoralf Skolem Awards were conferred this year to reward CADE papers that have passed the test of time by being most influential papers in the field:

CADE-20 (2005) Nominal techniques in Isabelle/HOL by Christian Urban and Christine Tasson: The first paper showing how to use nominal techniques to deal with bound variables in higher-order theorem provers.

CADE-14 (1997) SATO: An Efficient Propositional Prover by Hantao Zhang: For its seminal contribution to the design and implementation of novel techniques, including lazy data structures and clever Boolean constraint propagation that caused a step change in the area and deeply influenced later systems.

CADE-8 (1986) Commutation, Transformation, and Termination by Leo Bachmair and Nachum Dershowitz: For laying the foundations of today's termination theorem-proving techniques.

CADE-0-1 (1968 and 1975) The mathematical language AUTOMATH, its usage, and some of its extensions by N.G. de Bruijn: For his landmark and remarkable contribution to the design and implementation of higher-order theorem provers.

Also, several students received Woody Bledsoe Travel Awards, thus named to remember the late Woody Bledsoe, funded by CADE Inc. to sponsor student participation.

Many people contributed to making CADE-25 a success. We are very grateful to the members of the Program Committee and the external reviewers for carefully reviewing and evaluating papers. CADE-25 would not have been possible without the dedicated work of the Organizing Committee, headed by Conference Chair Christoph Benzmüller. Many thanks also go to Workshop, Tutorial, and Competition Co-chairs Jasmin Blanchette and Andrew Reynolds, and to Publicity and Web Chair Julian Röder. On behalf of the Program Committee, we also thank all the invited speakers for their contribution to the success of this jubilee edition. We also acknowledge the important contributions of the workshop organizers, tutorial speakers, competition organizers, and poster event organizer. Thanks also go to Andrei Voronkov and the development team of the EasyChair system. Last, but not least, we thank all authors who submitted papers to CADE-25 and all participants of the conference.

CADE-25 received support from many organizations. On behalf of all organizers, we thank the German Research Foundation, DFG, for supporting the special session, and the European Coordinating Committee for Artificial Intelligence (ECCAI) for supporting the invited talk given by Ulrich Furbach. We also gratefully acknowledge support from Freie Universität Berlin, the *Artificial Intelligence Journal,* and Microsoft Research.

May 2015   Amy P. Felty
                                                                       Aart Middeldorp

Affiliated Events

Workshops

- Bridging: Bridging the Gap Between Human and Automated Reasoning, organized by Ulrich Furbach, Natarajan Shankar, Marco Ragni, and Steffen Hölldobler
- DT: 29. Jahrestreffen der GI-Fachgruppe Deduktionssysteme, organized by Christoph Benzmüller, Matthias Horbach, Alexander Steen, and Max Wisniewski
- HOL4: HOL4 Workshop, organized by Ramana Kumar
- IWC: Fourth International Workshop on Confluence, organized by Takahito Aoto and Ashish Tiwari
- LFMTP: International Workshop on Logical Frameworks and Meta-Languages: Theory and Practice, organized by Kaustuv Chaudhuri and Iliano Cervesato
- PxTP: Workshop on Proof eXchange for Theorem Proving, organized by Cezary Kaliszyk and Andrei Paskevich
- QUANTIFY: Second International Workshop on Quantification, organized by Hubie Chen, Florian Lonsing, and Martina Seidl
- Vampire: The Vampire Workshop, organized by Laura Kovacs and Andrei Voronkov

Tutorials

- Abella: Reasoning About Computational Systems Using Abella, given by Kaustuv Chaudhuri and Gopalan Nadathur
- Beluga: Programming Proofs About Formal Systems, given by Brigitte Pientka
- CPROVER: From Programs to Logic: The CPROVER Verification Tools, given by Daniel Kroening, Martin Brain, and Peter Schrammel
- Isabelle: Isabelle Tutorial, given by Makarius Wenzel
- KeY: The Sequent Calculus of the KeY Tool, given by Reiner Hähnle and Peter Schmitt
- Lean: Lean Theorem Prover: A Tutorial, given by Leonardo de Moura, Soonho Kong, Jeremy Avigad, and Floris van Doorn
- Superposition: 25th Anniversary of Superposition: Status and Future, given by Stephan Schulz and Christoph Weidenbach

Competitions

- CoCo: The Fourth Confluence Competition, organized by Takahito Aoto, Nao Hirokawa, Julian Nagele, Naoki Nishida, and Harald Zankl
- CASC: The CADE ATP System Competition, organized by Geoff Sutcliffe
- termCOMP: Termination Competition, organized by Johannes Waldmann and Stefan von der Krone

Poster Events

- EPS: The CADE-25 Taskforce Towards an Encyclopedia of Proof Systems, organized by Bruno Woltzenlogel Paleo

Organization

Program Chairs

Amy Felty   University of Ottawa, Canada
Aart Middeldorp   University of Innsbruck, Austria

Program Committee

Carlos Areces   Universidad Nacional de Córdoba, Argentina
Alessandro Armando   University of Genova, Italy
Christoph Benzmüller   Freie Universität Berlin, Germany
Josh Berdine   Microsoft Research, UK
Jasmin Blanchette   Inria Nancy and LORIA, France
Marta Cialdea Mayer   Università di Roma Tre, Italy
Stephanie Delaune   CNRS, France
Gilles Dowek   Inria, France
Amy Felty   University of Ottawa, Canada
Reiner Hähnle   Technical University of Darmstadt, Germany
Stefan Hetzl   Vienna University of Technology, Austria
Marijn Heule   The University of Texas at Austin, USA
Nao Hirokawa   JAIST, Japan
Ullrich Hustadt   University of Liverpool, UK
Deepak Kapur   University of New Mexico, USA
Gerwin Klein   NICTA and UNSW, Australia
Laura Kovacs   Chalmers University of Technology, Sweden
Carsten Lutz   Universität Bremen, Germany
Assia Mahboubi   Inria, France
Aart Middeldorp   University of Innsbruck, Austria
Albert Oliveras   Technical University of Catalonia, Spain
Nicolas Peltier   CNRS, France
Brigitte Pientka   McGill University, Canada
Ruzica Piskac   Yale University, USA
André Platzer   Carnegie Mellon University, USA
Andrew Reynolds   EPFL Lausanne, Switzerland
Christophe Ringeissen   LORIA-Inria, France
Renate A. Schmidt   University of Manchester, UK
Stephan Schulz   DHBW Stuttgart, Germany
Georg Struth   University of Sheffield, UK
Geoff Sutcliffe   University of Miami, USA

Alwen Tiu   Nanyang Technological University, Singapore
Freek Wiedijk   Radboud University Nijmegen, The Netherlands

Conference Chair

Christoph Benzmüller   Freie Universität Berlin, Germany

Workshop, Tutorial, and Competition Co-chairs

Jasmin Blanchette   Inria Nancy and LORIA, France
Andrew Reynolds   EPFL Lausanne, Switzerland

Publicity and Web Chair

Julian Röder   Freie Universität Berlin, Germany

Additional Reviewers

Albarghouthi, Aws	Galmiche, Didier	Klebanov, Vladimir
Alberti, Francesco	Gao, Sicun	Klein, Dominik
Andronick, June	Ghilardi, Silvio	Kneuss, Etienne
Avanzini, Martin	Ghorbal, Khalil	Koenighofer, Robert
Balbiani, Philippe	Gieseke, Fabian	Kontchakov, Roman
Bard, Gregory	Giesl, Jürgen	Krishna, Siddharth
Basold, Henning	Gimenez, Stéphane	Kyas, Marcel
Baumgartner, Peter	Goré, Rajeev	Lange, Martin
Bjørner, Nikolaj	Griggio, Alberto	Le Berre, Daniel
Bonacina, Maria Paola	Gupta, Ashutosh	Lee, Matias David
Boy de La Tour, Thierry	Habermehl, Peter	Lewis, Corey
Bubel, Richard	Hansen, Peter	Limongelli, Carla
Cerrito, Serenella	Henriques, David	Lombardi, Henric
Chaudhuri, Kaustuv	Hermant, Olivier	Martins, João G.
Cohen, Cyril	Hladik, Jan	Matichuk, Daniel
David, Amélie	Horbach, Matthias	McMillan, Ken
Demri, Stéphane	Huisman, Marieke	Murray, Toby
Dima, Catalin	Ibanez-Garcia,	Müller, Andreas
Dimitrova, Rayna	Yazmin Angelica	Nadathur, Gopalan
van Ditmarsch, Hans	Iosif, Radu	Nalon, Cláudia
Dragan, Ioan	Jeannin, Jean-Baptiste	Niksic, Filip
Echahed, Rachid	Johansson, Moa	de Nivelle, Hans
Echenim, Mnacho	Jovanović, Dejan	Orbe, Ezequiel
Emmi, Michael	Junttila, Tommi	Papacchini, Fabio
Felgenhauer, Bertram	Kaliszyk, Cezary	Park, Sungwoo
Fontaine, Pascal	Kaminski, Mark	Paskevich, Andrei
Fuhs, Carsten	King, Tim	Peñaloza, Rafael

2015 Thoralf Skolem Award Committee

Board of Trustees of CADE Inc.

Board of the Association for Automated Reasoning

Jasmin Blanchette   Inria Nancy and LORIA, France
(Newsletter Editor)
Pascal Fontaine (CADE)   University of Lorraine and LORIA, France
Martin Giese (Secretary)   University of Oslo, Norway
Hans Jürgen Ohlbach   LMU Munich, Germany
(Vice-President)
Renate Schmidt (CADE)   University of Manchester, UK
Larry Wos (President)   Argonne National Laboratory, USA

Sponsors

The CADE conference series is sponsored by CADE Inc., a sub-corporation of the Association for Automated Reasoning. In addition, CADE-25 gratefully acknowledges support from Freie Universität Berlin, the *Artificial Intelligence Journal,* Microsoft Research, the German Research Foundation, DFG, and the European Coordinating Committee for Artificial Intelligence.

Abstracts of Invited Talks

The first three abstracts are for invited talks given in the Special Session on the Past, Present, and Future of Automated Deduction. The next three are for those given during the main conference. These are followed by three abstracts describing the competitions held at CADE-25.

History and Prospects for First-Order Automated Deduction

David A. Plaisted

352 Sitterson Hall
Department of Computer Science, UNC Chapel Hill
Chapel Hill, NC, 27599-3175, USA
http://www.cs.unc.edu/~plaisted

On the fiftieth anniversary of the appearance of Robinson's resolution paper [1], it is appropriate to consider the history and status of theorem proving, as well as its possible future directions. Here we discuss the history of first-order theorem proving both before and after 1965, with some personal reflections. We then generalize model-based reasoning to first-order provers, and discuss what it means for a prover to be goal sensitive. We also present a way to analyze asymptotically the size of the search space of a first-order prover in terms of the size of a minimal unsatisfiable set of ground instances of a set of first-order clauses.

Reference

1. Robinson, J.: A machine-oriented logic based on the resolution principle. J. ACM **12**(1), 23–41 (1965)

On the Role of Proof Theory
in Automated Deduction

Frank Pfenning

Carnegie Mellon University, USA

Since the seminal work by Gentzen, who developed both natural deduction and the sequent calculus, there has been a line of research concerned with discovering deep structural properties of proofs in order to control the search space in theorem proving. This is particularly important in non-classical logics where traditional model-theoretic techniques may be more difficult to apply. We will walk through some of the key developments, starting with cut elimination and identity expansion, followed by focusing, polarization, and the separation of judgments and propositions. These concepts have been surprisingly robust, applicable to many non-classical logics, to the extent that one may consider them a litmus test on whether a set of rules or axioms form a coherent logic. We illustrate how each of these ideas affect proof search. In some cases, proofs are sufficiently restricted so that proof search can be seen as a fundamental computational mechanism, giving rise to logic programming.

Stumbling Around in the Dark: Lessons from Everyday Mathematics

Ursula Martin

University of Oxford, UK
Ursula.Martin@cs.ox.ac.uk

The growing use of the internet for collaboration, and of numeric and symbolic software to perform calculations it is impossible to do by hand, not only augment the capabilities of mathematicians, but also afford new ways of observing what they do. In this essay we look at four case studies to see what we can learn about the everyday practice of mathematics: the *polymath* experiments for the collaborative production of mathematics, which tell us about mathematicians attitudes to working together in public; the *minipolymath* experiments in the same vein, from which we can examine in finer grained detail the kinds of activities that go on in developing a proof; the mathematical questions and answers in *math overflow,* which tell us about mathematical-research-in-the-small; and finally the role of computer algebra, in particular the GAP system, in the production of mathematics. We conclude with perspectives on the role of computational logic.

Automated Reasoning in the Wild

Ulrich Furbach, Björn Pelzer, and Claudia Schon

Universität Koblenz-Landau, Germany
{uli, bpelzer, schon}@uni-koblenz.de

This paper discusses the use of first order automated reasoning in question answering and cognitive computing. For this the natural language question answering project LogAnswer is briefly depicted and the challenges faced therein are addressed. This includes a treatment of query relaxation, web-services, large knowledge bases and co-operative answering. In a second part a bridge to human reasoning as it is investigated in cognitive psychology is constructed by using standard deontic logic.

Work supported by DFG FU 263/15-1 'Ratiolog'.

The Herbrand Manifesto

Thinking Inside the Box

Michael Genesereth and Eric J.Y. Kao

Computer Science Department
Stanford University, USA
genesereth@stanford.edu
erickao@cs.stanford.edu

The traditional semantics for (first-order) relational logic (sometimes called *Tarskian* semantics) is based on the notion of interpretations of constants in terms of objects external to the logic. *Herbrand* semantics is an alternative that is based on truth assignments for ground sentences without reference to external objects. Herbrand semantics is simpler and more intuitive than Tarskian semantics; and, consequently, it is easier to teach and learn.

Moreover, it is more expressive than Tarskian semantics. For example, while it is not possible to finitely axiomatize natural number arithmetic completely with Tarskian semantics, this can be done easily with Herbrand semantics. Herbrand semantics even enables us to define the least fixed-point model of a stratified logic program without any special constructs.

The downside is a loss of some familiar logical properties, such as compactness and proof-theoretic completeness. However, there is no loss of inferential power—anything that can be deduced according to Tarskian semantics can also be deduced according to Herbrand semantics.

Based on these results, we argue that there is value in using Herbrand semantics for relational logic in place of Tarskian semantics. It alleviates many of the current problems with relational logic and ultimately may foster a wider use of relational logic in human reasoning and computer applications. To this end, we have already taught several sessions of the computational logic course at Stanford and a popular MOOC using Herbrand semantics, with encouraging results in both cases.

Automating Leibniz's Theory of Concepts

Jesse Alama[1], Paul E. Oppenheimer[2], and Edward N. Zalta[2]

[1] Vienna University of Technology, Vienna, Austria
alama@logic.at
[2] Stanford University, Stanford, USA
{paul.oppenheimer,zalta}@stanford.edu

Our computational metaphysics group describes its use of automated reasoning tools to study Leibniz's theory of concepts. We start with a reconstruction of Leibniz's theory within the theory of abstract objects (henceforth 'object theory'). Leibniz's theory of concepts, under this reconstruction, has a nonmodal algebra of concepts, a concept-containment theory of truth, and a modal metaphysics of complete individual concepts. We show how the object-theoretic reconstruction of these components of Leibniz's theory can be represented for investigation by means of automated theorem provers and finite model builders. The fundamental theorem of Leibniz's theory is derived using these tools.

Confluence Competition 2015

Takahito Aoto[1], Nao Hirokawa[2], Julian Nagele[3],
Naoki Nishida[4], and Harald Zankl[3]

[1] Tohoku University, Japan
[2] JAIST, Japan
[3] University of Innsbruck, Austria
[4] Nagoya University, Japan

Confluence is one of the central properties of rewriting. Our competition aims to foster the development of techniques for proving/disproving confluence of various formalisms of rewriting automatically. We explain the background and setup of the 4th Confluence Competition.

The CADE-25 ATP System Competition CASC-25

Geoff Sutcliffe

University of Miami, USA

The CADE ATP System Competition (CASC) is an annual evaluation of fully automatic Automated Theorem Proving (ATP) systems for classical logic the world championship for such systems. One purpose of CASC is to provide a public evaluation of the relative capabilities of ATP systems. Additionally, CASC aims to stimulate ATP research, motivate development and implementation of robust ATP systems that are useful and easily deployed in applications, provide an inspiring environment for personal interaction between ATP researchers, and expose ATP systems within and beyond the ATP community. Fulfillment of these objectives provides insight and stimulus for the development of more powerful ATP systems, leading to increased and more effective use.

CASC-25 was held on 4th August 2015 as part of the 25th International Conference on Automated Deduction (CADE-25), run on computers supplied by the StarExec project. The CASC-25 web site provides access to all systems and competition resources: http://www.tptp.org/CASC/25.

CASC is run in divisions according to problem and system characteristics. For CASC-25 the divisions were:

- THF: **T**yped **H**igher-order **F**orm theorems (axioms with a provable conjecture).
- THN: **T**yped **H**igher-order form **N**on-theorems (axioms with a countersatisfiable conjecture, and satisfiable axiom sets). This division was new for CASC-25.
- TFA: **T**yped **F**irst-order with **A**rithmetic theorems (axioms with a provable conjecture).
- TFN: **T**yped **F**irst-order with arithmetic **N**on-theorems (axioms with a countersatisfiable conjecture, and satisfiable axiom sets). This division was new for CASC-25.
- FOF: **F**irst-**O**rder **F**orm theorems (axioms with a provable conjecture).
- FNT: **F**irst-order form syntactically non-propositional **N**on-**T**heorems (axioms with a countersatisfiable conjecture, and satisfiable axiom sets).
- EPR: **E**ffectively **PR**opositional clause normal form (non-)theorems.
- LTB: First-order form theorems (axioms with a provable conjecture) from **L**arge **T**heories, presented in **B**atches with a shared time limit.

Problems for CASC are taken from the TPTP Problem Library. The TPTP version used for CASC is released after the competition, so that new problems have not been seen by the entrants. The THF, TFA, FOF, FNT, and LTB divisions were ranked according to the number of problems solved with an acceptable proof/model output. The THN, TFN, and EPR divisions were ranked according to the number of problems solved, but not necessarily accompanied by a proof or model. Ties are broken according to the average time over problems solved. Division winners are announced and prizes are awarded.

The design and organization of CASC has evolved over the years to a sophisticated state. Decisions made for CASC (alongside the TPTP, and the ES* series of workshops) have influenced the direction of development in ATP for classical logic. CASC-25 was the 20th edition of CASC, and it is interesting to look back on some of the key decisions that have helped bring ATP to its current state.

- CASC-13, 1996: The first CASC stimulated research towards robust, fully automatic systems that take only logical formulae as input. It increased the visibility of systems and developers, and rewarded implementation efforts.
- CASC-14, 1997: Introduced the SAT division, stimulating the development of model finding systems for CNF.
- CASC-15, 1998: Introduced the FOF division, starting the slow demise of CNF to becoming just the "assembly language" of ATP.
- CASC-16, 1999: Changes to the problem selection motivated the development of techniques for automatic tuning of ATP systems' search parameters.
- CASC-JC, 2001: Introduced ranking based on proof output, starting the the trend towards ATP systems that efficiently output proofs and models. Introduced the EPR division, stimulating the development of specialized techniques for this important subclass of problems.
- CASC-20, 2005: Required systems to develop builtin equality reasoning, by removing the equality axioms from the TPTP problems.
- CASC-J3, 2006: The FOF division was promoted as the most important, stimulating development of ATP systems for full first-order logic.
- CASC-21, 2007: Introduced the FNT division, further stimulating the development of model finding systems.
- CASC-J4, 2008: Introduced the LTB division, leading to the development of techniques for automatically dealing with very large axiom sets.
- CASC-J5, 2010: Introduced the THF division, stimulating development of ATP systems for higher-order logic.
- CASC-23, 2011: Introduced the TFA division, stimulating development of ATP systems for full first-order logic with arithmetic.
- CASC-J6, 2012: Otter replaced by Prover9 as the "fixed-point" in the FOF division, demonstrating the progress in ATP.
- CASC-24, 2013: Removed the CNF division, confirming the demise of CNF.
- CASC-J7, 2014: Required use of the SZS ontology, so the ATP systems unambiguously report what they have established about the input.
- CASC-25, 2015: Introduced the THN and TFN divisions, stimulating development of model finding for the TFA and THF logics.

The ongoing success and utility of CASC depends on ongoing contributions of problems to the TPTP. The automated reasoning community is encouraged to continue making contributions of all types of problems.

Termination Competition (termCOMP 2015)

Jürgen Giesl[1], Frédéric Mesnard[2], Albert Rubio[3],
René Thiemann[4], and Johannes Waldmann[5]

[1] RWTH Aachen University, Germany
[2] Université de la Réunion, France
[3] Universitat Politècnica de Catalunya - BarcelonaTech, Spain
[4] Universität Innsbruck, Austria
[5] HTWK Leipzig, Germany

The termination competition focuses on automated termination analysis for all kinds of programming paradigms, including categories for term rewriting, imperative programming, logic programming, and functional programming. Moreover, the competition also features categories for automated complexity analysis. In all categories, the competition also welcomes the participation of tools providing certified proofs. The goal of the termination competition is to demonstrate the power of the leading tools in each of these areas.

F. Giesl—This author is supported by the Deutsche Forschungsgemeinschaft (DFG) grant GI 274/6-1.
A. Rubio—This author is supported by the Spanish MINECO under the grant TIN2013-45732- C4-3-P (project DAMAS).
R. Thiemann—This author is supported by the Austrian Science Fund (FWF) project Y757.

Contents

Unification

SAT/SMT

Past, Present and Future of Automated Deduction

Part II Models and Patterns of Adapted Education

History and Prospects for First-Order Automated Deduction

David A. Plaisted[✉]

Department of Computer Science, UNC Chapel Hill,
352 Sitterson Hall, Chapel Hill, NC 27599-3175, USA
plaisted@cs.unc.edu
http://www.cs.unc.edu/~plaisted

Abstract. On the fiftieth anniversary of the appearance of Robinson's resolution paper [57], it is appropriate to consider the history and status of theorem proving, as well as its possible future directions. Here we discuss the history of first-order theorem proving both before and after 1965, with some personal reflections. We then generalize model-based reasoning to first-order provers, and discuss what it means for a prover to be goal sensitive. We also present a way to analyze asymptotically the size of the search space of a first-order prover in terms of the size of a minimal unsatisfiable set of ground instances of a set of first-order clauses.

Keywords: First-order logic · Resolution · Theorem proving · Instance-based methods · Model-based reasoning · Goal-sensitivity · Search space sizes · Term rewriting · Complexity

1 Introduction and General Comments

This presentation concentrates on first-order logic. Sometimes it seems as if the development of first-order logic provers will be swallowed up by so many other logics and areas of inquiry in automated deduction, but first-order logic is really central to the field and there are probably significant advances yet to be made. In addition, one can expect that methods that are good for first-order logic will also help to design higher order logic provers.

This presentation will emphasize my personal experiences. I will talk about the past history of the field, and also present evaluations of where we are now and possible directions for the future. It is a pity that Bill McCune, Harald Ganzinger, Greg Nelson, and Mark Stickel are not here to give their insights into the history of the field.

1.1 Search Space Issues

Before talking about the history of the field, let us step back and ask what we are attempting to accomplish, in a general sense, and what is theoretically possible.

© Springer International Publishing Switzerland 2015
A.P. Felty and A. Middeldorp (Eds.): CADE-25, LNAI 9195, pp. 3–28, 2015.
DOI: 10.1007/978-3-319-21401-6_1

Jürg Nievergelt once gave me a depressing view of theorem proving work. He felt that different methods just explored different portions of the search space, so the point was to find the right method to prove a particular theorem. However, this does not tell the whole story. The search space can be reduced, such as by first-order resolution over propositional deduction, and DPLL over propositional resolution. Also, hardware verification tools typically use DPLL (with CDCL) and not propositional resolution, as another indication that a reduction in search space has been achieved. Normal forms in term rewriting also reduce the search space. All the different forms of a term ($x * 1$, $x * 1 * 1$ et cetera) reduce to the same thing. The Knuth-Bendix method gets even more reductions and therefore is even more efficient at reducing the search space in many cases. The Gröbner Basis method also reduces the search space, but needs more axioms to be able to apply. Some results in mathematics may have been derived precisely to reduce the search space in proving various kinds of theorems. Also, the efficiency of the basic operations of a prover matters, as Stickel showed with his prolog-technology theorem prover [63], and others have continued to demonstrate.

Concerning search space, there are some basic questions to be answered, such as: What fraction of first-order formulas can be decided by decision procedures? Does this question have meaning? The subset is constantly increasing.

2 Pre-resolution

The predicate calculus was originally developed by Frege [70]. He considered predicates like "is happy" as signifying a function $H()$ of one variable that maps its argument x to a truth value $H(x)$, either true or false.

Hilbert wanted to provide a secure foundation for mathematics including a formalization of mathematics and an algorithm for deciding the truth or falsity of any mathematical statement. This is known as Hilbert's program [67].

Gödel showed that Hilbert's program was impossible, in its obvious interpretation [69]. He showed that any sufficiently powerful consistent formal system is incomplete. In terms of Turing machines, this result can be understood as follows: For any sound and effective system F that can formalize Turing computations, there will be some Turing machine M that fails to halt on blank tape, but this fact cannot be shown in F. In fact, such a machine M can be constructed from F. These results apply to any consistent effective extension of Peano arithmetic, for example. Thus it is not possible to formalize all of mathematics in a computable way, and any attempt at such a formalism will omit some true mathematical statements.

The ATP community has inherited Hilbert's program to some extent, in attempting to prove and decide what can be proved and decided, but of course there can be no recursive time bound on proving theorems, because of the undecidability of first-order logic.

However, it is still possible to write theorem provers and attempt to improve their efficiency, even if not all true statements are provable. Herbrand's theorem [40] gives a method to test a formula in first-order logic by successively testing propositional formulas for validity. Herbrand's theorem is of major importance in software developed for theorem proving by computer.

Gilmore's method [25] was an early attempt to implement Herbrand's theorem. Another early approach was presented by Davis and Putnam [18]. The linked conjunct method [16] was still another early method that attempted to guide the instantiation of clauses to prove unsatisfiability. An early Wos paper [68] mentions Gilmore's method but states that Davis and Putnam's method applied to sets of propositional clauses is much more efficient. The Wos paper also states that resolution can reduce the combinatorial explosion in Davis and Putnam's method by a factor in excess of 10^{50}. However, with faster propositional calculus implementations and by enumerating ground terms in a different way, this figure can possibly be reduced, as it will be shown later.

3 Early Post-resolution

Unification and resolution as presented by Robinson in 1965 [57] were the beginning of the modern era of theorem proving. Theorems could be proved that were significantly harder than those obtainable previously.

3.1 The Argonne Group

The Argonne group was the first group to devote serious effort to implementing Robinson's resolution rule. Here are some of the earliest theorems that they proved [68]:

In an associative system with left and right solutions, there is right identity element.

In an associative system with an identity element, if the square of every element is the identity, the system is commutative.

In a group, if the square of every element is the identity, the group is commutative.

For these proofs, the associativity of multiplication was represented by the following axioms.

$\neg P(x, y, u) \lor \neg P(y, z, v) \lor \neg P(u, z, w) \lor P(x, v, w)$

$\neg P(x, y, u) \lor \neg P(y, z, v) \lor \neg P(x, v, w) \lor P(u, z, w)$

Also, $P(x, x, e)$ was used to mean that the square of every element is the identity. Thus multiplication was represented in a relational manner. This reduces the need for explicit equational reasoning.

The terms paramodulation and demodulation and the associated concepts were developed by the Argonne group. They also developed the set of support strategy. Hyper-resolution and P_1 deduction, on the other hand, were developed [56] by Robinson.

The Argonne prover was initially very slow. Finally McCune took the matter into his own hands and rewrote the entire prover in C, producing Otter [38], which was much faster and very easy to use.

3.2   Other Early Work

Maslov's method [71] appeared at about the same time as resolution. There were also many early refinements to resolution such as ancestry filter form, model elimination [36], semantic resolution, locking resolution, and merging. These refinements were an attempt to improve the performance of resolution, and they did help to some extent. There were also non-resolution methods such as the connection method of Bibel [6] and Andrews' matings method [1], which are now viewed as similar or identical to each other. At this time, the classic text of Chang and Lee [13] appeared, which is still helpful. The Pelletier problems [43] were often used to test theorem provers. There are some excellent collections [10,59,60] of early papers in automated deduction.

3.3   AI and Theorem Proving

There was initial enthusiasm for resolution in the artificial intelligence community; for example, it was the basis of Cordell Green's QA3 system [26]. Soon afterwards there was disenchantment with resolution and with uniform methods in general, and an emphasis on expert systems instead, which could perform at a human level in a narrow area. In fact, general systems were termed weak methods, and narrowly defined but more capable systems were termed strong methods by the AI community. Today ATP seems to be one tool in AI's toolkit, though not a solution to every problem. Formal logic still has a place in AI, such as in the situation calculus and in non-monotonic logic.

3.4   Personal Experiences

The book Computers and Thought [21] had an early influence on me. I saw the potential of computers for augmenting and simulating human intelligence. As an undergraduate I spent a summer working for MIT's Project MAC and got an early exposure to artificial intelligence research in this way. Near the end of my graduate school education, Vaughan Pratt explained the concept of NP-completeness to me. This led to some early papers on this topic; I also did some work on algorithms, partially motivated by Ed Reingold. However, the P vs. NP problem seemed too hard, and algorithms research did not seem to have any general, unifying features; also, there were already many bright and capable people in the field.

During my graduate studies Dave Luckham suggested that I study methods for equational theorem proving. The seminal paper of Knuth and Bendix [32] had only recently appeared. I developed something like unfailing completion and was about to write a thesis about it. Here is a paragraph from a thesis draft of June 10, 1974:

> We show that if there is an equational derivation over all-positive set
> S of equations, between $r_1\alpha$ and $r_2\alpha$ for some terms r_1 and r_2 and
> some substitution α, then an instance of $x \neq x$ is derivable from

$S \cup \{r_1 \neq r_2\}$ using the simplification strategy. The proof is approximately by induction on the complexity of the equational derivation between $r_1\alpha$ and $r_2\alpha$, where the complexity of an equational derivation remains to be defined. The proof is somewhat complicated and it could be that a simpler proof exists. The complexity of an equational derivation must be defined carefully so that it has the required properties. The following definitions help to state the definition of the complexity ordering on equational derivations:

My original Ph.D. thesis draft dealt, among other things, with strategies for equality in theorem proving, but then a departmental report by Dallas Lankford [34] appeared in which it seemed that he had already obtained a similar result about equality. So I switched to another topic. It turned out that Lankford did not have the result he had claimed, but hoped to find a proof for it in a subsequent departmental report, which never appeared. The full development of unfailing completion did not appear until several years later [2]. My original thesis draft also included some material on abstraction, which I later developed into a few papers.

I almost met Gerald Peterson at a conference but was diverted by talking to someone else. Had the two of us met, it might have led to some good work in equational theorem proving methods.

My early work on the path of subterms ordering [44,45] was done in 1978 at about the same time as Dershowitz' recursive path ordering [19]. These orderings are similar, but the recursive path ordering is simpler. Since then there have been many developments related to the recursive path ordering. Dershowitz in addition gave a general method for showing that orderings on terms are well-founded. Unfortunately, my work was sent to the wrong community for refereeing and was rejected, coming out instead as a technical report. This report led Claude Kirchner to enter the field of term rewriting systems; he did not read the proofs, but looked at the examples in the report.

As for first-order theorem proving, when I started graduate school, it seemed that resolution and its refinements, such as ancestry filter form and locking resolution, plus model elimination, were the only games in town. I didn't like resolution initially, but the concept is actually very natural. The idea is that if two clauses share a literal, then in some cases the shared literal can be eliminated and the remaining portions of the clauses can be joined together into another clause. Thus the eliminated literal is a "bridge concept" to relate two other connected sets of concepts. Thus two concepts related to the same concept are also related to each other, which is a natural idea.

I was greatly influenced by Bledsoe's work [8] in which he showed that some reasonably simple set theory problems could not be solved easily by resolution. He also did some early work on semantics [3,9], which also impressed me.

An early argument against resolution concerning the exponential number of clauses that can be generated by the clause form translation, was overcome by the structure-preserving translations [48]. My earliest paper about this for first-order logic may be the one by Greenbaum et al. [27] in CADE.

In the early days, we tested our provers on common problems such as Schubert's Steamroller, the Zebra problem, and similar problems, which seemed like personal friends. There was a sense of achievement when a proof was found. Today with the massive TPTP problem set [64], the art of testing has greatly advanced, but perhaps something has also been lost.

My work in theorem proving progressed through a number of strategies, finding deficiencies in each one. In 1974 I had implemented a back chaining resolution prover based on semantic trees with variables. Vaughan Pratt gave me a problem to show that if in a group the square of every element is the identity, then the group is commutative. My prover had a lot of trouble with this problem, and this example showed me that such a back chaining approach was not the way to go.

During my sabbatical at SRI I implemented a forward chaining resolution prover, but it had trouble with Pelletier's non-obviousness problem, which convinced me that this also was not the right approach. At the University of Illinois, Steve Greenbaum implemented the Violet prover. We put a lot of work into including abstraction, which did not turn out to be helpful. However, the basic prover was fairly efficient compared with resolution provers of that time. One of the main ideas was to resolve the pair of clauses whose sum of sizes was as small as possible. We also had some advanced data structures for term searching and for unification. This prover still merits some additional work.

Instance-Based Methods. After my attempt at a resolution prover at SRI, the next idea was to extend Prolog's back chaining strategy for Horn clauses to full first-order logic. This led to the simplified problem reduction format [47] and the modified problem reduction format [51], but still I was not satisfied with them. Eventually I decided that what was needed was DPLL-style search [17] in first-order logic. This led to work on instance based methods [4], leading to a sequence of provers including clause linking [35], semantic hyper-linking [14], the replacement rule theorem prover [42], and ordered semantic hyper-linking [52]. None of these were implemented with highly efficient data structures except for OSHL, which Hao Xu later implemented for his Ph.D. thesis with an inference rate often approaching 10,000 inferences per second.

Other instance-based methods also appeared, including Billon's disconnection calculus [7] implemented in the DCTP theorem prover [61], Equinox [15], and Inst-Gen [33]. The linked conjunct method [16] was an early instance-based method, and SATCHMO [37] can also be seen in this light, though it had to enumerate ground terms in some cases. One might consider some versions of hyper-tableaux as instance-based as well. The disconnection calculus, Inst-Gen, and DCTP are all somewhat in the style of clause linking, while OSHL and Equinox are more in the style of SATCHMO, though OSHL permits semantic guidance to select ground intances. Perhaps it would be worthwhile to reimplement SATCHMO with a very high inference rate. It appears that instance-based methods have even been incorporated in Vampire to some extent, showing their increasing importance in the field. Instance-based methods appear to perform particularly well on function-free clause sets, and these have some important applications.

4   Late Post-resolution

For other and later developments in first-order theorem proving, the two volume collection by Robinson and Voronkov [55] is a good source. After the initial developments surrounding resolution, there continued to be advances in refining resolution strategies, in refining paramodulation and rewriting, including basic paramodulation, in instance-based methods, in incorporating special axioms into first-order provers, in decision procedures for specialized theories, and in efficient data structures. The CADE system competition has become an important event and a significant test of various provers. The lean theorem provers [41] are noteworthy for their performance in a very compact prover due to the similarity of Prolog and first-order logic. Major provers including Vampire [54], E [58], and Spass [66] have become increasingly effective. The use of *strategy selection* has greatly helped major provers today, including Vampire and E.

5   Comments on Resolution

Resolution initiated the modern era of theorem proving in 1965. The computation of most general unifiers avoids the necessity to enumerate all propositional instances of first-order formulas.

Resolution not only uses most general unifiers, but with paramodulation and demodulation is also easily extendible to equality and rewriting. This is a good combination and may explain the persistence of resolution in theorem proving despite its inefficiency on non-Horn propositional problems. But is resolution a global maximum or a local maximum as a strategy? Is it possible to go beyond it? Perhaps we need to try to go beyond resolution and supplement it with other approaches in order to obtain truly powerful provers.

A resolution prover is like a prolific but not very well organized mathematician filling notebooks with trivial deductions, with no overall sense of where he is going. Once in a while he stumbles on something interesting.

What does a large set of clauses generated by resolution, mean? How is it making progress towards a proof? It is difficult to make any sense of tens of thousands of clauses in memory. Maybe culling small ground instances of derived clauses and finding their models by DPLL would give some insight into what is happening and how the models are being restricted.

Resolution is entirely syntactic; there is no semantics involved, though semantics can be introduced in semantic variants of resolution. Human mathematicians use semantics such as groups for group theory theorems. Perhaps our provers also should use more semantic information.

Even the most efficient propositional provers are benefited by conflict-driven clause learning (CDCL), which is essentially resolution. This shows that resolution is not going to disappear.

6   Propositional Calculus and SMT

One of the unexpected developments in theorem proving is the increasing efficiency of propositional provers, which are even used to solve problems in other domains by translating them into propositional logic and then using a propositional satisfiability procedure. Also, propositional provers are now often used for model checking applications, in the bounded model checking approach. We now at least have some understanding of the complexity of propositional satisfiability from the theory of NP completeness. It appears, assuming that P is not equal to NP, that satisfiability is exponential in the worst case. How is it then that propositional provers can be so efficient in practice? Part of the reason is the so-called satisfiability threshold [20]; for problems with a large ratio of clauses to literals, DPLL is likely to finish quickly because the search tree will be small. For clauses with a small ratio of clauses to literals, DPLL is likely to find a model quickly. The hard problems tend to be those in the middle. The fastest propositional provers use not only DPLL but also CDCL, which helps them avoid repetitive parts of the search by learning the reason for various conflicts. It would be helpful to have something like this in first-order provers, too.

Another recent development is the increasing effectiveness of provers based on satisfiability modulo theories (SMT) and their applications. What's the next step beyond SMT to include more of first-order logic and decision procedures while maintaining the propositional efficiency of DPLL? Equinox achieved respectable performance in a possibly complete theorem proving method by combining an OSHL style prover and DPLL, and dealt with equality by congruence closure. Is it possible to extend this approach to more specialized decision procedures, and thereby obtain a way to extend SMT to a complete first-order strategy?

7   Equality and Term Rewriting Systems

Much more could be said about term rewriting systems, completion, and Musser's inductionless induction [39]. I was amazed at the way one could prove inductive theorems by term rewriting system completion. Lankford had many pioneering papers in term-rewriting systems that unfortunately did not get published. Early termination techniques by Iturriaga [29] were pioneering but largely superceded by the recursive path orderings, which were a tremendous advance in termination, though earlier orderings are still significant. The survey of rewriting by Huet and Oppen [28] impressed me with the potential of term-rewriting techniques. Equational unification methods including AC unification [62] have had a tremendous impact as well. Termination techniques using the dependency pair ordering [24] are another significant development. Of course, one could also mention conditional term rewriting, higher order rewriting, rigid E-unification, and the Waldmeister prover [22].

8 Discussion of Prover Features

Now we turn our attention from the history of theorem proving to its possible future. Here are some features that are desirable in a theorem prover, but perhaps no current prover has all of them at the same time. Perhaps if this combination of features could be achieved it would lead to the next major advance in theorem proving. Perhaps the field has been struggling to get all these features in one prover. If one has a strategy with all of these features, then it would seem to be similar to human approaches to theorem proving. The inference rate might be slow, so such an approach might not be helpful for small, easy problems, but for harder problems, and with very long running times, the strategy might be more effective.

8.1 First-Order

The first feature that is desired is that the logic should be first (or higher) order; We are interested in first-order methods.

8.2 Model-Based Reasoning with Backtracking

The second feature is that it is desirable to have DPLL style model-based search and backtracking over models in a first-order prover. Model-based search with backtracking is what gives DPLL its efficiency. We discuss how one might generalize DPLL to obtain a first-order model-based method, that is, a method that involves the search for a model with backtracking as in DPLL, but generalized to first-order logic.

A survey paper [11] considers various model-based theorem proving strategies, and shows how the term model-based reasoning can mean many different things. Here, we consider model-based reasoning in the following sense. This presentation is partly inspired by point set topology.

We assume that initially a formula S (possibly a set of clauses) is given and it is desired to show that S is unsatisfiable. We define abstract first-order model-based methods $AMB(Ext, Inf)$ for demonstrating unsatisfiability of S, consisting of an extension method Ext and an inference method Inf that satisfy certain properties.

In general, an AMB method generates a sequence of interpretations I and formulas W contradicting the interpretations I. It is also convenient to specify a partial interpretation J with each I such that W also contradicts J. A method $AMB(Ext, Inf)$ is *complete* if for any unsatisfiable S, a formula which is a contradiction is eventually derived.

More specifically, let $LC(S)$ be a set of logical consequences of S such that $LC(S)$ is recursively enumerable. Also, for every interpretation I let $\mathcal{P}(I)$ be a set of partial interpretations, such that all elements J of $\mathcal{P}(I)$ are subsets of I. (An interpretation I is viewed as a set of literals that are satisfied by I, and similarly for partial interpretations J. Thus if a formula W contradicts J and $J \subseteq I$ then W contradicts I also).

Define a *conflict triple* for S to be a triple (I, J, W) such that $J \in \mathcal{P}(I)$, $W \in LC(S)$, and W contradicts J. If S is unsatisfiable, then for any interpretation I there must exist $J \in \mathcal{P}(I)$ and $W \in LC(S)$ such that (I, J, W) is a conflict triple.

A *conflict triple sequence* for S is a sequence (I_i, J_i, W_i) of conflict triples for S; the sequence is *productive* if no I_i has a previous J_k as a subset, for $k < i$.

A set $\{(I_i, J_i, W_i)\}$ of conflict triples is *refutational* if for every interpretation I there is a J_i such that J_i is a subset of I. This means that every I is contradicted by some W_i, which implies that S is unsatisfiable.

The *initial conflict triple sequence* is the empty sequence; after that a sequence is constructed using the extension operation Ext to add triples to the end of the sequence one by one. Given a sequence $T_1, T_2, \ldots, T_n$ of conflict triples, $Ext((T_1, T_2, \ldots, T_n))$ is another conflict triple. A (finite) *extension sequence* is either empty or of the form $T_1, T_2, \ldots, T_n, Ext((T_1, T_2, \ldots, T_n))$ where $T_1, T_2, \ldots, T_n$ is an extension sequence. Also, an infinite sequence is an extension sequence if all its finite prefixes are extension sequences. The extension method Ext can *fail*, in which case it is not possible to extend the sequence.

The extension method Ext must satisfy the following properties: (1) If $(I, J, W) = Ext((T_1, T_2, \ldots, T_n))$ where $T_i = (I_i, J_i, W_i)$ then I cannot have any J_i in the sequence as a subset. (2) Some pair (J, W) for I with J in $\mathcal{P}(I)$ is chosen fairly in the sense that in any infinite extension sequence (I_i, J_i, W_i), if for infinitely many I_i, (I_i, J, W) is a conflict triple, then eventually some conflict triple (I', J', W') with $J' \subseteq J$ will be chosen by extension and inserted in the sequence. (3) For any conflict triple (I_i, J_i, W_i) in an extension sequence, and any W in $LC(S)$, it is decidable whether W contradicts J_i. (4) If S is unsatisfiable, then the extension method fails if and only if the set of conflict triples in the sequence is refutational.

Now, any extension sequence will be productive by property (1).

For DPLL, W_i is a clause that contradicts some prefix of I_i, and J_i would be some such prefix. This means that J_i is always a finite interpretation for DPLL. However, for first-order logic J_i could be infinite, and might be defined for example on all ground instances of a finite set of possibly non-ground literals. The productive property is guaranteed in DPLL by the backtracking mechanism for DPLL.

Convergence: Say a sequence I_i of interpretations *converges to* I' if for all $J \in \mathcal{P}(I')$, all but finitely many elements I_i have $J \in \mathcal{P}(I_i)$.

An extension method Ext is *convergent* if for any infinite extension sequence (I_i, J_i, W_i), there is an interpretation I' and an infinite subsequence I_{n_i} of I_i such that I_{n_i} converges to I'.

Termination: If the extension method Ext is convergent and S is unsatisfiable, then there can be no infinite extension sequence.

Proof. Suppose (I_i, J_i, W_i) is an infinite extension sequence constructed by Ext. Because Ext is convergent, there will be an infinite subsequence I_{n_i} that converges to some interpretation I'. If sequence I_{n_i} of interpretations converges to I' and S is unsatisfiable, then because S is unsatisfiable there is a $J \in \mathcal{P}(I')$ and a $W \in LC(S)$ such that W contradicts J, and thus by definition of convergence all but finitely many I_{n_i} have $J \in \mathcal{P}(I_{n_i})$ so that (J, W) can also be chosen for infinitely many I_{n_i}. Hence (J', W') will eventually be chosen by fairness for some $J' \subseteq J$, and some W', so that some conflict triple (I_{n_i}, J', W') will be chosen for some n_i. This permits only finitely many elements I_{n_k} for $k > i$ to be chosen by productivity and because by definition of convergence, $J \in \mathcal{P}(I_{n_k})$ also for all but finitely many n_k. Thus there can be no infinite productive conflict triple sequence constructed by fair repeated extension, so the sequence has to stop.

The inference method Inf of AMB is *complete* if it is computable and the following holds:

For any refutational set $\{(I_1, J_1, W_1), \dots, (I_n, J_n, W_n)\}$ of conflict triples having more than one element, it is possible to apply the inference method Inf to two elements I_i and I_j in the set producing another conflict triple (I', J', W') such that $J' \subseteq J_i$ and $J' \subseteq J_j$, hence W' contradicts both J_i and J_j.

If the extension process stops, then AMB repeatedly applies the inference method Inf to the set of conflict triples in the extension sequence to replace conflict triples for I_i and I_j by a conflict triple for I', producing a smaller refutational set, and this operation is repeated until the set of conflict triples has only one element (I, J, W). Then all interpretations are contradicted by W. This W is then itself a contradiction (false), and its proof is a demonstration that S is unsatisfiable. In DPLL, this can correspond to CDCL deriving the empty clause.

Corollary: If the extension method Ext is convergent and the inference method Inf is complete then $AMB(Ext, Inf)$ is complete.

Proof: If S is unsatisfiable then eventually the extension method Ext stops with a finite extension sequence. The set of conflict triples in this sequence is refutational. Then the inference method Inf can be applied to this set repeatedly until a contradiction is derived.

In practice, inferences may be performed as the conflict triple sequence is generated even before a refutational set is obtained.

For a *strong* model-based system we assume that W_i is the universal closure of a formula W_i' and require that J_i should satisfy the universal closure of the negation of W_i', and similarly for results of inferences.

By a *model-based method* we mean either something that fits into the AMB formalism or something that is in the same style even if it doesn't exactly fit the formalism. Model Evolution [5] is one of the few strategies that is first-order and appears to be model-based in this sense. Perhaps it is a strong method. OSHL is model-based, but the clauses C_j are always ground clauses. Other instance-based methods may also be model-based in this sense.

8.3   Goal Sensitivity

The third feature that is desired in a prover is that inferences should be restricted to those that are related to the particular theorem and not just to general axioms. For clause form provers, if one wants to prove a theorem R from a set A of axioms, then one typically converts $A \wedge \neg R$ to clause form, obtaining a set S of clauses that is the union of clauses T from A and U from $\neg R$. In a goal sensitive method, clauses U from the negation of the theorem are typically considered to be relevant initially. Then the proof search is restricted so that clauses from T, and resolvents of input clauses, are only used if they are in some sense related to clauses from U. This means that when trying to prove a theorem, only the clauses that are related to the particular theorem R are used. Generally A will be satisfiable, so that T will also be satisfiable. Let I be a model of T. Then if S is unsatisfiable, only clauses from U will contradict I. Thus one may consider that only clauses that contradict such an I are be relevant initially. Such an I will typically be a nontrivial model, that is, not obtained simply by choosing truth values of predicate symbols in a certain way.

There are various ways to decide which derived formulas are relevant for a proof. One approach is to assign each formula a *relevance attribute*. The attribute can be true, indicating that the formula is relevant, or false, indicating that the formula is not relevant. These attributes are assigned initially so that only formulas related to the particular theorem are relevant. An inference is considered to be *relevant* if at least one of the hypotheses used in the inference is relevant. After each inference, the relevance attributes of formulas involved in the inference are updated. The conclusion of a relevant inference is always relevant, and the relevance attributes of the non-relevant hypotheses are changed from false to true. Some relevance strategies may also assign a numerical relevance distance attribute to formulas, indicating how relevant they are. A method is *goal sensitive* if it is a relevance strategy, that is, it only performs relevant inferences. Such a strategy only generates relevant formulas.

Thus the result of an inference rule such as resolution, applied to a relevant clause and another clause, is also relevant. Operations other than resolution, such as instantiation, can also create a relevant instance $C\Theta$ of a clause C if they unify a literal of C with the complement of a literal of a relevant clause D.

Say a literal is *relevant* if it is a literal of a relevant clause.

Relevance can be guaranteed for AMB-style model-based search methods with nontrivial semantics assuming that S is a set of first-order clauses, the initial interpretation I_1 is chosen to be a model of T, the clauses that contradict I_1 are the relevant clauses initially, the method generates a sequence C_i (W_i) of clauses that are logical consequences of S, and each C_i is either (a) a resolvent with a relevant parent, (b) a clause $C\Theta$ obtained by unifying some literal of a clause C with the complement of a relevant literal, or (c) a clause that contradicts the starting interpretation I_1. In such a case it is easy to show that all C_i are relevant. Choosing a semantics I_1 that models the general axioms T will guarantee that the first contradiction clause C_1 found is an instance of the negation of the theorem, and is therefore relevant.

A nontrivial semantics may be necessary for this, because there may not be a trivial model (choosing only truth values of predicate symbols) of T. The Gelernter prover [23] is an example of a prover using nontrivial semantics essentially for Horn clauses. A nontrivial interpretation I_1 can be represented by a procedure to test if a first-order formula or clause is satisfied by I_1. It is also necessary to represent other interpretations that arise from I_1 later in the search procedure.

For equational proofs of an equation $s = t$ from a set E of equations, if one can complete E then applying rewriting and narrowing (paramodulation) to s and t using the completed E suffices for completeness; $s = t$ is a logical consequence of E iff s and t rewrite to the same term. Thus such rewriting proofs are automatically goal sensitive, assuming that the theorem $s = t$ is selected to be relevant initially. Such goal-sensitivity is a tremendous advantage. If unfailing completion is used, the completion steps may not be relevant, but steps involving rewriting and narrowing of s and t will be relevant.

There are also methods [30, 46, 50, 65] that compute at the start which clauses and even which instances of clauses are closely related to the particular theorem, so that proof strategies can concentrate on such clauses. These methods typically compute a relevance distance d of each clause from the particular theorem, with smaller distances indicating clauses that are more closely related to the particular theorem. Then these methods compute a set $R_d(S)$ of clauses and instances of clauses of S such that all clauses in $R_d(S)$ are at relevance distance d or less. This set $R_d(S)$ is computed so that it is unsatisfiable if there is an unsatisfiable set of ground instances of S of cardinality d or less. Typically $R_d(S)$ is computable from S in polynomial time, or at worst in exponential time, and $R_d(S) \subseteq R_{d+1}(S)$ for all $d \geq 0$.

8.4   Importance of Goal-Sensitivity

Why should goal sensitivity help a theorem prover? This question will now be considered. There seems to be some disagreement in the deduction community about the importance of goal sensitivity, and also about which methods are goal sensitive. One report [53] states that goal sensitive strategies tend to do better, especially on large axiom sets. If a method has no goal sensitivity, and there are many general axioms, this means that when proving a theorem T from a set A of axioms, most inferences do not depend on T at all. Thus for various theorems $T_1, T_2, \ldots$ proved from A, most inferences will be repeated. This does not seem reasonable. If one saves results between $T_1, T_2, \ldots$, then one has even more formulas to retain, with a possible storage issue. Also, the theorems T_i will be even more overwhelmed by these additional formulas than before.

Axiom sets commonly used in mathematics are studied because they describe interesting objects such as the integers. Theorems concerning these objects often assert that these objects have certain properties. Because these objects are seen as interesting, from these axioms one can typically derive a huge number of logical consequences. For example, consider the axioms of number theory and the axioms of arithmetic, and the theorems that can be derived from them. Therefore it is not feasible to just combine axioms when proving a theorem, because so many theorems can be shown from the axioms, so goal sensitivity is important in such cases. In addition, such axiom sets will generally not have trivial models that simply assign truth values to predicate symbols. Also, because such structures are often well understood (such as integers and sets), it may be possible to specify semantics for them operationally (in an effective manner) to achieve goal sensitivity.

For very large axiom sets, or axiom sets with many consequences, relevance methods are especially important. If there is no particular theorem, and one simply wants to test an axiom set A for satisfiability, then one can still apply relevance methods by choosing a known satisfiable subset B of A as the general axioms, and one can then consider $A - B$ as the particular theorem. This approach will at least avoid combining axioms of A in the search for a contradiction.

Perhaps the concepts of goal relevance and semantics could even help propositional provers.

8.5   Proof Confluence

Proof confluence is another desirable property of a theorem prover. It means that there is no backtracking, so that no step has to be fully undone. One never completely erases the results of a step. Proof convergence means that if S is unsatisfiable, then one will eventually find a refutation without backtracking. It is possible for a method to be proof confluent even if it backtracks over partial interpretations; one can consider DPLL with CDCL to be proof confluent, because it saves information from previous attempts to find a model, even though it backtracks over partial interpretations.

8.6   Evalution of Methods

There is some controversy over which methods have which properties, but here is an attempted list of some methods and their properties. We consider that a method that does not permit the deletion of instances of more general clauses is not fully first-order; perhaps one could call it half order.

General resolution is first-order, not goal sensitive, and not model-based in the *AMB* sense. Resolution with set of support is first-order, goal sensitive, and not model-based. Model elimination is first-order, goal sensitive with the proper choice of starting clause, and not model-based in the sense we are discussing. OSHL is model-based and goal sensitive, but not first-order. DPLL is model-based and can be goal sensitive with the proper starting model, but is not first-order. Clause linking can be goal sensitive but requires that instances of clauses be kept, and thus is not fully first-order. It is also not really model-based. DCTP and the disconnection calculus are related to clause linking and do not seem to be goal sensitive, though there is some controversy about this. They may be viewed as model-based. They are not fully first-order because they keep instances of more general clauses. Inst-Gen and clause linking are related. Inst-Gen uses an arbitrary model given by the satisfiability procedure. It does not appear to be goal sensitive. It is not fully first-order because it sometimes keeps instances of more general clauses. It has a partial model-based character. Model Evolution is not goal sensitive but is model-based and first-order. SATCHMO is not goal sensitive or first-order but it is model-based. We are not sure how to classify hyper-tableaux and their many variations.

Knuth-Bendix completion is purely syntactic but sometimes very effective. The same is true of DPLL.

It is desired to have a method that is first-order, goal sensitive, and model-based. Proof confluence is also important and desirable when achievable. Could it be that we haven't yet reached the starting point in ATP?

Maria Paola Bonacina and the author are working on an SGGS method [12] that has all these desirable properties.

Another way to evaluate provers is to run them on examples. Concerning this method of evaluation, should it be necessary to so highly engineer a strategy to evaluate its effectiveness or get it published? Shouldn't there be a way to test strategies independently of the degree of engineering? Perhaps by counting the number of inferences needed to get proofs? Possibly by analyzing the asymptotic efficiency of a strategy in general or on specific classes of problems?

9   More Search Space Discussion

How can we give a rigorous complexity theoretic answer to the question whether one theorem proving strategy is better than another? For example, is resolution for first-order logic better than enumerating propositional instances, and if so, by how much? We can give specific examples, but it would be better to have a more general method of evaluation.

Theoreticians are highly interested in the complexity of resolution on propositional calculus problems. Is there some way to interest them in its complexity for first-order logic?

Even though first-order validity is only partially decidable, one can still discuss the asymptotic complexity of various theorem proving methods in terms of the size of a minimal Herbrand set. The book [49] presented this idea, but perhaps in too complex a way. The approach and notation here are different.

9.1   Terminology

Define a Herbrand set for a set S of clauses to be a set T of ground instances of clauses in S such that T is unsatisfiable. Define the linear size s_{lin} of terms, literals, clauses, and clause sets by

$$s_{lin}(\neg L) = s_{lin}(L)$$
$$s_{lin}(P(t_1 \ldots t_n)) = s_{lin}(f(t_1 \ldots t_n)) = 1 + s_{lin}(t_1) + \cdots + s_{lin}(t_n)$$
$$s_{lin}(L_1 \vee \cdots \vee L_n) = s_{lin}(L_1) + \cdots + s_{lin}(L_n)$$
$$s_{lin}(C_1 \wedge \cdots \wedge C_n) = s_{lin}(C_1) + \cdots + s_{lin}(C_n)$$
$$s_{lin}(x) = s_{lin}(c) = 1 \text{ for variables } x \text{ and constants } c$$

Define the directed acyclic graph size $s_{dag}(L)$ in the same way as s_{lin} but on a different term representation, with pointers to previous occurrences of a subterm so that the second and succeeding occurrences of a subterm only count as size 1. For this size measure, we represent a literal L with integers giving pointers to previous occurrences of a subterm so that for example $P(f(x), 2)$

represents $P(f(x), f(x))$ and the 2 indicates that the subterm begins in position 2 of the term. The representation of a literal L in this way may not be unique, so that we define $s_{dag}(L)$ to be the length of the shortest such representation of L. We assume that pointers are assigned so that no subterm appears more than once in a literal except possibly for variables and constants. Then $s_{dag}(P(f(x), f(x))) = 4$ because this literal can be represented as $P(f(x), 2)$, but $s_{lin}(P(f(x), f(x))) = 5$. Of course, $s_{dag}(L) \leq s_{lin}(L)$ for all L. This definition extends to clauses and sets of clauses in the usual way. For clauses we assume that pointers can refer to subterms in other literals of the clause so that no subterm of size more than one appears more than once in a clause. This assumes some ordering of the literals in the clause. For sets of clauses, some ordering on the clauses in the set is assumed, so that pointers can refer to subterms in other clauses.

We define a *binary resolution* between two clauses C_1 and C_2 as a resolution in which a literal of C_1 is unified with the complement of a literal of C_2 using a most general unifier, and a *binary factoring* on a clause C as a factoring in which two literals of C are unified by a most general unifier. Then it turns out that if clause D is a binary resolvent of clauses C_1 and C_2, $s_{dag}(D) < s_{dag}(C_1) + s_{dag}(C_2)$ and if D is a binary factor of C, then $s_{dag}(D) < s_{dag}(C)$ because no new term structure is created during unification, but only new pointers to existing term structure are created, and at least one literal is deleted.

DPLL Work Bounds. Given a set S of propositional clauses and an atom (Boolean variable) P appearing in S let $S(P \leftarrow T)$ and $S(P \leftarrow F)$ be S with P replaced by T (true), F (false) respectively everywhere and the following simplifications applied repeatedly:

$$C \lor T \rightarrow T, \ C \lor F \rightarrow C, \ S \land T \rightarrow S, \ S \land F \rightarrow F, \ \neg T \rightarrow F, \ \neg F \rightarrow T$$

Assume that such simplifications have already been applied to S. Define $\mathcal{W}_{DPLL}(S)$ for propositional clause set S by

if S is T or S is F then $\mathcal{W}_{DPLL}(S) = 1$ else
let P be an atom appearing in S.
$\quad \mathcal{W}_{DPLL}(S) =$
$\qquad$ if $S(P \leftarrow T)$ and $S(P \leftarrow F)$ are satisfiable
$\qquad$ then $1 + max(\mathcal{W}_{DPLL}(S(P \leftarrow T)), \mathcal{W}_{DPLL}(S(P \leftarrow F)))$
$\qquad$ else $1 + \mathcal{W}_{DPLL}(S(P \leftarrow T)) + \mathcal{W}_{DPLL}(S(P \leftarrow F))$

To within a polynomial $\mathcal{W}_{DPLL}(S)$ is an upper bound on the work for DPLL on S, not even considering CDCL, because DPLL might first choose a truth value for the Boolean variable P causing S to become unsatisfiable. In general, if there are n distinct atoms in S then $\mathcal{W}_{DPLL}(S) \leq 2^{n+1} - 1$.

Define $\mathcal{W}'_{DPLL}(S)$ for propositional clause set S by

if S is T or S is F then $\mathcal{W}'_{DPLL}(S) = 1$ else
let P be an atom appearing in S.
$\mathcal{W}'_{DPLL}(S) =$
    if $S(P \leftarrow T)$ and $S(P \leftarrow F)$ are satisfiable
    then $1 + (\mathcal{W}'_{DPLL}(S(P \leftarrow T)) + \mathcal{W}'_{DPLL}(S(P \leftarrow F)))/2$
    else
    if $S(P \leftarrow T)$ and $S(P \leftarrow F)$ are unsatisfiable
    then $1 + \mathcal{W}'_{DPLL}(S(P \leftarrow T)) + \mathcal{W}'_{DPLL}(S(P \leftarrow F))$
    else
    if $S(P \leftarrow T)$ is satisfiable and $S(P \leftarrow F)$ is unsatisfiable
    then $1 + \mathcal{W}'_{DPLL}(S(P \leftarrow T)) + \mathcal{W}'_{DPLL}(S(P \leftarrow F))/2$
    else
    if $S(P \leftarrow T)$ is unsatisfiable and $S(P \leftarrow F)$ is satisfiable
    then $1 + \mathcal{W}'_{DPLL}(S(P \leftarrow F)) + \mathcal{W}'_{DPLL}(S(P \leftarrow T))/2$

This is to within a polynomial an upper bound on the average time taken by DPLL on S with a random choice of whether to do $P \leftarrow T$ or $P \leftarrow F$ first, because if $S(P \leftarrow F)$ or $S(P \leftarrow F)$ is satisfiable then the other case is omitted. We shall typically be interested in the worst case time for $\mathcal{W}$ and the average case time for $\mathcal{W}'$. If there are n distinct atoms in S, then in many cases it appears that $\mathcal{W}'_{DPLL}(S)$ is polynomial or even linear in n based on practical experience.

Finally, define $\mathcal{W}''_{DPLL}(S)$ for clause set S to be n, where there are n distinct atoms in S. In many cases the time taken by DPLL seems to be within a small polynomial of $\mathcal{W}''_{DPLL}(S)$.

Resolution Work Bound. For resolution, denote a binary resolution between clauses C_1 and C_2 with resolvent C as the triple (C_1, C_2, C). Define the complexity $\mathcal{W}_{RES}(C_1, C_2, C)$ of a binary resolution (C_1, C_2, C) as $s_{dag}(C_1) + s_{dag}(C_2)$. Denote a binary factoring operation on C_1 with result C as the pair (C_1, C). The complexity $\mathcal{W}_{FACT}(C_1, C)$ of a binary factoring operation on C_1 is $s_{dag}(C_1)$. If R is a set of binary resolutions and binary factorings then define $\mathcal{W}_{RES}(R)$ as the sum of $\mathcal{W}_{RES}(C_1, C_2, C)$ for all resolutions (C_1, C_2, C) in R plus the sum of $\mathcal{W}_{FACT}(C_1, C)$ for all binary factorings in R. Also, for a set S of clauses, let $\mathcal{R}(S)$ be the set of all binary resolutions (C_1, C_2, C) for clauses C_1, C_2 in S together with all binary factoring operations (C_1, C) for clauses C_1 in S.

9.2   Literal Size Bounds

Bound for DPLL. Now, we can apply DPLL to first-order logic crudely, by enumerating all ground instances of a set of clauses in some order and looking for propositional (i.e., ground) refutations using DPLL periodically. Assume that this version of DPLL is applied to first-order clause sets; we want to analyze its complexity and compare it to breadth-first resolution. Let $lit_{lin,d}(S)$ be the set of ground instances C of clauses in S such that for all literals L in C, $s_{lin}(L) \leq d$. We want to evaluate the worst case for $\mathcal{W}_{DPLL}(lit_{lin,d}(S))$ and also evaluate $\mathcal{W}''_{DPLL}(lit_{lin,d}(S))$.

Suppose in a Herbrand set T for S, there are c symbols (function, constant, and predicate symbols) in all. Then there are at most $n = c^d$ ground atoms of linear size d. Adding in smaller d gives $c^d + c^{d-1} + \ldots$ which, assuming $c \geq 2$, is at most $2c^d$, or $O(c^d)$. Then $\mathcal{W}_{DPLL}(lit_{lin,d}(S))$ is upper bounded by $2^{O(c^d)}$, but $\mathcal{W}''_{DPLL}$ is only $O(c^d)$.

Let $lit_{dag,d}(S)$ be the set of ground instances C of clauses in S such that for all literals L in C, $s_{dag}(L) \leq d$. We want to evaluate the worst case for $\mathcal{W}_{DPLL}(lit_{dag,d}(S))$ and also evaluate $\mathcal{W}''_{DPLL}(lit_{dag,d}(S))$.

Suppose in a Herbrand set T for S, there are c symbols (function, constant, and predicate symbols) in all. In ground atoms of size d, the integers pointing to subterms could be between 1 and d so that the number of symbols that can appear in each position is bounded by $c + d$. Then there are at most $n = (c + d)^d$ ground atoms of dag size d. Adding in smaller sizes gives not more than $(c + d)^d + (c + d)^{d-1} + \ldots$ which, assuming $c \geq 2$, is at most $2(c + d)^d$ or $O((c + d)^d)$. Then $\mathcal{W}_{DPLL}(lit_{dag,d}(S))$ is upper bounded by $2^{O((c+d)^d)}$, but $\mathcal{W}''_{DPLL}$ is only $O((c + d)^d)$.

Bound for Resolution. Now, how can resolution obtain all proofs that could be obtained by DPLL if all ground literals have linear size bounded by d? To do this, the most favorable restriction of resolution would be to save all clauses containing literals of linear size bounded by d. Such a size bounded resolution strategy would also obtain some proofs that DPLL could not obtain. What is the complexity of this approach, and how does it compare to the complexity of DPLL on such clause sets?

Say a resolution (C_1, C_2, C) or a factoring (C_1, C) is d-*size bounded* if all literals in C have maximum literal size s_{lin} bounded by d. Define Π_r on resolutions and factorings by $\Pi_r((C_1, C_2, C_3)) = C_3$ and $\Pi_r((C_1, C_2)) = C_2$. Extend Π_r to sets elementwise. Let $\mathcal{R}_{lin,d}(S)$ be the set of d-size bounded resolutions and factorings in $\mathcal{R}(S)$. Let $\mathcal{R}_{lin,d}^k(S)$ be defined by $\mathcal{R}_{lin,d}^0(S) = \phi$ and $\mathcal{R}_{lin,d}^{k+1}(S) = \mathcal{R}_{lin,d}(S \cup \Pi_r(\mathcal{R}_{lin,d}^k(S)))$ for $k \geq 0$. Let $\mathcal{R}_{lin,d}^*(S)$ be $\bigcup_{k \geq 0} \mathcal{R}_{lin,d}^k(S)$. Essentially all resolvents, resolvents of resolvents, binary factorings, et cetera are computed in which every step is d-size bounded. We want to upper bound $\mathcal{W}_{RES}(\mathcal{R}_{lin,d}^*(S))$. If $lit_{lin,d}(S)$ is unsatisfiable then a refutation can be obtained by lifting a ground resolution refutation, which means that all clauses with more than $2c^d$ literals or with literals of size greater than d may be deleted. It may be necessary to factor clauses before resolution, and to temporarily save resolvents with more than $2c^d$ literals, which will be reduced to not more than $2c^d$ literals by a sequence of factorings, but these factorings will be ignored in the analysis.

For a very weak bound on $\mathcal{W}_{RES}(\mathcal{R}_{lin,d}^*(S))$, consider the clauses generated by resolution that have atoms of size at most d, where these literals may contain variables. Let h be c^d, an upper bound on the number of ground atoms of size d. Smaller sizes may increase the number of ground atoms by a constant factor (at most 2), as before. Assume that all resolvents with an atom of linear size greater than d are deleted. Then for a retained clause C having at most $2h$ literals of size at most d, $s_{lin}(C) \leq 2dh$. How many such clauses are there, accounting for the occurrences of variables? Assuming variables are named by first appearance from left to right, for $2h$ atoms of size d there can be c symbols in the leftmost place (which is a predicate symbol) in clause C, $c + 1$ in the next place, and so on up to $c + d * h - 1$ symbols in the last place, so the total number of possible combinations is bounded by $(c + 2dh - 1)!/(c - 1)!$. This in turn is bounded by $(c + 2dh)^{2dh}$. How many clauses can there then be, ignoring subsumption deletion? There are up to h atoms and each can be positive, negative, or absent for $3^{2h}(c + 2dh - 1)!/(c - 1)!$ clauses. All pairs of clauses might resolve so that we might have the square of this many resolutions, as an upper bound on the number of resolutions. Each pair of clauses could resolve by binary resolution in possibly $4h^2$ ways for another factor of $4h^2$, and the work per resolution might be on the order of $4dh$, giving a bound of $(3^{2h}(c + 2dh - 1)!/(c - 1)!)^2 * 4h^2 * 4dh$ on $\mathcal{W}_{RES}(\mathcal{R}_{lin,d}(S))$. Each clause could also have $2h(2h - 1)/2$ binary factors, but this is a lower order contribution. Accounting for smaller size atoms would increase the bound by a small factor. The bound on number of resolutions is considerably larger than 9^{2h} and even 9^{2h} is much worse than the 2^{2h} bound for $\mathcal{W}_{DPLL}(lit_{lin,d}(S))$, and $\mathcal{W}''_{DPLL}(lit_{lin,d}(S))$ is even smaller. Of course, this bound on resolution is a very weak bound, and in practice resolution may perform much better than this.

Unit Resolution. A UR resolution is a sequence of resolutions of a non-unit clause C against enough unit clauses to remove all but one of the literals of C, producing another unit clause. This strategy is complete for Horn sets. Define $\mathcal{UR}^*_{lin,d}(S)$ similarly to $\mathcal{R}^*_{lin,d}(S)$ except that only UR resolutions are done. Factorings are not needed in this case. We want to upper bound $\mathcal{W}_{RES}(\mathcal{UR}^*_{lin,d}(S))$. If one does UR resolution, then the number of UR resolvents that are kept is bounded by the total number of units which considering occurrences of variables in the units is bounded by $c(c + 1) \ldots (c + d - 1)$ or $(c + d - 1)!/(c - 1)!$ for literals of size d. Considering smaller sizes the bound becomes twice as large. So if there are n 3-literal clauses then the total number of UR resolutions done is at most proportional to $3n(2(c + d - 1)!/(c - 1)!)^2$ because each 3-literal clause can resolve two of its three literals in three ways and there are n 3-literal clauses. No new 3-literal clauses are produced, so n is the number of 3 literal clauses in the input set S. This quantity $3n(2(c + d - 1)!/(c - 1)!)^2$ is bounded by $12n(c + d)^{2d}$ and this bound holds for arbitrarily deep refutations. Then $\mathcal{W}_{RES}(\mathcal{UR}^*_{lin,d}(S))$ is at most a factor of $4d$ larger than this to account for the work per resolution depending on the clause sizes. (One clause can have at most 3 literals and the other at most one literal, explaining the factor of 4). This bound is single exponential in d and is not much worse than $\mathcal{W}''_{DPLL}(lit_{lin,d}(S))$, and in fact this bound is much better than the worst case bound $\mathcal{W}_{DPLL}(lit_{lin,d}(S))$ for DPLL, even for deep UR refutations. Of course, UR resolution is not a complete strategy, but perhaps resolution gets much of its power from clause sets having UR refutations. Perhaps resolution not restricted to UR resolution also performs well if the proof can be obtained using only resolvents having 2 or 3 literals.

If instead one considers s_{dag} instead of s_{lin} for these results, the bounds would be a little different.

9.3   Bounded Depth Resolution Refutations

Bound for Resolution. Now for a more favorable resolution bound, consider clause sets S of n clauses having resolution refutations of depth bounded by D. Let $\mathcal{R}^k(S)$ be defined by $\mathcal{R}^0(S) = \phi$ and $\mathcal{R}^{k+1}(S) = \mathcal{R}(S \cup \Pi_r(\mathcal{R}^k(S)))$ for $k \geq 0$.

We want to bound $\mathcal{W}_{RES}(\mathcal{R}^D(S))$. Suppose S is a set of n 3 literal clauses. Suppose each clause has dag size at most s. Then two clauses having at most 3 literals can have at most 9 binary resolvents and a clause with at most 3 literals can have at most 3 binary factors. So the number of clauses generated at depth 1 is at most $p_1(n) = n(n-1)/2 * 9 + 3n$ and these clauses are of dag size at most $2s$. This is a polynomial $p_1(n)$ of degree 2 for the number of clauses. Then at depth 2 there will be at most $(p_1(n) + n)(p_1(n) + n - 1)/2 * 25 + 10p_1(n)$ clauses and factors because the clauses at depth 1 may have 5 literals each. This is a polynomial $p_2(n)$ of degree 4 for the number of clauses. The dag sizes of the clauses may be at most $4s$. In general at depth D there will be at most $p_D(n)$ clauses for a polynomial $p_D(n)$ which is of degree 2^D, and the dag sizes of the clauses are at most $2^D s$. So the total work $\mathcal{W}_{RES}(\mathcal{R}^D(S))$ is bounded by the sum of $p_i(n)2^i s$ for i less than or equal to D. This is still polynomial in n and s for a fixed D, which (as shown below) is much better than $\mathcal{W}_{DPLL}(lit_{dag,d}(S))$ for $d = 2^D s$. However, $p_i(n)$ is on the order of n^{2^D}, which is double exponential in D, and the work bound for resolution is larger than this.

If one restricts the depth D refutations to unit resolution refutations, then all the resolvents have only one or two literals so that the number of unit resolvents between two clauses is at most 3. This reduces the polynomials p_i and leads to a more favorable bound for unit resolution, perhaps explaining even more the advantage of resolution over DPLL if shallow unit resolution refutations are common.

Bound for DPLL. Now consider how one could use DPLL to obtain all refutations obtainable by depth D resolution. If there is a depth D bounded resolution refutation, then the dag literal size s_{dag} of a clause in a minimal Herbrand set can be bounded by $2^D s$. This is because at most 2^D copies of input clauses can be used in a refutation of depth D, and each copy has dag size at most s. When all unifications between these copies are done, corresponding to the structure of the refutation, a Herbrand set is obtained whose instances can have dag size at most $2^D s$. Consider now how one might test satisfiability for such clause sets with DPLL. Suppose ground clauses are enumerated in order of s_{dag} instead of s_{lin}. We want to bound $\mathcal{W}_{DPLL}(lit_{dag,d}(S))$ for $d = 2^D s$. With c symbols, there can be no more than $(c + d)^d$ ground atoms of this dag size; including smaller sizes increases the bound by at most a factor of 2, to $O((c + d)^d)$. Therefore the worst case bound for $\mathcal{W}_{DPLL}$ is $2^{O((c+d)^d)}$ which for a fixed D is double exponential in s, while resolution is polynomial in n and s for fixed D. Also, even $\mathcal{W}''_{DPLL}(lit_{dag,d}(S))$ is $O((c + d)^d)$ which is one exponential less than $\mathcal{W}_{DPLL}(lit_{dag,d}(S))$, but still much worse than resolution for constant D. Thus resolution is asymptotically much better than DPLL for depth D bounded refutations if D is small. It may be that such small depth refutations are common, explaining the advantage of resolution over DPLL on many clause sets.

9.4 Herbrand Set Size Bound

Another approach is to consider bounds based on the total size of a minimal Herbrand set for S. If one restricts consideration to Herbrand sets T having size $s_{dag}(T)$ bounded by H, then one can bound the dag size of the resolvents and factors by H as well without losing the refutation. If $s_{dag}(T)$ is bounded by H, then the number n of clauses in T is also bounded by H, as is the number of distinct atoms in H. By resolving each atom away in turn, the depth of a ground resolution refutation can also be bounded by H. Lifting this refutation to first-order may require binary factorings to merge literals that correspond to the same ground literal. For simplicity assume that the set of input clauses is closed under binary factorings. $H - 1$ factoring operations or less on each clause after each resolution except the last one suffice to merge non-ground literals that correspond to the same ground literal. With not more than $H - 1$ binary factoring operations after each resolution, the total refutation depth will be not more than H^2. Recalling the polynomials p_i, $p_{H^2}(n)$ is bounded by $p_{H^2}(H)$ which is on the order of $H^{2^{H^2}}$, double exponential in H. Perhaps this is a more natural bound than one phrased in terms of resolution itself. However, the resolution work bound $\mathcal{W}_{RES}(\mathcal{R}^{H^2}(S))$ is double exponential in H. Another approach is to count the total number of clauses that can be generated. Even assuming that the dag literal sizes of the clauses are bounded by H, one still has more than 3^{c^H} possible clauses, which is still double exponential in H. For DPLL, the dag literal size is bounded by H so that there can be $O((c + H)^H)$ atoms altogether and $\mathcal{W}_{DPLL}(lit_{dag,H}(S))$ is $2^{O((c+H)^H)}$ which is also double exponential in H, but $\mathcal{W}''_{DPLL}(lit_{dag,H}(S))$ is $O((c + H)^H)$ which is only single exponential in H. In this case DPLL looks better, assuming the refutation depths are not small.

If one considers instead Herbrand sets T with a fixed number H of distinct atoms as the size of T increases, then the bounds in terms of $s_{dag}(T)$ are similar to those for bounded depth resolution refutations, and resolution looks better than DPLL. This measure, not depending on the properties of resolution itself, also seems to be a more natural measure than the depth of a resolution refutation.

9.5 Summary

So, to sum up, if one bounds the maximum literal size in a Herbrand set, DPLL seems to have an advantage. If one bounds the size of the Herbrand set as a whole, DPLL still seems to have an advantage but not as much. For resolution refutations having a small depth (which corresponds to a small number of literals in a Herbrand set T), resolution looks better. For sets S having unit refutations, resolution looks much better. It depends then on the nature of the problems which method is best, according to this analysis; of course, the upper bounds are very weak.

This analysis suggests that resolution performs best on clause sets having shallow refutations or unit refutations, and for other clause sets, perhaps instance based strategies or even DPLL on ground instances is better. Because many typical theorems are Horn sets or nearly Horn sets and unit resolution is complete for Horn sets, resolution may perform well on such theorems. As evidence that DPLL can be faster than resolution, SATCHMO has much the flavor of DPLL applied to an enumeration of ground instances, and in its day it frequently beat the best resolution provers. Another evidence is the good performance of instance-based methods on the function-free fragment in the annual TPTP competitions, and of DPLL for hardware verification applications. One might say that resolution is good on easy problems, but for hard problems, other methods are needed. Of course, many of these easy problems were not easy before resolution came along.

This analysis also suggests to look for methods that are asymptotically efficient both for small Herbrand sets, for sets having shallow resolution refutations, for sets having unit refutations, and for sets of clauses in which the maximum literal size in a Herbrand set is bounded. Perhaps instance based strategies already have this property, or could be modified to have it, especially if supplemented with special rules for unit clauses. This analysis does not include equality, however, which is easy to include in a resolution prover but possibly more difficult for an instance-based strategy.

10   Additional Comments

Generating conjectures is also a fruitful area of research. This area is related to interpolation and automatic program verification, as well as being important for mathematical discovery in general. Model finding is another important area, which can be used to show that a formula is satisfiable. However, model finding is especially difficult, and for first-order logic is not even partially decidable. Another area of interest is using resolution provers as decision procedures for subsets of first-order logic [31]; Maslov's method [71] can also be used in this way.

As another issue, what can we do to improve the status of theorem proving in the general computer science community, as well as the funding situation? It appears from personal experience that more and more people outside of computer science are becoming interested in automated deduction.

Should we be thinking about the social consequences of theorem proving work and of artificial intelligence work in general? If compouters learn to prove hard theorems, will they replace humans? What needs to happen for computer theorem provers to be able to prove hard theorems often, without human interaction?

References

1. Andrews, P.: Theorem proving via general matings. J. ACM **28**, 193–214 (1981)
2. Bachmair, L., Dershowitz, N., Plaisted, D.: Completion without failure. In: Kaci, H.A., Nivat, M. (eds.) Resolution of Equations in Algebraic Structures Progress in Theoretical Computer Science. Rewriting Techniques, vol. 2, pp. 1–30. Academic Press, New York (1989)
3. Ballantyne, A., Bledsoe, W.: On generating and using examples in proof discovery. In: Hayes, J., Michie, D., Pao, Y.H. (eds.) Machine Intelligence, vol. 10, pp. 3–39. Ellis Horwood, Chichester (1982)
4. Baumgartner, P., Thorstensen, E.: Instance based methods - a brief overview. Ger. J. Artif. Intell. (KI) **24**(1), 35–42 (2010)
5. Baumgartner, P., Tinelli, C.: The model evolution calculus as a first-order DPLL method. Artif. Intell. **172**(4–5), 591–632 (2008)
6. Bibel, W.: Automated Theorem Proving. Vieweg, Braunschweig (1982)
7. Billon, J.P.: The disconnection method. In: Miglioli, P., Moscato, U., Ornaghi, M., Mundici, D. (eds.) TABLEAUX 1996. LNCS, vol. 1071, pp. 110–126. Springer, Heidelberg (1996)
8. Bledsoe, W.W.: Non-resolution theorem proving. Artif. Intell. **9**(1), 1–35 (1977)
9. Bledsoe, W.W.: Using examples to generate instantiations of set variables. In: Proceedings of the 8th International Joint Conference on Artificial Intelligence, pp. 892–901. William Kaufmann (1983)
10. Bledsoe, W.W., Loveland, D.W. (eds.): Automated Theorem Proving: After 25 Years. Contemporary mathematics. American Mathematical Society, Providence (1984)
11. Bonacina, M.P., Furbach, U., Sofronie-Stokkermans, V.: On first-order model-based reasoning. In: Martí-Oliet, N., Olveczky, P., Talcott, C. (eds.) Logic, Rewriting, and Concurrency: Essays in Honor of José Meseguer. LNCS, vol. 9200, Springer, Heidelberg (2015, to appear)
12. Bonacina, M.P., Plaisted, D.A.: SGGS theorem proving: an exposition. In: Konev, B., Moura, L.D., Schulz, S. (eds.) Proceedings of the 4th Workshop on Practical Aspects in Automated Reasoning, EasyChair Proceedings in Computing (2014, to appear)
13. Chang, C., Lee, R.: Symbolic Logic and Mechanical Theorem Proving. Academic Press, New York (1973)
14. Chu, H., Plaisted, D.: Semantically guided first-order theorem proving using hyperlinking. In: Bundy, A. (ed.) CADE 1994. LNCS, vol. 814, pp. 192–206. Springer, Heidelberg (1994)
15. Claessen, K.: Equinox, a new theorem prover for full first-order logic with equality (2005). Presented at Dagstuhl Seminar on Deduction and Applications
16. Davis, M.: Eliminating the irrelevant from machanical proofs. In: Proceedings of Symposia in Applied Mathematics, vol. 15, pp. 15–30 (1963)
17. Davis, M., Logemann, G., Loveland, D.: A machine program for theorem-proving. Commun. ACM **5**, 394–397 (1962)
18. Davis, M., Putnam, H.: A computing procedure for quantification theory. J. ACM **7**, 201–215 (1960)
19. Dershowitz, N.: Orderings for term-rewriting systems. Theor. Comput. Sci. **17**, 279–301 (1982)
20. Dubois, O.: Upper bounds on the satisfiability threshold. Theor. Comput. Sci. **265**(1–2), 187–197 (2001)

21. Feigenbaum, E.A., Feldman, J.: Computers and Thought. McGraw-Hill, New York (1963)
22. Gaillourdet, J.-M., Hillenbrand, T., Löchner, B., Spies, H.: The new WALDMEISTER loop at work. In: Baader, F. (ed.) CADE 2003. LNCS (LNAI), vol. 2741, pp. 317–321. Springer, Heidelberg (2003)
23. Gelernter, H., Hansen, J., Loveland, D.: Empirical explorations of the geometry theorem proving machine. In: Feigenbaum, E., Feldman, J. (eds.) Computers and Thought, pp. 153–167. McGraw-Hill, New York (1963)
24. Giesl, J., Schneider-Kamp, P., Thiemann, R.: *AProVE 1.2*: automatic termination proofs in the dependency pair framework. In: Furbach, U., Shankar, N. (eds.) IJCAR 2006. LNCS (LNAI), vol. 4130, pp. 281–286. Springer, Heidelberg (2006)
25. Gilmore, P.C.: A proof method for quantification theory. IBM J. Res. Dev. **4**(1), 28–35 (1960)
26. Green, C.: Application of theorem proving to problem solving. In: Proceedings of the 1st International Joint Conference on Artificial Intelligence, pp. 219–239. Morgan Kaufmann (1969)
27. Greenbaum, S., Nagasaka, A., O'Rorke, P., Plaisted, D.A.: Comparison of natural deduction and locking resolution implementations. In: Loveland, D. (ed.) 6th Conference on Automated Deduction. LNCS, vol. 138, pp. 159–171. Springer, Heidelberg (1982)
28. Huet, G., Oppen, D.C.: Equations and rewrite rules: a survey. In: Book, R. (ed.) Formal Language Theory: Perspectives and Open Problems, pp. 349–405. Academic Press, New York (1980)
29. Iturriaga, R.: Contributions to mechanical mathematics. Ph.D. thesis, Carnegie-Mellon University, Pittsburgh, Pennsylvania (1967)
30. Jefferson, S., Plaisted, D.: Implementation of an improved relevance criterion. In: Proceedings of the 1st Conference on Artificial Intelligence Applications, pp. 476–482. IEEE Computer Society Press (1984)
31. Joyner, W.: Resolution strategies as decision procedures. J. ACM **23**(1), 398–417 (1976)
32. Knuth, D., Bendix, P.: Simple word problems in universal algebras. In: Leech, J. (ed.) Computational Problems in Abstract Algebra, pp. 263–297. Pergamon Press, Oxford (1970)
33. Korovin, K., Sticksel, C.: iProver-eq: an instantiation-based theorem prover with equality. In: Giesl, J., Hähnle, R. (eds.) IJCAR 2010. LNCS, vol. 6173, pp. 196–202. Springer, Heidelberg (2010)
34. Lankford, D.: Canonical algebraic simplification in computational logic. Technical report, Memo ATP-25, University of Texas, Austin, Texas (1975)
35. Lee, S.J., Plaisted, D.: Eliminating duplication with the hyper-linking strategy. J. Autom. Reasoning **9**(1), 25–42 (1992)
36. Loveland, D.: A simplified format for the model elimination procedure. J. ACM **16**, 349–363 (1969)
37. Manthey, R., Bry, F.: SATCHMO: a theorem prover implemented in prolog. In: Lusk, E., Overbeek, R. (eds.) 9th International Conference on Automated Deduction. LNCS, vol. 310, pp. 415–434. Springer, Heidelberg (1988)
38. McCune, W.W.: Otter 3.0 reference manual and guide. Technical report, ANL-94/6, Argonne National Laboratory, Argonne, IL (1994)
39. Musser, D.: On proving inductive properties of abstract data types. In: Proceedings of the 7th ACM Symposium on Principles of Programming Languages, pp. 154–162. ACM Press (1980)

40. O'Connor, J.J., Robertson, E.F.: Jacques Herbrand. http://www-history.mcs. st-andrews.ac.uk/Biographies/Herbrand.html. Accessed March 2015
41. Otten, J.: *PleanCoP 2.0* and *ileanCoP 1.2*: high performance lean theorem proving in classical and intuitionistic logic (system descriptions). In: Armando, A., Baumgartner, P., Dowek, G. (eds.) IJCAR 2008. LNCS (LNAI), vol. 5195, pp. 283–291. Springer, Heidelberg (2008)
42. Paramasivam, M., Plaisted, D.: A replacement rule theorem prover. J. Autom. Reasoning 18(2), 221–226 (1997)
43. Pelletier, F.J.: Seventy-five problems for testing automatic theorem provers. J. Autom. Reasoning 2, 191–216 (1986)
44. Plaisted, D.: A recursively defined ordering for proving termination of term rewriting systems. Technical report, R-78-943, University of Illinois at Urbana-Champaign, Urbana, IL (1978)
45. Plaisted, D.: Well-founded orderings for proving termination of systems of rewrite rules. Technical report, R-78-932, University of Illinois at Urbana-Champaign, Urbana, IL (1978)
46. Plaisted, D.: An efficient relevance criterion for mechanical theorem proving. In: Proceedings of the 1st Annual National Conference on Artificial Intelligence, pp. 79–83. AAAI Press (1980)
47. Plaisted, D.: A simplified problem reduction format. Artif. Intell. 18(2), 227–261 (1982)
48. Plaisted, D., Greenbaum, S.: A structure-preserving clause form translation. J. Symb. Comput. 2, 293–304 (1986)
49. Plaisted, D., Zhu, Y.: The Efficiency of Theorem Proving Strategies: A Comparative and Asymptotic Analysis. Vieweg, Wiesbaden (1997)
50. Plaisted, D., Yahya, A.: A relevance restriction strategy for automated deduction. Artif. Intell. 144(1–2), 59–93 (2003)
51. Plaisted, D.A.: Non-Horn clause logic programming without contrapositives. J. Autom. Reasoning 4(3), 287–325 (1988)
52. Plaisted, D.A., Zhu, Y.: Ordered semantic hyper linking. J. Autom. Reasoning 25(3), 167–217 (2000)
53. Reif, W., Schellhorn, G.: Theorem proving in large theories. In: Bibel, W., Schmitt, P.H. (eds.) Automated Deduction - A Basis for Applications. Applied Logic Series, vol. 10, pp. 225–241. Springer, Heidelberg (1998)
54. Riazanov, A., Voronkov, A.: Vampire. In: Ganzinger, H. (ed.) CADE 1999. LNCS (LNAI), vol. 1632, pp. 292–296. Springer, Heidelberg (1999)
55. Robinson, A., Voronkov, A. (eds.): Handbook of Automated Reasoning. Elsevier, Amsterdam (2001)
56. Robinson, J.: Automatic deduction with hyper-resolution. Int. J. Comput. Math. 1(3), 227–234 (1965)
57. Robinson, J.: A machine-oriented logic based on the resolution principle. J. ACM 12(1), 23–41 (1965)
58. Schulz, S.: System description: E 1.8. In: McMillan, K., Middeldorp, A., Voronkov, A. (eds.) LPAR-19 2013. LNCS, vol. 8312, pp. 735–743. Springer, Heidelberg (2013)
59. Siekmann, J., Wrightson, G. (eds.): Automation of Reasoning 1: Classical Papers on Computational Logic 1957–1966. Symbolic Computation. Springer, Heidelberg (1983)
60. Siekmann, J., Wrightson, G. (eds.): Automation of Reasoning 2: Classical Papers on Computational Logic 1967–1970. Symbolic Computation. Springer, Heidelberg (1983)

61. Letz, R., Stenz, G.: DCTP - a disconnection calculus theorem prover - system abstract. In: Goré, R.P., Leitsch, A., Nipkow, T. (eds.) IJCAR 2001. LNCS (LNAI), vol. 2083, pp. 381–385. Springer, Heidelberg (2001)
62. Stickel, M.: A unification algorithm for associative-commutative functions. J. ACM **28**(3), 423–434 (1981)
63. Stickel, M.: A prolog technology theorem prover: implementation by an extended prolog compiler. J. Autom. Reasoning **4**(4), 353–380 (1988)
64. Sutcliffe, G.: The TPTP problem library and associated infrastructure - the FOF and CNF parts, v3.5.0. J. Autom. Reasoning **43**(4), 337–362 (2009)
65. Sutcliffe, G., Puzis, Y.: *SRASS* - a semantic relevance axiom selection system. In: Pfenning, F. (ed.) CADE 2007. LNCS (LNAI), vol. 4603, pp. 295–310. Springer, Heidelberg (2007)
66. Weidenbach, C., Dimova, D., Fietzke, A., Kumar, R., Suda, M., Wischnewski, P.: SPASS version 3.5. In: Schmidt, R.A. (ed.) CADE-22. LNCS, vol. 5663, pp. 140–145. Springer, Heidelberg (2009)
67. Wikipedia: Hilbert's program – Wikipedia, the free encyclopedia (2014). http://en.wikipedia.org/wiki/Hilbert'sprogram. Accessed March 2015
68. Wos, L., Carson, D., Robinson, G.: The unit preference strategy in theorem proving. In: Proceedings of the Fall Joint Computer Conference, Part I. AFIPS Conference Proceedings, vol. 26, pp. 615–621 (1964)
69. Zak, R.: Hilbert's program. In: Zalta, E. (ed.) The Stanford Encyclopedia of Philosophy, Spring 2015 edn. (2015). Accessed March 2015
70. Zalta, E.N.: Gottlob Frege. In: Zalta, E. (ed.) The Stanford Encyclopedia of Philosophy. Fall 2014 edn. (2014). Accessed March 2015
71. Zamov, N.: Maslov's inverse method and decidable classes. Ann. Pure Appl. Logic **42**, 165–194 (1989)

Stumbling Around in the Dark: Lessons from Everyday Mathematics

Ursula Martin[✉]

University of Oxford, Oxford, UK
Ursula.Martin@cs.ox.ac.uk

Abstract. The growing use of the internet for collaboration, and of numeric and symbolic software to perform calculations it is impossible to do by hand, not only augment the capabilities of mathematicians, but also afford new ways of observing what they do. In this essay we look at four case studies to see what we can learn about the everyday practice of mathematics: the *polymath* experiments for the collaborative production of mathematics, which tell us about mathematicians attitudes to working together in public; the *minipolymath* experiments in the same vein, from which we can examine in finer grained detail the kinds of activities that go on in developing a proof; the mathematical questions and answers in *math overflow*, which tell us about mathematical-research-in-the-small; and finally the role of computer algebra, in particular the GAP system, in the production of mathematics. We conclude with perspectives on the role of computational logic.

1 Introduction

The popular image of a mathematician is of a lone genius (probably young, male and addicted to coffee) having a brilliant idea that solves a very hard problem. This notion has been fuelled by books such as Hadamard's 'Psychology of invention in the mathematical field', based on interviews with forty or so mathematicians, and dwelling on an almost mystical process of creativity. Journalistic presentations of famous mathematicians continue the theme - for example picking out Andrew Wiles's remark on his proof of Fermat's conjecture "And sometimes I realized that nothing that had ever been done before was any use at all. Then I just had to find something completely new; it's a mystery where that comes from."

However Wiles also points out that inspiration is by no means the whole story, stressing the sheer slog of research mathematics:

> "I used to come up to my study, and start trying to find patterns. I tried doing calculations which explain some little piece of mathematics. I tried to fit it in with some previous broad conceptual understanding of some part of mathematics that would clarify the particular problem I was thinking about. Sometimes that would involve going and looking it up in a book to see how it's done there. Sometimes it was a question of modifying things a bit, doing a little extra calculation"

A.P. Felty and A. Middeldorp (Eds.): CADE-25, LNAI 9195, pp. 29–51, 2015.
DOI: 10.1007/978-3-319-21401-6_2

and draws attention to the lengthy periods of hard work between the moments of clarity:

> "Perhaps I can best describe my experience of doing mathematics in terms of a journey through a dark unexplored mansion. You enter the first room of the mansion and it's completely dark. You stumble around bumping into the furniture, but gradually you learn where each piece of furniture is. Finally, after six months or so, you find the light switch, you turn it on, and suddenly it's all illuminated. You can see exactly where you were. Then you move into the next room and spend another six months in the dark. So each of these breakthroughs, while sometimes they're momentary, sometimes over a period of a day or two, they are the culmination of – and couldn't exist without – the many months of stumbling around in the dark that precede them."

Cedric Villani also unpacks the myth in his recent book, which gives an account (incorporating emails, Manga comics, and his love of French cheese), of the work that won him the Field's medal, and gives a gripping picture the mathematician's days trying out ideas that don't quite work, or turn out to be wrong, or right but not useful, and of the exhilaration of "the miracle" when "everything seemed to fit together as if by magic". For Wiles, the process was essentially a solitary one, but for Villani "One of the greatest misconceptions about mathematics is that it's a solitary activity in which you work with your pen, alone, in a room. But in fact, it's a very social activity. You constantly seek inspiration in discussions and encounters and randomness and chance and so on."

Wiles and Villani both talk about the importance of a broad view of mathematics, not so much for precise and formalisable correspondences, but in the hope that ideas that have worked in one area will stimulate new ways of looking at another. Edward Frenkel in "Love and Math", his recent popular account his work on the Langlands Program, draws attention to the role of the past literature: "It often happens like this. One proves a theorem, others verify it, new advances in the field are made based on the new result, but the true understanding of its meaning might take years or decades".

In a lecture in 2012 mathematician Michael Atiyah (who also remarked in the 1990's that too much emphasis was placed on correctness, and mathematics needed a more "buccaneering" approach) pointed to the importance of errors in developing understanding: "I make mistakes all the time... I published a theorem in topology. I didn't know why the proof worked, I didn't understand why the theorem was true. This worried me. Years later we generalised it — we looked at not just finite groups, but Lie groups. By the time we'd built up a framework, the theorem was obvious. The original theorem was a special case of this. We got a beautiful theorem and proof."

The personal accounts of outstanding mathematicians are complemented by the insights of ethnographers into the workaday worlds of less exalted individuals. Barany and Mackenzie, using the methods and language of sociology to look more keenly at this process of "stumbling around in the dark", observe that "the formal rigor at the heart of mathematical order becomes indissociable from the 'chalk in hand' character of routine mathematical work. We call attention to the vast labor of decoding, translating, and transmaterializing official texts without which advanced mathematics could not proceed. More than that, we suggest that these putatively passive substrates of mathematical knowledge and practice instead embody potent resources and constraints that combine to shape mathematical research in innumerable ways."

Barany and Mackenzie observed mathematicians in their offices and in front of blackboards, but the growing use of the internet for collaboration, and numeric and symbolic software to perform calculations it is impossible to do by hand, not only augment the capabilities of mathematicians, in particular by enabling collaboration, but also provide new ways of observing what they do.

In this paper we look at four case studies to see what we can learn about the everyday practice of mathematics, so as to shed new light the process of "stumbling around in the dark".

Two of the case studies are rooted in the mathematical area of group theory. A group is, roughly speaking, the set of symmetries of an object, and the field emerged in the nineteenth century, through the systematic study of roots of equations triggered by the work of Galois. It continues to provide a surprising and challenging abstract domain which underlies other parts of mathematics, such as number theory and topology, with practical applications in areas such as cryptography and physics. Its greatest intellectual achievement is the classification of finite simple groups, the basic "building blocks" of all finite groups. Daniel Gorenstein, one of the prime movers in coordinating the effort, estimates the proof variously as occupying between 5,000 and 15,000 journal pages over 30 years. Sociologist Alma Steingart describes the endeavour as "the largest and most unwieldy mathematical collaboration in recent history", and points to the flexible notion of the idea of proof over the life of the collaboration (which explains the varied estimates of the length of the proof). She points out that the sheer volume of material meant that only one or two individuals were believed to have the knowledge to understand and check the proof, or to understand it well enough to fix the 'local errors' that were still believed to be present. The field has a well-developed tradition of computer support and online resources, in particular early heroic endeavours which constructed so called "sporadic" simple groups by constructing certain matrices over finite fields. Today widely used software such as GAP incorporates many specialist algorithms, and exhaustive online data resources, such as the ATLAS of data about simple groups, capture information about these complex objects.

Our four studies involve the *polymath* experiments for the collaborative production of mathematics, which tell us about mathematicians attitudes to working together in public; the *minipolymath* experiments in the same vein, from which we can examine in finer grained detail the kinds of activities that go on in producing a proof; the mathematical questions and answers in *mathoverflow* , which tell us about mathematical-research -in-the-small; and finally the role of computer algebra, in particular the GAP system, in the production of mathematics. We conclude with remarks on the role of computational logic.

2   The Power of Collaboration: *polymath*

Timothy Gowers was awarded a Fields Medal in 1998 for work combining functional analysis and combinatorics, in particular his proof of Szemerdi's theorem. Gowers has characterised himself as a problem-solver rather than a theory-builder, drawing attention to the importance of problem solvers and problem solving in understanding and developing broad connections and analogies between topics not yet amenable to precise unifying theories. He writes articulately on his blog about many topics connected with mathematics, education and open science, and used this forum to launch his experiments in online collaborative proof which he called *"polymath"*. In a blog post on 27th January 2009 he asked "Is massively collaborative mathematics possible", suggesting that "If a large group of mathematicians could connect their brains efficiently, they could perhaps solve problems very efficiently as well.". Ground rules were formulated, designed to encourage massively collaborative mathematics both in the sense of involving as many people as possible: "we welcome all visitors, regardless of mathematical level, to contribute to active polymath projects by commenting on the threads"; and having a high degree of interaction and rapid exchange of informal ideas: "It's OK for a mathematical thought to be tentative, incomplete, or even incorrect". and "An ideal polymath research comment should represent a 'quantum of progress'."

The post attracted 203 comments from around the globe, exploring philosophical and practical aspects of working together on a blog to solve problems, and a few days later Gowers launched the first experiment. The problem chosen was to find a new proof of the density version of the "Hales Jewett Theorem", replacing the previously known very technical proof with a more accessible combinatorial argument which, it was hoped, would also open the door to generalisations of the result. Over the next seven weeks, 27 people contributed around 800 comments - around 170,000 words in all - with the contributors ranging from high-school teacher Jason Dyer to Gowers's fellow Fields Medallist Terry Tao. On March 10, 2009 Gowers was able to announce a new combinatorial proof of the result, writing "If this were a conventional way of producing mathematics, then it would be premature to make such an announcement - one would wait until the proof was completely written up with every single i dotted and every t crossed - but this is blog maths and we're free to make up conventions as we go along."

The result was written up as a conventional journal paper, with the author given as "D H J Polymath" - identifying the actual contributors requires some detective work on the blog - and published on the arxiv in 2009, and in the Annals of Mathematics in 2012. The journal version explains the process "Before we start working towards the proof of the theorem, we would like briefly to mention that it was proved in a rather unusual "open source" way, which is why it is being published under a pseudonym. The work was carried out by several researchers, who wrote their thoughts, as they had them, in the form of blog comments. Anybody who wanted to could participate, and at all stages of the process the comments were fully open to anybody who was interested".

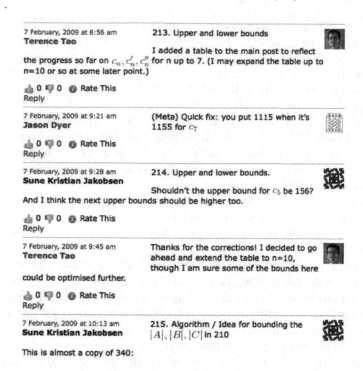

Fig. 1. An extract from the *polymath* blog for the proof of the Density Hales Jewett theorem

A typical extract from the blog (Fig. 1) shows the style of interaction. Participants, in line with the ground rules, were encouraged to present their ideas in an accessible way, to put forward partial ideas that might be wrong – "better to have had five stupid ideas than no ideas at all" – to test out ideas on other participants before doing substantial work on them, and to treat other participants with respect. As the volume of comments and ideas grew, it became apparent that the blog structure made it hard for readers to extract the thread of the argument and keep up with what was going on, without having to digest everything that had been previously posted, and in future experiments a leader took on the task of drawing together the threads from time to time, identifying the most appropriate next direction, and restarting the discussion with a substantial new blog post.

By 2015 there had been nine endeavours in the *polymath* sequence, and a number of others in similar style. Not all had achieved publishable results, with some petering out through lack of participation, but all have left the record of their partial achievements online for others to see and learn from - a marked contrast to partial proof attempts that would normally end up in a waste-basket.

The most recent experiment, *polymath* 8, was motivated by Yitang Zhang's proof of a result about bounded gaps between primes. The twin primes conjecture states that there are infinitely many pairs of primes that differ by 2: $3, 5, \ldots 11, 13$

and so on. Zhang proved that there is a number K such that infinitely many pairs of primes differ by at most K, and showed that K is less than 70,000,000. After various discussions on other blogs, Tao formally launched the project, to improve the bound on K, on 13th June 2013. The first part of the project, *polymath* 8a, concluded with a bound of 4,680, and a research paper, also put together collaboratively, appeared on the arxiv in February 2014. The second part, *polymath* 8b, combined this with techniques independently developed in parallel by James Maynard, to reach a bound of 246, with a research paper appearing on the arxiv in July 2014. The participants also used Tao's blog to seek input for a retrospective paper reflecting on the experience, which appeared in the arxiv in September 2014.

One immediate concern was the scoping of the enquiry so as to not to intimidate or hamper individuals working on their own on this hot topic: it was felt that this was more than countered by providing a resource for the mathematical community that would capture progress, and provide a way to pull together what would otherwise be many independent tweaks. The work was well suited to the *polymath* approach: the combination of Tao's leadership and the timeliness of the problem made it easy to recruit participants; the bound provided an obvious metric of progress and maintained momentum; and it naturally fell into five components forming what Tao called a "factory production line". The collaborative approach allowed people to learn new material, and brought rapid sharing of expertise, in particular knowledge of the literature, and access to computational and software skills.

Tao himself explained how he was drawn into the project by the ease of making simple improvements to Zhang's bound, even though it interrupted another big project (a piece of work that he expects to take some years), and summed up by saying "All in all, it was an exhausting and unpredictable experience, but also a highly thrilling and rewarding one." The time commitment was indeed intense - for example, a typical thread *"polymath* 8b, II: Optimising the variational problem and the sieve" started on 22 November 2013, and ran for just over two weeks until 8th December. The initial post by Tao runs to about 4000 words - it is followed by 129 posts, of which 36, or just under a third, are also by Tao.

Tao and other participants were motivated above all by the kudos of solving a high-profile problem, in a way that was unlikely had they worked individually, but also by the excitement of the project, and the enthusiasm of the participants and the wider community. They reported enjoying working in this new way, especially the opportunity to work alongside research stars, the friendliness of the other participants, and their tolerance of errors, and the way in which the problem itself and the *polymath* format provided the incentive of frequent incremental progress, in a way not typical of solo working.

Participants needed to balance the incentives for participation against other concerns. Chief among these was the time commitment: participants reported the need for intense concentration and focus, with some working on it at a "furious pace" for several months; some feeling that the time required to grasp everything that was happening on the blog make *polymath* collaborations more, rather than less, time consuming than traditional individual work or small-group collaboration; and some feeling that the fast pace was deterring participants whose working style was slower and more reflective.

Pure mathematicians typically produce one or two journal papers a year, so that, particularly for those who do not yet have established positions, there will be concerns that a substantial investment of time in *polymath* might damage their publication record. While such a time commitment would normally be worth the risk for a high-profile problem that has the likely reward of a good publication, the benefits are less clear-cut when the paper is authored under a group pseudonym (D H J Polymath), with the list of participants given in a linked wiki. As a participant remarked *"polymath* was a risk for those who did not have tenure". On the other hand, in a fast moving area, participants may feel that incorporating their ideas into the collective allows them to make a contribution that they would not have achieved with solo work, or that engaging in this way is better than being beaten to a result and getting no credit, especially if participation in a widely-read blog is already adding to their external reputation.

An additional risk for those worried about their reputation can be that mistakes are exposed for ever in a public forum: pre-tenure mathematician Pace Nielsen was surprised that people were "impressed with my bravery" and would advise considering this issue before taking part. Rising star James Maynard observed: "It was very unusual for me to work in such a large group and so publicly - one really needed to lose inhibitions and be willing to post ideas that were not fully formed (and potentially wrong!) online for everyone to see."

Those reading the *polymath* 8 sites went well beyond the experts - with an audience appreciating the chance to see how mathematics was done behind the scenes, or as Tao put it "How the sausage is made". At its height it was getting three thousand hits a day, and even readers who knew little mathematics reported the excitement of checking regularly and watching the bounds go down. All the members of a class of number theory students at a summer school on Bounded Gaps admitted to following *polymath*. Perhaps typical was Andrew Roberts, an undergraduate who thanked the organisers for such an educational resource and reported "reading the posts and following the 'leader-board' felt a lot like an academic spectator sport. It was surreal, a bit like watching a piece of history as it occurred. It made the mathematics feel much more alive and social, rather than just coming from a textbook. I don't think us undergrads often get the chance to peek behind closed doors and watch professional mathematicians 'in the wild' like this, so from a career standpoint, it was illuminating." David Roberts, an Australian educator who used *polymath* in his classes to show students how the things they were learning were being used in cutting-edge research, reported "For me personally it felt like being able to sneak into the garage and watch a high-performance engine being built up from scratch; something I could never do, but could appreciate the end result, and admire the process." The good manners remarked upon by the expert participants extended to less-well informed users, with questions and comments from non-experts generally getting a polite response, often from Tao himself, and a few more outlandish comments, such as claims of a simple proof, being ignored except for a plethora of down-votes.

The experiments have attracted widespread attention, in academia and beyond. Gowers had worked closely with physicist Michael Nielson in designing the wiki and blog structure to support *polymath*, and in an article in Nature in 2009 the pair reflected on its wider implications, a theme developed further in Nielsen's 2012 book, Reinventing Discovery, and picked up by researchers in social and computer science analysing the broader phenomenon of open science enabled by the internet.

Like the participants, the analysts remarked on the value of the *polymath* blogs for capturing the records of how mathematics is done, the kinds of thinking that goes into the production of a proof, such as experimenting with examples, computations and concepts, and showing the dead ends and blind alleys. As Gowers and Nielson put it, "Who would have guessed that the working record of a mathematical project would read like a thriller?"

Although *polymath* is often described as "crowdsourced science", the crowd is a remarkably small and expert one. The analogy has often been drawn with open source software projects - however these are typically organised in a much more modular and top down fashion than in possible in developing a mathematical proof, where many ideas and strands will be interwoven in a manner, as Nielsen comments, much more akin to a novel.

Research on collaboration and crowdsourcing carried out by psychologists, cognitive scientists and computer scientists helps explain the success of *polymath* and other attempts at open collaboration in mathematics. Mathematicians have well-established shared standards for exposition and argument, making it easy to resolve disputes. As the proof develops, the blog provides a shared cognitive space and short term working memory. The ground rules allow for a dynamic division of labour, and encourage a breakdown into smaller subtasks thus reducing barriers to entry and increasing diversity. At the same time, presenting the whole activity to readers, rather than in a more rigidly structured and compartmentalised way, allows more scope for serendipity and conversation across threads. Gowers gives an example of one contributor developing ideas in a domain which he was not familiar with (ergodic theory), and another who translated these ideas into one that he was familiar with (combinatorics), thus affecting his own line of reasoning.

A striking aspect of *polymath* is that senior figures in the field are prepared to try such a bold experiment, to think though clearly for themselves what the requirements are, and to take a "user centred" view of the design, based on their understanding of their own user community. For example it was suggested that participants might use a platform such as github, designed for collaborative working and version control, to make the final stage, collaborating on a paper, more straightforward. Tao responded "One thing I worry about is that if we use any form of technology more complicated than a blog comment box, we might lose some of the participants who might be turned off by the learning curve required."

These working records allow the analysis of what is involved in the creation of a proof, which we explore in the following section.

3   Examples, Conjectures, Concepts and Proofs:
minipolymath

The *minipolymath* series applied the *polymath* model to problems drawn from the International Mathematical Olympiad (IMO), a competition for national teams of high school students. Tao and Gowers are among successful Olympiad contestants who have gone on to win Fields medals. Using short but challenging high-school level problems allowed for a much greater range of participants, and for greater experimentation with the format, as problems typically took hours rather than months to be solved. This windmill-inspired problem was composed by Geoff Smith for the 2011 IMO, held in the Netherlands.

The *windmill* Problem. Let S be a finite set of at least two points in the plane. Assume that no three points of S are collinear. A *windmill* is a process that starts with a line l going through a single point $P \in S$. The line rotates clockwise about the pivot P until the first time that the line meets some other point Q belonging to S. This point Q takes over as the new pivot, and the line now rotates clockwise about Q, until it next meets a point of S. This process continues indefinitely.

Show that we can choose a point P in S and a line l going through P such that the resulting windmill uses each point of S as a pivot infinitely many times. (Tao, 8:00 pm)

Tao posted the problem as a *minipolymath* challenge at 8 pm on July 19th, 2011 a few days after the competition. Interest was immediate, and seventy four minutes later the participants had found a solution and by 9.50 pm, when Tao called a halt there were 147 comments on the blog, over 27 threads. To investigate this we developed a typology of comments as below (nine comments fell into two categories, and one fell into three). Figure 2 shows the proportion of each category.

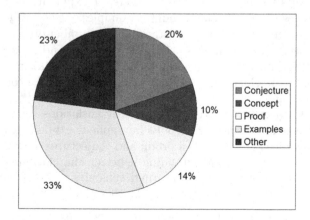

Fig. 2. The proportions of comments which concerned conjectures, concepts, proofs, examples, and other.

Exemplars of the Typology of Comments

Concept. Since the points are in general position, you could define "the wheel of p", w(p) to be radial sequence of all the other points p!=p around p. Then, every transition from a point p to q will "set the windmill in a particular spot" in q. This device tries to clarify that the new point in a windmill sequence depends (only) on the two previous points of the sequence. (Anonymous, 8:41 pm)

Example. If the points form a convex polygon, it is easy. (Anonymous, 8:08 pm)

Conjecture. One can start with any point (since every point of S should be pivot infinitely often), the direction of line that one starts with however matters! (Anonymous, 8:19 pm)

Conjecture. Perhaps even the line does not matter! Is it possible to prove that any point and any line will do? (Anonymous, 8:31 pm)

Proof. The first point and line P_0, l_0 cannot be chosen so that P_0 is on the boundary of the convex hull of S and l_0 picks out an adjacent point on the convex hull. Maybe the strategy should be to take out the convex hull of S from consideration; follow it up by induction on removing successive convex hulls. (Haggai Nuchi, 8:08 pm)

Example and Conjecture. Can someone give me *any* other example where the windmill cycles without visiting all the points? The only one I can come up with is: loop over the convex hull of S. (Srivatsan Narayanan, 9:08 pm)

Other. Got it! Kind of like a turn number in topology. Thanks! :) (Gal, 9:50 pm)

Analysing the typology in more depth we see that:

- *Examples* played a key role in the discussion, forming around a third of the comments. We saw some supporting examples to explain or justify conjectures, concepts and requests for clarification; but most of the examples concerned counter examples to the conjectures of others, or explanations to why these counter examples were not in fact, counter examples. As an IMO problem, the question was assumed to be correctly stated, but a number of 'counterexample' comments concerned participants attempts to understand it, with a number of people initially misled by the windmill analogy into thinking of the rotating line as a half-line, in which case the result does not hold, and a counterexample can indeed be found.
- *Conjectures.* Conjectures concerned possible translations to other domains which would provide results that could be applied; extensions of the initial problem; sub-conjectures towards a proof; and conjectures of properties of the windmill process aimed at understanding it better and clarifying thinking.
- *Concepts.* One class of concepts concerned analogies to everyday objects: as well as the somewhat misleading windmills, these included

  "We could perhaps consider "layers" of convex hulls (polygons) .. like peeling off an onion. If our line doesn't start at the "core" (innermost) polygon then I feel it'll get stuck in the upper layers and never reach the core." (Varun, 8:27 pm)

Notice the author's use of apostrophes to stress that this is an analogy. Other analogies were to related mathematical objects, so as to provide ideas or inspiration, rather than affording an exact translation so that results from the new domain could be immediately applied. An important development was the emergence of the idea of talking about the "direction" of a line, leading to the observation, important for the final proof, that the number of points on each side of the line stays constant throughout the windmill process. It was noticeable how new concepts rapidly spread among the participants, even before they were precisely pinned down, enabling communication.

- *Proof.* Twenty one comments concerned a proof. Fourteen were about possible proof strategies, one was clarification of the strategy, one was carrying out a plan and five were about identifying which properties were relevant to the proof. Three strategies were discussed: one by induction and two involving translation to analogous domains. Within this the proof itself occupies a mere 16 comments, by 7 participants (three comments are "Anonymous" but from the context appear to be the same person).
- *Other.* The 34 comments classified as "other" include clarification of duplicate comments, explanations of a claim, and friendly interjections. Some of these are mathematically interesting, guiding the direction of the discussion, while others are simply courtesy comments. All play an important role, along with smiley faces, exclamation marks, and so on, in creating and maintaining an environment which is friendly, collaborative, informal and polite.

4   Questions and Answers: *mathoverflow*

Discussion fora for research mathematics have evolved from the early newsnet newsgroups to modern systems based on the *stackexchange* architecture, which allow rapid informal interaction and problem solving. In three years *mathoverflow.net* has hosted 61,000 conversations and accumulated over 10,000 users, of whom about 500 are active in any month. The highly technical nature of research mathematics means that this is not currently an endeavour accessible to the public at large: a separate site *math.stackexchange.com* is a broader question and answer site "for people studying math at any level and professionals in related fields".

Within *mathoverflow* house rules give detailed guidance, and stress clarity, precision, and asking questions with a clear answer. Moderation is fairly tight, and some complain it constrains discussion. The design of such systems has been subject to considerable analysis by the designers and users, and *meta.mathoverflow* contains many reflective discussions. A key element of the success of the system is user ratings of questions and responses, which combine to form reputation ratings for users. These have been studied by psychologists

Tausczik and Pennebaker who concluded that *mathoverflow* reputations offline (assessed by numbers of papers published) and in *mathoverflow* were consistently and independently related to the *mathoverflow* ratings of authors' submissions, and that while more experienced contributors were more likely to be motivated by a desire to help others, all users were motivated by building their *mathoverflow* reputation.

We studied the mathematical content of *mathoverflow* questions and responses, choosing the subdomain of group theory so as to align with related work on GAP: at the time of writing (April 2015) around 3,500 of the *mathoverflow* questions are tagged "group theory", putting it in the top 5 topic-specific tags.

We analysed a sample of 100 questions drawn from April 2011 and July 2010 to obtain a spread and developed a typology:

Conjecture 36 % — asks if a mathematical statement is true. May ask directly "Is it true that" or ask under what circumstances a statement is true.

What is this 28 % — describes a mathematical object or phenomenon and asks what is known about it.

Example 14 % — asks for examples of a phenomenon or an object with particular properties.

Formula 5 % — ask for an explicit formula or computation technique.

Different Proof 5 % — asks if there is an alternative to a known proof. In particular, since our sample concerns the field of group theory, a number of questions concern whether a certain result can be proved without recourse to the classification of finite simple groups.

Reference 4 % — asks for a reference for something the questioner believes to be already in the literature.

Perplexed 3 % — ask for help in understanding a phenomenon or difficulty. A typical question in this area might concern why accounts from two different sources (for example Wikipedia and a published paper) seem to contradict each other.

Motivation 3 % — asks for motivation or background. A typical question might ask why something is true or interesting, or has been approached historically in a particular way.

Other 2 % — closed by moderators as out of scope, duplicates etc.

We also looked for broad phenomena in the structure of the successful responses. *mathoverflow* is very effective, with 90 % of our sample successful, in that they received responses that the questioner flagged as an "answer", of which 78 % were reasonable answers to the original question, and a further 12 % were partial or helpful responses that moved knowledge forward in some way. The high success rate suggests that, of the infinity of possible mathematical questions, questioners are becoming adept at choosing those for *mathoverflow* that are amenable to its approach. The questions and the answers build upon an assumption of a high level of shared background knowledge, perhaps at the level of a PhD in group theory.

The usual presentation of mathematics in research papers is in a standardised precise and rigorous style: for example, the response to a conjecture is either a counterexample, or a proof of a corresponding theorem, structured by means of intermediate definitions, theorems and proofs. By contrast, the typical response to a *mathoverflow* question, whatever the category, is a discussion presenting facts or short chains of inference that are relevant to the question, but may not answer it directly. The facts and inference steps are justified by reference to the literature, or to mathematical knowledge that the responder expects the other participants to have. Thus in modelling a *mathoverflow* discussion, we might think of each user as associated to a collection of facts and short inferences from them, with the outcome of the discussion being that combining the facts known to different users has allowed new inferences. Thus the power of *mathoverflow* comes from developing collective intelligence through sharing information and understanding.

In 56 % of the responses we found citations to the literature. This includes both finding papers that questioners were unaware of, and extracting results that are not explicit in the paper, but are straightforward (at least to experts), consequences of the material it contains. For example, the observation needed from the paper may be a consequence of an intermediate result, or a property of an example which was presented in the paper for other purposes. In 34 % of the responses, explicit examples of particular groups were given, as evidence for, or counter examples to, conjectures. The role of examples in mathematical practice, for example as evidence to refine conjectures, was explored by Lakatos: we return to this below.

In addition *mathoverflow* captures information known to individuals but not normally recorded in the research literature: for example unpublished material, motivation, explanations as to why particular approaches do not work or have been abandoned, and intuition about conjectures. The presentation is often speculative and informal, a style which would have no place in a research paper, reinforced by conversational devices that are accepting of error and invite challenge, such as "I may be wrong but...", "This isn't quite right, but roughly speaking...". Where errors are spotted, either by the person who made them or by others, the style is to politely accept and correct them: corrected errors of this kind were found in 37 % of our sample (we looked at *discussions* of error: we have no idea how many actual errors there are).

It is perhaps worth commenting on things that we did not see in our sample of technical questions tagged "group theory" in *mathoverflow*. In developing "new" mathematics considerable effort is put into the formation of new concepts and definitions: we saw little of this in *mathoverflow*, where questions are by and large focussed on extending or refining existing knowledge and theories. A preliminary scan suggests these are not present in other technical areas of *mathoverflow* either.

We see little serious disagreement in our *mathoverflow* sample: perhaps partly because of the effect of the "house rules", but also because of the style of discussion, which is based on evidence from the shared research background and knowledge of the participants: there is more debate in *meta.mathoverflow*, which has a broader range of non-technical questions about the development of the discipline and so on.

5   Everyday Calculation: GAP

GAP (Groups, Algorithms and Programming) is a substantial open-source computer algebra system, supporting research and teaching in computational group and semigroup theory, discrete mathematics, combinatorics and finite fields, and the applications of these techniques in areas such as cryptography and physics. It has been developed over the past 20 years or so by teams led initially from the University of Aachen, and currently from the University of St Andrews. According to google scholar it has been cited in around 3,500 research papers: the GAP making list has over a thousand members.

GAP provides well documented implementations of algorithms covering techniques for identifying, and computing in, finite and infinite groups defined by permutations, generators and relations, and matrices over finite fields. It also supports a variety of standard data-sets: for example the 52×10^{12} semigroups with up to 10 elements. It currently comprises over 0.6 million lines of code, with a further 1.1 million in over 100 contributed packages. Considerable effort is taken to ensure that GAP packages and datasets can be treated as objects in the scholarly ecosystem through establishing refereeing standards, citation criteria and so on.

Alongside the efforts one would expect of running a large open source project - a source code repository, mailing lists and archives, centralized testing services, issue tracker, release management, and a comprehensive website - the activity of the core GAP developers is driven by extending the power and reach of the system. Thus extensive efforts are being put into techniques for increasing the efficiency of algorithms handling matrices, permutations, finite fields and the like, for example by developing new data representations, and parallelising GAP over multicores in ways that do not increase complexity for the user. Extending the reach of GAP includes developing new theories and algorithms, and supporting these with high quality well-documented implementations: for example recent work has included devising computational methods for semigroups and new techniques for computing minimal polynomials for matrices over finite fields.

Research users of GAP typically use it to experiment with conjectures and theories. Whereas pencil and paper calculation restricts investigations to small and atypical groups, the ready availability in GAP of a plethora of examples, and the ease of computing with groups of large size, makes it possible to develop, explore and refine hypotheses, examples and possible counter examples, before proceeding to decide exactly what theorems to prove, and developing the proofs in a conventional journal paper. For example, we reviewed the 49 papers in google scholar which cited GAP Version 4.7. 5, 2014. A number of items were eliminated: duplicates; out of scope, such as lecture slides; and papers that did not appear to cite GAP, or cited it without mention in the text. The remaining 37 papers fell into six main groupings:

Explicit Computation as Part of a Proof, 25 % — These papers contained proofs that needed explicit and intricate calculation, carried out in GAP but difficult or impossible to do by hand. This arises particularly in theorems that depend on aspects of the classification of finite simple groups, or other results of a similar character, and hence require checking a statement for an explicitly given list of groups, each of which can be handled in GAP.

Examples and Counter examples, 25 % — These papers had used GAP to find or verify explicit examples of groups or other combinatorial objects: in some cases to illustrate a theorem, or as evidence for a conjecture; in others as counter-examples to a conjectured extension or variant of a result. Notice that GAP's built-in libraries of groups are often used to search for counter examples.

New Algorithms, 20 % — These were papers mainly devoted to the exposition of a new algorithm. In some cases these were supported by an explicit GAP implementation. In the rest the algorithm was more general than could be supported by GAP, but the paper contained a worked example, executed in GAP, for illustrative purposes.

Computations with Explicit Primes, 14 % — Groups whose number of elements is a power of a prime are the basic "building blocks" of finite group theory. As we have indicated, GAP can compute with fixed values of a prime number p, but is unable to handle statements of the form "For all primes p....". Many results of this form have generic proofs for "large enough" primes, while requiring a different proof for fixed small values of p, which can be computed by GAP, sometimes by making use of GAP's built in tables of particular families of groups. Thus for example Vaughan-Lee finds a formula for the number of groups of order p^8, with exponent p, where p is prime, $p > 7$. To complement this he computes a list of all $1,396,077$ groups of order 3^8, and these are made available in GAP.

Applications in Other Fields, 10 % — This included three papers in theoretical physics and gauge theory, all doing explicit computation using GAPs built in representations of Lie Algebras, and an example of how GAP has made group theory accessible to non-specialists. A further paper in education research presented symmetry and the Rubik's cube.

Other, 6 % — Two papers cited GAP as background material, one in describing how their own algorithmic approach went beyond it; and one mentioning that calculations in GAP had shown a claimed result in an earlier paper to be incorrect, and presenting a corrected statement and proof.

One factor encouraging the take-up of GAP in research is its widespread use in teaching mathematics, both at undergraduate level and as part of research training. For example a professor at Colorado State University in the USA writes "I have been using GAP for many years in my undergraduate and graduate classes in algebra and combinatorics [...]. I have found the system an indispensible tool for illustrating phenomena that are beyond simple pencil-and-paper methods [...]. It also has been most useful as a laboratory environment for students to investigate algebraic structures [...]. [T]his first-hand investigation gives students a much better understanding of what these algebraic structures are, and how their elements behave, than they would get by the traditional examples presented in a board-lecture situation." GAP enables an experimental approach, where students can explore examples and formulate and solve research questions of their own, developing their skills as mathematicians and building familiarity and confidence in using tools such as GAP later in their careers.

The examples above highlight the use of GAP in published mathematics research: supporting the traditional style of pure mathematics research paper through the use of computation as part of the proof of theorems; in the construction of examples and counter examples; and in algorithms research. As they are drawn from published papers they reflect what is documented in archival publications - much is omitted. Notice first that it would be unusual to have a proof that consisted entirely of a GAP computation; such a proof would probably not be considered deep enough to warrant journal publication, unless it was given context as part of a larger body of work which was felt to be significant. Thus in our sample one paper is devoted essentially entirely to computations of 9 pages of tables together with a narrative explanation of them: Harvey and Rayhaun build evidence for a connection between a particular modular form occurring in the work of Field's medal winner Borcherds, and the representation theory of the Thomson sporadic simple group. This is related to remarkable results linking sporadic simple groups, modular forms and conformal field theory, popularly labelled "Moonshine", which came about when Mackay observed in 1979 a connection between the "Monster" simple group and a certain modular form, through observing the role of the number 196883 in both.

However, sampling published papers to spot usage of GAP is also misleading, as it does not reflect significant and deep use of systems such as GAP in the process of doing mathematics, of exploring patterns, ideas and concepts, playing with examples and formulating and testing conjectures. We have heard, anecdotally, of mathematicians spending several months on calculations where the use of GAP was not even mentioned in the final paper: in one case a lengthy calculation involving the number 7 in GAP then informed published hand calculations which mimicked the computer calculation but with the 7 replaced by the variable p throughout; in another a lengthy computer search found a counterexample to a conjecture that could be readily described and shown to have the required properties without mentioning the computer search; in a third lengthy calculations in GAP were carried out to build evidence for a series of conjectures before time and effort was invested in a hand proof. Evidence for the use of computer algebra (though not GAP) in developing a proof can be drawn from the *polymath 8* project where several participants were using Maple, Mathematica and SAGE for experiment, calculation and search as a matter of course, alongside the main argument, from time to time reporting the evidence they had found, and comparing their results. The crowdsourcing approach clarified computational techniques and apparent variations in results, and provided added confidence in the approach.

The use of computer support in this way is in line with the manifesto of "Experimental Mathematics", ably presented in a series of books and papers by Bailey, Borwein and others. They articulate the possible uses of symbolic and numeric computation as:

(a) Gaining insight and intuition;
(b) Visualizing math principles;
(c) Discovering new relationships;
(d) Testing and especially falsifying conjectures;
(e) Exploring a possible result to see if it merits formal proof;
(f) Suggesting approaches for formal proof;
(g) Computing replacing lengthy hand derivations;
(h) Confirming analytically derived results.

The analysis of GAP papers above is consistent with the remark that while (a), (d) and (g) might all appear in current published papers, the rest are more likely to happen in the more speculative stage of the development of a proof. Bailey and Borwein argue that we need to think carefully about whether to allow a computation to be considered directly as a proof, and how to establish new standards for it to take its place in the literature. They go beyond the use of computation in support of traditional proof methodologies to assert:

> Robust, concrete and abstract, mathematical computation and inference on the scale now becoming possible should change the discourse about many matters mathematical. These include: what mathematics is, how we know something, how we persuade each other, what suffices as a proof, the infinite, mathematical discovery or invention, and other such issues.

While these are all worthy of debate, and indeed along with *polymath*, HOTT, and the work of Gonthier and Hales are stimulating increasing discussion in blogs and other scientific commentary, it is not clear that the practice of mathematics, as evidenced by mathematics publications, is yet changing. For example, a glance at the Volume 24, Issue 1, of the Journal of Experimental Mathematics, published in January 2015 finds 12 papers. All follow the standard mathematical style of presentation with theorems, proofs, examples and so on. Of these, 4 use computation as in (d) above - for testing and especially falsifying conjectures, by exhibiting a witness found through calculation or search. A further 5 are of type (a) and use computation to numerically evaluate or estimate a function, and hence conjecture an exact algebraic formula for it. Two start with computational experiments which stimulate a conjecture that all elements of a certain finite class of finite objects have a certain property, and then prove it by (computational) exhaustion: so these may be described as (a) followed by (d) One presents an algorithm plus a running example, so perhaps also (a). It would appear that all exhibit (g).

6   Learning from the Everyday

Gowers, Tao, Villani and Wiles are extraordinary mathematicians, which makes their reflections of the process of doing mathematics both fascinating and atypical. The case studies described above allow us to look at the everyday activities of more ordinary mathematicians. It also allows us to draw a number of conclusions related to the practice of mathematics, attitudes to innovations, and the possible deployment of computational logic systems.

Stumbling Around in the Dark. By looking at how mathematics is done - or as Tao puts it 'how the sausage is made' - we get a more detailed view of Wiles's 'stumbling around in the dark'. Our examples highlight the role of conjectures, concepts and examples in creating a proof. Interestingly, they provide an evidence base to challenge Lakatos's account of the development of proofs. To simplify somewhat, Lakatos presents a view of mathematical practice in which conjectures are subject to challenge through exhibiting examples, leading to modification of the hypothesis, or the concepts underlying it, with all the while progress towards a proof (of something, if not the original hypothesis) being maintained. While we certainly see this process at work in *polymath*, *minipolymath* and *mathoverflow* , this description suggests an all too tidy a view of the world, as we also see lines of enquiry abandoned because people get stuck, or make mistakes, or spot what might be a fruitful approach but lack the immediate resources of time or talent to address it, or judge that other activities are more worthwhile. Villani's playful account of the development of a proof is particularly insightful about this aspect of research, and we also observe numerous paths not taken or dismissed in the *polymath* and *minipolymath* problems. Likewise many of the *mathoverflow* questions demonstrate a general wish to understand a particular phenomenon or example, or find out what others know about it. rather than asking a precise question about its properties.

Examples and Computation. Our case studies exhibit a variety of ways in which examples are used: straightforwardly as part of an existence proof; in the Laktosian sense of a way to test and moderate hypotheses; to explain or clarify; or as a way of exploring what might be true or provable. Examples play an interesting role in collaborative endeavours: since the same example or phenomenon may occur in different areas of mathematics with different descriptions, sharing or asking for an example in *mathoverflow* may open up connections, or shed new light on a problem, or allow rapid interaction through a new researcher trying to understand how an unfamiliar concept apples to their own favourite family of examples. Computational methods allow construction, exploration, and retrieval of a much greater range of examples, and such techniques appear to be absorbed into the standard literature without comment, despite the well known limitations and non-reproducibility of computer algebra calculations.

Crowdsourcing, Leadership and the Strategic View. After a few iterations the *polymath* projects evolved to having a leader (most have been led by Tao) who took responsibility for overall guidance of the approach, drawing together the threads every so often to write a long blog post (with a view to it being part of the published paper), and setting the discussion on a new path. Perhaps most striking, and worthy of further study, is the strategic decisions that are made about which route to pursue in a complex landscape of possible proofs. While these are the result of the intuition and insight of extraordinary mathematicians, when the participants comment on this, we find judgements informed by what has worked in the past in similar situations, assessments of the relative difficulties or of the various approaches, or the likelihood that the approach will lead to something fruitful, even if it is not likely to solve the whole problem. Frenkel and Villani give similar insights in their books, with a frequent metaphor being that of searching for the proof as a journey, and the final proof as a road-map for the next explorer. Marcus du Sautoy, writes on proof as narrative: "The proof is the story of the trek and the map charting the coordinates of that journey". To continue the metaphor, collaboration enables new ways of exploration, to draw on different skills, and, crucially to share risks in a way that can make participants more adventurous in what they try out. In a section of *polymath* 8a for example, several participants are experimenting in an adventurous way with different computer algebra systems, and are joined by an expert in Maple who is able to transform and integrate the informal ideas and make rapid progress.

Institutional Factors in Innovation. All of our case studies show, in different ways, innovations in the practice of mathematics, through the use of machines to support collaboration, knowledge sharing, or calculations in support of a proof. What is noteworthy is that, however innovative the process, the outcomes of the activity remain unchanged: traditional papers in a traditional format in traditional journals, albeit with some of the elements executed by a machine. The reasons for this appear not to be any innate superiority of the format, indeed plenty have argued perceptively and plausibly for change, but the external drivers on research mathematicians. Research mathematicians are almost exclusively employed in the university system, either in the developed world, or in organisations in the developing world who are adopting similar norms and mechanisms, and are driven by the need to gather traditional indicators of esteem and recognition. The leaders of the field, such as Gowers, Tao, and Wiles, are perhaps best placed to resist these drivers, but are likewise aware of the pressures on younger colleagues - as evidence the discussions about authorship in *polymath*, and the advice not to spend too much time on it before tenure. Such pressures are active in other ways - for example publishing a so-called 'informalisation' of a formal proof in Homotopy Type theory in the high profile LICS conference, or in shaping decisions about how to spend ones time, so that, for example, the tactical goal of getting a paper written over a summer before teaching starts in the fall trumps loftier concerns.

Finally, what does this tell us about computational logic? We have described the "stumbling around in the dark" that currently seems a inevitable part of developing a proof, using as evidence the traces left by participants in collaborative activities on the web, and users of computer group theory systems. We have stressed the importance of a strategic view of proof, and the diversity and sharing of risk provided by collaboration. While we have not yet studied this in detail, we no of no evidence that the same is not true of developing large machine proofs. Understanding the collaborative development of human proofs should help us understand the collaborative development of machine proofs as well, and the best way to combine the two: Obua and Fleuriot's ProofPeer is making a start. Vladimir Voevodsky argues that computer proof will lead to a flowering of collaboration, as it enables trust between participants, who can rely on the machine to check each others work, and hence enables participants to take more risks, leading to much greater impact for activities like *polymath*.

At the same time, we see increasing recognition of the power of machine proof for mathematics: the work of Gonthier, Hales and Voevodsky to the fore. A recent triumph for SAT solving in mathematics was the discovery by Konev and Lisitsa in 2014 of a sequence of length 1160 giving the best possible bound on a solution to the Erdos discrepancy problem, resolving a question that had been partially solved in an earlier *polymath* discussion, which found a bound of 1124, which Gowers and others believed was best possible. Konev and Lisitsa write "The negative witness, that is, the DRUP unsatisfiability certificate, is probably one of longest proofs of a non-trivial mathematical result ever produced. Its gigantic size is comparable, for example, with the size of the whole Wikipedia, so one may have doubts about to which degree this can be accepted as a proof of a mathematical statement." It is an indication of how attitudes to computer proof have evolved since the more negative comments and concerns reported by Mackenzie 20 years before, that Gowers responded on his blog that "I personally am relaxed about huge computer proofs like this. It is conceivable that the authors made a mistake somewhere, but that is true of conventional proofs as well."

Our review of papers in group theory showed that they often contain significant amounts of detailed symbolic hand calculation of the kind that it would be straightforward to carry out in a proof-assistant, even though this is not current practice. Likewise machine assistance would surely confer some advantages in organising proofs that rely on complicated "minimum counterexample" arguments, a common pattern when considering finite simple groups. Similarly machine "book-keeping" would help in handling elaborate case-splits, as often occur in proofs of results about groups of order p^r, for all, or for all sufficiently large, primes, where the behaviour of different residue classes of $r \bmod p$ need to be considered. As Vaughan-Lee writes of the work mentioned above "all the proofs are traditional "hand" proofs, albeit with machine assistance with linear algebra and with adding, multiplying, and factoring polynomials. However the proofs involve a case by case analysis of hundreds of different cases, and although most of the cases are straightforward enough it is virtually impossible to avoid the occasional slip or transcription error." Since use of GAP is now accepted and routine in such papers, it is hard to see why use of a proof assistant could not be also.

In a panel discussion at the 2014 ceremonies for the Breakthrough Prize, the winners Simon Donaldson, Maxim Kontsevich, Jacob Lurie, Terence Tao and Richard Taylor addressed computer proof in various ways: asking for better search facilities (Tao), wondering if " Perhaps at some point we will write our papers not in LaTeX but instead directly in some formal mathematics system" (Tao), and remarking "I would like to see a computer proof verification system with an improved user interface, something that doesn't require 100 times as much time as to write down the proof. Can we expect, say in 25 years, widespread adoption of computer verified proofs?" (Lurie). Several speakers pointed to the length of time it can take for humans to be certain that a complex proof is true, and Kontsevich pointed out that "The refinement and cleaning up of earlier, more complicated stories is an important and undervalued contribution in mathematics.". This last point chimes with an observation made by Steingart on the classification of finite simple groups: the concern that the protagonists had that, with the leading figures in the field growing older, and few new recruits as other areas now seemed more exciting, the skills needed to understand these complex proofs and fix, if necessary, any local errors were being lost, and the proof risked being 'uninvented'. Perhaps ensuring that mathematics, once invented with such difficulty, does not become uninvented again, and that we don't forget how to read the map. is the greatest contribution computational logic can make to the field.

Acknowledgements. Ursula Martin acknowledges EPSRC support from EP/K040251. This essay acknowledges with thanks a continuing collaboration with Alison Pease, and incorporates material from two workshop papers which we wrote together.

Further Reading

Barany, M., Mackenzie, D.: Chalk: Materials and Concepts in Mathematics Research. Representation in Scientific Practice Revisited. MIT Press (2014)

Frenkel, E.: Love and Math: The Heart of Hidden Reality. Basic Books (2014)

Gorenstein, D.: Finite Simple Groups: An Introduction to their Classification. Plenum Press, New York (1982)

Hadamard, J.: The Psychology of Invention in the Mathematical Field. Princeton (1954)

Mackenzie, D.: Mechanizing Proof: Computing, Risk, and Trust. MIT Press (2001)

Steingart, A.: A group theory of group theory: collaborative mathematics and the 'uninvention' of a 1000-page proof. Soc. Stud. Sci. **42**, 185–213 (2014)

Villani, C.: Birth of a Theorem. Random House (2015)

Wiles, A.: Transcript of interview on PBS. www.pbs.org/wgbh/nova/physics/andrew-wiles-fermat.html

The Power of Collaboration: *polymath*

Gowers, T., Nielsen, M.: Massively collaborative mathematics. Nature **461**, 879–881 (2009)

The *polymath* Blog. polymathprojects.org

The *polymath* wiki. michaelnielsen.org/polymath1

"Is massively collaborative mathematics possible?", Gowers's Weblog, gowers.wordpress.com/2009/01/27/is-massively-collaborative-mathematics-possible

Nielsen, M.: Reinventing Discovery. Princeton (2012)

D H J Polymath: The 'bounded gaps between primes' polymath project: a retrospective analysis'. Newslett. Eur. Math. Soc. **94**, 13–23

Examples, Conjectures, Concepts and Proofs: *minipolymath*

Mini-polymath3 discussion. terrytao.wordpress.com/2011/07/19/mini-polymath 3-discussion-thread/

Minipolymath3 project. polymathprojects.org/2011/07/19/minipolymath3-project-2011-imo

Martin, U., Pease, A.: Seventy four minutes of mathematics: an analysis of the third Mini-Polymath project. In: Proceedings of AISB/IACAP 2012, Symposium on Mathematical Practice and Cognition II (2012)

Questions and Answers: *mathoverflow*

Martin, U., Pease, A.: What does mathoverflow tell us about the production of mathematics? arxiv.org/abs/1305.0904

Tausczik, Y.R., Pennebaker, J.W.: Participation in an online mathematics community: differentiating motivations to add. In: Proceedings CSCW 2012, pp. 207–216. ACM (2012)

Mendes Rodrigues, E., Milic-Frayling, N.: Socializing or knowledge sharing?: Characterizing social intent in community question answering. In: Proceedings CIKM 2009, pp. 1127–1136. ACM (2009)

Everyday Calculation: GAP

Vaughan-Lee, M.: Groups of order p^8 and exponent p. Int. J. Group Theor. Available Online from 28 June 2014

Bailey, D.H., Borwein, J.M.: Exploratory experimentation and computation. Not. Am. Math. Soc. **58**, 1410–1419 (2011)

GAP – Groups, Algorithms, and Programming, Version 4.7.7. The GAP Group (2015). www.gap-system.org

Harvey, J.A., Rayhaun, B.C.: Traces of Singular Moduli and Moonshine for the Thompson Group. arxiv.org/abs/1504.08179

Conclusions

Donaldson, S., Kontsevich, M., Lurie, J., Tao, T., Taylor, R.: Panel discussion at the 2014 award of the Breakthrough Prize. experimentalmath.info/blog/2014/11/breakthrough-prize-recipients-give-math-seminar-talks/

du Sautoy, M.: How mathematicians are storytellers and numbers are the characters. www.theguardian.com/books/2015/jan/23/mathematicians-storytellers-numbers-characters-marcus-du-sautoy

Lakatos, I.: Proofs and Refutations. CUP, Cambridge (1976)

Obua, S., Fleuriot, J., Scott, P., Aspinall, D.: ProofPeer: Collaborative Theorem Proving. arxiv.org/abs/1404.6186

Invited Talks

Automated Reasoning in the Wild

Ulrich Furbach[(✉)], Björn Pelzer, and Claudia Schon

Universität Koblenz-Landau, Koblenz, Germany
{uli,bpelzer,schon}@uni-koblenz.de

Abstract. This paper discusses the use of first order automated reasoning in question answering and cognitive computing. For this the natural language question answering project LogAnswer is briefly depicted and the challenges faced therein are addressed. This includes a treatment of query relaxation, web-services, large knowledge bases and co-operative answering. In a second part a bridge to human reasoning as it is investigated in cognitive psychology is constructed by using standard deontic logic.

1 Introduction

Automated reasoning as it is researched within the CADE and IJCAR community has always been aiming at applications; from the very beginning of AI, deduction and reasoning papers were published at the major AI conferences. In the last decades, however, the area split into various sub-disciplines, and unfortunately the connections between the parts began to loosen in some cases. For example, papers about common-sense reasoning or argumentation theory are published very rarely in CADE. And vice versa, there are fewer papers on automated reasoning in general AI conferences. This paper tries to demonstrate that it is worth investigating applications which do not allow the use of off–the–shelf theorem provers and it discusses the challenges which have to be faced.

It is the case that first and higher order automatic reasoning nowadays have proven to be very helpful in many application areas. Besides the classical domain of mathematics, the most important application certainly is program and software verification — an overview of the state of the art is contained in [3]. In most application areas the reasoning machinery is applied in such a way that correctness and completeness are guaranteed. In description logic applications, the quest for decidability is even more urgent, and thus there is a goal to push the expressiveness and to keep decidability at the same time.

In various different applications we developed and used our automated reasoning system *Hyper*, which is based on the hyper tableaux calculus with equality [2]. In most applications, it was not just the proof of the problem which was of interest, rather it was the model (or the representation of the model) which was returned by the prover. This model was used by the overarching software system to perform its computation. A typical example was the *Living Book*

Work supported by DFG FU 263/15-1 'Ratiolog'.

A.P. Felty and A. Middeldorp (Eds.): CADE-25, LNAI 9195, pp. 55–72, 2015.
DOI: 10.1007/978-3-319-21401-6_3

project, where a model of the prover run was used to construct a LATEX-file which could be compiled as an individual book according to the user specification [1]. One interesting challenge in this project was that on the one hand, we had huge amounts of formulae and in particular facts, which represented small LATEX-snippets which could be used to compose a book. On the other hand, there was a user, waiting for the response of the Living Book system according to her query. This aspect became much more crucial when we started using Hyper in the question answering system *LogAnswer* [6]. LogAnswer is an open domain natural language question answering system, which uses Hyper as one software component in the process of answering a user question – and of course there is an urgent need to answer quickly. In the following we will depict the architecture of the question answering system and in particular we will discuss why and how we sacrificed completeness of the reasoning mechanism.

Question answering became a prominent topic after the tremendous success of IBM's *Watson*-System in the Jeopardy quiz show. Watson succeeded in beating two human champions in this open domain quiz under very strong time constraints. Since then, IBM has been developing the Watson system further and also tailoring it to various application domains [5]. The keyword which turns the Jeopardy winning system into the basis of a business plan is *cognitive computing system*. Such a system is designed to learn and to interact with people in a way that the result could not be achieved either by humans or machine on their own. Of course, mastering *Big Data* also plays an important role—IBM's marketing slogan is "Artificial Intelligence meets Business Intelligence". Such a cognitive computing system has the following properties:

– Multiple knowledge formats have to be processed: formal knowledge like ontologies but also a broad variety of natural language sources, like textbooks, encyclopedias, newspapers and literary works.
– The different formats of knowledge also entail the necessity to work with different reasoning mechanisms, including information retrieval, automated deduction in formal logic, and probabilistic reasoning.
– The different parts and modules have to interact and cooperate very closely.
– The entire processing is time critical because of the interaction with humans.
– The system must be aware of its own state and accuracy in order to rank its outcomes.

Natural language question answering is obviously one example of cognitive computing as depicted above. To answer natural language questions, huge text corpora together with background knowledge given in various formats are used. The user interaction is rather simple: The user asks a natural language question and the system answers in natural language. In Sect. 2 natural language question answering as it is used in the LogAnswer system is briefly introduced. In Sect. 3 we discuss lessons learned from this and how this translates into requirements. The need for combining automated reasoning with findings from cognitive psychology is depicted and discussed in Sect. 4.

2   Deep Question Answering

One characteristic of open domain question answering is that the automated reasoning system, in our case the first-order logic theorem prover Hyper, is only one component of the overall system, whose parts all have to cooperate closely. The LogAnswer-System, as described in Fig. 1, basically consists of Hyper, an information retrieval module, of machine learning and of natural language answer generation. These modules cooperate, and in particular the machine learning generated decision trees are used for ranking of results. For this task, attributes from the information retrieval are used, e.g. number of keywords found or distance of keywords, together with attributes from the Hyper proofs. In other words, with the information retrieval procuring 200 answer candidates to be evaluated by Hyper, we end up with up to 200 different proofs, which then have to be compared and ranked in order to find the best one. From a proof theoretic viewpoint we could take the length of the proof or the number of formulae used therein into account. The problem, however, is that we rarely have a real proof. Let us explain this with the following example where we have a question Q and one of the 200 answer candidates C:

Q: Rudy Giuliani war Bürgermeister welcher US-Stadt?[1]
C: Hinter der Anklage stand der spätere Bürgermeister von New York, Rudolph Giuliani.[2]

The logical representation of both, generated by the linguistic module to capture the semantics of natural language text is:

$Q =$

$$\neg attr(X1, X2) \lor \neg attr(X1, X3) \lor \neg sub(X2, nachname.1.1)$$
$$\lor \neg val(X2, giuliani.0) \lor \neg sub(X3, vorname.1.1)$$
$$\lor \neg val(X3, rudy.0) \lor \neg sub(X1, buergermeister.1.1)$$
$$\lor \neg attch(FOCUS, X1) \lor \neg sub(FOCUS, us\text{-}stadt.1.1)$$

$C =$

$$hinter(c221, c210) \land sub(c220, nachname.1.1) \land val(c220, giuliani.0)$$
$$\land sub(c219, vorname.1.1) \land val(c219, rudolph.0) \land prop(c218, spaet.1.1)$$
$$\land attr(c218, c220) \land attr(c218, c219) \land sub(c218, buergermeister.1.1)$$
$$\land val(c216, new\text{-}york.0) \land sub(c216, name.1.1) \land sub(c215, stadt.1.1)$$
$$\land attch(c215, c218) \land attr(c215, c216) \land subs(c211, stehen.1.1)$$
$$\land loc(c211, c221) \land scar(c211, c218) \land sub(c210, anklage.1.1)$$

A proof attempt of this task fails because several subgoals of Q cannot be proved: $\neg sub(X3, vorname.1.1) \lor \neg val(X3, rudy.0)$ cannot be proved, because C only

---

[1] Rudy Giuliani was the mayor of which city in the USA?
[2] Responsible for the charges was the future mayor of New York, Rudolph Giuliani.

contains $\neg sub(c219, vorname.1.1) \vee \neg val(c219, rudolph.0)$ stating that the given name is *Rudolph* instead of *Rudy*. Instead of returning a *failure* the system simply skips these subgoals and tries to run the proof again with this modified goal. A similar effect happens when Hyper tries to "identify" $val(c216, new\text{-}york.0)$ with $sub(FOCUS, us\text{-}stadt.1.1)$. Without additional knowledge identifying *Rudy* with *Rudolph* and *New-York* with *US-Stadt* this proof is not possible. Hence the system skips part of the goal again. This second skip is different from the first because here a literal is deleted and remaining unproven which contains the variable *FOCUS*. This is a special variable whose instantiation is important to construct an answer from the proof – it basically represents the core of the answer. When Hyper finally manages to succeed, this is only possible after two skips of a literal, one containing the critical focus variable. The resulting proof must therefore be judged to have a low quality for the purpose of answer generation.

This example demonstrates two aspects:

- It is not always possible to find a proof for all of the subgoals; in many cases it is only possible to get partial proofs. This is even more dramatic if we take into account that time constraints are very tight. For one query we have 200 answer candidates, hence 200 proof tasks, each using clauses from the query, the answer candidate under consideration and the huge amount of background knowledge (which will be discussed later). This is why we have a time limit for every attempt to prove a subgoal. In most cases the failure for a subgoal is because of such a time-out.
- The partial proofs returned by Hyper are useful, its properties, like number of skipped literals, type of skipped literals and many more have to be used for a ranking of proofs. This ranking can be used among other attributes to evaluate the plausibility of the final answer. Another aspect of ranking is discussed in the Sect. 4.

To sum up, in this application we definitely sacrifice completeness for efficiency and plausibility.

3   Lessons for Automated Reasoning

In the previous section we very briefly depicted the LogAnswer system as an example for cognitive computing; for a more detailed description we refer to [6,8]. This section summarizes some lessons and requirements for automated reasoning which came up during this project.

It was already pointed out that completeness of the calculus was not exploited in this application. One reason was depicted, namely the need to skip literals, a process called *relaxation*, which became necessary because of time limitations. Another reason is that the original text from Wikipedia already contains incorrect or contradictory knowledge. And even if the text did not contain such wrongness, it could be very well introduced by the parser which reads and transforms the textual sources. The parser can fail at correctly resolving ambiguous or complex statements or peculiarities in the layout, for example at embedded figures

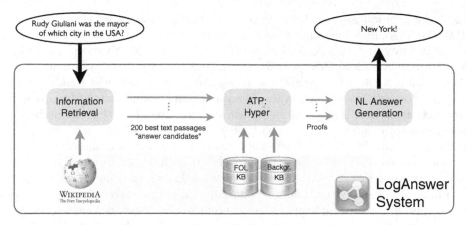

The knowledge source is the text corpus of the German Wikipedia which contains 12+ million natural language sentences. This text corpus is processed according to the query with usual shallow information retrieval techniques. The resulting text pieces are ranked with machine learning techniques and the 200 best are selected. For these answer candidates the semantic representation in first-order logic from *FOL KB*—which is computed beforehand for the entire text corpus — together with background knowledge *Backgr. KB* is fed into the automated reasoning system Hyper. For these 200 proofs a ranking taking into account various properties from the proofs is done and the 5 best are passed to the natural language answer generator.

Fig. 1. Overview of the LogAnswer-System

and tables. Besides this "incomplete embedding" of the Hyper prover there are more challenges for using automated reasoning in the wild, which are discussed in this section.

Background Knowledge

Answering a query on the basis of a given text is certainly only possible if background knowledge is available. These can be very simple things like basic knowledge about properties of temporal or spatial relations, but also much more sophisticated knowledge given that huge ontologies could be a source for background knowledge. In LogAnswer we have tried several different approaches. To get a feeling for the size of problems for automated theorem provers, one may visit the *TPTP* [24], which is meant as a representative and challenging collection of problems. In [18] we discussed size statistics of the TPTP, showing that most of the problems are not very large. Almost half of the problems have at most 50 clauses or formulae; indeed, the median amount is 52. Over 85 % of the problems have no more than 1,000 clauses or formulas, and only a few outliers exceed 100,000 or even a million. The latter large problems are not typical for automated theorem proving; they are included within the TPTP as a challenge— not because of their complicated structure, but because of their abnormal size.

And indeed, ontologies which might serve as background knowledge, like Cyc, SUMO or Yago, have amounts of formulae which are far beyond the size that can be handled by general purpose theorem provers. For example Yago contains more than 10 million entities (like persons, organizations, cities, etc.) and more than 120 million facts about these entities. ResearchCyc contains more than 500,000 concepts, forming an ontology in the domain of human consensus reality. Nearly 5,000,000 assertions (facts and rules) using more than 26,000 relations inter-relate, constrain, and, in effect, (partially) define the concepts. And even the smallest version of Cyc, OpenCyc, still contains more than 3 million formulae.

There are basically two different approaches to include background knowledge for the reasoning task: Firstly, one can include the knowledge directly into the set of formulae handled by the proof procedure and applying clever clause selection criteria during the proof attempt. The second possibility is to keep the knowledge external and query a special mechanism to reason with the knowledge base from within the proof procedure.

Clause Selection

Technically, some forms of clause selection have been used in automated reasoning for decades, as a theorem prover that naively evaluates all the available clauses quickly gets bogged down in an exploding number of inference results. The *set of support strategy* [29] changed this by dividing the clauses into sets, only some of which were allowed to act as premises. Implementations of this strategy, for example the *given-clause algorithm* of the prover *Otter* [17], also implemented selection heuristics to pick the most suitable clauses from the support set first. While such methods are complete, provided they are fair, their methods are still intended for sets that are small compared to the typical background knowledge, and they rely on all clauses being already stored in the prover in a complex, thoroughly analyzed form. Just storing the typical background knowledge in this manner will easily exceed hardware capacities — and this before the actual reasoning has even started.

Thus the growth of clause sets has lead to an increased interest in incomplete selection methods, which forego completeness in favour of selecting a set of manageable size at high speed, while still hoping to pick all the clauses actually needed for a proof. A plethora of methods exist: *Graph-based clustering methods* arrange symbols in a graph depending on how they occur in clauses, and then use this graph to form clause clusters. *Heuristic clustering methods* use a measure of closeness between clauses for clustering. More recently machine learning methods have been employed to learn selection criteria from known successful proofs, such as in the systems *MaLARea* [25] and *Flyspeck* [13].

Among the most successful methods in practice is the *SInE* algorithm from [11], a quick heuristic axiom selection algorithm that can operate on very shallow axiom representations which in principle do not even have to be in clause normal form. Hyper implements both the SInE algorithm and a slight modification, which addresses some weaknesses of SInE. SInE's main idea is that when rare symbols (with respect to the entire knowledge base) occur in a clause or axiom, then they are more relevant for this clause than its other symbols. The algorithm is goal-oriented: It selects those clauses from a clause set containing a rare symbol which also occurs in a given goal. This process is recursively continued with these selected symbols until a fixed point is reached. Of course, the necessary computation for selecting rare symbols can be done beforehand in a preprocessing step, which has the big advantage that it can be done on the pure clause or axiom set, before any other necessary preprocessing of the prover adds more structure, like pointers to index trees or other clause features. Thus, the axiom representations can remain very shallow at this point, as the selection method does not require any deeper structural axiom analysis beyond simply counting symbols.

When testing the SInE algorithm within the LogAnswer system we found it to be less suitable for our background knowledge than we initially expected from its general performance. When investigating this, we made an observation described in more detail in [18]: A knowledge base like ours, which uses a sparse inventory of predicate symbols for a large number of axioms, tends to throw off the rarity-based heuristics of SInE. As every predicate symbol is very common in the knowledge base, in particular in comparison to function symbols, essential clauses will often not get selected, as the heuristics considers them to be only relevant for some rare function symbols which are not requested by the goal. We will illustrate this phenomenon with an example:

$$Q : \leftarrow predator(x) \wedge mammal(x)$$
$$C_1 : predator(wolf) \leftarrow$$
$$C_2 : predator(eagle) \leftarrow$$
$$C_3 : mammal(wolf) \leftarrow$$
$$C_4 : mammal(cow) \leftarrow$$
$$C_5 : mammal(sheep) \leftarrow$$
$$C_6 : bird(eagle) \leftarrow$$
$$C_7 : bird(hummingbird) \leftarrow$$

The SInE algorithm will select axioms via the query predicate symbols $predator/1$ and $mammal/1$. For $predator/1$, it selects C_1 and C_2, as $predator/1$ is among the rarest symbols there—it occurs twice in the entire knowledge base. However, for $mammal/1$ no axioms are selected, as the $mammal/1$ symbol occurs three times in the knowledge base and is thus more common than the animal names in the unit clauses. Instead, the selection of C_2 triggers the selection of C_6 via $eagle$, as both $eagle$ and $bird/1$ are equally rare. Without any of C_3, C_4 and C_5, it is obviously impossible to prove the goal Q. A tolerance factor would help in this case but in real-world knowledge bases, the same predicate may be used to store thousands of facts, making their predicate so common as to render these facts inaccessible to SInE, unless the tolerance is increased so far that nearly the entire knowledge base is selected.

To deal with this issue, LogAnswer therefore also uses a modified version of SInE: Firstly it distinguishes between positive and negative occurrences and, secondly, it differentiates function and predicate symbols, such that a clause can be selected by a predicate symbol that is more common than its function symbols or vice versa. In the example above the Hyper axiom selection method would thus first select C_1 and C_2 for $predator/1$ and C_3, C_4 and C_5 for $mammal/1$. Unlike with SInE there is no selection of C_6 triggered by C_2 via $eagle$, because this symbol occurs only in positive literals in both clauses, while our selection method requires complementary occurrences. With the selection of C_1 and C_3, the query Q can now be proven. In general, our method selects more clauses than SInE does, even though the distinction between positive and negative occurrences helps to avoid some unnecessary selections made by SInE.

In [18] all this is described more carefully and more formal together with an extensive evaluation. To summarize this evaluation, we just mention that 1,805 LogAnwer problems have been tested with Hyper in modes which apply "no selection", "SInE selection" and the modified "Hyper selection". The main observations are: With "no selection", Hyper solves 95.7 % of the problems with an average of 0.99 s; with "SInE selection", it solves only 41.3 % but much faster, namely in 0.01 s; and finally, with "Hyper selection", it solves slightly more, namely 46,6 %, but it takes 0.06 s.

Altogether clause selection improves efficiency of the proof procedure for large knowledge bases significantly but at the same time completeness gets lost—again!

Web Services

The alternative to the method depicted above is to keep the knowledge external and query it from within the proof procedure. Recent years have seen the emergence of vast digital data sources on the web, some of which provide their content in a structured form, for example DBpedia.[3] Such easily machine-readable sources, which we will refer to as *web services* in this paper, have the advantage of offering large quantites of data that in principle could be used for reasoning without much preprocessing. At the same time, there is the problem that web connections may be slow, and also that web services usually only work well for very concrete queries — one has to know what one is looking for, the full data behind the web service is never available in its entirety at once.

A first attempt to employ web services in automatic reasoning is the *SPASS-XDB* system [23], where the theorem prover *SPASS* [28] is connected to a number of external data sources. The authors note that the technical issues of web services infringe upon completeness, but do not explore the matter much beyond this.

As web services contain information that could be valuable for question answering, naturally, in LogAnswer, we were also interested in an integration. For Hyper we decided to include the querying mechanism not just on the technical level, but also on the calculus level, by extending the Hyper tableau inference rule. This gave us a more thorough understanding of the theoretical possibilities

---

[3] http://dbpedia.org/.

and limitations of web services in the context of automated reasoning. The following description is taken from [19]. There are two viewpoints to this extension, one is dealing with the formal representation of external sources, which allows an analysis of the extended calculus, and the other is concerned with an efficient implementation. For the formal representation we assume a binary predicate $ext/2$ which is used to represent the relation $ext(q, a)$ between the request term q and the response term a. An external source is represented as a possibly infinite set of ground ext units; e.g.

$ext(weather\_service(weather('Stockholm', 27\text{-}06\text{-}2012)), 'cloudy') \leftarrow$
$ext(weather\_service(temperature('Stockholm', 27\text{-}06\text{-}2012)), '15°C') \leftarrow$
$ext(currency\_exchange\_service(eur, usd, 2.99, 27\text{-}06\text{-}2012), '\$3.74') \leftarrow$

In the formal treatment within the hyper extension rule, it is assumed that all the positive ground units from the external source are given in a set C^{ext} and unsatisfiability of a clause set C is now defined as *unsatisfiability with respect to C^{ext}*. The new hyper extension step with external access works mostly like the original, except that ext atoms from negative literals in the extending clause can also unify with units from the external source, provided the request terms in these atoms are ground before their respective unification. This latter condition is necessary as web services only respond to fully instantiated queries. In [19] this extended calculus is proven to be correct, but—of course—there is no guarantee of completeness. The following example illustrates the usage of train schedule webservice.

C_1: $at('CeBIT', 'Hanover') \leftarrow$
C_2: $next\_train\_to(Event, Time) \leftarrow$
$\qquad\qquad at(Event, ToCity),$
$\qquad\qquad ext(user\_location\_service, FromCity),$
$\qquad\qquad ext(next\_train\_finder\_service(FromCity, ToCity), Time)$
Q_3: $\leftarrow next\_train\_to('CeBIT', Time)$

Clause C_2 uses two external sources, one for giving the user's current location and another for train schedules. The two occurrences of the ext predicate are handled by the Hyper extension inference rule. Since the real data is provided by a web service, the proof procedure has no guarantee concerning response time or even availability of the service. Therefore, we implemented a proxy service which handles the request called by an ext predicate. This proxy answers synchronously to the prover, although the web service answers asynchronously. The proxy has a cache in which previously answered requests are stored and it can answer either with *wait*, *failed* or with the *response*. In case of a *wait* the proof procedure can continue its derivation and can check later whether a response has been received. The *failed* case is handled like a non-matching in C^{ext} and a *response* is handled as a set of unifying unit clauses.

Cooperative Answers

In natural language question answering it is often the case that no correct answer can be given but the answer to a slightly modified question could be helpful—or, even worse, the correct answer may not be the desired one (e.g. *"Do you know what time it is?"*). This problem is approached by cooperative answering.

The following example is taken from [18] again. Consider the following question Q together with the sentence S that was retrieved from the KB as possibly relevant:

Q: What is the weight of the "Maus" ("Mouse") tank?
S: At 188 tons the "Maus" is the heaviest armoured fighting vehicle ever built.

While S does not mention any tank, tanks are a subclass of vehicles, and given S we could use abduction to form the hypothesis that the vehicle *'Maus'* is a tank and then answer the question. For this, the *ext* predicate together with the web service from the previous subsection can be used:

Let $\mathcal{C}^{ext}$ be an external source containing positive ground *ext* units of the form $ext(subclass\_of(c), d) \leftarrow$, which is the external source conforming representation of the subclass relationship $subclass\_of(c, d)$ between two concept identifiers c and d. Consider the query $Q: \leftarrow is\_a('Maus', tank), has\_weight('Maus', x)$ and the rest of the clause set based on S, such that the first part of the query cannot be proven. We now obtain the abductive relaxation supporting clause set $\mathcal{C}^{ar}$ from $\mathcal{C}$ by adding two clauses as follows.

Q^{ar}: $relaxed\_answer(rlx('Maus', x_1), rlx(tank, x_2), rlx('Maus', x_3)) \leftarrow$

$\qquad is\_a(x_1, x_2), has\_weight(x_3, x), ext(subclass\_of('Maus'), x_1),$
$\qquad\qquad\qquad ext(subclass\_of(tank), x_2), ext(subclass\_of('Maus'), x_3)$
C^{rs}: $ext(subclass\_of(x), x) \leftarrow$

Instead of relaxing Q by simply skipping it as described in Sect. 2, we can use the generalized literals in the modified query Q^{ar}, namely $is\_a(x_1, x_2)$ and $has\_weight(x_3, x)$, together with the request predicates $ext(subclass\_of(tank), x_2)$ and $ext(subclass\_of('Maus'), x_3)$, which finally results in a unit clause:

$C_3 : relaxed\_answer(rlx('Maus', 'Maus'), rlx(tank, vehicle), rlx('Maus', 'Maus')) \leftarrow$

In a system like LogAnswer this relaxed answer clause can be used to cooperatively answer the question *"What is the weight of the 'Maus' tank?"* with *"188t, if by 'tank' you mean 'vehicle'"*. The user must then judge whether this particular relaxation was acceptable.

In addition to the result from Sect. 2, the need to give up the completeness of the calculus, we discussed in this section that it is important to include external sources of data and knowledge and that cooperation with the person who asked the question is mandatory.

4 Common Sense and Cognitive Science

In the previous sections, we argued that in open domain question answering the reasoning process has to master a lot of non-logical challenges. This is also typical for the more general class of problems which can be subsumed under the keyword of *cognitive computing*. This paradigm was depicted in the introduction and question answering was introduced as an instance of cognitive computing. An important feature is that humans and the system have to cooperate aiming at solutions of tasks which could not be solved by humans or machines alone.

For such a cooperation it is important to model within the automated reasoning system the way humans do their reasoning. Indeed, there is the field of *common sense reasoning* which aims at bringing aspects from everyday reasoning into the field; circumscription, Bayesian reasoning and answer set programming certainly are examples for this. More recently there is increasing interest in modelling results from cognitive psychology about reasoning in logical systems.

One interesting observation is that humans are able to solve certain problems very fast and correctly, while other problems appear to be very difficult and error-prone, even though they have a very similar logical structure. A famous example is the so-called *Wason selection task* [27], which has been investigated in psychology since the 1960s. An extensive treatment from a logical perspective can be found in [7, 22]. Automated reasoning might be able to learn a lot from such interdisciplinary studies. In a keynote talk at the previous CADE, Natarjan Shankar even coined a new name for this research – *Cueology* [21].

Another observation from psychology is that humans use weird but useful inference rules. Examples for this are investigated in the context of the so-called *suppression task* [4]: Assume the following statements:

If she has an essay to write, she will study late in the library.
She has an essay to write.

In an experiment, persons are asked to draw a valid conclusion out of these premisses. It turned out that 98 % of the test persons conclude correctly that the following statement is a consequence of the above presented statements.

She will study late in the library.

This shows that in such a setting modus ponens a is very natural rule of deduction. If an additional statement is given, namely

If she has some textbooks to read, she will study late in the library.

this does not change the percentage of correct answers. Obviously this additional conditional is understood as an alternative. And indeed, we can transform the two conditionals

$$essay\_to\_write \rightarrow study\_late$$
$$textbooks\_to\_read \rightarrow study\_late$$

equivalently into a single one, where the premiss is a disjunction:

$$essay\_to\_write \lor textbooks\_to\_read \rightarrow study\_late$$

If, however, as an additional premiss

If the library stays open, she will study late in the library.

or as a formula *library_open* $\rightarrow$ *study_late* is added, only 38 % draw the correct conclusion, although modus ponens is applicable in this case as well. People are understanding this additional conditional not as an alternative but as an additional premiss.

When trying to simulate human reasoning with the help of automated reasoning, there are different approaches: In [14] it is logic programming, in [12] abductive logic programming is used and in [22] 3-valued logic is a favourite. In [7] we used deontic logic for modelling different kinds of human reasoning, e.g. the Wason selection task, but also the suppression task is modelled therein. There also is a transformation of deontic logic into a decidable fragment of first-order logic, which can be decided by the Hyper prover.

Standard deontic logic (SDL) is obtained from the well-known modal logic K by adding the seriality axiom D:

$$D : \quad \Box P \rightarrow \Diamond P$$

In this logic, the $\Box$-operator is interpreted as "it is obligatory that" and the $\Diamond$ as "it is permitted that". The $\Diamond$-operator can be defined by the following equivalence:

$$\Diamond P \equiv \neg \Box \neg P$$

The additional axiom D: $\Box P \rightarrow \Diamond P$ in SDL states that if a formula has to hold in all reachable worlds, then there exists such a world. With the deontic reading of $\Box$ and $\Diamond$ this means: Whenever the formula P ought to be true, then there exists a world where it holds. In consequence, there is always a world which is ideal in the sense that all the norms formulated by the 'ought to be'-operator hold.

SDL can be used in a natural way to describe knowledge about norms or licenses. The use of conditionals for expressing rules which should be considered as norms seems straightforward, but holds some subtle difficulties. If we want to express that *if P then Q* is a norm, an obvious solution would be to use

$$\Box(P \rightarrow Q)$$

which reads *it is obligatory that Q holds if P holds*. An alternative would be

$$P \rightarrow \Box Q$$

meaning *if P holds, it is obligatory that Q holds*. In [26] there is a careful discussion which of these two possibilities should be used for conditional norms.

The first one has severe disadvantages. The most obvious disadvantage is, that P together with $\Box(P \to Q)$ does not imply $\Box Q$. This is why we prefer the latter method, where the $\Box$ operator is in the conclusion of the conditional. For a more detailed discussion of such aspects we refer to [9].

Besides understanding and interacting with human-like reasoning mechanisms it is also important that the reasoning system *knows* about the relevance of its results. In the previous sections it was already demonstrated that different kinds of relaxation during the derivation gives us an attribute for computing relevance (more relaxations = less relevant). As an important candidate mechanism for further refinements we also considered *defeasible reasoning*. Certain knowledge is assumed to be *defeasible*, while other parts are *strict* knowledge which is specified by *contingent facts* (e.g. "Tom is an emu") and *general rules* holding in all possible worlds without exception (e.g. "emus do not fly"). Strict knowledge is always preferred to knowledge depending also on *defeasible rules* (e.g. "birds *normally* fly"). Hence, it is possible to compare two or more different derivations with respect to the clauses and rules, which are used. In [8] this approach is discussed in the context of comparing proofs during question answering. Deontic Logic turns out to be helpful also for this because rules expressing norms can be seen as defeasible rules and such the entire approach of defeasible reasoning could be applicable for normative reasoning as well.

A Benchmark Suite

Most sub-disciplines of automated reasoning use more or less well-established benchmark suits for evaluating concepts and systems. In some areas there are even competitions which directly allow comparison of systems. In common-sense reasoning there also is a certain effort for establishing benchmarks; there is the Choice of Plausible Alternatives (COPA) in [20] or the Winograd Schema Challenge from [15]. Both benchmark sets are based on natural language, such that they are not easily accessible by most common-sense reasoning systems. This is different in the Triangle Choice of Plausible Alternatives (Triangle-COPA) [16], which is a set of currently 100 problems for common-sense reasoning, which are presented not only in natural language but also as a logical description.

The setting of the problems in Triangle-COPA resembles the setting in the famous Heider-Simmel film. Figure 2 depicts a screenshot[4] from the film which was shown to students. After the presentation of the film, the students were asked to narrate what they observed in the film. Nearly all students offered a story where they interpreted the geometrical shapes as persons interacting with each other. The fact that humans interpret even such a restricted setting based on common-sense theories is reflected in the Triangle-COPA benchmarks as well.

Each problem consists of a logical description of a situation and a question. Furthermore, two possible answers are given in logic. The task is to determine which of the two answers is the correct one. Both the situation description as

---

[4] Screenshot available at: http://orphanfilmsymposium.blogspot.com/2008/05/national-science-foundation-grants.html (retrieved: 21st of april 2015).

Fig. 2. In 1944, Heider and Simmel conducted a famous study [10], where they presented a 90 sec film to undergrad students. This film showed simple geometric shapes, a little triange (*lt*), a big triangle (*bt*) and a circle (*c*), moving in and around a rectangle with a little part which opened and closed.

well as the possible answers are formalized in first-order logic using a controlled vocabulary. Therefore, natural language processing is not necessary to solve the problems. In order to facilitate the focus on automated common-sense reasoning, the domain of the problems is very restricted.

Like in the setting used in the Heider-Simmel film, the Triangle-COPA problems describe a sequence of interactions between a circle, a big triangle, a little triangle, a box, and a door. This sequence is described with the help of a fixed, rather restricted vocabulary. This vocabulary consists of 1-character actions like *shake* or *run*, 2-character actions like *approach*, spacial relations like *enter*, relations for assertions about time and negations, abstract actions like *attack*, mental actions like *agree*, emotions like *afraid* and social relationships like *friend*. Furthermore, the two possible answers to the question corresponding to the interpretations of the sequence is given in the same vocabulary. The question is not part of the formalization. The following set of facts gives an example for such a sequence given in the Triangle-COPA vocabulary:

$$friend(e1, bt, lt).$$
$$enter(e2, bt).$$
$$approach(e3, c, lt).$$
$$shake(e4, lt).$$
$$attack(e5, c, lt).$$
$$leave(e6, bt).$$
$$seq(e1, e2, e3, e4, e5, e6, e7).$$

It describes the situation, in which the little triangle and the big triangle are friends. The big triangle enters the room and sees the circle approaching the little triangle. The little triangle starts shaking and it is attacked by the circle. In the next event, the big triangle leaves the room.

A possible question would be: How does the little triangle feel? Possible answers, given as facts in the Triangle-COPA vocabulary, could be:

1. *dissapointed*(*e7*, *lt*).
2. *excited*(*e7*, *lt*).

Many emotions in everyday life can be explained with unmet expectations. The husband not bringing flowers on the wedding anniversary and the friend arriving delayed to a date are only two examples where unmet expectations cause negative feelings. The example given above can explained with the help of unmet expectations as well. The big triangle and the little triangle are friends. The circle attacks the little triangle. Normally one should defend a friend who is attacked by someone. This is why the little triangle expects the big triangle to hurry to its defence. In the described situation, however, the big triangle does not defend its friend but leaves the room. Therefore, the little triangle is disappointed.

It is possible to model unmet expectations with the help of deontic logic. Normative statements are used to model expected behavior. In our example, the fact that one should defend friends if they are attacked can be modelled by a set of deontic logic formulae. This set of deontic formula is the set of ground instances of the following formula:

$$friend(E, X, Y) \wedge attack(E', Z, X) \wedge after(E, E') \rightarrow \Box defend(E', Y, X). \quad (1)$$

where *after* is a transitive predicate, stating that one event occurs after another. *after*(*e1*, *e2*) means that event *e2* occurs after *e1*. Since formula (1) contains variables, it is not a SDL formula. However, we use it as an abbreviation for its set of ground instances. The ground instance interesting for our example is:

$$friend(e1, bt, lt) \wedge attack(e5, c, lt) \wedge after(e1, e5) \rightarrow \Box defend(e5, bt, lt). \quad (2)$$

With the help of formula (2), it is possible to derive that the big triangle ought to defend the little triangle in event *e5*.

Ground instances of the following formula can be used to deduce that someone is disappointed if he ought to be defended by someone but in fact is not defended:[5]

$$(\Box defend(E, X, Y) \wedge \bigwedge_{\substack{\forall E' \\ after(E, E')}} \neg defend(E', X, Y)) \rightarrow (\bigwedge_{\substack{\forall E'' \\ after(E, E'')}} disappointed(E'', Y)).$$

$$(3)$$

---

[5] We are aware that this formalization is too strong since it causes someone only to be disappointed if he is never defended in the future. However, in the Triangle-COPA, the future only consists of a very small number of events and therefore this formalization is sufficient for our purposes.

The negation sign occurring in the formula is negation as failure. A ground instance interesting for our example is:

$$(\Box defend(e5, bt, lt) \land \neg defend(e6, bt, lt) \land \neg defend(e7, bt, lt)) \to$$
$$(disappointed(e6, lt) \land disappointed(e7, lt)). \tag{4}$$

With the help of this formula, it is possible to deduce that the little triangle is disappointed in $e6$ and $e7$. Referring to the question "How does the little triangle feel?" formulated before, we can use the derived $disappointed(e7, lt)$ to show that the first alternative given is the correct one. First experiments using this approach with the Hyper theorem prover yield promising results. Of course, it is not desirable to formalize all rules manually. Rules like (3) can be generated automatically by formalizing a metarule stating that: whenever x and y are friends and y is obliged to do something for x but does not act according to his obligation, x is disappointed. This metarule can then be instantiated by the respective obligation. Furthermore, like in the case of deep question answering, it is desirable to integrate background knowledge. Ontologies like OpenCyc could be used to include information about the different emotions and how they are connected.

5   Conclusion

This paper discusses the use of first order automated reasoning in question answering and cognitive computing. We demonstrated that a first challenge faced therein is to give up completeness property of the proof procedure. This is necessary because of query relaxation which needs to be done in order to get results. We furthermore demonstrated how to connect a reasoning system to web services and to large knowledge bases, which also enables cooperative question answering. In a second part a bridge to human reasoning as it is investigated in cognitive psychology is constructed by using standard deontic logic. We briefly gave examples for experiments from cognitive science and we demonstrated how deontic logic can be used to tackle benchmarks from common-sense reasoning. All this was done in a rather informal manner, trying to open areas of research instead of already offering complete solutions.

References

1. Baumgartner, P., Furbach, U., Groß-Hardt, M., Sinner, A.: Living book - deduction, slicing, and interaction. J. Autom. Reasoning **32**(3), 259–286 (2004)
2. Baumgartner, P., Furbach, U., Pelzer, B.: Hyper tableaux with equality. In: Pfenning, F. (ed.) CADE 2007. LNCS (LNAI), vol. 4603, pp. 492–507. Springer, Heidelberg (2007)
3. Beckert, B., Hähnle, R.: Reasoning and verification: state of the art and current trends. IEEE Intell. Syst. **29**(1), 20–29 (2014)
4. Byrne, R.M.J.: Suppressing valid inferences with conditionals. Cognition **31**(1), 61–83 (1989)

5. Ferrucci, D., Levas, A., Bagchi, S., Gondek, D., Mueller, E.T.: Watson: beyond jeopardy!. Artif. Intell. **199–200**, 93–105 (2013)
6. Furbach, U., Glöckner, I., Pelzer, B.: An application of automated reasoning in natural language question answering. AI Commun. **23**(2–3), 241–265 (2010)
7. Furbach, U., Schon, C.: Deontic logic for human reasoning. In: Eiter, T., Strass, H., Truszczyński, M., Woltran, S. (eds.) Advances in Knowledge Representation. LNCS, vol. 9060, pp. 63–80. Springer, Heidelberg (2015)
8. Furbach, U., Schon, C., Stolzenburg, F., Weis, K.-H., Wirth, C.-P.: The RatioLog Project - Rational Extensions of Logical Reasoning. ArXiv e-prints, March 2015
9. Gabbay, D., Horty, J., Parent, X., van der Meyden, R., van der Torre, L. (eds.): Handbook of Deontic Logic and Normative Systems. College Publications, London (2013)
10. Heider, F., Simmel, M.: An experimental study of apparent behavior. Am. J. Psychol. **57**(2), 243–259 (1944)
11. Hoder, K., Voronkov, A.: Sine qua non for large theory reasoning. In: Bjørner, N., Sofronie-Stokkermans, V. (eds.) CADE 2011. LNCS, vol. 6803, pp. 299–314. Springer, Heidelberg (2011)
12. Hölldobler, S., Philipp, T., Wernhard, C.: An abductive model for human reasoning. In: AAAI Spring Symposium: Logical Formalizations of Commonsense Reasoning (2011)
13. Kaliszyk, C., Urban, J.: Learning-assisted automated reasoning with Flyspeck. J. Autom. Reasoning **53**(2), 173–213 (2014)
14. Kowalski, R.: Computational Logic and Human Thinking: How to be Artificially Intelligent. Cambridge University Press, Cambridge (2011)
15. Levesque, H.J.: The winograd schema challenge. In: LogicalFormalizations of Commonsense Reasoning, Papers from the 2011 AAAI Spring Symposium, Technical Report SS-11-06, Stanford, California, USA, 21–23 March 2011. AAAI (2011)
16. Maslan, N., Roemmele, M., Gordon, A.S.: One hundred challenge problems for logical formalizations of commonsense psychology. In: Twelfth International Symposium on Logical Formalizations of Commonsense Reasoning, Stanford, CA (2015)
17. McCune, W.: OTTER 3.3 Reference Manual. Argonne National Laboratory, Argonne, Illinois (2003)
18. Pelzer, B.: Automated Reasoning Embedded in Question Answering. Ph.D. thesis, University of Koblenz (2013)
19. Pelzer, B.: Automated theorem proving with web services. In: Timm, I.J., Thimm, M. (eds.) KI 2013. LNCS, vol. 8077, pp. 152–163. Springer, Heidelberg (2013)
20. Roemmele, M., Bejan, C.A., Gordon, A.S.: Choice of plausible alternatives: an evaluation of commonsense causal reasoning. In: Logical Formalizations of Commonsense Reasoning,Papers from the 2011 AAAI Spring Symposium, Technical Report SS-11-06, Stanford, California, USA, 21–23 March 2011. AAAI (2011)
21. Shankar, N.: Automated reasoning, fast and slow. In: Bonacina, M.P. (ed.) CADE 2013. LNCS, vol. 7898, pp. 145–161. Springer, Heidelberg (2013)
22. Stenning, K., Van Lambalgen, M.: Human Reasoning and Cognitive Science. MIT Press, Cambridge (2008)
23. Suda, M., Sutcliffe, G., Wischnewski, P., Lamotte-Schubert, M., de Melo, G.: External sources of axioms in automated theorem proving. In: Mertsching, B., Hund, M., Aziz, Z. (eds.) KI 2009. LNCS, vol. 5803, pp. 281–288. Springer, Heidelberg (2009)
24. Sutcliffe, G.: The TPTP problem library and associated infrastructure: the FOF and CNF parts, v3.5.0. J. Autom. Reasoning **43**(4), 337–362 (2009)

25. Urban, J., Sutcliffe, G., Pudlák, P., Vyskočil, J.: Malarea SG1 - machine learner for automated reasoning with semantic guidance. In: Armando, A., Baumgartner, P., Dowek, G. (eds.) IJCAR 2008. LNCS (LNAI), vol. 5195, pp. 441–456. Springer, Heidelberg (2008)
26. von Kutschera, F.: Einführung in die Logik der Normen. Werte und Entscheidungen, Alber (1973)
27. Wason, P.C.: Reasoning about a rule. Q. J. Exp. Psychol. **20**(3), 273–281 (1968)
28. Weidenbach, C., Schmidt, R.A., Hillenbrand, T., Rusev, R., Topic, D.: System sescription: SPASS version 3.0. In: Pfenning, F. (ed.) CADE 2007. LNCS (LNAI), vol. 4603, pp. 514–520. Springer, Heidelberg (2007)
29. Wos, L., Overbeek, R., Lusk, E., Boyle, J.: Automated Reasoning: Introduction and Applications. Prentice-Hall, Englewood Cliffs (1984)

Automating Leibniz's Theory of Concepts

Jesse Alama[1], Paul E. Oppenheimer[2], and Edward N. Zalta[2(✉)]

[1] Vienna University of Technology, Vienna, Austria
alama@logic.at
[2] Stanford University, Stanford, USA
{paul.oppenheimer,zalta}@stanford.edu

Abstract. Our computational metaphysics group describes its use of automated reasoning tools to study Leibniz's theory of concepts. We start with a reconstruction of Leibniz's theory within the theory of abstract objects (henceforth 'object theory'). Leibniz's theory of concepts, under this reconstruction, has a non-modal algebra of concepts, a concept-containment theory of truth, and a modal metaphysics of complete individual concepts. We show how the object-theoretic reconstruction of these components of Leibniz's theory can be represented for investigation by means of automated theorem provers and finite model builders. The fundamental theorem of Leibniz's theory is derived using these tools.

Keywords: Computational philosophy · Theory exploration · Leibniz · Theory of concepts · Automated reasoning · Theorem prover · Abstract objects · Containment theory of truth · Modal metaphysics · Complete individual concepts · Finite model · Modal logic · Model theory · Representation · Higher-order logic

1 Introduction

The computational metaphysics group at Stanford University's Metaphysics Research Lab has been engaged in a project of implementing object theory, i.e., the axiomatic theory of abstract objects [19,20], in a first-order (with identity) automated reasoning environment. The first efforts [4] established that PROVER9 could be used to represent and derive the theorems that govern possible worlds [22] and Platonic Forms [17]. Our focus over the past several years has been to represent and derive the theorems in [24], a paper that shows how to apply object theory to derive Leibniz's non-modal 'calculus' of concepts, his containment theory of truth, and his modal metaphysics of 'complete individual concepts' (defined below). Leibniz's theory of concepts is still interesting today, for several reasons. The calculus of concepts was one of the first axiomatizations of semi-lattices; his containment theory of truth anticipated work on generalized quantifiers [12]; and the modal metaphysics of complete individual concepts shows how to reconcile Lewis's 'counterpart' interpretation of quantified modal logic [11] with the standard (Kripke) interpretation [8]. Though these features

© Springer International Publishing Switzerland 2015
A.P. Felty and A. Middeldorp (Eds.): CADE-25, LNAI 9195, pp. 73–97, 2015.
DOI: 10.1007/978-3-319-21401-6_4

were explored in detail in [24], we rehearse the basics below, when we sketch the background of our work with automated reasoning tools.

In this paper, we describe not only our results of using E, VAMPIRE, and PARADOX to automate the theory of concepts, but also the obstacles, insights, and other interesting issues that arose during the course of our investigations. The use of automated deduction tools allowed us to find out interesting things about representing richer logics in FOL= (first-order logic with identity), and about the reasoning needed to derive Leibniz's results. In what follows, we focus on the core principles of Leibniz's theory of concepts, and we make no attempt to derive or explain the many other ambitious theories Leibniz attempted to develop (such as his theodicy for explaining the presence of evil, or his plan for world peace by dissolving the ideological case for religious wars).

2   Overview of Object Theory and Two Applications

2.1   The Basics of Object Theory

Object theory [19,20,22,24] is an axiom system formalized in a syntactically second-order modal predicate calculus in which there is a primitive 1-place predicate $E!$ ('concreteness'). (Identity is not primitive; see below). The system uses complex terms of two kinds, namely, definite descriptions and λ-expressions, where the latter denote relations rather than functions. The distinguishing feature of object theory is that the language uses two kinds of atomic formulas:

- $F^n x_1 \ldots x_n$ (for $n \geq 0$)
    These are the atomic ('exemplification') formulas of standard FOL. When $n \geq 1$, these are read as "$x_1, \ldots, x_n$ exemplify F^n" and when $n = 0$, as "F^0 is true".
- xF^1
    These are new, monadic atomic ('encoding') formulas. These are read as "x encodes F^1" and we henceforth drop the superscript on the F^1.

In what follows, we'll often substitute the variables $x, y, z, \ldots$ for $x_1, x_2, \ldots$ and say that they range over *objects* or *individuals*, while the variables $F^n, G^n, \ldots$ range over n-place *relations* (recall that the language of object theory is second-order).

Encoding formulas are best explained by first introducing two defined predicates used in object theory: *being ordinary* ('$O!$') and *being abstract* ('$A!$'). An ordinary object is one that is possibly concrete (i.e., $O!x =_{df} \Diamond E!x$), whereas an abstract object is one that couldn't be concrete (i.e., $A!x =_{df} \neg\Diamond E!x$). Note that these definitions partition the domain of objects.

Intuitively, ordinary objects are the kinds of things we might encounter in experience. They only *exemplify* their properties, and the standard formulas of the classical predicate calculus are sufficient to represent claims about which properties and relations ordinary objects exemplify or stand in.

But abstract objects aren't given in experience; nor is there a Platonic heaven out there containing abstract objects waiting to be discovered. Instead, abstract objects are identified by the properties by which we conceive of them. For example, mathematical objects are abstract objects; the only way we can get information about them is by way of our theories of them. We use encoding formulas to indicate the properties F by which we theoretically conceive of an abstract object x. For example, where κ is a uniquely defined object term in some mathematical theory T, object theory identifies κ as the abstract object that encodes all and only the (mathematical) properties attributed to κ in T (either by assumption or by proof). Though mathematical objects are identified by their *encoded* properties, they also *exemplify* non-mathematical properties. So whereas the number 0 of Peano Number Theory encodes such mathematical properties as $[\lambda x \ x < 1]$, $[\lambda x \ \forall y (y + x = y)]$, etc., it exemplifies non-mathematical properties, such as *being thought about by mathematician z, being abstract, not being a building*, etc. The first of the latter group is contingently exemplified by 0 (depending on the mathematician), while the second and third are necessarily exemplified by 0. However, the properties encoded by 0 *constitute* 0, since they are even more important to the identity of 0 than its necessarily exemplified properties.

We mentioned that there are two kinds of complex terms, definite descriptions and λ-expressions. Descriptions denote individuals, while the n-place λ-expressions denote n-place relations. Definite descriptions have the form $\imath x\phi$, and are to be read: the x in fact such that ϕ. In the modal contexts of object theory, these terms are interpreted rigidly (i.e., semantically, $\imath x\phi$ denotes the unique object that satisfies ϕ, if there is one, at the actual world of the model). λ-expressions have the form $[\lambda x_1 \ldots x_n \phi]$. The principles of α-, β-, and η-conversion for λ-expressions are assumed as axioms, though β-conversion is taken to be an equivalence and not an equation. It is important to note that λ-expressions obey the restriction that ϕ have no encoding subformulas. This is to avoid a Russell-style paradox.[1] As previously mentioned, λ-expressions are to be understood relationally in object theory, not functionally. That is, $[\lambda x_1 \ldots x_n \ \phi]$ doesn't denote an n-ary function, but rather an n-place relation, i.e., an element of a primitive domain of n-place relations.[2]

---

[1] If we were to allow a predicate of the form $[\lambda x \ \exists F (xF \ \& \ \neg Fx)]$, then an abstract object that encodes such a property would exemplify the property if and only if it doesn't. The paradox is avoided by banishing encoding from λ-expressions.

[2] Note that since λ-expressions may not contain encoding subformulas, the comprehension principle for relations derivable from β-Conversion becomes similarly restricted. β-Conversion asserts that $[\lambda x_1 \ldots x_n \ \phi]y_1 \ldots y_n \equiv \phi^{y_1,\ldots,y_n}_{x_1,\ldots,x_n}$. We can universally generalize on each of the y_is to obtain:

$$\forall y_1 \ldots \forall y_n ([\lambda x_1 \ldots x_n \ \phi]y_1 \ldots y_n \equiv \phi^{y_1,\ldots,y_n}_{x_1,\ldots,x_n})$$

Then we apply the Rule of Necessitation and existential generalization to obtain:

$$\exists F^n \Box \forall y_1 \ldots \forall y_n (F^n y_1 \ldots y_n \equiv \phi),$$ provided F^n doesn't occur free in ϕ and ϕ has no encoding subformulas.

This comprehension principle doesn't guarantee that there are any relations definable in terms of encoding predications.

The two main principles governing encoding predications and abstract objects are a comprehension principle and an identity principle for objects. The comprehension principle asserts: for any formula ϕ that places a condition on properties, there is an abstract object that encodes all and only the properties F satisfying (in Tarski's sense) ϕ. The comprehension principle is formalized as a schema:

$$\exists x(A!x \,\&\, \forall F(xF \equiv \phi)), \text{ for any formula } \phi \text{ in which } x \text{ doesn't occur free}$$

Here are some instances:

$$\exists x(A!x \,\&\, \forall F(xF \equiv Fa))$$
$$\exists x(A!x \,\&\, \forall F(xF \equiv F = R \lor F = S))$$
$$\exists x(A!x \,\&\, \forall F(xF \equiv \text{In Peano Number Theory}, F0))$$

These respectively assert the existence of an abstract object that: (a) encodes just the properties exemplified by object a; (b) encodes just the properties R and S, and (c) encodes just the properties attributed to 0 in Peano Number Theory.[3]

Identity is not primitive in object theory. Rather, it is defined for both objects and relations. Since the definition of relation identity doesn't play a role in what follows, we discuss only the definition of identity for objects. We define: $x = y$ iff either x and y are ordinary objects that necessarily exemplify the same properties or x and y are abstract objects that necessarily encode the same properties, i.e.,

$$x = y \;=_{df}\; (O!x \,\&\, O!y \,\&\, \Box\forall F(Fx \equiv Fy)) \lor (A!x \,\&\, A!y \,\&\, \Box\forall F(xF \equiv yF))$$

Thus, if we know that objects x and y are abstract, we have to show that they necessarily encode the same properties to show that they are identical. With identity for objects and relations defined, object theory adds an axiom schema for substitution of identicals.

In addition to the above principles, one other principle is added to the standard principles of second-order quantified modal logic, namely, the claim that if x possibly encodes a property F it necessarily encodes F:

$$\Diamond xF \to \Box xF$$

Thus, encoding predications are not relative to any circumstance; under the standard interpretation of the modal operators, this principle guarantees that if an encoding statement is true at any possible world, it is true at every possible world. By contrast, the truth of exemplification statements (and complex statements containing them) can vary from world to world.

---

[3] The informal construction "In Peano Number Theory, $F0$" can be analyzed in object theory as well. A theory is analyzed as an abstract object that encodes propositions p by encoding the propositional properties of the form $[\lambda y\, p]$ (read this predicate as: being such that p). Then we can define "In theory T, $F0$" as $T[\lambda y\, F0]$, i.e., as T encodes the property *being such that 0 exemplifies F*. This analysis applies to any other mathematical individual κ, mathematical theory T, and constructions of the form "In theory T, $F\kappa$". For a full discussion of the analysis of mathematics within object theory, see [13].

2.2   Application to Possible Worlds

Object theory has been applied in a variety of ways. Our present focus is on Leibniz's theory of concepts, which includes a non-modal calculus of concepts, the concept containment theory of truth, and a modal metaphysics of concepts. However, to represent the latter, we must explain how possible worlds are analyzed in object theory. Though possible worlds (since [7]) are usually taken as semantically-primitive entities and used to formulate truth conditions for modal statements (as we did above, in explaining the axiom $\Diamond xF \rightarrow \Box xF$), object theory takes a different approach. Though the language of object theory includes modal operators, it uses these operators to *define* possible worlds as abstract objects of a certain kind and *derive* the main principles governing worlds from object-theoretic axioms. This, it is claimed, justifies the use of possible worlds when doing modal semantics (including in the modal semantics of object theory). So we will often contrast the possible worlds definable in object theory with the 'semantically primitive possible worlds' used in standard modal semantics that we sometimes reference. In what follows, we first rehearse the object-theoretic analysis of possible worlds and then rehearse the theory of Leibnizian concepts.

Possible worlds are defined as situations, where a *situation* is any abstract object that encodes only propositional properties of the form $[\lambda y\, p]$:

$$Situation(x) \;=_{df}\; A!x \,\&\, \forall F(xF \rightarrow \exists p(F = [\lambda y\, p]))$$

When x is a situation, we say that x *makes* proposition p *true* (or p is *true in* x), written $x \models p$, just in case x encodes *being such that* p:

$$x \models p \;=_{df}\; Situation(x) \,\&\, x[\lambda y\, p]$$

Then, we define a *possible world* to be any situation that *might* be such that it encodes all and only the true propositions [22]:

$$World(x) \;=_{df}\; Situation(x) \,\&\, \Diamond \forall p((x \models p) \equiv p)$$

If we then say that an object x is *maximal* just in case x is a situation and, for every proposition p, either p is true in x or $\neg p$ is true in x, i.e.,

$$Maximal(x) \;=_{df}\; Situation(x) \,\&\, \forall p((x \models p) \vee (x \models \neg p))$$

then it follows that every possible world is maximal:

$$\forall x(World(x) \rightarrow Maximal(x))$$

Moreover, let us say that an object x is *actual* just in case x is a situation such that every proposition true in x is true, i.e.,

$$Actual(x) \;=_{df}\; Situation(x) \,\&\, \forall p((x \models p) \rightarrow p)$$

It then follows that there is a unique actual world. That is, where $\exists! x \phi$ asserts that there is a unique x such that ϕ:

$$\exists! x(World(x) \,\&\, Actual(x))$$

The fundamental theorems of world theory are also provable, namely, that p is necessarily true if and only if p is true in all possible worlds, and p is possibly true if and only if p is true in some possible world [22]:

$$\Box p \equiv \forall x (World(x) \rightarrow x \models p)$$
$$\Diamond p \equiv \exists x (World(x) \,\&\, x \models p)$$

These theorems play an important role in the analysis of Leibniz's modal metaphysics of concepts, in which he asserts that if an object x exemplifies a property F but might not have, then not only does the *individual concept* of x *contain* the *general concept* of F, but there is a *counterpart* of the concept of x that doesn't contain the concept of F and that *appears at* some other possible world. One of our main goals is to represent and prove this claim within an automated reasoning environment.

2.3　Application to Leibniz's Theory of Concepts

As mentioned previously on several occasions, Leibniz's theory of concepts has three components: a non-modal calculus of concepts, the containment theory of truth and a modal metaphysics of concepts. We can integrate and unify all three facets of Leibniz's work by deriving the main theorems of each within object theory. In the remainder of this section, we shall use the variables x, y, z as restricted variables that range just over abstract objects.

Leibniz's Non-modal Calculus of Concepts. The first step of this integration is to recognize that abstract objects serve as a good analysis of Leibnizian concepts generally, so that we may define:

x *is a Leibnizian concept* ('$C!x$') $=_{df} A!x$

Since Leibnizian concepts are abstract objects, we immediately obtain the first three theorems of Leibniz's [9], namely, that identity for concepts is reflexive, symmetric, and transitive. This is derivable from the definition of identity for abstract objects (see Sect. 2.1).[4]

　The two final key definitions that yield, as theorems, the axioms of Leibniz's calculus of concepts [9] are: concept summation ($\oplus$) and concept inclusion ($\preceq$):

$$x \oplus y \;=_{df}\; \imath z(C!z \,\&\, \forall F(zF \equiv xF \vee yF))$$
$$x \preceq y \;=_{df}\; \forall F(xF \rightarrow yF)$$

From these two definitions, it follows that $\oplus$ is an idempotent, commutative and associative operation on the concepts, and that $\preceq$ is a reflexive, anti-symmetric

---

[4] It is an easy logical theorem that $\Box \forall F(xF \equiv xF)$. But then, when x is a concept, it is abstract, and so it follows from the definition of identity that $x = x$. Using the principle of substitution of identicals, we can then derive the symmetry and transitivity of identity for abstract objects.

and transitive condition [24].[5] Thus, if we think of concept summation as a *join* operation, Leibniz's 'calculus' of concepts is in effect a semi-lattice. Though we have not pursued the matter, the semi-lattice can be extended to a lattice by introducing a *meet* operation $x \otimes y$, i.e., concept multiplication, by replacing the disjunction sign in the definition of $x \oplus y$ with an ampersand.

Here are some other key theorems derivable from the above theory of concepts (see [24], Theorems 25–27):

$$x \preceq y \equiv \exists z(x \oplus z = y)$$
$$x \preceq y \equiv x \oplus y = y$$
$$x \oplus y = y \equiv \exists z(x \oplus z = y)$$

Finally, Leibniz's notion of *concept containment* is just the converse of *concept inclusion*:

$$x \succeq y =_{df} y \preceq x$$

Thus, one can prove theorems analogous to the above that are stated in terms of concept containment instead of concept inclusion.

We have not yet brought automated reasoning tools to bear on the above object-theoretic *reconstruction* of Leibniz's algebra of concepts. Instead, the focus of our investigations was on the work described in the remainder of this section. However, in a separate project using PROVER9, we verified *versions* of the above theorems in which $\oplus$ and $\preceq$ were taken as primitive and axiomatized instead of defined as in object theory.[6]

Leibniz's Containment Theory of Truth. Though Leibnizian concepts are identified generally as abstract objects, special kinds of Leibnizian concepts can be defined. For example, there are general concepts of properties (e.g., the concept of being a king, etc.) and concepts of individuals (e.g., the concept of Alexander the Great). Both play a role in Leibniz's containment theory of truth.

First, we define "the concept of F" ('c_F') as follows:

$$c_F =_{df} \imath x(C!x \,\&\, \forall G(xG \equiv F \Rightarrow G))$$

In this definition, $F \Rightarrow G$ is itself defined as necessary implication: $\Box \forall x(Fx \rightarrow Gx)$. Thus, c_F is the concept that encodes exactly the properties that are necessarily implied by the property F.

Next, we define 'the concept of individual u' ('c_u'), where u is a restricted variable ranging over ordinary individuals, as follows:

$$c_u =_{df} \imath x(C!x \,\&\, \forall F(xF \equiv Fu))$$

---

[5] Note we say *condition* here rather than *relation* because the definition of $\preceq$ has encoding subformulas and, as such, $\preceq$ is not guaranteed to be a relation.

[6] See http://mally.stanford.edu/cm/leibniz/ for a description of this work and links to all the input and output files.

For example, the concept of Alexander (c_a) is the concept that encodes exactly the properties Alexander (a) exemplifies. Note that in this example, Alexander gets correlated with a concept that contains his properties. If we restate this using the concepts of simple set theory: the proper name 'a' is correlated with a set of properties. This recalls the treatment of proper names as generalized quantifiers [12].[7]

The definitions of c_F and c_u put us into a position to represent Leibniz's containment theory of truth. Leibniz asserts that a simple predication 'Alexander is king' ('Ka') is to be analyzed as: the concept Alexander contains the concept of being a king.[8] Where c_K is the concept of being a king and c_a is the concept of Alexander, we can represent Leibniz's analysis as:

$$c_a \succeq c_K$$

The equivalence of Ka and Leibniz's analysis $c_a \succeq c_K$ is *derivable* in object theory, since it is a general theorem that for any ordinary object u and property F:

Theorem 38, [24]:

$$Fu \equiv c_u \succeq c_F$$

It is important to note that Theorem 38 is an example of a theorem that isn't a necessary truth. The reason is easy to see if we take possible worlds as semantically primitive and speak in terms of the classical semantics of modal logic: the formula Fu can change in truth value from world to world, but the formula $c_u \succeq c_F$ uses a term, c_u, that is rigidly defined in terms of what is true at the actual world. c_u encodes all and only the properties that u in fact exemplifies. Hence, if u is in fact F (i.e., Fu is true at the actual world w_0) and there is a world w_1 where u fails to be F, then the left side of Theorem 38 is false at w_1

---

[7] Indeed, if we define 'the concept of every person' as the concept that encodes exactly the properties F such that every person exemplifies F, then the containment theory of truth described below will offer a unified 'subject-predicate' analysis of the sentences 'Alexander is happy' and 'Every person is happy'. On the containment theory of truth, the former is true because the concept of Alexander contains the concept of being happy, while the latter is true because the concept of every person contains the concept of being happy.

[8] Leibniz asserts his containment theory of truth in the following passage taken from the translation in [15], 18–19 (the source is [3], 51):

> ...every true universal affirmative categorical proposition simply shows some connection between predicate and subject (a *direct* connexion, which is what is always meant here). This connexion is, that the predicate is said to be in the subject, or to be contained in the subject; either absolutely and regarded in itself, or at any rate, in some instance; i.e., that the subject is said to contain the predicate in a stated fashion. This is to say that the concept of the subject, either in itself or with some addition, involves the concept of the predicate.

The translator titled the fragment from which this passage is taken as 'Elements of a Calculus'.

while the right side of Theorem 38 is true at w_1 (c_u is the object that encodes all the properties that u exemplifies at w_0, and since u does in fact exemplify F at w_0, one can show that c_u contains c_F).[9] The fact that the proof of Theorem 38 rests on a contingency actually works in Leibniz's favor: he introduced the notion of a hypothetical necessity in response to (his contemporary) Antoine Arnauld, who charged that he mistakenly analyzed the contingent claim Ka in terms of the necessary claim $c_a \succeq c_K$. Theorem 38 is indeed a kind of hypothetical necessity, if we understood that to mean that its proof depends on a contingency and that we can detach and then prove $c_a \succeq c_K$ only given the contingent Ka as a premise.

The above facts about Theorem 38 can be traced back to an interesting feature of object theory, namely, that rigid definite descriptions are governed by a logical axiom that fails to be a necessary truth.[10] When one includes such a rigidifying operator that is semantically interpreted with respect to the facts at the actual world of the model, then one must stipulate: the Rule of Necessitation may not be applied to any axiom governing the operator having the form of a conditional in which a (potentially contingent) formula ϕ appears on one side of the conditional in a non-rigid context and appears on the other side of the conditional in a rigid context. Nor can the Rule of Necessitation be allowed to apply to any theorem derived from such axioms. Note that the actuality operator is similar to the rigid definite description operator in this regard; for a full discussion of logical truths that aren't necessary, see [21]. As we shall see, this issue won't surface when we represent definite descriptions, since primitive descriptions will be eliminated under the standard Russellian analysis.

Leibniz's Modal Metaphysics of Concepts. Finally, we reach the most compelling ideas in Leibniz's metaphysics, in which he uses the containment theory of truth to analyze modal predications and, in the course of doing so, relates concepts of individuals and possible worlds. In various passages, Leibniz talks about possible individuals.[11] However, it is generally thought that Leibniz's

---

[9] It is interesting to note that for some properties G, the proof that c_u encodes G will rest on a contingency (namely, when G is a property contingently exemplified by u), but the proof that c_F encodes G doesn't. That's because it is provable that if $F \Rightarrow G$ then $\Box(F \Rightarrow G)$.

[10] Instead of stating this logical axiom in its full generality, here is an example of an instance:

$$F\imath x Gx \equiv \exists x(Gx \,\&\, \forall y(Gy \rightarrow y{=}x) \,\&\, Fx)$$

This is a version of Russell's analysis of definite description, first described in [18]. If the description $\imath x Gx$ rigidly denotes the object that is uniquely G at the actual world (assuming there is one), then the above principle will fail to be necessarily true if there is a unique G at the actual world that is F at a world w_1, but where nothing is G at w_1 or where two distinct things are G.

[11] See, for example, the *Theodicy* ([10], 371 = [5], vi, 363), where he talks about the 'several Sextuses', and in a letter to Hessen-Rheinfels, where he talks about the 'many possible Adams' ([16], 51 = [5], ii, 20).

containment theory of truth was designed to replace talk of individuals having properties with talk of containment holding between concepts. Consequently, most commentators believe that we should interpret Leibniz's references to possible individuals as references to *concepts of individuals*. Leibniz does after all say that a concept of an individual may *appear* at a (unique) possible world.

To represent these ideas, we continue to use u as a restricted variable over ordinary individuals and use w as a restricted variable ranging over possible worlds (i.e., the worlds defined in an Sect. 2.2). We then define:

$$RealizesAt(u, x, w) =_{df} \forall F((w \models Fu) \equiv xF)$$
$$AppearsAt(x, w) =_{df} \exists u RealizesAt(u, x, w)$$
$$IndividualConcept(x) =_{df} \exists w AppearsAt(x, w)$$

From these definitions, it follows that every individual concept appears at a unique world ([24], Theorem 31):

$$IndividualConcept(x) \rightarrow \exists! w AppearsAt(x, w)$$

Moreover, not only is there a concept of the individual Alexander (which we've defined as c_a), but for each possible world w, there is a concept of the individual of Alexander at w, c_a^w, which encodes exactly the properties F that Alexander exemplifies at w. Thus, we may define:

$$c_u^w =_{df} \iota x(C!x \,\&\, \forall F(xF \equiv w \models Fu))$$

Where w_α is the actual world, it is easy to show that c_a is identical to $c_a^{w_\alpha}$ (i.e., the concept of Alexander is identical to the concept of Alexander at the actual world). Moreover, one can show that for any ordinary individual u, c_u is an individual concept, and that for any ordinary individual u and world w, c_u^w is an individual concept:

$$\forall u\, IndividualConcept(c_u)$$
$$\forall u, w\, IndividualConcept(c_u^w)$$

These facts put us in a position to see that both the Kripkean [8] and Lewisian [11] interpretation of possible objects can exist side-by-side (though Lewis's possible individuals are represented at the level of concepts). Kripke believes that a modal claim such as "Obama might have had a son" is true because

there is a possible world where *Obama himself* has a son.

By contrast, Lewis takes this modal claim to be true because

there is a possible world where *a counterpart of Obama* has a son.

The precise Leibnizian picture we've developed enables us to show how Kripke's view holds with respect to ordinary individuals, while Lewis's view holds with respect to Leibnizian complete individual concepts.

To see that Kripke's view holds with respect to ordinary individuals, we need only observe that it is Obama himself who fails to have a son in our world but who has a son at some other world. However, to see why Lewis's view holds with respect to concepts of ordinary individuals, we must first partition the concepts of ordinary individuals into groups of counterparts. Let (italic, non-bold) c, c' range over individual concepts. Then we may define:

$$Counterparts(c, c') =_{df} \exists u \exists w_1 \exists w_2 (c = \boldsymbol{c}_u^{w_1} \,\&\, c' = \boldsymbol{c}_u^{w_2})$$

In other words, individual concepts c and c' are counterparts whenever there is an ordinary object u and worlds w_1 and w_2 such that c is the concept of u-at-w_1 and c' is the concept of u-at-w_2. So, if w' is a world where Obama does have a son, $\boldsymbol{c}_o^{w'}$ is a counterpart of the concept of Obama ($\boldsymbol{c}_o$), given that $\boldsymbol{c}_o = \boldsymbol{c}_o^{w_\alpha}$. Obama, $\boldsymbol{w}_\alpha$ and w' are thus witnesses to the definition of $Counterparts(\boldsymbol{c}_o, \boldsymbol{c}_o^{w'})$.

Now, as we saw above, individual concepts appear at a unique world. So they are, in some sense, world-bound individuals. Thus, we obtain Lewis-style truth conditions for modal claims in the domain of Leibnizian individual concepts, given the following *fundamental theorem* (applied to Alexander):

If Alexander is king but might not have been, then:

(a) the concept of Alexander contains the concept of being a king, and
(b) some individual concept that is a counterpart to the concept of Alexander fails to contain the concept of being a king and *appears at* some non-actual possible world.

Formally and generally, where u is any ordinary object, c is any individual concept, and F is any property, we have:

Theorem 40a [24]:
$(Fu \,\&\, \Diamond \neg Fu) \rightarrow [\boldsymbol{c}_u \succeq \boldsymbol{c}_F \,\&$
$\quad \exists c(Counterparts(c, \boldsymbol{c}_u) \,\&\, c \not\succeq \boldsymbol{c}_F \,\&\, \exists w(w \neq \boldsymbol{w}_\alpha \,\&\, Appears(c, w)))]$

Similarly, we have:

If Obama doesn't have a son but might have, then:

(a) the concept of Obama fails to contains the concept of having a son, and
(b) some individual concept that is a counterpart to the concept of Obama contains the concept of having a son and *appears at* some non-actual possible world.

Formally and more generally, this becomes:

Theorem 40b [24]:
$(\neg Fu \,\&\, \Diamond Fu) \rightarrow [\boldsymbol{c}_u \not\succeq \boldsymbol{c}_F \,\&$
$\quad \exists c(Counterparts(c, \boldsymbol{c}_u) \,\&\, c \succeq \boldsymbol{c}_F \,\&\, \exists w(w \neq \boldsymbol{w}_\alpha \,\&\, Appears(c, w)))]$

The main goal of our efforts to implement Leibniz's modal metaphysics computationally was to obtain a proof of the above two theorems using an automated reasoning engine. In what follows, we'll work our way to an understanding of our computational implementation of Theorem 40a.

3   Summary of Our Representational Techniques

The fundamental idea behind our work in implementing object theory in an automated reasoning environment is this: instead of building a customized theorem prover that understands the syntax of object theory, we use the language of standard theorem provers to represent the *first-order truth conditions* of the statements of object theory. The first order truth conditions can be understood in terms of the object theory's natural semantics and model theory. This is entirely appropriate because the minimal models of object theory reveal that despite its second-order syntax, it has general Henkin models [6]. We did not employ a mechanical procedure for translating the formulas of object theory into TPTP syntax, but rather carried it out by hand. In constructing the translations, we adopted various conventions, some of which are discussed below.

Our basic convention was to translate the second-order quantified modal syntax of object theory into $\mathsf{FOL}_=$, supplemented with predicates that sort individuals into four domains: objects, properties, propositions, and points. Our representation can be written directly in TPTP syntax without further processing. In what follows, we summarize the techniques we developed in order to produce TPTP problem files for the theorems in [24].

3.1   Representing Second-Order Syntax Using First-Order Syntax

The most important first step of the process is to recognize that in the semantics of object theory, 1-place properties have an exemplification extension among objects and that this extension varies from possible world to possible world. However, since possible worlds are going to be one of the targets of object-theoretic analysis, we call the semantically-primitive possible worlds *points*. Moreover, we refer to 0-place relations as *propositions*. Thus to translate modal claims involving the individual variables $x, y, z, \ldots$, property variables $F^1, G^1, \ldots$ and propositional variables $F^0, G^0, \ldots$ (which we write using $P, Q, \ldots$), we introduce the following basic sorts:

```
object(X)
property(F)
proposition(P)
point(D)
```

Moreover, since the simple and complex predications in object theory take place in a modal language, we adopted the convention of introducing an extra argument place in primitive or defined conditions, which relativizes them with respect to a point D. When translating explicitly modal claims, that extra argument place can be bound by a quantifier over points.

Whereas uppercase D is a variable ranging over points, we represent a basic (non-modal) predication of the form Fx using a named point:

```
ex1_wrt(F,X,d)
```

Here, d is the semantically primitive 'actual world' that serves as the distinguished element of the domain of possible worlds found in classical modal semantics. In general, then, since a formula like Fx can appear within a modal context, we represent it with the primitive condition ex1_wrt(F,X,D), which has an argument place for point D. Note that we also add as an axiom the *right-handed sorting* rule that asserts that if ex1_wrt(F,X,D), then F is a property, X is an object, and D is a point:

```
fof(sort_ex1_wrt,type,
  (! [F,X,D] : (ex1_wrt(F,X,D) =>
  (property(F) & object(X) & point(D)))))).
```

This, in effect, tells the theorem prover that we're primarily interested in models in which the arguments of the relation ex1_wrt are entities of the appropriate sorts. Such right-handed sorting rules allow us to represent facts that come for free in a second-order language.

Next, we represent a basic modal predication of the form $\Box Fx$ as:

```
(! [D] : (point(D) => ex1_wrt(F,X,D)))
```

Possibility claims are represented in a similar way, using existential quantifications over points.

To represent the primitive predicate '$E!x$' and the defined predicates '$O!x$' and '$A!x$', we introduced the property constants e, o and a. Just as in object theory, e is primitive. But o and a are defined as (cf. the definitions described in Sect. 2.1):

```
fof(o,definition,
  (! [X,D] : ((object(X) & point(D)) => (ex1_wrt(o,X,D) <=>
  (? [D2] : (point(D2) & ex1_wrt(e,X,D2))))))))).
```

```
fof(a,definition,
  (! [X,D] : ((object(X) & point(D)) => (ex1_wrt(a,X,D) <=>
  ~(? [D2] : (point(D2) & ex1_wrt(e,X,D2))))))))).
```

Now to introduce the constant c to denote the property of *being a concept*, we asserted that the property of being a concept is identical to the property of being abstract, i.e.,

```
fof(being_a_concept_is_being_abstract,axiom,c=a).
```

The above techniques are an important first step towards solving the problem of representing the object-theoretic definitions of the various metaphysical kinds used by Leibniz, such as concepts of individuals, concepts of properties, possible worlds, etc.

3.2   Representing the Two Modes of Predication

Now the distinguishing feature of the language of object theory is that it has two fundamental modes of predication. In addition to predications of the form Fx

(familiar from standard FOL), there are also predications of the form xF. We represent the latter as enc_wrt(X,F,D), and require the following right-handed sorting rule:

```
fof(sort_enc_wrt,type,
  (! [X,F,D] : (enc_wrt(X,F,D) =>
  (object(X) & property(F) & point(D))))).
```

The formula enc_wrt(X,F,D) will appear in several of the definitions that are given below.

3.3   Representing Identity Claims

Recall the disjunctive definition of object-theoretic identity $x = y$ in Sect. 2.1. To represent that definition, we developed two preliminary definitions, one for the identity of ordinary objects (o_equal_wrt) and one for the identity of abstract objects (a_equal_wrt).[12] We then represented $x = y$ in terms of the general notion object_equal_wrt(X,Y,D) by stipulating that object_equal_wrt(X,Y,D) holds if and only if either o_equal_wrt(X,Y,D) or a_equal_wrt(X,Y,D), as follows:

```
fof(object_equal_wrt,definition,
  (! [X,Y,D] : ((object(X) & object(Y) & point(D)) =>
  (object_equal_wrt(X,Y,D) <=>
  (o_equal_wrt(X,Y,D) | a_equal_wrt(X,Y,D))))))).
```

Finally, we then connected general identity object_equal_wrt(X,Y,D) with the built-in equality of the reasoning system. This *bridge principle* asserts:

```
fof(object_equal_wrt_implies_identity,theorem,
  (! [X,Y] : ((object(X) & object(Y)) =>
  (? [D] : (point(D) & object_equal_wrt(X,Y,D)) =>
  X = Y)))).
```

That is, if objects X and Y are object_equal at some point D, then they are identical. This definition suffices because it is a theorem that any objects o_equal

---

[12] The definition of o_equal_wrt is:

```
fof(o_equal_wrt,definition,
  (! [X,Y,D] : ((object(X) & object(Y) & point(D)) =>
  (o_equal_wrt(X,Y,D) <=> (ex1_wrt(o,X,D) & ex1_wrt(o,Y,D) &
  (! [D2] : (point(D2) => (! [F] : (property(F) =>
  (ex1_wrt(F,X,D2) <=> ex1_wrt(F,Y,D2)))))))))))).
```

This says that X and Y are o_equal with respect to point D just in case X and Y are both ordinary objects and at every point D2, they exemplify the same properties. A similar definition defines: X and Y are a_equal with respect to point D just in case X and Y are both abstract objects and at every point D2, they encode the same properties.

at some point are o_equal at every point, and also a theorem that any objects a_equal at some point are a_equal at every point.

With the above bridge principle in place, inferences about object-theoretic identity can be drawn by the automated reasoning engine using system equality (e.g., via demodulation).

3.4 Representing Definite Descriptions

Here are two examples of how we represent definite descriptions. Recall that we introduced the term c_u to denote *the* concept c that encodes all and only the properties that u in fact exemplifies, and we introduced the term c_F to denote *the* concept c that encodes all and only the properties necessarily implied by F. Consider first how we represent c_u. We begin by introducing the relational condition concept_of_individual_wrt(X,U,D), which holds just in case U is an ordinary object and X is a concept (i.e., abstract object) that encodes exactly the properties F such that U exemplifies F:

```
fof(concept_of_individual_wrt,definition,
  (! [X,U,D] : ((object(X) & object(U) & point(D)) =>
  (concept_of_individual_wrt(X,U,D) <=> (ex1_wrt(c,X,D) &
  ex1_wrt(o,U,D) & (! [F] : (property(F) =>
  (enc_wrt(X,F,D) <=> ex1_wrt(F,U,D)))))))))).
```

We then introduce is_the_concept_of_individual_wrt(X,U,D) as holding whenever X is a concept of individual U with respect to point D and anything Z that is a concept of individual U with respect to point D is object_equal to X:

```
fof(is_the_concept_of_individual_wrt,definition,
  (! [X,U,D] : ((object(X) & object(U) & point(D)) =>
  (is_the_concept_of_individual_wrt(X,U,D) <=>
  (concept_of_individual_wrt(X,U,D) &
  (! [Z] : (concept_of_individual_wrt(Z,U,D) =>
  object_equal_wrt(Z,X,D)))))))).
```

Here X corresponds to c_u when is_the_concept_of_individual_wrt(X,U,d).

Consider second how we represent c_F. We begin by introducing the relational condition concept_of_wrt(Y,F,D), which holds just in case Y is a concept that encodes just the properties necessarily implied by F. Formally:

```
fof(concept_of_wrt,definition,
  (! [Y,F,D] : ((object(Y) & property(F) & point(D)) =>
  (concept_of_wrt(Y,F,D) <=>
  (ex1_wrt(c,Y,D) & (! [G] : (property(G) =>
  (enc_wrt(Y,G,D) <=> implies_wrt(F,G,D)))))))))).
```

Then we introduce is_the_concept_of_wrt(Y,F,D) as holding whenever Y is a concept of property F at point D and anything Z that is a concept of F at D is object_equal to Y:

```
fof(is_the_concept_of_wrt,definition,
 (! [Y,F,D] : ((object(Y) & property(F) & point(D)) =>
 (is_the_concept_of_wrt(Y,F,D) <=>
 (concept_of_wrt(Y,F,D) & (! [Z] : (object(Z) =>
 (concept_of_wrt(Z,F,D) => object_equal_wrt(Z,Y,D)))))))))).
```

Thus, Y corresponds to c_F when is_the_concept_of_wrt(Y,F,d).[13] All of the above definitions play an important role in the statement and proof of the fundamental theorem of Leibniz's modal theory of concepts.

3.5 Representing λ-Expressions

As an example of how we represented λ-expressions, consider $[\lambda z\ Py]$, which denotes the property: being a z such that y exemplifies P. We discussed such λ-expressions in footnote 3 and at the beginning of Sect. 2.2 (on world theory). To properly understand these expressions, note that, in object theory, it is a theorem that for every property P and object y, there exists a proposition Py. Moreover, for each such proposition, object theory's comprehension principle for properties asserts (cf. footnote 2):

$$\exists F\Box\forall x(Fx \equiv Py)$$

i.e., there is a property F such that, necessarily, an object x exemplifies F if and only if Py. We use the λ-expression $[\lambda z\ Py]$ to denote such a property. It obeys the principle:

$$\Box([\lambda z\ Py]x \equiv (Py)_z^x)$$

However, the variable z bound by the λ in $[\lambda z\ Py]$ is vacuously bound since it doesn't appear in Py. So $(Py)_z^x$ (i.e., the result of substituting x for z in Py) is just the formula Py. Hence we have: $\Box([\lambda z\ Py]x \equiv Py)$, i.e., necessarily, x exemplifies *being such that* Py if and only if Py.

We represented these facts as follows:

```
fof(existence_proposition_plug1,axiom,
 (! [X,F] : ((object(X) & property(F)) =>
 (? [P] : (proposition(P) & plug1(P,F,X))))))).
```

```
fof(proposition_plug1_truth,definition,
 (! [X,F,P] : ((object(X) & property(F) & proposition(P)) =>
 (plug1(P,F,X) => (! [D] : (point(D) =>
 (true_wrt(P,D) <=> ex1_wrt(F,X,D)))))))))).
```

---

[13] It is important to note here that, in contrast to the concept of an individual, we need not have linked Y to the concept of F at the distinguished point d, given what we said in footnote 9.

```
fof(existence_vac,axiom,
 (! [P] : (proposition(P) =>
  (? [Q] : (property(Q) & is_being_such_that(Q,P)))))).
```

```
fof(truth_wrt_vac,axiom,
 (! [P,Q] : ((proposition(P) & property(Q)) =>
  (is_being_such_that(Q,P) =>
   (! [D,X] : ((point(D) & object(X)) =>
   (ex1_wrt(Q,X,D) <=> true_wrt(P,D))))))))).
```

The first asserts that for any property F and object X, there is a proposition P obtained by *plugging* X into F, where the truth conditions for plugging are defined by the second principle as: if P is the proposition obtained by plugging X into F, then for every point D, P is true with respect to D whenever X exemplifies F with respect to D. The third asserts that for every proposition P, there is a property Q such that Q is being such that P. The fourth asserts that if Q is being such that P, then for any point D and object X, X exemplifies Q at D if and only if P is true at D.

4   Representing the Fundamental Theorem

We now apply the techniques just summarized to the representation of one of the two fundamental theorems described at the end of Sect. 2, namely, Theorem 40a. The antecedent of Theorem 40a is:

$$Fu \,\&\, \Diamond \neg Fu \tag{A}$$

Since the variable 'u' is a restricted variable ranging over ordinary objects, we represent the first conjunct as:

```
ex1_wrt(o,U,d) & ex1_wrt(F,U,d),
```

where o is the property of being ordinary, F is a variable ranging over properties, U is a variable ranging over objects, and d is the distinguished point. By using sortal predicates to make everything explicit, the conjunction (A) can be represented as follows:

```
object(U) & property(F) & ex1_wrt(o,U,d) & ex1_wrt(F,U,d) &
  (? [D] : (point(D) & ~ex1_wrt(F,U,D)))
```
$$\tag{$\overline{\text{A}}$}$$

In other words, the antecedent of Theorem 40a becomes:

(If) U is an object, F is a property, U exemplifies being ordinary at d, U exemplifies F at d, and there is a point D such that U fails to exemplify F at D, (then) ...

Now the first conjunct of the consequent of Theorem 40a is:

$$c_u \succeq c_F \tag{B}$$

This is not a theorem of object theory, though it follows from the facts that $Fu \equiv c_u \succeq c_F$ (referenced earlier as Theorem 38a) and the premise Fu. Hence it follows from the antecedent of Theorem 40a. If we recall the earlier definitions of c_u, c_F and $\succeq$, we can represent this clause as follows, in which c_u is represented by X, c_F is represented by Y and $\succeq$ is represented by contains_wrt(X,Y,d):

> (? [X,Y] : object(X) & object(Y) & ex1_wrt(c,X,d) &
> ex1_wrt(c,Y,d) & is_the_concept_of_individual_wrt(X,U,d) &
> is_the_concept_of_wrt(Y,F,d) & contains_wrt(X,Y,d))   ($\overline{\mathrm{B}}$)

The first four conjuncts of ($\overline{\mathrm{B}}$) tell us that X and Y are both objects and, in particular, both concepts. The fifth and sixth conjuncts of ($\overline{\mathrm{B}}$), i.e.,

> is_the_concept_of_individual_wrt(X,U,d)
> is_the_concept_of_wrt(Y,F,d)

were defined in Sect. 3.4. We therefore know that the X and Y asserted to exist in ($\overline{\mathrm{B}}$) corresponds to c_u and c_F, respectively, in the language of object theory.

The final conjunct of ($\overline{\mathrm{B}}$) is contains_wrt(X,Y,d). This asserts that object X contains object Y at point d, where this is defined this as:

> fof(contains_wrt,definition,
> (! [X,Y,D] : ((object(X) & object(Y) & point(D)) =>
> (contains_wrt(Y,X,D) <=> included_in_wrt(X,Y,D)))))).

Here, included_in_wrt(X,Y,D) is defined as you might expect given our discussion of $\succeq$ in Sect. 2.3.[14]

Finally, we turn to the second conjunct of the consequent of Theorem 40a. It asserts:

$$\exists c(\mathit{Counterparts}(c, c_u) \,\&\, c \not\succeq c_F \,\&\, \exists w(w \neq w_\alpha \,\&\, \mathit{Appears}(c, w)))$$

Given that we already know X represents c_u and Y represents c_F, this becomes represented as follows:

> (? [Z] : (object(Z) & ex1_wrt(c,Z,d) & counterparts_wrt(Z,X,d) &
> ~contains_wrt(Z,Y,d) & (? [A,W] : (object(A) & object(W) &
> is_the_actual_world_wrt(A,d) & world_wrt(W,d) &
> ~equal_wrt(W,A,d) & appears_in_wrt(Z,W,d)))))).

We can now represent Theorem40a:

---

[14] The definition is:

> fof(included_in_wrt,definition,
> (! [X,Y,D] : ((object(X) & object(Y) & point(D)) =>
> (included_in_wrt(X,Y,D) <=>
> (ex1_wrt(c,X,D) & ex1_wrt(c,Y,D) &
> (! [F] : (property(F) => (enc_wrt(X,F,D) => enc_wrt(Y,F,D)))))))))).

```
fof(theorem_40a,conjecture,
 (! [U,F] : ((object(U) & property(F)) =>
 ((ex1_wrt(o,U,d) & ex1_wrt(F,U,d) &
 (? [D] : (point(D) & ~ex1_wrt(F,U,D)))) =>
 (? [X,Y] : (object(X) & object(Y) & ex1_wrt(c,X,d) &
 ex1_wrt(c,Y,d) & is_the_concept_of_individual_wrt(X,U,d) &
 is_the_concept_of_wrt(Y,F,d) & contains_wrt(X,Y,d) &
 (? [Z] : (object(Z) & ex1_wrt(c,Z,d) &
 counterparts_wrt(Z,X,d) & ~contains_wrt(Z,Y,d) &
 (? [A,W] : (object(A) & object(W) &
 is_the_actual_world_wrt(A,d) & world_wrt(W,d) &
 ~equal_wrt(W,A,d) & appears_in_wrt(Z,W,d))))))))))).
```

Theorem 40b has a similar representation. The problem files for both Theorem40a and Theorem40b are available online.[15] A web page containing the links to all the relevant files is also available.[16]

5   Techniques for Speeding up the Workflow

Our work consisted of developing a theory (i.e., a structured collection of theorems and definitions) in TPTP notation. Faced with the task of adding premises to a TPTP file for some conjecture, we generally used [24] as a guide. To formalize (proofs of) theorems, we proceeded in the usual naive way. If the original proof referred to a previous theorem, we included it in the TPTP file. Similarly, whenever defined notions appeared in a conjecture or premise, we included their definitions in the TPTP file. Our workflow can intuitively be understood as taking a kind of poor man's closure operation: first we looked at the primitive and defined notions that appeared in the conjecture to be proved and then we kept adding to the file those axioms, definitions, and previous theorems governing the primitive and defined notions that we thought would be needed to yield a proof of the conjecture.

To facilitate constructing this "closure", we wrote a script that inspects a TPTP file and reports on the predicate symbols and function symbols appearing in it. (The script is essentially just a front end to the standard GetSymbols tool distributed with TPTP). The procedure is embarrassingly simple, but it prevented us from wasting time trying to diagnose a countersatisfiable conjecture that failed because of lack of information about one of the notions in the conjecture. The script thus highlighted in red those symbols that occur exactly once in the problem (i.e., *hapax legomena*). This is a quick check for whether there is a gap in the problem because, intuitively, a predicate or function that occurs exactly once is a red flag. Of course, such a heuristic is far from complete. In certain situations, a problem may well be solvable while having symbols that appear exactly once. Equality counts as an undefined binary predicate symbol

---

[15] See http://mally.stanford.edu/cm/concepts/theorem40a.p and theorem40b.p.
[16] See http://mally.stanford.edu/cm/concepts/.

from our program's point of view, and at times we worked with problems where a single equation was present in the problem. At other times, it is acceptable for there to be primitive (undefined) notions that by chance do occur exactly once.

Another useful program we found ourselves in need of and developed was a tool for running multiple theorem provers and extracting the sets of premises used in the proofs (TSTP/TPTP derivations). It is quite interesting to see that different theorem provers "react" differently to one and the same theorem proving problem. After relying on the theorem provers to tell us which premises they used in a proof, we systematically tried removing premises and testing for the existence of a proof or a countermodel using the reduced set of premises. What we found is that our premise sets were almost always bigger than necessary. At times, a surprisingly large number of premises could be cut. We were often delightfully puzzled into rethinking our initial proof because we had expected that certain lemmas or definitions could not be removed. In order to be very clear about the power of the axioms, we wanted the extra insight that came from minimizing the premise sets.

The above tools, developed to prune unneeded concepts and premises from the proof of a given conjecture, were bundled together to make the TIPI program, which is written in Perl and available online.[17] TIPI was constructed as a catch-all tool for our formalization project. TIPI has proved of value in other formalization projects as well [1,2].

When we first started our project, we produced a separate input file for each theorem in [24]. This works fine as long as one doesn't end up having to go back and redo theorems when representational improvements are discovered. In the course of working on theorem-proving problems, we often had to make on-the-fly adjustments to our representations. If one regards our theory as a kind of tree of dependencies, in which formulas depend (either by definition or by derivation) on other formulas, such on-the-fly changes can quickly lead to confusion. One may confidently think that a small change to a formula ϕ makes no difference to the provability of a theorem ψ that depends it, only to be shown wrong by a countermodel. In general, *any* change to a formula reopens the question of whether some other dependent formula is provable; indeed, any change to a prior axiom, definition, or theorem, requires checking all the theorems that depended on that changed formula. Once dozens of theorems are involved, one has to ensure that axioms, definitions, and theorems are kept synchronized across multiple files. We sometimes were satisfied that a problem was solved only to have to revisit the problem when we discovered we could correct or improve the formulation some principle.

This problem of dependence is of course not unique to theorem proving; it is clearly an old, well-recognized issue in software engineering generally. In our case, we designed a suite of makefiles to help keep ourselves honest about the status of our theory as we made changes to it. The solution we arrived at is to regenerate problems from a master file. Each formula capable of generating a problem, that is, each theorem or lemma, is annotated with the axioms, definitions, sorts, theorems,

---

[17] https://github.com/jessealama/tipi and http://arxiv.org/abs/1204.0901.

and lemmas that are used in its derivation. The TPTP problem files are generated automatically from a master file containing the latest version of the dependencies.

6   Observations

6.1   What We've Learned

Although our computational study of [24] didn't reveal any errors of reasoning, we did come away from the research with some new insights about the implementation of object theory using automated reasoning tools. One of the interesting things we learned concerned the demands that our representational methods placed on the definition of notions from object theory. Originally, we thought that it might help cut down the search space for proofs if we prefaced each definiendum with an antecedent that both sorted the variables and also introduced any restrictions on the variables. After all, reasoning with restricted variables eliminates inference steps and thus potential errors of reasoning. So, for example, situations are definable in object theory as abstract objects of a certain kind. We wondered whether proof search would be more efficient with this restriction, i.e., if we defined the world-relative condition situation_wrt(X,D) only for those objects X known to be abstract, as follows:

```
fof(situation_wrt,definition,
  (! [X,D] : ((object(X) & point(D)) => (ex1_wrt(a,X,D) =>
  (situation_wrt(X,D) <=> (! [F] : (property(F) => (enc_wrt(X,F,D) =>
  (? [P] : (proposition(P) & is_being_such_that(F,P)))))))))))).
```

However, in the end, we discovered that our provers do better if we don't use restricted variables when introducing the definiendum. The more general way of formulating such definitions is to introduce the definiendum as soon as the variables in the argument places are sorted. On that method, the above definition becomes:

```
fof(situation_wrt,definition,
  (! [X,D] : ((object(X) & point(D)) => (situation_wrt(X,D) <=>
  (ex1_wrt(a,X,D) & (! [F] : (property(F) => (enc_wrt(X,F,D) =>
  (? [P] : (proposition(P) & is_being_such_that(F,P))))))))))).
```

Thus, we adhered to the following format for introducing an n-place condition Definiendum(X1,...,Xn):

```
(! [X1,...,Xn]: ((sort1(X1) & ... & sortn(Xn)) =>
(Definiendum(X1,...,Xn) <=> ...X1...Xn...))).
```

Another interesting question that arose was when to formulate our theorems in their most general modal form, i.e., as necessary truths (prefaced by a universal quantifier over all points), as opposed to formulating them as non-modal facts that hold just of the distinguished point d. In many modal systems, this question doesn't arise since every theorem is a necessary truth. But object theory allows for reasoning with contingent premises and with a contingent axiom governing rigid definite descriptions. We noted earlier that the presence of rigid definite descriptions in certain contexts is a tip-off that it may be inappropriate to use the Rule of Necessitation (see especially the discussion following Theorem 38). One has to keep track of theorems that are proved with a contingent premise or that depend on the contingent axiom governing rigid definite descriptions.

In Sect. 2.3, we introduced Theorem 38, which asserts that an ordinary object u exemplifies F if and only if the concept of individual u contains the concept of property F. This theorem is a key part of the proof of Theorem 40a. It is important to recognize that Theorem 38 should be proved only in the following form (as a fact about the distinguished point d):

```
fof(theorem38,theorem,
  (! [U,F] : ((object(U) & property(F)) => (ex1_wrt(o,U,d) =>
  (? [Y,Z] : (object(Y) & object(Z) & ex1_wrt(c,Y,d) &
  ex1_wrt(c,Z,d) & is_the_concept_of_individual_wrt(Y,U,d) &
  is_the_concept_of_wrt(Z,F,d) &
  (ex1_wrt(F,U,d) <=> contains_wrt(Y,Z,d)))))))).
```

and *not* in the following form (as a fact about every point D):

```
fof(theorem38,theorem,
  (! [U,F] : ((object(U) & property(F)) => (! [D] : (point(D) =>
  (ex1_wrt(o,U,D) => (? [Y,Z] : (object(Y) & object(Z) &
  ex1_wrt(c,Y,D) & ex1_wrt(c,Z,D) &
  is_the_concept_of_individual_wrt(Y,U,D) &
  is_the_concept_of_wrt(Z,F,D) &
  (ex1_wrt(F,U,D) <=> contains_wrt(Y,Z,D))))))))))).
```

When representing object theoretic claims, one always has to ask: is this provably true only with respect to the distinguished point d or is it provable for every point D? Of course, one must take care not to get confused by the fact that *possible worlds* are defined in object theory, and so we can express claims that have both variables ranging over defined possible worlds as well as variables ranging over the primitive sort `point`. The question of when to represent a theorem as a necessary truth affects only those claims involving the modal operator 'necessarily', not claims about possible worlds *per se*.[18]

---

[18] In general, we adopted the policy of proving necessitations of theorems only when they were required for the proof of another theorem. For example, the following necessary truth was needed for the proof of Theorem 40a:

```
fof(uniqueness_of_concept_of_individual_in_wrt,lemma,
  (! [D] : (point(D) => (![X,Y,U,W] : ((object(X) & object(Y) &
  object(U) & object(W)) => (world_wrt(W,D) =>
  ((concept_of_individual_in_wrt(X,U,W,D) &
  concept_of_individual_in_wrt(Y,U,W,D)) =>
  a_equal_wrt(X,Y,D)))))))).
```

This asserts that for every point D, if X and Y are both concepts of the individual U with respect to D, then X and Y are identical abstract objects with respect to D.

An interesting point emerged about representing object theory's two main axiom schemata. Basically, we adopted the expedient of representing only the instances of the schemata that we needed as a premise to prove a conjecture. For example, as noted earlier, the main comprehension schema for abstract objects asserts:

$$\exists x(A!x \;\&\; \forall F(xF \equiv \phi)), \text{ provided } x \text{ doesn't occur free in } \phi$$

Since there is no way to represent schemata in first-order syntax, our policy was this: if any theorem that required the *existence* of an abstract object given by some instance of the above schema, then we formulated the particular instance as a premise. So, for example, if a theorem required the existence of the concept of individual u, we would represent the following instance:

$$\exists x(A!x \;\&\; \forall F(xF \equiv Fu))$$

Then, given the definition of the concept of individual u, we would be assured that the domain contained such a concept.

We also followed this procedure to address the problem of representing the β-conversion schema. Since we don't have a general way of representing all the various different λ-abstracts in $\mathsf{FOL}_=$, we simply had to manually represent various λ-abstracts and axiomatize them as needed. Thankfully, few instances were needed to complete the formalization.

We see two possible ways to represent these two axiom schemata in full generality. One is to reason syntactically about the formulas allowed in instances of the schema. The other is to formulate them in third-order logic, analogously to the way in which the induction axiom for arithmetic can be formulated as a single second-order axiom rather than as a schema for generating first-order axioms.

6.2   Future Work

At the time of writing, we haven't yet proved every lemma upon which Theorems 40a and 40b depend; only a few remain. Once the work is complete, we hope to put the theorems into a form that can be submitted to the TPTP Library. This will require an additional step of determining which of the axioms, definitions, lemmas, sorting principles, etc., constitute the core part of the theory. Once we identify the core part of the theory, we can look for a model of all of the key principles upon which Theorems 40a and 40b depend. We've found models of the premise sets in each of the separate input files for the theorems we've proved thus far, but until the work is complete, we won't be in a position to identify the core group of principles that require a consistency check. As of release 6.1.0 of the TPTP Library, there is a new section devoted to philosophy. We plan to submit this theory for inclusion in that section.

One long-term goal of our project is to identify desiderata for the design and implementation of a customized, native prover for object theory. A customized prover would allow us to input formulas that more closely resemble those of object theory. Furthermore, customized theorem provers and model builders might be able to recognize subformulas, recognize which formulas have no encoding subformulas, generate instances of the comprehension schema for relations, and generally be more attuned to the special features of object theory. Reasoning in object theory is more structured than simply throwing a set of formulas at a theorem prover and looking for a refutation. There are dependencies such as the dependence of definitions on their justifying theorems, and the restriction on the Rule of Necessitation to formulas that do not depend on contingent assumptions. This makes the definition of provability in object theory more subtle, which understandably complicates the implementation of any system that tries to be faithful to it.

However, there may be obstacles to developing a native theorem prover for object theory. If the work in [14] is correct, there is a feature of object theory that suggests it may be difficult to adapt those existing reasoning engines which are based on some form of functional type theory. For the discussion in [14] established that object theory (a) contains formulas that neither are terms themselves nor can be converted to terms by λ-abstraction, and therefore (b) involves reasoning that seems to be capturable only in the logic of relational rather than functional type theory. Consequently, if existing automated reasoning engines depend essentially on some form of functional type theory to define and navigate the search space for finding proofs, then it may be that new methods (e.g., ones that work in a relational type-theoretic environment and not just in a functional type-theoretic environment) will have to be incorporated into the design and implementation of a customized prover for object theory.

Finally, if new methods and tools are developed to make the process go more quickly and smoothly, it should be easier to investigate object theory's Frege-style derivation of the Dedekind-Peano axioms for arithmetic [23].

Acknowledgments. The second and third authors would like to thank Branden Fitelson for collaborating on an earlier automated reasoning project and making us aware of the advances in theorem-proving technology.

References

1. Alama, J.: Complete independence of an axiom system for central translations. Note di Matematica **33**, 133–142 (2013)
2. Alama, J.: The simplest axiom system for hyperbolic geometry revisited, again. Stud. Logica **102**(3), 609–615 (2014)
3. Couturat, L. (ed.): Opuscules et fragments inédits de Leibniz. F. Alcan, Paris (1903)
4. Fitelson, B., Zalta, E.: Steps toward a computational metaphysics. J. Philos. Logic **36**(2), 227–247 (2007)
5. Gerhardt, C.I. (ed.): Die Philosophischen Schriften von Gottfried Wilhelm Leibniz, vol. i–vii. Weidmann, Berlin (1875–1990)
6. Henkin, L.: Completeness in the theory of types. J. Symb. **15**(2), 81–91 (1950)
7. Kripke, S.: A completeness theorem in modal logic. J. Symb. Logic **24**(1), 1–14 (1959)

8. Kripke, S.: Semantical considerations on modal logic. Acta Philos. Fennica **16**, 83–94 (1963)
9. Leibniz, G.W.: A study in the calculus of real addition. In: Parkinson, G. (ed.) Leibniz Logical Papers, pp. 131–144. Clarendon, Oxford (1996)
10. Leibniz, G.W.: Theodicy. Yale University Press, New Haven (1952)
11. Lewis, D.: Counterpart theory and quantified modal logic. J. Philos. **54**(5), 113–126 (1968)
12. Montague, R.: The proper treatment of quantification in ordinary English. In: Hintikka, K.J.J., Moravcsik, J.M.E., Suppes, P. (eds.) Approaches to Natural Language, pp. 221–242. D. Reidel, Dordrecht (1973)
13. Nodelman, U., Zalta, E.: Foundations for mathematical structuralism. Mind **123**(489), 39–78 (2014)
14. Oppenheimer, P., Zalta, E.: Relations versus functions at the foundations of logic: type-theoretic considerations. J. Logic Comput. **21**, 351–374 (2011)
15. Parkinson, G. (ed.): Leibniz: Logical Papers. Clarendon, Oxford (1966)
16. Parkinson, G. (ed.): Leibniz: Philosophical Writings. Dent & Sons, London (1973)
17. Pelletier, F., Zalta, E.: How to say goodbye to the third man. Noûs **34**(2), 165–202 (2000)
18. Russell, B.: On denoting. Mind **14**, 479–493 (1905)
19. Zalta, E.: Abstract Objects: An Introduction to Axiomatic Metaphysics. D. Reidel, Dordrecht (1983)
20. Zalta, E.: Intensional Logic and the Metaphysics of Intentionality. MIT Press, Cambridge (1988)
21. Zalta, E.: Logical and analytic truths that are not necessary. J. Philos. **85**(2), 57–74 (1988)
22. Zalta, E.: Twenty-five basic theorems in situation and world theory. J. Philos. Logic **22**(4), 385–428 (1993)
23. Zalta, E.: Natural numbers and natural cardinals as abstract objects: a partial reconstruction of Frege's Grundgesetze in object theory. J. Philos. Logic **28**(6), 619–660 (1999)
24. Zalta, E.: A (Leibnizian) theory of concepts. Philosophiegeschichte logische Analyse/Logical Anal. Hist. Philos. **3**, 137–183 (2000)

Competition Descriptions

Confluence Competition 2015

Takahito Aoto[1]([✉]), Nao Hirokawa[2], Julian Nagele[3], Naoki Nishida[4], and Harald Zankl[3]

[1] Tohoku University, Sendai, Japan
aoto@nue.riec.tohoku.ac.jp
[2] JAIST, Nomi, Japan
hirokawa@jaist.ac.jp
[3] University of Innsbruck, Innsbruck, Austria
{Julian.Nagele,Harald.Zankl}@uibk.ac.at
[4] Nagoya University, Nagoyo, Japan
nishida@is.nagoya-u.ac.jp

Abstract. Confluence is one of the central properties of rewriting. Our competition aims to foster the development of techniques for proving/disproving confluence of various formalisms of rewriting automatically. We explain the background and setup of the 4th Confluence Competition.

1 Introduction

Confluence (Fig. 1) provides a general notion of determinism and has been conceived as one of the central properties of rewriting [1]. Confluence has been investigated in many formalisms of rewriting such as first-order rewriting, lambda-calculi, higher-order rewriting, constrained rewriting, conditional rewriting, etc. More precisely, a rewrite system is a set of rewrite rules, and for each rewrite system $\mathcal{R}$, rewrite steps $s \to_{\mathcal{R}} t$ are associated. A rewrite system $\mathcal{R}$ is said to be confluent if for any $t_1 \;_{\mathcal{R}}\overset{*}{\leftarrow} t_0 \overset{*}{\to}_{\mathcal{R}} t_2$ there exists t_3 such that $t_1 \overset{*}{\to}_{\mathcal{R}} t_3 \;_{\mathcal{R}}\overset{*}{\leftarrow} t_2$, where $\overset{*}{\to}_{\mathcal{R}}$ is the reflexive transitive closure of $\to_{\mathcal{R}}$. The notions of rewrite rules, associated rewrite steps, and terms to be rewritten vary from one formalism to another. Confluence is also related to many important properties of rewriting such as the unique normal form property, ground confluence, etc.

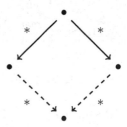

Fig. 1. Confluence

The task of our competition is to foster the development of techniques for proving/disproving confluence automatically by setting up a dedicated and fair competition among confluence proving/disproving tools. The 4th Confluence Competition (CoCo 2015)[1] runs live during the *4th International Workshop on Confluence (IWC 2015)* collocated with the *25th International Conference on Automated Deduction (CADE-25)* in Berlin, Germany.

---

[1] http://coco.nue.riec.tohoku.ac.jp.

© Springer International Publishing Switzerland 2015
A.P. Felty and A. Middeldorp (Eds.): CADE-25, LNAI 9195, pp. 101–104, 2015.
DOI: 10.1007/978-3-319-21401-6_5

2   Categories

Since different formalisms capture different confluence problems and techniques for confluence proving, the competition is separated into several categories. Categories are divided into *competition* categories and *demonstration* categories.

Demonstration categories are a novelty of CoCo 2015. These categories are one-time events for demonstrating new attempts and/or merits of particular tools. Demonstration categories can be requested until 2 months prior to the competition.

In contrast to demonstration categories, competition categories are not only run in a single competition but also in future editions of the confluence competition. Competition categories can be requested until 6 months prior to the competition, in order to allow organizers to make a decision on the framework and semantics of the rewriting formalism and the input format of the problems. The following 4 competition categories are run in CoCo 2015. See Fig. 2 for examples of the different rewriting formalisms.

$$\text{TRS:} \quad \left\{ \begin{array}{ll} +(0, y) \quad \rightarrow y, & \text{sum(nil)} \quad \rightarrow 0 \\ +(\mathsf{s}(x), y) \rightarrow \mathsf{s}(+(x, y)), & \text{sum}(\mathsf{cons}(x, ys)) \rightarrow +(x, \mathsf{sum}(ys)) \end{array} \right\}$$

$$\text{CTRS:} \quad \left\{ \begin{array}{l} +(0, y) \quad \rightarrow y \\ +(\mathsf{s}(x), y) \rightarrow \mathsf{s}(+(x, y)) \\ \text{fib}(0) \quad \rightarrow \mathsf{pair}(\mathsf{s}(0), 0) \\ \text{fib}(\mathsf{s}(x)) \quad \rightarrow \mathsf{pair}(w, y) \Leftarrow \mathsf{fib}(x) = \mathsf{pair}(y, z), +(y, z) = w \end{array} \right\}$$

$$\text{HRS:} \quad \left\{ \begin{array}{ll} \mathsf{map} \ (\lambda n. \ f \ n) \ \mathsf{nil} & \rightarrow \mathsf{nil} \\ \mathsf{map} \ (\lambda n. \ f \ n) \ (\mathsf{cons} \ x \ xs) \rightarrow \mathsf{cons} \ (f \ x) \ (\mathsf{map} \ (\lambda n. \ f \ n) \ xs) \end{array} \right\}$$

Fig. 2. Three different formalisms of rewrite systems in CoCo 2015.

TRS category. This is a category for first-order term rewrite systems. The framework of first-order term rewrite systems is most fundamental in the theory of term rewriting (e.g. [1]).

CTRS category. This is a category for *conditional* term rewriting. Conditional term rewriting allows to deal with conditions whose evaluation is defined recursively using the rewrite relation. Incorporation of conditional expressions is fundamental from the point of views of universal algebra (quasi-variety) and of functional programming. Depending on the interpretation of the conditions, 3 condition types are considered—namely, semi-equational, join and oriented types. We refer to the textbook [3] for details.

HRS category. Many expressive formal systems such as systems of predicate logics, λ-calculi, process calculi, etc. need variable binding. Higher-order rewriting is a framework that extends first-order term rewriting by a binding mechanism. Various formalisms of higher-order rewriting have been proposed in the literature. This category deals with one of the most classical frameworks of higher-order rewriting, namely *higher-order rewriting systems* [2].

CPF category. This category is for the certification of confluence proofs based on interactive theorem provers. Here confluence tools must produce machine-checkable proofs which are checked by trustable certifiers in a second step.

	number of tools	number of tool authors	categories
CoCo 2012	4	8	TRS/CPF
CoCo 2013	4	10	TRS/CPF
CoCo 2014	7	15	TRS/CTRS/CPF

Fig. 3. Statistics in the previous competitions.

In Fig. 3, we list statistics and categories of the previous competitions. The HRS category has been incorporated for the first time in CoCo 2015.

3   Problems and Evaluation Process

We maintain a database of *confluence problems* (Cops), dealing with the three rewriting formalisms reflected in the competition categories. The community can submit problems prior to the competition. For the competition, only problems from the literature are considered, where this family collects examples from the literature (articles, papers, technical notes, and so on) dealing with confluence, avoiding test examples generated automatically or tend to have a similar structure. The actual problem sets for the competition are selected randomly from these problems considering the time balance. For the demonstration categories, the participants are requested to prepare the problems for the competition.

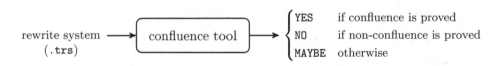

Fig. 4. Input and output scheme of a confluence tool.

Figure 4 shows the input and output scheme required for tool participants. In the competition a set of confluence problems is submitted to each participating tool. Each tool is supposed to answer whether the given rewrite system is confluent or not. Tools must be able to run on the designated execution platform and read problems as input. The output of the tools must contain an answer in the first line followed by some proof argument understandable for human experts. Valid answers are YES (the input is confluent) and NO (the input is not confluent). Any other answer (such as MAYBE is interpreted as the tool could not determine the status of the input. The timeout for each problem is set to 60 seconds for all categories. Every problem in the CTRS category is classified by the pair of a condition type (oriented, join, semi-equational), and a type of CTRS (type 1, 2, 3, or 4). A tool for the CTRS category should output UNSUPPORTED in place of YES/NO for input CTRSs in classes that the tool does not support. The unsupported classes of each tool must be declared at the time of registration. For the CPF category, each participant should output certifiable proofs as well as the result of certification: YES if confluence proof is certified and NO if non-confluence proof is certified.

The score is computed in percent of solved vs. supported problems (i.e. number of YES/NO answers vs. total number of UNSUPPORTED answers). In case of a draw there might be more winners. The tool with the maximal score wins. An answer is plausible if it was not falsified (automatically or manually). A tool with at least one non-plausible answer cannot be a winner.

4   Competition Platform and LiveView

The competition runs on a dedicated high-end cross-community competition platform *StarExec* [4]. The progress of the live competition is shared with the audience visually through the *LiveView* tool which interacts with StarExec. A screenshot of the LiveView is shown in Fig. 5.

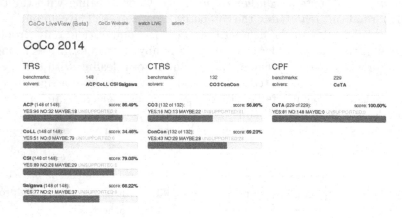

Fig. 5. LiveView of CoCo 2014.

References

1. Baader, F., Nipkow, T.: Term Rewriting and All That. Cambridge University Press, Cambridge (1998)
2. Mayr, R., Nipkow, T.: Higher-order rewrite systems and their confluence. Theor. Comput. Sci. **192**(1), 3–29 (1998)
3. Ohlebusch, E.: Advanced Topics in Term Rewriting Systems. Springer, New York (2002)
4. Stump, A., Sutcliffe, G., Tinelli, C.: StarExec: a cross-community infrastructure for logic solving. In: Demri, S., Kapur, D., Weidenbach, C. (eds.) IJCAR 2014. LNCS, vol. 8562, pp. 367–373. Springer, Heidelberg (2014)

Termination Competition (termCOMP 2015)

Jürgen Giesl[1], Frédéric Mesnard[2], Albert Rubio[3]([✉]),
René Thiemann[4], and Johannes Waldmann[5]

[1] RWTH Aachen University, Aachen, Germany
[2] Université de la Réunion, Saint Denis, France
[3] Universitat Politècnica de Catalunya - BarcelonaTech, Barcelona, Spain
albert@cs.upc.edu
[4] Universität Innsbruck, Innsbruck, Austria
[5] HTWK Leipzig, Leipzig, Germany

Abstract. The termination competition focuses on automated termination analysis for all kinds of programming paradigms, including categories for term rewriting, imperative programming, logic programming, and functional programming. Moreover, the competition also features categories for automated complexity analysis. In all categories, the competition also welcomes the participation of tools providing certified proofs. The goal of the termination competition is to demonstrate the power of the leading tools in each of these areas.

1 Introduction

The termination competition has been organized annually since 2004. There are usually between 10 and 20 participating termination/complexity/certification tools. Recent competitions were executed live during the main conferences of the field (at VSL 2014, RDP 2013, IJCAR 2012, RTA 2011, and FLoC 2010).

2 Competition Categories

2.1 Termination Analysis

The termination competition features numerous categories for different forms of languages. These languages can be classified into real programming languages (Sect. 2.1.2) and languages based on rewriting and/or transition systems (Sect. 2.1.1). Termination of such languages is also of great practical interest, because they are often used as *back-end* languages. More precisely, one can prove termination of programs by first translating them into such a back-end language automatically and by analyzing termination of the resulting rewrite or transition system afterwards.

J. Giesl—This author is supported by the Deutsche Forschungsgemeinschaft (DFG) grant GI 274/6-1.
A. Rubio—This author is supported by the Spanish MINECO under the grant TIN2013-45732-C4-3-P (project DAMAS).
R. Thiemann—This author is supported by the Austrian Science Fund (FWF) project Y757.

© Springer International Publishing Switzerland 2015
A.P. Felty and A. Middeldorp (Eds.): CADE-25, LNAI 9195, pp. 105–108, 2015.
DOI: 10.1007/978-3-319-21401-6_6

2.1.1 Rewriting and Transition Systems

There are several categories for termination analysis of different variants of term rewriting. This includes classical term rewriting, conditional term rewriting, term rewriting under specific strategies (innermost rewriting, outermost rewriting, and context-sensitive rewriting), string rewriting (where all function symbols are unary), relative term or string rewriting (where one has to prove that certain rules cannot be used infinitely often), term rewriting modulo equations, and higher-order rewriting.

In 2014, the competition also had categories for systems with built-in integers for the first time. More precisely, there was a category for term rewriting extended by integers and a category for integer transition systems (which do not feature terms, and where one has to prove absence of infinite runs originating in designated start states).

2.1.2 Programming Languages

The termination competition has categories for termination of programs in several languages from different paradigms. This includes functional languages (Haskell), logic languages (Prolog), and imperative languages. While a category for termination of Java programs has already been part of the competition since 2009, since 2014 there is also a category for the analysis of C programs.

2.2 Complexity Analysis

Since 2008, the termination competition has categories for asymptotic worst-case complexity analysis of term rewriting. Here, one tries to find an upper bound on the function that maps any natural number n to the length of the longest possible derivation starting with a term of size n or less. In the competition, different forms of complexity are investigated, depending on whether one regards full or innermost rewriting. Moreover, these forms of complexity differ in the shape of the possible start terms. For *derivational complexity*, one allows arbitrary start terms. In contrast, for *runtime complexity*, one only allows start terms of the form $f(t_1, \ldots, t_n)$, where a defined symbol f (i.e., an "algorithm") is applied to "data objects" $t_1, \ldots, t_n$ (i.e., the terms t_i may not contain any defined symbols). So runtime complexity corresponds to the notion of complexity typically used for programs.

2.3 Certified Categories

It regularly occurred during previous competitions that bugs of tools have been detected by conflicting answers. However, even if there are no conflicting answers there is the potential of wrong answers. To this end, the termination competition provides a certification option. If enabled, tools must generate their proofs in a machine-readable and fully specified *certification problem format*. These proofs will then be validated by certifiers whose soundness has to be justified, e.g., by a machine-checked soundness proof of the certifier itself, or via on-the-fly generation of proof scripts for proof-assistants like Coq, Isabelle, or PVS.

The certification option is currently supported for most categories on first-order term rewriting, for both termination and complexity analysis.

3   Termination Problem Data Base

The *Termination Problem Data Base* (TPDB) is the collection of all the examples used in the competition. Its structure is closely related to the categories in the competition. Each example in the TPDB is sufficiently specified to precisely determine a Boolean answer for termination (or an optimal answer for complexity). For instance, although we aim to detect duplicates and eliminate them from the TPDB (modulo renaming and order of rewrite rules), the data base may contain two examples with the same program which differ in their evaluation strategy or in the set of start terms. These details are important in the competition, where the tools are asked to investigate the termination and complexity behavior for exactly the given evaluation strategy and initial terms. Although there is a unique correct answer for each example, these answers are not stored in the TPDB and might even be unknown. For instance, the TPDB also contains Collatz' open termination problem of the "$3n + 1$" function.

New problems for the TPDB can be submitted at any time and will be added after a short reviewing process of the steering committee. This steering committee consists of representatives of the participating research groups. It is in charge of strategic decisions for the competition and its future. Currently, the examples in the TPDB are distributed as follows w.r.t. their source languages: term rewriting (10755), Haskell (1671), integer transition systems (953), Java (859), Prolog (492), and C (480).

4   Running the Competition

Here is a brief description of the rules of the competition:

- For termination tools: given an input program from the TPDB, try to determine whether it terminates or not within a given time limit. Positive and negative answers are equally scored when determining the winner of a category.
- For complexity tools: try to figure out the worst-case complexity of an input program from the TPDB within a given time limit in big-O notation. Here, the scoring depends on the precision of the answer in comparison to the answers of the other competing tools.
- For certifiers: try to check as many machine-readable termination/complexity proofs as possible.

Both termination and complexity tools must provide a human- or machine-readable proof in addition to their answer. The input problems and tools are partitioned w.r.t. the categories presented in Sect. 2. A category is only scheduled in the competition if there are at least two participating tools for that category. Other categories may be scheduled for demonstration purposes.

From 2004 to 2007, the competition was hosted by the University of Paris-Sud, France. From 2008 to 2013, the competition was hosted by the University of Innsbruck, Austria. In 2014, the competition was run for the first time on the StarExec platform (https://www.starexec.org/) at the University of Iowa, USA, while results were presented on the web front-end star-exec-presenter (see Fig. 1) running at HTWK Leipzig, Germany. The same infrastructure will be used for the 2015 competition.

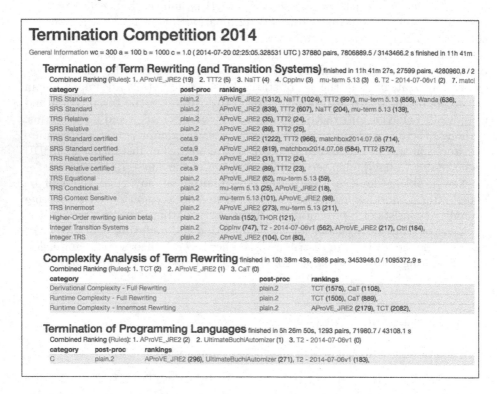

Fig. 1. The web front-end star-exec-presenter summarizing the 2014 competition

In order to run the competition within the duration of a conference, in the last years only a subset of termination problems from the TPDB was selected for each competition. Separate "full runs" of all tools on all TPDB problems were also executed, which took around a week. In 2014, StarExec provided enough computing power to execute a full run in 12 hours. Time-out per problem was 5 min. A total of 377880 problem/tool pairs were executed using $8 \cdot 10^6$ seconds (2200 hours) CPU time and running on almost 200 nodes in parallel.

For further details, we refer to the main web site of the termination competition (http://termination-portal.org/wiki/Termination_Competition).

Rewriting

Non-E-Overlapping, Weakly Shallow, and Non-Collapsing TRSs are Confluent

Masahiko Sakai[1]($\boxtimes$), Michio Oyamaguchi[1], and Mizuhito Ogawa[2]

[1] Graduate School of Information Science, Nagoya University,
Furo-cho, Chikusa-ku, Nagoya 464-8603, Japan
sakai@is.nagoya-u.ac.jp, oyamaguchi@za.ztv.ne.jp
[2] Japan Advanced Institute of Science and Technology,
1-1 Asahidai, Nomi, Ishikawa 923-1292, Japan
mizuhito@jaist.ac.jp

Abstract. A term is *weakly shallow* if each defined function symbol occurs either at the root or in the ground subterms, and a term rewriting system is weakly shallow if both sides of a rewrite rule are weakly shallow. This paper proves that non-E-overlapping, weakly-shallow, and non-collapsing term rewriting systems are confluent by extending *reduction graph* techniques in our previous work [19] with *towers of expansions*.

1 Introduction

Confluence of term rewriting systems (TRSs) is undecidable, even for flat TRSs [11] or length-two string rewrite systems [18]. Two decidable subclasses are known: right-linear and shallow TRSs by tree automata techniques [2] and terminating TRSs by resolving to finite search [8]. Many sufficient conditions have been proposed, and they are classified into two categories.

- Local confluence for terminating TRSs [8]. It was extended to TRSs with relative termination [5,7]. Another criterion comes with the decomposition to linear and terminating non-linear TRSs [10]. It requires conditions for the existence of well-founded *ranking*.
- Peak elimination with an explicit well-founded measure. Lots of works explore left-linear TRSs under the non-overlapping condition and its extensions [6,14–17,21]. For non-linear TRSs, there are quite few works [3,22] under the non-E-overlapping condition (which coincides with non-overlapping if left-linear) and additional restrictions that allow to define such measures.

We have proposed a different methodology, called a *reduction graph* [19], and shown that *"weakly non-overlapping, shallow, and non-collapsing TRSs are confluent"*. An original idea comes from observing that, when non-E-overlapping, peak-elimination uses only *"copies"* of reductions in an original rewrite sequences. Thus, if we focus on terms appearing in peak elimination, they

---

The results without proofs are orally presented at IWC 2014 [20].

M. Ogawa—This work is supported by JSPS KAKENHI Grant Number 25540003.

A.P. Felty and A. Middeldorp (Eds.): CADE-25, LNAI 9195, pp. 111–126, 2015.
DOI: 10.1007/978-3-319-21401-6_7

are finitely many. We regard a rewrite relation over these terms as a directed graph, and construct a confluent directed acyclic graph (DAG) in a bottom-up manner, in which the shallowness assumption works. The keys are, such a DAG always has a unique normal form (if it is finite), and convergence is preserved if we add an arbitrary reduction starting from a normal form. Our reduction graph technique is carefully designed to preserve both acyclicity and finiteness.

This paper introduces the notion of *towers of expansions*, which extends a reduction graph by adding terms and edges expanded with function symbols in an on-demand way, and shows that "*weakly shallow, non-E-overlapping, and non-collapsing TRSs are confluent*". A term is weakly shallow if each defined function symbol appears either at the root or in the ground subterms, and a TRS is weakly shallow if the both sides of rules are weakly shallow. It is worth mentioning:

- A Turing machine is simulated by a weakly shallow TRS [9] (see Remark 1), and many decision problems, such as the word problem, termination and confluence, are undecidable [12]. Note that the word problem is decidable for shallow TRSs [1]. The fact distinguishes these classes.
- The non-E-overlapping property is undecidable for weakly shallow TRSs [12]. A decidable sufficient condition is *strongly non-overlapping*, where a TRS is *strongly non-overlapping* if its linearization is non-overlapping [13]. Here, these conditions are the same when left-linear.
- Our result gives a new criterion for confluence provers of TRSs. For instance,

$$\{d(x,x) \rightarrow h(x), f(x) \rightarrow d(x, f(c)), c \rightarrow f(c), h(x) \rightarrow h(g(x))\}$$

is shown to be confluent only by ours.

Remark 1. Let Q, Σ and Γ ($\supseteq \Sigma$) be finite sets of states, input symbols and tape symbols of a Turing machine M, respectively. Let $\delta : Q \times \Gamma \rightarrow Q \times \Gamma \times \{\text{left}, \text{right}\}$ be the transition function of M. Each configuration $a_1 \cdots a_i q a_{i+1} \cdots a_n \in \Gamma^+ Q \Gamma^+$ (where $q \in Q$) is represented by a term $q(a_i \cdots a_1(\$), a_{i+1} \cdots a_n(\$))$ where arities of function symbols q, a_j ($1 \leq j \leq n$) and $\$ $ are 2, 1 and 0, respectively. The corresponding TRS R_M consists of rewriting rules below:

$$\begin{aligned} q(x, a(y)) &\rightarrow p(b(x), y) && \text{if } \delta(q, a) = (p, b, \text{right}), \\ q(a'(x), a(y)) &\rightarrow p(x, a'(b(y))) && \text{if } \delta(q, a) = (p, b, \text{left}) \end{aligned}$$

2 Preliminaries

2.1 Abstract Reduction System

For a binary relation $\rightarrow$, we use $\leftarrow$, $\leftrightarrow$, $\rightarrow^+$ and $\rightarrow^*$ for the inverse relation, the symmetric closure, the transitive closure, and the reflexive and transitive closure of $\rightarrow$, respectively. We use $\cdot$ for the composition operation of two relations.

An *abstract reduction system* (ARS) is a directed graph $G = \langle V, \rightarrow \rangle$ with reduction $\rightarrow \subseteq V \times V$. If $(u, v) \in \rightarrow$, we write it as $u \rightarrow v$. An element u of V

is *($\rightarrow$-)normal* if there exists no $v \in V$ with $u \rightarrow v$. We sometimes call a normal element a *normal form*. For subsets V' and V'' of V, $\rightarrow|_{V' \times V''} = \rightarrow \cap (V' \times V'')$.

Let $G = \langle V, \rightarrow \rangle$ be an ARS. We say G is *finite* if V is finite, *confluent* if $\leftarrow^* \cdot \rightarrow^* \subseteq \rightarrow^* \cdot \leftarrow^*$, *Church-Rosser (CR)* if $\leftrightarrow^* \subseteq \rightarrow^* \cdot \leftarrow^*$, and *terminating* if it does not admit an infinite reduction sequence from a term. G is *convergent* if it is confluent and terminating. Note that confluence and CR are equivalent.

We refer standard terminology in graphs. Let $G = \langle V, \rightarrow \rangle$ and $G' = \langle V', \rightarrow' \rangle$ be ARSs. We use $V_{G'}$ and $\rightarrow_{G'}$ to denote V' and $\rightarrow'$, respectively. An edge $v \rightarrow u$ is an *outgoing-edge* of v and an *incoming-edge* of u, and v is the *initial vertex* of $\rightarrow$. A vertex v is $\rightarrow$-*normal* if it has no outgoing-edges. The union of graphs is defined as $G \cup G' = \langle V \cup V', \rightarrow \cup \rightarrow' \rangle$. We say

- G is *connected* if $(u, v) \in \leftrightarrow^*$ for each $u, v \in V$.
- G' *includes* G, denoted by $G' \supseteq G$, if $V' \supseteq V$ and $\rightarrow' \supseteq \rightarrow$.
- G' *weakly subsumes* G, denoted by $G' \sqsupseteq G$, if $V' \supseteq V$ and $\leftrightarrow'^* \supseteq \rightarrow$.
- G' *conservatively extends* G, if $V' \supseteq V$ and $\leftrightarrow'^*|_{V \times V} = \leftrightarrow^*$.

The weak subsumption relation $\sqsupseteq$ is transitive.

2.2 Term Rewriting System

Let F be a finite set of function symbols, and X be an enumerable set of variables with $F \cap X = \emptyset$. $T(F, X)$ denotes the set of terms constructed from F and X and $\mathrm{Var}(t)$ denotes the set of variables occurring in a term t. A *ground* term is a term in $T(F, \emptyset)$. The set of positions in t is $\mathrm{Pos}(t)$, and the *root* position is ε. For $p \in \mathrm{Pos}(t)$, the subterm of t at position p is denoted by $t|_p$. The root symbol of t is $\mathrm{root}(t)$, and the set of positions in t whose symbols are in S is denoted by $\mathrm{Pos}_S(t) = \{p \mid \mathrm{root}(t|_p) \in S\}$. The term obtained from t by replacing its subterm at position p with s is denoted by $t[s]_p$. The *size* $|t|$ of a term t is $|\mathrm{Pos}(t)|$. As notational convention, we use s, t, u, v, w for terms, x, y for variables, a, b, c, f, g for function symbols, p, q for positions, and σ, θ for substitutions.

We define $\mathrm{sub}(t)$ as $\mathrm{sub}(x) = \emptyset$ and $\mathrm{sub}(t) = \{t_1, \ldots, t_n\}$ if $t = f(t_1, \ldots, t_n)$. A *rewrite rule* is a pair (ℓ, r) of terms such that $\ell \notin X$ and $\mathrm{Var}(\ell) \supseteq \mathrm{Var}(r)$. We write it $\ell \rightarrow r$. A *term rewriting system* (TRS) is a finite set R of rewrite rules. The *rewrite relation* of R on $T(F, X)$ is denoted by $\xrightarrow{R}$. We sometimes write $s \xrightarrow[R]{p} t$ to indicate the *rewrite step* at the position p. Let $s \xrightarrow[R]{p} t$. It is a *top reduction* if $p = \varepsilon$. Otherwise, it is an *inner reduction*, written as $s \xrightarrow[R]{\varepsilon <} t$.

Given a TRS R, the set D of *defined symbols* is $\{\mathrm{root}(\ell) \mid \ell \rightarrow r \in R\}$. The set C of *constructor symbols* is $F \setminus D$. For $T \subseteq T(F, X)$ and $f \in F$, we use $T|_f$ to denote $\{s \in T \mid \mathrm{root}(s) = f\}$. For a subset F' of F, we use $T|_{F'}$ to denote the union $\cup_{f \in F'} T|_f$.

A *constructor term* is a term in $T(C, X)$, and a *semi-constructor term* is a term in which defined function symbols appear only in the ground subterms. A term is *shallow* if the length $|p|$ is 0 or 1 for every position p of variables in the term. A *weakly shallow term* is a term in which defined function symbols appear only either at the root or in the ground subterms (i.e., $p \neq \varepsilon$ and $\mathrm{root}(s|_p) \in D$ imply that $s|_p$ is ground). Note that every shallow term is weakly shallow.

A rewrite rule $\ell \to r$ is *weakly shallow* if ℓ and r are weakly shallow, and *collapsing* if r is a variable. A TRS is *weakly shallow* if each rewrite rule is weakly shallow. A TRS is *non-collapsing* if it contains no collapsing rules.

Example 2. A TRS R_1 is weakly shallow and non-collapsing.

$$R_1 = \{f(x,x) \to a,\ f(x,g(x)) \to b,\ c \to g(c)\}\ [6]$$

Let $\ell_1 \to r_1$ and $\ell_2 \to r_2$ be rewrite rules in a TRS R. Let p be a position in ℓ_1 such that $\ell_1|_p$ is not a variable. If there exist substitutions θ_1, θ_2 such that $\ell_1|_p\theta_1 = \ell_2\theta_2$ (resp. $\ell_1|_p\theta_1 \overset{\varepsilon\leq}{\underset{R}{\leftrightarrow}}{}^* \ell_2\theta_2$), we say that the two rules are *overlapping* (resp. *E-overlapping*), except that $p = \varepsilon$ and the two rules are identical (up to renaming variables). A TRS R is *overlapping* (resp. *E-overlapping*) if it contains a pair of overlapping (resp. E-overlapping) rules. Note that TRS R_1 in Example 2 is E-overlapping since $f(c,c) \overset{\varepsilon\leq}{\underset{R}{\leftrightarrow}}{}^* f(c,g(c))$.

3   Extensions of Convergent Abstract Reduction Systems

This section describes a transformation system from a finite ARS to obtain a convergent (i.e., terminating and confluent) ARS that preserves the connectivity.

Let $G = \langle V, \to \rangle$ be an ARS. If G is finite and convergent, then we use a function $\downarrow_G$ (called the choice mapping) that takes an element of V and returns the normal form [19]. We also use $v\downarrow_G$ instead of $\downarrow_G(v)$.

Definition 3. *For ARSs $G_1 = \langle V_1, \to_1 \rangle$ and $G_2 = \langle V_2, \to_2 \rangle$, we say that $G_1 \cup G_2$ is the* hierarchical combination *of G_2 with G_1, denoted by $G_1 > G_2$, if $\to_1\ \subseteq (V_1 \setminus V_2) \times V_1$.*

Proposition 4. *$G_1 > G_2$ is terminating if both G_1 and G_1 are so.*

Lemma 5. *Let $G_1 > G_2$ be a confluent and hierarchical combination of ARSs. If a confluent ARS G_3 weakly subsumes G_2 and $G_1 > G_3$ is a hierarchical combination, then $G_1 > G_3$ is confluent.*

Proof. We use $\langle V_i, \to_i \rangle$ to denote G_i. Let $\alpha : u' \overset{*}{\leftarrow}_{G_1 > G_3} u \overset{*}{\to}_{G_1 > G_3} u''$. If $u \in V_3$, only $\to_3$ appears in α, and hence $u' \to_3^* \cdot \leftarrow_3^* u''$ follows from the confluence of G_3. Otherwise, α is represented as $u' \leftarrow_3^* v' \leftarrow_1^* u \to_1^* v'' \to_3^* u''$. Since $v' \to_1^* w' \to_2^* \cdot \leftarrow_2^* w'' \leftarrow_1^* v''$ for some w' and w'' (from the confluence of $G_1 > G_2$) and $G_2 \sqsubseteq G_3$, we obtain $u' \leftarrow_3^* v' \to_1^* w' \leftrightarrow_3^* w'' \leftarrow_1^* v'' \to_3^* u''$. Since $G_1 > G_3$ is a hierarchical combination, $v' = w'$ if $v' \in V_3$, and $v' = u'$ otherwise. Hence, $u' \to_1^* \cdot \leftrightarrow_3^* w'$. Similarly either $v'' = w''$ or $v'' = u''$. Thus, $u' \to_1^* \cdot \leftrightarrow_3^* \cdot \leftarrow_1^* u''$. The confluence of G_3 gives $u' \to_1^* \cdot \to_3^* \cdot \leftarrow_3^* \cdot \leftarrow_1^* u''$, and $u' \to_{G_1 > G_3}^* \cdot \leftarrow_{G_1 > G_3}^* u''$.   □

In the sequel, we generalize properties of ARSs obtained in [19].

Definition 6. *Let $G = \langle V, \rightarrow \rangle$ be a convergent ARS. Let v, v' be vertices such that $v \neq v'$ and if $v \in V$ then v is $\rightarrow$-normal. Then G', denoted by $G \multimap (v \rightarrow v')$, is defined as follows (see Fig. 1):*

$$
\begin{cases}
\langle V \cup \{v'\}, \rightarrow \cup \{(v, v')\} \rangle & \text{if } v \in V \text{ and } v' \notin V & (1) \\
\langle V, \rightarrow \cup \{(v, v')\} \rangle & \text{if } v, v' \in V \text{ and } v' \not\leftrightarrow^* v & (2) \\
\langle V, \rightarrow \setminus \{(v', v'') \mid v' \rightarrow v''\} \cup \{(v, v')\} \rangle & \text{if } v, v' \in V \text{ and } v' \leftrightarrow^* v & (3) \\
\langle V \cup \{v, v'\}, \rightarrow \cup \{(v, v')\} \rangle & \text{if } v \notin V & (4)
\end{cases}
$$

Note that v' becomes a normal form of G' when the first or the third transformation is applied.

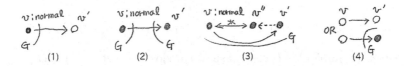

Fig. 1. Adding an edge to a convergent ARS

Proposition 7. *For a convergent ARS G, the ARS $G' = G \multimap (v \rightarrow v')$ is convergent, and satisfies $G' \sqsupseteq G$.*

We represent $G \multimap (v_0 \rightarrow v_1) \multimap (v_1 \rightarrow v_2) \multimap \cdots \multimap (v_{n-1} \rightarrow v_n)$ as $G \multimap (v_0 \rightarrow v_1 \rightarrow \cdots \rightarrow v_n)$ (if Definition 6 can be repeatedly applied).

Proposition 8. *Let $G = \langle V, \rightarrow \rangle$ be a convergent ARS. Let $v_0, v_1, \ldots, v_n$ satisfy $v_i \neq v_j$ (for $i \neq j$), and one of the following conditions:*

(1) $v_0 \in V$, v_0 is $\rightarrow$-normal, and $v_i \in V$ implies $v_i \leftrightarrow^* v_0$ for each $i(< n)$,
(2) $v_0, \cdots, v_{n-1} \notin V$.

Then, $G' = G \multimap (v_0 \rightarrow v_1 \rightarrow \cdots \rightarrow v_n)$ is well-defined and convergent, and $G' \sqsupseteq G$ holds.

4   Reduction Graphs

From now on, we fix C and D as the sets of constructors and defined function symbols for a TRS R, respectively. We assume that there exists a constructor with a positive arity in C, otherwise all weakly shallow terms are shallow.

4.1   Reduction Graphs and Monotonic Extension

Definition 9 ([19]). *An ARS $G = \langle V, \rightarrow \rangle$ is an R-reduction graph if V is a finite subset of $\mathrm{T}(F, X)$ and $\rightarrow \subseteq \underset{R}{\rightarrow}$.*

For an R-reduction graph $G = \langle V, \to \rangle$, *inner-edges*, *strict inner-edges*, and *top-edges* are given by $\overset{\varepsilon\leq}{\to} = \to \cap \overset{\varepsilon\leq}{\underset{R}{\to}}$, $\overset{\neq\varepsilon}{\to} = \to \setminus \overset{\varepsilon}{\underset{R}{\to}}$, and $\overset{\varepsilon}{\to} = \to \cap \overset{\varepsilon}{\underset{R}{\to}}$, respectively. We use $G^{\varepsilon\leq}$, $G^{\neq\varepsilon}$, and G^{ε} to denote $\langle V, \overset{\varepsilon\leq}{\to} \rangle$, $\langle V, \overset{\neq\varepsilon}{\to} \rangle$, and $\langle V, \overset{\varepsilon}{\to} \rangle$, respectively. Remark that for $R = \{a \to b, f(x) \to f(b)\}$ $V = \{f(a), f(b)\}$, and $G = \langle V, \{(f(a), f(b))\} \rangle$, we have $G^{\varepsilon\leq} = G^{\varepsilon} = G$ and $G^{\neq\varepsilon} = \langle V, \emptyset \rangle$.

For an R-reduction graph $G = \langle V, \to \rangle$ and $F' \subseteq F$, we represent $G|_{F'} = \langle V, \to|_{F'} \rangle$ where $\to|_{F'} = \to|_{V|_{F'} \times V}$. Note that $\to|_C = \to|_{V|_C \times V|_C}$ and $\to = \to|_D \cup \to|_{V|_C \times V|_C}$.

Definition 10. *Let $G = \langle V, \to \rangle$ be an R-reduction graph. The* direct-subterm reduction-graph $\mathrm{sub}(G)$ *of G is $\langle \mathrm{sub}(V), \mathrm{sub}(\to) \rangle$ where*

$$
\begin{cases}
\mathrm{sub}(V) = \bigcup_{t \in V} \mathrm{sub}(t) \\
\mathrm{sub}(\to) = \{(s_i, t_i) \mid f(s_1, \ldots, s_n) \overset{\varepsilon\leq}{\to} f(t_1, \ldots, t_n),\ s_i \neq t_i,\ 1 \leq i \leq n\}.
\end{cases}
$$

An R-reduction graph $G = \langle V, \to \rangle$ is subterm-closed *if $\mathrm{sub}(G^{\neq\varepsilon}) \sqsubseteq G$.*

Lemma 11. *Let $G = \langle V, \to \rangle$ be a subterm-closed R-reduction graph. Assume that (1) $s[t]_p \leftrightarrow^* s[t']_p$, and (2) for any $p' < p$, if $(s[t]_p)|_{p'} \leftrightarrow^* (s[t']_p)|_{p'}$ then $(s[t]_p)|_{p'} \overset{\neq\varepsilon}{\leftrightarrow}^* (s[t']_p)|_{p'}$. Then $t \leftrightarrow^* t'$.*

Proof. By induction on $|p|$. If $p = \varepsilon$, trivial. Let $p = iq$ and $s = f(s_1, \ldots, s_n)$. Since $s[t]_p \overset{\neq\varepsilon}{\leftrightarrow}^* s[t']_p$ from the assumptions, the subterm-closed property of G implies $s_i[t]_q \leftrightarrow^* s_i[t']_q$. Hence, $t \leftrightarrow^* t'$ holds by induction hypothesis. $\square$

Definition 12. *For a set F' ($\subseteq F$) and an R-reduction graph $G = \langle V, \to \rangle$, the F'-monotonic extension $M_{F'}(G) = \langle V_1, \to_1 \rangle$ is*

$$
\begin{cases}
V_1 = \{f(s_1, \ldots, s_n) \mid f \in F',\ s_1, \ldots, s_n \in V\}, \\
\to_1 = \{(f(\cdots s \cdots), f(\cdots t \cdots)) \in V_1 \times V_1 \mid s \to t\}.
\end{cases}
$$

Example 13. As a running example, we use the following TRS, which is non-E-overlapping, non-collapsing, and weakly shallow with $C = \{g\}$ and $D = \{c, f\}$:

$$
R_2 = \{f(x, g(x)) \to g^3(x),\ c \to g(c)\}.
$$

Consider a subterm-closed R_2-reduction graph $G = \langle \{c, g(c), g^2(c)\}, \{(c, g(c))\} \rangle$. In the sequel, we use a simple representation of graphs as $G = \{c \to g(c),\ g^2(c)\}$. The C-monotonic extension $M_C(G)$ of G is $M_C(G) = \{g(c) \to g^2(c),\ g^3(c)\}$.

Proposition 14. *Let $M_{F'}(G) = \langle V', \to' \rangle$ be the F'-monotonic extension of an R-reduction graph $G = \langle V, \to \rangle$. Then,*

(1) if G is terminating (resp. confluent), then $M_{F'}(G)$ is.

(2) If G is subterm-closed, then for $u, v \in V|_{F'}$, we have (a) $u, v \in V'$, and (b) $u \overset{\neq\varepsilon}{\to} v$ implies $u \leftrightarrow'^* v$.

(3) $\mathrm{sub}(M_{F'}(G)) \subseteq G$ if F' contains a function symbol with a positive arity.

4.2   Constructor Expansion

Definition 15. *For a subterm-closed R-reduction graph G, a constructor expansion $\overline{M_C}(G)$ is the hierarchical combination $G|_D \gg M_C(G)$ $(= G|_D \cup M_C(G))$. The k-times application of $\overline{M_C}$ to G is denoted by $\overline{M_C}^k(G)$.*

Example 16. For G in Example 13, the constructor expansions $\overline{M_C}^i(G)$ of G $(i = 1, 3)$ are

$$\overline{M_C}(G) = \{c \to g(c) \to g^2(c), \; g^3(c)\},$$
$$\overline{M_C}^3(G) = \{c \to g(c) \to g^2(c) \to g^3(c) \to g^4(c), \; g^5(c)\}.$$

Lemma 17. *Let G be a subterm-closed R-reduction graph. Then,*

(1) $\mathrm{sub}(\overline{M_C}(G)^{\neq\varepsilon}) \sqsubseteq G$, *and*
(2) $\to_{G^{\neq\varepsilon}} \subseteq \leftrightarrow^*_{M_F(G)}$, *that is, $G \sqsubseteq G^\varepsilon \cup M_F(G)$,*

Proof. Let $G = \langle V, \to \rangle$. We refer $M_C(G)$ by $G' = \langle V', \to' \rangle$. Thus, for $v \in V'$, $\mathrm{root}(v) \in C$. Note that $\overline{M_C}(G) = G|_D \gg M_C(G) = \langle V' \cup V, \to' \cup \to|_{V|_D \times V} \rangle$.

(1) Due to $\mathrm{sub}(\overline{M_C}(G)^{\neq\varepsilon}) = \mathrm{sub}(G^{\neq\varepsilon}|_D) \cup \mathrm{sub}(M_C(G))$, it is enough to show $\mathrm{sub}(G^{\neq\varepsilon}|_D) \sqsubseteq G$ and $\mathrm{sub}(M_C(G)) \sqsubseteq G$. The former follows from the fact that $\mathrm{sub}(G^{\neq\varepsilon}|_D) \subseteq \mathrm{sub}(G^{\neq\varepsilon})$ and G is subterm-closed. The latter follows from $\mathrm{sub}(M_C(G)) \subseteq G$.
(2) Obvious from Proposition 14 (2).   □

Lemma 18. *For a subterm-closed R-reduction graph G,*

(1) $G \sqsubseteq \overline{M_C}(G)$,
(2) $\overline{M_C}(G)$ *is subterm-closed, and*
(3) $\overline{M_C}(G)$ *is convergent if G is convergent.*

Proof. Let $G = \langle V, \to \rangle$. Note that $\overline{M_C}(G) = (G|_D \gg M_C(G)) = \langle V \cup V_{M_C(G)}, \to|_D \cup \to_{M_C(G)} \rangle$.

(1) Since $\to|_{V|_C \times V|_C} \subseteq \xrightarrow{\neq\varepsilon}_G$, we have $\to|_{V|_C \times V|_C} \subseteq \leftrightarrow^*_{M_C(G)}$ (by Proposition 14 (2)), so that $G \sqsubseteq \overline{M_C}(G)$.
(2) By Lemma 17 (1), $\mathrm{sub}(\overline{M_C}(G)^{\neq\varepsilon}) \sqsubseteq G$. Combining this with $G \sqsubseteq \overline{M_C}(G)$, we obtain $\mathrm{sub}(\overline{M_C}(G)^{\neq\varepsilon}) \sqsubseteq \overline{M_C}(G)$. Thus, $\overline{M_C}(G)$ is subterm-closed.
(3) If we show $G' = \langle V|_C, \to|_{V|_C \times V|_C} \rangle \sqsubseteq M_C(G)$, the confluence of $\overline{M_C}(G) = G|_D \gg M_C(G)$ follows from Lemma 5, since $G = G|_D \gg G'$ and $M_C(G)$ is confluent by Proposition 14 (1). Since G is subterm-closed, we have $V|_C \subseteq V_{M_C(G)}$ and $\to|_{V|_C \times V|_C} \subseteq \leftrightarrow^*_{M_C(G)}$ by Proposition 14 (2). Hence, $G' \sqsubseteq M_C(G)$. The termination of $\overline{M_C}(G)$ follows from Proposition 4, since $G|_D$ and $M_C(G)$ are terminating.   □

Corollary 19. *For a subterm-closed R-reduction graph G and $k \geq 0$, we have:*

(1) $G \sqsubseteq \overline{M_C}^k(G)$.

(2) $\overline{M_C}^k(G)$ is subterm-closed.

(3) $\overline{M_C}^k(G)$ is convergent, if G is convergent.

Remark 20. When an R-reduction graph G is subterm-closed, we observe that $\leftrightarrow_{\overline{M_C}^k(G)}^* = \leftrightarrow_{G \cup M_C(G) \cup \cdots \cup M_C^k(G)}^*$ from $\to_{G|_C} \subseteq \leftrightarrow_{M_C(G)}^*$ by Proposition 14 (2).

Proposition 21. Let G be a subterm-closed R-reduction graph. Then, $\overline{M_C}^k(G) \sqsubseteq \overline{M_C}^m(G)$ for $m > k \geq 0$.

Proof. By $\overline{M_C}^m(G) = \overline{M_C}^{m-k}(\overline{M_C}^k(G))$ and Corollary 19 (1) and (2). $\qquad \square$

5 Tower of Constructor Expansions

From now on, let G be a convergent and subterm-closed R-reduction graph. We call $M_F(\overline{M_C}^i(G))$ a *tower of constructor expansions* of G for $i \geq 0$. We use $G_{2_i} = \langle V_{2_i}, \to_{2_i} \rangle$ to denote $M_F(\overline{M_C}^i(G))$.

5.1 Enriching Reduction Graph

We show that there exists a convergent R-reduction graph G_1 with $M_F(G) \sqsubseteq G_1$ such that G_{2_i} is a conservative extension of G_1 for large enough i.

Lemma 22. *For a convergent and subterm-closed R-reduction graph G, there exist k (≥ 0) and an R-reduction graph G_1 satisfying the following conditions.*

(i) G_1 is convergent, and consists of inner-edges.

(ii) $G_1 \sqsubseteq G_{2_k}$.

(iii) $u \leftrightarrow_{2_i}^* v$ implies $u \leftrightarrow_1^* v$ for each $u, v \in V_1$ and i (≥ 0).

(iv) $M_F(G) \sqsubseteq G_1$.

Proof. Let $G_1 := M_F(G)$ and $k := 0$. We define a condition (iii)' as "(iii) holds for all i ($< k$)". Initially, (i) holds by Proposition 14 (1) since G is convergent. (ii) and (iv) hold from $G_1 = M_F(G) = G_{2_0}$, and (iii)' holds from $k = 0$.

We transform G_1 so that (i), (ii), (iii)' and (iv) are preserved and the number $|V_1/\leftrightarrow_1^*|$ of connected components of G_1 decreases. This transformation $(G_1, k) \vdash (G_1', k')$ continues until (iii) eventually holds, since $|V_1/\leftrightarrow_1^*|$ is finite.

For current G_1 and k, we assume that (i), (ii), (iii)' and (iv) hold. If G_1 fails (iii), there exist i with $i \geq k$ and $u, v \in V_1$ such that $u \neq v$ and $(u, v) \in \leftrightarrow_{2_i}^* \setminus \leftrightarrow_1^*$. We choose such k' as the least i. Remark that G_1 is convergent from (i), and $G_{2_{k'}}$ is convergent from Corollary 19 (3) and Proposition 14 (1). Let $\downarrow_1$ and $\downarrow_{2_{k'}}$ be the choice mappings of G_1 and $G_{2_{k'}}$, respectively. Since $G_1 \sqsubseteq G_{2_{k'}}$ from (ii) and Proposition 21, we have $(u\downarrow_1, v\downarrow_1) \in \leftrightarrow_{2_{k'}}^*$ and $u\downarrow_1 \neq v\downarrow_1$. From the convergence of $G_{2_{k'}}$, we have

$$\begin{cases} u\downarrow_1 = u_0 \to_{2_{k'}} u_1 \to_{2_{k'}} \cdots \to_{2_{k'}} u_{n'} \to_{2_{k'}} \cdots \to_{2_{k'}} u_n = (u\downarrow_1)\downarrow_{2_{k'}} \\ \qquad\qquad\qquad\qquad\qquad\qquad \| \qquad\qquad\qquad\qquad\quad \| \\ v\downarrow_1 = v_0 \to_{2_{k'}} v_1 \to_{2_{k'}} \cdots \to_{2_{k'}} v_{m'} \to_{2_{k'}} \cdots \to_{2_{k'}} v_m = (v\downarrow_1)\downarrow_{2_{k'}} \end{cases}$$

where (n', m') is the smallest pair under the lexicographic ordering such that $u_{n'} = v_{m'}$. Note that u_j's and v_j's do not necessarily belong to V_1. We define a transformation $(G_1, k) \vdash (G_1', k')$ with G_1' to be

$$
\begin{cases}
G_1 \multimap (u_0 \to \cdots \to u_j) & \text{if there exists (the smallest) } j \text{ such that} \\
& \quad 0 < j \leq n', u_j \in V_1, \text{ and } u_j \not\leftrightarrow_1^* u \\
G_1 \multimap (v_0 \to \cdots \to v_{j'}) & \text{if there exists (the smallest) } j' \text{ such that} \\
& \quad 0 < j' \leq m', v_{j'} \in V_1, \text{and } v_{j'} \not\leftrightarrow_1^* v \\
G_1 \multimap (u_0 \to \cdots \to u_{n'}) \multimap (v_0 \to \cdots \to v_{m'}) & \text{otherwise.}
\end{cases}
$$

Since the condition (1) of Proposition 8 holds, (i) is preserved. From $G_1 \sqsubseteq G_1'$ (iv) holds, and (ii) $G_1' \sqsubseteq G_{2_{k'}}$ by Proposition 21. If $k' = k$, (iii)' does not change. If $k' > k$, then $u \leftrightarrow_{2_i}^* v$ implies $u \leftrightarrow_1^* v$ for i with $k \leq i < k'$, since we chose k' as the least. Hence (iii)' holds. In either case, $|V_1/\leftrightarrow_1^*|$ decreases. □

Example 23. For G in Example 13, Lemma 22 starts from $M_F(G)$, which is displayed by the solid edges in Fig. 2. G_1 is constructed by augmenting the dashed edges with $k = 1$.

$$
\begin{array}{cccc}
c & f(c,c) & \to \quad f(g(c),c) & f(g^2(c),c) \\
& \downarrow & \downarrow & \downarrow \\
g(c) & f(c,g(c)) & \to \quad f(g(c),g(c)) & \dashrightarrow f(g^2(c),g(c)) \\
\downarrow & & & \vdots \\
g^2(c) & f(c,g^2(c)) & \to \quad f(g(c),g^2(c)) & \dashrightarrow f(g^2(c),g^2(c)) \\
\vdots & & & \\
g^3(c) & & &
\end{array}
$$

Fig. 2. G_1 constructed by Lemma 22 from G in Example 13

Corollary 24. *Assume that $G_1 = \langle V_1, \to_1 \rangle$ and h (≥ 0) satisfy the conditions i) to iv) in Lemma 22. Let $v_0, v_1, \ldots, v_n$ satisfy $v_j \neq v_{j'}$ for $j \neq j'$ and $v_{j-1} (\leftrightarrow_{2_k}^* \cap \xrightarrow{\varepsilon \leq}{R}) v_j$ for $1 \leq j \leq n$. If either (1) $v_0 \in V_1$ and v_0 is $\to_1$-normal, or (2) $v_0, \cdots, v_{n-1} \notin V_1$ and $v_n \in V_1$, then the conditions (i) to (iv) hold for $G_{1'} = G_1 \multimap (v_0 \to v_1 \to \cdots \to v_n)$ and $k' = \max(k, h)$.*

Proof. For (1), from (iii) of G_1, $v_j \in V_1$ implies $v_j \leftrightarrow_1^* v_0$. For either case, from (i) and (iv) of G_1 and Proposition 8, $G_{1'}$ satisfies (i) and (iv). Since $v_{j-1} \leftrightarrow_{2_k}^* v_j$, $G_{1'}$ immediately satisfies (ii). Since $v_0 \in V_1$ or $v_n \in V_1$, $G_{1'}$ satisfies (iii). □

5.2 Properties of Tower of Expansions on Weakly Shallow Systems

Lemma 25. *Let R be a non-E-overlapping and weakly shallow TRS. Let $G = \langle V, \to \rangle$ be a convergent and subterm-closed R-reduction graph, and let $\ell \to r \in R$.*

(1) *If* $\ell\sigma \leftrightarrow^*_{2_i} \ell\theta$, *then* $x\sigma \leftrightarrow^*_{\overline{M_C}^i(G)} x\theta$ *for each variable* $x \in \mathrm{Var}(\ell)$.

(2) *For a weakly shallow term* s *with* $s \notin X$, *assume that* $x\sigma \leftrightarrow^*_{\overline{M_C}^i(G)} x\theta$ *for each variable* $x \in \mathrm{Var}(s)$. *If* $s\sigma \in V_{2_i}$, *then* $s\sigma \leftrightarrow^*_{2_k} s\theta$ *for some* $k \ (\geq i)$.

(3) *If* $\ell\sigma \leftrightarrow^*_{2_i} u$, *then there exist a substitution* θ *and* $k \ (\geq i)$ *such that* $u \ (\xrightarrow{\varepsilon\leq}_R \cap \leftrightarrow^*_{2_k})^* \ell\theta$ *and* $x\sigma \rightarrow^*_{\overline{M_C}^i(G)} x\theta$ *for each variable* $x \in \mathrm{Var}(\ell)$.

Proof. Note that G_{2_i} is convergent by Corollary 19 (3) and Proposition 14 (1).

(1) Let $\ell = f(\ell_1, \ldots, \ell_n)$. For each j $(1 \leq j \leq n)$, $\ell_j\sigma \leftrightarrow^*_{\overline{M_C}^i(G)} \ell_j\theta$. Since $\overline{M_C}^i(G)$ is convergent by Corollary 19 (3), there exists v_j such that $\ell_j\sigma \rightarrow^*_{\overline{M_C}^i(G)} v_j \leftarrow^*_{\overline{M_C}^i(G)} \ell_j\theta$. Since $\overline{M_C}^i(G)$ is subterm-closed by Corollary 19 (2) and ℓ_j is semi-constructor, we have $x\sigma \leftrightarrow^*_{\overline{M_C}^i(G)} x\theta$ for every $x \in \mathrm{Var}(\ell)$ by Lemma 11.

(2) First, we show that for a semi-constructor term t if $t\sigma \in V_{\overline{M_C}^i(G)}$, there exists $k \ (\geq i)$ such that $t\sigma \leftrightarrow^*_{\overline{M_C}^k(G)} t\theta$ by induction on the structure of t. If t is either a variable or a ground term, immediate. Otherwise, let $t = f(t_1, \ldots, t_n)$ for $f \in C$. Since $\overline{M_C}^i(G)$ is subterm-closed, $t_j\sigma \in V_{\overline{M_C}^i(G)}$ for each j. Hence, induction hypothesis ensures $t_j\sigma \leftrightarrow^*_{\overline{M_C}^{k_j}(G)} t_j\theta$ for some $k_j \geq i$. Since $M_C(\overline{M_C}^i(G)) \subseteq \overline{M_C}^{i+1}(G)$ and Proposition 21, we have $t\sigma \leftrightarrow^*_{\overline{M_C}^k(G)} t\theta$ for $k = 1 + \max\{k_1, \ldots, k_n\}$.

We show the statement (2). Since $s \notin X$, s is represented as $f(s_1, \ldots, s_n)$ where each s_i is a semi-constructor term in $V_{\overline{M_C}^i(G)}$. Since there exists $k \ (\geq i)$ such that $s_j\sigma \leftrightarrow^*_{\overline{M_C}^k(G)} s_j\theta$, we have $s\sigma \leftrightarrow^*_{M_F(\overline{M_C}^k(G))} s\theta$.

(3) Since G_{2_i} is convergent, there exists v with $\ell\sigma \rightarrow^*_{2_i} v \leftarrow^*_{2_i} u$. Here, $u \rightarrow^*_{2_i} v$ and $\ell\sigma \rightarrow^*_{2_i} v$ imply $u \ (\rightarrow_{2_i} \cap \xrightarrow{\varepsilon\leq}_R)^* v$ and $\ell\sigma \ (\rightarrow_{2_i} \cap \xrightarrow{\varepsilon\leq}_R)^* v$, respectively. Since R is non-E-overlapping, $\ell\sigma \rightarrow^*_{2_i} v$ has no reductions at $\mathrm{Pos}_F(\ell)$. By a similar argument to that of (1), we have $\ell|_p\sigma \leftrightarrow^*_{\overline{M_C}^i(G)} v|_p$ for each $p \in \mathrm{Pos}_X(\ell)$.

Let $x \in \mathrm{Var}(\ell)$. Since $\overline{M_C}^i(G)$ is convergent from Corollary 19 (3), we have $x\sigma = \ell\sigma|_p \rightarrow^*_{\overline{M_C}^i(G)} x\theta \leftarrow^*_{\overline{M_C}^i(G)} v|_p$ for each $p \in \mathrm{Pos}_{\{x\}}(\ell)$ by taking θ as $x\theta = x\sigma\downarrow_{\overline{M_C}^i(G)}$. Since ℓ is weakly shallow, by repeating (2) to each step in $v|_p \rightarrow^*_{\overline{M_C}^i(G)} x\theta$, there exists k with $v \leftrightarrow^*_{2_k} \ell\theta$. We have $u \ (\xrightarrow{\varepsilon\leq}_R \cap \leftrightarrow^*_{2_k})^* v \ (\xrightarrow{\varepsilon\leq}_R \cap \leftrightarrow^*_{2_k})^* \ell\theta$ by Proposition 21.   □

6 Bottom-Up Construction of Convergent Reduction Graph

From now on, we assume that a TRS R is non-E-overlapping, non-collapsing, and weakly shallow. We show that R is confluent by giving a transformation of any R-reduction graph G_0 (possibly) containing a divergence into a convergent and subterm-closed R-reduction graph G_4 with $G_0 \sqsubseteq G_4$. The non-collapsing condition is used only in Lemma 27. Note that non-overlapping is not enough to ensure confluence as R_1 in Example 2. Now, we see an overview by an example.

Example 26. Consider R_2 in Example 13. Given $G_0 = \{f(g(c), c) \leftarrow f(c, c) \rightarrow f(c, g(c)) \xrightarrow{\varepsilon} g^3(c)\}$, we firstly take the subterm graph $\mathrm{sub}(G_0)$ and apply the transformation on it recursively to obtain a convergent and subterm-closed reduction graph G. In the example case, $\mathrm{sub}(G_0)$ happens to be equal to G in Example 13, and already satisfies the conditions. Secondly, we apply Lemma 22 on $M_F(G)$ and obtain G_1 in Example 2. As the next steps, we will merge the top edges T_1 in $G_0 \cup G$ into G_1, where $T_1 = \{f(c, g(c)) \xrightarrow{\varepsilon} g^3(c), c \xrightarrow{\varepsilon} g(c)\}$. Note that top edges in G is necessary for subterm-closedness. The union $G_1 \cup T_1$ is not, however, confluent in general. Thirdly, we remove unnecessary edges from T_1 by Lemma 27, and obtain T (in the example $T = T_1$). Finally, by Lemma 28, we transform edges in T into S with modifying G_1 into $G_{1'}$ so that $G_4 = G_{1'}|_D \cup S \cup M_C(\overline{M_C}^{k'}(G))$ is confluent ($k' \geq k$). The resultant reduction graph G_4 is shown in Fig. 3, where the dashed edges are in S and some garbage vertices are not presented. (See Example 30 for details of the final step).

$$
\begin{array}{ccccccc}
c & & f(c,c) & \rightarrow & f(g(c),c) & & f(g^2(c),c) \\
\vdots & & \downarrow & & \downarrow & & \downarrow \\
g(c) & & f(c,g(c)) & \rightarrow & f(g(c),g(c)) & \rightarrow & f(g^2(c),g(c)) \\
\downarrow & & & & & & \downarrow \\
g^2(c) & & f(c,g^2(c)) & \rightarrow & f(g(c),g^2(c)) & \rightarrow & f(g^2(c),g^2(c)) \\
\downarrow & & & & & & \downarrow \\
g^3(c) & \rightarrow & g^4(c) & \rightarrow & g^5(c) & \dashleftarrow & f(g^2(c),g^3(c))
\end{array}
$$

Fig. 3. G_4 constructed by Lemma 29 from G_0 in Example 26

6.1   Removing Redundant Edges and Merging Components

For R-reduction graphs $G_1 = \langle V_1, \rightarrow_1 \rangle$ and $T_1 = \langle V_1, \rightarrow_{T_1} \rangle$, the *component graph* (denoted by T_1/G_1) of T_1 with G_1 is the graph $\langle \mathcal{V}, \rightarrow_{\mathcal{V}} \rangle$ having connected components of G_1 as vertices and $\rightarrow_{T_1}$ as edges such that

$$
\mathcal{V} = \{[v]_{\leftrightarrow_1^*} \mid v \in V_1\}, \quad \rightarrow_{\mathcal{V}} = \{([u]_{\leftrightarrow_1^*}, [v]_{\leftrightarrow_1^*}) \mid (u, v) \in \rightarrow_{T_1}\}.
$$

Lemma 27. *Let* $G_1 = \langle V_1, \rightarrow_1 \rangle$ *be an R-reduction graph obtained from Lemma 22, and let* $T_1 = \langle V_1, \rightarrow_{T_1} \rangle$ *be an R-reduction graph with* $\rightarrow_{T_1} = \xrightarrow{\varepsilon}_{T_1}$. *Then, there exists a subgraph* $T = \langle V_1, \rightarrow_T \rangle$ *of* T_1 *with* $\rightarrow_T \subseteq \rightarrow_{T_1}$ *that satisfies the following conditions.*

(1) $(\leftrightarrow_1 \cup \leftrightarrow_{T_1})^* = (\leftrightarrow_1 \cup \leftrightarrow_T)^*$.
(2) The component graph T/G_1 *is acyclic in which each vertex has at most one outgoing-edge.*

Proof. We transform the component graph T_1/G_1 by removing edges in cycles and duplicated edges so that preserving its connectivity. This results in an acyclic directed subgraph $T = \langle V_1, \to_T \rangle$ without multiple edges.

Suppose some vertex in T/G_1 has more than one outgoing-edges, say $\ell\sigma \to_T r\sigma$ and $\ell'\theta \to_T r'\theta$, where $\ell\sigma \leftrightarrow_1^* \ell'\theta$, $r\sigma, r\theta \in V_1$ and $\ell \to r, \ell' \to r' \in R$. Since R is non-E-overlapping, we have $\ell = \ell'$ and $r = r'$. By the condition (ii) of Lemma 22, $\ell\sigma \leftrightarrow_{2_k}^* \ell\theta$ holds. Since R is non-collapsing, Lemma 25 (1) and (2) ensure $r\sigma \leftrightarrow_{2_j}^* r\theta$ for some j ($\geq k$). By the condition (iii) of Lemma 22, $r\sigma \leftrightarrow_1^* r\theta$. These edges duplicate, contradicting to the assumption.   □

In Lemma 27, if $\to_T$ is not empty, there exists a vertex of T/G_1 that has outgoing-edges, but no incoming-edges. We call such an outgoing-edge a *source edge*. Lemma 28 converts T to S in a source to sink order (by repeatedly choosing source edges) such that, for each edge in S, the initial vertex is $\to_1$-normal.

Lemma 28. *Let G_1, S, and T be R-reduction graphs, where G_1 and k satisfy the conditions (i) to (iv) of Lemma 22. Assume that the following conditions hold.*

(v) $V_S = V_T = V_{G_1}$, $\to_S = \xrightarrow{\varepsilon}_S$, $\to_T = \xrightarrow{\varepsilon}_T$, and $\to_S \cap \to_T = \emptyset$.
(vi) The component graph $(S \cup T)/G_1$ is acyclic, where outgoing-edges are at most one for each vertex. Moreover, if $[u]_{\leftrightarrow_1^*}$ has an incoming-edge in T/G_1 then it has no outgoing-edges in S/G_1.
(vii) u is $\to_1$-normal and $u \not\leftrightarrow_1^* v$ for each $(u, v) \in \to_S$.

When $\to_T \neq \emptyset$, there exists a conversion $(S, T, G_1, k) \vdash (S', T', G_{1'}, k')$ that preserves the conditions (i) to (iv) of Lemma 22, and conditions (v) to (vii), and satisfies the following conditions (1) to (3).

(1) $G_{1'}$ is a conservative extension of G_1.
(2) $(\leftrightarrow_T \cup \leftrightarrow_S)^* \subseteq (\leftrightarrow_{T'} \cup \leftrightarrow_{S'} \cup \leftrightarrow_{1'})^*$.
(3) $|\to_T| > |\to_{T'}|$

Proof. We design $\vdash$ as sequential applications of $\vdash_\ell$, $\vdash_r$, and $\vdash_e$ in this order. We choose a source edge $(\ell\sigma, r\sigma)$ (of T/G_1) from T. We will construct a substitution θ such that $(\ell\sigma){\downarrow_1} (\xrightarrow{\varepsilon\leq}_R \cap \leftrightarrow_{2_{k'}}^*)^* \ell\theta$ and $(r\sigma){\downarrow_1} (\xrightarrow{\varepsilon\leq}_R \cap \leftrightarrow_{2_{k'}}^*)^* \cdot (\xrightarrow{\varepsilon\leq}_R \cap \leftrightarrow_{2_{k'}}^*)^* r\theta$ for enough large k'. The former sequence is added to G_1 by $\vdash_\ell$, the latter is added to G_1 by $\vdash_r$, and $\vdash_e$ removes $(\ell\sigma, r\sigma)$ from T and adds $(l\theta, r\theta)$ to S.

We have $\ell\sigma \to_1^* (\ell\sigma){\downarrow_1}$ by i), and $\ell\sigma \leftrightarrow_{2_k}^* (\ell\sigma){\downarrow_1}$ by ii). From Lemma 25 (3), there are $k^\ell \geq k$ and a substitution θ such that $x\sigma \xrightarrow{*}_{MC^k(G)} x\theta$ for each $x \in$ Var(ℓ), $(\ell\sigma){\downarrow_1} = u_0 \xrightarrow{\varepsilon\leq}_R u_1 \xrightarrow{\varepsilon\leq}_R \cdots \xrightarrow{\varepsilon\leq}_R u_n = \ell\theta$, and $u_{j-1} \leftrightarrow_{2_{k^\ell}}^* u_j$ for each $j (\leq n)$.

$(\vdash_\ell)$ We define $(S, T, G_1, k) \vdash_\ell (S, T, G_{1^\ell}, k^\ell)$ by $G_{1^\ell} = G_1 \multimap (u_0 \to \cdots \to u_n)$ to satisfy $(\ell\sigma){\downarrow_1} \leftrightarrow_{1^\ell}^* \ell\theta$ such that $\ell\theta$ is G_{1^ℓ}-normal. Since u_0 is $\to_1$-normal, the case (1) of Corollary 24 holds, so that $\vdash_\ell$ preserves (i) to (iv) for G_{1^ℓ} and k^ℓ. (1) and (2) are immediate. From (1), (vi) is preserved. Since $[\ell\sigma]_{\leftrightarrow_1^*}$ does not have outgoing edges in S by (vi), (vii) is preserved.

($\vdash_r$) We define $(S, T, G_{1^\ell}, k^\ell) \vdash_r (S, T, G_{1'}, k')$. Let $G_{1^\ell} = \langle V_{1^\ell}, \to_{1^\ell} \rangle$. Since $x\sigma \leftrightarrow^*_{\overline{M_C}^{k^\ell}(G)} x\theta$ by Proposition 21 and $r\sigma \in V_{2_{k^\ell}}$, we obtain $r\sigma \leftrightarrow^*_{2_{k'}} r\theta$ for some $k' \geq k^\ell$ by Lemma 25 (2). We construct $G_{1'}$ to satisfy $(r\sigma){\downarrow}_{1^\ell} \leftrightarrow^*_{1'} r\theta$. Since the confluence of $G_{2_{k'}}$ follows from Corollary 19 (3) and Proposition 14 (1), we have the following sequences.

$$\begin{cases} (r\sigma){\downarrow}_{1^\ell} = u_0 \to_{2_{k'}} u_1 \to_{2_{k'}} \cdots \to_{2_{k'}} u_n = v, \\ r\theta = v_0 \to_{2_{k'}} v_1 \to_{2_{k'}} \cdots \to_{2_{k'}} v_m = v, \end{cases}$$

where we choose the least n satisfying $u_n = v_m$. There are two cases according to the second sequence.

(a) If $v_i \in V_{1^\ell}$ for some i, we choose i as the least. If $i = 0$, then $G_{1'} = G_{1^\ell}$. Otherwise, let $G_{1'} := G_{1^\ell} \multimap (v_0 \to v_1 \to \cdots \to v_i)$. Since G_{1^ℓ} satisfies the case (2) of Corollary 24, $\vdash_r$ preserves (i) to (iv). Since $u_0 \leftrightarrow^*_{2_{k'}} v_i$ and $u_0, v_i \in V_{1^\ell}$, $u_0 \leftrightarrow^*_{1^\ell} v_i$ by (iii). Thus, $(r\sigma){\downarrow}_{1^\ell} \leftrightarrow^*_{1'} r\theta$.

(b) Otherwise (i.e., $v_i \notin V_{1^\ell}$ for each i), let

$$\begin{cases} G_{1''} := G_{1^\ell} \multimap (u_0 \to u_1 \to \cdots \to u_n) \\ G_{1'} := G_{1''} \multimap (v_0 \to v_1 \to \cdots \to v_m). \end{cases}$$

Since u_0 is G_{1^ℓ}-normal and $u_j \in V_{1^\ell}$ implies $u_0 \leftrightarrow^*_{1^\ell} u_j$ (by iii) of G_{1^ℓ}, $G_{1''}$ and k' satisfy (i) to (iv) by Corollary 24. Let $G_{1''} = \langle V_{1''}, \to_{1''} \rangle$. Since $v_i \notin V_{1''}$ for each i ($< m$) and $v_m = u_n = v \in V_{1''}$, $G_{1'}$ and k' also satisfy (i) to (iv) by Corollary 24. By construction, $(r\sigma){\downarrow}_{1^\ell} \leftrightarrow^*_{1'} r\theta$ holds.

Since S and T do not change, $\vdash_r$ keeps (v), (1), and (2). Lastly, (vi) and (vii) follows from (1).

($\vdash_e$) We define $(S, T, G_{1'}, k') \vdash_e (S', T', G_{1'}, k')$, where $V_{S'} = V_{G_{1'}}$, $V_{T'} = V_{G_{1'}}$, $\to_{S'} = \to_S \cup \{(\ell\theta, r\theta)\}$, and $\to_{T'} = \to_T \setminus \{(\ell\sigma, r\sigma)\}$. Since $(\ell\sigma, r\sigma)$ is a source edge of T/G_1, $\vdash_e$ preserves (vi). Conditions (i) to (v), (1) and (3) are trivial. Since $\ell\sigma \leftrightarrow_{G_{1'}} (\ell\sigma){\downarrow}_1 \leftrightarrow_{G_{1'}} \ell\theta \to_{S'} r\theta \leftrightarrow_{G_{1'}} (r\sigma){\downarrow}_{1^\ell} \leftrightarrow_{G_{1'}} r\sigma$ implies $(\ell\sigma, r\sigma) \in \leftrightarrow^*_{S' \cup G_{1'}}$, we have (2). (vii) holds from (vi).   □

6.2   Construction of a Convergent and Subterm-Closed Graph

Lemma 29. *Let $G_0 = \langle V_0, \to_0 \rangle$ be an R-reduction graph. Then, there exists a convergent and subterm-closed R-reduction graph G_4 with $G_0 \sqsubseteq G_4$.*

Proof. By induction on the sum of the size of terms in V_0, i.e., $\Sigma_{v \in V_0} |v|$. If G_0 has no vertex, we set $G_4 = G_0$, which is the base case. Otherwise, by induction hypothesis, we obtain a convergent and subterm-closed R-reduction graph G with $\mathrm{sub}(G_0) \sqsubseteq G$. We refer to the conditions (i) to (vii) in Lemma 28.

Let $G_1 = \langle V_1, \to_1 \rangle$ and k be as in Lemma 22. Let T be obtained from G_1 and $T_1 = \langle V_1, \to_{G^\varepsilon} \cup \to_{G_5^\varepsilon} \rangle$ by applying Lemma 27.

Let $S = \langle V_1, \emptyset \rangle$. For G_1 and k, (i) to (iv) hold by Lemma 22. (vi) holds by Lemma 27 (2) and $\to_S = \emptyset$, and (vii) trivially holds. Starting from (S, T, G_1, k), we repeatedly apply $\vdash$ (in Lemma 28), which moves edges in T to S until $\to_T = \emptyset$.

Finally, we obtain $(S', \langle V_{1'}, \emptyset \rangle, G_{1'}, k')$ that satisfies (i) to (vii) and (1) to (3) in Lemma 28, where $G_{1'} = \langle V_{1'}, \to_{1'} \rangle$ and $V_{S'} = V_{1'}$. From Lemmas 27 and 28 (1) and (2), $(\leftrightarrow_1 \cup \leftrightarrow_{G^\varepsilon} \cup \leftrightarrow_{G_0^\varepsilon})^* = (\leftrightarrow_1 \cup \leftrightarrow_T)^* \subseteq (\leftrightarrow_{1'} \cup \leftrightarrow_{S'})^*$. Note that $G_{1'}$ is convergent by (i).

Let $G_3 = \langle V_3, \to_3 \rangle$ be $S' \cup G_{1'}$. This is obtained by repeatedly extending $G_{1'}$ by $G_{1'} \multimap (u \to v)$ for each $(u, v) \in \to_{S'}$, since in each step (vii) is preserved; u is $\to_{1'}$-normal and $u \not\leftrightarrow_{1'}^* v$. Thus, the convergence of G_3 follows from Proposition 7.

We show $G_0 \sqsubseteq G_3$. Since $G_0^\varepsilon \subseteq T_1 \sqsubseteq G_1 \cup T \sqsubseteq G_{1'} \cup S'$ (by Lemmas 27 and 28) and $M_F(\mathrm{sub}(G_0)) \sqsubseteq M_F(G) \sqsubseteq G_1 \sqsubseteq G_{1'}$ (by $\mathrm{sub}(G_0) \sqsubseteq G$ and iv)), $G_0 \subseteq G_0^\varepsilon \cup M_F(\mathrm{sub}(G_0)) \sqsubseteq S' \cup G_{1'} = G_3$.

Let $G_4 = \langle V_4, \to_4 \rangle$ be given by $G_4 := G_3|_D \rhd M_C(\overline{M_C}^{k'}(G))$. We show $G_0 \sqsubseteq G_4$ by showing $G_3 \sqsubseteq G_4$. Since $G_{1'} \sqsubseteq G_{2_{k'}}$ by (ii) where $G_{2_{k'}}$ contains no top edges, we have $V_{1'}|_C \subseteq V_{2_{k'}}|_C$ and $\to_{1'}|_C \subseteq (\leftrightarrow_{2_{k'}}|_C)^*$. Since $\to_{2_{k'}}|_C = \to_{M_C(\overline{M_C}^{k'}(G))}$, we have $G_{1'}|_C \sqsubseteq \langle V_{1'}, \emptyset \rangle \cup M_C(\overline{M_C}^{k'}(G))$. Thus, $G_{1'} = G_{1'}|_D \cup G_{1'}|_C \sqsubseteq G_{1'}|_D \cup M_C(\overline{M_C}^{k'}(G))$. By $S' = S'|_D$, we have $G_3 = S' \cup G_{1'} \sqsubseteq S'|_D \cup G_{1'}|_D \cup M_C(\overline{M_C}^{k'}(G)) = G_4$.

Now, our goal is to show that G_4 is convergent and subterm-closed. The convergence of $G_4 = G_3|_D \rhd M_C(\overline{M_C}^{k'}(G))$ is reduced to that of $G_3 = G_3|_D \rhd \langle V_3|_C, \to_3|_C \rangle$ by Proposition 4 and Lemma 5. Their requirements are satisfied from $\langle V_3|_C, \to_3|_C \rangle = \langle V_{1'}|_C, \to_{1'}|_C \rangle \sqsubseteq M_C(\overline{M_C}^{k'}(G))$ by (ii) and the convergence of $M_C(\overline{M_C}^{k'}(G))$ by Corollary 19 (3) and Proposition 14 (1).

We will prove that G_4 is subterm-closed by showing $\mathrm{sub}(G_4^{\neq \varepsilon}) \sqsubseteq \overline{M_C}^{k'}(G)$ and $\overline{M_C}^{k'}(G) \sqsubseteq G_4$. Note that $\mathrm{sub}(G_4^{\neq \varepsilon}) = \mathrm{sub}((S'|_D)^{\neq \varepsilon} \cup (G_{1'}|_D)^{\neq \varepsilon} \cup (M_C(\overline{M_C}^{k'}(G)))^{\neq \varepsilon}) \subseteq \mathrm{sub}(S'^{\neq \varepsilon}) \cup \mathrm{sub}(G_{1'}|_D) \cup \overline{M_C}^{k'}(G)$. We have $\mathrm{sub}(S'^{\neq \varepsilon}) = \langle \mathrm{sub}(V_{1'}), \emptyset \rangle$. Since $G_{2_{k'}}$ has no top edges and $G_{1'} \sqsubseteq G_{2_{k'}}$ by (ii), $\mathrm{sub}(G_{1'}) \sqsubseteq \mathrm{sub}(G_{2_{k'}}) = \mathrm{sub}(M_F(\overline{M_C}^{k'}(G))) \subseteq \overline{M_C}^{k'}(G)$. Thus, $\mathrm{sub}(G_4^{\neq \varepsilon}) \sqsubseteq \overline{M_C}^{k'}(G)$.

It remains to show $\overline{M_C}^{k'}(G) \sqsubseteq G_4$, which is reduced to $G|_D \sqsubseteq G_4$ from $\overline{M_C}^{k'}(G) = G|_D \cup M_C(\overline{M_C}^{k'-1}(G))$, $M_C(\overline{M_C}^{k'}(G)) \subseteq G_4$, and Proposition 21. Since $G|_D \subseteq G \sqsubseteq G^\varepsilon \cup M_F(G)$ by Lemma 17 (2), it is sufficient to show that $G^\varepsilon \sqsubseteq G_4$ and $M_F(G) \sqsubseteq G_4$.

Obviously, $M_F(G) \sqsubseteq G_{1'} \subseteq G_3 \sqsubseteq G_4$ holds, since $M_F(G) \sqsubseteq G_{1'}$ by (iv). We show $G^\varepsilon \sqsubseteq G_4$. Since $V_G \subseteq V_{M_F(G)}$ by Proposition 14 (2), we have $V_{G^\varepsilon} = V_G \subseteq V_{M_F(G)} \subseteq V_{1'} \subseteq V_3 \subseteq V_4$. By Lemmas 27 (1) and 28 (2), $\to_{G^\varepsilon} \subseteq (\leftrightarrow_{G_{1'}} \cup \leftrightarrow_{S'})^*$ holds, and by (ii) we have $\to_{G_{1'}|_C} \subseteq \leftrightarrow_{M_C(\overline{M_C}^{k'}(G))}^*$. Hence, $\to_{G^\varepsilon} \subseteq (\leftrightarrow_{G_{1'}|_D} \cup \leftrightarrow_{S'} \cup \leftrightarrow_{M_C(\overline{M_C}^{k'}(G))})^* = \leftrightarrow_{G_4}^*$. Therefore G_4 is subterm-closed. □

Example 30. Let us consider applying Lemma 29 on G_1 and T in Example 26, where $k = 1$. The edge $c \to g(c)$ in T is simply moved to S. For the edge $f(c, g(c)) \to g^3(c)$ in T, $\vdash_\ell$ adds $f(g^2(c), g^2(c)) \to f(g^2(c), g^3(c))$ to G_1. $\vdash_r$ adds $g^3(c) \to g^4(c) \to g^5(c)$ to G_1 and increases k to 3. $\vdash_e$ adds $f(g^2(c), g^3(c)) \to$

$g^5(c)$ to S. Since $M_C(\overline{M_C}^3(G))$ is $\{g(c) \to g^2(c) \to \cdots \to g^4(c) \to g^5(c), \ g^6(c)\}$, $G_4 = (S \cup G_1|_D) > M_C(\overline{M_C}^3(G))$ is as in Fig. 3.

Theorem 31. *Non-E-overlapping, weakly shallow, and non-collapsing TRSs are confluent.*

Proof. Let $u \leftarrow^*_R s \to^*_R t$. We obtain G_4 by applying Lemma 29 to an R-reduction graph G_0 consisting of the sequence. By $G_0 \sqsubseteq G_4$ and the convergence of G_4, $u\downarrow_{G_4} = t\downarrow_{G_4}$. Thus we have $u \to^*_R s' \leftarrow^*_R t$ for some s'.   □

Corollary 32. *Strongly non-overlapping, weakly shallow, and non-collapsing TRSs are confluent.*

7 Conclusion

This paper extends the reduction graph technique [19] and has shown that *non-E-overlapping, weakly shallow, and non-collapsing TRSs are confluent.*

We think that the *non-collapsing* condition can be dropped by refining the reduction graph techniques. A further step will be to relax the *weakly shallow* to the *almost weakly shallow* condition, which allows at most one occurrence of a defined function symbol in each path from the root to a variable.

References

1. Comon, H., Haberstrau, M., Jouannaud, J.-P.: Syntacticness, cycle-syntacticness, and shallow theories. Inf. Comput. **111**, 154–191 (1994)
2. Godoy, G., Tiwari, A.: Confluence of shallow right-linear rewrite systems. In: Ong, L. (ed.) CSL 2005. LNCS, vol. 3634, pp. 541–556. Springer, Heidelberg (2005)
3. Gomi, H., Oyamaguchi, M., Ohta, Y.: On the church-rosser property of root-E-overlapping and strongly depth-preserving term rewriting systems. IPS J **39**(4), 992–1005 (1998)
4. Gramlich, B.: Confluence without termination via parallel critical pairs. In: Kirchner, H. (ed.) CAAP 1996. LNCS, vol. 1059, pp. 211–225. Springer, Heidelberg (1996)
5. Hirokawa, N., Middeldorp, A.: Decreasing diagrams and relative termination. J. Autom. Reason. **47**(4), 481–501 (2011)
6. Huet, G.: Confluent reductions: abstract properties and applications to term rewriting systems. J. ACM **27**, 797–821 (1980)
7. Klein, D., Hirokawa, N.: Confluence of non-left-linear TRSs via relative termination. In: Bjørner, N., Voronkov, A. (eds.) LPAR-18 2012. LNCS, vol. 7180, pp. 258–273. Springer, Heidelberg (2012)
8. Knuth, D.E., Bendix, P.B.: Simple word problems in universal algebras. In: Leech, J. (ed.) Computational Problems in Abstract Algebra, pp. 263–297 (1970)
9. Klop, J.W.: Term Rewriting Systems, in Handbook of Logic in Computer Science, vol. 2. Oxford University Press, New York (1993)
10. Liu, J., Dershowitz, N., Jouannaud, J.-P.: Confluence by critical pair analysis. In: Dowek, G. (ed.) RTA-TLCA 2014. LNCS, vol. 8560, pp. 287–302. Springer, Heidelberg (2014)

11. Mitsuhashi, I., Oyamaguch, M., Jacquemard, F.: The confluence problem for flat TRSs. In: Calmet, J., Ida, T., Wang, D. (eds.) AISC 2006. LNCS (LNAI), vol. 4120, pp. 68–81. Springer, Heidelberg (2006)
12. Mitsuhashi, I., Oyamaguchi, M., Matsuura, K.: On the E-overlapping property of Wweak monadic TRSs. IPS J. **53**(10), 2313–2327 (2012). (in Japanese)
13. Ogawa, M., Ono, S.: On the uniquely converging property of nonlinear term rewriting systems. Technical report of IEICE, COMP89-7, pp. 61–70 (1989)
14. Okui, S.: Simultaneous critical pairs and church-rosser property. In: Nipkow, T. (ed.) RTA 1998. LNCS, vol. 1379, pp. 2–16. Springer, Heidelberg (1998)
15. van Oostrom, V.: Development closed critical pairs. In: Dowek, G., Heering, J., Meinke, K., Möller, B. (eds.) HOA 1995. LNCS, vol. 1074, pp. 185–200. Springer, Heidelberg (1996)
16. Oyamaguchi, M., Ohta, Y.: A new parallel closed condition for Church-Rosser of left-linear term rewriting systems. In: Comon, H. (ed.) RTA 1997. LNCS, vol. 1232, pp. 187–201. Springer, Heidelberg (1997)
17. Rosen, B.K.: Tree-manipulating systems and Church-Rosser theorems. J. ACM **20**, 160–187 (1973)
18. Sakai, M., Wang, Y.: Undecidable properties on length-two string rewriting systems. ENTCS **204**, 53–69 (2008)
19. Sakai, M., Ogawa, M.: Weakly-non-overlapping non-collapsing shallow term rewriting systems are confluent. Inf. Process. Lett. **110**, 810–814 (2010)
20. Sakai, M., Oyamaguchi, M., Ogawa, M.: Non-E-overlapping and weakly shallow TRSs are confluent (extended abstract). IWC **34–38**, 2014 (2014)
21. Toyama, Y.: Commutativity of term rewriting systems. In: Programming of Future Generation Computer II, pp. 393–407 (1988)
22. Toyama, Y., Oyamaguchi, M.: Church-Rosser property and unique normal form property of non-duplicating term rewriting systems. In: Lindenstrauss, N., Dershowitz, N. (eds.) CTRS 1994. LNCS, vol. 968, pp. 316–331. Springer, Heidelberg (1995)

CoLL: A Confluence Tool for Left-Linear Term Rewrite Systems

Kiraku Shintani and Nao Hirokawa[✉]

School of Information Science, JAIST, Nomi, Japan
{s1310032,hirokawa}@jaist.ac.jp

Abstract. We present a confluence tool for left-linear term rewrite systems. The tool proves confluence by using Hindley's commutation theorem together with three commutation criteria, including Church-Rosser modulo associative and/or commutative theories. Despite a small number of its techniques, experiments show that the tool is comparable to recent powerful confluence tools.

1 Introduction

In this paper we present the new confluence tool CoLL for left-linear term rewrite systems (TRSs). The tool has two distinctive features. One is use of Jouannaud and Kirchner's theorem for the Church-Rosser modulo property. Our tool supports rewriting modulo associativity *and/or* commutativity rules. Another notable feature is that confluence is proved by commutation criteria only. By using Hindley's Commutation Theorem [1] confluence is proved via commutation of subsystems of an input TRS. In addition to them, CoLL implements a simple transformation technique that eliminates redundant rewrite rules.

The remaining part of the paper is organized as follows: In Sect. 3 we discuss how to use the Church-Rosser modulo theorem for associativity and/or commutativity rules. Commutation criteria and decomposition techniques supported in the tool are described in Sect. 4. In Sect. 5 we report experimental results to assess effectiveness of the presented techniques and the tool. The final section describes related work and concluding remarks. The tool is available at:

http://www.jaist.ac.jp/project/saigawa/coll/

2 Preliminaries

We assume familiarity of term rewriting and unification theory [2]. We recall here only some notions for rewriting and rewriting modulo. We consider terms built from a signature $\mathcal{F}$ and a set $\mathcal{V}$ of variables. We write $s \rhd t$ if t is a proper subterm of s.

Supported by JSPS KAKENHI Grant Number 25730004 and Core to Core Program.

A.P. Felty and A. Middeldorp (Eds.): CADE-25, LNAI 9195, pp. 127–136, 2015.
DOI: 10.1007/978-3-319-21401-6_8

Commutation. Let $\mathcal{R}$ and $\mathcal{S}$ be TRSs. We say that $\mathcal{R}$ and $\mathcal{S}$ commute if $_{\mathcal{R}}{\leftarrow}^{*} \cdot \rightarrow_{\mathcal{S}}^{*} \subseteq \rightarrow_{\mathcal{S}}^{*} \cdot _{\mathcal{R}}{\leftarrow}^{*}$. Confluence of $\mathcal{R}$ is equivalent to *self-commutation* of $\mathcal{R}$, i.e., commutation of $\mathcal{R}$ and $\mathcal{R}$. The relation $\rightarrow_{\mathcal{S}}^{*} \cdot \rightarrow_{\mathcal{R}} \cdot \rightarrow_{\mathcal{S}}^{*}$ is called the *relative step* of $\mathcal{R}$ over $\mathcal{S}$, and denoted by $\rightarrow_{\mathcal{R}/\mathcal{S}}$. We say that $\mathcal{R}/\mathcal{S}$ is terminating if $\rightarrow_{\mathcal{R}/\mathcal{S}}$ is terminating.

Multi-steps. The *multi-step* $\multimap_{\mathcal{R}}$ of a TRS $\mathcal{R}$ is inductively defined on terms as follows:

1. $x \multimap_{\mathcal{R}} x$ for all $x \in \mathcal{V}$,
2. $f(s_1, \ldots, s_n) \multimap_{\mathcal{R}} f(t_1, \ldots, t_n)$ if $s_i \multimap_{\mathcal{R}} t_i$ for all $1 \leqslant i \leqslant n$, and
3. $\ell\sigma \multimap_{\mathcal{R}} r\tau$ if $\ell \rightarrow r \in \mathcal{R}$ and σ and τ are substitutions that $x\sigma \multimap_{\mathcal{R}} x\tau$ for all variables x.

Rewriting Modulo. Let $\mathcal{R}$ and $\mathcal{E}$ be TRSs. The rewrite step $\rightarrow_{\mathcal{R},\mathcal{E}}$ of $\mathcal{R}$ *modulo theory* $\mathcal{E}$ is defined as follows: $s \rightarrow_{\mathcal{R},\mathcal{E}} t$ if If $s|_p \leftrightarrow_{\mathcal{E}}^{*} \ell\sigma$ and $t = s[r\sigma]_p$ for some position $p \in \mathcal{P}os_{\mathcal{F}}(s)$, rule $\ell \rightarrow r \in \mathcal{R}$, and substitution σ. The relation $\rightarrow_{\mathcal{R},\mathcal{E}}$ is *Church-Rosser modulo* $\mathcal{E}$, denoted by $\mathsf{CR}(\mathcal{R},\mathcal{E})$, if $\leftrightarrow_{\mathcal{R}\cup\mathcal{E}}^{*} \subseteq \rightarrow_{\mathcal{R},\mathcal{E}}^{*} \cdot \leftrightarrow_{\mathcal{E}}^{*} \cdot _{\mathcal{R},\mathcal{E}}{\leftarrow}^{*}$. Let $\mathcal{F}_{\mathsf{A}}$, $\mathcal{F}_{\mathsf{C}}$, and $\mathcal{F}_{\mathsf{AC}}$ be pairwise disjoint sets of binary function symbols. We define the three theories A (associativity), C (commutativity), and AC as:

$$\mathsf{A} = \{f(f(x,y),z) \rightarrow f(x,f(y,z)), \ f(x,f(y,z)) \rightarrow f(f(x,y),z) \mid f \in \mathcal{F}_{\mathsf{A}}\}$$
$$\mathsf{C} = \{f(x,y) \rightarrow f(y,x) \mid f \in \mathcal{F}_{\mathsf{C}}\}$$
$$\mathsf{AC} = \{f(f(x,y),z) \rightarrow f(x,f(y,z)), \ f(x,y) \rightarrow f(y,x) \mid f \in \mathcal{F}_{\mathsf{AC}}\}$$

Critical Pairs. Conditions for confluence are often based on the notion of critical pairs. We denote by $\mathcal{U}_{\mathcal{E}}(s \approx t)$ a fixed complete set of $\mathcal{E}$-unifiers for terms s and t. Let $\ell_1 \rightarrow r_1$ be a rule in a TRS $\mathcal{R}$ and $\ell_2 \rightarrow r_2$ a variant of a rule in a TRS $\mathcal{S}$ with $\mathcal{V}ar(\ell_1) \cap \mathcal{V}ar(\ell_2) = \varnothing$. When $p \in \mathcal{P}os_{\mathcal{F}}(\ell_2)$ and $\sigma \in \mathcal{U}_{\mathcal{E}}(\ell_1 \approx \ell_2|_p)$, the pair $(\ell_2\sigma[r_1\sigma]_p, r_2\sigma)$ is called an *$\mathcal{E}$-extended critical pair* (or simply *$\mathcal{E}$-critical pair*) of $\mathcal{R}$ on $\mathcal{S}$, and written $\ell_2\sigma[r_1\sigma]_p \ {}_{\mathcal{R},\mathcal{E}}{\leftarrow}\!\times\!\rightarrow_{\mathcal{S}} r_2\sigma$.

3   Confluence via Church-Rosser Modulo

In this section we explain how the next theorem by Jouannaud and Kirchner [3] is used for confluence analysis. Especially, we discuss how to deal with associativity and/or commutativity rules.

Theorem 1. *Let $\mathcal{R}$ and $\mathcal{E}$ be TRSs that $\mathcal{R}/\mathcal{E}$ is terminating and $\rhd \cdot \leftrightarrow_{\mathcal{E}}^{*}$ is well-founded. Then, $\mathsf{CR}(\mathcal{R},\mathcal{E})$ iff $_{\mathcal{R},\mathcal{E}}{\leftarrow}\!\times\!\rightarrow_{\mathcal{R}\cup\mathcal{E}\cup\mathcal{E}^{-1}} \subseteq \rightarrow_{\mathcal{R},\mathcal{E}}^{*} \cdot \leftrightarrow_{\mathcal{E}}^{*} \cdot _{\mathcal{R},\mathcal{E}}{\leftarrow}$.* $\square$

We use the next left-linear TRS $\mathcal{R}_1$ to illustrate problems that arise when employing Theorem 1:

$$1: 0 + x \rightarrow x \qquad 2: x + (y + z) \rightarrow (x + y) + z \qquad 3: (x + y) + z \rightarrow x + (y + z)$$

Let $\mathcal{F}_{\mathsf{A}} = \{+\}$. We may assume $\mathsf{A} = \{2, 3\}$. An idea here is proving $\mathsf{CR}(\{1\}, \mathsf{A})$ to conclude confluence of $\mathcal{R}_1$. The next trivial lemma validates this idea. We call a TRS $\mathcal{E}$ *reversible* if $\rightarrow_{\mathcal{E}} \subseteq {}_{\mathcal{E}}^{*}{\leftarrow}$ holds.

Lemma 2. *Suppose $\mathcal{E}$ is reversible. If* $\mathsf{CR}(\mathcal{R}, \mathcal{E})$ *then* $\mathcal{R} \cup \mathcal{E}$ *is confluent.*   □

Reversibility of A and well-foundedness of $\rhd \cdot \leftrightarrow^*_{\{2,3\}}$ are trivial. Termination of $\{1\}/\mathsf{A}$ can be shown by AC-RPO [4]. Therefore, it remains to test joinability of extended critical pairs to apply Theorem 1.

3.1   Associative Unification

How to compute A-critical pairs? Plotkin [5] introduced a procedure that enumerates a minimal complete set of A-unifiers. It is well-known that a minimal complete set need not to be finite, and thus the procedure may not terminate. In fact there is a one-rule TRS that admits infinitely many A-critical pairs. Probably this is one of main reasons that existing confluence tools do not support Theorem 1 for associativity theory. However, as observed in [6], a minimal complete set resulting from the procedure is finite whenever an input equality is a pair of linear terms that share no variables. Therefore, for every left-linear TRS one can safely use Plotkin's procedure to compute their A-critical pairs.

We present a simple variant of Plotkin's procedure [5,7] specialized for our setting. Let S and T be sets of substitutions. We abbreviate the set $\{\sigma\tau \mid \sigma \in S \text{ and } \tau \in T\}$ to ST. Given a term t, we write $t\!\downarrow_{\mathsf{A}'}$ for the normal form of t with respect to A'. Here A' stands for the confluent and terminating TRS $\{f(f(x,y), z) \to f(x, f(y, z)) \mid f \in \mathcal{F}_{\mathsf{A}}\}$.

Definition 3. *Let s and t be terms. The function $\langle s \approx t \rangle$ is inductively defined as follows:*

$$\langle s \approx t \rangle = \begin{cases} \{\{s \mapsto t\}\} & \text{if } s \in \mathcal{V} \\ \{\{t \mapsto s\}\} & \text{if } s \notin \mathcal{V} \text{ and } t \in \mathcal{V} \\ A_1 \cdots A_n \cup A_{s,t} \cup A_{t,s} & \text{if } s = f(s_1, \ldots, s_n) \text{ and } t = f(t_1, \ldots, t_n) \\ \varnothing & \text{otherwise} \end{cases}$$

where,

$$A_i = \langle s_i \approx t_i \rangle, \qquad A_{s,t} = \begin{cases} \{\{s_1 \mapsto f(t_1, s_1)\}\}\langle s \approx t_2 \rangle & \text{if } (*) \\ \varnothing & \text{otherwise} \end{cases}$$

and $(*)$ stands for $s = f(s_1, s_2)$, $t = f(t_1, t_2)$, $f \in \mathcal{F}_{\mathsf{A}}$, and $s_1 \in \mathcal{V}$.

Theorem 4. *Let s and t be linear terms with $\mathcal{V}ar(s) \cap \mathcal{V}ar(t) = \varnothing$. The set $\langle s\!\downarrow_{\mathsf{A}'} \approx t\!\downarrow_{\mathsf{A}'} \rangle$ is a finite complete set of their A-unifiers.*   □

We illustrate the use of the theorem. Let $s = 0 + x$, $t = (x' + y') + z'$, and $\mathcal{F}_{\mathsf{A}} = \{+\}$. A complete set of A-unifiers for the terms is computed as follows:

$$\langle s\!\downarrow_{\mathsf{A}'} \approx t\!\downarrow_{\mathsf{A}'} \rangle = \langle 0 + x \approx x' + (y' + z') \rangle$$
$$= (\langle 0 \approx x' \rangle \langle x \approx y' + z' \rangle) \cup (\{\{x' \mapsto 0 + x'\}\}\langle x' + (y' + z') \approx x \rangle)$$
$$= \{\{x' \mapsto 0, x \mapsto y' + z'\}, \{x' \mapsto 0 + x', x \mapsto x' + (y' + z')\}\}$$

This set induces the A-critical pairs:

$$y' + z' \quad {}_{\{1\},A} \leftarrow \bowtie \rightarrow_{\{3\}} \quad 0 + (y' + z')$$
$$x' + (y' + z') \quad {}_{\{1\},A} \leftarrow \bowtie \rightarrow_{\{3\}} \quad (0 + x') + (y' + z')$$

Both of the right-hand sides reduce to the corresponding left-hand sides by the rewriting modulo step $\rightarrow_{\{1\},A}$. What about the other critical pairs?

3.2 Coherence

Consider the A-critical pair:

$$x + z \quad {}_{\{1\},A} \leftarrow \bowtie \rightarrow_{\{2\}} \quad (x + 0) + z$$

Contrary to our intention, $(x + 0) + z \rightarrow_{\{1\},A} x + z$ does not hold, and thus $\mathsf{CR}(\{1\}, A)$ is refuted by Theorem 1. This undesired incapability of rewriting modulo is known as the coherence problem [3,8].

Definition 5. *A pair $(\mathcal{R}, \mathcal{E})$ is strongly coherent if $\leftrightarrow_{\mathcal{E}}^* \cdot \rightarrow_{\mathcal{R},\mathcal{E}} \subseteq \rightarrow_{\mathcal{R},\mathcal{E}} \cdot \leftrightarrow_{\mathcal{E}}^*$.*

Lemma 6. *Suppose $\mathcal{E}$ is reversible and $(\mathcal{R}, \mathcal{E})$ is strongly coherent. If $\mathsf{CR}(\mathcal{R}, \mathcal{E})$ then $\mathcal{R} \cup \mathcal{E}$ is confluent, and vice versa.* □

While the strong coherence property always holds for rewriting modulo C, rewriting modulo A and AC rarely satisfy the property. This can be overcome by *extending* a rewrite system. Since an extension for AC is known [3,8], here we consider an A-extension of a TRS.

Definition 7. *Let $\mathcal{R}$ be a TRS. The A-extended TRS $\mathsf{Ext}_A(\mathcal{R})$ consists of*

$$\ell \rightarrow r \qquad f(\ell, x) \rightarrow f(r, x) \qquad f(x, f(\ell, y)) \rightarrow f(x, f(r, y))$$
$$f(x, \ell) \rightarrow f(x, r)$$

for all rules $\ell \rightarrow r \in \mathcal{R}$ with $f = \mathrm{root}(\ell) \in \mathcal{F}_A$. Here x and y are fresh variables not in ℓ.

Lemma 8. *The pair $(\mathsf{Ext}_A(\mathcal{R}), A)$ is strongly coherent and $\rightarrow_{\mathsf{Ext}_A(\mathcal{R})} = \rightarrow_{\mathcal{R}}$.*

Proof. From the inclusion $\rightarrow_A \cdot \rightarrow_{\mathsf{Ext}_A(\mathcal{R}),A} \subseteq \rightarrow_{\mathsf{Ext}_A(\mathcal{R}),A} \cdot \rightarrow_A^*$ the first claim follows. Since $\rightarrow_{\mathcal{R}}$ is closed under contexts, the second claim is trivial.   □

The TRS $\mathsf{Ext}_A(\{1\})$ consists of the four rules:

$$0 + x \rightarrow x \qquad\qquad w + (0 + x) \rightarrow w + x$$
$$(0 + x) + y \rightarrow x + y \qquad w + ((0 + x) + y) \rightarrow w + (x + y)$$

As the extended TRS contains all original rules, we have again the previous A-critical pair:

$$x + z \quad {}_{\{1\},A} \leftarrow \bowtie \rightarrow_{\{2\}} \quad (x + 0) + z$$

Since $(x + 0) + z \rightarrow_{\mathsf{Ext}_A(\{1\}),A} x + z$ holds, the pair is joinable. Similarly, one can verify that all other A-critical pairs are joinable. Therefore, $\mathsf{CR}(\mathsf{Ext}_A(\{1\}), A)$ is concluded by Theorem 1. Finally, confluence of $\mathcal{R}_1$ is established.

3.3 Commutative Unification

Commutative unification also benefits from left-linearity. We define $\mathcal{U}_{\mathcal{E}}^{\mathsf{C}}(s \approx t)$ as $\{\mu \mid s \multimap_{\mathsf{C}} s'$ and $\mu \in \mathcal{U}_{\mathcal{E}}(s' \approx t)$ for some $s'\}$.

Lemma 9. *Suppose* C *and* $\mathcal{E} \cup \mathcal{E}^{-1}$ *commute. If* $\mathsf{Var}(s) \cap \mathsf{Var}(t) = \varnothing$ *and* s *is linear then* $\mathcal{U}_{\mathcal{E}}^{\mathsf{C}}(s \approx t)$ *is a complete set of* $\mathsf{C} \cup \mathcal{E}$-*unifiers for* s *and* t.

Proof. Since it is trivial that $\mathcal{U}_{\mathcal{E}}^{\mathsf{C}}(s \approx t)$ consists of $\mathcal{E} \cup \mathsf{C}$-unifiers, we only show completeness of the set. Let $s\sigma \leftrightarrow^{*}_{\mathsf{C} \cup \mathcal{E}} t\sigma$. One can show $(\leftrightarrow_{\mathcal{E}} \cup \rightarrow_{\mathsf{C}})^{*} \subseteq {}^{*}_{\mathsf{C}}{\leftarrow} \cdot \leftrightarrow^{*}_{\mathcal{E}}$ by using induction, commutation, and the reversibility of C. Thus, $s\sigma \;{}^{*}_{\mathsf{C}}{\leftarrow}\; u \leftrightarrow^{*}_{\mathcal{E}} t\sigma$ for some u. Since $\multimap_{\mathsf{C}} = \leftrightarrow^{*}_{\mathsf{C}}$ holds, $s\sigma \multimap_{\mathsf{C}} u$. Due to the linearity of s there are s' and σ' such that $u = s'\sigma'$, $s \multimap_{\mathsf{C}} s'$, and $x\sigma \multimap_{\mathsf{C}} x\sigma'$ for all variables x. We define the substitution μ as follows:

$$\mu = \{x \mapsto x\sigma \mid x \in \mathsf{Var}(s)\} \cup \{x \mapsto x\sigma' \mid x \in \mathsf{Var}(t)\}$$

Since s and t share no variables, μ is well-defined. By the definition we obtain $s'\mu \leftrightarrow^{*}_{\mathcal{E}} t\mu$. $\qquad\square$

Since A and C are left-linear TRSs that share no function symbols, their commutation can be proved (by using e.g. Theorem 11 in the next section). So Lemma 9 gives a way to compute $\mathsf{A} \cup \mathsf{C}$-critical pairs.

Example 10. Consider the left-linear TRS $\mathcal{R}_2$ with $\mathcal{F}_{\mathsf{A}} = \{*\}$ and $\mathcal{F}_{\mathsf{C}} = \{\mathsf{eq}\}$:

1: $\mathsf{eq}(\mathsf{a}, \mathsf{a}) \to \mathsf{T}$ 3: $\mathsf{eq}(\mathsf{a} * x, y * \mathsf{a}) \to \mathsf{eq}(x, y)$ 5: $(x * y) * z \to x * (y * z)$
2: $\mathsf{eq}(\mathsf{a}, x * y) \to \mathsf{F}$ 4: $\mathsf{eq}(x, y) \to \mathsf{eq}(y, x)$ 6: $x * (y * z) \to (x * y) * z$

Let $\mathcal{R} = \{1, 2, 3\}$ and $\mathcal{E} = \{4, 5, 6\}$. Note that $\mathcal{E} = \mathsf{C} \cup \mathsf{A}$. It is sufficient to show $\mathsf{CR}(\mathsf{Ext}_{\mathsf{A}}(\mathcal{R}), \mathcal{E})$. We can use AC-RPO to prove termination of $\mathcal{R}/\mathcal{E}$, which is equivalent to that of $\mathsf{Ext}_{\mathsf{A}}(\mathcal{R})/\mathcal{E}$ due to the identity in Lemma 6. Let $s = \mathsf{eq}(\mathsf{a} * x, y * \mathsf{a})$ and $t = \mathsf{eq}(\mathsf{a} * x', y' * \mathsf{a})$. A complete set of their $\mathsf{A} \cup \mathsf{C}$-unifiers is:

$$\mathcal{U}_{\mathsf{A}}^{\mathsf{C}}(s \approx t) = \langle s \approx t \rangle \cup \langle \mathsf{eq}(y * \mathsf{a}, \mathsf{a} * x) \approx \mathsf{eq}(\mathsf{a} * x', y' * \mathsf{a}) \rangle$$

$$= \left\{ \begin{array}{llll} \{x \mapsto x', & y \mapsto y'\}, & & \\ \{x \mapsto \mathsf{a}, & y \mapsto \mathsf{a}, & x' \mapsto \mathsf{a}, & y' \mapsto \mathsf{a}\} \\ \{x \mapsto y' * \mathsf{a}, & y \mapsto \mathsf{a} * y, & x' \mapsto y * \mathsf{a}, & y' \mapsto \mathsf{a} * y'\} \end{array} \right\}$$

In this way we can compute complete sets to induce the set of all $\mathcal{E}$-critical pairs. Since all pairs are joinable, $\mathsf{CR}(\mathcal{R}, \mathcal{E})$ is concluded.

4 Commutation

4.1 Commutation Criteria

Our tool employs three commutation criteria. The first commutation criterion is the *development closedness theorem* [9–12].

Theorem 11 (Development Closedness). *Left-linear TRSs $\mathcal{R}$ and $\mathcal{S}$ commute if the inclusions $\mathcal{R}{\leftarrow}{\Join}{\rightarrow}\mathcal{S} \subseteq {\multimap}\mathcal{S}$ and $\mathcal{R}{\leftarrow}{\ltimes}{\rightarrow}\mathcal{S} \subseteq {\rightarrow}_{\mathcal{S}}^{*} \cdot {\leftarrow}{\circ}_{\mathcal{R}}$ hold.* □

The second criterion is the commutation version of the confluence criterion based on rule labeling with weight function [13, 14].

Definition 12. *A weight function w is a function from $\mathcal{F}$ to $\mathbb{N}$. The weight $w(C)$ of a context C is defined as follows:*

$$w(C) = \begin{cases} 0 & \text{if } C = \Box \\ w(f) + w(C') & \text{if } C = f(t_1, \ldots, C', \ldots, t_n) \text{ with a context } C' \end{cases}$$

The weight is admissible *for a TRS $\mathcal{R}$ if*

$$\{w(C) \mid C[x] = \ell\} \geqslant^{\mathrm{mul}} \{w(C) \mid C[x] = r\}$$

holds for all $\ell \to r \in \mathcal{R}$ and $x \in \mathsf{Var}(r)$. Here $\geqslant^{\mathrm{mul}}$ stands for the multiset extension of the standard order $>$ on $\mathbb{N}$ (see e.g. [2]). A rule labeling ϕ for a TRS $\mathcal{R}$ is a function from $\mathcal{R}$ to $\mathbb{N}$. The labeled step $\xrightarrow{\alpha}_{\mathcal{R}}$ is defined as follows: $s \to_{\mathcal{R},(k,m)} t$ if there are a rule $\ell \to r$, a context C, and a substitution σ such that $s = C[\ell\sigma]$, $t = C[r\sigma]$, and $\alpha = (w(C), \phi(\ell \to r))$.

In the next theorem we use the following abbreviations for labeled steps:

$$\xrightarrow{I} = \bigcup_{\alpha \in I} \xrightarrow{\alpha} \qquad {\Upsilon}\alpha = \{\beta \in I \mid \alpha \succ \beta\} \qquad {\Upsilon}\alpha\beta = ({\Upsilon}\alpha) \cup ({\Upsilon}\beta)$$

where, $\succ$ stands for the lexicographic order on $\mathbb{N} \times \mathbb{N}$.

Theorem 13 (Rule Labeling). *Left-linear TRSs $\mathcal{R}$ and $\mathcal{S}$ commute if there are an admissible weight function w and a rule labeling ϕ for $\mathcal{R} \cup \mathcal{S}$ such that*

$$(\mathcal{R}{\xleftarrow{\alpha}}{\Join}{\xrightarrow{\beta}}\mathcal{S}) \cup (\mathcal{R}{\xleftarrow{\alpha}}{\ltimes}{\xrightarrow{\beta}}\mathcal{S}) \subseteq \xrightarrow{{\Upsilon}\alpha}_{\mathcal{S}}^{*} \cdot \xrightarrow{\beta}_{\mathcal{S}}^{=} \cdot \xrightarrow{{\Upsilon}\alpha\beta}_{\mathcal{S}}^{*} \cdot {}_{\mathcal{R}}^{*}{\xleftarrow{{\Upsilon}\alpha\beta}} \cdot {}_{\mathcal{R}}^{=}{\xleftarrow{\alpha}} \cdot {}_{\mathcal{R}}^{*}{\xleftarrow{{\Upsilon}\beta}}$$

for all pairs $\alpha, \beta \in \mathbb{N} \times \mathbb{N}$. □

The final criterion is a trivial adaptation of Theorem 1 to the commutation property, integrating Lemmata 6, 8, and 9.

Theorem 14 (Church-Rosser Modulo). *Let $\mathcal{R}, \mathcal{S}$ be left-linear TRSs and $\mathcal{E} \in \{\mathsf{A}, \mathsf{AC}\}$ such that $\mathcal{R}/\mathcal{E}'$ is terminating for $\mathcal{E}' = \mathcal{E} \cup \mathsf{C}$. The TRSs $\mathcal{R} \cup \mathcal{E}'$ and $\mathcal{S} \cup \mathcal{E}'$ commute if and only if the inclusion holds:*

$$(\mathcal{R}',\mathcal{E}'{\leftarrow}{\Join}{\rightarrow}\mathcal{S}'\cup\mathcal{E}') \cup (\mathcal{R}'\cup\mathcal{E}'{\leftarrow}{\ltimes}{\rightarrow}\mathcal{S}',\mathcal{E}') \subseteq \xrightarrow{}_{\mathcal{S}',\mathcal{E}'}^{*} \cdot {\leftrightarrow}_{\mathcal{E}'}^{*} \cdot {}_{\mathcal{R}',\mathcal{E}'}^{*}{\leftarrow}$$

Here $\mathcal{R}' = \mathsf{Ext}_{\mathcal{E}}(\mathcal{R})$ and $\mathcal{S}' = \mathsf{Ext}_{\mathcal{E}}(\mathcal{S})$. □

Note that our tool uses the algorithm in [15] for AC unification and flattened term representation for overcoming the coherence problem of AC-rewriting. Since we use the dedicated algorithms for A and AC unification, currently we cannot employ Theorem 1 with $\mathcal{E} = \mathsf{A} \cup \mathsf{AC}$.

4.2 Commutation Theorem

The next theorem is known as Hindley's Commutation Theorem [1].

Theorem 15 (Commutation Theorem). *If $\xrightarrow{\alpha}$ and $\xrightarrow{\beta}$ commute for all $\alpha \in I$ and $\beta \in J$ then $\xrightarrow{I}$ and $\xrightarrow{J}$ commute.* $\square$

Example 16. Consider the left-linear TRS $\mathcal{R}_3$:

1:	$0 \times y \to \mathsf{nil}$	5:	$\mathsf{nil} \mathbin{+\!\!+} x \to x$
2:	$\mathsf{s}(x) \times y \to y \mathbin{+\!\!+} (x \times y)$	6:	$x \mathbin{+\!\!+} \mathsf{nil} \to x$
3:	$\mathsf{hd}(\mathsf{c}(x)) \to x$	7:	$x \mathbin{+\!\!+} (y \mathbin{+\!\!+} z) \to (x \mathbin{+\!\!+} y) \mathbin{+\!\!+} z$
4:	$\mathsf{hd}(\mathsf{c}(x) \mathbin{+\!\!+} y) \to x$	8:	$(x \mathbin{+\!\!+} y) \mathbin{+\!\!+} z \to x \mathbin{+\!\!+} (y \mathbin{+\!\!+} z)$
		9:	$\mathsf{from}(x) \to x : \mathsf{from}(\mathsf{s}(x))$

By using the Commutation Theorem we show self-commutation of $\mathcal{R}_3$:

(i) Self-commutation of $\{1, \ldots, 8\}$ follows from Theorem 14.
(ii) Commutation of $\{1, \ldots, 8\}$ and $\{9\}$ follows from Theorem 11.
(iii) Self-commutation of $\{9\}$ is proved by Theorem 11.

Hence, $\mathcal{R}_3$ is confluent.

It is a non-trivial task to find suitable commuting subsystems from an exponential number of candidates. In order to address the problem we introduce a decomposition method based on *composability*, which was introduced by Ohlebusch [16]. Let $\mathcal{R}$ be a TRS. We write $\mathcal{F}_\mathcal{R}$, $\mathcal{D}_\mathcal{R}$, and $\mathcal{C}_\mathcal{R}$ for the following sets:

$$\mathcal{F}_\mathcal{R} = \bigcup_{\ell \to r \in \mathcal{R}} \mathcal{F}\mathsf{un}(\ell) \cup \mathcal{F}\mathsf{un}(r) \qquad \mathcal{D}_\mathcal{R} = \{\mathsf{root}(\ell) \mid \ell \to r \in \mathcal{R}\} \qquad \mathcal{C}_\mathcal{R} = \mathcal{F}_\mathcal{R} \setminus \mathcal{D}_\mathcal{R}$$

Definition 17. *We say that TRSs $\mathcal{R}$ and $\mathcal{S}$ are* composable *if $\mathcal{C}_\mathcal{R} \cap \mathcal{D}_\mathcal{S} = \mathcal{C}_\mathcal{S} \cap \mathcal{D}_\mathcal{R} = \varnothing$ and $\{\ell \to r \in \mathcal{R} \cup \mathcal{S} \mid \mathsf{root}(\ell) \in \mathcal{D}_\mathcal{R} \cup \mathcal{D}_\mathcal{S}\} \subseteq \mathcal{R} \cap \mathcal{S}$.*

Ohlebusch [16] posed the following question.

Question 18. Are left-linear composable TRSs $\mathcal{R}$ and $\mathcal{S}$ confluent if and only if $\mathcal{R} \cup \mathcal{S}$ is confluent?

Although the question still remains open, the following variation is valid.

Theorem 19. *Commuting composable TRSs $\mathcal{R}$ and $\mathcal{S}$ are confluent if and only if $\mathcal{R} \cup \mathcal{S}$ is confluent.* $\square$

Example 20. Recall the TRS $\mathcal{R}_3$ from Example 16. The TRS is the union of the three commuting composable subsystems: $\{1, 2, 5, 6, 7, 8\}$, $\{3, 4, 5, 6, 7, 8\}$, and $\{9\}$. Confluence of each subsystem can be proved in the same method used in the previous example. Hence, $\mathcal{R}_3$ is confluent.

5   Implementation

The confluence tool CoLL consists of about 5,000 lines of OCaml code. Given an input TRS, the tool first performs the next trivial *redundant rule elimination*.

Theorem 21. *Let $\mathcal{R}$ and $\mathcal{S}$ be TRSs with $\mathcal{S} \subseteq \to_{\mathcal{R}}^*$. The TRS $\mathcal{R} \cup \mathcal{S}$ is confluent if and only if $\mathcal{R}$ is confluent.*   □

Example 22. We illustrate the elimination technique with a small example taken from the Confluence Problem Database (Cops)[1]. Consider the TRS:

$$1\colon f(x) \to g(x, f(x)) \qquad 2\colon f(f(f(f(x)))) \to f(f(f(g(x, f(x)))))$$

Since $\{2\} \subseteq \to_{\{1\}}^*$ holds, we eliminate the redundant rule 2. Confluence of the simplified system $\{1\}$ is easily shown by Theorem 11. Note that CoLL cannot prove confluence without using the elimination technique.

Next, the tool employs Theorem 19 to split the simplified TRS into commuting composable subsystems $\mathcal{R}_1, \ldots, \mathcal{R}_n$. For each subsystem $\mathcal{R}_i$ the tool performs the non-confluence test of [17, Lemma 1]. If non-confluence is detected, the tool outputs NO (non-confluence is proved). Otherwise, the tool uses the Commutation Theorem together with the three commutation criteria (Theorems 11, 13, and 14) to determine self-commutation of $\mathcal{R}_i$. Suitable commuting subsystems are searched by enumeration. It outputs YES (confluence is proved) if all of $\mathcal{R}_1, \ldots, \mathcal{R}_n$ are confluent. Concerning automation, we employed AC-RPO for checking termination of $\mathcal{R}/\mathcal{E}$ automatically. Automation of Theorem 13 is based on the SAT encoding technique of [18].

We tested the presented techniques on 188 left-linear TRSs in Cops Nos. 1–425, where we ruled out duplicated problems.[2] The tests were run on a PC equipped with an Intel Core i7-4500U CPU with 1.8 GHz and 3.8 GB of RAM using a time-out of 120 sec. For the sake of comparison we also ran the tools that participated in the 3rd Confluence Competition: ACP v0.5 [9], CSI v0.4.1 [17], and Saigawa v1.7[3]. The first table in Fig. 1 summarizes the results. The first three indicate the results of each commutation criterion without using the Commutation Theorem. The second table indicates the results of individual theories for Theorem 14. The row 'all three' in the first table is the summation of their results, and 'all with elimination' is the same but the elimination technique is enabled. The row CoLL corresponds to the strategy stated above. On our problem set, all confluence proofs by Saigawa are covered by CoLL. The results of CoLL, ACP, and CSI are incomparable.

---

[1] http://cops.uibk.ac.at/.

[2] All problems and results are available at the tool website (see the URL in Section 1).

[3] http://www.jaist.ac.jp/project/saigawa/.

	YES	NO	timeout	$\mathcal{E}$	YES	NO	timeout
Church-Rosser modulo	93	8	1	$\varnothing$	18	8	0
development closed	17	0	0	A	24	0	0
rule labeling	58	0	26	C	42	8	0
all three	125	8	–	AC	64	8	1
all with elimination	136	9	–	C ⊎ AC	88	8	1
CoLL	137	16	21	A ⊎ C ⊎ AC	93	8	1
ACP	134	41	0				
CSI	118	38	11				
Saigawa	105	16	17				

Fig. 1. Experimental results.

6 Conclusion

We presented the new confluence tool CoLL for left-linear TRSs, which proves confluence via commutation. Our primary contribution is automation of Jouannaud and Kirchner's Church-Rosser modulo criterion for associativity and/or commutativity theory, where left-linearity is exploited in several ways.

We briefly compare CoLL with existing confluence tools. CRC 3 [19] is a powerful Church-Rosser checker for Maude and supports the Church-Rosser modulo theorem for any combination of associativity, commutativity, and/or identity theories, except associativity theory. When handling TRSs that contain reversible rules, ACP [9] employs *reduction-preserving completion* [20]. This method effectively works for C and AC rules, but not for associativity rules. ACP and CSI [17] employ layer-preserving decomposition [16] to split a TRS into subsystems. The technique is incomparable to Theorem 19. If Question 18 is affirmatively solved, it generalizes the two techniques for the class of left-linear TRSs. Finally, CoLL was designed for a complement of Saigawa. The two tools will be merged in the next version.

As future work we plan to investigate whether Theorem 19 can be generalized to cover a subclass of *hierarchical combination* [16]. Another interesting direction is the modularity of the commutation property. Since confluence is a modular property [21], it is closed under *signature extension*. Contrary to our expectation, (even local) commutation is *not* signature extensible. Consider the TRSs $\mathcal{R}$ and $\mathcal{S}$ over the signature $\mathcal{F} = \{f^{(2)}, a^{(0)}, b^{(0)}\}$:

$$\mathcal{R} = \{\, a \to b \,\} \qquad \mathcal{S} = \left\{ \begin{array}{ll} f(x, x) \to b, & f(a, x) \to b, \quad f(x, a) \to b \\ f(b, x) \to b, & f(x, b) \to b \end{array} \right\}$$

Since $C[t] \to_{\mathcal{S}}^* b$ holds for all contexts C and $t \in \{a, b\}$, we obtain the strong commutation $_{\mathcal{R}}\!\leftarrow \cdot \to_{\mathcal{S}} \subseteq \to_{\mathcal{S}}^* \cdot \overset{=}{_{\mathcal{R}}}\!\leftarrow$, which entails commutation of $\mathcal{R}$ and $\mathcal{S}$. However, if one extends the signature to $\mathcal{F} \cup \{g^{(1)}\}$, the local peak $f(g(b), g(a))\ _{\mathcal{R}}\!\leftarrow$ $f(g(a), g(a)) \to_{\mathcal{S}} b$ no longer commutes. We conjecture that (local) commutation is closed under signature extension for left-linear TRSs.

Acknowledgements. We are grateful for the detailed comments of the anonymous reviewers, which helped us to improve the presentation.

References

1. Hindley, J.R.: The Church-Rosser Property and a Result in Combinatory Logic. Ph.D. thesis, University of Newcastle-upon-Tyne (1964)
2. Baader, F., Nipkow, T.: Term Rewriting and All That. Cambridge University Press, Cambridge (1998)
3. Jouannaud, J.P., Kirchner, H.: Completion of a set of rules modulo a set of equations. SIAM J. Comput. **15**(4), 1155–1194 (1986)
4. Rubio, A.: A fully syntactic AC-RPO. Inf. Comput. **178**(2), 515–533 (2002)
5. Plotkin, G.: Building in equational theories. Mach. Intell. **7**, 73–90 (1972)
6. Schulz, K.: Word unification and transformation of generalized equations. In: Abdulrab, H., Pecuchet, J.-P. (eds.) IWWERT 1991. LNCS, vol. 677, pp. 150–176. Springer, Heidelberg (1993)
7. Schmidt, R.A.: E-Unification for subsystems of $S4$. In: Nipkow, T. (ed.) RTA 1998. LNCS, vol. 1379, pp. 106–120. Springer, Heidelberg (1998)
8. Peterson, G., Stickel, M.: Complete sets of reductions for some equational theories. J. ACM **28**(2), 233–264 (1981)
9. Aoto, T., Yoshida, J., Toyama, Y.: Proving confluence of term rewriting systems automatically. In: Treinen, R. (ed.) RTA 2009. LNCS, vol. 5595, pp. 93–102. Springer, Heidelberg (2009)
10. Huet, G.: Confluent reductions: abstract properties and applications to term rewriting systems. J. ACM **27**(4), 797–821 (1980)
11. Toyama, Y.: Commutativity of term rewriting systems. In: Fuchi, K., Kott, L. (eds.) Programming of Future Generation Computers II, pp. 393–407. North-Holland, Amsterdam (1988)
12. van Oostrom, V.: Developing developments. Theoret. Comput. Sci. **175**(1), 159–181 (1997)
13. van Oostrom, V.: Confluence by decreasing diagrams. In: Voronkov, A. (ed.) RTA 2008. LNCS, vol. 5117, pp. 306–320. Springer, Heidelberg (2008)
14. Aoto, T.: Automated confluence proof by decreasing diagrams based on rule-labelling. In: Lynch, C. (ed.) RTA 2010, LIPIcs, vol. 6, pp. 7–16 (2010)
15. Pottier, L.: Minimal solutions of linear diophantine systems: bounds and algorithms. In: Book, R.V. (ed.) Rewriting Techniques and Applications. LNCS, vol. 488, pp. 162–173. Springer, Heidelberg (1991)
16. Ohlebusch, E.: Modular Properties of Composable Term Rewriting Systems. Ph.D. thesis, Universität Bielefeld (1994)
17. Zankl, H., Felgenhauer, B., Middeldorp, A.: CSI – a confluence tool. In: Bjørner, N., Sofronie-Stokkermans, V. (eds.) CADE-23. LNCS, vol. 6803, pp. 499–505. Springer, Heidelberg (2011)
18. Hirokawa, N., Middeldorp, A.: Decreasing diagrams and relative termination. J. Autom. Reason. **47**(4), 481–501 (2011)
19. Durán, F., Meseguer, J.: A Church-Rosser checker tool for conditional order-sorted equational maude specifications. In: Ölveczky, P.C. (ed.) WRLA 2010. LNCS, vol. 6381, pp. 69–85. Springer, Heidelberg (2010)
20. Aoto, T., Toyama, Y.: A reduction-preserving completion for proving confluence of non-terminating term rewriting systems. LMCS **8**(1), 1–29 (2012)
21. Toyama, Y.: On the Church-Rosser property for the direct sum of term rewriting systems. J. ACM **34**(1), 128–143 (1987)

Term Rewriting with Prefix Context
Constraints and Bottom-Up Strategies

Florent Jacquemard[1][✉], Yoshiharu Kojima[2], and Masahiko Sakai[2]

[1] INRIA and Ircam, 1 Place Igor Stravinsky, 75004 Paris, France
florent.jacquemard@inria.fr
[2] Graduate School of Information Science, Nagoya University,
Furo-cho, Chikusa-ku, Nagoya 464-8603, Japan
kojima@trs.cm.is.nagoya-u.ac.jp, sakai@is.nagoya-u.ac.jp

Abstract. We consider an extension of term rewriting rules with context constraints restricting the application of rewriting to positions whose prefix (*i.e.* the sequence of symbols from the rewrite position up to the root) belongs to a given regular language. This approach, well studied in string rewriting, is similar to node selection mechanisms in XML transformation languages, and also generalizes the context-sensitive rewriting. The systems defined this way are called prefix constrained TRS (*p*CTRS), and we study the decidability of reachability of regular tree model checking and the preservation of regularity for some subclasses. The two latter properties hold for linear and right-shallow standard TRS but not anymore when adding context constraints. We show that these properties can be restored by restricting derivations to bottom-up ones, and moreover that it implies that left-linear and right-ground *p*CTRS preserve regularity and have a decidable regular model checking problem.

1 Introduction

Term rewriting systems (TRS) are a rule-based computation model for the definition of ranked trees (*terms*) transformations. In the context of formal verification, they can be used to model the dynamics of a system whose configurations are represented by terms. The rewrite relation represents the transitions between configurations. For instance, functional programs manipulating structured data values with pattern matching can be described by rewrite rules [13] such that the rewriting relation represents the program evaluation. This approach can also be applied to distributed algorithms or imperative programs [2] modifying some parts of tree shaped data structures in place, while leaving the rest unchanged.

Regular model checking (RMC) [1] is a useful approach for the automatic reachability and flow analysis of programs or systems modeled by TRS. This technique works by constructing an automaton-based finite representation of the set of reachable configurations of the system analyzed, and uses this representation to detect possible erroneous reachable configurations. Tree automata

---

Y. Kojima—Currently working at TOSHIBA CORPORATION.

A.P. Felty and A. Middeldorp (Eds.): CADE-25, LNAI 9195, pp. 137–151, 2015.
DOI: 10.1007/978-3-319-21401-6_9

(TA [3]) appear to be appropriate for this purpose. A sufficient condition for the decision of RMC is the effective *preservation of regularity*: given a TRS $\mathcal{R}$ satisfying some restrictions, and a TA recognizing a set of terms L_{in} which represents initial configurations, can we compute a TA recognizing the rewrite closure of L_{in} by $\mathcal{R}$, *i.e.* the set of reachable configurations? *Static type checking* of XML transformations can sometimes be solved with similar techniques (see *e.g.* [16]).

Standard TRSs are a Turing-complete low-level formalism with a simple definition by pattern matching and subterm replacement: one rewrite rule can be applied at any position in a term, provided that the left-hand-side of the rule matches the subterm at this position in the term. For instance, a rule with left-hand-side $a(x)$ can be applied at positions labelled by a.

For some applications, one may need to add *context conditions* for the application of rewriting, for instance: rename the label a into b at some position π in a term with the rewrite rule $a(x) \to b(x)$, provided that there exist more than one occurrence of b above π. This is analogous to XML node selection in *e.g.* XQuery update[1] expressed by languages such as XPath. Of course, context conditions can be encoded with additional rewrite rules but this way, small programs or systems will have complex TRS representations, making the modeling process tedious and error prone, and the verification with RMC complicated.

The goal of this paper is to study an expressive extension of TRS with context conditions, which eases modeling, while preserving decidability of RMC under restrictions. More precisely, we study a class called *p*CTRS (prefix controlled TRS) where term rewriting rules are extended with conditions restricting the application of rewriting to positions π whose path (i.e. sequence of symbols and directions from the root down to rewrite position π) belongs to a given regular language. Such context constraints have been studied intensively for string rewriting [4,20] but very few results are known in the case of terms.

First, we show that regularity preservation does not hold and RMC or reachability become undecidable with prefix constraints, already for rewrite systems with strong restrictions such as linearity and flatness of left or right hand sides of rules (Sect. 3.1) which are known to ensure the preservation of regularity in the case of unconstrained TRS [17]. We consider next a natural restriction ensuring effective regularity preservation by bottom-up derivations [6] for linear and right-shallow *p*CTRS (Sect. 4). Considering bottom-up strategy is quite natural in the context of the applications mentioned above. Left-linear and right-ground *p*CTRS enforce bottom-up derivations (Sect. 5), and hence effectively preserve regularity.

2   Preliminaries

Terms. We use the standard notations for terms and positions, see [15]. A *signature* Σ is a finite set of function symbols with fixed arity. We denote the arity of $f \in \Sigma$ as $ar(f)$ and the maximal arity of a symbol of Σ as $max(\Sigma)$. Given an infinite set $\mathcal{X}$ of variables, the set of terms built over Σ and $\mathcal{X}$ is

---

[1] http://www.w3.org/TR/xquery-update-10/.

denoted $\mathcal{T}(\Sigma, \mathcal{X})$, and the subset $\mathcal{T}(\Sigma, \emptyset)$ of *ground* terms is denoted $\mathcal{T}(\Sigma)$. The set of variables occurring in a term $t \in \mathcal{T}(\Sigma, \mathcal{X})$ is denoted $var(t)$. A signature is called *unary* (resp. *strictly unary*) if all its symbols have arity at most 1 (resp. arity exactly 1). In the following, given a strictly unary signature Σ, a string $a_1 a_2 \ldots a_n \in \Sigma^*$ is represented by the term $a_1(a_2(\ldots a_n(x)))$, where $x \in \mathcal{X}$.

A term $t \in \mathcal{T}(\Sigma, \mathcal{X})$ can be seen as a function from its set of positions $\mathcal{P}os(t)$ into $\Sigma \cup \mathcal{X}$. Positions in terms are denoted by sequences of natural numbers, ε is the empty sequence (root position), and $\pi \cdot \pi'$ denotes the concatenation of positions π and π'. The concatenation is naturally extended to sets of positions. The *subterm* of t at position π is denoted $t|_\pi$ defined by $t|_\varepsilon = t$ and $f(t_1, \ldots, t_m)|_{i \cdot \pi} = t_i|_\pi$. The *size* $\|t\|$ of a term t is the cardinality of $\mathcal{P}os(t)$. We write $|s|$ for the length of a finite sequence s. The *depth* of a symbol that occurs in a term at a position π is $|\pi|$. Note that for a string s and its associated term representation t (over a strictly unary signature), $|s| = \|t\| - 1$. A term t is *linear* if no variable occurs more than once in t, *flat* if its depth is at most one and *shallow* if every variable of $var(t)$ occurs at depth at most one in t.

A *substitution* is a mapping from variables of $\mathcal{X}$ into terms of $\mathcal{T}(\Sigma, \mathcal{X})$. It is called *grounding for* $V \subseteq \mathcal{X}$ if the codomain of the restriction $\sigma|_V$ is a set of ground terms. The application of a substitution σ to a term t is denoted as $t\sigma$.

A *context* is a term $C \in \mathcal{T}(\Sigma, \mathcal{X})$ with one distinguished variable x_C occurs exactly once in C. Given a context C and one terms $t \in \mathcal{T}(\Sigma, \mathcal{X})$, we write $C[t]_\pi$ to denote $C\sigma$, where σ is the substitution associating t to x_C and π is the (unique) position of x_C in C. The notation $s = C[t]_\pi$ may also be used to emphasize that $s|_\pi$ is t.

Controlled Term Rewriting Systems. We propose a formalism that strictly extends standard term rewriting systems by forcing, for every rewrite position π in a term t, the path in t from the root into π to belong to a given regular language. For this purpose we use a notion of path carrying both the labels (in Σ) and directions (in $1 .. \max(\Sigma)$). More precisely, let $\mathcal{D}ir(\Sigma) = \{\langle g, i \rangle \mid g \in \Sigma, 0 < i \le ar(g)\}$; we associate with a ground term $t = g(t_1, \ldots, t_{ar(g)}) \in \mathcal{T}(\Sigma)$ and a position $\pi \in \mathcal{P}os(t)$, a *path* in $\mathcal{D}ir(\Sigma)$ defined recursively by

$$path(g(t_1, \ldots, t_{ar(g)}), \varepsilon) = \varepsilon,$$
$$path(g(t_1, \ldots, t_{ar(g)}), i \cdot \pi) = \langle g, i \rangle \cdot path(t_i, \pi) \text{ (with } 1 \le i \le ar(g)).$$

In the case of unary signatures, we may omit the direction (which is always 1) from the path notation, *i.e.* we write g instead of $\langle g, 1 \rangle$.

A *prefix controlled term rewriting system* (pCTRS) over a signature Σ is a finite set $\mathcal{R}$ of prefix controlled *rewrite rules* of the form $L : \ell \to r$, where $L \subseteq \mathcal{D}ir(\Sigma)^*$ is a regular language over $\mathcal{D}ir(\Sigma)$, $\ell \in \mathcal{T}(\Sigma, \mathcal{X}) \setminus \mathcal{X}$ (the left-hand side, or *lhs*), and $r \in \mathcal{T}(\Sigma, var(\ell))$ (the right-hand side, or *rhs*). We use a finite automaton $\mathcal{A}_L$ or a regular expression to present the regular language L.

A term t is rewritten to t' in one step by a pCTRS $\mathcal{R}$, denoted by $t \xrightarrow{\mathcal{R}} t'$, if there exist a controlled rewrite rule $L : \ell \to r \in \mathcal{R}$, a position $\pi \in \mathcal{P}os(t)$ such that $path(t, \pi) \in L$, and a substitution σ such that $t|_\pi = \ell\sigma$ and $t' = t[r\sigma]_\pi$. The reflexive and transitive closure of $\xrightarrow{\mathcal{R}}$ is denoted $t_1 \xrightarrow{*}{\mathcal{R}} t_n$, which we call *a derivation* by $\mathcal{R}$. The size of a *p*CTRS rule $L : \ell \to r$ is the sum of the sizes of the given automaton $\mathcal{A}_L$ defining the control language L, the lhs ℓ and the rhs r. The size $\|\mathcal{R}\|$ of a *p*CTRS $\mathcal{R}$ is the sum of the sizes of its rules.

A controlled rewrite rule $L : \ell \to r$ is *ground, flat, linear, shallow* if ℓ and r are so. It is *right-flat, etc* (resp. *left-flat*) if r (resp. ℓ) is. It is *collapsing* if $r \in \mathcal{X}$, and otherwise *non-collapsing*. A *p*CTRS is *flat, etc* if all its rules are so.

Example 1. Let us consider the *p*CTRS

$$\mathcal{R} = \left\{ \langle h, 1 \rangle^{+} : a \to c, \ \langle g, 2 \rangle^{+} : b \to d, \ \left(\langle h, 1 \rangle \mid \langle h, 2 \rangle \right)^{*} : h(x, y) \to g(x, y, b) \right\}.$$

The rewriting $h(a, b) \to h(c, b)$ is possible with the first rule of $\mathcal{R}$, and $h(a, b) \to g(a, b, b) \to g(a, d, b)$ with the third and then the second rule of $\mathcal{R}$, but neither $g(a, b, b) \to g(c, b, b)$ nor $g(a, b, b) \to g(a, b, d)$ are possible because of their control languages. ◇

Related Work: TRS with Context Constraints. Standard (uncontrolled) TRSs [15] are particular cases of *p*CTRSs with rules of the form $\mathcal{D}ir(\Sigma)^{*} : \ell \to r$. Rewrite systems with context constraints expressed with regular languages have been studied in the case of string rewriting, see [20], and also [4] for the case of conditional context-free (string) grammars.

In [11], we studied a class called CntTRS more general than *p*CTRS. The context constraints in *p*CTRS are specified, for each rewrite rule, by a selection automaton which defines a set of positions in a term based on tree automata computations. Reachability is undecidable for ground CntTRS, whereas we show here that it is decidable for left-linear and right-ground *p*CTRS (Sect. 5).

Under the *context-sensitive rewriting* [10] the rewrite positions are selected according to priorities on the evaluation of arguments of function symbols. More precisely, let us call CS TRS over Σ a pair $\langle \mathcal{R}, \mu \rangle$ made of an uncontrolled TRS $\mathcal{R}$ over Σ and a mapping μ associating to every symbol of Σ the subset of the indexes of its argument that can be rewritten. It means that the positions selected for rewriting in a term $f(t_1, \dots, t_n)$ are defined recursively as the root position and all the positions selected in every t_i such that $i \in \mu(f)$. In the above definition, a *path* in $\mathcal{D}ir(\Sigma)$ contains information both of the symbols and directions. It follows that CS TRSs are particular case of *p*CTRSs.

Proposition 1. *For all CS TRS $\langle \mathcal{R}, \mu \rangle$ over Σ, there exists a pCTRS $\mathcal{R}'$ over Σ such that the rewrite relations defined by $\langle \mathcal{R}, \mu \rangle$ and $\mathcal{R}'$ coincide.*

Consequently, the results below for *p*CTRSs (Corollary 10) extend to CS TRS.

Automata. A finite (string) automaton (FSA) $\mathcal{B}$ over an alphabet Γ with state set P is presented as a tuple $\langle P, p_0, G, \Theta \rangle$ where $p_0 \in P$ is the initial state (denoted $init(\mathcal{B})$) and $G \subseteq P$ (denoted $final(\mathcal{B})$) and the set of transitions $\Theta \subseteq P \times \Gamma \times P$. A transition $\langle p, a, p' \rangle \in \Theta$ is denoted $p \xrightarrow{a} p'$. The size of $\mathcal{B}$ is $\|\mathcal{B}\| = 3 * |\Theta|$.

A *tree automaton* (TA) $\mathcal{A}$ over a signature Σ is a tuple $\langle Q, F, \Delta \rangle$ where Q is a finite set of nullary state symbols, disjoint from Σ, $F \subseteq Q$ is the subset of final states and Δ is a set of transition rules of the form: $g(q_1, \ldots, q_{ar(g)}) \to q$, or $q_1 \to q$ (ε-transition) where $q_1, \ldots, q_{ar(g)}, q \in Q$. Sometimes, the components of a TA $\mathcal{A}$ are written with $\mathcal{A}$ as subscript, like in $Q_{\mathcal{A}}$ to indicate that Q is the state set of $\mathcal{A}$. The size of the transition $g(q_1, \ldots, q_{ar(g)}) \to q$ (resp. ε-transition) is $ar(g)+2$ (resp. 2), and the size $\|\mathcal{A}\|$ of $\mathcal{A}$ is the sum of the sizes of its transitions.

The transition set of a TA $\mathcal{A}$ over Σ is an (uncontrolled) ground TRS, hence we can define a TA transition from $s \in \mathcal{T}(\Sigma \cup Q_{\mathcal{A}})$ into $t \in \mathcal{T}(\Sigma \cup Q_{\mathcal{A}})$ as a rewrite step, denoted $s \xrightarrow[\mathcal{A}]{} t$. The *language* $\mathcal{L}(\mathcal{A}, q)$ of $\mathcal{A}$ in the state $q \in Q_{\mathcal{A}}$ is the set of terms $t \in \mathcal{T}(\Sigma)$ such that $t \xrightarrow[\mathcal{A}]{*} q$. A TA $\mathcal{A}$ is called *clean* if for all $q \in Q_{\mathcal{A}}$, $\mathcal{L}(\mathcal{A}, q) \neq \emptyset$. The language of $\mathcal{A}$ is $\mathcal{L}(\mathcal{A}) = \bigcup_{q \in F_{\mathcal{A}}} \mathcal{L}(\mathcal{A}, q)$. A set of terms $L \subseteq \mathcal{T}(\Sigma)$ is called *regular* if it is the language of a TA.

Regular (tree) languages are effectively closed by intersection, union and complement. The problems of emptiness (given a TA $\mathcal{A}$, does it hold that $\mathcal{L}(\mathcal{A}) = \emptyset$?) and membership (given a TA $\mathcal{A}$ and a ground term t, does it hold that $t \in \mathcal{L}(\mathcal{A})$?) are decidable in deterministic time respectively linear and quadratic.

Rewrite Closure and Decision Problems. The *rewrite closure* of a set of ground terms L by a pCTRS $\mathcal{R}$ is $\mathcal{R}^*(L) = \{t \mid \exists s \in L, s \xrightarrow[\mathcal{R}]{*} t\}$. *Reachability* is the problem to decide, given two terms $s, t \in \mathcal{T}(\Sigma, \mathcal{X})$ and a pCTRS $\mathcal{R}$ whether $s \xrightarrow[\mathcal{R}]{*} t$. *Regular model checking* (RMC) is the problem to decide, given two regular tree languages L_{in} and L_{err} and a pCTRS $\mathcal{R}$ whether $\mathcal{R}^*(L_{\mathsf{in}}) \cap L_{\mathsf{err}} = \emptyset$. Note that non-reachability corresponds to the particular case where $L_{\mathsf{in}} = \{s\}$ and $L_{\mathsf{err}} = \{t\}$. The name RMC is coined after state exploration techniques for checking safety properties. In this setting, L_{in} and L_{err} represent (possibly infinite) sets of initial, respectively error, states. This problem is also related to the problem of *typechecking* tree transformations, see *e.g.* [16].

3 Regularity Preservation for pCTRSs

A pCTRS $\mathcal{R}$ is said to *preserve regularity* if for every regular language $L \subseteq \mathcal{T}(\Sigma)$, the closure $\mathcal{R}^*(L)$ is regular. The preservation is *effective* if moreover a TA recognizing $\mathcal{R}^*(L)$ can be constructed. Thanks to the closure and decidability properties of TAs, the effective preservation is a sufficient condition for RMC.

3.1 Linear and Flat pCTRSs

Every linear and right-flat (uncontrolled) TRS effectively preserves regularity [17]. This property does not hold when adding prefix control.

Proposition 2. *Linear and flat pCTRSs do not preserve regularity.*

Proof. Let us consider the unary signature $\Sigma = \{a, a', b, b', c, d, \bot\}$ where $\bot$ has arity 0 and all other symbols have arity 1, and the linear and flat pCTRS $\mathcal{R}$ over Σ containing the 4 following rules

$$c^* : c(x) \to a'(x), \quad c^*a'a^*b^* : d(x) \to b'(x),$$
$$c^* : a'(x) \to a(x), \quad c^*a^*b^* : b'(x) \to b(x).$$

For the sake of readability, given a string $w \in (\Sigma \setminus \{\perp\})^*$, we simply write below w for the term $w\sigma_0$ where σ_0 is the substitution associating $\perp$ to the (single) variable of the term representing w. The intersection of the regular term set a^*b^* and the rewrite closure of c^*d^* by $\mathcal{R}$ is $\{a^n b^m \mid n \geq m\}$, which is context free (CF) and not regular. Indeed, the control language $c^*a'a^*b^*$ imposes a pairing between rewritings of d into b and rewritings of c into a: for each rewriting of d into b', there must have been one (and only one) earlier rewriting of c into a', as illustrated by the following rewrite sequence $ccdd \xrightarrow{\mathcal{R}} ca'dd \xrightarrow{\mathcal{R}} ca'b'd \xrightarrow{\mathcal{R}} cab'd \xrightarrow{\mathcal{R}} cabd \xrightarrow{\mathcal{R}} a'abd \xrightarrow{\mathcal{R}} a'abb' \xrightarrow{\mathcal{R}} aabb' \xrightarrow{\mathcal{R}} aabb.$ □

We can generalize the principle of the construction of Proposition 2, in order to build a linear and flat pCTRS producing a rewrite closure of the form $\{a^n b^m c^p \mid n \geq m \geq p\}$ (after intersection with the regular language $a^*b^*c^*$), starting from a regular set of the form $d^*e^*f^*$ and using a flat pCTRS. Since the produced language is context-sensitive (CS), it follows that there is no hope for a polynomial time decision procedure (congruence closure like) for the decision of reachability for linear and flat pCTRS.

3.2 Left-(linear and Flat) pCTRSs

Let us consider the situation where the linearity and flatness restrictions apply only to left-hand-side of rewrite rules. In the literature, (uncontrolled) TRSs with such syntactical restrictions are called *inverse-monadic*. They also have the same expressiveness as production rules of CF Tree Grammars.

When restricting to strictly unary signatures, these TRSs correspond to string rewriting rules with lhs of length exactly one. It is folklore knowledge that this kind of string rewriting systems transform CF languages into CF languages. This result generalizes to trees (see *e.g.* [11]). It follows that reachability and RMC are decidable for left-(linear and flat) TRSs. This does not hold when extending the expressiveness with prefix control, even in the case of strings. This is a direct consequence of the following lemma, based on a transformation of CS grammars into Pentonnen normal form [18].

Lemma 3. *For every CS (resp. recursively enumerable (RE)) language L over a strictly unary signature Σ, there exists a linear, left-flat and non-collapsing (resp. linear and left-flat) pCTRS $\mathcal{R}$ over an extended strictly unary signature $\Sigma' \supset \Sigma$ such that $L = \mathcal{R}^*(\{s\}) \cap \mathcal{T}(\Sigma)$ for some term $s \in \mathcal{T}(\Sigma', \mathcal{X})$.*

Proof. Assume that $L \subseteq \Sigma^*$ is a CS language, and let $\mathcal{G} = \langle \mathcal{N}, \Sigma, S, P \rangle$ be a CS grammar generating L, with non-terminal set $\mathcal{N}$, set of terminals Σ, $S \in \mathcal{N}$, and let $\Sigma' = \Sigma \cup \mathcal{N}$ (where the symbols of $\mathcal{N}$ are unary). We can assume that the production rules of $\mathcal{G}$ are in Pentonnen normal form [18]: $AB \to AC$, $A \to BC$, $A \to a$ where $A, B, C \in \mathcal{N}$ and $a \in \Sigma$. Transforming any CS grammar into a grammar of this form can be done in PTIME. It follows that L is the intersection between $\mathcal{T}(\Sigma)$ and the rewrite closure of $\{S(x)\}$ by the linear, left-flat and non-collapsing pCTRS $\mathcal{R}$ simulating the production rules of $\mathcal{G}$, as described in Fig. 1. Note that the size of $\mathcal{R}$ is linear in the size of $\mathcal{G}$.

$\mathcal{G}$	$\mathcal{R}$
$A \;\; \to BC$	$(\mathcal{N} \cup \Sigma)^* : A(x) \to B(C(x))$
$AB \to AC$	$(\mathcal{N} \cup \Sigma)^*A : B(x) \to C(x)$
$A \;\; \to a$	$(\mathcal{N} \cup \Sigma)^* : A(x) \to a(x)$
$A \;\; \to \varepsilon$	$(\mathcal{N} \cup \Sigma)^* : A(x) \to x$

Fig. 1. Construction of a linear CF pCTRS for the proof of Proposition 4.

Every RE language can be generated by a CS grammar as above, completed with some deleting rules of the form $A \to \varepsilon$. It corresponds to the collapsing rewrite rule $A(x) \to x$ (last line of Fig. 1). $\qquad\qquad\square$

Proposition 4. *Over strictly unary signatures, (i) reachability is undecidable for linear and left-flat pCTRSs, and (ii) reachability is PSPACE-complete and regular model checking is undecidable for linear, left-flat, non-collapsing pCTRSs.*

Proof. The undecidability of the reachability problem for linear and left-flat pCTRSs (Claim (i)) follows from Lemma 3, and undecidability of the membership problem of RE languages.

For Claim (ii), the PSPACE-hardness of the reachability problem and undecidability of RMC for linear, left-flat, non-collapsing pCTRSs follow from Lemma 3 and, respectively, the PSPACE-completeness of the membership problem and undecidability of emptiness problem for CS languages.

The PSPACE upper bound for reachability follows immediately from the fact that for a left-flat and non-collapsing pCTRS $\mathcal{R}$ over a strictly unary signature, the size of every rhs of rule of $\mathcal{R}$ is larger or equal to the size of the corresponding lhs. Hence, $s \xrightarrow[\mathcal{R}]{*} t$ can be checked by a backward exploration of the ancestors of t *wrt* $\mathcal{R}$, and they are all smaller than or equal to t. $\qquad\square$

To sum up, (unconstrained) TRSs with syntactic restrictions of flatness and linearity benefit good results of regularity preservation and decision, but these results are lost when adding prefix constraints. The reason is that these constraints permit to test the context of rewrite positions and therefore simulate computations of Turing Machines (Proposition 4(i)) or Linear Bounded Automata (Proposition 4(ii) and remark after Proposition 2). We can observe that in the simulations, it is important to rewrite alternatively in two directions, top-down and bottom-up (see for instance the rewrite sequence presented in the proof of Proposition 2). A key property of regularity preservation results such as [14] is that in every step $C[t] \xrightarrow[\mathcal{R}]{} D[t]$, either all redexes in t are preserved or all are inactivated. For an ordinary step of a left-linear right-shallow pCTRS, such a property does not hold in general, because a reduction in context C may activate a redex in t. However, if we restrict to bottom-up rewriting, the above properties are recovered for linear and right-shallow pCTRSs. We show this in the next sections, and show consequently regularity preservation and decision results for right-ground pCTRSs (Sect. 5) pCTRSs.

4   Bottom-Up Rewrite Strategy

We show in this section that when we restrict to bottom-up derivations [6], the preservation of regularity holds for linear and right-shallow pCTRSs.

4.1   Definition

We define a bottom-up derivation on terms by introducing a bottom-up marked rewriting on marked terms, where the latter is called *weakly* bottom-up in [6]. Following the definition of [6], we use a marked copy of the signature $\overline{\Sigma} = \{\bar{g} \mid g \in \Sigma\}$. A *marked term* is a term in $\mathcal{T}(\Sigma \cup \overline{\Sigma}, \mathcal{X})$. Given a term of $t \in \mathcal{T}(\Sigma, \mathcal{X})$, we use the notation $\bar{t}$ to represent a marked term in $\mathcal{T}(\Sigma \cup \overline{\Sigma}, \mathcal{X})$ associated with t in a way that t is obtained from $\bar{t}$ by replacing each symbol $\bar{g}$ by g. Moreover, $\tilde{t}$ denotes the unique marked term associated with t which belongs to $\mathcal{T}(\overline{\Sigma} \cup \mathcal{X})$. This notation is extended to contexts and substitutions as expected.

The *bottom-up marked rewriting* relation for a pCTRS $\mathcal{R}$ is defined as

$$\overline{C}[\overline{\ell\sigma}]_\pi \xrightarrow[\mathcal{R}]{bu} \overline{C}[r\,\tilde{\sigma}]_\pi$$

for a context $\overline{C}$ if $L : \ell \to r \in \mathcal{R}$, $path(C, \pi) \in L$ and the root symbol of $\overline{\ell}$ is in Σ (the other symbols may be marked or not). We say that the derivation $s \xrightarrow[\mathcal{R}]{*} t$ on terms is *bottom-up* if there exist a marking $\bar{t}$ and a bottom-up marked rewriting sequence $s \xrightarrow[\mathcal{R}]{bu,*} \bar{t}$. In this case, we write $s \Rightarrow_\mathcal{R}^{bu} t$. Note that the derivation $\Rightarrow_\mathcal{R}^{bu}$ on terms is not transitive.

Example 2. Let $\mathcal{R}$ contain the two following prefix controlled rules $\varepsilon : h(x) \to g(x)$ and $\langle h, 1 \rangle : a \to b$. For the controlled rewrite derivation $h(a) \xrightarrow[\mathcal{R}]{*} g(b)$, there exists a bottom-up marked rewriting sequence $h(a) \xrightarrow[\mathcal{R}]{bu} h(b) \xrightarrow[\mathcal{R}]{bu} g(\bar{b})$. Thus the former sequence is bottom-up, *i.e.* $h(a) \Rightarrow_\mathcal{R}^{bu} g(b)$.   ◇

Example 3. Let $\mathcal{R} = \{\varepsilon : h(x) \to g(x), \langle g, 1 \rangle : a \to b\}$. The controlled rewrite derivation $h(a) \xrightarrow[\mathcal{R}]{*} g(b)$ is not bottom-up. Indeed, following the above definition of the bottom-up marked rewriting, we have $h(a) \xrightarrow[\mathcal{R}]{bu} g(\bar{a})$ but we do not have $g(\bar{a}) \xrightarrow[\mathcal{R}]{bu} g(\bar{b})$ because $\bar{a}$ is not in Σ.   ◇

Related Rewrite Strategies. The notion of bottom-up derivations was firstly introduced as the basic narrowing [15]. The bottom-up marked rewriting BU of [6] (that we shall call BU [6] to avoid confusions) is defined with integer marking. It is more general than the above bottom-up marked rewriting, the latter being roughly the restriction of BU [6] using 1 marker.

In [6], a result of regularity preservation is proved for the subclass of linear TRSs such that every rewrite derivation can be simulated by a BU [6] rewrite derivation (such TRSs are called BU). It is shown in [6] that linear and right-flat TRSs are BU. We have seen (Proposition 2) that with prefix control, regularity is not preserved by linear and right-flat pCTRSs. However, we will prove in the next section that regularity is preserved by linear and right-flat pCTRSs when restricting to bottom-up derivations.

There have been studies on regularity preservation, or the decidability of RMC, for (unconstrained) term rewriting under other strategies. It is show in [14] that regularity is preserved by rewriting with linear and right-shallow TRS under the *context-sensitive* strategy. The *innermost* rewriting $\xrightarrow[\mathcal{R}]{in}$ [15] corresponds to the *call by value* computation for programming languages, where arguments are fully evaluated before the function application. More precisely, a rewrite rule can be applied to a subterm at position π if all the proper subterms at children positions of π are normal forms. It is easily shown that $\xrightarrow[\mathcal{R}]{in,*} \subseteq \Rightarrow_{\mathcal{R}}^{bu}$, where the relation is proper for most of TRSs. Regularity preservation have been shown for innermost rewriting with linear right-shallow term rewriting systems [14], and with constructor based systems with additional restrictions [19].

The *one-pass leaf-started derivation* $\Rightarrow_{\mathcal{R}}^{1pls}$ [9] is defined using an auxiliary symbol which acts as a token passed from leaves to root. It is shown in [9] that RMC is decidable for left-linear TRS with one-pass leaf-started rewriting (but regularity is not necessarily preserved). It is shown in [5] that regularity is preserved for *one IO rewrite pass* [7] (denoted $\Rightarrow_{\mathcal{R}}^{IO}$ for its reflexive extension) by linear TRS $\mathcal{R}$. It can be observed that $\Rightarrow_{\mathcal{R}}^{1pls} \subseteq \Rightarrow_{\mathcal{R}}^{IO} \subseteq \Rightarrow_{\mathcal{R}}^{bu}$, where the relations are proper for most of TRSs.

To our knowledge, our approach of studying closure under bottom-up derivations for rewrite rules with context constraints is original.

4.2   Tree Automata Completion

We show now that linear and right-shallow pCTRSs effectively preserve regularity when used with bottom-up rewriting. For this purpose we use a procedure completing a given TA $\mathcal{A}$ with respect to a given pCTRS $\mathcal{R}$. Assuming *wlog* that the initial TA $\mathcal{A}$ is clean and is given without ε-transitions, we complete it first into a TA $\mathcal{A}'$, with one new state $\underline{v}$ for every ground subterm v of a *rhs* of $\mathcal{R}$ and with appropriate transitions such that $\mathcal{L}(\mathcal{A}', \underline{v}) = \{v\}$. Note that $\mathcal{A}'$ can also be assumed to be clean and without ε-transitions.

For each rule $L : \ell \to r$ in $\mathcal{R}$, we assume given an FSA $\mathcal{C}_L$ over $\mathcal{D}ir(\Sigma)$ recognizing L. The respective state sets of all these FSAs are assumed disjoint. We define an automaton $\mathcal{C}_0 = \langle 2^P, \mathcal{D}ir(\Sigma), S, 2^G, \Theta \rangle$ that simulates all the control automata $\mathcal{C}_L$ as follows: P (resp. G) is the union of all the state sets (resp. final state sets) of the $\mathcal{C}_L$'s, S is the set of all the initial states of $\mathcal{C}_L$'s, and Θ contains all the transitions of the form $s \xrightarrow{\langle g,i \rangle} s'$ where $s' = \{p' \in P \mid \exists p \in s\ \exists L : \ell \to r \in \mathcal{R} \text{ s.t. } p \xrightarrow{\langle g,i \rangle} p' \text{ is a transition of } \mathcal{C}_L\}$. Note that $\mathcal{C}_0$ is deterministic.

Let $\mathcal{A}_0 = \langle Q, F, \Delta_0 \rangle$ where $Q = Q_{\mathcal{A}'} \times 2^P$, $F = \{\langle q, S \rangle \mid q \in F_{\mathcal{A}'}\}$, and Δ_0 is the set of transitions of the form:

$$g(\langle q_1, s_1 \rangle, \ldots, \langle q_m, s_m \rangle) \to \langle q_0, s_0 \rangle$$

with $g \in \Sigma$ and such that $g(q_1, \ldots, q_m) \to q_0$ is a transition of $\mathcal{A}'$, and $s_0 \xrightarrow{\langle g,i \rangle} s_i \in \Theta$ for all i with $1 \leq i \leq m$. Intuitively, a term $C[\ell\sigma] \in \mathcal{A}$ and the displayed redex $\ell\sigma$ is reducible by a rule $L : \ell \to r$, if and only if there exists a transition $C[\ell\sigma] \xrightarrow{*}{}_{\mathcal{A}_0} C[\langle q_0, s_0 \rangle] \xrightarrow{*}{}_{\mathcal{A}_0} \langle q, s \rangle \in F$ such that $s_0 \cap final(\mathcal{C}_L) \neq \emptyset$.

We show now how to complete an automaton $\mathcal{A}_k = \langle Q, F, \Delta_k \rangle$, for $k \geq 0$, into $\mathcal{A}_{k+1} = \langle Q, F, \Delta_{k+1} \rangle$, in order to simulate one bottom-up rewrite step with $\mathcal{R}$. At each construction step $k \geq 0$, we construct Δ_{k+1} by adding rules.

(Rules 1). For all $L : \ell \to g(r_1, \ldots, r_m)$ in $\mathcal{R}$, with $m \geq 0$, for all substitutions θ from $\mathcal{X}$ into Q grounding for $var(\ell)$, such that $\ell\theta \xrightarrow{*}{}_{\mathcal{A}_k} \langle q_0, s_0 \rangle$, the last step of this derivation is not an ε-transition, and $s_0 \cap final(\mathcal{C}_L) \neq \emptyset$, we add to Δ_k all the following rules:

$$g\left(\langle q_1, s_1 \rangle, \ldots, \langle q_m, s_m \rangle\right) \to \langle q_0, s_0 \rangle$$

such that for all $1 \leq j \leq m$, if r_j is a variable then $\langle q_j, s_j \rangle = r_j\theta$, otherwise, $q_j = r_j$, and $s_0 \xrightarrow{\langle g,j \rangle} s_j \in \Theta$.

(Rules 2). For all $L : \ell \to x$ in $\mathcal{R}$ with $x \in var(\ell)$, for all substitutions θ from $\mathcal{X}$ into Q grounding for $var(\ell)$ such that $\ell\theta \xrightarrow{*}{}_{\mathcal{A}_k} \langle q_0, s_0 \rangle$, and $s_0 \cap final(\mathcal{C}_L) \neq \emptyset$, we add to Δ_k the following rule:

$$x\theta \to \langle q_0, s_0 \rangle$$

The completion terminates with a fixpoint Δ_k, and the TA $\mathcal{A}^* = \langle Q, F, \Delta_k \rangle$ recognizes the bottom-up rewrite closure of $\mathcal{L}(\mathcal{A})$ by $\mathcal{R}$.

Example 4. Let us consider the pCTRS of Example 2, and two FSA describing the control languages of its two rules: the first FSA has one state v_0, both initial and final, and no transitions, and the second FSA has two states v_1 (initial) and v_2 (final) and one transition $v_1 \xrightarrow{\langle h,1 \rangle} v_2$. Let the initial $\mathcal{A}$ recognize the singleton language $\{h(a)\}$, with the two transitions $a \to q_a$ and $h(q_a) \to q$ (q is the only final state). The automaton $\mathcal{A}_0$ contains the following transitions:

$$a \to \langle q_a, s \rangle \text{ for all } s \subseteq \{v_0, v_1, v_2\},$$
$$h\left(\langle q_a, \{v_2\} \rangle\right) \to \langle q, s \rangle \quad \text{for all } s \text{ with } \{v_1\} \subseteq s \subseteq \{v_0, v_1, v_2\},$$
$$h\left(\langle q_a, \emptyset \rangle\right) \to \langle q, s \rangle \quad \text{for all } s \subseteq \{v_0, v_2\}.$$

The completion process adds the following transitions to $\mathcal{A}^*$, by the case (Rules 1):

$$b \to \langle q_a, s \rangle \text{ for all } s \text{ with } \{v_2\} \subseteq s \subseteq \{v_0, v_1, v_2\},$$
$$g\left(\langle q_a, \{v_2\} \rangle\right) \to \langle q, s \rangle \quad \text{for all } s \text{ with } \{v_0, v_1\} \subseteq s \subseteq \{v_0, v_1, v_2\},$$
$$g\left(\langle q_a, \emptyset \rangle\right) \to \langle q, s \rangle \quad \text{for all } s \text{ with } \{v_0\} \subseteq s \subseteq \{v_0, v_2\}.$$

The only final state of $\mathcal{A}^*$ is $\langle q, \{v_0, v_1\} \rangle$. Then $g(a)$ and $g(b)$ which are both in the bottom-up closure of $h(a)$ are recognized by $\mathcal{A}^*$ with the derivation $g(a) \xrightarrow{}{}_{\mathcal{A}_0}$ $g\left(\langle q_a, \{v_2\} \rangle\right) \xrightarrow{}{}_{\mathcal{A}^*} \langle q, \{v_0, v_1\} \rangle$ and $g(b) \xrightarrow{}{}_{\mathcal{A}^*} g\left(\langle q_a, \{v_2\} \rangle\right) \xrightarrow{}{}_{\mathcal{A}^*} \langle q, \{v_0, v_1\} \rangle$. ◇

Example 5. With the pCTRS of Example 3, we have a first FSA for control identical to the one-state FSA of Example 4 and a second one with two states v_1 (initial) and v_2 (final) and one transition $v_1 \xrightarrow{\langle g,1 \rangle} v_2$. Let us consider the same initial automaton $\mathcal{A}$ as in Example 4. The automaton $\mathcal{A}_0$ contains now the transitions: $a \to \langle q_a, s \rangle$ and $h(\langle q_a, \emptyset \rangle) \to \langle q, s \rangle$ for all $s \subseteq \{v_0, v_1, v_2\}$ (q is final). We obtain the following additional transitions in $\mathcal{A}^*$ by the case (Rules 1):

$$b \to \langle q_a, s \rangle \quad \text{for all } s \text{ with } \{v_2\} \subseteq s \subseteq \{v_0, v_1, v_2\},$$
$$g(\langle q_a, \emptyset \rangle) \to \langle q, s \rangle \quad \text{for all } s \text{ with } \{v_0\} \subseteq s \subseteq \{v_0, v_1, v_2\}.$$

The term $g(a)$ is in the bottom-up rewrite closure of $h(a)$ by $\mathcal{R}$. It is recognized by $\mathcal{A}^*$ with the following derivation $g(a) \xrightarrow{\mathcal{A}_0} g(\langle q_a, \emptyset \rangle) \xrightarrow{\mathcal{A}^*} \langle q, \{v_0, v_1\} \rangle$. The term $h(b)$ is not in the bottom-up rewrite closure of $h(a)$ by $\mathcal{R}$; it holds that $h(b) \xrightarrow{\mathcal{A}^*} h(\langle q_a, s \rangle)$ if $v_2 \in s$, but $\mathcal{A}^*$ cannot compute from such configurations. The situation is similar for $g(b)$, which is neither in the bottom-up rewrite closure of $h(a)$ by $\mathcal{R}$ (see Example 3). $\diamond$

Note that the number of states of the automaton $\mathcal{A}^*$ is exponential in the number of states of the control automata used in the definition of $\mathcal{R}$, and polynomial in $|Q_{\mathcal{A}}|$. When we do not account the size of control automata in the evaluation of the size of $\mathcal{R}$, then the size of $\mathcal{A}^*$ is polynomial and the above construction is PTIME (as well as the decidability results in the next corollary).

Theorem 5. *Given a TA $\mathcal{A}$ and a linear and right-shallow pCTRS $\mathcal{R}$ over Σ, one can construct in EXPTIME a TA over Σ recognizing the bottom-up rewrite closure of $\mathcal{L}(\mathcal{A})$ by $\mathcal{R}$, and whose size is exponential in the size of $\mathcal{A}$ and $\mathcal{R}$.*

In the rest of the section we prove the theorem, by establishing the correctness and completeness of the construction of $\mathcal{A}^*$. For this purpose, we use a relation defined as $t \xrightarrow[\mathcal{R},s]{\pi} t'$, with $\pi \in \mathcal{P}os(t)$ and $s \subseteq P$, iff there exist $L : \ell \to r \in \mathcal{R}$ and a substitution σ such that $t|_\pi = \ell\sigma$, $t' = t[r\sigma]_\pi$, $s \xrightarrow[\Theta]{path(t,\pi)} s'$ and s' contains a final state of C_L. The suffix π in $\xrightarrow[\mathcal{R},s]{\pi}$ might be dropped. We associate to this relation its bottom-up marked counterpart $\xrightarrow[\mathcal{R},s]{bu}$ as above. Note that $\xrightarrow{\mathcal{R}} = \xrightarrow{\mathcal{R},S}$ and $\xrightarrow[\mathcal{R}]{bu} = \xrightarrow[\mathcal{R},S]{bu}$. The next lemma follows immediately from the definition of the relation $\xrightarrow[\mathcal{R},s]{}$.

Lemma 6. *For all $u,t \in \mathcal{T}(\Sigma)$, $i \cdot \pi \in \mathcal{P}os(u)$, and $s \subseteq P$, $u \xrightarrow[\mathcal{R},s]{i \cdot \pi} t$ iff there exist $s_i \subseteq P$ and $g \in \Sigma \cup \overline{\Sigma}$ such that $u = g(u_1, \ldots, u_m)$, $t = g(t_1, \ldots, t_m)$, $u_i \xrightarrow[\mathcal{R},s_i]{\pi} t_i$, and $s \xrightarrow[\Theta]{\langle g,i \rangle} s_i$.*

The next lemma follows from the construction of $\mathcal{A}_0$, as $\mathcal{A}_0$ embeds both $\mathcal{A}'$ and the control automata C_L in the first, resp. second, components of its states.

Lemma 7. *For all $t \in \mathcal{T}(\Sigma)$,*

  i. *if $t \xrightarrow[\mathcal{A}]{*} q$, then for all $s \subseteq P$ there exists $\bar{t}$ such that $\bar{t} \xrightarrow[\mathcal{A}_0]{*} \langle q, s \rangle$.*
  ii. *if $\bar{t} \xrightarrow[\mathcal{A}_0]{*} \langle q, s \rangle$, then $t \xrightarrow[\mathcal{A}]{*} q$.*

The correctness of the construction, *i.e.* the inclusion of $L(\mathcal{A}^*)$ in the bottom-up closure of $L(\mathcal{A})$ by $\mathcal{R}$, results from the following lemma.

Lemma 8. *For all $t \in \mathcal{T}(\Sigma)$ and all state $\langle q, s \rangle \in Q$ such that $t \xrightarrow[\mathcal{A}^*]{*} \langle q, s \rangle$, there exist u and $\bar{t}$ such that (i) $u \xrightarrow[\mathcal{A}_0]{*} \langle q, s \rangle$ and (ii) $u \xrightarrow[\mathcal{R},s]{bu,*} \bar{t}$. Moreover, if the last step in $t \xrightarrow[\mathcal{A}^*]{*} \langle q, s \rangle$ is not an ε-transition, then the top symbol of $\bar{t}$ is in Σ.*

Proof. Let the *index* of a transition rule γ of $\mathcal{A}^*$ be 0 if γ is a transition of $\mathcal{A}_0$ and otherwise, the minimal $k > 0$ such that γ is a transition of $\mathcal{A}_k$ and not a transition of $\mathcal{A}_{k-1}$. We do a proof by induction on the multiset of the indexes of transition rules of $\mathcal{A}^*$ used in the derivation $t \xrightarrow[\mathcal{A}^*]{*} \langle q, s \rangle$, which we call ρ.

We illustrate only an interesting case where the rule γ used in its last step is not an ε-transition and is nor in Δ_0, *i.e.* we assume that the derivation ρ has the following form for $k > 0$:

$$\rho: \ t = g(t_1, \ldots, t_m) \xrightarrow[\mathcal{A}^*]{*} g(\langle q_1, s_1 \rangle, \ldots, \langle q_m, s_m \rangle) \xrightarrow[\mathcal{A}_k]{} \langle q, s \rangle \qquad (\rho_1)$$

This means that $t_j \xrightarrow[\mathcal{A}^*]{*} \langle q_j, s_j \rangle$ for all j with $1 \leq j \leq m$, and then by induction hypothesis, there exist u_j and $\bar{t}_j$ such that (i_0) $u_j \xrightarrow[\mathcal{A}_0]{*} \langle q_j, s_j \rangle$, and (ii_0) $u_j \xrightarrow[\mathcal{R},s_j]{bu,*} \bar{t}_j$.

In this case γ has been added by the case (Rules 1) of the construction, because there exist a rewrite rule $L : \ell \to r \in \mathcal{R}$, a substitution θ from $\mathcal{X}$ into Q, grounding for $var(\ell)$, such that $\ell\theta \xrightarrow[\mathcal{A}_{k-1}]{*} \langle q, s \rangle$, the last step of this derivation is not an ε-transition, and $s \cap final(\mathcal{C}_L) \neq \emptyset$. Moreover, letting $r = g(r_1, \ldots, r_m)$, it holds that for all $1 \leq j \leq m$, if r_j is a variable then $\langle q_j, s_j \rangle = r_j\theta$, and otherwise, $q_j = r_j$ and $s \xrightarrow{\langle g,j \rangle} s_j \in \Theta$.

Without loss of generality, let us assume that for some i, $r_1, \ldots, r_i$ are ground terms and $r_{i+1}, \ldots, r_m$ are distinct variables (remember that $\mathcal{R}$ is linear and right-shallow). Let us now construct a substitution σ from $\mathcal{X}$ into $\mathcal{T}(\Sigma, \mathcal{X})$, grounding for $var(\ell)$. For each $x \in var(\ell) \cap var(r)$, there exists $i + 1 \leq j \leq m$ such that $x = r_j$, and we let $x\sigma = t_j$. For each $x \in var(\ell) \setminus var(r)$, we let $x\sigma$ be an arbitrary ground term in $L(\mathcal{A}_0, x\theta)$ (such a term exists by assumption that $\mathcal{A}_0$ is clean). One can check, using (ρ_1), the construction of γ, and the linearity of the rewrite rules of $\mathcal{R}$, that $\ell\sigma \xrightarrow[\mathcal{A}^*]{*} \ell\theta \xrightarrow[\mathcal{A}_{k-1}]{*} \langle q, s \rangle$, where the last step is not an ε-transition. This derivation is strictly smaller that ρ wrt the induction ordering. Thus, by induction hypothesis, there exist u and $\overline{\ell\sigma}$ such that (i_1) $u \xrightarrow[\mathcal{A}_0]{*} \langle q, s \rangle$ and (ii_1) $u \xrightarrow[\mathcal{R},s]{bu,*} \overline{\ell\sigma}$, and moreover, the top symbol of $\overline{\ell\sigma}$ is in Σ.

By construction of γ, $s \cap final(\mathcal{C}_L) \neq \emptyset$, hence, using the rule $L : \ell \to r \in \mathcal{R}$, it holds that: $\overline{\ell\sigma} \xrightarrow[\mathcal{R},s]{bu} g(r_1, \ldots, r_i, \widetilde{t_{i+1}}, \ldots, \widetilde{t_m})$. Moreover, for all j with $1 \leq j \leq i$, it holds by construction of γ that $q_j = r_j$, hence (i_0) and Lemma 7(ii) imply that $u_j = r_j$, hence $r_j \xrightarrow[\mathcal{R},s_j]{bu,*} \bar{t}_j$ by (ii_0). Using Lemma 6, it follows that

$$u \xrightarrow[\mathcal{R},s]{bu,*} \overline{\ell\sigma} \xrightarrow[\mathcal{R},s]{bu} g(r_1, \ldots, r_i, \widetilde{t_{i+1}}, \ldots, \widetilde{t_m}) \xrightarrow[\mathcal{R},s]{bu,*} g(\overline{t_1}, \ldots, \overline{t_i}, \widetilde{t_{i+1}}, \ldots, \widetilde{t_m}).$$

Letting $\bar{t} = g(\overline{t_1}, \ldots, \overline{t_i}, \widetilde{t_{i+1}}, \ldots, \widetilde{t_m})$, we can conclude for (ii) in this case. Note that the top symbol of $\bar{t}$ is in Σ. $\square$

The following lemma implies the completeness of the construction of $\mathcal{A}^*$.

Lemma 9. *For all $u \in \mathcal{T}(\Sigma)$, $\bar{t} \in \mathcal{T}(\Sigma \cup \overline{\Sigma})$, state $\langle q, s \rangle \in Q$, if $u \xrightarrow[\mathcal{R},s]{bu,*} \bar{t}$, and $u \xrightarrow[\mathcal{A}_0]{*} \langle q, s \rangle$, then $t \xrightarrow[\mathcal{A}^*]{*} \langle q, s \rangle$. Moreover, if the top symbol of $\bar{t}$ is in Σ, then the last step in $t \xrightarrow[\mathcal{A}^*]{*} \langle q, s \rangle$ is not an ε-transition.*

Proof. We do a proof by induction on the lexical combination of the length of the derivation $u \xrightarrow[\mathcal{R},s]{bu,*} \bar{t}$ and the structure of u.

We illustrate only an interesting case that some rewrite steps are performed at the root position, and the last rewrite step performed at the root position involves a non-collapsing rule.

Since $u \xrightarrow[\mathcal{R},s]{bu,*} \bar{t}$ is a bottom-up marked rewriting, no earlier derivation is performed at the root position with a collapsing rule (because the root symbol of every redex in a bottom-up marked derivation must be in Σ), and the top symbol of $\bar{t}$ is in Σ. We can write the rewrite sequence as follows:

$$u \xrightarrow[\mathcal{R},s]{bu,*} \overline{\ell\sigma} \xrightarrow[\mathcal{R},s]{bu} g(r_1, \ldots, r_m)\widetilde{\sigma} \xrightarrow[\mathcal{R},s]{bu,*} \bar{t}$$

where $L : \ell \to g(r_1, \ldots, r_m) \in \mathcal{R}$ and $\varepsilon \in L$. Without loss of generality, we assume that $r_1, \ldots, r_i$ are ground terms and $r_{i+1}, \ldots, r_m$ are variables for some $i \leq m$. By induction hypothesis, it holds that $\ell\sigma \xrightarrow[\mathcal{A}^*]{*} \langle q, s \rangle$ and the last step of this derivation is not an ε-transition. This rewrite sequence can be decomposed into $\ell\sigma \xrightarrow[\mathcal{A}^*]{*} \ell\theta \xrightarrow[\mathcal{A}^*]{*} \langle q, s \rangle$ where θ is a substitution from $\mathcal{X}$ into Q, grounding for $var(\ell)$. Note that we use the assumption that $\mathcal{R}$ is linear in order to construct this θ. Moreover, $s \cap final(\mathcal{C}_L) \neq \emptyset$. Then from the construction case (Rules 1), $\mathcal{A}^*$ contains the transition rule $g(\langle q_1, s_1 \rangle, \ldots, \langle q_m, s_m \rangle) \to \langle q, s \rangle$ where

– for all j with $1 \leq j \leq i$, $q_j = r_j$,
– for all j with $i < j \leq m$, $\langle q_j, s_j \rangle = r_j\theta$,
– for all j with $1 \leq j \leq m$, $s \xrightarrow[\Theta]{\langle g, j \rangle} s_j$.

For all j with $1 \leq j \leq i$, $r_j \xrightarrow[\mathcal{A}_0]{*} \langle q_j, s_j \rangle$ by the construction of $q_j = r_j$ and Lemma 7(i), and for all j with $i \leq j \leq m$, $r_j\sigma \xrightarrow[\mathcal{A}^*]{*} r_j\theta$. Therefore

$$t = g(r_1, \ldots, r_i, r_{i+1}\sigma \ldots, r_m\sigma) \xrightarrow[\mathcal{A}_0]{*} g(\langle q_1, s_1 \rangle, \ldots, \langle q_i, s_i \rangle, r_{i+1}\sigma \ldots, r_m\sigma)$$
$$\xrightarrow[\mathcal{A}^*]{*} g(\langle q_1, s_1 \rangle, \ldots, \langle q_i, s_i \rangle, \langle q_{i+1}, s_{i+1} \rangle, \ldots, \langle q_m, s_m \rangle) \xrightarrow[\mathcal{A}_0]{*} \langle q, s \rangle.$$

Hence $t = x\sigma \xrightarrow[\mathcal{A}^*]{*} x\theta \xrightarrow[\mathcal{A}^*]{*} \langle q, s \rangle$. $\square$

Corollary 10. *Reachability and RMC wrt. bottom-up rewriting are decidable in EXPTIME for linear and right-shallow pCTRSs.*

Proof. Given a linear and right-shallow pCTRS $\mathcal{R}$, the reachability problem $s \xrightarrow[\mathcal{R}]{bu,*} t$ wrt bottom-up rewriting is equivalent to $t \in L(\mathcal{A}^*)$ where $\mathcal{A}^*$ is the TA constructed from a TA recognizing $\{s\}$ as in Theorem 5. The RMC problem $\mathcal{R}^*(L_{\text{in}}) \cap L_{\text{err}} = \emptyset$, wrt bottom-up rewriting, is equivalent to $L(\mathcal{A}^*_{\text{in}}) \cap L_{\text{err}} = \emptyset$, where $\mathcal{A}^*_{\text{in}}$ is the TA constructed from a TA recognizing L_{in} as in Theorem 5.

Both problems can be decided in PTIME in the size of $\mathcal{A}^*$ and t on one hand and $\mathcal{A}^*_{\text{in}}$ and a TA recognizing L_{err} on the other hand. $\square$

5 Left-Linear and Right-Ground pCTRSs

It can be observed that every rewrite sequence with a right-ground pCTRS is bottom-up. Hence the following corollary immediately follows.

Corollary 11. *Given a TA $\mathcal{A}$ and a left-linear and right-ground pCTRS $\mathcal{R}$ over Σ, one can construct in EXPTIME a TA over Σ recognizing the rewrite closure of $\mathcal{L}(\mathcal{A})$ by $\mathcal{R}$, and whose size is exponential in the size of $\mathcal{A}$ and $\mathcal{R}$. Reachability and RMC are decidable for left-linear and right-ground pCTRSs.*

The following proposition establishes a lower bound for the construction.

Proposition 12. *Reachability is PSPACE-hard for ground pCTRSs.*

Proof. We make a reduction of the intersection emptiness problem for regular string languages. Let $L_1, \ldots, L_n (n \geq 2)$ be regular languages, and let

$$\mathcal{R} = \{\Sigma^* : \sharp_1 \to a(\sharp_1) \mid a \in \Sigma\} \cup \{L_i : \sharp_i \to \sharp_{i+1} \mid 1 \leq i \leq n-1\}$$
$$\cup \{L_n : \sharp_n \to \flat\} \qquad\qquad \cup \{\Sigma^* : a(\flat) \to \flat \mid a \in \Sigma\}$$

It can be easily checked that $\sharp_1 \xrightarrow{*}_{\mathcal{R}} \flat$ iff $L_1 \cap \cdots \cap L_n \neq \emptyset$. $\qquad\square$

Given a linear and right-shallow (uncontrolled) TRS $\mathcal{R}$, for every rewrite sequence $s \xrightarrow{*}_{\mathcal{R}} t$ there exists a bottom-up rewrite sequence $s \xrightarrow{*}_{\mathcal{R}} t$ [6]. This is however not the case in presence of prefix control, since linear and right-shallow pCTRSs do not preserve regularity (Proposition 2).

6 Conclusion

This work could be extended in several directions. A question is whether the results of Sect. 4 still hold when weakening the linearity restriction into right-linearity. Note that regularity preservation has been established for right-linear and right-shallow (unconstrained) TRSs in [17]. An alternative approach might be to construct automata recognizing regular over-approximating of the closures, for larger classes of pCTRS, like in [8]. The completion algorithms of this paper terminate because the shallowness of the rhs ensure that no new state needs to be added to the automata; other automata completion methods [8] accept non-shallow rhs and thus need to normalize the new transitions by adding new states, they ensure their termination by merging states, at the cost of precision.

Finally, a difficult problem is the generalization of the problems presented in this paper to unranked tree rewriting [12], where variables are instantiated by forests (i.e. finite sequences of trees) instead of terms.

References

1. Abdulla, P.A., Jonsson, B., Mahata, P., d'Orso, J.: Regular tree model checking. In: Brinksma, E., Larsen, K.G. (eds.) CAV 2002. LNCS, vol. 2404, pp. 555–568. Springer, Heidelberg (2002)

2. Bouajjani, A., Habermehl, P., Rogalewicz, A., Vojnar, T.: Abstract regular tree model checking of complex dynamic data structures. In: Yi, K. (ed.) SAS 2006. LNCS, vol. 4134, pp. 52–70. Springer, Heidelberg (2006)
3. Comon, H., Dauchet, M., Gilleron, R., Jacquemard, F., Löding, C., Lugiez, D., Tison, S., Tommasi, M.: Tree Automata Techniques and Applications (2007). http://tata.gforge.inria.fr
4. Dassow, J., Paun, G., Salomaa, A.: Grammars with controlled derivations. In: Rozenberg, G., Salomaa, A. (eds.) Handbook of Formal Languages, vol. 2, pp. 101–154. Springer, Heidelberg (1997)
5. Dauchet, M., De Comite, F.: A Gap Between linear and non-linear term-rewriting systems. In: Lescanne, P. (ed.) Rewriting Techniques and Applications. LNCS, vol. 256, pp. 95–104. Springer, Heidelberg (1987)
6. Durand, I., Sénizergues, G.: Bottom-up rewriting is inverse recognizability preserving. In: Baader, F. (ed.) RTA 2007. LNCS, vol. 4533, pp. 107–121. Springer, Heidelberg (2007)
7. Engelfriet, J., Schmidt, E.M.: IO and OI. II. J. Comput. Syst. Sci. **16**(1), 67–99 (1978)
8. Feuillade, G., Genet, T., Tong, V.V.T.: Reachability analysis over term rewriting systems. J. Autom. Reasoning **33**(3–4), 341–383 (2004)
9. Fülöp, Z., Jurvanen, E., Steinby, M., Vágvölgyi, S.: On one-pass term rewriting. Acta Cybernetica **14**(1), 83–98 (1999)
10. Futatsugi, K., Goguen, J.A., Jouannaud, J.-P., Meseguer, J.: Principles of OBJ2. In: Proceedings 12th ACM SIGACT-SIGPLAN Symposium on Principles of Programming Languages (POPL), pp. 52–66 (1985)
11. Jacquemard, F., Kojima, Y., Sakai, M.: Controlled term rewriting. In: Tinelli, C., Sofronie-Stokkermans, V. (eds.) FroCoS 2011. LNCS, vol. 6989, pp. 179–194. springer, Heidelberg (2011)
12. Jacquemard, F., Rusinowitch, M.: Rewrite closure and CF hedge automata. In: Dediu, A.-H., Martín-Vide, C., Truthe, B. (eds.) LATA 2013. LNCS, vol. 7810, pp. 371–382. Springer, Heidelberg (2013)
13. Jones, N.D., Andersen, N.: Flow analysis of lazy higher-order functional programs. Theor. Comput. Sci. **375**(1–3), 120–136 (2007)
14. Kojima, Y., Sakai, M.: Innermost reachability and context sensitive reachability properties are decidable for linear right-shallow term rewriting systems. In: Voronkov, A. (ed.) RTA 2008. LNCS, vol. 5117, pp. 187–201. Springer, Heidelberg (2008)
15. de Vrijer, R., Bezem, M., Klop, J.W. (eds.): Term Rewriting Systems by Terese. Cambridge Tracts in TCS, vol. 55. Cambridge University Press, Cambridge (2003)
16. Milo, T., Suciu, D., Vianu, V.: Typechecking for XML transformers. J. Comput. Syst. Sci. **66**(1), 66–97 (2003)
17. Nagaya, T., Toyama, Y.: Decidability for left-linear growing term rewriting systems. Inf. Comput. **178**(2), 499–514 (2002)
18. Penttonen, M.: One-sided and two-sided context in formal grammars. Inf. Control **25**, 371–392 (1974)
19. Réty, P., Vuotto, J.: Tree Automata for Rewrite Strategies. J. Symbolic Comput. **40**, 749–794 (2005)
20. Sénizergues, G.: Some decision problems about controlled rewriting systems. Theor. Comput. Sci. **71**(3), 281–346 (1990)

Encoding Dependency Pair Techniques
and Control Strategies for Maximal Completion

Haruhiko Sato[1][(✉)] and Sarah Winkler[2]

[1] Graduate School of Information Science and Technology, Hokkaido University,
Sapporo, Japan
[2] Institute of Computer Science, University of Innsbruck, Innsbruck, Austria
haru@complex.ist.hokudai.ac.jp, sarah.winkler@uibk.ac.at

Abstract. This paper describes two advancements of SAT-based Knuth-Bendix completion as implemented in Maxcomp. (1) Termination techniques using the dependency pair framework are encoded as satisfiability problems, including dependency graph and reduction pair processors. (2) Instead of relying on pure maximal completion, different SAT-encoded control strategies are exploited.

Experiments show that these developments let Maxcomp improve over other automatic completion tools, and produce novel complete systems.

Keywords: Term rewriting · Completion · SAT encoding · Dependency pairs

1 Introduction

Recently, some impressive progress was been achieved by exploiting SAT/SMT solvers in theorem proving [6]. Maximal completion is a simple yet highly efficient Knuth-Bendix completion approach which relies on MaxSAT solving [5]. It is hence inherently limited to compute complete term rewrite systems (TRSs) whose termination can be expressed as a SAT problem. The maximal completion tool Maxcomp restricts to LPO and KBO, which naturally narrows the range of possible completions. For instance, in the following presentation of CGE_2 the last equation cannot be oriented:

$$e \cdot x \approx x \qquad f(x \cdot y) \approx f(x) \cdot f(y) \qquad x \cdot (y \cdot z) \approx (x \cdot y) \cdot z$$
$$i(x) \cdot x \approx e \qquad g(x \cdot y) \approx g(x) \cdot g(y) \qquad f(x) \cdot g(y) \approx g(y) \cdot f(x)$$

In general, Maxcomp cannot complete CGE problems, which describe commuting group endomorphisms as occurring in the theory of uninterpreted functions [10]. Another potential limitation of Maxcomp is given by the fact that its exclusive search strategy is to orient as many equations as possible.

This paper presents two advancements of Maxcomp. (1) Our abstract framework for SMT encodings allows to first switch from a termination problem to a

---

This research was supported by the Austrian Science Fund project I963.

A.P. Felty and A. Middeldorp (Eds.): CADE-25, LNAI 9195, pp. 152–162, 2015.
DOI: 10.1007/978-3-319-21401-6_10

dependency pair (DP) problem and subsequently apply an arbitrary sequential combination of dependency pair processors. We give encodings for different estimations of the dependency graph (DG), and show how to apply reduction pair processors in this context. Though encoding termination of a TRS as a satisfiability problem has become common practice, to the best of our knowledge all previous encodings restrict to a specific reduction order or interpretations into a particular domain. (2) The original version of Maxcomp always tried to generate a complete TRS by orienting as many equations as possible. However, this control strategy is not always optimal to guide the proof search. We devised satisfiability encodings for a number of alternative control strategies, and compared them experimentally.

Our results show that these enhancements allow Maxcomp to not only complete CGE problems but in general improve over previous automatic completion tools. Though we described preliminary results on DP encodings in [7], our recent work on control strategies greatly enhanced the tool's power and scalability.

The remainder of this paper is structured as follows. Section 2 collects some preliminaries before our encodings for dependency pair techniques are outlined in Sect. 3. Section 4 presents the developed control strategies. Some further implementation issues are described in Sect. 5, and experimental results are presented in Sect. 6.

2 Preliminaries

We assume familiarity with term rewriting [1]. Knuth-Bendix completion aims to transform an equational system (ES) $\mathcal{E}_0$ into a TRS $\mathcal{R}$ which is complete for $\mathcal{E}_0$, i.e., terminating, confluent and equivalent to $\mathcal{E}_0$. The set of critical pairs $\mathrm{CP}(\ell_1 \to r_1, \ell_2 \to r_2)$ denotes all equations $\ell_2\sigma[r_1\sigma]_p \approx r_2\sigma$ such that p is a function symbol position in ℓ_2, $\ell_2|_p$ and ℓ_1 are unifiable with mgu σ, and if $p = \epsilon$ then the two rules are not variants. We write $\mathrm{CP}(\mathcal{R})$ for the set of critical pairs among rules from a TRS $\mathcal{R}$. The relation $\downarrow_\mathcal{R}$ denotes $\to_\mathcal{R}^* \cdot {}_\mathcal{R}^*\!\leftarrow$. We also write $(s \approx t)\downarrow_\mathcal{R}$ for an equation $s' \approx t'$ such that s' and t' are some $\mathcal{R}$-normal forms of s and t, respectively, and mean the natural extension to sets of equations $\mathcal{E}$ when writing $\mathcal{E}\downarrow_\mathcal{R}$. For an ES $\mathcal{E}$ we write $\widetilde{\mathcal{E}}$ to denote the set of all equations $\ell \approx r$ such that $\ell \approx r \in \mathcal{E} \cup \mathcal{E}^{-1}$ and $\ell \to r$ is a valid rewrite rule.

Maximal completion is a simple completion approach based on MaxSAT solving. For an input ES $\mathcal{E}_0$, it tries to compute $\Phi_{\mathcal{E}_0}(\mathcal{E}_0)$ as follows:

Definition 1. *Let $\mathcal{E}$ be a fixed ES. For any ES $\mathcal{C}$, $\Phi_\mathcal{E}$ is defined by*

$$\Phi_\mathcal{E}(\mathcal{C}) = \begin{cases} \mathcal{R} & \textit{if } \mathcal{E} \cup \mathrm{CP}(\mathcal{R}) \subseteq \downarrow_\mathcal{R} \quad \textit{for some } \mathcal{R} \in \mathfrak{R}(\mathcal{C}) \\ \Phi_\mathcal{E}(\mathcal{C} \cup S(\mathcal{C})) & \textit{otherwise} \end{cases} \tag{1}$$

where $\mathfrak{R}(\mathcal{C})$ consists of terminating TRSs $\mathcal{R}$ such that $\mathcal{R} \subseteq \widetilde{\mathcal{C}}$, and $S(\mathcal{C}) \subseteq \leftrightarrow_\mathcal{C}^*$.

Theorem 1 ([5])**.** *The TRS $\Phi_{\mathcal{E}_0}(\mathcal{E}_0)$ is complete for $\mathcal{E}_0$ if it is defined.*

In the maximal completion tool Maxcomp, $\mathfrak{R}(\mathcal{C})$ is computed by maximizing the number of satisfied clauses in $\bigvee_{s \approx t \in \mathcal{C}} [s > t] \vee [t > s]$, subject to the side constraints implied by the SAT/SMT encoding $[\cdot > \cdot]$ of some reduction order $>$, and $S(\mathcal{C})$ is a subset of $\bigcup_{\mathcal{R} \in \mathfrak{R}(\mathcal{C})} (\mathrm{CP}(\mathcal{R}) \cup \mathcal{E}_0){\downarrow}_{\mathcal{R}}$.

In this paper we use the dependency pair (DP) framework to show termination of TRSs [4]. A DP problem is a pair of two TRSs $(\mathcal{P}, \mathcal{R})$, it is finite if it does not admit an infinite chain. A DP processor Proc is a function which maps a DP problem to either a set of DP problems or "no". It is sound if a DP problem d is finite whenever $\mathrm{Proc}(d) = \{d_1, \ldots, d_n\}$ and all of d_i are finite.

For an ES $\mathcal{C}$, let the set of dependency pair candidates $\mathrm{DPC}(\mathcal{C})$ be all rules $F(t_1, \ldots, t_n) \to G(u_1, \ldots, u_n)$ such that $f(t_1, \ldots, t_n) \approx r \in \widetilde{\mathcal{C}}$, $r \trianglerighteq g(u_1, \ldots, u_n)$ but $\ell \not\trianglerighteq g(u_1, \ldots, u_n)$, and F, G are fresh function symbols.

3   Encodings

We first illustrate the idea of our encodings by means of a simple example.

Example 1. Suppose that the current set of equations contains a potential rule $\alpha\colon \mathsf{aa} \to \mathsf{ba}$. (To enhance readability we here use string notation and write aa to denote the term $\mathsf{a}(\mathsf{a}(x))$, etc.). Note that this rule gives rise to the dependency pair $\beta\colon \mathsf{Aa} \to \mathsf{Ba}$ if b is a defined symbol. Rule α gives rise to the following constraints:

$$S_\alpha^0 \to W_\alpha^1 \wedge X_\mathsf{a}^{\mathrm{def}} \wedge (X_\mathsf{b}^{\mathrm{def}} \to S_\beta^1) \tag{a}$$

$$S_\beta^1 \to [\mathsf{Aa} \geqslant^{\mathsf{poly}} \mathsf{Ba}] \wedge (\neg[\mathsf{Aa} >^{\mathsf{poly}} \mathsf{Ba}] \to S_\beta^2) \tag{b}$$

$$S_\beta^2 \to [\mathsf{Aa} \geqslant^{\mathsf{lpo}} \mathsf{Ba}] \wedge (\neg[\mathsf{Aa} >^{\mathsf{lpo}} \mathsf{Ba}] \to S_\beta^3)$$

$$W_\alpha^1 \to [\mathsf{aa} \geqslant^{\mathsf{poly}} \mathsf{ba}] \wedge (\neg[\mathsf{aa} >^{\mathsf{poly}} \mathsf{ab}] \to W_\alpha^2) \tag{c}$$

$$W_\alpha^2 \to [\mathsf{aa} \geqslant^{\mathsf{lpo}} \mathsf{ba}]$$

$$\neg S_\beta^3 \tag{d}$$

Here the boolean variables $S_\alpha^0, W_\alpha^1, \ldots, W_\alpha^3, S_\beta^1, \ldots, S_\beta^3$ express strict/weak orientation of α and β in different proof stages, and $X_\mathsf{a}^{\mathrm{def}}, X_\mathsf{b}^{\mathrm{def}}$ express whether a and b are defined. The constraint (a) triggers the DP β and "moves" rule α to the weak component. Constraint (b) expresses that if the DP β is not oriented it remains to be considered, both for stage 2 and 3. Constraint (c) ensures that rule α is weakly oriented. Since monotonic polynomial interpretations allow for rule removal, α can be removed after stage 2 if it was strictly oriented. Finally, (d) demands that the DP β needs no consideration after stage 3.

The following paragraphs transfer standard notions of the DP framework to our satisfiability setting.

Definition 2. *A* DP problem encoding *is a tuple* $\mathcal{D} = (\mathcal{S}, \mathcal{W}, \varphi)$ *consisting of two sets of boolean variables* $\mathcal{S} = \{S_{\ell \to r} \mid \ell \to r \in \mathcal{P}\}$ *and* $\mathcal{W} = \{W_{\ell \to r} \mid \ell \to r \in \mathcal{R}\}$ *for TRSs* $\mathcal{P}$ *and* $\mathcal{R}$, *and a formula* φ. *We call an assignment* α *finite for a DP problem encoding* $(\mathcal{S}, \mathcal{W}, \varphi)$ *if* $\alpha(\varphi) = \top$ *and the DP problem* $(\mathcal{P}_\alpha^\mathcal{S}, \mathcal{R}_\alpha^\mathcal{W})$ *is finite, given by the TRSs* $\mathcal{P}_\alpha^\mathcal{S} = \{\ell \to r \mid S_{\ell \to r} \in \mathcal{S}, \alpha(S_{\ell \to r}) = \top\}$ *and* $\mathcal{R}_\alpha^\mathcal{W} = \{\ell \to r \mid W_{\ell \to r} \in \mathcal{W}, \alpha(W_{\ell \to r}) = \top\}$.

Definition 3. *A* DP processor encoding Proc *maps a DP problem encoding* $\mathcal{D} = (\mathcal{S}, \mathcal{W}, \varphi)$ *to a finite set of DP problem encodings* $\mathrm{Proc}(\mathcal{D}) = \{\mathcal{D}_1, \dots, \mathcal{D}_n\}$. *The encoding* Proc *is sound if for any* $\mathcal{D}$ *such that* $\mathrm{Proc}(\mathcal{D}) = \{\mathcal{D}_1, \dots, \mathcal{D}_n\}$ *and any assignment* α *that is finite for all* $\mathcal{D}_i$, *it also holds that* α *is finite for* $\mathcal{D}$.

Definition 4. *The set of* initial variables *for an ES* $\mathcal{C}$ *is* $\mathcal{I}_\mathcal{C} = \{I_{\ell \to r} \mid \ell \approx r \in \widetilde{\mathcal{C}}\}$. *For an ES* $\mathcal{C}$ *the* initial DP problem encoding *is given by* $\mathcal{D}_\mathcal{C} = (\mathcal{S}, \mathcal{W}, \varphi)$ *where* $\mathcal{S} = \{S_{\ell \to r} \mid \ell \to r \in \mathrm{DPC}(\mathcal{C})\}$, $\mathcal{W} = \{W_{\ell \to r} \mid \ell \approx r \in \widetilde{\mathcal{C}}\}$ *and*

$$\varphi = \bigwedge_{\ell \approx r \in \widetilde{\mathcal{C}}} I_{\ell \to r} \to \left(W_{\ell \to r} \wedge X_{\mathrm{root}(\ell)}^{def} \wedge \bigwedge_{s \to t \in \mathrm{DPC}(\ell \to r)} X_{\mathrm{root}(t)}^{def} \to S_{s \to t} \right)$$

Lemma 1. *Let* $\mathcal{C}$ *be an ES. Suppose there is a tree whose nodes are DP problem encodings satisfying the following conditions:*

- *The root is the initial DP problem encoding* $\mathcal{D}_\mathcal{C}$.
- *For every non-leaf node* $\mathcal{D}$ *with children* $\mathcal{D}_1, \dots, \mathcal{D}_n$ *there is a sound processor encoding* Proc *such that* $\mathrm{Proc}(\mathcal{D}) = \{\mathcal{D}_1, \dots, \mathcal{D}_n\}$.

Let the leaves be $\{(\mathcal{S}_i, \mathcal{W}_i, \varphi_i) \mid 1 \leq i \leq k\}$. *If an assignment* α *satisfies*

$$\bigwedge_{i=1}^k \varphi_i \wedge \bigwedge_{s \to t \in \mathcal{S}_i} \neg S_{s \to t}$$

then the TRS $\mathcal{R} = \{\ell \to r \mid \ell \approx r \in \widetilde{\mathcal{C}}, \alpha(I_{\ell \to r}) = \top\}$ *is terminating.*

Proof. By induction on the tree structure, α is finite for all nodes. Termination of $\mathcal{R}$ follows from finiteness of α for the root $\mathcal{D}_\mathcal{C}$. □

Definition 5 (Reduction Pair Processor). *Let* $(>, \geqslant)$ *be a reduction pair and* π *an argument filtering, with satisfiability encodings* $[\cdot \geqslant_\pi \cdot]$ *and* $[\cdot >_\pi \cdot]$

A DP problem encoding $(\mathcal{S}, \mathcal{W}, \varphi)$ *is mapped to* $\{(\mathcal{S}', \mathcal{W}', \varphi \wedge T_S \wedge T_W)\}$ *where* $\mathcal{S}' = \{S'_{\ell \to r} \mid S_{\ell \to r} \in \mathcal{S}\}$, $\mathcal{W}' = \{W'_{\ell \to r} \mid W_{\ell \to r} \in \mathcal{W}\}$, *and*

$$T_S = \bigwedge_{S_{\ell \to r} \in \mathcal{S}} S_{\ell \to r} \to [\ell \geqslant_\pi r] \wedge (\neg[\ell >_\pi r] \to S'_{\ell \to r})$$

$$T_W = \bigwedge_{W_{\ell \to r} \in \mathcal{W}} W_{\ell \to r} \to W'_{\ell \to r} \wedge [\ell \geqslant_\pi r]$$

Here all boolean variables in $\mathcal{S}'$ and $\mathcal{W}'$ are assumed to be fresh. Concrete encodings $[\cdot \geqslant_\pi \cdot]$ and $[\cdot >_\pi \cdot]$ for LPO/RPO, KBO as well as reduction orders given by polynomial and matrix interpretations—also with argument filterings and usable rules—are well-studied, see for instance [3, 8, 13–15].

Note that Definition 5 can easily be modified to admit rule removal by adding clauses $(\neg[\ell >_\pi r] \to W'_{\ell \to r})$ to the conjunction defining T_W, similar as for T_S.

Definition 6 (Dependency Graph Processor). *A DP problem encoding of the form* $(\mathcal{S}, \mathcal{W}, \varphi)$ *is mapped to* $\{(\mathcal{S}', \mathcal{W}', \psi)\}$ *such that* $\mathcal{S}' = \{S'_{\ell \to r} \mid S_{\ell \to r} \in \mathcal{S}\}$, $\mathcal{W}' = \{W'_{\ell \to r} \mid S_{\ell \to r} \in \mathcal{S}\} \cup \{W'_{\ell \to r} \mid W_{\ell \to r} \in \mathcal{W}\}$, *and* $\psi = \varphi \wedge T_S \wedge T_W$ *where*

$$T_S = \bigwedge_{S_p, S_{p'} \in \mathcal{S}} S_p \wedge S_{p'} \wedge [p \Rightarrow p'] \wedge \neg S'_p \wedge \neg S'_{p'} \to X_p^{\mathrm{w}} > X_{p'}^{\mathrm{w}}$$

$$T_W = \bigwedge_{S_{\ell \to r} \in \mathcal{S}} S_{\ell \to r} \to W'_{\ell \to r} \wedge \bigwedge_{W_{\ell \to r} \in \mathcal{W}} W_{\ell \to r} \to W'_{\ell \to r}$$

Here T_S encodes cycle analysis of the graph in the sense that a cycle $p_1 \Rightarrow p_2 \Rightarrow \cdots \Rightarrow p_n \Rightarrow p_1$ issues the unsatisfiable constraint $X_{p_1}^{\mathrm{w}} > X_{p_2}^{\mathrm{w}} > \cdots > X_{p_n}^{\mathrm{w}} > X_{p_1}^{\mathrm{w}}$. For the formula $[s \to t \Rightarrow u \to v]$ encoding the presence of an edge from $s \to t$ to $u \to v$ one can simply use $\top$ if $\mathsf{root}(t) = \mathsf{root}(u)$ and $\bot$ otherwise. (We also experimented with an encoding in terms of the unifiability between $\mathsf{REN}(\mathsf{CAP}(t))$ and u, but due to reasons of space do not present it here).

The above encoding does not allow to use different orderings in SCCs, in contrast to what is commonly done in termination provers. However, it can be modified to consider SCCs by mapping a problem encoding to k independent problem encodings.

Definition 7 (Dependency Graph Processor with k SCCs). *A DP problem encoding* $\mathcal{D} = (\mathcal{S}, \mathcal{W}, \varphi)$ *is mapped to* $\{\mathcal{D}_i\}_{1 \le i \le k} = \{(\mathcal{S}_i, \mathcal{W}_i, \psi_i)\}_{1 \le i \le k}$ *where* $\mathcal{S}_i = \{S_{i, \ell \to r} \mid S_{\ell \to r} \in \mathcal{S}\}$, $\mathcal{W}_i = \{W_{i, \ell \to r} \mid S_{\ell \to r} \in \mathcal{S}\} \cup \{W_{i, \ell \to r} \mid W_{\ell \to r} \in \mathcal{W}\}$, $\psi_i = \varphi \wedge T_{\mathrm{scc}}(k) \wedge T_S(i) \wedge T_W(i)$, *and*

$$T_{\mathrm{scc}}(k) = \bigwedge_{S_p \in \mathcal{S}} 1 \le X_p^{\mathrm{scc}} \le k \wedge \bigwedge_{S_p, S_{p'} \in \mathcal{S}} S_p \wedge S_{p'} \wedge [p \Rightarrow p'] \to X_{p,p'}^{\Rightarrow} \wedge X_p^{\mathrm{scc}} \ge X_{p'}^{\mathrm{scc}}$$

$$T_S(i) = \bigwedge_{S_p, S_{p'} \in \mathcal{S}} X_{p,p'}^{\Rightarrow} \wedge X_p^{\mathrm{scc}} = i \wedge X_{p'}^{\mathrm{scc}} = i \wedge \neg S_{i,p} \wedge \neg S_{i,p'} \to X_p^{\mathrm{w}} > X_{p'}^{\mathrm{w}}$$

$$T_W(i) = \bigwedge_{S_p \in \mathcal{S}} S_p \wedge X_p^{\mathrm{scc}} = i \to W_{i,p} \wedge \bigwedge_{W_p \in \mathcal{W}} W_p \to W_{i,p}$$

Here $X_{p_1, p_2}^{\Rightarrow}$ is a boolean variable encoding the presence of both DPs p_1 and p_2 as well as an edge from p_1 to p_2, and X_p^{scc} is an integer variable assigning an SCC number to a DP p. Hence $T_{\mathrm{scc}}(k)$ encodes the separation of the graph into at most k SCCs, and $T_S(i), T_W(i)$ encode conditions to orient the ith SCC.

Soundness of all the above encodings can be shown by relating them to their processor counterparts [4], but we omit the proofs here due to lack of space.

4   Control Strategies

In its original version, Maxcomp generated terminating TRSs $\mathfrak{R}(\mathcal{C})$ by orienting as many equations in $\mathcal{C}$ as possible. This was motivated by the following observation: whenever a TRS $\mathcal{R}$ is complete for $\mathcal{E}_0$, then any terminating TRS $\mathcal{R}'$ satisfying $\mathcal{R} \subseteq \mathcal{R}' \subseteq \leftrightarrow^*_{\mathcal{E}_0}$ is complete for $\mathcal{E}_0$ as well. However, this choice of $\mathfrak{R}(\mathcal{C})$ has drawbacks in the case where the selected TRS is *not* yet complete: In case of multiple possibilities, the search is not guided towards "more useful" TRSs. Moreover, the chosen TRSs are large such that critical pair generation and normalization tend to be inefficient.

We therefore experimented with different components of control strategies which can be combined in a variety of ways. The following desirable properties of TRSs $\mathcal{R} \in \mathfrak{R}(\mathcal{C})$ constitute the basis of the below definitions, reflecting the aim to eventually derive a complete TRS for the axioms $\mathcal{E}_0$.

(1) All nontrivial equations in $\mathcal{C}$ should be reducible by $\mathcal{R}$.
(2) The axioms $\mathcal{E}_0$ should be derivable from $\mathcal{R}$.
(3) Preferably, the critical pairs of $\mathcal{R}$ should be joinable.

We use sets of constraints cs and mc, where constraints in cs have to be always satisfied, whereas the number of satisfied constraints from mc is to be maximized. To determine cs and mc, the following options c and mc are considered:

$$\text{c} ::= \text{Red} \mid \text{Comp} \qquad \text{mc} ::= \text{None} \mid \text{MaxRed} \mid \text{CPRed} \mid \text{Oriented} \mid \text{NotOriented}$$

Here Red ensures property (1) by demanding that $\varphi_{red}(\mathcal{C})$ is satisfied.

$$\varphi_{red}(\mathcal{C}) = \bigwedge_{s \approx t \in \mathcal{C}} \varphi_{red}(s \approx t) \qquad \varphi_{red}(s \approx t) = \bigvee \{ I_{\ell \to r} \mid \ell \to r \in \widetilde{\mathcal{C}} \text{ reduces } s \text{ or } t \}$$

Option Comp ensures property (2) by demanding that $\varphi_{comp}(\mathcal{C})$ is satisfied. To that end, every equation $\ell \approx r \in \mathcal{C}$ is associated with a fresh boolean variable $E_{\ell \approx r}$ and a fresh integer variable $w_{\ell \approx r}$.

$$\varphi_{comp}(\mathcal{C}) = \bigwedge_{\ell \approx r \in \mathcal{E}_0} E_{\ell \approx r} \wedge \bigwedge_{\ell \approx r \in \mathcal{C}} E_{\ell \approx r} \to \ell = r \vee I_{\ell \to r} \vee I_{r \to \ell} \vee \varphi_{\leftrightarrow}(\ell \approx r)$$

$$\varphi_{\leftrightarrow}(\ell \approx r) = \bigvee_{\mathcal{E} \in \mathfrak{D}_{\ell \approx r}} \bigwedge_{e' \in \mathcal{E}} E_{e'} \wedge w_{\ell \approx r} > w_{e'}$$

Here $\mathfrak{D}_{\ell \approx r}$ consists of ESs $\mathcal{E} \subseteq \mathcal{C}$ satisfying $\ell \leftrightarrow^*_{\mathcal{E}} r$, and $w_{\ell \approx r}$ avoids cyclic dependencies among equations by requiring that equations are only derived from equations associated with smaller values. Suitable sets $\mathfrak{D}_e$ can be collected when rewriting equations: Whenever an equation e is simplified to e' using rules $\mathcal{R}$, $\{e'\} \cup \mathcal{R}$ is added to $\mathfrak{D}_e$, and $\{e\} \cup \mathcal{R}$ is added to $\mathfrak{D}_{e'}$.

Concerning options for mc, None requires nothing, MaxRed maximizes the number of clauses $\varphi_{red}(s \approx t)$ for $s \approx t \in \mathcal{C}$, Oriented maximizes the number of oriented equations in $\mathcal{C}$, and NotOriented maximizes the number of unoriented equations in $\mathcal{C}$. The option CPRed tries to reduce as many critical pairs of $\mathcal{R} \in \mathfrak{R}(\mathcal{C})$ as possible by maximizing the number of satisfied clauses in $\varphi_{CPred}(\mathcal{C})$. Here $\mathcal{R}(\mathcal{C})$ consists of all rewrite rules $\ell \to r$ such that $\ell \approx r \in \widetilde{\mathcal{C}}$.

$$\varphi_{CPred}(\mathcal{C}) = \{I_{r_1} \wedge I_{r_2} \to \varphi_{red}(s \approx t) \mid r_1, r_2 \in \mathcal{R}(\mathcal{C}) \text{ and } s \approx t \in \mathrm{CP}(r_1, r_2)\}$$

5   Implementation

We next describe some further implementation details of our extension of Maxcomp, which will in the sequel be referred to as MaxcompDP. The general layout of Maxcomp based on the control loop shown in Fig. 1 was kept. As input parameters, the function `maxcomp` obtains a set of equations $\mathcal{C}$ and an overall strategy $\mathcal{S}$ (described below). The ES $\mathcal{C}$ is initialized with $\mathcal{E}_0$. In each recursive call, the `max_k` function tries to find k terminating TRSs $\mathfrak{R}(\mathcal{C})$ according to the strategy $\mathcal{S}$, and it returns a possibly modified strategy $\mathcal{S}'$. For each $\mathcal{R} \in \mathfrak{R}(\mathcal{C})$, if $\mathcal{R}$ is confluent and joins $\mathcal{E}_0$ then `maxcomp` succeeds; otherwise n new equations are selected and added to $\mathcal{C}$. In order to find $\mathfrak{R}(\mathcal{C})$, `max_k` uses (MAX)SAT calls to Yices [2]. In some important aspects MaxcompDP deviates from Maxcomp, the next paragraphs describe these changes.

```
function maxcomp(C, S)
    R(C), S' := max_k(C, S, k)
    for all R ∈ R(C)
        if CP(R) ∪ E₀ ⊆↓_R then R
        else C := C ∪ select(n, (CP(R) ∪ C)↓_R)
    maxcomp(C, S')
```

Fig. 1. Main control function.

Termination Strategies. Termination of TRSs $\mathfrak{R}(\mathcal{C})$ is encoded according to a certain termination strategy. Besides LPO and KBO as used in Maxcomp, MaxcompDP now also provides as base orders simple linear polynomial interpretations of the shape $x_1 + \ldots + x_n + c$ and the instance MPOL of the weighted path order [13]. We implemented the DP techniques presented in Sect. 3, and also support argument filterings for LPO and KBO. Overall, termination strategies can thus be constructed according to the following grammar:

$$o ::= \mathsf{LPO} \mid \mathsf{KBO} \mid \mathsf{MPol} \mid \mathsf{LPol} \qquad t ::= os \mid \mathsf{DP}(os) \mid \mathsf{DG}(os) \mid \mathsf{DGk}(int, os)$$

where os abbreviates o *list*, which is interpreted as lexicographic combination of the associated reduction orders/reduction pairs. DP switches to a DP problem using Definition 4, while DG and DGk use DG encodings according to Definitions 6 and 7, respectively. In the sequel we consider the strategies $t_{lpo} := [\mathsf{LPO}]$, $t_{dp} := \mathsf{DP}([\mathsf{LPol}, \mathsf{LPO}])$, $t_{dg} := \mathsf{DG}([\mathsf{LPol}, \mathsf{LPO}])$ and $t_{dg2} := \mathsf{DGk}(2, [\mathsf{LPol}, \mathsf{LPO}])$.

Overall Strategies. An overall strategy S for the `max_k` function combines termination and control strategies and has the type $(t, c \; set, mc) \; list$. Such a strategy is used as follows: If S is the empty list, `max_k` fails. If S is a non-empty list $(t, cs, mc) :: S'$, `max_k` tries to find a TRS $\mathcal{R}_i$ by satisfying the constraints t and cs, and maximizing the satisfied constraints of mc, such that $\mathcal{R}_i$ is different from $\mathcal{R}_1, \ldots, \mathcal{R}_{i-1}$. If `max_k` can find k TRSs in this way, it returns $\mathcal{R}_1, \ldots, \mathcal{R}_k$ and S. If it fails to do so for $\mathcal{R}_i$, it tries to find the remaining TRSs using strategy S', and returns S'. This allows to change to a more appropriate termination and/or control strategy if the current one can, e.g., not orient sufficiently many equations. We experimented with different strategies such as $S_{red} := [(t, \{\mathsf{Red}\})]$, $S_{comp} := [(t, \{\mathsf{Red}, \mathsf{Comp}\})]$, $S_{CPred} := [(t, \{\mathsf{Red}\}, \mathsf{CPRed})]$, $S_{maxcomp} := [(t, \varnothing, \mathsf{Oriented})]$, $S_{notOriented} := [(t, \{\mathsf{Red}, \mathsf{Comp}\}, \mathsf{NotOriented})]$ for different termination strategies t. Here we write (t, cs) for (t, cs, None). The strategy $S_{maxcomp}$ corresponds to the original Maxcomp approach. For a termination strategy t, $s_{full}(t)$ denotes $(t, \{\mathsf{Red}, \mathsf{Comp}\}, \mathsf{CPRed})$, and $S_{full}(t)$ denotes $[s_{full}(t)]$. The S_{auto} strategy turned out particularly useful, it is defined by $[s_{full}(t_{lpo}), s_{full}(t_{dp}), (t_{lpo}, \{\mathsf{Comp}\}, \mathsf{MaxRed})]$.

The number k has considerable impact; MaxcompDP lets the user control it by an input parameter. By default, $k = 6$ in the first two recursive calls and $k = 2$ afterwards. The rationale behind this choice is that considering a wide variety of orientations in the beginning of a run reduces the risk of getting stuck with an initial, possibly unfortunate orientation.

Selection of New Equations. The `select` function in Fig. 1 plays the role of $S(\mathcal{C})$ from Definition 1. Maxcomp by default selected up to 7 equations of size at most 20 from the set $\mathrm{CP}(\mathcal{R}) \downarrow_\mathcal{R}$ (since it is practically infeasible to add all critical pairs). In contrast, MaxcompDP does not only add critical pairs of a TRS $\mathcal{R}$ to $\mathcal{C}$ but also reduced equations. Therefore the n smallest equations from the set $(\mathrm{CP}(\mathcal{R}) \cup \mathcal{C}) \downarrow_\mathcal{R}$ are selected, without inducing a size bound. The number n can be controlled by the user, by default $n = 12$.

Incremental Termination Checks. The formulas obtained with our termination encodings easily grow large. However, though in the course of a completion run many satisfiability checks are required, the termination constraint issued for a specific rule does not change. We hence use Yices in an incremental way: whenever a new equation $\ell \approx r$ gives rise to a potential rule $\ell \to r$, its termination constraint $[\ell > r]$ is computed and added to the context of Yices. To find a terminating TRS, we temporarily add the constraints cs and mc according to the current control strategy but backtrack after the SAT check. This allows to use the same Yices context throughout the completion run, and issue termination constraints only once per rule (though new constraints need to be computed if the termination strategy changes).

MaxcompDP as well as all experimental results are available from

http://cl-informatik.uibk.ac.at/software/maxcompdp

6   Experiments

Table 1 summarizes our experimental results for the test bed comprising 115 equational systems from the distribution of mkbTT [12], run on a system equipped with an Intel Core i7 with four cores of 2.1GHz each and 7.5 GB of memory. Each ES was given a time limit of 600 seconds, timeouts are marked ∞. The rows labeled (1)–(4) correspond to MaxcompDP using S_{full} with different termination strategies, and (5) applies the automatic mode S_{auto} as described in Sect. 5. Rows (DP1)–(DP5) use t_{dp} within different control strategies, and (LPO1) combines t_{lpo} with $S_{maxcomp}$. All runs used the default values for k and n. Finally, we compare with other completion tools that are automatic in that no reduction order is required as input, namely Maxcomp, mkbTT, KBCV [9], and Slothrop [11].

Column # lists the number of successful completions, the next column gives the average time for a completion in seconds. Columns (a)–(d) show the results for some selected systems, namely CGE_2, CGE_5, proofreduction, and equiv_proofs.

The DP strategy (2) allows to successfully complete problems (a)–(d), which cannot be completed using LPO or KBO. However, some other systems are lost, compared to the setting using LPO. Typically, these problems require many iterations and/or give rise to many equations. Also, the average time compared to (1) is multiplied. Settings (3) and (4) require more encoding effort such that completion takes even more time than for setting (2). However, the tradeoff between the more complex encoding and the gain in power turns out more beneficial for setting (4), which can complete the same number of problems as (2) but more

Table 1. Experimental Results.

		#	avg. time	(a)	(b)	(c)	(d)
(1)	$S_{full}(t_{lpo})$	81	2.2	∞	∞	∞	∞
(2)	$S_{full}(t_{dp})$	89	33.5	17.1	79.5	5.2	3.1
(3)	$S_{full}(t_{dg})$	86	37.3	18.5	155.5	5.7	3.1
(4)	$S_{full}(t_{dg2})$	89	41.0	12.3	254.0	13.2	6.5
(5)	S_{auto}	97	11.6	4.1	104.4	3.6	1.5
(DP1)	$S_{maxcomp}$	56	4.8	157.7	∞	7.5	3.9
(DP2)	S_{red}	81	31.8	568.4	∞	9.8	2.1
(DP3)	S_{comp}	87	45.9	15.5	302.7	3.4	2.2
(DP4)	S_{CPred}	90	29.7	17.1	273.5	9.2	3.4
(DP5)	$S_{notOriented}$	85	15.9	3.6	15.9	20.6	3.8
(LPO1)	$S_{maxcomp}$	77	13.5	∞	∞	∞	∞
	Maxcomp	87	3.8	∞	∞	∞	∞
	mkbTT	85	40.1	33.5	∞	7.3	237.9
	KBCV	88	12.4	∞	∞	∞	∞
	Slothrop	76	65.8	∞	∞	209.4	12.1

than (3). Overall S_{auto} proved to be most powerful since it can often be efficient by applying LPO, but also switch to a more sophisticated strategy in case of unorientable equations.

Concerning control strategies, a comparison of (1) with (LPO1) and (2) with (DP1) suggests that $S_{maxcomp}$ is by far more suited for plain reduction orders than for complex DP strategies. One reason for that might be that powerful DP strategies can orient more equations such that $S_{maxcomp}$ gives rise to even larger TRSs. But we also observed that $S_{maxcomp}$ with DPs prefers unfortunate orientations in presence of group theory, which occurs in many problems.

All of S_{red}, S_{comp} and S_{CPred} positively influence the number of completed systems (though at the price of lower efficiency). In the S_{auto} setting, their combination was most successful.

As Table 1 shows, MaxcompDP with strategy S_{auto} can complete more systems than any other automatic completion tool , although the tools are incomparable in the sense that for each tool there is an ES that it can complete, but no other tools can. It manages to complete CGE_5 in 104.4 seconds, whereas for mkbTT it was a major effort requiring more than 35000 seconds. Moreover, MaxcompDP can also complete CGE_6 and CGE_7 in 307 and 362 seconds, respectively (the latter using $n = 18$, though). No other tool could complete these ESs so far.

References

1. Baader, F., Nipkow, T.: Term Rewriting and All That. Cambridge University Press, Cambridge (1998)
2. Dutertre, B., de Moura, L.: A fast linear-arithmetic solver for DPLL(T). In: Ball, T., Jones, R.B. (eds.) CAV 2006. LNCS, vol. 4144, pp. 81–94. Springer, Heidelberg (2006)
3. Endrullis, J., Waldmann, J., Zantema, H.: Matrix interpretations for proving termination of term rewriting. JAR **40**(2–3), 195–220 (2008)
4. Giesl, J., Thiemann, R., Schneider-Kamp, P.: The dependency pair framework: combining techniques for automated termination proofs. In: Baader, F., Voronkov, A. (eds.) LPAR 2004. LNCS (LNAI), vol. 3452, pp. 301–331. Springer, Heidelberg (2005)
5. Klein, D., Hirokawa, N.: Maximal completion. In: RTA, vol. 10 of LIPIcs, pp. 71–80 (2011)
6. Korovin, K.: Inst-Gen – a modular approach to instantiation-based automated reasoning. In: Voronkov, A., Weidenbach, C. (eds.) Programming Logics. LNCS, vol. 7797, pp. 239–270. Springer, Heidelberg (2013)
7. Sato, H., Winkler, S.: A satisfiability encoding of dependency pair techniques for maximal completion. In: WST (2014)
8. Schneider-Kamp, P., Thiemann, R., Annov, E., Codish, M., Giesl, J.: Proving termination using recursive path orders and SAT solving. In: Konev, B., Wolter, F. (eds.) FroCos 2007. LNCS (LNAI), vol. 4720, pp. 267–282. Springer, Heidelberg (2007)
9. Sternagel, T., Zankl, H.: KBCV – Knuth-Bendix Completion Visualizer. In: Gramlich, B., Miller, D., Sattler, U. (eds.) IJCAR 2012. LNCS, vol. 7364, pp. 530–536. Springer, Heidelberg (2012)

10. Stump, A., Löchner, B.: Knuth-Bendix completion of theories of commuting group endomorphisms. IPL **98**(5), 195–198 (2006)
11. Wehrman, I., Stump, A., Westbrook, E.: Slothrop: knuth-bendix completion with a modern termination checker. In: Pfenning, F. (ed.) RTA 2006. LNCS, vol. 4098, pp. 287–296. Springer, Heidelberg (2006)
12. Winkler, S., Sato, H., Middeldorp, A., Kurihara, M.: Multi-completion with termination tools. JAR **50**(3), 317–354 (2013)
13. Yamada, A., Kusakari, K., Sakabe, T.: A unified ordering for termination proving. Science of Computer Programming (2014). doi:10.1016/j.scico.2014.07.009
14. Zankl, H., Hirokawa, N., Middeldorp, A.: Constraints for argument filterings. In: van Leeuwen, J., Italiano, G.F., van der Hoek, W., Meinel, C., Sack, H., Plášil, F. (eds.) SOFSEM 2007. LNCS, vol. 4362, pp. 579–590. Springer, Heidelberg (2007)
15. Zankl, H., Hirokawa, N., Middeldorp, A.: KBO orientability. JAR **43**(2), 173–201 (2009)

Reducing Relative Termination to Dependency Pair Problems

José Iborra[1], Naoki Nishida[1], Germán Vidal[2], and Akihisa Yamada[3]([✉])

[1] Graduate School of Information Science, Nagoya University, Nagoya, Japan
[2] MiST, DSIC, Universitat Politècnica de València, Valencia, Spain
[3] Institute of Computer Science, University of Innsbruck, Innsbruck, Austria
`akihisa.yamada@uibk.ac.at`

Abstract. Relative termination, a generalized notion of termination, has been used in a number of different contexts like proving the confluence of rewrite systems or analyzing the termination of narrowing. In this paper, we introduce a new technique to prove relative termination by reducing it to dependency pair problems. To the best of our knowledge, this is the first significant contribution to Problem #106 of the RTA List of Open Problems. The practical significance of our method is illustrated by means of an experimental evaluation.

1 Introduction

Proving that a program terminates is a fundamental problem that has been extensively studied in almost all programming paradigms. For *term rewrite systems (TRSs)*, termination analysis has attracted considerable attention (see, e.g., the survey of Zantema [31] and the termination portal[1]), and various automated termination provers for TRSs have been developed, e.g. AProVE [9], $\mathsf{T_TT_2}$ [20], and NaTT [28]. Among them the *dependency pair (DP) method* [2,12] and its successor the *DP framework* [10] became a modern standard.

Termination of a TRS is usually checked for all possible reduction sequences. In some cases, however, one is interested in proving a generalized notion, *relative termination* [7,17]. Roughly speaking, a TRS $\mathcal{R}$ is relatively terminating w.r.t. another TRS $\mathcal{B}$ (that here we call the *base*), when any infinite reduction using both systems contains only a finite number of steps given with rules from $\mathcal{R}$. For instance, consider the following base:

$$\mathcal{B}_{\mathsf{comlist}} = \{\mathsf{cons}(x, \mathsf{cons}(y, ys)) \to \mathsf{cons}(y, \mathsf{cons}(x, ys))\}$$

specifying a property for commutative lists (i.e., that the order of elements is irrelevant). Termination of operations on commutative lists, described by a TRS

German Vidal is partially supported by the EU (FEDER) and the Spanish *Ministerio de Economía y Competitividad* under grant TIN2013-44742-C4-1-R and by the *Generalitat Valenciana* under grant PROMETEOII2015/013. Akihisa Yamada is supported by the Austrian Science Fund (FWF): Y757.

[1] Available from URL http://www.termination-portal.org/.

© Springer International Publishing Switzerland 2015
A.P. Felty and A. Middeldorp (Eds.): CADE-25, LNAI 9195, pp. 163–178, 2015.
DOI: 10.1007/978-3-319-21401-6_11

$\mathcal{R}$, can be analyzed as the relative termination of $\mathcal{R}$ w.r.t. $\mathcal{B}_{\mathsf{comlist}}$. Note also that the base $\mathcal{B}_{\mathsf{comlist}}$ is clearly non-terminating.

Relative termination has been used in various contexts: proving confluence of a rewrite system [7,13]; liveness properties in the presence of fairness [18]; and termination of narrowing [15,24,27], an extension of rewriting to deal with non-ground terms (see, e.g., [14]). Moreover, analyzing relative termination can also be useful for other purposes, like dealing with random values or considering rewrite systems with so called *extra-variables* (i.e., variables that occur in the right-hand side of a rule but not in the corresponding left-hand side). For instance, the following base $\mathcal{B}_{\mathsf{rand}}$ specifies a random number generator:

$$\mathcal{B}_{\mathsf{rand}} = \{\mathsf{rand}(x) \to x, \ \mathsf{rand}(x) \to \mathsf{rand}(\mathsf{s}(x))\}$$

We have $\mathsf{rand}(0) \to^*_{\mathcal{B}_{\mathsf{rand}}} \mathsf{s}^n(0)$ for arbitrary $n \in \mathbb{N}$. Now consider

$$\mathcal{R}_{\mathsf{quot}} = \{ \qquad x - 0 \to x, \qquad \mathsf{s}(x) - \mathsf{s}(y) \to x - y, \\ \mathsf{quot}(0, \mathsf{s}(y)) \to 0, \quad \mathsf{quot}(\mathsf{s}(x), \mathsf{s}(y)) \to \mathsf{s}(\mathsf{quot}(x - y, \mathsf{s}(y)))\}$$

from [2]. Termination of $\mathcal{R}_{\mathsf{quot}}$ can be shown using the DP method [2]. However, it is unknown if $\mathcal{R}_{\mathsf{quot}}$ is relatively terminating w.r.t. $\mathcal{B}_{\mathsf{rand}}$ using previously known techniques. Note also that it seems not so obvious, since $\mathcal{R}_{\mathsf{quot}}$ is not relatively terminating w.r.t. the following similar variant $\mathcal{B}_{\mathsf{gen}}$:

$$\mathcal{B}_{\mathsf{gen}} = \{\mathsf{gen} \to 0, \ \mathsf{gen} \to \mathsf{s}(\mathsf{gen})\}$$

which is considered in the context of termination of narrowing [15,24,27]. Indeed, we can construct the following infinite reduction sequence using $\mathcal{B}_{\mathsf{gen}}$:

$$\mathsf{s}(\mathsf{gen}) - \mathsf{s}(\mathsf{gen}) \to_{\mathcal{R}_{\mathsf{quot}}} \mathsf{gen} - \mathsf{gen} \to^*_{\mathcal{B}_{\mathsf{gen}}} \mathsf{s}(\mathsf{gen}) - \mathsf{s}(\mathsf{gen}) \to_{\mathcal{R}_{\mathsf{quot}}} \cdots$$

We expect that a similar technique can also be used to deal with TRSs with extra-variables. In principle, these systems are always non-terminating, since extra-variables can be replaced by any term. However, one can still consider an interesting termination property: is the system terminating if the extra-variables can only be instantiated with terms built from a restricted signature? Consider, e.g., the following TRS from [23]:

$$\mathcal{R} = \{\mathsf{f}(x, 0) \to \mathsf{s}(x), \mathsf{g}(x) \to \mathsf{h}(x, y), \mathsf{h}(0, x) \to \mathsf{f}(x, x), \mathsf{a} \to \mathsf{b}\}$$

This system is clearly non-terminating due to the extra variable in the second rewrite rule. However, by assuming that y can only take values built from constructor symbols (e.g., natural numbers), one can reformulate these rewrite rules as follows: $\mathcal{R}' = \{\mathsf{f}(x, 0) \to \mathsf{s}(x), \mathsf{g}(x) \to \mathsf{h}(x, \mathsf{gen}), \mathsf{h}(0, x) \to \mathsf{f}(x, x), \mathsf{a} \to \mathsf{b}\}$, using $\mathcal{B}_{\mathsf{gen}}$ above. Obviously, $\mathcal{R}' \cup \mathcal{B}_{\mathsf{gen}}$ is still non-terminating since $\mathcal{B}_{\mathsf{gen}}$ is non-terminating. Nevertheless, one can still prove relative termination of $\mathcal{R}'$ w.r.t. $\mathcal{B}_{\mathsf{gen}}$, which is an interesting property since one can ensure terminating derivations by using an appropriate heuristics to instantiate extra-variables.

Another interesting application of relative termination w.r.t. $\mathcal{B}_{\mathsf{rand}}$ is to generate *test cases*. For example, in the QuickCheck technique, lists over, e.g., natural numbers are generated at random. Assume f and g are defined externally by a TRS $\mathcal{R}_{fg}$, and consider the TRS $\mathcal{R}_{test}$ consisting of the following rules:

$$\mathsf{rands}(0, y) \to \mathsf{done}(y) \qquad \mathsf{rands}(\mathsf{s}(x), y) \to \mathsf{rands}(x, \mathsf{cons}(\mathsf{rand}(0), y))$$
$$\mathsf{tests}(0) \to \mathsf{true} \qquad \mathsf{tests}(\mathsf{s}(x)) \to \mathsf{test}(\mathsf{rands}(\mathsf{rand}(0), \mathsf{nil})) \wedge \mathsf{tests}(x)$$
$$\mathsf{eq}(x, x) \to \mathsf{true} \qquad \mathsf{test}(\mathsf{done}(y)) \to \mathsf{eq}(f(y), g(y))$$

where lists are built from nil and cons. Execution of $\mathsf{tests}(\mathsf{s}^n(0))$ tests the equivalence between f and g by feeding them random inputs n times. Even when f and g are defined by $f(x) \to x$ and $g(x) \to x$, AProVE fails to prove relative termination of $\mathcal{R}_{test} \cup \mathcal{R}_{fg}$ w.r.t. $\mathcal{B}_{\mathsf{rand}}$.

In this paper, we present a new technique for proving relative termination by reducing it to the finiteness of dependency pair problems. To the best of our knowledge, we provide the first significant contribution to Problem #106 of the *RTA List of Open Problems*:[2] "Can we use the dependency pair method to prove relative termination?" We implemented the proposed method in the termination tool NaTT[3] and showed its significance through experiments. Using results of this paper and [29], NaTT can prove relative termination of $\mathcal{R}_{\mathsf{quot}}$ w.r.t. $\mathcal{B}_{\mathsf{rand}}$, and relative termination of $\mathcal{R}_{test} \cup \mathcal{R}_{fg}$ w.r.t. $\mathcal{B}_{\mathsf{rand}}$ for e.g., naive and tail recursive definitions of summation as f and g.

This paper is organized as follows. In Sect. 2, we briefly review some notions and notations of term rewriting. In Sects. 3–5, we present our main contributions for reducing relative termination to a dependency pair problem. Moreover, some subtle features about minimality are discussed in Sect. 6. Then, Sect. 7 describes implementation issues and presents selected results from an experimental evaluation. Finally, Sect. 8 compares our technique with some related work, and Sect. 9 concludes and points out some directions for future research. Missing proofs of technical results can be found in the appendix.

2   Preliminaries

We assume some familiarity with basic concepts and notations of term rewriting. We refer the reader to, e.g., [4] for further details.

A *signature* $\mathcal{F}$ is a set of function symbols. Given a set of *variables* $\mathcal{V}$ with $\mathcal{F} \cap \mathcal{V} = \emptyset$, we denote the domain of *terms* by $\mathcal{T}(\mathcal{F}, \mathcal{V})$. We use $\mathsf{f}, \mathsf{g}, \ldots$ to denote function symbols and $x, y, \ldots$ to denote variables. The *root symbol* of a term $t = f(t_1, \ldots, t_n)$ is f and denoted by $\mathsf{root}(t)$. We assume an extra fresh constant $\square$ called a *hole*. Then, $C \in \mathcal{T}(\mathcal{F} \cup \{\square\}, \mathcal{V})$ is called a *context* on $\mathcal{F}$. We use the notation $C[\,]$ for the context containing one hole, and if $t \in \mathcal{T}(\mathcal{F}, \mathcal{V})$, then $C[t]$ denotes the result of placing t in the hole of $C[\,]$.

---

[2] http://www.win.tue.nl/rtaloop/.
[3] Available at http://www.trs.cm.is.nagoya-u.ac.jp/NaTT/.

A *position* p in a term t is represented by a finite sequence of natural numbers, where ϵ denotes the root position. We let $t|_p$ denote the *subterm* of t at position p, and $t[s]_p$ the result of replacing the subterm $t|_p$ by the term s. We denote by $s \unrhd t$ that t is a subterm of s, and by $s \rhd t$ that it is a *proper* subterm.

$\mathcal{V}ar(t)$ denotes the set of variables appearing in t. A *substitution* is a mapping $\sigma : \mathcal{V} \to \mathcal{T}(\mathcal{F}, \mathcal{V})$, which is extended to a morphism from $\mathcal{T}(\mathcal{F}, \mathcal{V})$ to $\mathcal{T}(\mathcal{F}, \mathcal{V})$ in a natural way. We denote the application of a substitution σ to a term t by $t\sigma$.

A *rewrite rule* $l \to r$ is a pair of terms such that $l \notin \mathcal{V}$ and $\mathcal{V}ar(l) \supseteq \mathcal{V}ar(r)$. The terms l and r are called the left-hand side and the right-hand side of the rule, respectively. A *term rewriting system (TRS)* is a set of rewrite rules. Given a TRS $\mathcal{R}$, we write $\mathcal{F}_\mathcal{R}$ for the set of function symbols appearing in $\mathcal{R}$, $\mathcal{D}_\mathcal{R}$ for the set of the *defined* symbols, i.e., the root symbols of the left-hand sides of the rules, and $\mathcal{C}_\mathcal{R}$ for the set of *constructors*; $\mathcal{C}_\mathcal{R} = \mathcal{F}_\mathcal{R} \setminus \mathcal{D}_\mathcal{R}$.

For a TRS $\mathcal{R}$, we define the associated *rewrite relation* $\to_\mathcal{R}$ as follows: given terms $s, t \in \mathcal{T}(\mathcal{F}, \mathcal{V})$, $s \to_\mathcal{R} t$ holds iff there exist a position p in s, a rewrite rule $l \to r \in \mathcal{R}$ and a substitution σ with $s|_p = l\sigma$ and $t = s[r\sigma]_p$; the rewrite step is often denoted by $s \xrightarrow{p}_\mathcal{R} t$ to make the rewritten position explicit, and $s \xrightarrow{>\epsilon}_\mathcal{R} t$ if the position is strictly below the root. Given a binary relation $\to$, we denote by $\to^+$ the transitive closure of $\to$ and by $\to^*$ its reflexive and transitive closure.

Now we recall the formal definition of relative termination:

Definition 1 (Relative Termination [17]). Given two TRSs $\mathcal{R}$ and $\mathcal{B}$, we define the relation $\to_{\mathcal{R}/\mathcal{B}}$ as $\to_\mathcal{B}^* \cdot \to_\mathcal{R} \cdot \to_\mathcal{B}^*$. We say that $\mathcal{R}$ *relatively terminates* *w.r.t.* $\mathcal{B}$, or simply that $\mathcal{R}/\mathcal{B}$ is terminating, if the relation $\to_{\mathcal{R}/\mathcal{B}}$ is terminating. We say that a term t is $\mathcal{R}/\mathcal{B}$-*nonterminating* if it starts an infinite $\to_{\mathcal{R}/\mathcal{B}}$ derivation, and $\mathcal{R}/\mathcal{B}$-*terminating* otherwise.

In other words, $\mathcal{R}/\mathcal{B}$ is terminating if every (possibly infinite) $\to_{\mathcal{R} \cup \mathcal{B}}$ derivation contains only finitely many $\to_\mathcal{R}$ steps. Note that sequences of $\to_\mathcal{B}$ steps are "collapsed" and seen as a single $\to_{\mathcal{R}/\mathcal{B}}$ step. Hence, an infinite $\to_{\mathcal{R}/\mathcal{B}}$ derivation must contain an infinite number of $\to_\mathcal{R}$ steps and thus only finite $\to_\mathcal{B}$ subderivations.

The Dependency Pair Framework. The dependency pair (DP) framework [2,10] enables analyzing cyclic dependencies between rewrite rules, and has become one of the most popular approaches to proving termination in term rewriting. Indeed, it underlies virtually all modern termination tools for TRSs.

Let us briefly recall the fundamentals of the DP framework. Here, we consider that the signature $\mathcal{F}$ is implicitly extended with fresh function symbols $f^\sharp$ for each defined function $f \in \mathcal{D}_\mathcal{R}$. Also, given a term $t = f(\bar{t})$ with $f \in \mathcal{D}_\mathcal{R}$, we let $t^\sharp$ denote $f^\sharp(\bar{t})$. Here, $\bar{t}$ is an abbreviation for $t_1, \ldots, t_n$ for an appropriate n. If $l \to r \in \mathcal{R}$ and t is a subterm of r with a defined root symbol, then the rule $l^\sharp \to t^\sharp$ is a *dependency pair* of $\mathcal{R}$. The set of all dependency pairs of $\mathcal{R}$ is denoted by $\mathsf{DP}(\mathcal{R})$. Note that $\mathsf{DP}(\mathcal{R})$ is also a TRS.

A key ingredient in this framework is the notion of a *chain*, which informally represents a sequence of calls that can occur during a reduction. In the following, $\mathcal{P}$ will often denote a set of dependency pairs. A $(\mathcal{P}, \mathcal{R})$-*chain* (*à la* [12]) is a possibly infinite rewrite sequence $s_1 \xrightarrow{\epsilon}_{\mathcal{P}} t_1 \xrightarrow{*}_{\mathcal{R}} s_2 \xrightarrow{\epsilon}_{\mathcal{P}} t_2 \xrightarrow{*}_{\mathcal{R}} \cdot \xrightarrow{\epsilon}_{\mathcal{P}} \cdots$. The basic result from [2] is then that a TRS $\mathcal{R}$ is terminating iff there is no infinite $(\mathsf{DP}(\mathcal{R}), \mathcal{R})$-chain. In order to check absence of infinite chains, the DP framework introduces the notions of a *DP problem*. A DP problem is just a pair $(\mathcal{P}, \mathcal{R})$, and is called *finite* if there is no infinite $(\mathcal{P}, \mathcal{R})$-chain. To prove DP problems finite, we can use several techniques implemented in current termination tools, e.g., AProVE, $\mathsf{T_TT_2}$, and NaTT.

3   Relative Termination as a Dependency Pair Problem

Let us start with some basic conditions in terms of dependency pair problems. First, it is folklore that, given two TRSs $\mathcal{R}$ and $\mathcal{B}$, termination of $\mathcal{R} \cup \mathcal{B}$ implies the relative termination of $\mathcal{R}$ w.r.t. $\mathcal{B}$. Therefore, an obvious *sufficient* condition for relative termination can be stated as follows:

Proposition 1. *Let $\mathcal{R}$ and $\mathcal{B}$ be TRSs. $\mathcal{R}/\mathcal{B}$ is terminating if the DP problem $(\mathsf{DP}(\mathcal{R} \cup \mathcal{B}), \mathcal{R} \cup \mathcal{B})$ is finite.*

Observe that $\mathsf{DP}(\mathcal{R}) \cup \mathsf{DP}(\mathcal{B}) \subseteq \mathsf{DP}(\mathcal{R} \cup \mathcal{B})$ but $\mathsf{DP}(\mathcal{R} \cup \mathcal{B}) = \mathsf{DP}(\mathcal{R}) \cup \mathsf{DP}(\mathcal{B})$ is not true when there are shared symbols.

On the other hand, using a proof technique from the standard DP framework, we can easily prove the following *necessary* condition for relative termination.

Proposition 2. *Let $\mathcal{R}$ and $\mathcal{B}$ be TRSs. If $\mathcal{R}/\mathcal{B}$ is terminating, then the DP problem $(\mathsf{DP}(\mathcal{R}), \mathcal{R} \cup \mathcal{B})$ is finite.*

Now, we aim at finding more precise characterizations of relative termination in terms of DP problems. First we need some auxiliary definitions and results.

Definition 2 (Order Pair). *We say that $(\succsim, \succ)$ is a (well-founded) order pair on carrier A if $\succsim$ is a quasi-ordering on A, $\succ$ is a (well-founded) strict order on A, and $\succsim$ and $\succ$ are compatible (i.e., $\succsim \circ \succ \circ \succsim \subseteq \succ$).*

The *multiset extension* of an order pair $(\succsim, \succ)$ on A is the order pair $(\succsim^{\mathrm{mul}}, \succ^{\mathrm{mul}})$ on multisets over A which is defined as follows: $X \succsim^{\mathrm{mul}} Y$ if X and Y are written $X = X' \uplus \{x_1, \ldots, x_n\}$ and $Y = Y' \uplus \{y_1, \ldots, y_n\}$—where "$\uplus$" denotes union on multisets—such that

- $\forall y \in Y'. \exists x \in X'. x \succ y$, and
- $\forall i \in \{1, \ldots, n\}. x_i \succsim y_i$.

We have $X \succ^{\mathrm{mul}} Y$ if it also holds that $X' \neq \emptyset$. It is shown that the multiset extension of a well-founded order pair is also a well-founded order pair [26].

In the following, we will consider a particular order pair:

Definition 3. *For two TRSs $\mathcal{R}$ and $\mathcal{B}$, the pair $(\gtrsim_{\mathcal{R}/\mathcal{B}}, \succ_{\mathcal{R}/\mathcal{B}})$ of relations on terms is defined as follows:* $\gtrsim_{\mathcal{R}/\mathcal{B}} = (\rightarrow_{\mathcal{R} \cup \mathcal{B}} \cup \rhd)^*$ *and* $\succ_{\mathcal{R}/\mathcal{B}} = (\rightarrow_{\mathcal{R}/\mathcal{B}} \cup \rhd)^+$.

The relations $\gtrsim_{\mathcal{R}/\mathcal{B}}$ and $\succ_{\mathcal{R}/\mathcal{B}}$ enjoy the following key property:

Lemma 1. *For two TRSs $\mathcal{R}$ and $\mathcal{B}$, $(\gtrsim_{\mathcal{R}/\mathcal{B}}, \succ_{\mathcal{R}/\mathcal{B}})$ is a well-founded order pair on $\mathcal{R}/\mathcal{B}$-terminating terms.*

Proof. For an $\mathcal{R}/\mathcal{B}$-terminating term t, $t \rightarrow_{\mathcal{R} \cup \mathcal{B}} t'$ implies that t' is also $\mathcal{R}/\mathcal{B}$-terminating. Furthermore, $t \rhd t'$ implies that t' is also $\mathcal{R}/\mathcal{B}$-terminating. Using these facts, the required properties are straightforward from the definition. □

We now introduce the multisets that we will use to prove our main results.

Definition 4. *Let $\mathcal{R}$ be a TRS and t a term. The multiset $\nabla_{\mathcal{R}}(t)$ of maximal $\mathcal{R}$-defined subterms of t is defined as follows:*

- $\nabla_{\mathcal{R}}(x) = \emptyset$ *if* $x \in \mathcal{V}$,
- $\nabla_{\mathcal{R}}(f(t_1, \ldots, t_n)) = \nabla_{\mathcal{R}}(t_1) \cup \cdots \cup \nabla_{\mathcal{R}}(t_n)$ *if* $f \notin \mathcal{D}_{\mathcal{R}}$, *and*
- $\nabla_{\mathcal{R}}(f(t_1, \ldots, t_n)) = \{f(t_1, \ldots, t_n)\}$ *if* $f \in \mathcal{D}_{\mathcal{R}}$.

An essential property is that an $\rightarrow_{\mathcal{R}}$ reduction step corresponds to a decrease in $\succ_{\mathcal{R}/\mathcal{B}}^{mul}$, which is stated as follows:

Lemma 2. *Let $\mathcal{R}$ and $\mathcal{B}$ be TRSs. If $s \rightarrow_{\mathcal{R}} t$ then $\nabla_{\mathcal{R}}(s) \succ_{\mathcal{R}/\mathcal{B}}^{mul} \nabla_{\mathcal{R}}(t)$.*

Proof. Let $s \xrightarrow{p}_{\mathcal{R}} t$ and q be the shortest prefix of p such that $\mathsf{root}(s|_q) \in \mathcal{D}_{\mathcal{R}}$, that is, $\nabla_{\mathcal{R}}(s) = \nabla_{\mathcal{R}}(s[\,]_q) \cup \{s|_q\}$. Note that q always exists since $\mathsf{root}(s|_p) \in \mathcal{D}_{\mathcal{R}}$. We distinguish the following cases:

- Suppose that $q < p$. Since $\mathsf{root}(s|_q) = \mathsf{root}(t|_q) \in \mathcal{D}_{\mathcal{R}}$, we have $\nabla_{\mathcal{R}}(t) = \nabla_{\mathcal{R}}(s[\,]_q) \cup \{t|_q\}$. Trivially $s|_q \succ_{\mathcal{R}/\mathcal{B}} t|_q$, and thus $\nabla_{\mathcal{R}}(s) \succ_{\mathcal{R}/\mathcal{B}}^{mul} \nabla_{\mathcal{R}}(t)$.
- Suppose that $p = q$. We have $\nabla_{\mathcal{R}}(t) = \nabla_{\mathcal{R}}(s[\,]_p) \cup \nabla_{\mathcal{R}}(t|_p)$. For every $t' \in \nabla_{\mathcal{R}}(t|_p)$, we have $s|_p \rightarrow_{\mathcal{R}} t|_p \rhd t'$ and thus $s|_p \succ_{\mathcal{R}/\mathcal{B}} t'$. We conclude $\nabla_{\mathcal{R}}(s) \succ_{\mathcal{R}/\mathcal{B}} \nabla_{\mathcal{R}}(t)$. □

Unfortunately, a $\rightarrow_{\mathcal{B}}$ reduction does not generally imply a weak decrease in $\gtrsim_{\mathcal{R}/\mathcal{B}}^{mul}$ without further conditions. Hence, we introduce the following notion:

Definition 5 ($\mathcal{R}/\mathcal{B}$ Weak-Decreasing). *Let $\mathcal{R}$ and $\mathcal{B}$ be TRSs. We say that $\mathcal{R}/\mathcal{B}$ is weak-decreasing if $t \rightarrow_{\mathcal{B}} t'$ implies $\nabla_{\mathcal{R}}(t) \gtrsim_{\mathcal{R}/\mathcal{B}}^{mul} \nabla_{\mathcal{R}}(t')$.*

Keen readers may notice that $\mathcal{R}/\mathcal{B}$ weak-decreasingness is somewhat related to the notion of a *rank non-increasing* TRS from [22]. Intuitively speaking, given two disjoint signatures, the rank of a term is given by the number of nested functions from different sets. E.g., given signatures $\mathcal{F}_1 = \{f, a\}$ and $\mathcal{F}_2 = \{g\}$, the term $f(f(a))$ has rank 1, while $f(f(g(a)))$ has rank 3. A TRS $\mathcal{R}$ is then called rank non-increasing if $t \rightarrow_{\mathcal{R}} t'$ implies that the rank of t is equal or greater than the rank of t'. The following example illustrates the difference between our notion of $\mathcal{R}/\mathcal{B}$ weak-decreasingness and that of rank non-increasingness:

Example 1. Consider the two TRSs $\mathcal{R} = \{a \to b\}$ and $\mathcal{B} = \{b \to a\}$. Clearly, $\mathcal{R} \cup \mathcal{B}$ is rank non-increasing (there are no nested functions, so the rank is always 1). On the other hand, $\mathcal{R}/\mathcal{B}$ is not weak-decreasing since $b \to_{\mathcal{B}} a$ but $\nabla_{\mathcal{R}}(b) = \{\,\} \not\succeq_{\mathcal{R}/\mathcal{B}}^{mul} \{a\} = \nabla_{\mathcal{R}}(a)$. Not also that $\mathcal{R}/\mathcal{B}$ is not terminating.

Now we can show the following result using Lemmas 1 and 2, and the fact that the multiset extension preserves well-foundedness.

Lemma 3. *Let $\mathcal{R}$ and $\mathcal{B}$ be TRSs such that $\mathcal{R}/\mathcal{B}$ is weak-decreasing. A term s is $\mathcal{R}/\mathcal{B}$-terminating if all elements in $\nabla_{\mathcal{R}}(s)$ are $\mathcal{R}/\mathcal{B}$-terminating.*

Now we recall the notion of minimal (*nonterminating*) terms. We say that an $\mathcal{R}/\mathcal{B}$-nonterminating term is *minimal* if all its proper subterms are $\mathcal{R}/\mathcal{B}$-terminating. It is clear that any $\mathcal{R}/\mathcal{B}$-nonterminating term has some minimal $\mathcal{R}/\mathcal{B}$-nonterminating subterm.

Lemma 4. *Let $\mathcal{R}$ and $\mathcal{B}$ be TRSs such that $\mathcal{R}/\mathcal{B}$ is weak-decreasing. If t is a minimal $\mathcal{R}/\mathcal{B}$-nonterminating term, then $\mathsf{root}(t) \in \mathcal{D}_{\mathcal{R}}$.*

Proof. We prove the claim by contradiction. Consider a minimal $\mathcal{R}/\mathcal{B}$-nonterminating term s such that $\mathsf{root}(s) \notin \mathcal{D}_{\mathcal{R}}$. Since $\mathsf{root}(s) \notin \mathcal{D}_{\mathcal{R}}$, all elements in $\nabla_{\mathcal{R}}(s)$ are proper subterms of s, which are $\mathcal{R}/\mathcal{B}$-terminating due to minimality. Lemma 3 implies that s is $\mathcal{R}/\mathcal{B}$-terminating, hence we have a contradiction.

Using the previous results, we can state the following sufficient condition which states that relative termination of $\mathcal{R}$ w.r.t. $\mathcal{B}$ coincides with the finiteness of the DP problem $(\mathsf{DP}(\mathcal{R}), \mathcal{R} \cup \mathcal{B})$, even if $\mathcal{B}$ is non-terminating. To facilitate the following discussion, besides $\mathcal{R}/\mathcal{B}$ weak-decreasingness we further impose that $\mathcal{R}$ and $\mathcal{B}$ share no defined symbol, *i.e.*, $\mathcal{D}_{\mathcal{R}} \cap \mathcal{D}_{\mathcal{B}} = \emptyset$. This condition will be relaxed in the later development. The proof of the following theorem is analogous to the standard dependency pair proof scheme.

Theorem 1. *Let $\mathcal{R}$ and $\mathcal{B}$ be TRSs such that $\mathcal{R}/\mathcal{B}$ is weak-decreasing and $\mathcal{D}_{\mathcal{R}} \cap \mathcal{D}_{\mathcal{B}} = \emptyset$. Then, $\mathcal{R}/\mathcal{B}$ is terminating iff the DP problem $(\mathsf{DP}(\mathcal{R}), \mathcal{R} \cup \mathcal{B})$ is finite.*

Theorem 1 is not yet applicable in practice; whether $\mathcal{R}/\mathcal{B}$ is weak-decreasing or not is obviously undecidable in general. Thus in the next section, we provide decidable syntactic conditions to ensure $\mathcal{R}/\mathcal{B}$ weak-decreasingness.

4   Syntactic Conditions for Weak-Decreasingness

In this section, we focus on finding syntactic and decidable conditions that ensure $\mathcal{R}/\mathcal{B}$ weak-decreasingness. For this purpose, this time we require $\mathcal{B}$ to be *non-duplicating*, i.e., no variable has more occurrences in the right-hand side of a rule than in its left-hand side, together with the following condition:

Definition 6 (Dominance). *We say that a TRS $\mathcal{R}$ dominates a TRS $\mathcal{B}$ iff the right-hand sides of all rules in $\mathcal{B}$ contain no symbol from $\mathcal{D}_{\mathcal{R}}$.*

Before proving that the above two conditions ensure $\mathcal{R}/\mathcal{B}$ weak-decreasingness, we state an auxiliary result. Let $\mathcal{MV}\mathrm{ar}(s)$ denote the multiset of variables occurring in a term s. The following lemma can easily be proved.

Lemma 5. *Let $\mathcal{R}$ and $\mathcal{B}$ be TRSs such that $\mathcal{R}$ dominates $\mathcal{B}$. For every term t and substitution σ, $\nabla_{\mathcal{R}}(t\sigma) \succsim_{\mathcal{R}/\mathcal{B}}^{\mathrm{mul}} \biguplus_{x \in \mathcal{MV}\mathrm{ar}(t)} \nabla_{\mathcal{R}}(x\sigma)$.*

The following lemma is the key result of this section:

Lemma 6. *Let $\mathcal{R}$ and $\mathcal{B}$ be TRSs such that $\mathcal{R}$ dominates $\mathcal{B}$ and $\mathcal{B}$ is non-duplicating. Then $\mathcal{R}/\mathcal{B}$ is weak-decreasing.*

Proof. We prove that $t \xrightarrow{p}_{\mathcal{B}} t'$ implies $\nabla_{\mathcal{R}}(t) \succsim_{\mathcal{R}/\mathcal{B}}^{\mathrm{mul}} \nabla_{\mathcal{R}}(t')$ for arbitrary terms t and t' and a position p. We distinguish the following two cases:

– First, assume that p has a prefix q such that $t|_q \in \nabla_{\mathcal{R}}(t)$. Then, we have

$$\nabla_{\mathcal{R}}(t) = \nabla_{\mathcal{R}}(t[\,]_q) \cup \{t|_q\} \qquad \text{and} \qquad \nabla_{\mathcal{R}}(t') = \nabla_{\mathcal{R}}(t[\,]_q) \cup \{t'|_q\}$$

Since $t|_q \rightarrow_{\mathcal{B}} t'|_q$, we have $t|_q \succsim_{\mathcal{R}/\mathcal{B}}^{\mathrm{mul}} t'|_q$ and thus $\nabla_{\mathcal{R}}(t) \succsim_{\mathcal{R}/\mathcal{B}}^{\mathrm{mul}} \nabla_{\mathcal{R}}(t')$.
– Assume now that $t_q \notin \nabla_{\mathcal{R}}(t)$ for any prefix q of p. Let $l \rightarrow r \in \mathcal{B}$, $t|_p = l\sigma$, and $t' = t[r\sigma]_p$. In this case, we have

$$\nabla_{\mathcal{R}}(t) = \nabla_{\mathcal{R}}(t[\,]_p) \cup \nabla_{\mathcal{R}}(l\sigma) \qquad \text{and} \qquad \nabla_{\mathcal{R}}(t') = \nabla_{\mathcal{R}}(t[\,]_p) \cup \nabla_{\mathcal{R}}(r\sigma)$$

From Lemma 5, we have $\nabla_{\mathcal{R}}(l\sigma) \succsim_{\mathcal{R}/\mathcal{B}}^{\mathrm{mul}} \bigcup_{x \in \mathcal{MV}\mathrm{ar}(l)} \nabla_{\mathcal{R}}(x\sigma)$. Since $\mathcal{R}$ dominates $\mathcal{B}$, r cannot contain symbols from $\mathcal{D}_{\mathcal{R}}$. Therefore,

$$\nabla_{\mathcal{R}}(l\sigma) \succsim_{\mathcal{R}/\mathcal{B}}^{\mathrm{mul}} \bigcup_{x \in \mathcal{MV}\mathrm{ar}(l)} \nabla_{\mathcal{R}}(x\sigma) \quad \text{and} \quad \nabla_{\mathcal{R}}(r\sigma) = \bigcup_{x \in \mathcal{MV}\mathrm{ar}(r)} \nabla_{\mathcal{R}}(x\sigma)$$

Since $\mathcal{B}$ is non-duplicating, we have $\mathcal{MV}\mathrm{ar}(l) \supseteq \mathcal{MV}\mathrm{ar}(r)$ and thus $\nabla_{\mathcal{R}}(l\sigma) \supseteq \nabla_{\mathcal{R}}(r\sigma)$. Therefore, we conclude that $\nabla_{\mathcal{R}}(t) \succsim_{\mathcal{R}/\mathcal{B}}^{\mathrm{mul}} \nabla_{\mathcal{R}}(t')$. $\qquad \square$

Finally, the following result is a direct consequence of Theorem 1 and Lemma 6.

Corollary 1. *Let $\mathcal{R}$ and $\mathcal{B}$ be TRSs such that $\mathcal{R}$ dominates $\mathcal{B}$, $\mathcal{B}$ is non-duplicating, and $\mathcal{D}_{\mathcal{R}} \cap \mathcal{D}_{\mathcal{B}} = \emptyset$. Then, $\mathcal{R}/\mathcal{B}$ is terminating iff the DP problem $(\mathsf{DP}(\mathcal{R}), \mathcal{R} \cup \mathcal{B})$ is finite.*

The following simple example illustrates that Corollary 1 indeed advances the state-of-the-art in proving relative termination.

Example 2. Consider the following two TRSs:

$$\mathcal{R} = \{\mathsf{g}(\mathsf{s}(x), y) \rightarrow \mathsf{g}(\mathsf{f}(x, y), y)\} \qquad \mathcal{B} = \{\mathsf{f}(x, y) \rightarrow x, \ \mathsf{f}(x, y) \rightarrow \mathsf{f}(x, \mathsf{s}(y))\}$$

Since they satisfy the conditions of Corollary 1, we obtain the DP problem $(\mathsf{DP}(\mathcal{R}), \mathcal{R} \cup \mathcal{B})$, where $\mathsf{DP}(\mathcal{R}) = \{\mathsf{g}^{\sharp}(\mathsf{s}(x), y) \rightarrow \mathsf{g}^{\sharp}(\mathsf{f}(x, y), y)\}$. The DP problem can be proved finite using classic techniques, e.g. polynomial interpretation $\mathcal{P}ol$ such that $\mathsf{f}_{\mathcal{P}ol}(x, y) = x$. On the other hand, all the tools we know that support relative termination, namely AProVE (ver. 2014), $\mathsf{T_TT_2}$ (ver. 1.15), Jambox (ver. 2006) [6], and TPA (ver. 1.1) [19], fail on this problem.

The dominance condition and the non-duplication condition are indeed necessary for Corollary 1 to hold. It is clear that the former condition is necessary from Example 1, which violates the dominance condition. For the latter condition, the following example illustrates that it is also necessary.

Example 3. Consider the following two TRSs:

$$\mathcal{R} = \{a \to b\} \qquad\qquad \mathcal{B} = \{f(x) \to c(x, f(x))\}$$

We have the following infinite $\to_{\mathcal{R}/\mathcal{B}}$-derivation:

$$f(a) \to_{\mathcal{B}} c(a, f(a)) \to_{\mathcal{R}} c(b, f(a)) \to_{\mathcal{B}} c(b, c(a, f(a))) \to_{\mathcal{R}} c(b, c(b, f(a))) \to_{\mathcal{B}} \cdots$$

However, there is no infinite $(\mathsf{DP}(\mathcal{R}), \mathcal{R} \cup \mathcal{B})$-chain since $\mathsf{DP}(\mathcal{R}) = \emptyset$. Note that this is a counterexample against [15, Theorem 5].

5   Improving Applicability

In contrast to dominance and non-duplication, the condition $\mathcal{D}_{\mathcal{R}} \cap \mathcal{D}_{\mathcal{B}} = \emptyset$ is not necessary. In order to show that this is indeed the case, let us recall the following result from [7]:

Proposition 3. *Let $\mathcal{R}$, $\mathcal{B}'$ and $\mathcal{B}''$ be TRSs. Then, $(\mathcal{R} \cup \mathcal{B}')/\mathcal{B}''$ is terminating iff both $\mathcal{R}/(\mathcal{B}' \cup \mathcal{B}'')$ and $\mathcal{B}'/\mathcal{B}''$ are terminating.*

Therefore, we have the following corollary in our context:

Corollary 2. *Let $\mathcal{R}$ and $\mathcal{B}$ be TRSs with $\mathcal{B} = \mathcal{B}' \cup \mathcal{B}''$. If $(\mathcal{R} \cup \mathcal{B}')/\mathcal{B}''$ is terminating, then $\mathcal{R}/\mathcal{B}$ is terminating.*

Now we state the first theorem of this section.

Theorem 2. *Let $\mathcal{R}$ and $\mathcal{B}$ be TRSs such that $\mathcal{R}$ dominates $\mathcal{B}$ and $\mathcal{B}$ is non-duplicating. If the DP problem $(\mathsf{DP}(\mathcal{R}), \mathcal{R} \cup \mathcal{B})$ is finite then $\mathcal{R}/\mathcal{B}$ is terminating.*

Proof. Let $\mathcal{B}'$ be the set of rules in $\mathcal{B}$ that define $\mathcal{D}_{\mathcal{R}}$ symbols, i.e., $\mathcal{B}' = \{l \to r \in \mathcal{B} \mid \mathsf{root}(l) \in \mathcal{D}_{\mathcal{R}}\}$, and let $\mathcal{B}'' = \mathcal{B} \setminus \mathcal{B}'$. Since the right-hand sides of $\mathcal{B}'$ rules cannot contain symbols from $\mathcal{D}_{\mathcal{R}}$ ($= \mathcal{D}_{\mathcal{R} \cup \mathcal{B}'}$), we have $\mathsf{DP}(\mathcal{R} \cup \mathcal{B}') = \mathsf{DP}(\mathcal{R})$.

Now, observe that $\mathcal{R} \cup \mathcal{B}'$ dominates $\mathcal{B}''$, $\mathcal{D}_{\mathcal{R} \cup \mathcal{B}'} \cap \mathcal{D}_{\mathcal{B}''} = \emptyset$, and $\mathcal{B}''$ is non-duplicating. Thus, Corollary 1 implies the relative termination of $\mathcal{R} \cup \mathcal{B}'$ w.r.t. $\mathcal{B}''$ and Corollary 2 implies the relative termination of $\mathcal{R}$ w.r.t. $\mathcal{B}$.   □

Unfortunately, the remaining two, namely the dominance and non-duplication conditions, might be too restrictive in practice. For instance, only six out of 44 examples in the relative TRS category of the TPDB satisfy both conditions.

Luckily, we can employ again Corollary 2 to relax the conditions. Consider TRSs $\mathcal{R}$ and $\mathcal{B}$ such that we want to prove that $\mathcal{R}/\mathcal{B}$ is terminating but the conditions of Theorem 2 do not hold. Then, we might still find a partition $\mathcal{B} = \mathcal{B}' \uplus \mathcal{B}''$ such that $\mathcal{R} \cup \mathcal{B}'$ and $\mathcal{B}''$ satisfy the conditions.

If we succeed, then by Theorem 2 and Corollary 2, we have that $\mathcal{R}/\mathcal{B}$ is terminating (i.e., by Theorem 2, $(\mathcal{R} \cup \mathcal{B}')/\mathcal{B}''$ is terminating and, by Corollary 2, $\mathcal{R}/(\mathcal{B}' \cup \mathcal{B}'')$ is also terminating with $\mathcal{B}' \cup \mathcal{B}'' = \mathcal{B}$).

Corollary 3. *Let $\mathcal{R}$ and $\mathcal{B}$ be TRSs. If $\mathcal{B}$ is split into $\mathcal{B} = \mathcal{B}' \uplus \mathcal{B}''$ such that (1) $\mathcal{B}''$ is non-duplicating, (2) $\mathcal{R} \cup \mathcal{B}'$ dominates $\mathcal{B}''$, and (3) the DP problem $(\mathsf{DP}(\mathcal{R} \cup \mathcal{B}'), \mathcal{R} \cup \mathcal{B})$ is finite, then $\mathcal{R}/\mathcal{B}$ is terminating.*

Example 4. Consider the following TRSs $\mathcal{R}$ and $\mathcal{B}$:

$$\mathcal{R} = \{\mathsf{a} \to \mathsf{b}\} \qquad \mathcal{B} = \{\mathsf{f}(\mathsf{s}(x)) \to \mathsf{c}(x, \mathsf{f}(x)), \ \mathsf{c}(x, \mathsf{c}(y, z)) \to \mathsf{c}(y, \mathsf{c}(x, z))\}$$

The first rule of $\mathcal{B}$ is duplicating, and hence Theorem 2 does not apply. However, we can split $\mathcal{B}$ into the following TRSs $\mathcal{B}'$ and $\mathcal{B}''$:

$$\mathcal{B}' = \{\mathsf{f}(\mathsf{s}(x)) \to \mathsf{c}(x, \mathsf{f}(x))\} \qquad \mathcal{B}'' = \{\mathsf{c}(x, \mathsf{c}(y, z)) \to \mathsf{c}(y, \mathsf{c}(x, z))\}$$

so that Corollary 3 applies. Now, we have $\mathsf{DP}(\mathcal{R} \cup \mathcal{B}') = \{\mathsf{f}^\sharp(\mathsf{s}(x)) \to \mathsf{f}^\sharp(x)\}$, whose finiteness can be proved using standard techniques.

Corollary 3 requires the rules in $\mathcal{B}$ that are duplicating or violate the dominance condition to be relatively terminating w.r.t. other rules in $\mathcal{B}$. This is not overly restrictive, as shown by the following two examples.

Example 5. Consider again the TRS $\mathcal{B}$ of Example 3, which is duplicating and nonterminating. We can construct an infinite $\to_{\mathcal{R}/\mathcal{B}}$-reduction as in Example 3 for any nonempty TRS $\mathcal{R}$; thus, any nonempty TRS is not relatively terminating w.r.t. $\mathcal{B}$.

Example 6. Consider the two TRSs $\mathcal{R} = \{\mathsf{a} \to \mathsf{b}\}$ and $\mathcal{B} = \{\mathsf{d} \to \mathsf{c}(\mathsf{a}, \mathsf{d})\}$. Note that $\mathcal{R}$ does not dominate $\mathcal{B}$. The DP problem $(\mathsf{DP}(\mathcal{R}), \mathcal{R} \cup \mathcal{B}) = (\emptyset, \mathcal{R} \cup \mathcal{B})$ is trivially finite. However, $\mathcal{R}$ is not relatively terminating w.r.t. $\mathcal{B}$, as the following infinite derivation exists:

$$\mathsf{d} \to_\mathcal{B} \mathsf{c}(\mathsf{a}, \mathsf{d}) \to_\mathcal{R} \mathsf{c}(\mathsf{b}, \mathsf{d}) \to_\mathcal{B} \mathsf{c}(\mathsf{b}, \mathsf{c}(\mathsf{a}, \mathsf{d})) \to_\mathcal{R} \mathsf{c}(\mathsf{b}, \mathsf{c}(\mathsf{b}, \mathsf{d})) \to_\mathcal{B} \cdots$$

6   Relative Termination and Minimality

A DP chain $s_1^\sharp \xrightarrow{\epsilon}_\mathcal{P} t_1^\sharp \to^*_\mathcal{R} s_2^\sharp \xrightarrow{\epsilon}_\mathcal{P} t_2^\sharp \to^*_\mathcal{R} \cdots$ is said to be *minimal* if every $t_i^\sharp$ is terminating w.r.t. $\mathcal{R}$. It is well-known that absence of infinite minimal $(\mathsf{DP}(\mathcal{R}), \mathcal{R})$-chains implies absence of infinite $(\mathsf{DP}(\mathcal{R}), \mathcal{R})$-chains and thus termination of $\mathcal{R}$. A couple of techniques, namely *usable rules* and the *subterm criterion* have been proposed to prove absence of infinite minimal chains [12].

Unfortunately, for the DP problems produced by our relative termination criteria, the minimality property cannot be assumed. Therefore, usable rules and subterm criterion do not apply in general.

Example 7. Consider the TRSs $\mathcal{R} = \{\mathsf{f}(\mathsf{s}(x)) \to \mathsf{f}(x)\}$ and $\mathcal{B} = \{\mathsf{inf} \to \mathsf{s}(\mathsf{inf})\}$. Theorem 2 yields the DP problem $(\{\mathsf{f}^\sharp(\mathsf{s}(x)) \to \mathsf{f}^\sharp(x)\}, \mathcal{R} \cup \mathcal{B})$, which satisfies the subterm criterion in the argument of $\mathsf{f}^\sharp$. Moreover, since no rule is *usable* from the dependency pair $\mathsf{f}^\sharp(\mathsf{s}(x)) \to \mathsf{f}^\sharp(x)$, the usable rule technique would yield the DP problem $(\{\mathsf{f}^\sharp(\mathsf{s}(x)) \to \mathsf{f}^\sharp(x)\}, \emptyset)$, which any standard technique proves finite. However, $\mathcal{R}/\mathcal{B}$ is not terminating as the following infinite reduction exists:

$$\mathsf{f}(\mathsf{s}(\mathsf{inf})) \to_\mathcal{R} \mathsf{f}(\mathsf{inf}) \to_\mathcal{B} \mathsf{f}(\mathsf{s}(\mathsf{inf})) \to_\mathcal{R} \mathsf{f}(\mathsf{inf}) \to_\mathcal{B} \cdots$$

Nonetheless, we show that both the subterm criterion and usable rules are still applicable when $\mathcal{B}$ satisfies the following condition:

Definition 7 (Quasi-Termination [5]). *We say that a TRS $\mathcal{R}$ is quasi-terminating iff the set $\{t \mid s \to_{\mathcal{R}}^* t\}$ is finite for every term s.*

Now we naturally extend the notion of minimality to relative termination.

Definition 8 (Relative DP Problem). *A relative DP problem is a triple of TRSs, written $(\mathcal{P}, \mathcal{R}/\mathcal{B})$. A $(\mathcal{P}, \mathcal{R}/\mathcal{B})$-chain is a possibly infinite sequence*

$$s_1 \xrightarrow{\epsilon}_{\mathcal{P}} t_1 \xrightarrow{>\epsilon}{}^*_{\mathcal{R}\cup\mathcal{B}} s_2 \xrightarrow{\epsilon}_{\mathcal{P}} t_2 \xrightarrow{>\epsilon}{}^*_{\mathcal{R}\cup\mathcal{B}} \cdots$$

and is called minimal *if every t_i is $\mathcal{R}/\mathcal{B}$-terminating. The relative DP problem is* minimally finite *if it admits no infinite minimal chain.*

Clearly, finiteness of $(\mathsf{DP}(\mathcal{R}), \mathcal{R}/\mathcal{B})$ is equivalent to that of $(\mathsf{DP}(\mathcal{R}), \mathcal{R} \cup \mathcal{B})$. Hence our previous results hold as well for the corresponding relative DP problems.

When the base $\mathcal{B}$ is quasi-terminating, we can apply the subterm criterion. A *simple projection* π assigns each n-ary symbol $f^\sharp$ an argument position $i \in \{1, \ldots, n\}$. For a term $t^\sharp = f^\sharp(t_1, \ldots, t_n)$ and $i = \pi(f^\sharp)$, we denote t_i by $\pi(t^\sharp)$. For a relation $\sqsupset$ on terms, $\sqsupset^\pi$ is defined as follows: $s \sqsupset^\pi t$ iff $\pi(s) \sqsupset \pi(t)$.

Theorem 3 (Relative Subterm Criterion). *Let $\mathcal{B}$ be a quasi-terminating TRS, $(\mathcal{P}, \mathcal{R}/\mathcal{B})$ a relative DP problem and π a simple projection such that $\mathcal{P} \subseteq \trianglerighteq^\pi$. Then, $(\mathcal{P}, \mathcal{R}/\mathcal{B})$ is finite if $(\mathcal{P} \setminus \triangleright^\pi, \mathcal{R}/\mathcal{B})$ is finite.*

The proof of the above theorem mimics that of [12, Theorem 11], but here we need the relative termination of $\triangleright/\mathcal{B}$.

The quasi-termination condition also enables the usable rules technique.

Theorem 4 (Relative Usable Rules). *If $\mathcal{B}$ is quasi-terminating, then the usable rule argument can be applied to the relative DP problem $(\mathcal{P}, \mathcal{R}/\mathcal{B})$.*

Proof (Sketch). The proof basically follows the standard case of [12, Theorem 29]. Note however that we require $\mathcal{B}$ to be quasi-terminating, in order for the interpretation $I_{\mathcal{G}}$ to be well-defined for all $\mathcal{R}/\mathcal{B}$-terminating terms. $\qquad\square$

It is well-known that, unfortunately, the quasi-termination condition is undecidable [5]. In our implementation, we only use a trivial sufficient condition, *size-non-increasingness*. We admit that this is quite restrictive, and thus leave it for future work to find more useful syntactic condition for this purpose.

From Example 7, it is clear that the usable rule argument does not apply to the rules in $\mathcal{B}$ if they are not quasi-terminating. Nonetheless, we conjecture that the usable rule argument may be still applicable to the rules in $\mathcal{R}$.

7   Experimental Evaluation

A key technique for proving finiteness of DP problems are *reduction pairs* [2]: A *reduction pair* is a well-founded order pair $(\gtrsim, >)$ on terms such that $\gtrsim$ is closed on contexts and substitutions, and $>$ is closed on substitutions.

Proposition 4 ([2,10]). *Let $(\mathcal{P}, \mathcal{R})$ be a DP problem and $(\gtrsim, >)$ a reduction pair such that $\mathcal{P} \cup \mathcal{R} \subseteq \gtrsim$. The DP problem $(\mathcal{P}, \mathcal{R})$ is finite iff $(\mathcal{P} \setminus >, \mathcal{R})$ is.*

In the experiments, we use the following reduction pairs:

- *polynomial interpretations* with negative constants [2,11,21],
- the *lexicographic path order* [16],
- the *weighted path order* with partial status [29], and
- (2- or 3-dimensional) *matrix interpretations* [6].

Geser [7] proposed a technique to reduce relative termination of TRSs to relative termination of simpler TRSs. This technique is incorporated into the DP framework for proving standard termination, as *rule removal processors* [10]. We say a reduction pair $(\gtrsim, >)$ is *monotone* if $>$ is closed under contexts.

Proposition 5 (Relative Rule Removal Processor).   *Let $\mathcal{R}$ and $\mathcal{B}$ be TRSs, and $(\gtrsim, >)$ a monotone reduction pair such that $\mathcal{R} \cup \mathcal{B} \subseteq \gtrsim$. Then $\mathcal{R}$ is relatively terminating w.r.t. $\mathcal{B}$ if and only if $\mathcal{R} \setminus >$ is relatively terminating w.r.t. $\mathcal{B} \setminus >$.*

For monotone reduction pairs, we use polynomial and matrix interpretations with top-left elements of coefficients being at least 1 [6].

We implemented our technique into the termination prover NaTT (ver.1.2). In the following, we show the significance of our technique through an experimental evaluation. The experiments[4] were run on a server equipped with a quad-core Intel Xeon E5-3407v2 processor running at a clock rate of 2.40GHz and 32GB of main memory. NaTT uses z3 4.3.2[5] as a back-end SMT solver.

The first test set consists of the 44 examples in the "TRS Relative" category of the *termination problem database* (TPDB) 9.0.[6] The results are presented in the left half of Table 1. In the first two rows, we directly apply Theorem 2 and Corollary 3, and then apply the aforementioned reduction pairs. We observe that they are of limited applicability on the TPDB set of problems due to the non-duplication and dominance conditions. Nonetheless, Corollary 3 could prove relative termination of two problems[7] which no tools participating in the *termination competition 2014* were able to prove. For comparison, we include results for rule removal processors by matrix interpretations in the third row.

We also prepared 44 examples by extending examples of [3] with the random number generator $\mathcal{B}_{\mathsf{rand}}$ or the commutative list specification $\mathcal{B}_{\mathsf{comlist}}$. The results

---

[4] Details are available at http://www.trs.cm.is.nagoya-u.ac.jp/papers/CADE2015.

[5] Available at http://z3.codeplex.com/.

[6] Available at http://termination-portal.org/wiki/TPDB.

[7] For one of the two problems, the union is terminating.

Table 1. Experiments

Method	TPDB relative (44)				AG01+relative (44)			
	Yes	Maybe	T.O	Time	Yes	Maybe	T.O	Time
Theorem 2	4	40	0	1.21	29	15	0	5.07
Corollary 3	6	28	0	37.03	29	15	0	5.08
Proposition 5	23	17	4	406.01	9	35	0	8.19
Proposition 5 + Corollary 3	25	11	8	505.94	35	9	0	12.70
AProVE	27	(no: 8)	9	756.66	14	0	30	1959.91

are presented in the right half of Table 1. In these examples, the power of our method should be clear. Theorem 2 is already able to prove relative termination of 29 examples, while AProVE succeeds only in 14 examples.

The DP framework allows combining termination proving techniques. In the fourth row, we combine the rule removal processors and the technique presented in this paper. This combination indeed boosts the power of NaTT; e.g., by combining Proposition 5 and Corollary 3, NaTT can prove relative termination for a total of 60 examples (out of 88), while AProVE can only prove it for 41 examples.[8] Therefore, we can conclude that our technique improves the state-or-the-art methods for proving relative termination.

8 Related Work

One of the most comprehensive works on relative termination is Geser's PhD thesis [7]. One of the main results in this work is formulated in Proposition 5 in the previous section. A similar technique has been used, e.g., to prove confluence in [13]. Of course dependency pairs are not considered in [7] since it was introduced almost a decade later. Dependency pairs are considered in [6], but they are mainly used to prove termination of a TRS $\mathcal{R}$ by proving the termination of $DP(\mathcal{R})/\mathcal{R}$, which is quite a different purpose from ours.

Giesl and Kapur [8] adapted the dependency pair method for proving termination of *equational rewriting*, a special case of relative termination where the base is symmetric ($\mathcal{B} = \mathcal{B}^{-1}$). For more specific *associative-commutative (AC)* rewriting, a number of papers exist (e.g., [1]). The key technique behind them is to compute an extension of $\mathcal{R}$ w.r.t. the considered equations. This operation allows symbols in $\mathcal{B}$ (e.g., AC symbols) to be defined also in $\mathcal{R}$, and hence no counterpart of the dominance condition is required. However, such extensions are computable only for certain equations (e.g., AC), and thus they are not appropriate in our setting, where an arbitrary base $\mathcal{B}$ is considered.

The closer approach is [15], where the main aim was proving termination of narrowing [14] by proving relative termination of a corresponding rewrite relation, similarly to [24, 27]. In [15], a first attempt to reduce relative termination

---

[8] For four examples, AProVE proved relative termination but NaTT failed. There AProVE used *semantic labeling* [30], which is currently not implemented in NaTT.

to a DP problem is made by requiring $\mathcal{R}$ and $\mathcal{B}$ to form a so called *hierarchical combination (HC)* [25], i.e., $\mathcal{D}_{\mathcal{R}} \cap \mathcal{F}_{\mathcal{B}} = \emptyset$. Unfortunately, we found that [15, Theorem 5] was incorrect since requiring $\mathcal{B}$ to be non-duplicating is also necessary. In fact, Example 3 is a counterexample to [15, Theorem 5]. The present paper corrects and significantly extends [15]; namely, all results in Sects. 3, 5 and 6 are new, and those in Sect. 4 correct and extend the previous result of [15]. Note also that the HC condition of [15] is a special case of our dominance condition. Moreover, we developed an implementation that allowed us to experimentally verify that our technique indeed pays off in practice.

9   Conclusion

In this paper, we have introduced a new approach to proving relative termination by reducing it to DP problems. The relevance of such a result should be clear, since it allows one to prove relative termination by reusing many existing techniques and tools for proving termination within the DP framework. Indeed, such an approach was included in the RTA List of Open Problems (Problem #106). To the best of our knowledge, this work makes the first significant contribution to positively answering this problem. Moreover, as shown in Sect. 7, our method is competitive w.r.t. state-of-the-art provers for the problems in TPDB, and is clearly superior for examples including the generation of random values or the simulation of extra-variables, as discussed in Sect. 1.

As future work, we plan to improve the precision of our technique by extending the DP framework to be more suitable for proving relative termination. We will also continue the research on finding less restrictive conditions on $\mathcal{R}$ and $\mathcal{B}$ so that the technique becomes more widely applicable.

Acknowledgement. We would like to thank Nao Hirokawa and the anonymous reviewers for their helpful comments and suggestions in early stages of this work.

References

1. Alarcón, B., Lucas, S., Meseguer, J.: A dependency pair framework for $A \vee C$-termination. In: Ölveczky, P.C. (ed.) WRLA 2010. LNCS, vol. 6381, pp. 35–51. Springer, Heidelberg (2010)
2. Arts, T., Giesl, J.: Termination of term rewriting using dependency pairs. Theor. Comput. Sci. **236**(1–2), 133–178 (2000)
3. Arts, T., Giesl, J.: A collection of examples for termination of term rewriting using dependency pairs. Technical report AIB-2001-09, RWTH Aachen (2001)
4. Baader, F., Nipkow, T.: Term Rewriting and All That. Cambridge University Press, Cambridge (1998)
5. Dershowitz, N.: Termination of rewriting. J. Symb. Comput. **3**(1&2), 69–115 (1987)
6. Endrullis, J., Waldmann, J., Zantema, H.: Matrix interpretations for proving termination of term rewriting. J. Autom. Reasoning **40**(2–3), 195–220 (2008)
7. Geser, A.: Relative termination. Dissertation, Fakultät für Mathematik und Informatik, Universität Passau, Germany (1990)

8. Giesl, J., Kapur, D.: Dependency pairs for equational rewriting. In: Middeldorp, A. (ed.) RTA 2001. LNCS, vol. 2051, pp. 93–107. Springer, Heidelberg (2001)
9. Giesl, J., Schneider-Kamp, P., Thiemann, R.: *AProVE 1.2*: automatic termination proofs in the dependency pair framework. In: Furbach, U., Shankar, N. (eds.) IJCAR 2006. LNCS (LNAI), vol. 4130, pp. 281–286. Springer, Heidelberg (2006)
10. Giesl, J., Thiemann, R., Schneider-Kamp, P., Falke, S.: Mechanizing and improving dependency pairs. J. Autom. Reasoning **37**(3), 155–203 (2006)
11. Hirokawa, N., Middeldorp, A.: Polynomial interpretations with negative coefficients. In: Buchberger, B., Campbell, J. (eds.) AISC 2004. LNCS (LNAI), vol. 3249, pp. 185–198. Springer, Heidelberg (2004)
12. Hirokawa, N., Middeldorp, A.: Dependency pairs revisited. In: van Oostrom, V. (ed.) RTA 2004. LNCS, vol. 3091, pp. 249–268. Springer, Heidelberg (2004)
13. Hirokawa, N., Middeldorp, A.: Decreasing diagrams and relative termination. J. Autom. Reasoning **47**(4), 481–501 (2011)
14. Hullot, J.M.: Canonical forms and unification. CADE-5. LNCS, vol. 87, pp. 318–334. Springer, Heidelberg (1980)
15. Iborra, J., Nishida, N., Vidal, G.: Goal-directed and relative dependency pairs for proving the termination of narrowing. In: De Schreye, D. (ed.) LOPSTR 2009. LNCS, vol. 6037, pp. 52–66. Springer, Heidelberg (2010)
16. Kamin, S., Lévy, J.J.: Two generalizations of the recursive path ordering (1980, unpublished note)
17. Klop, J.W.: Term rewriting systems: a tutorial. Bull. Eur. Assoc. Theor. Comput. Sci. **32**, 143–183 (1987)
18. Koprowski, A., Zantema, H.: Proving liveness with fairness using rewriting. In: Gramlich, B. (ed.) FroCos 2005. LNCS (LNAI), vol. 3717, pp. 232–247. Springer, Heidelberg (2005)
19. Koprowski, A.: TPA: termination proved automatically. In: Pfenning, F. (ed.) RTA 2006. LNCS, vol. 4098, pp. 257–266. Springer, Heidelberg (2006)
20. Korp, M., Sternagel, C., Zankl, H., Middeldorp, A.: Tyrolean termination tool 2. In: Treinen, R. (ed.) RTA 2009. LNCS, vol. 5595, pp. 295–304. Springer, Heidelberg (2009)
21. Lankford, D.: Canonical algebraic simplification in computational logic. Technical report ATP-25, University of Texas (1975)
22. Liu, J., Dershowitz, N., Jouannaud, J.-P.: Confluence by critical pair analysis. In: Dowek, G. (ed.) RTA-TLCA 2014. LNCS, vol. 8560, pp. 287–302. Springer, Heidelberg (2014)
23. Nishida, N., Sakai, M., Sakabe, T.: Narrowing-based simulation of term rewriting systems with extra variables. ENTCS **86**(3), 52–69 (2003)
24. Nishida, N., Vidal, G.: Termination of narrowing via termination of rewriting. Appl. Algebra Eng. Commun. Comput. **21**(3), 177–225 (2010)
25. Ohlebusch, E.: Advanced Topics in Term Rewriting. Springer-Verlag, London (2002)
26. Thiemann, R., Allais, G., Nagele, J.: On the formalization of termination techniques based on multiset orderings. In: RTA 2012. LIPIcs, vol. 15, pp. 339–354. Schloss Dagstuhl - Leibniz-Zentrum für Informatik (2012)
27. Vidal, G.: Termination of narrowing in left-linear constructor systems. In: Garrigue, J., Hermenegildo, M.V. (eds.) FLOPS 2008. LNCS, vol. 4989, pp. 113–129. Springer, Heidelberg (2008)
28. Yamada, A., Kusakari, K., Sakabe, T.: Nagoya termination tool. In: Dowek, G. (ed.) RTA-TLCA 2014. LNCS, vol. 8560, pp. 466–475. Springer, Heidelberg (2014)

29. Yamada, A., Kusakari, K., Sakabe, T.: A unified ordering for termination proving. Sci. Comput. Program. (2014). doi:10.1016/j.scico.2014.07.009
30. Zantema, H.: Termination of term rewriting by semantic labelling. Fundamenta Informaticae **24**(1/2), 89–105 (1995)
31. Zantema, H.: Termination. In: Bezem, M., Klop, J.W., de Vrijer, R. (eds.) Term Rewriting Systems. Cambridge Tracts in Theoretical Computer Science, vol. 55, pp. 181–259. Cambridge University Press, Cambridge (2003)

Decision Procedures

Decidability of Univariate Real Algebra
with Predicates for Rational and Integer Powers

Grant Olney Passmore$^{(\boxtimes)}$

Aesthetic Integration, London and Clare Hall,
University of Cambridge, Cambridge, England
grant.passmore@cl.cam.ac.uk

Abstract. We prove decidability of univariate real algebra extended
with predicates for rational and integer powers, i.e., "$x^n \in \mathbb{Q}$" and
"$x^n \in \mathbb{Z}$." Our decision procedure combines computation over real algebraic cells with the rational root theorem and witness construction via
algebraic number density arguments.

1 Introduction

From the perspective of decidability, the reals stand in stark contrast to the
rationals and integers. While the elementary arithmetical theories of the integers
and rationals are undecidable, the corresponding theory of the reals is decidable
and admits quantifier elimination. The immense utility real algebraic reasoning
finds within the mathematical sciences continues to motivate significant progress
towards practical automatic proof procedures for the reals.

However, in mathematical practice, we are often faced with problems involving a combination of nonlinear statements over the reals, rationals and integers.
Consider the existence and irrationality of $\sqrt{2}$, expressed in a language with
variables implicitly ranging over $\mathbb{R}$:

$$\exists x(x \geq 0 \wedge x^2 = 2) \ \wedge \ \neg\exists x(x \in \mathbb{Q} \wedge x \geq 0 \wedge x^2 = 2)$$

Though easy to prove by hand this sentence has never to our knowledge been
placed within a broader decidable theory so that, e.g., the existence and irrationality of solutions to any univariate real algebra problem can be decided automatically. This $\sqrt{2}$ example is relevant to the theorem proving community as its
formalisation has been used as a benchmark for comparing proof assistants [21].
It would be useful if such proofs were fully automatic.

In this paper, we prove decidability of univariate real algebra extended with
predicates for rational and integer powers. This guarantees we can always decide
sentences like the above, and many more besides. For example, the following
conjectures are decided by our method in a fraction of a second:

$$\forall x(x^3 \in \mathbb{Z} \wedge x^5 \notin \mathbb{Z} \ \Rightarrow \ x \notin \mathbb{Q})$$
$$\exists x(x^2 \in \mathbb{Q} \wedge x \notin \mathbb{Q} \wedge x^5 + 1 > 20)$$
$$\forall x(x^2 \notin \mathbb{Q} \Rightarrow x \notin \mathbb{Q})$$
$$\exists x(x \notin \mathbb{Q} \wedge x^2 \in \mathbb{Z} \wedge 3x^4 + 2x + 1 > 5 \wedge 4x^3 + 1 < 2)$$

© Springer International Publishing Switzerland 2015
A.P. Felty and A. Middeldorp (Eds.): CADE-25, LNAI 9195, pp. 181–196, 2015.
DOI: 10.1007/978-3-319-21401-6_12

2   Preliminaries

We assume a basic grounding in commutative algebra. We do not however assume exposure to real algebraic geometry and give a high-level treatment of the relevant foundations.

The theory of real closed fields (RCF) is $Th(\langle \mathbb{R}, +, -, \times, <, 0, 1 \rangle)$, the collection of all true sentences of the reals in the elementary language of ordered rings. RCF is complete, decidable and admits effective elimination of quantifiers [3].

A real algebraic number is a real number that is a root of a (non-zero) univariate polynomial with integer coefficients.

The real algebraic numbers,

$$\mathbb{R}_{alg} = \{x \in \mathbb{R} \mid \exists p \neq 0 \in \mathbb{Z}[x] \text{ s.t. } p(x) = 0\},$$

form a computable subfield (a computable sub-RCF) of $\mathbb{R}$. Indeed, $\mathbb{R}_{alg}$ embeds isomorphically into every RCF. The field operations of $\mathbb{R}_{alg}$ are performed on computable representations of field elements. The *minimal polynomial* of $\alpha \in \mathbb{R}_{alg}$ is the unique monic $p \in \mathbb{Q}[x]$ of least degree s.t. $p(\alpha) = 0$. The degree of an algebraic number is the degree of its minimal polynomial.

An element $\alpha \in \mathbb{R}_{alg}$ can be represented by two pieces of data: (i) a polynomial $p(x) \in \mathbb{Z}[x]$ s.t. $p(\alpha) = 0$, and (ii) an identifier specifying which root of $p(x)$ is denoted by α. A *root-triple* representation is often used where α is "pinned down" among the roots of $p(x)$ by an interval with rational endpoints:

$$\langle p(x) \in \mathbb{Z}[x], q_1, q_2 \in \mathbb{Q} \rangle \text{ s.t. } p(\alpha) = 0 \ \wedge \ \#\{r \in [q_1, q_2] \mid p(r) = 0\} = 1.$$

The process of *root isolation* is a key component of computing over $\mathbb{R}_{alg}$. Given a polynomial $p \in \mathbb{Z}[x]$ with k unique real roots, root isolation computes a sequence of disjoint real intervals with rational endpoints $I_1, \ldots, I_k$ s.t. each I_j contains precisely one real root of p. Much work has been done on efficient root isolation. Common approaches include those based on Sturm's Theorem and Descartes' Rule of Signs [4,12,19]. Sturm's Theorem also plays a key role in computing the sign of a polynomial evaluated at a real algebraic number.

Given representations of $\alpha, \beta \in \mathbb{R}_{alg}$, there are two main approaches to performing the field operations, i.e., for computing representations of α^{-1}, $\alpha + \beta$, $\alpha\beta$, etc. Both approaches rely on root isolation. The first approach uses bivariate resultants to compute representation polynomials [12]. The second approach uses a recursive representation of real algebraic numbers through an explicit treatment of field towers and does not require computing resultants [13,17]. Computing α^n (which plays a key role in our decision procedure) can in general be done by repeated squaring, requiring on the order of $\log n$ real algebraic number multiplications. More sophisticated methods for α^n are also available [6].

The Intermediate Value Theorem (IVT) holds over every RCF. Armed with machinery for computing the sign of a polynomial $p(x) \in \mathbb{Z}[x]$ at a real algebraic point $\alpha \in \mathbb{R}_{alg}$, the combination of IVT and root isolation can be used as the basis of a decision method for univariate real algebra.

Consider

$$\varphi(x) = \left[\bigwedge_{i=1}^{k_1} \bigvee_{j=1}^{k_2} (p_{ij}(x) \odot_{ij} 0) \right] \quad \text{s.t. } p_{ij} \in \mathbb{Z}[x], \ \odot_{ij} \in \{<, \leq, =, \geq, >\}.$$

We can decide the satisfiability of φ over $\mathbb{R}$, i.e., whether or not

$$\langle \mathbb{R}, +, -, \times, <, 0, 1 \rangle \models \exists x(\varphi(x))$$

in the following manner:

- Let $P = \prod_{ij} p_{ij} \in \mathbb{Z}[x]$, the product of all polynomials appearing in φ.
- Let $\alpha_1 < \ldots < \alpha_k \in \mathbb{R}_{alg}$ be all distinct real roots of P.
- Then, the roots α_i partition $\mathbb{R}$ into finitely many connected components:

$$\mathbb{R} = {]-\infty, \alpha_1[} \cup [\alpha_1] \cup {]\alpha_1, \alpha_2[} \cup \cdots \cup {]\alpha_{k-1}, \alpha_k[} \cup [\alpha_k] \cup {]\alpha_k, +\infty[}.$$

- By IVT, the sign of each polynomial p_{ij} appearing in φ is invariant over any component of the partitioning.
- Thus, we can simply select one sample point from each component of the partitioning and obtain a sequence of $2k + 1$ real algebraic points $S = \{r_1, \ldots, r_{k+1}\} \subset \mathbb{R}_{alg}$ s.t.

$$\langle \mathbb{R}, +, -, \times, <, 0, 1 \rangle \models \exists x(\varphi(x)) \quad \Longleftrightarrow \quad \bigvee_{i=1}^{2k+1} \varphi(r_i).$$

Now $\exists x(\varphi(x))$ can be decided simply by evaluating $\varphi(x)$ at finitely many real algebraic points. The partitioning of $\mathbb{R}$ constructed above is called an *algebraic decomposition* induced by P (equivalently, by the polynomials p_{ij}).

3   Decision Procedure

Our decision procedure extends the IVT-based method for univariate real algebra with means to handle predicates expressing the rationality and integrality of powers of the variable of the formula, i.e., $(x^n \in \mathbb{Q})$ and $(x^n \in \mathbb{Z})$. As will be made clear (cf. Sect. 5), the restriction of these predicates to powers of the variable is important: The method would fail if we allowed more general polynomials $p(x) \in \mathbb{Z}[x]$ to appear in constraints of the form $(p(x) \in \mathbb{Q})$.

Formally, we work over the univariate language of ordered rings $\mathcal{L}$ extended with infinitely many predicate symbols of one real variable:

$$(x \in \mathbb{Q}), (x^2 \in \mathbb{Q}), (x^3 \in \mathbb{Q}), \ldots \quad \text{and} \quad (x \in \mathbb{Z}), (x^2 \in \mathbb{Z}), (x^3 \in \mathbb{Z}), \ldots .$$

We use $\mathcal{L}_{\mathbb{QZ}}$ to mean the resulting extended language and $\mathcal{L}_{\mathbb{Q}}$ (resp. $\mathcal{L}_{\mathbb{Z}}$) to mean $\mathcal{L}$ extended only with the rationality (resp. integrality) predicates.

We present a method to decide the satisfiability of quantifier-free $\mathcal{L}_{\mathbb{Q}\mathbb{Z}}$ formulas over $\mathbb{R}$. It suffices to consider $\mathcal{L}_{\mathbb{Q}\mathbb{Z}}$ formulas of the form

$$\varphi(x) \wedge \Gamma(x)$$

where $\varphi \in \mathcal{L}$ is a formula of univariate real algebra and

$$\Gamma = \Gamma_{\mathbb{Q}} \wedge \Gamma_{\mathbb{Z}}$$

s.t.

$$\Gamma_{\mathbb{Q}} = \left[\bigwedge_{i=1}^{k_1} (x^{w_1(i)} \in \mathbb{Q}) \wedge \bigwedge_{i=1}^{k_2} (x^{w_2(i)} \notin \mathbb{Q}) \right]$$

and

$$\Gamma_{\mathbb{Z}} = \left[\bigwedge_{i=1}^{k_3} (x^{w_3(i)} \in \mathbb{Z}) \wedge \bigwedge_{i=1}^{k_4} (x^{w_4(i)} \notin \mathbb{Z}) \right].$$

Informed by the IVT-based method for univariate real algebra, we can reduce this $\mathcal{L}_{\mathbb{Q}\mathbb{Z}}$ decision problem to an even more restricted one. Crucial to this reduction is treating the connected components of an algebraic decomposition as "first class" objects, rather than only computing with single sample points selected from them. We call such components *r-cells*.

Definition 1 (r-cell). *An r-cell is a connected component of $\mathbb{R}$ of one of the following four forms (with $\alpha, \beta \in \mathbb{R}_{alg}$): (i) $[\alpha]$, (ii) $]-\infty, \alpha[$ s.t. $\alpha \leq 0$, (iii) $]\alpha, \beta[$ s.t. $0 \leq \alpha < \beta$ or $\alpha < \beta \leq 0$, (iv) $]\alpha, +\infty[$ s.t. $\alpha \geq 0$.*

Observe that the only r-cell containing zero is the singleton (type (i)) r-cell $[0]$. Note that r-cells of type (i) are 0-dimensional subsets of $\mathbb{R}$ while r-cells of types (ii)-(iv) are 1-dimensional. We call these 0-cells and 1-cells, resp. An algebraic decomposition can always be transformed into an *r-cell decomposition* by splitting any 1-cell containing zero into three parts.

Given $\Phi(x) = \varphi(x) \wedge \Gamma(x)$, we must decide whether or not $\mathbb{R}$ contains any point x s.t. $\Phi(x)$ holds. To do so, we will first compute an r-cell decomposition of $\mathbb{R}$ induced by the polynomials of φ. Let $c_1, \ldots, c_k$ be these r-cells. Then by IVT, the truth of φ is invariant within each c_i. Note, however, that the truth of Γ may vary over each c_i. Let C be the result of filtering out all r-cells c_i that falsify φ:

$$C = \{c_i \mid \exists r \in c_i(\varphi(r)), \ 1 \leq i \leq k\}.$$

This can be done by evaluating φ at a single sample point drawn from each c_i. If $C = \emptyset$, then Φ is clearly unsatisfiable over $\mathbb{R}$. Otherwise, C is a non-empty collection of r-cells over which φ is satisfied. To decide Φ, we need only to decide whether or not Γ is satisfied over any $c \in C$.

We present a method to do so. We first develop a method to decide rationality constraints over an r-cell. We then lift the method to handle general combinations of rationality and integrality constraints.

3.1 Deciding Rationality Constraints

Given a system of rationality constraints $\Gamma_{\mathbb{Q}}$ and an r-cell c, we need a method to decide whether or not $\Gamma_{\mathbb{Q}}$ is satisfied over c. To accomplish this, we will extract a system of *degree constraints* from $\Gamma_{\mathbb{Q}}$ and give a method to decide if c contains a real algebraic number satisfying them.

We must however take care of the following issue: If we prove there exists no algebraic real in c satisfying $\Gamma_{\mathbb{Q}}$, how do we know there exists no *transcendental* real in c satisfying $\Gamma_{\mathbb{Q}}$ as well? That is, in the presence of rationality constraints, can we still transfer results from $\mathbb{R}_{alg}$ to $\mathbb{R}$ as a whole? We answer this question in the affirmative by proving a suitable transfer principle (cf. Theorem 2).

It turns out we need essentially two methods for deciding $\Gamma_{\mathbb{Q}}$ over c: One method for 0-cells and another for 1-cells. We begin with the 1-cell case.

1-Cells. To construct our system of degree constraints, we shall utilise a fundamental property relating the degree of a "binomial root" real algebraic number to the rationality of its powers. We employ a result on the density of real algebraic numbers to show that any consistent system of degree constraints gives rise to a real algebraic solution in a 1-cell. We then prove completeness of the method and a transfer principle enabling us to lift results from $\mathbb{R}_{alg}$ to $\mathbb{R}$.

Lemma 1 (Minimal binomials). *Let $\alpha \in \mathbb{R}_{alg}$ s.t. $\alpha^n \in \mathbb{Q}$ for some $n \in \mathbb{N}$. Then, the minimal polynomial for α over $\mathbb{Q}[x]$ is a binomial of the form $x^d - q$.*

Proof. Let $k \in \mathbb{N}$ be the least power s.t. $\alpha^k \in \mathbb{Q}$. We shall prove that $p(x) = x^k - \alpha^k \in \mathbb{Q}[x]$ is the minimal polynomial for α. Assume $p(x)$ is reducible over $\mathbb{Q}[x]$. Observe that $p(x) = \prod_{i=1}^{k}(x - \alpha\zeta^i)$ where ζ is a kth root of unity. As $p(x)$ is reducible, it must have a nontrivial factor $f(x) = \prod_{i=1}^{m}(x - \alpha\zeta^{s_i}) \in \mathbb{Q}[x]$ with $m < k$ and $s_i \in \mathbb{N}$. But then $(\alpha^m \prod_{i=1}^{m} \zeta^{s_i}) \in \mathbb{Q}$, and since α is real, we must have $\alpha^m \in \mathbb{Q}$. But $m < k$. Contradiction. Thus, as $p(x) = x^k - \alpha^k$ is irreducible and monic, it is the minimal polynomial for α over $\mathbb{Q}[x]$. □

Lemma 2 (Binomial algebraic degree and divisibility). *Let $\alpha \in \mathbb{R}_{alg}$ s.t. α is a root of some $x^k - q \in \mathbb{Q}[x]$. Let $n \in \mathbb{N}$. Then,*

$$(\alpha^n \in \mathbb{Q}) \iff deg(\alpha) \mid n.$$

Proof. Let $d = deg(\alpha)$. ($\Leftarrow$) By Lemma 1, $\alpha^d \in \mathbb{Q}$. But, as $d \mid n$, we have $\alpha^n = (\alpha^d)^k$ for some $k \in \mathbb{N}$. Thus, $\alpha^n \in \mathbb{Q}$. ($\Rightarrow$) We use the method of infinite descent. Consider $\alpha^n = q \in \mathbb{Q}$. Then, $x^n - q$ has α as a root, and thus $d \leq n$. Assume $d \nmid n$. It follows that $d < n$, $gcd(d, n) = 1$, $q = \alpha^d \alpha^{n-d}$ and $gcd(d, n - d) = 1$. As $\alpha^d \in \mathbb{Q}$, we have $\alpha^{n-d} = \frac{q}{\alpha^d} \in \mathbb{Q}$. Note $n - d < n$. But then $\alpha^{n-d} \in \mathbb{Q}$ s.t. $d \nmid n - d$, and we can continue this process ad infinitum. Contradiction. □

Let $c \subset \mathbb{R}$ be a 1-cell and $\Gamma_{\mathbb{Q}}$ a system of rationality constraints s.t.

$$\Gamma_{\mathbb{Q}} = \left[\bigwedge_{i=1}^{k_1}(x^{w_1(i)} \in \mathbb{Q}) \land \bigwedge_{i=1}^{k_2}(x^{w_2(i)} \notin \mathbb{Q}) \right].$$

To $\Gamma_\mathbb{Q}$, we associate a system of *degree constraints* $\mathcal{D}(\Gamma_\mathbb{Q})$ as follows:

$$\mathcal{D}(\Gamma_\mathbb{Q}) = \left[\bigwedge_{i=1}^{k_1} (d \mid w_1(i)) \wedge \bigwedge_{i=1}^{k_2} (d \nmid w_2(i)) \right].$$

Note that each $w_j(i)$ is a concrete natural number. Thus, $\mathcal{D}(\Gamma_\mathbb{Q})$ is a system of arithmetical constraints with a single free variable d. We shall prove that $\Gamma_\mathbb{Q}$ is satisfied over c iff $\mathcal{D}(\Gamma_\mathbb{Q})$ is consistent over $\mathbb{N}$, i.e., iff

$$\exists d \in \mathbb{N} \text{ s.t. } \mathcal{D}(\Gamma_\mathbb{Q})(d).$$

We proceed in two steps. First, we prove that $\Gamma_\mathbb{Q}$ is satisfied by a *real algebraic number* in c iff $\mathcal{D}(\Gamma_\mathbb{Q})$ is satisfied over $\mathbb{N}$. Next, we show that this result can be lifted to $\mathbb{R}$ as a whole, i.e., that $\Gamma_\mathbb{Q}$ is satisfied over c (by any real, be it algebraic or transcendental) iff $\mathcal{D}(\Gamma_\mathbb{Q})$ is satisfied over $\mathbb{N}$.

These results elucidate a deep *homogeneity* of $\mathbb{R}$. Intuitively, $\mathbb{R}$ is so saturated with real algebraic numbers that, given any open interval $I \subset \mathbb{R}$, the only way I can fail to contain an algebraic number satisfying $\Gamma_\mathbb{Q}$ is if the *purely arithmetical facts* induced by $\Gamma_\mathbb{Q}$ (via Lemma 2) are mutually inconsistent over $\mathbb{N}$. Moreover, from the perspective of rationality constraints, transcendental elements cannot be distinguished from algebraic ones. To prove these results, we shall need to understand a bit about the density of real algebraic numbers of arbitrary degree.

Lemma 3 (Density of ratios of primes). *Given $a < b \in \mathbb{R}$, there exists $\frac{p}{q} \in]a, b[$ s.t. $|p| \neq |q|$ are both prime.*

Proof. A straightforward application of the Prime Number Theorem.

Lemma 4 (Density of real algebraic numbers of degree n). *Let $a < b \in \mathbb{R}$ and $n \in \mathbb{N}$. Then, $\exists \alpha \in \mathbb{R}_{alg}$ s.t. $a < \alpha < b$ and $deg(\alpha) = n$ and $\alpha^n \in \mathbb{Q}$.*

Proof. We construct an irreducible $p(x) = x^n - q \in \mathbb{Q}[x]$ s.t. $a < \sqrt[n]{q} < b$. Then, $\alpha = \sqrt[n]{q}$ will suffice. WLOG, assume $a > 0$. Let Q be a rational in $]a, b[$. Let $f : \mathbb{R}^+ \to \mathbb{R}$ be the nth-root function, i.e., $f(r) = \sqrt[n]{r}$. Consider $Q^n \in \mathbb{Q}$. By continuity of f, $\exists \epsilon > 0$ s.t. $f(]Q^n - \epsilon, Q^n + \epsilon[) \subset]a, b[$. For each rational $q \in]Q^n - \epsilon, Q^n + \epsilon[$, we thus have $a < f(q) < b$ with $f(q)$ algebraic, as $(f(q))^n - q = 0$. To prove the theorem, we must choose q s.t. $deg(f(q)) = n$. It suffices to find $q \in]Q^n - \epsilon, Q^n + \epsilon[$ s.t. $p(x) = x^n - q$ is irreducible over $\mathbb{Q}[x]$. By Lemma 3, we can choose $q = \frac{q_1}{q_2} \in]Q^n - \epsilon, Q^n + \epsilon[$ s.t. $q_1 \neq q_2$ are both prime. By Eisenstein's criterion, $q_2 x^n - q_1$ is irreducible over $\mathbb{Q}[x]$. Thus, $x^n - \frac{q_1}{q_2}$ is irreducible and $\alpha = \sqrt[n]{\frac{q_1}{q_2}}$ completes the proof. □

With Lemma 4 in hand, it is not hard to see that $\Gamma_\mathbb{Q}$ is satisfied by a *real algebraic number* in a 1-cell c iff $\mathcal{D}(\Gamma_\mathbb{Q})$ is satisfied over $\mathbb{N}$.

Theorem 1 (1-cell arithmetical reduction: algebraic case). *Let $\Gamma_\mathbb{Q}$ be a system of rationality constraints and $c \subseteq \mathbb{R}$ a 1-cell. Then, $\Gamma_\mathbb{Q}$ is satisfiable over c by a real algebraic number iff $\mathcal{D}(\Gamma_\mathbb{Q})$ is satisfiable over $\mathbb{N}$.*

Proof. ($\Rightarrow$) *Let* $\alpha \in (c \cap \mathbb{R}_{alg})$ *satisfy* $\Gamma_{\mathbb{Q}}$. *Then, by Lemma 2,* $d = deg(\alpha)$ *satisfies* $\mathcal{D}(\Gamma_{\mathbb{Q}})$. ($\Leftarrow$) *Let* $d \in \mathbb{N}$ *satisfy* $\mathcal{D}(\Gamma_{\mathbb{Q}})$. *Then, by Lemma 2, any algebraic* $\alpha \in c$ *s.t.* $deg(\alpha) = d$ *will satisfy* $\Gamma_{\mathbb{Q}}$. *But, by Lemma 4, such an* α *must exist in* c. $\square$

Thus, we have reduced the satisfiability of $\Gamma_{\mathbb{Q}}$ by real algebraic numbers present in a 1-cell c to the satisfiability of $\mathcal{D}(\Gamma_{\mathbb{Q}})$ over $\mathbb{N}$. However, we must still attend to the possibility that $\Gamma_{\mathbb{Q}}$ could be satisfied by a *transcendental* element in c without being satisfied by an algebraic element in c. Let us now prove that this scenario is impossible. In fact, we will prove this for both the 0 and 1-dimensional cases.

Theorem 2 (Rationality constraints transfer principle). *Let* $\Gamma_{\mathbb{Q}}$ *be a system of rationality constraints and* c *an r-cell. Then, it is impossible for* $\Gamma_{\mathbb{Q}}$ *to be satisfied by a transcendental real in* c *without also being satisfied by an algebraic real in* c.

Proof. Let $\Gamma_{\mathbb{Q}} = \left[\bigwedge_{i=1}^{k_1}(x^{w_1(i)} \in \mathbb{Q}) \wedge \bigwedge_{i=1}^{k_2}(x^{w_2(i)} \notin \mathbb{Q}) \right]$. *If* c *is a 0-cell, then* c *contains no transcendental elements, so the theorem holds. Consider* c *a 1-cell. We examine the structure of* $\Gamma_{\mathbb{Q}}$. *If* $k_1 > 0$, *i.e.,* $\Gamma_{\mathbb{Q}}$ *contains at least one positive rationality constraint, then* $\Gamma_{\mathbb{Q}}$ *cannot be satisfied by any transcendental element, and the theorem holds. Thus, we are left to consider* $\Gamma_{\mathbb{Q}} = \bigwedge_{i=1}^{k_2}(x^{w_2(i)} \notin \mathbb{Q})$ *s.t.* $\Gamma_{\mathbb{Q}}$ *is satisfied by a transcendental element in* c. *Let* $m = max(w_2(1), \ldots, w_2(k_2))$. *Then,* $\Gamma_{\mathbb{Q}}$ *will be satisfied by any* $\alpha \in \mathbb{R}_{alg}$ *s.t.* $deg(\alpha) > m$. *But by Lemma 4,* c *must contain an algebraic* α *s.t.* $deg(\alpha) = m+1$.
$\square$

In addition to giving us a complete method for deciding the satisfiability of systems of rationality constraints over 1-cells, the combination of Theorem 2 and the completeness of the theory of real closed fields tells us something of a fundamental model-theoretic nature:

Corollary 1 (Transfer principle for $\mathcal{L}_{\mathbb{Q}}$). *Given* $\phi \in \mathcal{L}_{\mathbb{Q}}$,

$$\langle \mathbb{R}, +, \times, <, (x^n \in \mathbb{Q})_{n \in \mathbb{N}}, 0, 1 \rangle \models \phi \iff \langle \mathbb{R}_{alg}, +, \times, <, (x^n \in \mathbb{Q})_{n \in \mathbb{N}}, 0, 1 \rangle \models \phi.$$

That is, extending the language $\mathcal{L}$ to include rationality constraints ($\mathcal{L}_{\mathbb{Q}}$) still guarantees a sound transfer of results from $\mathbb{R}_{alg}$ to $\mathbb{R}$.

Finally, let us put the pieces together and prove our main theorem for 1-cells.

Theorem 3 (1-cell arithmetical reduction: general case) *Let* $\Gamma_{\mathbb{Q}}$ *be a system of rationality constraints and* $c \subseteq \mathbb{R}$ *a 1-cell. Then,* $\Gamma_{\mathbb{Q}}$ *is satisfiable over* c *iff* $\mathcal{D}(\Gamma_{\mathbb{Q}})$ *is satisfiable over* $\mathbb{N}$.

Proof. Immediate by Theorems 1 and 2. $\square$

Thus, to decide if $\Gamma_{\mathbb{Q}}$ is satisfied over a 1-cell c, we need only check the consistency of $\mathcal{D}(\Gamma_{\mathbb{Q}})$ over $\mathbb{N}$. It is easy to derive an algorithm for doing so. Consider $\mathcal{D}(\Gamma_{\mathbb{Q}})$ s.t.

$$\mathcal{D}(\Gamma_{\mathbb{Q}}) = \left[\bigwedge_{i=1}^{k_1}(d \mid w_1(i)) \wedge \bigwedge_{i=1}^{k_2}(d \nmid w_2(i)) \right].$$

If $k_1 = 0$, then $d = max(w_2(1), \ldots, w_2(k_2)) + 1$ satisfies $\mathcal{D}(\Gamma_{\mathbb{Q}})$. If $k_2 = 0$, then $d = 1$ satisfies $\mathcal{D}(\Gamma_{\mathbb{Q}})$. Finally, if $k_1 > 0$ and $k_2 > 0$, then $m = min(w_1(1), \ldots, w_1(k_1))$ gives us an upper bound on all d satisfying $\mathcal{D}(\Gamma_{\mathbb{Q}})$. Thus, we need only search for such a d from 1 to m. For efficiency, we can augment this bounded search by various cheap sufficient conditions for recognising inconsistencies in $\mathcal{D}(\Gamma_{\mathbb{Q}})$.

0-Cells. When deciding rationality constraints over r-cells of the form $[\alpha]$, we will need to decide, when given some $j \in \mathbb{N}$, whether or not $\alpha^j \in \mathbb{Q}$. Recall that a root-triple for α^j can be computed from a root-triple for α (cf. Sect. 2). A key component for deciding a system of rationality constraints over a 0-cell is then an algorithm for deciding whether or not a given real algebraic number $\beta = \alpha^j$ is rational. Naively, one might try to solve this problem in the following way:

> Given β presented as a root-triple $\langle p \in \mathbb{Z}[x], l, u \rangle$, fully factor p over $\mathbb{Q}[x]$. Then, $\beta \in \mathbb{Q}$ iff the factorisation of p contains a linear factor of the form $(x - q)$ with $q \in [l, u]$.

From the perspective of theorem proving, the problem with this approach is that it is difficult in general to establish the "completeness" of a factorisation. While it is easy to verify that the product of a collection of factors equals the original polynomial, it can be very challenging (without direct appeal to the functional correctness of an implemented factorisation algorithm) to prove that a given polynomial is irreducible, i.e., that it cannot be factored any further. Indeed, deep results in algebraic number theory are used even to classify the irreducible factors of binomials [8]. Moreover, univariate factorisation can be computationally expensive, especially when one is only after rational roots.

We would like the steps in our proofs to be as clear and obvious as possible, and to minimise the burden of formalising our procedure as a tactic in a proof assistant. Thus, we shall go a different route. To decide whether or not a given α is rational, we apply a simple but powerful result from high school mathematics:

Theorem 4 (Rational roots). Let $p(x) = \sum_{i=0}^{n} a_n x^n \in \mathbb{Z}[x] \setminus \{0\}$. If $\frac{a}{b} \in \mathbb{Q}$ s.t. $p(q) = 0$ and $gcd(a, b) = 1$, then $a \mid a_0$ and $b \mid a_n$.

Proof. A straightforward application of Gauss's lemma.

Given Theorem 4, we can decide the rationality of α simply by enumerating potential rational roots $q_1, \ldots, q_k$ and checking by evaluation whether any q_i satisfies $(l \leq q_i \leq r \wedge p(q_i) = 0)$. Then, to decide whether α satisfies a given system of rationality constraints, e.g., $\Gamma_{\mathbb{Q}} = \left[(x^2 \in \mathbb{Q}) \wedge (x \notin \mathbb{Q}) \right]$, we first compute a root-triple representation for α^2 and then test α and α^2 for rationality as described. This process clearly always terminates. To make this more efficient when faced with many potential rational roots, we can combine (i) dividing our polynomial p by $(x - q)$ whenever q is realised to be a rational root, and (ii) various cheap irreducibility criteria over $\mathbb{Q}[x]$ for recognising when a polynomial has no linear factors over $\mathbb{Q}[x]$ and thus has no rational roots.

3.2   Deciding Integrality Constraints

Integrality Constraints Over an Unbounded 1-Cell. WLOG let $c =]\alpha, +\infty[$ with $\alpha \geq 0$. Consider $\Gamma = \Gamma_{\mathbb{Q}} \wedge \Gamma_{\mathbb{Z}}$ with

$$\Gamma_{\mathbb{Z}} = \left[\bigwedge_{i=1}^{k_3} (x^{w_3(i)} \in \mathbb{Z}) \wedge \bigwedge_{i=1}^{k_4} (x^{w_4(i)} \notin \mathbb{Z}) \right].$$

We use the notation $\phi : \Gamma$ to mean that the constraint ϕ is present as a conjunct in Γ. It is convenient to also view Γ as a set. Let $\overline{\Gamma}$ denote the closure of Γ under the following saturation rules:

1. $(x^n \notin \mathbb{Q}) : \overline{\Gamma} \rightarrow (x^n \notin \mathbb{Z}) : \overline{\Gamma}$
2. $(x^n \in \mathbb{Z}) : \overline{\Gamma} \rightarrow (x^n \in \mathbb{Q}) : \overline{\Gamma}$
3. $(x^n \in \mathbb{Z}) : \overline{\Gamma} \wedge (x^m \notin \mathbb{Z}) : \overline{\Gamma} \rightarrow (x \notin \mathbb{Q}) : \overline{\Gamma}$
4. $(x^n \in \mathbb{Z}) : \overline{\Gamma} \wedge (x^m \in \mathbb{Q}) : \overline{\Gamma} \rightarrow (x^m \in \mathbb{Z}) : \overline{\Gamma}$
5. $(x^n \in \mathbb{Z}) : \overline{\Gamma} \wedge (x^m \notin \mathbb{Z}) : \overline{\Gamma} \rightarrow (x^m \notin \mathbb{Q}) : \overline{\Gamma}$

This saturation process is clearly finite. The soundness of rules 1 and 2 is obvious. The soundness of rules 3-5 is easily verified by the following lemmata.

Lemma 5 (Soundness: rule 3). $(x^n \in \mathbb{Z}) \wedge (x^m \notin \mathbb{Z}) \rightarrow (x \notin \mathbb{Q})$

Proof. Since $x^m \notin \mathbb{Z}$, we know $x \notin \mathbb{Z}$. Suppose $x \in \mathbb{Q}$. Then $x = \frac{a}{b}$ s.t. $gcd(a, b) = 1$. Thus, $a^n = x^n b^n$. Thus, $b \mid a$. Recall $gcd(a, b) = 1$. So, $b = 1$. But then $x = a \in \mathbb{Z}$. Contradiction.   □

Lemma 6 (Soundness: rule 4). $(x^n \in \mathbb{Z}) \wedge (x^m \in \mathbb{Q}) \rightarrow (x^m \in \mathbb{Z})$

Proof. Let $d = deg(x)$. By Lemma 2, $d \mid n$ and $d \mid m$. If $d = n$, then $x^m = (x^n)^k$ for some $k \in \mathbb{N}$ and thus $x^m \in \mathbb{Z}$. Otherwise, $d < n$. Let $x^d = \frac{a}{b} \in \mathbb{Q}$ s.t. $gcd(a, b) = 1$. Thus, $x^n = (x^d)^k = \frac{a^k}{b^k} \in \mathbb{Z}$ for some $k \in \mathbb{N}$. But then $b = 1$, and thus $x^d \in \mathbb{Z}$. So, as $d \mid m$, $x^m \in \mathbb{Z}$ as well.   □

Lemma 7 (Soundness: rule 5). $(x^n \in \mathbb{Z}) \wedge (x^m \notin \mathbb{Z}) \rightarrow (x^m \notin \mathbb{Q})$

Proof. Assume $(x^n \in \mathbb{Z})$ and $(x^m \notin \mathbb{Z})$ but $(x^m \in \mathbb{Q})$. But then $(x^m \in \mathbb{Z})$ by rule 4. Contradiction.   □

Let us now prove that these rules[1] are *complete* for deciding the satisfiability of systems of rationality and integrality constraints over unbounded 1-cells. Let $\overline{\Gamma}_{\mathbb{Q}}$ (resp. $\overline{\Gamma}_{\mathbb{Z}}$) denote the collection of rationality (resp. integrality) constraints present in $\overline{\Gamma}$. Intuitively, we shall exploit the following observation: The construction of $\overline{\Gamma}$ projects all information pertaining to the *consistency* of the combined rationality and integrality constraints of Γ onto $\overline{\Gamma}_{\mathbb{Q}}$. Then, if $\overline{\Gamma}_{\mathbb{Q}}$ is consistent, i.e., $\exists d \in \mathbb{N}$ satisfying $\mathcal{D}(\overline{\Gamma}_{\mathbb{Q}})$, this will impose a strict correspondence between $\overline{\Gamma}_{\mathbb{Q}}$ and $\overline{\Gamma}_{\mathbb{Z}}$. From this correspondence and a least d witnessing $\mathcal{D}(\overline{\Gamma}_{\mathbb{Q}})$, we can construct an algebraic real satisfying Γ.

---

[1] In fact, the completeness proof shows that rule 3 is logically unnecessary. Nevertheless, we find its inclusion in the saturation process useful in practice.

Lemma 8 ($\overline{\Gamma}_{\mathbb{Q}}$-$\overline{\Gamma}_{\mathbb{Z}}$ correspondence). *If $\Gamma_{\mathbb{Z}}$ contains at least one positive integrality constraint, then*

$$\forall m \in \mathbb{N} \left[(x^m \in \mathbb{Q}) : \overline{\Gamma} \iff (x^m \in \mathbb{Z}) : \overline{\Gamma} \right]$$

and

$$\forall m \in \mathbb{N} \left[(x^m \notin \mathbb{Q}) : \overline{\Gamma} \iff (x^m \notin \mathbb{Z}) : \overline{\Gamma} \right].$$

Proof. Let us call the first conjunct A and the second B. ($A \Rightarrow$) As $\Gamma_{\mathbb{Z}}$ contains at least one positive integrality constraint, rule 4 guarantees $(x^m \in \mathbb{Z}) : \overline{\Gamma}$. ($A \Leftarrow$) Immediate by rule 2. ($B \Rightarrow$) Immediate by rule 1. ($B \Leftarrow$) As $\Gamma_{\mathbb{Z}}$ contains at least one positive integrality constraint, rule 5 guarantees $(x^m \notin \mathbb{Q}) : \overline{\Gamma}$. □

Theorem 5 (Completeness of Γ-saturation method). *Let $\Gamma = \Gamma_{\mathbb{Q}} \wedge \Gamma_{\mathbb{Z}}$ be a system of rationality and integrality constraints, and $c \subseteq \mathbb{R}$ an unbounded 1-cell. Then, $\mathcal{D}(\overline{\Gamma}_{\mathbb{Q}})$ is consistent over $\mathbb{N}$ iff Γ is consistent over c.*

Proof. ($\Leftarrow$) Immediate by Theorem 3 and the soundness of our saturation rules. ($\Rightarrow$) We proceed by cases.

[Case 1: Γ contains no positive rationality constraint]: Then, by Lemma 8 and the consistency of $\mathcal{D}(\overline{\Gamma}_{\mathbb{Q}})$, $\Gamma_{\mathbb{Z}}$ must contain no positive integrality constraints. But then it is consistent with Γ that every power of x listed in Γ be irrational. Let $k \in \mathbb{N}$ be the largest power s.t. x^k appears in a constraint in Γ. Then, by Lemma 2, any $\alpha \in c$ s.t. $\deg(\alpha) > k$ will satisfy Γ. By Lemma 4, we can always find such an α in c, e.g., we can select $\alpha \in c$ s.t. $\deg(\alpha) = k+1$.

[Case 2: Γ contains a positive rationality constraint but no positive integrality constraints]: By the consistency of $\mathcal{D}(\overline{\Gamma}_{\mathbb{Q}})$, it is consistent with Γ for every power of x listed in Γ to be non-integral. Let $d \in \mathbb{N}$ be the least natural number satisfying $\mathcal{D}(\Gamma_{\mathbb{Q}})$. Then, we can satisfy Γ with an α s.t. $\deg(\alpha) = d$ with $\alpha^{dk} \in (\mathbb{Q} \setminus \mathbb{Z})$ for each x^{dk} appearing in a constraint in Γ. By Lemma 4, we know such an α is present in c of the form $\alpha = \sqrt[d]{\frac{p}{q}}$ for primes $p \neq q$.

[Case 3: Γ contains both positive rationality and integrality constraints] By Lemma 8, the rows of $\overline{\Gamma}_{\mathbb{Q}}$ and $\overline{\Gamma}_{\mathbb{Z}}$ are in perfect correspondence. Let $d \in \mathbb{N}$ be the least natural number satisfying $\mathcal{D}(\overline{\Gamma}_{\mathbb{Q}})$. Since $\overline{\Gamma}_{\mathbb{Q}}$ is consistent, we can satisfy Γ by finding an $\alpha \in c$ s.t. $\alpha^{dk} \in \mathbb{Z}$ for every x^{dk} appearing in a constraint in Γ. Recall c is unbounded towards $+\infty$. Thus, c contains infinitely many primes p s.t. $\sqrt[d]{p} \in c$. Let $p \in c$ be such a prime. Then, $x^d - p \in \mathbb{Q}[x]$ is irreducible by Eisenstein's criterion. Thus, $\sqrt[d]{p} \in c$ and satisfies Γ. □

Integrality Constraints Over a Bounded 1-Cell. Let us now consider the satisfiability of $\Gamma = \Gamma_{\mathbb{Q}} \wedge \Gamma_{\mathbb{Z}}$ over a bounded 1-cell $c \subset \mathbb{R}$. Given the results of the last section, it is easy to see that if $\mathcal{D}(\overline{\Gamma}_{\mathbb{Q}})$ is unsatisfiable over $\mathbb{N}$, then Γ is unsatisfiable over c. However, as Γ is bounded on both sides, it is possible for $\mathcal{D}(\overline{\Gamma}_{\mathbb{Q}})$ to be satisfiable over $\mathbb{N}$ while Γ is unsatisfiable over c. That is, provided $\mathcal{D}(\overline{\Gamma}_{\mathbb{Q}})$ is consistent over $\mathbb{N}$, we must find a way to determine if c actually contains some α s.t. $\Gamma(\alpha)$ holds. Afterall, even with $\overline{\Gamma}_{\mathbb{Q}}$ satisfied over c, it is possible that c itself is not "wide enough" to satisfy the integrality constraints $\overline{\Gamma}_{\mathbb{Z}}$.

WLOG, let $c =]\alpha, \beta[$ s.t. $0 \leq \alpha < \beta \in \mathbb{R}_{alg}$. Let $\mathcal{D}(\overline{\Gamma}_{\mathbb{Q}})$ be satisfied by $d \in \mathbb{N}$. If Γ contains no positive integrality constraints, then we can reason as we did in the proof of Theorem 5 to show Γ is satisfied over c. The difficulty arises when a positive constraint $(x^k \in \mathbb{Z})$ appears in $\Gamma_{\mathbb{Z}}$. We can solve this case as follows.

Theorem 6 (Satisfiability over a bounded 1-cell). *Let $\Gamma_{\mathbb{Z}}$ contain at least one positive integrality constraint. Let $\mathcal{D}(\overline{\Gamma}_{\mathbb{Q}})$ be satisfiable over $\mathbb{N}$ with $d \in \mathbb{N}$ the least witness. Let $c =]\alpha, \beta[$ s.t. $0 \leq \alpha < \beta \in \mathbb{R}_{alg}$. Then, Γ is satisfiable over c iff $\exists z \in \left(]\alpha^d, \beta^d[\cap\mathbb{Z}\right)$ s.t. $x^d - z \in \mathbb{Z}[x]$ is irreducible over $\mathbb{Q}[x]$.*

Proof. $(\Rightarrow)$ *Assume Γ is satisfied by $\alpha \in c$. Then, by soundness of $\overline{\Gamma}$ saturation, $\overline{\Gamma}$ is satisfied by α as well. By Lemma 8, $(x^d \in \mathbb{Z}) : \overline{\Gamma}$. Moreover, d is the least natural number with this property. As $0 \leq \alpha < \beta$, $\{r^d \mid r \in c\} =]\alpha^d, \beta^d[$. Thus, as Γ is satisfied by $\alpha \in c$, there must exist an integer $z \in]\alpha^d, \beta^d[$ s.t. $\deg(\sqrt[d]{z}) = d$. But then by uniqueness of minimal polynomials, $x^d - z$ is irreducible over $\mathbb{Q}[x]$. $(\Leftarrow)$ Assume $z \in \left(]\alpha^d, \beta^d[\cap\mathbb{Z}\right)$ s.t. $x^d - z$ is irreducible over $\mathbb{Q}[x]$. Let $\gamma = \sqrt[d]{z}$ and note that $\gamma \in]\alpha, \beta[$. By Lemma 2, $\deg(\gamma) = d$. Thus, $\overline{\Gamma}_{\mathbb{Q}}$ is satisfied by γ. As $\gamma^d \in \mathbb{Z}$, it follows by Lemma 8 that Γ is satisfied by γ as well.* □

By Eisenstein's criterion, we obtain a useful corollary.

Corollary 2. *Let $\mathcal{D}(\overline{\Gamma}_{\mathbb{Q}})$ be satisfiable with $d \in \mathbb{N}$ the least natural number witness. Let $c =]\alpha, \beta[$ s.t. $0 \leq \alpha < \beta \in \mathbb{R}_{alg}$. Then, Γ is satisfiable over c if $\exists p \in]\alpha^d, \beta^d[$ s.t. p is prime.*

These results give us a simple algorithm to decide satisfiability of Γ over c: If $\mathcal{D}(\overline{\Gamma}_{\mathbb{Q}})$ is unsatisfiable over $\mathbb{N}$, then Γ is unsatisfiable. Otherwise, let $d \in \mathbb{N}$ be the minimal solution to $\mathcal{D}(\overline{\Gamma}_{\mathbb{Q}})$. Gather all integers $\{z_1, \ldots, z_k\}$ in $I =]\alpha^d, \beta^d[$. If any z_i is prime, Γ is satisfied over c. Otherwise, for each z_i, form the real algebraic number $\sqrt[d]{z_i}$ and check by evaluation if it satisfies Γ. By Theorem 6, Γ is satisfiable over c iff one of the $\sqrt[d]{z_i} \in c$ satisfies this process.

Integrality Constraints Over a 0-Cell. Finally, we consider the case of $\Gamma = \Gamma_{\mathbb{Q}} \wedge \Gamma_{\mathbb{Z}}$ over a 0-cell $[\alpha]$. Clearly, Γ is satisfied over c iff Γ is satisfied at α. By the soundness of Γ-saturation, if $\mathcal{D}(\overline{\Gamma}_{\mathbb{Q}})$ is unsatisfiable over $\mathbb{N}$, then Γ is unsatisfiable over c. Thus, we first form $\overline{\Gamma}$ and check satisfiability of $\mathcal{D}(\overline{\Gamma}_{\mathbb{Q}})$ over $\mathbb{N}$. Provided it is satisfiable, we then check $\Gamma(x \mapsto \alpha)$ by evaluation.

4   Examples

We have implemented[2] our decision method in a special version of the Meti-Tarski theorem prover [15]. We do not use any of the proof search mechanisms of MetiTarski, but rather its parsing and first-order formula data structures.

---

[2] The implementation of our procedure, including computations over r-cells, Γ-saturation and the proof output routines can be found in the RCF/ modules in the MetiTarski source code at http://metitarski.googlecode.com/.

In the examples that follow, all output (including the prose and LaTeX formatting) has been generated automatically by our implementation of the method.

4.1 Example 1

Let us decide $\exists x(\varphi(x) \wedge \Gamma(x))$, where

$$\varphi = (x^2 - 2 = 0) \quad \text{and} \quad \Gamma = (x \in \mathbb{Q}).$$

We first compute $\overline{\Gamma}$, the closure of Γ under the saturation rules:

$$\overline{\Gamma} = (x \in \mathbb{Q}).$$

Observe $\mathcal{D}(\overline{\Gamma}_{\mathbb{Q}})$ is satisfied (minimally) by $d = 1$.

We next compute an r-cell decomposition of $\mathbb{R}$ induced by φ, yielding:

1. $]-\infty, Root(x^2 - 2, [-2, -1/3])[$,
2. $[Root(x^2 - 2, [-2, -1/3])]$,
3. $]Root(x^2 - 2, [-2, -1/3]), 0[$,
4. $[0]$,
5. $]0, Root(x^2 - 2, [1/3, 2])[$,
6. $[Root(x^2 - 2, [1/3, 2])]$,
7. $]Root(x^2 - 2, [1/3, 2]), +\infty[$.

By IVT, φ has constant truth value over each such r-cell. Only two r-cells in the decomposition satisfy φ:

$$[Root(x^2 - 2, [-2, -1/3])], [Root(x^2 - 2, [1/3, 2])].$$

Let us now see if any of these r-cells satisfy Γ.

1. We check if $[Root(x^2 - 2, [-2, -1/3])]$ satisfies Γ.
   (a) Evaluating $(\alpha \in \mathbb{Q})$ for $\alpha = Root(x^2 - 2, [-2, -1/3])$. We shall determine the numerical type of α. Let $p(x) = x^2 - 2$. By RRT and the root interval, we reduce the set of possible rational values for α to $\{-1, -2\}$. But none of these are roots of $p(x)$. Thus, $\alpha \in (\mathbb{R} \setminus \mathbb{Q})$.
   So, the r-cell does not satisfy Γ.
2. We check if $[Root(x^2 - 2, [1/3, 2])]$ satisfies Γ.
   (b) Evaluating $(\alpha \in \mathbb{Q})$ for $\alpha = Root(x^2 - 2, [1/3, 2])$. We shall determine the numerical type of α. Let $p(x) = x^2 - 2$. By RRT and the root interval, we reduce the set of possible rational values for α to $\{1, 2\}$. But none of these are roots of $p(x)$. Thus, $\alpha \in (\mathbb{R} \setminus \mathbb{Q})$.
   So, the r-cell does not satisfy Γ.

Thus, as all r-cells have been ruled out, the conjecture is false.   □

4.2   Example 2

Let us decide $\exists x(\varphi(x) \wedge \Gamma(x))$, where

$$\varphi = True \ \text{ and } \ \Gamma = (x^3 \in \mathbb{Z}) \wedge (x^5 \notin \mathbb{Z}) \wedge (x \in \mathbb{Q}).$$

We first compute $\overline{\Gamma}$, the closure of Γ under the saturation rules:

$$\overline{\Gamma} = (x \notin \mathbb{Z}) \wedge (x \in \mathbb{Z}) \wedge (x \notin \mathbb{Q}) \wedge (x \in \mathbb{Q}) \wedge (x^3 \in \mathbb{Q}) \wedge (x^3 \in \mathbb{Z}) \wedge (x^5 \notin \mathbb{Q}) \wedge (x^5 \notin \mathbb{Z}).$$

But, $\overline{\Gamma}$ is obviously inconsistent. Thus, the conjecture is false.   □

4.3   Example 3

Let us decide $\exists x(\varphi(x) \wedge \Gamma(x))$, where

$$\varphi = ((x^3 - 7 > 3) \wedge (x^2 + x + 1 < 50)) \ \text{ and } \ \Gamma = (x^2 \notin \mathbb{Q}) \wedge (x^3 \in \mathbb{Z}).$$

We first compute $\overline{\Gamma}$, the closure of Γ under the saturation rules:

$$\overline{\Gamma} = (x^2 \notin \mathbb{Q}) \wedge (x^2 \notin \mathbb{Z}) \wedge (x^3 \in \mathbb{Z}) \wedge (x^3 \in \mathbb{Q}).$$

Observe $\mathcal{D}(\overline{\Gamma}_{\mathbb{Q}})$ is satisfied (minimally) by $d = 3$.

We next compute an r-cell decomposition of $\mathbb{R}$ induced by φ, yielding:

1. $]-\infty, Root(x^2 + x - 49, [-8, -1/50])[$,
2. $[Root(x^2 + x - 49, [-8, -1/50])]$,
3. $]Root(x^2 + x - 49, [-8, -1/50]), 0[$,
4. $[0]$,
5. $]0, Root(x^3 - 10, [57/44, 5/2])[$,
6. $[Root(x^3 - 10, [57/44, 5/2])]$,
7. $]Root(x^3 - 10, [57/44, 5/2]), Root(x^2 + x - 49, [401/100, 8])[$,
8. $[Root(x^2 + x - 49, [401/100, 8])]$,
9. $]Root(x^2 + x - 49, [401/100, 8]), +\infty[$.

By IVT, φ has constant truth value over each such r-cell. Only one r-cell in the decomposition satisfies φ:

$$]Root(x^3 - 10, [57/44, 5/2]), Root(x^2 + x - 49, [401/100, 8])[.$$

Let us now see if any of these r-cells satisfy Γ.

1. We check if $]Root(x^3 - 10, [57/44, 5/2]), Root(x^2 + x - 49, [401/100, 8])[$ satisfies Γ. Call the boundaries of this r-cell L and U. As Γ contains a positive integrality constraint and $d = 3$, any satisfying witness in this r-cell must be of the form $\sqrt[3]{z}$ for z an integer in $]L^3, U^3[$. The set of integers in question is $Z = \{z \in \mathbb{Z} \mid 11 \leq z \leq 276\}$, containing 266 members. We shall examine $\sqrt[3]{z}$ for each $z \in Z$ in turn.
   (a) Evaluating $(\alpha^2 \notin \mathbb{Q})$ for $\alpha = Root(x^3 - 11, [1/12, 11])$. Observe $\alpha^2 = Root(x^3 - 121, [1/144, 121])$. We shall determine the numerical type of α^2. Let $p(x) = x^3 - 121$. By RRT and the root interval, we reduce the set of possible rational values for α^2 to $\{1, 11, 121\}$. But none of these are roots of $p(x)$. Thus, $\alpha^2 \in (\mathbb{R} \setminus \mathbb{Q})$.

(b) Evaluating $(\alpha^3 \in \mathbb{Z})$ for $\alpha = Root(x^3 - 11, [1/12, 11])$. Observe $\alpha^3 = Root(x^3 - 1331, [1/1728, 1331])$. We shall determine the numerical type of α^3. Let $p(x) = x^3 - 1331$. By RRT and the root interval, we reduce the set of possible rational values for α^3 to $\{1, 11, 121, 1331\}$. Thus, we see $\alpha^3 = 11 \in \mathbb{Z}$.

Witness found: $Root(x^3 - 11, [1/12, 11])$. So, the r-cell does satisfy Γ.

Thus, the conjecture is true. $\square$

5   Discussion and Related Work

Let us describe some related results that help put our work into context.

- The existence of rational or integer solutions to univariate polynomial equations over $\mathbb{Q}[x]$ has long been known to be decidable. The best known algorithms are based on univariate factorisation via lattice reduction [7].
- Due to Weispfenning, the theory of linear, multivariate mixed real-integer arithmetic is known to be decidable and admit quantifier elimination [20].
- Due to van den Dries, the theory of real closed fields extended with a predicate for powers of two is known to be decidable [5]. Avigad and Yin have given a syntactic decidability proof for this theory, establishing a non-elementary upper bound for eliminating a block of quantifiers [2].
- Due to Davis, Putnam, Robinson and Matiyasevich, the $\exists^3$ nonlinear, equational theories of arithmetic over $\mathbb{N}$ and $\mathbb{Z}$ are known to be undecidable ("Hilbert's Tenth Problem" and reductions of its negative solution) [11].
- The decidability of the $\exists^2$ nonlinear, equational theories of arithmetic over $\mathbb{N}$ and $\mathbb{Z}$ is open.
- Due to Poonen, the $\forall^2\exists^7$ theory of nonlinear arithmetic over $\mathbb{Q}$ is known to be undecidable [16]. This is an improvement of Julia Robinson's original undecidability proof of $Th(\mathbb{Q})$ via a $\forall^2\exists^7\forall^6$ definition of $\mathbb{Z}$ over $\mathbb{Q}$ [18].
- Due to Koenigsmann, the $\forall^{418}$ and $\forall^1\exists^{1109}$ theories of nonlinear arithmetic over $\mathbb{Q}$ are known to be undecidable, via explicit definitions of $\mathbb{Z}$ over $\mathbb{Q}$ [9,10].
- The decidability of the $\exists^k$ equational nonlinear theory of arithmetic over $\mathbb{Q}$ is open for $k > 1$ ("Hilbert's Tenth Problem over $\mathbb{Q}$").

Our present result — the decidability of the nonlinear, univariate theory of the reals extended with predicates for rational and integer powers — fills a gap somewhere between the positive result on linear, multivariate mixed real-integer arithmetic, and the negative result for Hilbert's Tenth Problem in three variables.

Next, we would like to turn our decision method into a verified proof procedure within a proof assistant. The deepest result needed is the Prime Number Theorem (PNT). As Avigad et al. have formalised a proof of PNT within Isabelle/HOL [1], we are hopeful that a verified version of our procedure can be built in Isabelle/HOL [14] in the near future. To this end, it is useful to observe that PNT is not needed by the restriction of our method to deciding the rationality of real algebraic numbers like $\sqrt{2}$ and $\sqrt{3} + \sqrt{5}$. Thus, a simpler tactic could be constructed for this fragment.

Finally, we hope to extend the method to allow constraints of the form $(p(x) \in \mathbb{Q})$ for more general polynomials $p(x) \in \mathbb{Z}[x]$. The key difficulty lies with Lemma 2. This crucial property relating the degree of an algebraic number to the rationality of its powers applies to "binomial root" algebraic numbers, but not to algebraic numbers in general. For example, consider $\alpha = \sqrt{2} + \sqrt[4]{2}$. Then, the minimal polynomial of α over $\mathbb{Q}[x]$ is $x^4 - 4x^2 - 8x + 2$, but $\alpha^4 \notin \mathbb{Q}$. Thus, in the presence of richer forms of rationality and integrality constraints, our degree constraint reasoning is no longer sufficient. We expect to need more powerful tools from algebraic number theory to extend the method in this way.

6 Conclusion

We have established decidability of univariate real algebra extended with predicates for rational and integer powers. Our decision procedure combines computations over real algebraic cells with the rational root theorem and results on the density of real algebraic numbers. We have implemented the method, instrumenting it to produce readable proofs. In the future, we hope to extend our result to richer systems of rationality and integrality constraints, and to construct a verified version of the procedure within a proof assistant.

Acknowledgements. We thank Jeremy Avigad, Wenda Li, Larry Paulson, András Salamon and the anonymous referees for their helpful comments.

References

1. Avigad, J., Donnelly, K., Gray, D., Raff, P.: A formally verified proof of the prime number theorem. ACM Trans. Comp. Logic vol. 9(1), Article No. 2 (2007)
2. Avigad, J., Yin, Y.: Quantifier elimination for the reals with a predicate for the powers of two. Theor. Comput. Sci. **370**, 1–3 (2007)
3. Basu, S., Pollack, R., Roy, M.F.: Algorithms in Real Algebraic Geometry. Springer, Secaucus (2006)
4. Collins, G.E., Akritas, A.G.: Polynomial real root isolation using Descarte's rule of signs. In: ACM Symposium on Symbolic and Algebraic computation. ACM (1976)
5. van den Dries, L.: The field of reals with a predicate for the powers of two. Manuscr. Math. **54**(1–2), 187–195 (1985)
6. Hirvensalo, M., Karhumäki, J., Rabinovich, A.: Computing partial information out of intractable: powers of algebraic numbers as an example. J. Number Theor. **130**(2), 232–253 (2010)
7. van Hoeij, M.: Factoring polynomials and the knapsack problem. J. Number Theor. **95**(2), 167–189 (2002)
8. Hollmann, H.: Factorisation of $x^n - q$ over Q. Acta Arith. **45**(4), 329–335 (1986)
9. Koenigsmann, J.: Defining $\mathbb{Z}$ in $\mathbb{Q}$. Annals of Mathematics, To appear (2015)
10. Koenigsmann, J.: Personal communication (2015)
11. Matiyasevich, Y.: Hilbert's Tenth Problem. MIT Press, Cambridge (1993)
12. Mishra, B.: Algorithmic Algebra. Springer, New York (1993)

13. de Moura, L., Passmore, G.O.: Computation in real closed infinitesimal and transcendental extensions of the rationals. In: Bonacina, M.P. (ed.) CADE 2013. LNCS, vol. 7898, pp. 178–192. Springer, Heidelberg (2013)
14. Paulson, L.C.: Isabelle: A Generic Theorem Prover, vol. 828. Springer, New York (1994)
15. Paulson, L.C.: MetiTarski: past and future. In: Beringer, L., Felty, A. (eds.) ITP 2012. LNCS, vol. 7406, pp. 1–10. Springer, Heidelberg (2012)
16. Poonen, B.: Characterizing integers among rational numbers with a universal-existential formula. Am. J. Math. **131**(3), 675–682 (2009)
17. Rioboo, R.: Towards faster real algebraic numbers. J. Sym. Comp. **36**(3–4), 513–533 (2003)
18. Robinson, J.: Definability and Decision Problems in Arithmetic. Ph.D. thesis, University of California, Berkeley (1948)
19. Uspensky, J.V.: Theory of Equations. McGraw-Hill, New York (1948)
20. Weispfenning, V.: Mixed real-integer linear quantifier elimination. In: ISSAC 1999, New York, NY, USA (1999)
21. Wiedijk, F.: The Seventeen Provers of the World. Springer, New York (2006)

A Decision Procedure for (Co)datatypes in SMT Solvers

Andrew Reynolds[1]($\boxtimes$) and Jasmin Christian Blanchette[2,3]

[1] École Polytechnique Fédérale de Lausanne (EPFL), Lausanne, Switzerland
andrew.j.reynolds@gmail.com
[2] Inria Nancy and LORIA, Villers-lès-Nancy, France
jasmin.blanchette@inria.fr
[3] Max-Planck-Institut für Informatik, Saarbrücken, Germany

Abstract. We present a decision procedure that combines reasoning about datatypes and codatatypes. The dual of the acyclicity rule for datatypes is a uniqueness rule that identifies observationally equal codatatype values, including cyclic values. The procedure decides universal problems and is composable via the Nelson–Oppen method. It has been implemented in CVC4, a state-of-the-art SMT solver. An evaluation based on problems generated from theories developed with Isabelle demonstrates the potential of the procedure.

1 Introduction

Freely generated algebraic datatypes are ubiquitous in functional programs and formal specifications. They are especially useful to represent finite data structures in computer science applications but also arise in formalized mathematics. They can be implemented efficiently and enjoy properties that can be exploited in automated reasoners.

To represent infinite objects, a natural choice is to turn to coalgebraic datatypes, or *codatatypes*, the non-well-founded dual of algebraic *datatypes*. Despite their reputation for being esoteric, codatatypes have a role to play in computer science. The verified C compiler CompCert [13], the verified Java compiler JinjaThreads [14], and the formalized Java memory model [15] all depend on codatatypes to capture infinite processes.

Codatatypes are freely generated by their constructors, but in contrast with datatypes, infinite constructor terms are also legitimate values for codatatypes (Sect. 2). Intuitively, the values of a codatatype consist of all well-typed finite and infinite ground constructor terms, and only those. As a simple example, the coalgebraic specification

$$\textbf{codatatype } enat = \mathsf{Z} \mid \mathsf{S}(enat)$$

introduces a type that models the natural numbers Z, $\mathsf{S}(\mathsf{Z})$, $\mathsf{S}(\mathsf{S}(\mathsf{Z}))$, ..., in Peano notation but extended with an infinite value $\infty = \mathsf{S}(\mathsf{S}(\mathsf{S}(\ldots)))$. The equation $\mathsf{S}(\infty) \approx \infty$ holds as expected, because both sides expand to the infinite

---

In memoriam Morgan Deters 1979–2015.

© Springer International Publishing Switzerland 2015
A.P. Felty and A. Middeldorp (Eds.): CADE-25, LNAI 9195, pp. 197–213, 2015.
DOI: 10.1007/978-3-319-21401-6_13

term $S(S(S(\ldots)))$, which uniquely identifies ∞. Compared with the conventional definition **datatype** $enat = Z \mid S(enat) \mid \mathsf{Infty}$, the codatatype avoids one case by unifying the infinite and finite nonzero cases.

Datatypes and codatatypes are an integral part of many proof assistants, including Agda, Coq, Isabelle, Matita, and PVS. In recent years, datatypes have emerged in a few automatic theorem provers as well. The SMT-LIB format, implemented by most SMT solvers, has been extended with a syntax for datatypes. In this paper, we introduce a unified decision procedure for universal problems involving datatypes and codatatypes in combination (Sect. 3). The procedure is described abstractly as a calculus and is composable via the Nelson–Oppen method [18]. It generalizes the procedure by Barrett et al. [2], which covers only datatypes. Detailed proofs are included in a report [20].

Datatypes and codatatypes share many properties, so it makes sense to consider them together. There are, however, at least three important differences. First, *codatatypes need not be well-founded*. For example, the type **codatatype** $stream_\tau = SCons(\tau, stream_\tau)$ of infinite sequences or streams over an element type τ is allowed, even though it has no base case. Second, *a uniqueness rule takes the place of the acyclicity rule of datatypes*. Cyclic constraints such as $x \approx S(x)$ are unsatisfiable for datatypes, thanks to an acyclicity rule, but satisfiable for codatatypes. For the latter, a uniqueness rule ensures that two values having the same infinite expansion must be equal; from $x \approx S(x)$ and $y \approx S(y)$, it deduces $x \approx y$. These two rules are needed to ensure completeness (solution soundness) on universal problems. They cannot be expressed as finite axiomatizations, so they naturally belong in a decision procedure. Third, *it must be possible to express cyclic (regular) values as closed terms and to enumerate them*. This is necessary both for finite model finding [22] and for theory combinations. The μ-binder notation associates a name with a subterm; it is used to represent cyclic values in the generated models. For example, the μ-term $SCons(1, \mu s.\ SCons(0, SCons(9, s)))$ stands for the lasso-shaped sequence $1, 0, 9, 0, 9, 0, 9, \ldots$.

Our procedure is implemented in the SMT solver CVC4 [1] as a combination of rewriting and a theory solver (Sect. 4). It consists of about 2000 lines of C++ code, most of which are shared between datatypes and codatatypes. The code is integrated in the development version of the solver and is expected to be part of the CVC4 1.5 release. An evaluation on problems generated from Isabelle theories using the Sledgehammer tool [3] demonstrates the usefulness of the approach (Sect. 5).

Barrett et al. [2] provide a good account of related work on datatypes as of 2007, in addition to describing their implementation in CVC3. Since then, datatypes have been added not only to CVC4 (a complete rewrite of CVC3) but also to the SMT solver Z3 [17] and a SPASS-like superposition prover [27]. Closely related are the automatic structural induction in both kinds of provers [9,21], the (co)datatype and (co)induction support in Dafny [12], and the (semi-)decision procedures for datatypes implemented in Leon [26] and RADA [19]. Datatypes are supported by the higher-order model finder Refute [28].

Its successor, Nitpick [4], can also generate models involving cyclic codatatype values. Cyclic values have been studied extensively under the heading of regular or rational trees—see Carayol and Morvan [5] and Djellou et al. [6] for recent work. The μ-notation is inspired by the μ-calculus [11].

Conventions. Our setting is a monomorphic (or many-sorted) first-order logic. A signature $\Sigma = (\mathcal{Y}, \mathcal{F})$ consists of a set of types $\mathcal{Y}$ and a set of function symbols $\mathcal{F}$. Types are atomic sorts and interpreted as nonempty domains. The set $\mathcal{Y}$ must contain a distinguished type *bool* interpreted as the set of truth values. Names starting with an uppercase letter are reserved for constructors. With each function symbol f is associated a list of argument types $\tau_1, \ldots, \tau_n$ (for $n \geq 0$) and a return type τ, written f : $\tau_1 \times \cdots \times \tau_n \to \tau$. The set $\mathcal{F}$ must at least contain true, false : *bool*, interpreted as truth values. The only predicate is equality ($\approx$). The notation t^τ stands for a term t of type τ. When applied to terms, the symbol = denotes syntactic equality.

2 (Co)datatypes

We fix a signature $\Sigma = (\mathcal{Y}, \mathcal{F})$. The types are partitioned into $\mathcal{Y} = \mathcal{Y}_{\mathrm{dt}} \uplus \mathcal{Y}_{\mathrm{codt}} \uplus \mathcal{Y}_{\mathrm{ord}}$, where $\mathcal{Y}_{\mathrm{dt}}$ are the *datatypes*, $\mathcal{Y}_{\mathrm{codt}}$ are the *codatatypes*, and $\mathcal{Y}_{\mathrm{ord}}$ are the *ordinary types* (which can be interpreted or not). The function symbols are partitioned into $\mathcal{F} = \mathcal{F}_{\mathrm{ctr}} \uplus \mathcal{F}_{\mathrm{sel}}$, where $\mathcal{F}_{\mathrm{ctr}}$ are the *constructors* and $\mathcal{F}_{\mathrm{sel}}$ are the *selectors*. There is no need to consider further function symbols because they can be abstracted away as variables when combining theories. Σ-terms are standard first-order terms over Σ, without μ-binders.

In an SMT problem, the signature is typically given by specifying first the uninterpreted types in any order, then the (co)datatypes with their constructors and selectors in groups of mutually (co)recursive groups of (co)datatypes, and finally any other function symbols. Each (co)datatype specification consists of ℓ mutually recursive types that are either all datatypes or all codatatypes. Nested (co)recursion and datatype–codatatype mixtures fall outside this fragment.

Each (co)datatype δ is equipped with $m \geq 1$ constructors, and each constructor for δ takes zero or more arguments and returns a δ value. The argument types must be either ordinary, among the already known (co)datatypes, or among the (co)datatypes being introduced. To every argument corresponds a selector. The names for the (co)datatypes, the constructors, and the selectors must be fresh. Schematically:

$$\textbf{(co)datatype } \delta_1 = \mathsf{C}_{11}(\left[\mathsf{s}_{11}^1{:}\right] \tau_{11}^1, \ldots, \left[\mathsf{s}_{11}^{n_{11}}{:}\right] \tau_{11}^{n_{11}}) \mid \cdots \mid \mathsf{C}_{1m_1}(\ldots)$$
$$\vdots$$
$$\textbf{and } \delta_\ell = \mathsf{C}_{\ell 1}(\ldots) \mid \cdots \mid \mathsf{C}_{\ell m_\ell}(\ldots)$$

with $\mathsf{C}_{ij} : \tau_{ij}^1 \times \cdots \times \tau_{ij}^{n_{ij}} \to \delta_i$ and $\mathsf{s}_{ij}^k : \delta_i \to \tau_{ij}^k$. Defaults are assumed for the selector names if they are omitted. The δ constructors and selectors are denoted by $\mathcal{F}_{\mathrm{ctr}}^\delta$ and $\mathcal{F}_{\mathrm{sel}}^\delta$. For types with several constructors, it is useful to provide discriminators $d_{ij} : \delta_i \to bool$. Instead of extending $\mathcal{F}$, we let $d_{ij}(t)$ be an abbreviation for $t \approx C_{ij}(s_{ij}^1(t), \ldots, s_{ij}^{n_{ij}}(t))$.

A type δ depends on another type ε if ε is the type of an argument to one of δ's constructors. Semantically, a set of types is *mutually (co)recursive* if and only if the associated dependency graph is strongly connected. A type is *(co)recursive* if it belongs to such a set of types. Non(co)recursive type specifications such as **datatype** $option_\tau = $ None | Some(τ) are permitted.

One way to characterize datatypes is as the initial model of the selector–constructor equations [2]. A related semantic view of datatypes is as initial algebras. Codatatypes are then defined dually as final coalgebras [24]. The datatypes are generated by their constructors, whereas the codatatypes are viewed through their selectors.

Datatypes and codatatypes share many basic properties:

$$\begin{aligned}
\text{Distinctness:} &\quad C_{ij}(\bar{x}) \not\approx C_{ij'}(\bar{y}) \quad \text{if } j \neq j' \\
\text{Injectivity:} &\quad C_{ij}(x_1, \ldots, x_{n_{ij}}) \approx C_{ij}(y_1, \ldots, y_{n_{ij}}) \longrightarrow x_k \approx y_k \\
\text{Exhaustiveness:} &\quad d_{i1}(x) \vee \cdots \vee d_{im_i}(x) \\
\text{Selection:} &\quad s_{ij}^k(C_{ij}(x_1, \ldots, x_{n_{ij}})) \approx x_k
\end{aligned}$$

Datatypes are additionally characterized by an induction principle. The principle ensures that the interpretation of datatypes is standard. For the natural numbers constructed from Z and S, induction prohibits models that contain cyclic values—e.g., an n such that $n \approx S(n)$—or even infinite acyclic values $S(S(\ldots))$.

For codatatypes, the dual notion is called coinduction. This axiom encodes a form of extensionality: Two values that yield the same observations must be equal, where the observations are made through selectors and discriminators. In addition, codatatypes are guaranteed to contain all values corresponding to infinite ground constructor terms.

Given a signature Σ, $\mathcal{DC}$ refers to the *theory of datatypes and codatatypes*, which defines a class of Σ-interpretations $\mathcal{J}$, namely the ones that satisfy the properties mentioned in this section, including (co)induction. The interpretations in $\mathcal{J}$ share the same interpretation for constructor terms and correctly applied selector terms (up to isomorphism) but may differ on variables and wrongly applied selector terms. A formula is $\mathcal{DC}$-*satisfiable* if there exists an interpretation in $\mathcal{J}$ that satisfies it. For deciding universal formulas, induction can be replaced by the acyclicity axiom schema, which states that constructor terms cannot be equal to any of their proper subterms [2]. Dually, coinduction can be replaced by the uniqueness schema, which asserts that codatatype values are fully characterized by their expansion [24, Theorem 8.1, $2 \Leftrightarrow 5$].

Some codatatypes are so degenerate as to be finite even though they have infinite values. A simple example is **codatatype** $a = $ A(a), whose unique value is $\mu a.$ A(a). Other specimens are *stream*$_{unit}$ and **codatatype** $b = $ B(b, c, b, *unit*) **and** $c = $ C(a, b, c), where *unit* is a datatype with the single constructor Unity : *unit*. We call such types *corecursive singletons*. For the decision procedure, it will be crucial to detect these. A type may also be a corecursive singleton only in some models. If the example above is altered to make *unit* an uninterpreted type, b and c will be singletons precisely when *unit* is interpreted as a singleton. Fortunately, it is easy to characterize this degenerate case.

Lemma 1. *Let δ be a corecursive codatatype. For any interpretation in $\mathcal{J}$, the domain interpreting δ is either infinite or a singleton. In the latter case, δ necessarily has a single constructor, whose arguments have types that are interpreted as singleton domains.*

3 The Decision Procedure

Given a fixed signature Σ, the decision procedure for the universal theory of (co)datatypes determines the $\mathcal{DC}$-satisfiability of finite sets E of literals: equalities and disequalities between Σ-terms, whose variables are interpreted existentially. The procedure is formulated as a tableau-like calculus. Proving a universal quantifier-free conjecture is reduced to showing that its negation is unsatisfiable. The presentation is inspired by Barrett et al. [2] but higher-level, using unoriented equations instead of oriented ones.

To simplify the presentation, we make a few assumptions about Σ. First, all codatatypes are corecursive. This is reasonable because noncorecursive codatatypes can be seen as nonrecursive datatypes. Second, all ordinary types have infinite cardinality. Without quantifiers, the constraints E cannot entail an upper bound on the cardinality of any uninterpreted type, so it is safe to consider these types infinite. As for ordinary types interpreted finitely by other theories (e.g., bit vectors), each interpreted type having finite cardinality n can be viewed as a datatype with n nullary constructors [2].

A derivation rule can be applied to E if the preconditions are met. The conclusion either specifies equalities to be added to E or is $\bot$ (contradiction). One rule has multiple conclusions, denoting branching. An application of a rule is *redundant* if one of its non-$\bot$ conclusions leaves E unchanged. A *derivation tree* is a tree whose nodes are finite sets of equalities, such that child nodes are obtained by a nonredundant application of a derivation rule to the parent. A derivation tree is *closed* if all of its leaf nodes are $\bot$. A node is *saturated* if no nonredundant instance of a rule can be applied to it.

The calculus consists of three sets of rules, given in Figs. 1 to 3, corresponding to three phases. The first phase computes the bidirectional closure of E. The second phase makes inferences based on acyclicity (for datatypes) and uniqueness (for codatatypes). The third phase performs case distinctions on constructors for various terms occurring in E. The rules belonging to a phase have priority over those of subsequent phases. The rules are applied until the derivation tree is closed or all leaf nodes are saturated.

Phase 1: Computing the Bidirectional Closure (Fig. 1). In conjunction with Refl, Sym, and Trans, the Cong rule computes the congruence (upward) closure, whereas the Inject and Clash rules compute the unification (downward) closure. For unification, equalities are inferred based on the injectivity of constructors by

Inject, and failures to unify equated terms are recognized by Clash. The Conflict rule recognizes when an equality and its negation both occur in E, in which case E has no model.

Let $\mathcal{T}(E)$ denote the set of Σ-terms occurring in E. At the end of the first phase, E induces an equivalence relation over $\mathcal{T}(E)$ such that two terms t and u are equivalent if and only if $t \approx u \in E$. Thus, we can regard E as a set of equivalence classes of terms. For a term $t \in \mathcal{T}(E)$, we write $[t]$ to denote the equivalence class of t in E.

Phase 2: Applying Acyclicity and Uniqueness (Fig. 2). We describe the rules in this phase in terms of a mapping $\mathcal{A}$ that associates with each equivalence class a μ-term as its representative.

Formally, μ-*terms* are defined recursively as being either a variable x or an applied constructor $\mu x.\, \mathsf{C}(\bar{t})$ for some $\mathsf{C} \in \mathcal{F}_{\mathrm{ctr}}$ and μ-terms $\bar{t}$ of the expected types. The variable x need not occur free in the μ-binder's body, in which case the binder can be omitted. $\mathrm{FV}(t)$ denotes the set of free variables occurring in the μ-term t. A μ-term is *closed* if it contains no free variables. It is *cyclic* if it contains a bound variable. The α-*equivalence* relation $t =_\alpha u$ indicates that the μ-terms t and u are syntactically equivalent for some capture-avoiding renaming of μ-bound variables. Two μ-terms can denote the same value despite being α-disequivalent—e.g., $\mu x.\, \mathsf{S}(x) \neq_\alpha \mu y.\, \mathsf{S}(\mathsf{S}(y))$.

The μ-term $\mathcal{A}[t^\tau]$ describes a class of τ values that t and other members of t's equivalence class can take in models of E. When τ is a datatype, a cyclic μ-term describes an infeasible class of values.

The mapping $\mathcal{A}$ is defined as follows. With each equivalence class $[u^\tau]$, we associate a fresh variable $\tilde{u}^\tau$ and set $\mathcal{A}[u] := \tilde{u}$, that is to say there are initially no constraints on the values for any equivalence class $[u]$. The mapping $\mathcal{A}$ is refined by applying the following unfolding rule exhaustively:

$$\frac{\tilde{u} \in \mathrm{FV}(\mathcal{A}) \quad \mathsf{C}(t_1,\ldots,t_n) \in [u] \quad \mathsf{C} \in \mathcal{F}_{\mathrm{ctr}}}{\mathcal{A} := \mathcal{A}[\tilde{u} \mapsto \mu\tilde{u}.\, \mathsf{C}(\widetilde{t_1},\ldots,\widetilde{t_n})]}$$

$\mathrm{FV}(\mathcal{A})$ denotes the set of free variables occurring in $\mathcal{A}$'s range, and $\mathcal{A}[x \mapsto t]$ denotes the *variable-capturing* substitution of t for x in $\mathcal{A}$'s range. It is easy to see that the height of terms produced as a result of the unfolding is bounded by the number of equivalence classes of E, and thus the construction of $\mathcal{A}$ will terminate.

Example 1. Suppose that E contains four distinct equivalence classes $[w]$, $[x]$, $[y]$, and $[z]$ such that $\mathsf{C}(w,y) \in [x]$ and $\mathsf{C}(z,x) \in [y]$ for some $\mathsf{C} \in \mathcal{F}_{\mathrm{ctr}}$. A possible sequence of unfolding steps is given below, omitting trivial entries such as $[w] \mapsto \tilde{w}$.

1. Unfold $\tilde{x}$: $\mathcal{A} = \{[x] \mapsto \mu\tilde{x}.\, \mathsf{C}(\tilde{w}, \tilde{y})\}$
2. Unfold $\tilde{y}$: $\mathcal{A} = \{[x] \mapsto \mu\tilde{x}.\, \mathsf{C}(\tilde{w}, \mu\tilde{y}.\, \mathsf{C}(\tilde{z}, \tilde{x})), [y] \mapsto \mu\tilde{y}.\, \mathsf{C}(\tilde{z}, \tilde{x})\}$
3. Unfold $\tilde{x}$: $\mathcal{A} = \{[x] \mapsto \mu\tilde{x}.\, \mathsf{C}(\tilde{w}, \mu\tilde{y}.\, \mathsf{C}(\tilde{z}, \tilde{x})), [y] \mapsto \mu\tilde{y}.\, \mathsf{C}(\tilde{z}, \mu\tilde{x}.\, \mathsf{C}(\tilde{w}, \tilde{y}))\}$

$$\frac{t \in \mathcal{T}(E)}{E := E, t \approx t} \; \text{Refl} \qquad \frac{t \approx u \in E}{E := E, u \approx t} \; \text{Sym} \qquad \frac{s \approx t, \, t \approx u \in E}{E := E, s \approx u} \; \text{Trans}$$

$$\frac{\bar{t} \approx \bar{u} \in E \quad \mathsf{f}(\bar{t}), \mathsf{f}(\bar{u}) \in \mathcal{T}(E)}{E := E, \mathsf{f}(\bar{t}) \approx \mathsf{f}(\bar{u})} \; \text{Cong} \qquad \frac{t \approx u, \, t \not\approx u \in E}{\bot} \; \text{Conflict}$$

$$\frac{\mathsf{C}(\bar{t}) \approx \mathsf{C}(\bar{u}) \in E}{E := E, \bar{t} \approx \bar{u}} \; \text{Inject} \qquad \frac{\mathsf{C}(\bar{t}) \approx \mathsf{D}(\bar{u}) \in E \quad \mathsf{C} \neq \mathsf{D}}{\bot} \; \text{Clash}$$

Fig. 1. Derivation rules for bidirectional closure

$$\frac{\delta \in \mathcal{Y}_{\mathrm{dt}} \quad \mathcal{A}[t^\delta] = \mu x. \, u \quad x \in \mathrm{FV}(u)}{\bot} \; \text{Acyclic} \qquad \frac{\delta \in \mathcal{Y}_{\mathrm{codt}} \quad \mathcal{A}[t^\delta] =_\alpha \mathcal{A}[u^\delta]}{E := E, t \approx u} \; \text{Unique}$$

Fig. 2. Derivation rules for acyclicity and uniqueness

$$\frac{\begin{array}{c} t^\delta \in \mathcal{T}(E) \quad \mathcal{F}^\delta_{\mathrm{ctr}} = \{\mathsf{C}_1, \ldots, \mathsf{C}_m\} \\ \left(\mathsf{s}(t) \in \mathcal{T}(E) \text{ and } \mathsf{s} \in \mathcal{F}^\delta_{\mathrm{sel}} \right) \text{ or } \left(\delta \in \mathcal{Y}_{\mathrm{dt}} \text{ and } \delta \text{ is finite} \right) \end{array}}{E := E, t \approx \mathsf{C}_1\big(\mathsf{s}_1^1(t), \ldots, \mathsf{s}_1^{n_1}(t)\big) \quad \cdots \quad E := E, t \approx \mathsf{C}_m\big(\mathsf{s}_m^1(t), \ldots, \mathsf{s}_m^{n_m}(t)\big)} \; \text{Split}$$

$$\frac{t^\delta, u^\delta \in \mathcal{T}(E) \quad \delta \in \mathcal{Y}_{\mathrm{codt}} \quad \delta \text{ is a singleton}}{E := E, t \approx u} \; \text{Single}$$

Fig. 3. Derivation rules for branching

The resulting $\mathcal{A}$ indicates that the values for x and y in models of E must be of the forms $\mathsf{C}(\widetilde{w}, \mathsf{C}(\widetilde{z}, \mathsf{C}(\widetilde{w}, \mathsf{C}(\widetilde{z}, \ldots))))$ and $\mathsf{C}(\widetilde{z}, \mathsf{C}(\widetilde{w}, \mathsf{C}(\widetilde{z}, \mathsf{C}(\widetilde{w}, \ldots))))$, respectively. □

Given the mapping $\mathcal{A}$, the Acyclic and Unique rules work as follows. For acyclicity, if $[t]$ is a datatype equivalence class whose values $\mathcal{A}[t] = \mu x. \, u$ are cyclic (as expressed by $x \in \mathrm{FV}(u)$), then E is $\mathcal{DC}$-unsatisfiable. For uniqueness, if $[t]$, $[u]$ are two codatatype equivalence classes whose values $\mathcal{A}[t], \mathcal{A}[u]$ are α-equivalent, then $t \approx u$. Comparison for α-equivalence may seem too restrictive, since $\mu x. \, \mathsf{S}(x)$ and $\mu y. \, \mathsf{S}(\mathsf{S}(y))$ specify the same value despite being α-disequivalent, but the rule will make progress by discovering that the subterm $\mathsf{S}(y)$ of $\mu y. \, \mathsf{S}(\mathsf{S}(y))$ must be equal to the entire term.

Example 2. Let $E = \{x \approx \mathsf{S}(x), \, y \approx \mathsf{S}(\mathsf{S}(y))\}$. After phase 1, the equivalence classes are $\{x, \mathsf{S}(x)\}$, $\{y, \mathsf{S}(\mathsf{S}(y))\}$, and $\{\mathsf{S}(y)\}$. Constructing $\mathcal{A}$ yields

$$\mathcal{A}[x] = \mu \widetilde{x}. \, \mathsf{S}(\widetilde{x}) \qquad \mathcal{A}[y] = \mu \widetilde{y}. \, \mathsf{S}(\widetilde{\mu\mathsf{S}(y)}. \, \mathsf{S}(\widetilde{y})) \qquad \mathcal{A}[\mathsf{S}(y)] = \mu \widetilde{\mathsf{S}(y)}. \, \mathsf{S}(\mu \widetilde{y}. \, \mathsf{S}(\widetilde{\mathsf{S}(y)}))$$

Since $\mathcal{A}[y] =_\alpha \mathcal{A}[\mathsf{S}(y)]$, the Unique rule applies to derive $y \approx \mathsf{S}(y)$. At this point, phase 1 is activated again, yielding $\{x, \mathsf{S}(x)\}$ and $\{y, \mathsf{S}(y), \mathsf{S}(\mathsf{S}(y))\}$. The mapping $\mathcal{A}$ is updated accordingly: $\mathcal{A}[y] = \mu \widetilde{y}. \, \mathsf{S}(\widetilde{y})$. Since $\mathcal{A}[x] =_\alpha \mathcal{A}[y]$, Unique can finally derive $x \approx y$. □

Phase 3: Branching (Fig. 3). If a selector is applied to a term t, or if t's type is a finite datatype, t's equivalence class must contain a δ constructor term. This is enforced in the third phase by the Split rule. Another rule, Single, focuses on the degenerate case where two terms have the same corecursive singleton type and are therefore equal. Both Split's finiteness assumption and Single's singleton constraint can be evaluated statically based on a recursive computation of the cardinalities of the constructors' argument types.

Correctness. Correctness means that if there exists a closed derivation tree with root node E, then E is $\mathcal{DC}$-unsatisfiable; and if there exists a derivation tree with root node E that contains a saturated node, then E is $\mathcal{DC}$-satisfiable.

Theorem 2 (Termination). *All derivation trees are finite.*

Proof. Consider a derivation tree with root node E. Let $D \subseteq \mathcal{T}(E)$ be the set of terms whose types are finite datatypes, and let $S \subseteq \mathcal{T}(E)$ be the set of terms occurring as arguments to selectors. For each term $t \in D$, let $S_t^0 = \{t\}$ and $S_t^{i+1} = S_t^i \cup \{\mathsf{s}(u) \mid u^\delta \in S_t^i, \delta \in \mathcal{Y}_{\mathrm{dt}}, |\delta| \text{ is finite}, \mathsf{s} \in \mathcal{F}_{\mathrm{sel}}^\delta\}$, and let S_t^∞ be the limit of this sequence. This is a finite set for each t, because all chains of selectors applied to t are finite. Let S^∞ be the union of all sets S_t^∞ where $t \in D$, and let $\mathcal{T}^\infty(E)$ be the set of subterms of $E \cup \{\mathsf{C}_j(\mathsf{s}_j^1(t), \ldots, \mathsf{s}_j^{n_j}(t)) \mid t^\delta \in S \cup S^\infty, \mathsf{C}_j \in \mathcal{F}_{\mathrm{ctr}}^\delta\}$. In a derivation tree with root node E, it can be shown by induction on the rules of the calculus that each non-root node F is such that $\mathcal{T}(F) \subseteq \mathcal{T}^\infty(E)$, and hence contains an equality between two terms from $\mathcal{T}^\infty(E)$ not occurring in its parent node. Thus, the depth of a branch in a derivation tree with root node E is at most $|\mathcal{T}^\infty(E)|^2$, which is finite since $\mathcal{T}^\infty(E)$ is finite.   □

Theorem 3 (Refutation Soundness). *If there exists a closed derivation tree with root node E, then E is $\mathcal{DC}$-unsatisfiable.*

Proof. The proof is by structural induction on the derivation tree with root node E. If the tree is an application of Conflict, Clash, or Acyclic, then E is $\mathcal{DC}$-unsatisfiable. For Conflict, this is a consequence of equality reasoning. For Clash, this is a consequence of distinctness. For Acyclic, the construction of $\mathcal{A}$ indicates that the class of values that term t can take in models of E is infeasible. If the child nodes of E are closed derivation trees whose root nodes are the result of applying Split on t^δ, by the induction hypothesis $E \cup t \approx \mathsf{C}_j(\mathsf{s}_j^1(t), \ldots, \mathsf{s}_j^{n_j}(t))$ is $\mathcal{DC}$-unsatisfiable for each $\mathsf{C}_j \in \mathcal{F}_{\mathrm{ctr}}^\delta$. Since by exhaustiveness, all models of $\mathcal{DC}$ entail exactly one $t \approx \mathsf{C}_j(\mathsf{s}_j^1(t), \ldots, \mathsf{s}_j^{n_j}(t))$, E is $\mathcal{DC}$-unsatisfiable. Otherwise, the child node of E is a closed derivation tree whose root node $E \cup t \approx u$ is obtained by applying one of the rules Refl, Sym, Trans, Cong, Inject, Unique, or Single. In all these cases, $E \vdash_{\mathcal{DC}} t \approx u$. For Refl, Sym, Trans, Cong, this is a consequence of equality reasoning. For Inject, this is a consequence of injectivity. For Unique, the construction of $\mathcal{A}$ indicates that the values of t and u are equivalent in all models of E. For Single, t and u must have the same value since the cardinality of their type is one. By the induction hypothesis, $E \cup t \approx u$ is $\mathcal{DC}$-unsatisfiable and thus E is $\mathcal{DC}$-unsatisfiable.   □

It remains to show the converse of the previous theorem: If a derivation tree with root node E contains a saturated node, then E is $\mathcal{DC}$-satisfiable. The proof relies on a specific interpretation $\mathcal{J}$ that satisfies E.

First, we define the set of interpretations of the theory $\mathcal{DC}$, which requires custom terminology concerning μ-terms. Given a μ-term t with subterm u, the *expansion of u with respect to t* is the μ-term $\langle u \rangle_t^{\emptyset}$, abbreviated to $\langle u \rangle_t$, as returned by the function

$$\langle x \rangle_t^B = \begin{cases} x & \text{if } x \in B \\ \mu x.\, \mathsf{C}(\langle \bar{u} \rangle_t^{B \uplus \{x\}}) & \text{if } \mu x.\, \mathsf{C}(\bar{u}) \text{ binds this occurrence of } x \notin B \text{ in } t \end{cases}$$

$$\langle \mu x.\, \mathsf{C}(\bar{u}) \rangle_t^B = \begin{cases} x & \text{if } x \in B \\ \mu x.\, \mathsf{C}(\langle \bar{u} \rangle_t^{B \uplus \{x\}}) & \text{otherwise} \end{cases}$$

The recursion will eventually terminate because each recursion adds one bound variable to B and there are finitely many distinct bound variables in a μ-term. Intuitively, the expansion of a subterm is a stand-alone μ-term that denotes the same value as the original subterm—e.g., $\langle \mu y.\, \mathsf{D}(x) \rangle_{\mu x.\, \mathsf{C}(\mu y.\, \mathsf{D}(x))} = \mu y.\, \mathsf{D}(\mu x.\, \mathsf{C}(y))$.

The μ-term u is a *self-similar subterm* of t if u is a proper subterm of t, t and u are of the forms $\mu x.\, \mathsf{C}(t_1, \ldots, t_n)$ and $\mu y.\, \mathsf{C}(u_1, \ldots, u_n)$, and $\langle t_k \rangle_t =_\alpha \langle u_k \rangle_t$ for all k. The μ-term t is *normal* if it does not contain self-similar subterms and all of its proper subterms are also normal. Thus, $t = \mu x.\, \mathsf{C}(\mu y.\, \mathsf{C}(y))$ is not normal because $\mu y.\, \mathsf{C}(y)$ is a self-similar subterm of t. Their arguments have the same expansion with respect to t: $\langle \mu y.\, \mathsf{C}(y) \rangle_t = \mu y.\, \mathsf{C}(\langle y \rangle_t^{\{y\}}) = \mu y.\, \mathsf{C}(y)$ is α-equivalent to $\langle y \rangle_t = \mu y.\, \mathsf{C}(\langle y \rangle_t^{\{y\}}) = \mu y.\, \mathsf{C}(y)$. The term $u = \mu x.\, \mathsf{C}(\mu y.\, \mathsf{C}(x))$ is also not normal, since $\mu y.\, \mathsf{C}(x)$ is a self-similar subterm of u, noting that $\langle \mu y.\, \mathsf{C}(x) \rangle_u = \mu y.\, \mathsf{C}(\langle x \rangle_u^{\{y\}}) = \mu y.\, \mathsf{C}(\langle \mu x.\, \mathsf{C}(\mu y.\, \mathsf{C}(x)) \rangle_u^{\{y\}}) = \mu y.\, \mathsf{C}(\mu x.\, \mathsf{C}(\langle \mu y.\, \mathsf{C}(x) \rangle_u^{\{x,y\}})) = \mu y.\, \mathsf{C}(\mu x.\, \mathsf{C}(y))$ is α-equivalent to $\langle x \rangle_u = u$.

For any μ-term t of the form $\mu x.\, \mathsf{C}(\bar{u})$, its *normal form* $\lfloor t \rfloor$ is obtained by replacing all of the self-similar subterms of t with x and by recursively normalizing the other subterms. For variables, $\lfloor x \rfloor = x$. Thus, $\lfloor \mu x.\, \mathsf{C}(\mu y.\, \mathsf{C}(x)) \rfloor = \mu x.\, \mathsf{C}(x)$.

We now define the class of interpretations for $\mathcal{DC}$. $\mathcal{J}(\tau)$ denotes the interpretation type τ in $\mathcal{J}$—that is, a nonempty set of domain elements for that type. $\mathcal{J}(\mathsf{f})$ denotes the interpretation of a function f in $\mathcal{J}$. If $\mathsf{f} : \tau_1 \times \cdots \times \tau_n \to \tau$, then $\mathcal{J}(\mathsf{f})$ is a total function from $\mathcal{J}(\tau_1) \times \cdots \times \mathcal{J}(\tau_n)$ to $\mathcal{J}(\tau)$. All types are interpreted as sets of μ-terms, but only values of types in $\mathcal{Y}_{\mathrm{codt}}$ may contain cycles.

Definition 4 (Normal Interpretation). An interpretation $\mathcal{J}$ is *normal* if these conditions are met:

1. For each type τ, $\mathcal{J}(\tau)$ includes a maximal set of closed normal μ-terms of that type that are unique up to α-equivalence and acyclic if $\tau \notin \mathcal{Y}_{\mathrm{codt}}$.
2. For each constructor term $\mathsf{C}(\bar{t})$ of type τ, $\mathcal{J}(\mathsf{C})(\mathcal{J}(\bar{t}))$ is the value in $\mathcal{J}(\tau)$ that is α-equivalent to $\lfloor \mu x.\, \mathsf{C}(\mathcal{J}(\bar{t})) \rfloor$, where x is fresh.

3. For each selector term $s_j^k(t)$ of type τ, if $\mathcal{J}(t)$ is $\mu x.\ C_j(\bar{u})$, then $\mathcal{J}(s_j^k)(\mathcal{J}(t))$ is the value in $\mathcal{J}(\tau)$ that is α-equivalent to $\langle u_k \rangle_{\mathcal{J}(t)}$.

Not all normal interpretations are models of codatatypes, because models must contain all possible infinite terms, not only cyclic ones. However, acyclic infinite values are uninteresting to us, and for quantifier-free formulas it is trivial to extend any normal interpretation with extra domain elements to obtain a genuine model if desired.

When constructing a model $\mathcal{J}$ of E, it remains only to specify how $\mathcal{J}$ interprets wrongly applied selector terms and variables. For the latter, this will be based on the mapping $\mathcal{A}$ computed in phase 2 of the calculus.

First, we need the following definitions. We write $t =_\alpha^x u$ if μ-terms t and u are syntactically equivalent for some renaming that avoids capturing any variable other than x. For example, $\mu x.\ D(x) =_\alpha^y \mu x.\ D(y)$ (by renaming y to x), $\mu x.\ C(x,x) =_\alpha^x \mu y.\ C(x,y)$, and $\mu x.\ C(z,x) =_\alpha^z \mu y.\ C(z,y)$, but $\mu x.\ D(x) \neq_\alpha^x \mu x.\ D(y)$ and $\mu x.\ C(x,x) \neq_\alpha^y \mu y.\ C(x,y)$. For a variable x^τ and a normal interpretation $\mathcal{J}$, we let $\mathcal{V}_{\mathcal{J}}^x(\mathcal{A})$ denote the set consisting of all values $v \in \mathcal{J}(\tau)$ such that $v =_\alpha^x \langle u \rangle_t$ for some subterm u of a term t occurring in the range of $\mathcal{A}$. This set describes shapes of terms to avoid when assigning a μ-term to x.

The *completion* $\mathcal{A}^\star$ of $\mathcal{A}$ for a normal interpretation $\mathcal{J}$ assigns values from $\mathcal{J}$ to unassigned variables in the domain of $\mathcal{A}$. We construct $\mathcal{A}^\star$ by initially setting $\mathcal{A}^\star := \lfloor \mathcal{A} \rfloor$ and by exhaustively applying the following rule:

$$\frac{\widetilde{x}^\tau \in \mathrm{FV}(\mathcal{A}^\star) \quad \mu\widetilde{x}.\ t =_\alpha v \quad v \in \mathcal{J}(\tau) \quad v \notin \mathcal{V}_{\mathcal{J}}^{\widetilde{x}}(\mathcal{A}^\star)}{\mathcal{A}^\star := \lfloor \mathcal{A}^\star[\widetilde{x} \mapsto \mu\widetilde{x}.\ t] \rfloor}$$

Given an unassigned variable in $\mathcal{A}^\star$, this rule assigns it a fresh value—one that does not occur in $\mathcal{V}_{\mathcal{J}}^{\widetilde{x}}(\mathcal{A}^\star)$ modulo α-equivalence—excluding not only existing terms in the range of $\mathcal{A}^\star$ but also terms that could emerge as a result of the update. Since this update removes one variable from $\mathrm{FV}(\mathcal{A}^\star)$ and does not add any variables to $\mathrm{FV}(\mathcal{A}^\star)$, the process eventually terminates. We normalize all terms in the range of $\mathcal{A}^\star$ at each step.

To ensure disequality literals are satisfied by an interpretation based on $\mathcal{A}^\star$, it suffices that $\mathcal{A}^\star$ is injective modulo α-equivalence. This invariant holds initially, and the last precondition in the above rule ensures that it is maintained. The set $\mathcal{V}_{\mathcal{J}}^{\widetilde{x}}(\mathcal{A}^\star)$ is an overapproximation of the values that, when assigned to $\widetilde{x}$, will cause values in the range of $\mathcal{A}^\star$ to become α-equivalent. For infinite codatatypes, it is always possible to find fresh values v because $\mathcal{V}_{\mathcal{J}}^{\widetilde{x}}(\mathcal{A}^\star)$ is a finite set.

Example 3. Let δ be a codatatype with the constructors $C, D, E : \delta \to \delta$. Let E be the set $\{u \approx C(z),\ v \approx D(z),\ w \approx E(y),\ x \approx C(v),\ z \not\approx v\}$. After applying the calculus to saturation on E, the mapping $\mathcal{A}$ is as follows:

$$\begin{array}{lll}
\mathcal{A}[u] = \mu\widetilde{u}.C(\widetilde{z}) & \mathcal{A}[w] = \mu\widetilde{w}.E(\widetilde{y}) & \mathcal{A}[y] = \widetilde{y} \\
\mathcal{A}[v] = \mu\widetilde{v}.D(\widetilde{z}) & \mathcal{A}[x] = \mu\widetilde{x}.C(\mu\widetilde{v}.\ D(\widetilde{z})) & \mathcal{A}[z] = \widetilde{z}
\end{array}$$

To construct a completion $\mathcal{A}^\star$, we must choose values for $\widetilde{y}$ and $\widetilde{z}$, which are free in $\mathcal{A}$. Modulo α-equivalence, $\mathcal{V}_{\mathcal{J}}^{\widetilde{z}}(\mathcal{A}) = \{\mu a.\ \mathsf{C}(a),\ \mu a.\ \mathsf{D}(a),\ \mu a.\ \mathsf{C}(\mathsf{D}(a)),$ $\mathsf{C}(\mu a.\ \mathsf{D}(a))\}$. Now consider a normal interpretation $\mathcal{J}$ that evaluates variables in E based on $\mathcal{A}$: $\mathcal{J}(u) = \mathcal{A}[u]$, $\mathcal{J}(v) = \mathcal{A}[v]$, and so on. Assigning a value for $\mathcal{A}[z]$ that is α-equivalent to a value in $\mathcal{V}_{\mathcal{J}}^{\widetilde{z}}(\mathcal{A})$ may cause values in the range of $\mathcal{A}$ to become α-equivalent, which in turn may cause E to be falsified by $\mathcal{J}$. For example, assign $\mu\widetilde{z}.\ \mathsf{D}(\widetilde{z})$ for $\widetilde{z}$. After the substitution, $\mathcal{A}[v] = \mu\widetilde{v}.\ \mathsf{D}(\mu\widetilde{z}.\ \mathsf{D}(\widetilde{z}))$, which has normal form $\mu\widetilde{v}.\ \mathsf{D}(\widetilde{v})$, which is α-equivalent to $\mu\widetilde{z}.\ \mathsf{D}(\widetilde{z})$. However, this contradicts the disequality $z \not\approx v$ in E. On the other hand, if the value assigned to $\widetilde{z}$ is fresh, the values in the range of $\mathcal{A}$ remain α-disequivalent. We can assign a value such as $\mu\widetilde{z}.\ \mathsf{E}(\widetilde{z})$, $\mu\widetilde{z}.\ \mathsf{D}(\mathsf{C}(\widetilde{z}))$, or $\mu\widetilde{z}.\ \mathsf{C}(\mathsf{C}(\mathsf{D}(\widetilde{z})))$ to $\widetilde{z}$. □

In the following lemma about $\mathcal{A}^\star$, $\mathrm{Var}(t) = \begin{cases} t & \text{if } t \text{ is a variable} \\ x & \text{if } t \text{ is of the form } \mu x.\ u. \end{cases}$

Lemma 5. *If $\mathcal{A}$ is constructed for a saturated set E and $\mathcal{A}^\star$ is a completion of $\mathcal{A}$ for a normal interpretation $\mathcal{J}$, the following properties hold:*

(1) $\mathcal{A}^\star[x^\tau]$ *is α-equivalent to a value in $\mathcal{J}(\tau)$.*
(2) $\mathcal{A}^\star[x] = \langle t \rangle_{\mathcal{A}^\star[y]}$ *for all subterms t of $\mathcal{A}^\star[y]$ with $\mathrm{Var}(t) = \widetilde{x}$.*
(3) $\mathcal{A}^\star[x] =_\alpha \mathcal{A}^\star[y]$ *if and only if $[x] = [y]$.*

Intuitively, this lemma states three properties of $\mathcal{A}^\star$ that ensure a normal interpretation $\mathcal{J}$ can be constructed that satisfies E. Property (1) states that the values in the range of $\mathcal{A}^\star$ are α-equivalent to a value in normal interpretation. This means they are closed, normal, and acyclic when required. Property (2) states that the interpretation of all subterms in the range of $\mathcal{A}^\star$ depends on its associated variable only. In other words, the interpretation of a subterm t where $\mathrm{Var}(t) = \widetilde{x}$ is equal to $\mathcal{A}^\star[x]$, independently of the context. Property (3) states that $\mathcal{A}^\star$ is injective (modulo α-equivalence), which ensures that distinct values are assigned to distinct equivalence classes.

Theorem 6 (Solution Soundness). *If there exists a derivation tree with root node E containing a saturated node, then E is $\mathcal{DC}$-satisfiable.*

Proof. Let F be a saturated node in a derivation tree with root node E. We consider a normal interpretation $\mathcal{J}$ that interprets wrongly applied selectors based on equality information in F and that interprets the variables of F based on the completion $\mathcal{A}^\star$. For the variables, let $\mathcal{J}(x^\tau)$ be the value in $\mathcal{J}(\tau)$ that is α-equivalent with $\mathcal{A}^\star[x]$ for each variable $x \in \mathcal{T}(F)$, which by Lemma 5(1) is guaranteed to exist.

We first show that $\mathcal{J}$ satisfies all equalities $t_1 \approx t_2 \in F$. To achieve this, we show by structural induction on t^τ that $\mathcal{J}(t) =_\alpha \mathcal{A}^\star[t]$ for all terms $t \in \mathcal{T}(F)$, which implies $\mathcal{J} \models t_1 \approx t_2$ since $\mathcal{J}$ is normal.

If t is a variable, then $\mathcal{J}(t) =_\alpha \mathcal{A}^\star[t]$ by construction.

If t is a constructor term of the form $\mathsf{C}(u_1, \ldots, u_n)$, then $\mathcal{J}(t)$ is α-equivalent with $\lfloor \mu x.\ \mathsf{C}(\mathcal{J}(u_1), \ldots, \mathcal{J}(u_n)) \rfloor$ for some fresh x, which by the induction hypothesis is α-equivalent with $\lfloor \mu x.\ \mathsf{C}(\mathcal{A}^\star[u_1], \ldots, \mathcal{A}^\star[u_n]) \rfloor$. Call this term t'. Since

Inject and Clash do not apply to F, by the construction of $\mathcal{A}^\star$ we have that $\mathcal{A}^\star[t]$ is a term of the form $\mu t.\ \mathsf{C}(w_1, \ldots, w_n)$ where $\mathrm{Var}(w_i) = \widetilde{u}_i$ for each i. Thus by Lemma 5(2), $\langle w_i \rangle_{\mathcal{A}^\star[t]} = \mathcal{A}^\star[u_i]$. For each i, let $u_i{}'$ be the ith argument of t'. Clearly, $\langle u_i{}' \rangle_{t'} =_\alpha \mathcal{A}^\star[u_i]$. Thus, $\langle u_i{}' \rangle_{t'} =_\alpha \langle w_i \rangle_{\mathcal{A}^\star[t]}$. Thus, $\mathcal{I}(t) =_\alpha t' =_\alpha \mathcal{A}^\star[t]$, and we have $\mathcal{I}(t) =_\alpha \mathcal{A}^\star[t]$.

If t is a selector term $\mathsf{s}_j^k(u)$, since Split does not apply to F, $[u]$ must contain a term of the form $\mathsf{C}_{j'}(\mathsf{s}_{j'}^1(u), \ldots, \mathsf{s}_{j'}^n(u))$ for some j'. Since Inject and Clash are not applicable, by construction $\mathcal{A}^\star[u]$ must be of the form $\mu \widetilde{u}.\ \mathsf{C}_{j'}(w_1, \ldots, w_n)$, where $\mathrm{Var}(w_i) = \mathsf{s}_{j'}^i(u)$ for each i, and thus by Lemma 5(2), $\langle w_i \rangle_{\mathcal{A}^\star[u]} = \mathcal{A}^\star[\mathsf{s}_{j'}^i(u)]$. If $j = j'$, then $\mathcal{I}(t)$ is α-equivalent with $\langle w_k \rangle_{\mathcal{A}^\star[u]}$, which is equal to $\mathcal{A}^\star[\mathsf{s}_j^k(u)] = \mathcal{A}^\star[t]$. If $j \neq j'$, since Cong does not apply, any term of the form $\mathsf{s}_j^k(u')$ not occurring in $[t]$ is such that $[u] \neq [u']$. By the induction hypothesis and Lemma 5(3), $\mathcal{I}(u) \neq \mathcal{I}(u')$ for all such u, u'. Thus, we may interpret $\mathcal{I}(\mathsf{s}_j^k)(\mathcal{I}(u))$ as the value in $\mathcal{I}(\tau)$ that is α-equivalent with $\mathcal{A}^\star[t]$.

We now show that all disequalities in F are satisfied by $\mathcal{I}$. Assume $t \not\approx u \in F$. Since Conflict does not apply, $t \approx u \notin F$ and thus $[t]$ and $[u]$ are distinct. Since $\mathcal{I}(t) =_\alpha \mathcal{A}^\star[t]$ and $\mathcal{I}(u) =_\alpha \mathcal{A}^\star[u]$, by Lemma 5(3), $\mathcal{I}(t) \neq \mathcal{I}(u)$, and thus $\mathcal{I} \vDash t \not\approx u$.

Since by assumption F contains only equalities and disequalities, we have $\mathcal{I} \vDash F$, and since $E \subseteq F$, we conclude that $\mathcal{I} \vDash E$.   □

By Theorems 2, 3, and 6, the calculus is sound and complete for the universal theory of (co)datatypes. We can rightly call it a decision procedure for that theory. The proof of solution soundness is constructive in that it provides a method for constructing a model for a saturated configuration, by means of the mapping $\mathcal{A}^\star$.

4   Implementation as a Theory Solver in CVC4

The decision procedure was presented at a high level of abstraction, omitting quite a few details. This section describes the main aspects of the implementation within the SMT solver CVC4: the integration into CDCL(T) [7], the extension to quantified formulas, and some of the optimizations.

The decision procedure is implemented as a theory solver of CVC4, that is, a specialized solver for determining the satisfiability of conjunctions of literals for its theory. Given a theory $T = T_1 \cup \cdots \cup T_n$ and a set of input clauses F in conjunctive normal form, the CDCL(T) procedure incrementally builds partial assignments of truth values to the atoms of F such that no clause in F is falsified. We can regard such a partial assignment as a set M of true literals. By a variant of the Nelson–Oppen method [8,18], each T_i-solver takes as input the union M_i of (1) the purified form of T_i-literals occurring in M, where fresh variables replace terms containing symbols not belonging to T_i; (2) additional (dis)equalities between variables of types not belonging to T_i. Each T_i-solver either reports that a subset C of M_i is T_i-unsatisfiable, in which case $\neg C$ is added to F, adds a clause to F, or does nothing. When M is a complete assignment for F, a theory solver can choose to do nothing only if M_i is indeed T_i-satisfiable.

Assume E is initially the set M_i described above. With each equality $t \approx u$ added to E, we associate a set of equalities from M_i that together entail $t \approx u$, which we call its *explanation*. Similarly, each $\mathcal{A}[x]$ is assigned an explanation—that is, a set of equalities from M_i that entail that the values of $[x]$ in models of E are of the form $\mathcal{A}[x]$. For example, if $x \approx \mathsf{C}(x) \in M_i$, then $x \approx \mathsf{C}(x)$ is an explanation for $\mathcal{A}[x] = \mu \widetilde{x}.\ \mathsf{C}(\widetilde{x})$. The rules of the calculus are implemented as follows. For all rules with conclusion $\bot$, we report the union of the explanations for all premises is $\mathcal{DC}$-unsatisfiable. For Split, we add the exhaustiveness clause $t \approx \mathsf{C}_1\big(\mathsf{s}_1^1(t), \ldots, \mathsf{s}_1^{n_1}(t)\big) \vee \cdots \vee t \approx \mathsf{C}_m\big(\mathsf{s}_m^1(t), \ldots, \mathsf{s}_m^{n_m}(t)\big)$ to F. Decisions on which branch to take are thus performed externally by the SAT solver. All other rules add equalities to the internal state of the theory solver. The rules in phase 1 are performed eagerly—that is, for partial satisfying assignments M—while the rules in phases 2 and 3 are performed only for complete satisfying assignments M.

Before constructing a model for F, the theory solver constructs neither μ-terms nor the mapping $\mathcal{A}$. Instead, $\mathcal{A}$ is computed implicitly by traversing the equivalence classes of E during phase 2. To detect whether Acyclic applies, the procedure considers each equivalence class $[t]$ containing a datatype constructor $\mathsf{C}(t_1, \ldots, t_n)$. It visits $[t_1], \ldots, [t_n]$ and all constructor arguments in these equivalence classes recursively. If while doing so it returns to $[t]$, it deduces that Acyclic applies. To recognize when the precondition of Unique holds, the procedure considers the set S of all codatatype equivalence classes. It simultaneously visits the equivalence classes of arguments of constructor terms in each equivalence class in S, while partitioning S into $S_1, \ldots, S_n$ based on the top-most symbol of constructor terms in these equivalence classes and the equivalence of their arguments of ordinary types. It then partitions each set recursively. If the resulting partition contains a set S_i containing distinct terms u and v, it deduces that Unique applies to u and v.

While the decision procedure is restricted to universal conjectures, in practice we often want to solve problems that feature universal axioms and existential conjectures. Many SMT solvers, including CVC4, can reason about quantified formulas using incomplete instantiation-based methods [16,23]. These methods extend naturally to formulas involving datatypes and codatatypes.

However, the presence of quantified formulas poses an additional challenge in the context of (co)datatypes. Quantified formulas may entail an upper bound on the cardinality of an uninterpreted type u. When assuming that u has infinite cardinality, the calculus presented in Sect. 3 is incomplete since it may fail to recognize cases where Split and Single should be applied. This does not impact the correctness of the procedure in this setting, since the solver is already incomplete in the presence of quantified formulas. Nonetheless, two techniques help increase the precision of the solver. First, we can apply Split to datatype terms whose cardinality depends on the finiteness of uninterpreted types. Second, we can conditionally apply Single to codatatype terms that may have cardinality one. For example, the *stream$_u$* codatatype has cardinality one precisely when u has cardinality one. If there exist two equivalence classes $[s]$ and $[t]$ for this type, the implementation adds the clause $(\exists x\, y^u.\ x \not\approx y) \vee s \approx t$ to F.

The implementation of the decision procedure uses several optimizations following the lines of Barrett et al. [2]. Discriminators are part of the signature and not abbreviations. This requires extending the decision procedure with several rules, which apply uniformly to datatypes and codatatypes. This approach often leads to better performance because it introduces terms less eagerly to $\mathcal{T}(E)$. Selectors are collapsed eagerly: If $s_j^k(t) \in \mathcal{T}(E)$ and $t = C_j(u_1, \ldots, u_n)$, the solver directly adds $s_j^k(t) \approx u_k$ to E, whereas the presented calculus would apply Split and Inject before adding this equality. To reduce the number of unique literals considered by the calculus, we compute a normal form for literals as a preprocessing step. In particular, we replace $u \approx t$ by $t \approx u$ if t is smaller than u with respect to some term ordering, replace $C_j(\bar{t}) \approx C_{j'}(\bar{u})$ with $\bot$ when $j \neq j'$, replace all selector terms of the form $s_j^k(C_j(t_1, \ldots, t_n))$ by t_k, and replace occurrences of discriminators $d_j(C_{j'}(\bar{t}))$ by $\top$ or $\bot$ based on whether $j = j'$.

As Barrett et al. [2] observed for their procedure, it is both theoretically and empirically beneficial to delay applications of Split as long as possible. Similarly, Acyclic and Unique are fairly expensive because they require traversing the equivalence classes, which is why they are part of phase 2.

5   Evaluation on Isabelle Problems

To evaluate the decision procedure, we generated benchmark problems from existing Isabelle formalizations using Sledgehammer [3]. We included all the theory files from the Isabelle distribution (Distro, 879 goals) and the *Archive of Formal Proofs* (AFP, 2974 goals) [10] that define codatatypes falling within the supported fragment. We added two unpublished theories about Bird and Stern–Brocot trees by Peter Gammie and Andreas Lochbihler (G&L, 317 goals). To exercise the datatype support, theories about lists and trees were added to the first two benchmark sets. The theories were selected before conducting any experiments. The experimental data are available online.[1]

For each goal in each theory, Sledgehammer selected about 16 lemmas, which were monomorphized and translated to SMT-LIB 2 along with the goal. The resulting problem was given to the development version of CVC4 and to Z3 4.3.2 for comparison, each running for up to 60 s on the StarExec cluster [25]. Problems not involving any (co)datatypes were left out. Due to the lack of machinery for reconstructing inferences about (co)datatypes in Isabelle, the solvers are trusted as oracles. The development version of CVC4 was run on each problem several times, with the support for datatypes and codatatypes either enabled or disabled. The contributions of the acyclicity and uniqueness rules were also measured. Even when the decision procedure is disabled, the problems may contain basic lemmas about constructors and selectors, allowing some (co)datatype reasoning.

The results are summarized in Table 1. The decision procedure makes a difference across all three benchmark suites. It accounts for an overall success rate increase of over 5 %, which is quite significant. The raw evaluation data also suggest that the theoretically stronger decision procedures almost completely

---

[1] http://lara.epfl.ch/~reynolds/CADE2015-cdt/.

Table 1. Number of solved goals for the three benchmark suites

	Distro		AFP		G&L		Overall	
	CVC4	Z3	CVC4	Z3	CVC4	Z3	CVC4	Z3
No (co)datatypes	221	209	775	777	52	51	1048	1037
Datatypes without Acyclic	227	–	780	–	52	–	1059	–
Full datatypes	227	213	786	791	52	51	1065	1055
Codatatypes without Unique	222	–	804	–	56	–	1082	–
Full codatatypes	223	–	804	–	**59**	–	1086	–
Full (co)datatypes	**229**	–	**815**	–	**59**	–	**1103**	–

subsume the weaker ones in practice: We encountered only one goal (out of 4170) that was solved by a configuration of CVC4 and unsolved by a configuration of CVC4 with more features enabled.

Moreover, every aspect of the procedure, including the more expensive rules, make a contribution. Three proofs were found thanks to the acyclicity rule and four required uniqueness. Among the latter, three are simple proofs of the form **by** *coinduction auto* in Isabelle. The fourth proof, by Gammie and Lochbihler, is more elaborate:

> **lemma** *num_mod_den_uniq*: $x = $ Node 0 num $x \implies x = $ num_mod_den
> **proof** (*coinduction arbitrary*: x *rule*: *tree.coinduct_strong*)
>   **case** (*Eq_tree x*) **show** *?case*
>   **by** (*subst* (1 2 3 4) *Eq_tree*) (*simp add*: *eqTrueI*[OF *Eq_tree*])
> **qed**

where num_mod_den is defined as num_mod_den = Node 0 num num_mod_den.

6   Conclusion

We introduced a decision procedure for the universal theory of datatypes and codatatypes. Our main contribution has been the support for codatatypes. Both the metatheory and the implementation in CVC4 rely on μ-terms to represent cyclic values. Although this aspect is primarily motivated by codatatypes, it enables a uniform account of datatypes and codatatypes—in particular, the acyclicity rule for datatypes exploits μ-terms to detect cycles. The empirical results on Isabelle benchmarks confirm that CVC4's new capabilities improve the state of the art.

This work is part of a wider program that aims at enriching automatic provers with high-level features and at reducing the gap between automatic and interactive theorem proving. As future work, it would be useful to implement proof reconstruction for (co)datatype inferences in Isabelle. CVC4's finite model finding capabilities [22] could also be interfaced for generating counterexamples in proof assistants; in this context, the acyclicity and uniqueness rules are crucial to exclude spurious countermodels. Finally, it might be worthwhile to extend SMT solvers with dedicated support for (co)recursion.

Acknowledgment. We owe a great debt to the development team of CVC4, including Clark Barrett and Cesare Tinelli, and in particular Morgan Deters, who jointly with the first author developed the initial version of the theory solver for datatypes in CVC4. Our present and former bosses, Viktor Kuncak, Stephan Merz, Tobias Nipkow, Cesare Tinelli, and Christoph Weidenbach, have either encouraged the research on codatatypes or at least benevolently tolerated it, both of which we are thankful for. Peter Gammie and Andreas Lochbihler provided useful benchmarks. Andrei Popescu helped clarify our thoughts regarding codatatypes and indicated related work. Dmitriy Traytel took part in discussions about degenerate codatatypes. Pascal Fontaine, Andreas Lochbihler, Andrei Popescu, Christophe Ringeissen, Mark Summerfield, Dmitriy Traytel, and the anonymous reviewers suggested many textual improvements. The second author's research was partially supported by the Deutsche Forschungsgemeinschaft project "Den Hammer härten" (grant NI 491/14-1) and the Inria technological development action "Contre-exemples Utilisables par Isabelle et Coq" (CUIC).

References

1. Barrett, C., Conway, C.L., Deters, M., Hadarean, L., Jovanovic, D., King, T., Reynolds, A., Tinelli, C.: CVC4. In: Gopalakrishnan, G., Qadeer, S. (eds.) CAV 2011. LNCS, vol. 6806, pp. 171–177. Springer, Heidelberg (2011)
2. Barrett, C., Shikanian, I., Tinelli, C.: An abstract decision procedure for satisfiability in the theory of inductive data types. J. Satisf. Boolean Model. Comput. **3**, 21–46 (2007)
3. Blanchette, J.C., Böhme, S., Paulson, L.C.: Extending Sledgehammer with SMT solvers. J. Autom. Reasoning **51**(1), 109–128 (2013)
4. Blanchette, J.C., Nipkow, T.: Nitpick: a counterexample generator for higher-order logic based on a relational model finder. In: Kaufmann, M., Paulson, L.C. (eds.) ITP 2010. LNCS, vol. 6172, pp. 131–146. Springer, Heidelberg (2010)
5. Carayol, A., Morvan, C.: On rational trees. In: Ésik, Z. (ed.) CSL 2006. LNCS, vol. 4207, pp. 225–239. Springer, Heidelberg (2006)
6. Djelloul, K., Dao, T., Frühwirth, T.W.: Theory of finite or infinite trees revisited. Theor. Pract. Log. Prog. **8**(4), 431–489 (2008)
7. Ganzinger, H., Hagen, G., Nieuwenhuis, R., Oliveras, A., Tinelli, C.: DPLL(T): fast decision procedures. In: Alur, R., Peled, D.A. (eds.) CAV 2004. LNCS, vol. 3114, pp. 175–188. Springer, Heidelberg (2004)
8. Jovanović, D., Barrett, C.: Sharing is caring: combination of theories. In: Tinelli, C., Sofronie-Stokkermans, V. (eds.) FroCoS 2011. LNCS, vol. 6989, pp. 195–210. Springer, Heidelberg (2011)
9. Kersani, A., Peltier, N.: Combining superposition and induction: a practical realization. In: Fontaine, P., Ringeissen, C., Schmidt, R.A. (eds.) FroCoS 2013. LNCS, vol. 8152, pp. 7–22. Springer, Heidelberg (2013)
10. Klein, G., Nipkow, T., Paulson, L. (eds.): Archive of Formal Proofs. http://afp.sf.net/
11. Kozen, D.: Results on the propositional μ-calculus. Theor. Comput. Sci. **27**, 333–354 (1983)

12. Leino, K.R.M., Moskal, M.: Co-induction simply. In: Jones, C., Pihlajasaari, P., Sun, J. (eds.) FM 2014. LNCS, vol. 8442, pp. 382–398. Springer, Heidelberg (2014)
13. Leroy, X.: A formally verified compiler back-end. J. Autom. Reasoning **43**(4), 363–446 (2009)
14. Lochbihler, A.: Verifying a compiler for java threads. In: Gordon, A.D. (ed.) ESOP 2010. LNCS, vol. 6012, pp. 427–447. Springer, Heidelberg (2010)
15. Lochbihler, A.: Making the java memory model safe. ACM Trans. Program. Lang. Syst. **35**(4), 12:1–12:65 (2014)
16. de Moura, L., Bjørner, N.S.: Efficient E-matching for smt solvers. In: Pfenning, F. (ed.) CADE 2007. LNCS (LNAI), vol. 4603, pp. 183–198. Springer, Heidelberg (2007)
17. de Moura, L., Bjørner, N.S.: Z3: an efficient SMT solver. In: Ramakrishnan, C.R., Rehof, J. (eds.) TACAS 2008. LNCS, vol. 4963, pp. 337–340. Springer, Heidelberg (2008)
18. Nelson, G., Oppen, D.C.: Simplification by cooperating decision procedures. ACM Trans. Program. Lang. Syst. **1**(2), 245–257 (1979)
19. Pham, T., Whalen, M.W.: RADA: a tool for reasoning about algebraic data types with abstractions. In: Meyer, B., Baresi, L., Mezini, M. (eds.) ESEC/FSE 2013, pp. 611–614. ACM (2013)
20. Reynolds, A., Blanchette, J.C.: A decision procedure for (co)datatypes in SMT solvers. Technical report (2015). http://lara.epfl.ch/reynolds/CADE2015-cdt/
21. Reynolds, A., Kuncak, V.: Induction for SMT solvers. In: D'Souza, D., Lal, A., Larsen, K.G. (eds.) VMCAI 2015. LNCS, vol. 8931, pp. 80–98. Springer, Heidelberg (2015)
22. Reynolds, A., Tinelli, C., Goel, A., Krstić, S., Deters, M., Barrett, C.: Quantifier instantiation techniques for finite model finding in SMT. In: Bonacina, M.P. (ed.) CADE 2013. LNCS, vol. 7898, pp. 377–391. Springer, Heidelberg (2013)
23. Reynolds, A., Tinelli, C., de Moura, L.: Finding conflicting instances of quantified formulas in SMT. In: FMCAD 2014. pp. 195–202. IEEE (2014)
24. Rutten, J.J.M.M.: Universal coalgebra—a theory of systems. Theor. Comput. Sci. **249**, 3–80 (2000)
25. Stump, A., Sutcliffe, G., Tinelli, C.: StarExec: a cross-community infrastructure for logic solving. In: Demri, S., Kapur, D., Weidenbach, C. (eds.) IJCAR 2014. LNCS, vol. 8562, pp. 367–373. Springer, Heidelberg (2014)
26. Suter, P., Köksal, A.S., Kuncak, V.: Satisfiability modulo recursive programs. In: Yahav, E. (ed.) Static Analysis. LNCS, vol. 6887, pp. 298–315. Springer, Heidelberg (2011)
27. Wand, D.: Polymorphic+typeclass superposition. In: de Moura, L., Konev, B., Schulz, S. (eds.) PAAR 2014 (2014)
28. Weber, T.: SAT-Based Finite Model Generation for Higher-Order Logic. Ph.D. Thesis, Technische Universität München (2008)

Deciding ATL* Satisfiability by Tableaux

Amélie David[(✉)]

Laboratoire IBISC, Université D'Évry Val-d'Essonne,
EA 4526 23 Bd de France, 91037 Évry Cedex, France
adavid@ibisc.univ-evry.fr

Abstract. We propose a tableau-based decision procedure for the full
Alternating-time Temporal Logic ATL*. We extend our procedure for
ATL+ in order to deal with nesting of temporal operators. As a side
effect, we obtain a new and conceptually simple tableau method for
CTL*. The worst case complexity of our procedure is 3EXPTIME, which
is suboptimal compared to the 2EXPTIME complexity of the problem. However our method is human-readable and easily implementable.
A web application and binaries for our procedure are available at http://
atila.ibisc.univ-evry.fr/tableau_ATL_star/.

Keywords: Alternating-time temporal logic · ATL* · Automated theorem prover · Satisfiability · Tableaux

1   Introduction

The logic ATL* is the full version of the Alternating-time Temporal Logic introduced in [1] in order to describe open systems, that is systems that can interact
with their environment. Thus ATL and ATL* are the multi-agent versions of the
branching-time temporal logics CTL and CTL*. Indeed, in ATL and ATL* the
environment is modelled by an extra agent e interfering with the system components (the remaining agents) who need to succeed their task no matter how
e responds to their actions. ATL* is an important extension of ATL and ATL+
(an intermediate logic between ATL and ATL*) since it allows one to express
useful properties such as fairness constraints. Such properties can be expressed
only if nesting of temporal operators is possible, which is not the case in ATL
and ATL+. It is worth noting that ATL+ only permits Boolean combination of
unnested temporal operators.

   ' The problem studied in this paper is about deciding the satisfiability of ATL*
formulae. Models for ATL* formulae are directed graphs called *concurrent game
structures* where transitions between two states depend on the chosen action
of each agent. In general, there exists two ways for deciding the satisfiability:
using automata, as done in [8] or using tableaux; as we do here. In this paper,

---

All proofs of lemmas, propositions and theorems, as well as complete examples can
be found in the version with appendices at https://www.ibisc.univ-evry.fr/~adavid/
fichiers/cade15_tableaux_atl_star_long.pdf.

© Springer International Publishing Switzerland 2015
A.P. Felty and A. Middeldorp (Eds.): CADE-25, LNAI 9195, pp. 214–228, 2015.
DOI: 10.1007/978-3-319-21401-6_14

we propose the first tableau-based decision procedure for ATL*, as well as the first implementation of a decision procedure for ATL*, which is also the first implementation of a tableau-based decision procedure for ATL$^+$. We extend our procedure for ATL$^+$ [2] following the natural basic idea: separate present and future. However, this extension is not trivial since the separation into present and future is more subtle than for ATL$^+$ and needs to keep track of path formulae so as to be able to check eventualities. We think that our tableau-based decision procedure for ATL* is easy to understand and therefore also provides a new tableau-based decision procedure for CTL* which is conceptually simple. We prove that our procedure runs in at most 3EXPTIME, which is suboptimal (the optimal worst case complexity has been proved to be 2EXPTIME in [8]). However, we do not know of any specific cases where our procedure runs in 3EXPTIME, which leaves the possibility that it is optimal, after all.

This paper is organized as follows: Sect. 2 gives the syntax and semantics for ATL*. The general structure of the tableau-based decision procedure for ATL* that we propose can be found in Sect. 3 and details of the procedure in Sects. 4, 5 and 6. Theorems about soundness, completeness and complexity are given in Sect. 7 with their sketch of proof. The Sect. 8 is about the implementation of our procedure. The paper ends with some concluding remarks indicating also some possible directions of future research.

2 Syntax and Semantics of ATL*

ATL* can be seen as an extension of the *computational tree logic* (CTL*) [5] where the path quantifiers E – there exists a path – and A – for all paths – are replaced by $\langle\!\langle A \rangle\!\rangle$ and $[\![A]\!]$ where A is a coalition of agents. Intuitively $\langle\!\langle A \rangle\!\rangle \Phi$ means "There exists a strategy for the coalition A such that, no matter which strategy the remaining agents follow, Φ holds". On the other hand, $[\![A]\!]\Phi$ means "For all strategies of the coalition A, there exists a strategy of the remaining agents such that Φ holds". Also, whereas *transition systems* or *Kripke structures* are used in order to evaluate CTL* formulae, *concurrent game models* (CGM), whose definition is given in the Sect. 2.2 are used to evaluate ATL* formulae.

2.1 Syntax of ATL*

Before giving the syntax of ATL*, we recall that, as for CTL* or LTL, $\bigcirc$, $\square$ and $\mathcal{U}$ mean "*Next*", "*Always*" and "*Until*" respectively. In this paper, we give the syntax in negation normal form over a fixed set $\mathbb{P}$ of atomic propositions and primitive temporal operators $\bigcirc$ "*Next*", $\square$ "*Always*" and $\mathcal{U}$ "*Until*". The syntax of ATL* in negation normal form is defined as follows:

$$\text{State formulae:} \varphi := l \mid (\varphi \vee \varphi) \mid (\varphi \wedge \varphi) \mid \langle\!\langle A \rangle\!\rangle \Phi \mid [\![A]\!]\Phi \tag{1}$$

$$\text{Path formulae:} \Phi := \varphi \mid \bigcirc\Phi \mid \square\Phi \mid (\Phi\mathcal{U}\Phi) \mid (\Phi \vee \Phi) \mid (\Phi \wedge \Phi) \tag{2}$$

where $l \in \mathbb{P} \cup \{\neg p \mid p \in \mathbb{P}\}$ is a literal, A is a fixed finite set of agents and $A \subseteq \mathbb{A}$ is a coalition. Note that $\top := p \vee \neg p$, $\bot := \neg\top$, $\neg\langle\!\langle A \rangle\!\rangle\Phi := [\![A]\!]\neg\Phi$.

The temporal operator *"Sometimes"* $\Diamond$ can be defined as $\Diamond\varphi := \top \mathcal{U}\varphi$ and the temporal operator *"Release"* as $\psi \mathcal{R}\varphi := \Box\varphi \vee \varphi\mathcal{U}(\varphi \wedge \psi)$. When unnecessary, parentheses can be omitted.

In this paper, we use φ, ψ, η to denote arbitrary state formulae and Φ, Ψ to denote path formulae. By an ATL* formula we will mean by default a *state* formula of ATL*.

2.2   Concurrent Game Models

As for ATL or ATL$^+$, ATL* formulae are evaluated over *concurrent game models*. Concurrent games models are transition systems where each transition to a unique successor state results from the combination of actions chosen by all the agents (components and/or environment) of the system.

Notation: Given a set X, $\mathcal{P}(X)$ denotes the power set of X.

Definition 1 (Concurrent game model and structure). *A concurrent game model (in short CGM) is a tuple* $\mathcal{M} = (\mathbb{A}, \mathbb{S}, \{\mathsf{Act}_\mathsf{a}\}_{\mathsf{a}\in\mathbb{A}}, \{\mathsf{act}_\mathsf{a}\}_{\mathsf{a}\in\mathbb{A}}, \mathsf{out}, \mathbb{P}, \mathsf{L})$ *where*

- $\mathbb{A} = \{1, \ldots, k\}$ *is a finite, non-empty set of* players (agents),
- $\mathbb{S}$ *is a non-empty set of* states,
- *for each agent* $\mathsf{a} \in \mathbb{A}$, Act_a *is a non-empty set of* actions.
  For any coalition $A \subseteq \mathbb{A}$ *we denote* $\mathsf{Act}_A := \prod_{\mathsf{a}\in A} \mathsf{Act}_\mathsf{a}$ *and use* σ_A *to denote a tuple from* Act_A. *In particular,* $\mathsf{Act}_\mathbb{A}$ *is the set of all possible* action vectors *in* $\mathcal{M}$.
- *for each agent* $\mathsf{a} \in \mathbb{A}$, $\mathsf{act}_\mathsf{a} : \mathbb{S} \to \mathcal{P}(\mathsf{Act}_\mathsf{a}) \setminus \{\emptyset\}$ *defines for each state* s *the actions available to* a *at* s,
- out *is a transition function assigning to every state* $s \in \mathbb{S}$ *and every action vector* $\sigma_\mathbb{A} = \{\sigma_1, \ldots, \sigma_k\} \in \mathsf{Act}_\mathbb{A}$ *a state* $\mathsf{out}(s, \sigma_\mathbb{A}) \in \mathbb{S}$ *that results from* s *if every agent* $\mathsf{a} \in \mathbb{A}$ *plays action* σ_a, *where* $\sigma_\mathsf{a} \in \mathsf{act}_\mathsf{a}(s)$ *for every* $\mathsf{a} \in \mathbb{A}$.
- $\mathbb{P}$ *is a non-empty set of atomic propositions.*
- $\mathsf{L} : \mathbb{S} \to \mathcal{P}(\mathbb{P})$ *is a labelling function.*

The sub-tuple $\mathcal{S} = (\mathbb{A}, \mathbb{S}, \{\mathsf{Act}_\mathsf{a}\}_{\mathsf{a}\in\mathbb{A}}, \{\mathsf{act}_\mathsf{a}\}_{\mathsf{a}\in\mathbb{A}}, \mathsf{out})$ *is called a* concurrent game structure *(CGS).*

2.3   Semantics of ATL*

In order to give the semantics of ATL*, we use the following notions. Although they are the same as those in [2], we recall them here to make the paper self-contained.

Computations. A *play*, or *computation*, is an infinite sequence $s_0 s_1 s_2 \cdots \in \mathbb{S}^\omega$ of states such that for each $i \geq 0$ there exists an action vector $\sigma_\mathbb{A} = \langle \sigma_1, \ldots, \sigma_k \rangle$ such that $\mathsf{out}(s_i, \sigma_\mathbb{A}) = s_{i+1}$. A *history* is a finite prefix of a play. We denote by $\mathsf{Plays}_\mathcal{M}$ and $\mathsf{Hist}_\mathcal{M}$ respectively the set of plays and set of histories in $\mathcal{M}$.

For a state $s \in \mathbb{S}$, we use $\mathsf{Plays}_{\mathcal{M}}(s)$ and $\mathsf{Hist}_{\mathcal{M}}(s)$ as the set of plays and set of histories with initial state s. Given a sequence of states λ, we denote by λ_0 its initial state, by λ_i its $(i+1)$th state, by $\lambda_{\leq i}$ the prefix $\lambda_0 \dots \lambda_i$ of λ and by $\lambda_{\geq i}$ the suffix $\lambda_i \lambda_{i+1} \dots$ of λ. When $\lambda = \lambda_0 \dots \lambda_\ell$ is finite, we say that it has length ℓ and write $|\lambda| = \ell$. Also, we set $last(\lambda) = \lambda_\ell$.

Strategies. A *strategy* for an agent a in $\mathcal{M}$ is a mapping $F_{\mathsf{a}} : \mathsf{Hist}_{\mathcal{M}} \to \mathsf{Act}_{\mathsf{a}}$ such that for all histories $h \in \mathsf{Hist}_{\mathcal{M}}$, we have $F_{\mathsf{a}}(h) \in act_{\mathsf{a}}(last(h))$. This kind of strategies is also known as "perfect recall" strategies. We denote by $\mathsf{Strat}_{\mathcal{M}}(\mathsf{a})$ the set of strategies of agent a. A collective strategy of a coalition $\mathsf{A} \subseteq \mathbb{A}$ is a tuple $(F_{\mathsf{a}})_{\mathsf{a} \in \mathsf{A}}$ of strategies, one for each agent in A. We denote by $\mathsf{Strat}_{\mathcal{M}}(\mathsf{A})$ the set of collective strategies of coalition A. A play $\lambda \in \mathsf{Plays}_{\mathcal{M}}$ is consistent with a collective strategy $F_{\mathsf{A}} \in \mathsf{Strat}_{\mathcal{M}}(\mathsf{A})$ if for every $i \geq 0$ there exists an action vector $\sigma_{\mathbb{A}} = \langle \sigma_1, \dots, \sigma_k \rangle$ such that $out(\lambda_i, \sigma_{\mathbb{A}}) = \lambda_{i+1}$ and $\sigma_{\mathsf{a}} = F_{\mathsf{a}}(\lambda_{\leq i})$ for all $\mathsf{a} \in \mathsf{A}$. The set of plays with initial state s that are consistent with F_{A} is denoted $\mathsf{Plays}_{\mathcal{M}}(s, F_{\mathsf{A}})$. For any coalition $\mathsf{A} \subseteq \mathbb{A}$ and a given state $s \in \mathbb{S}$ in a given CGM $\mathcal{M}$, an A *-co-action* at s in $\mathcal{M}$ is a mapping $\mathsf{Act}^{\mathsf{c}}_{\mathsf{A}} : \mathsf{Act}_{\mathsf{A}} \to \mathsf{Act}_{\mathbb{A} \setminus \mathsf{A}}$ that assigns to every collective action of A at the state s a collective action at s for the complementary coalition $\mathbb{A} \setminus \mathsf{A}$. Likewise, an A *-co-strategy* in $\mathcal{M}$ is a mapping $F^{\mathsf{c}}_{\mathsf{A}} : \mathsf{Strat}_{\mathcal{M}}(\mathsf{A}) \times \mathsf{Hist}_{\mathcal{M}} \to \mathsf{Act}_{\mathbb{A} \setminus \mathsf{A}}$ that assigns to every collective strategy of A and every history h a collective action at $last(h)$ for $\mathbb{A} \setminus \mathsf{A}$, and $\mathsf{Plays}_{\mathcal{M}}(s, F^{\mathsf{c}}_{\mathsf{A}})$ is the set of plays with initial state s that are consistent with $F^{\mathsf{c}}_{\mathsf{A}}$.

Semantics. The semantics of ATL* is the same as the one of CTL* [5] (modulo CGM as intended interpretations) with the exception of the two following items:

- $\mathcal{M}, s \models \langle\!\langle \mathsf{A} \rangle\!\rangle \Phi$ iff there exists an A-strategy F_{A} such that, for all computations $\lambda \in \mathsf{Plays}_{\mathcal{M}}(s, F_{\mathsf{A}})$, $\mathcal{M}, \lambda \models \Phi$
- $\mathcal{M}, s \models [\![\mathsf{A}]\!] \Phi$ iff there exists an A-co-strategy $F^{\mathsf{c}}_{\mathsf{A}}$ such that, for all computations $\lambda \in \mathsf{Plays}_{\mathcal{M}}(s, F^{\mathsf{c}}_{\mathsf{A}})$, $\mathcal{M}, \lambda \models \Phi$

Valid, satisfiable and equivalent formulae in ATL* are defined as usual.

3   Tableau-Based Decision Procedure for ATL*

In this section, we give the general description of our tableau-based decision procedure for ATL* formulae. The different steps of the procedure are summarized in this section and Fig. 1 and then detailed in the next three sections.

From an initial formula η, the tableau-based decision procedure for ATL* that we propose attempts to build step-by-step a directed graph from which it is possible to extract a CGM for η. This attempt will lead to a failure if η is not satisfiable.

Nodes of that graph are labelled by sets of *state formulae* and are partitioned into two categories: *prestates* and *states*.

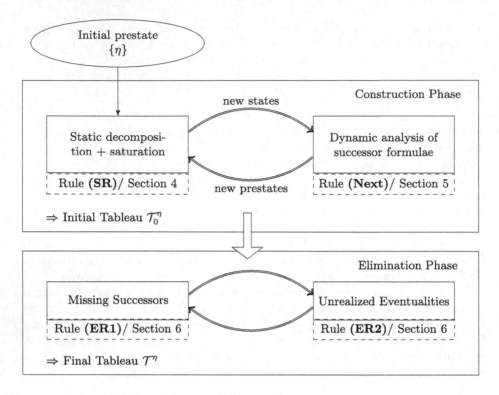

Fig. 1. Overview of the tableau-based decision procedure for ATL*

A prestate can be seen as a node where the information contained in its formulae is "implicit". When we decompose all the formulae of a prestate and saturate the prestate, we obtain one or several states as successor nodes. States have the particularity of containing formulae of the form $\langle\!\langle A \rangle\!\rangle \bigcirc \varphi$ or $[\![A]\!] \bigcirc \varphi$ from which it is possible to compute the next steps of the tableau creation. All prestates have states as successors and directed edges between them are of the form $\Longrightarrow$; on the other hand, all states have prestates as successors and directed edges between them are of the form $\xrightarrow{\sigma_{\mathbb{A}}}$ where $\sigma_{\mathbb{A}}$ is an action vector.

The procedure is in two phases: the *construction phase* and the *elimination phase*. First, we create an initial node, that is a prestate containing the initial formula η, and we construct the graph by expanding prestates into states via a rule called **(SR)** and by computing prestates from states with a rule called **(Next)**. The rule **(SR)** decomposes each ATL* formula of a prestate, and then saturates the prestate into new states. Explanation of rules **(SR)** and **(Next)** can be found in Sects. 4 and 5, respectively.

The procedure avoids creation of duplicated nodes (a form of loop check), which ensures termination of the procedure. The construction phase ends when no new states can be added to the graph. The graph obtained at the end of the construction phase is called the *initial tableau for η*, also noted $\mathcal{T}_0^\eta$.

The second phase of the procedure eliminates via the rule **(ER1)** all nodes with missing successors, that is prestates with no more successors at all or states with at least one missing action vector on its outcome edges. Also, by means of a rule called **(ER2)** it eliminates all states with "unrealized eventualities", that is states that cannot ensure that all the objectives it contains will be eventually fulfilled. The graph obtained at the end of the elimination phase of the procedure is called the *final tableau for* η, also noted $\mathcal{T}^\eta$. Explanation of rules **(ER1)** and **(ER2)** can be found in Sect. 6.

Our tableau-based decision procedure for ATL* deals with what [6] calls "tight satisfiability": the set $\mathbb{A}$ of agents involved in the tableau (and the CGMs it tries to build) is the set of agents present in the input formula.

4   Construction Phase: Decomposition and Saturation

Decomposition of ATL* Formulae All ATL* formulae can be partitioned into four categories: *primitive formulae, α-formulae, β-formulae* and *γ-formulae*. Primitive formulae correspond to the "simplest formulae" in the sense that they cannot be decomposed. These formulae are $\top, \bot$, the literals and all ATL* *successor formulae*, of the form $\langle\!\langle A \rangle\!\rangle \bigcirc \psi$ or $[\![A]\!] \bigcirc \psi$ where ψ is called the *successor component* of $\langle\!\langle A \rangle\!\rangle \bigcirc \psi$ or $[\![A]\!] \bigcirc \psi$ respectively. Every non-primitive formula must be decomposed into primitive formulae. α-formulae are of the form $\varphi \wedge \psi$ where φ and ψ are α-components while β-formulae are of the form $\varphi \vee \psi$ where φ and ψ are β-components. Their decomposition is classical. Other formulae, that is of the form $\langle\!\langle A \rangle\!\rangle \Phi$ or $[\![A]\!]\Phi$, where $\Phi \neq \bigcirc \psi$, are γ-formulae. This notion firstly introduced in [2] reveals quite useful also in the more expressive context of ATL*. Decomposition of these formulae is trickier than for α- and β-formulae. Indeed, we will need to extract all possibilities of truth encapsulated in γ-formula ξ, which concretely aims at defining one or several conjunctions of primitive formulae such that their disjunction is equivalent to the γ-formulae ξ (see lemma 1).

Decomposition of γ-Formulae. This subsection contains the heart of the decision procedure for ATL*, indeed the main difference with our decision procedure for ATL$^+$ lies in the treatment of γ-formulae. The first difficulty is that quantifiers $\langle\!\langle A \rangle\!\rangle$ or $[\![A]\!]$ cannot distribute over Boolean connectors as seen in [2]. An additional difficulty specific to ATL* is the fact that it is now necessary to also deal with nesting of temporal operators, resulting in a second level of recurrence when the temporal operators $\square$ and $\mathcal{U}$ are encountered in the decomposition function described below.

In temporal logics, e.g. LTL, the operator $\mathcal{U}$ is considered as an eventuality operator, that is an operator that promises to verify a given formula at some instant/state. When we write $\lambda \models \varphi_1 \mathcal{U} \varphi_2$, where φ_1 and φ_2 are state formulae, we mean that there is a state λ_i of the computation λ where φ_2 holds and φ_1 holds for all the states of λ preceding λ_i. So, once the property φ_2 is verified, we do not need to take care of φ_1, φ_2 and $\varphi_1 \mathcal{U} \varphi_2$ any more. We say that $\varphi_1 \mathcal{U} \varphi_2$ is *realized*. However, if φ_1 and φ_2 are path formulae, e.g. $\square \Phi_1$ and $\square \Phi_2$ respectively, state λ_i is such that from it Φ_2 must hold forever – we say that $\square \Phi_2$

is "initiated" at λ_i, in the sense that we start to make $\Box\Phi_2$ true at λ_i −, and for every computation $\lambda_{\geq j}$, where $j < i$, $\Box\Phi_1$ must hold. So Φ_1 has to be true forever, that is even after $\Box\Phi_2$ had been initiated. This explains the fact that at a possible state s the path formula $\varphi_1\mathcal{U}\varphi_2$ may become $\varphi_1\mathcal{U}\varphi_2 \wedge \varphi_1$ when φ_1 is a path formula and we postpone φ_2. Note that φ_1 is then also initiated at s. We now face the problem of memorizing the fact that a path formula Φ is initiated since path formulae cannot be stored directly in a state. That is why, during the decomposition of γ-formulae, we add a new set of path formulae linked to a γ-component and the current state.

The definition and general treatment of eventualities in our procedure are given in Sect. 6.

In order to decompose γ-formulae $\varphi = \langle\!\langle A \rangle\!\rangle\Phi$ or $\varphi = [\![A]\!]\Phi$, we analyse the path formula Φ in terms of present (current state) and future (next states). This analysis is done by a γ-*decomposition* function $\mathsf{dec} : \mathsf{ATL}_\mathsf{p}^* \to \mathcal{P}(\mathsf{ATL}_\mathsf{s}^* \times \mathsf{ATL}_\mathsf{p}^* \times \mathcal{P}(\mathsf{ATL}_\mathsf{p}^*))$ where $\mathsf{ATL}_\mathsf{p}^*$ is the set of ATL^* path formulae and $\mathsf{ATL}_\mathsf{s}^*$ is the set of ATL^* state formulae. Intuitively, the function dec assigns to the path formula Φ, a set of triples $\langle \psi, \Psi, S \rangle$ where ψ is a state formula true at the current state, Ψ is a path formula expressing what must be true at next states and S is the set of path formulae initiated at the current state during the γ-decomposition This set S will be used during the elimination phase to determine if eventualities are realized or not, see Sect. 6.

We first define two operators $\otimes$ and $\oplus$ between two sets $\mathfrak{S}_1$ and $\mathfrak{S}_2$ of triples.

⋆ $\mathfrak{S}_1 \otimes \mathfrak{S}_2 := \{\langle \psi_i \,\dot\wedge\, \psi_j, \Psi_i \,\dot\wedge\, \Psi_j, S_i \cup S_j \rangle \mid \langle\psi_i,\Psi_i,S_i\rangle \in \mathfrak{S}_1, \langle\psi_j,\Psi_j,S_j\rangle \in \mathfrak{S}_2\}$
⋆ $\mathfrak{S}_1 \oplus \mathfrak{S}_2 := \{\langle \psi_i \,\dot\wedge\, \psi_j, \Psi_i \,\dot\vee\, \Psi_j, S_i \cup S_j \rangle \mid \langle\psi_i,\Psi_i,S_i\rangle \in \mathfrak{S}_1, \langle\psi_j,\Psi_j,S_j\rangle \in \mathfrak{S}_2,$
$\Psi_i \neq \top, \Psi_j \neq \top\}$

The function dec is defined by induction on the structure of path formula Φ as follows:

⋆ $\mathsf{dec}(\varphi) = \{\langle\varphi, \top, \emptyset\rangle\}$ for any ATL^* state formula φ
⋆ $\mathsf{dec}(\bigcirc\Phi_1) = \{\langle\top, \Phi_1, \emptyset\rangle\}$ for any path formula Φ_1
⋆ $\mathsf{dec}(\Box\Phi_1) = \{\langle\top, \Box\Phi_1, \{\Phi_1\}\rangle\} \otimes \mathsf{dec}(\Phi_1)$
⋆ $\mathsf{dec}(\Phi_1\mathcal{U}\Phi_2) = (\{\langle\top, \Phi_1\mathcal{U}\Phi_2, \{\Phi_1\}\rangle\} \otimes \mathsf{dec}(\Phi_1)) \cup (\{\langle\top, \top, \{\Phi_2\}\rangle\} \otimes \mathsf{dec}(\Phi_2))$
⋆ $\mathsf{dec}(\Phi_1 \wedge \Phi_2) = \mathsf{dec}(\Phi_1) \otimes \mathsf{dec}(\Phi_2)$
⋆ $\mathsf{dec}(\Phi_1 \vee \Phi_2) = \mathsf{dec}(\Phi_1) \cup \mathsf{dec}(\Phi_2) \cup (\mathsf{dec}(\Phi_1) \oplus \mathsf{dec}(\Phi_2))$

Note that the definition of the function dec is based on the fixed-point equivalences of LTL [4]: $\Box\Psi \equiv \Psi \wedge \bigcirc\Box\Psi$ and $\Phi\mathcal{U}\Psi \equiv \Psi \vee (\Phi \wedge \bigcirc(\Phi\mathcal{U}\Psi))$.

The operators $\dot\wedge$ and $\dot\vee$ correspond respectively to the operators $\wedge$ and $\vee$ where the associativity, commutativity, idempotence and identity element properties are embedded in the syntax. The aim of both $\dot\wedge$ and $\dot\vee$ is to automatically transform resultant formulae in conjunctive normal form without redundancy, and therefore ensures the termination of our tableau-based decision procedure. For instance, when applying the function dec on $\Box\Diamond\Phi \wedge \Diamond\Phi$ we may obtain a path formula $\Box\Diamond\Phi \wedge \Diamond\Phi \wedge \Diamond\Phi$ and applying again the function dec on the so-obtained path formula will return $\Box\Diamond\Phi \wedge \Diamond\Phi \wedge \Diamond\Phi \wedge \Diamond\Phi$, and so on forever. Also when the formula is complicated with $\wedge$ and $\vee$ embedded in temporal operators, we may not be able to define which part of a path formula is identical to another one. We avoid these unwanted behaviours with our use of $\dot{\wedge}$ and $\dot{\vee}$ and the transformation of any new path formula in conjunctive normal form without redundancies.

Now, let $\zeta = \langle\!\langle A \rangle\!\rangle \Phi$ or $\zeta = [\![A]\!]\Phi$ be a γ-formula to be decomposed. Each triple $\langle \psi, \Psi, S \rangle \in \mathsf{dec}(\Phi)$ is then converted to a γ-*component* $\gamma_c(\psi, \Psi, S)$ as follows:

$$\gamma_c(\psi, \Psi, S) = \psi \quad \text{if } \Psi = \top \tag{3}$$

$$\gamma_c(\psi, \Psi, S) = \psi \wedge \langle\!\langle A \rangle\!\rangle \bigcirc \langle\!\langle A \rangle\!\rangle \Psi \quad \text{if } \zeta \text{ is of the form } \langle\!\langle A \rangle\!\rangle \Phi, \tag{4}$$

$$\gamma_c(\psi, \Psi, S) = \psi \wedge [\![A]\!] \bigcirc [\![A]\!] \Psi \quad \text{if } \zeta \text{ is of the form } [\![A]\!]\Phi \tag{5}$$

and a γ-set $\gamma_s(\psi, \Psi, S) = S$.

The following key lemma claims that every γ-formula is equivalent to the disjunction of its γ-components.

Lemma 1. *For any* ATL* γ-*formula* $\zeta = \langle\!\langle A \rangle\!\rangle \Phi$ *or* $\zeta = [\![A]\!]\Phi$

1. $\Phi \equiv \bigvee \{\psi \wedge \bigcirc \Psi \mid \langle \psi, \Psi, S \rangle \in \mathsf{dec}(\Phi)\}$
2. $\langle\!\langle A \rangle\!\rangle \Phi \equiv \bigvee \{\langle\!\langle A \rangle\!\rangle (\psi \wedge \bigcirc \Psi) \mid \langle \psi, \Psi, S \rangle \in \mathsf{dec}(\Phi)\}$, *and*
   $[\![A]\!]\Phi \equiv \bigvee \{[\![A]\!](\psi \wedge \bigcirc \Psi) \mid \langle \psi, \Psi, S \rangle \in \mathsf{dec}(\Phi)\}$
3. $\langle\!\langle A \rangle\!\rangle \Phi \equiv \bigvee \{\gamma_c(\psi, \Psi, S) \mid \langle \psi, \Psi, S \rangle \in \mathsf{dec}(\Phi)\}$

Example 1. (Decomposition of $\theta = \langle\!\langle 1 \rangle\!\rangle ((\Box \Diamond q \vee \Diamond r) \wedge (\Diamond q \vee \Diamond r)))$*).* First, we apply the decomposition function to the path formula $\Phi = (\Box \Diamond q \vee \Diamond r) \wedge (\Diamond q \vee \Diamond r)$, see Fig. 2. We recall that $\Diamond \varphi \equiv \top \, \mathcal{U} \varphi$. It is worth noting that p and r can be replaced by any state formulae without affecting the basic structure of the computation of the function dec.

Then, for instance, from the triple $\langle r, \Box \Diamond q \wedge \Diamond q, \{r, \Diamond q\} \rangle$ of $\mathsf{dec}(\Phi)$, we obtain the γ-component $\gamma_c(r, \Box \Diamond q \wedge \Diamond q, \{r, \Diamond q\}) = r \wedge \langle\!\langle 1 \rangle\!\rangle \bigcirc \langle\!\langle 1 \rangle\!\rangle (\Box \Diamond q \wedge \Diamond q)$ and the γ-set $\gamma_s(r, \Box \Diamond q \wedge \Diamond q, \{r, \Diamond q\}) = \{r, \Diamond q\}$; from the triple $\langle \top, \Box \Diamond q \wedge \Diamond q, \{\Diamond q\} \rangle$ we obtain $\gamma_c(\top, \Box \Diamond q \wedge \Diamond q, \{\Diamond q\}) = \langle\!\langle 1 \rangle\!\rangle \bigcirc \langle\!\langle 1 \rangle\!\rangle (\Box \Diamond q \wedge \Diamond q)$ and $\gamma_s(\top, \Box \Diamond q \wedge \Diamond q, \{\Diamond q\}) = \{\Diamond q\}$.

Closure. The closure $cl(\varphi)$ of an ATL* state formula φ is the least set of ATL* formulae such that $\varphi, \top, \bot \in cl(\varphi)$, and $cl(\varphi)$ is closed under taking successor, α-, β- and γ-components of φ. For any set of state formulae Γ we define

$$cl(\Gamma) = \bigcup \{cl(\psi) \mid \psi \in \Gamma\} \tag{6}$$

We denote by $|\psi|$ the length of ψ and by $||\Gamma||$ the cardinality of Γ.

Lemma 2. *For any* ATL* *state formula* φ, $||cl(\varphi)|| < 2^{2^{2|\varphi|}}$.

Sketch of proof. The double exponent, in the size of the φ, of the closure comes from the fact that, during decomposition of γ-formulae, path formulae are put in disjunctive normal form. We recall that this form is necessary to ensure the termination of our procedure.

$\text{dec}(\Phi) = \text{dec}(\Box\Diamond q \vee \Diamond r) \otimes \text{dec}(\Diamond q \vee \Diamond r)$
① $\text{dec}(\Box\Diamond q \vee \Diamond r) = \text{dec}(\Box\Diamond q) \cup \text{dec}(\Diamond r) \cup (\text{dec}(\Box\Diamond q) \oplus \text{dec}(\Diamond r))$

$$\text{dec}(\Phi) = \text{dec}(\Box\Diamond q \vee \Diamond r) \otimes \text{dec}(\Diamond q \vee \Diamond r)$$

① $\text{dec}(\Box\Diamond q \vee \Diamond r) = \text{dec}(\Box\Diamond q) \cup \text{dec}(\Diamond r) \cup (\text{dec}(\Box\Diamond q) \oplus \text{dec}(\Diamond r))$

❶ $\text{dec}(\Box\Diamond q)$
$= \{\langle \top, \Box\Diamond q, \{q\}\rangle\} \otimes \quad (\{\langle \top, \Diamond q, \emptyset\rangle\} \otimes \text{dec}(\top) \cup$
$\{\langle \top, \top, \{q\}\rangle\} \otimes \text{dec}(q))$
$= \{\langle \top, \Box\Diamond q, \{\Diamond q\}\rangle\} \otimes \{\langle \top, \Diamond q, \emptyset\rangle, \langle q, \top, \{q\}\rangle\}$
$= \{\langle \top, \Box\Diamond q \wedge \Diamond q, \{\top, \Diamond q\}\rangle, \langle q, \Box\Diamond q, \{\Diamond q, q\}\rangle\}$

❷ $\text{dec}(\Diamond q)$
$= \{\langle \top, \Diamond r, \emptyset\rangle, \langle r, \top, \{r\}\rangle\}$

① $\text{dec}(\Box\Diamond q \vee \Diamond r) = \{\langle \top, \Box\Diamond q \wedge \Diamond q, \{\Diamond q\}\rangle, \langle q, \Box\Diamond q, \{\Diamond q, q\}\rangle,$
$\langle \top, \Diamond r, \emptyset\rangle, \langle r, \top, \{r\}\rangle, \langle q, \Box\Diamond q \vee \Diamond r, \{\Diamond q, q\}\rangle,$
$\langle \top, (\Box\Diamond q \dot\wedge \Diamond q) \dot\vee \Diamond r \equiv (\Box\Diamond q \vee \Diamond r) \wedge (\Diamond q \vee \Diamond r), \{\Diamond q\}\rangle,$
$\}$

② $\text{dec}(\Diamond q \vee \Diamond r)$
$= \text{dec}(\Diamond q) \cup \text{dec}(\Diamond r) \cup (\text{dec}(\Diamond q) \oplus \text{dec}(\Diamond r))$
$= \{\langle \top, \Diamond q, \emptyset\rangle, \langle q, \top, \{q\}\rangle\} \cup \{\langle \top, \Diamond r, \emptyset\rangle, \langle r, \top, \{r\}\rangle\}$
$\cup \{\langle \top, \Diamond q \vee \Diamond r, \emptyset\rangle\}$
$= \{\langle \top, \Diamond q, \emptyset\rangle, \langle q, \top, \{q\}\rangle, \langle \top, \Diamond r, \emptyset\rangle, \langle r, \top, \{r\}\rangle,$
$\langle \top, \Diamond q \vee \Diamond r, \emptyset\rangle$

$\text{dec}(\Phi) = ① \otimes ② = \{\langle \top, \Box\Diamond q \dot\wedge \Diamond q \dot\wedge \Diamond q \equiv \Box\Diamond q \wedge \Diamond q, \{\Diamond q\}\rangle, \ldots ,$
$\langle \top, (\Box\Diamond q \dot\vee \Diamond r) \dot\wedge (\Diamond q \dot\vee \Diamond r) \dot\wedge \Diamond q, \{\Diamond q\}\rangle, \ldots$
$\langle r, \Box\Diamond q \wedge \Diamond q, \{r, \Diamond q\}\rangle, \langle \top, \Box\Diamond q \wedge \Diamond q, \{\Diamond q\}\rangle, \ldots\}$

Fig. 2. Function dec applied on the path formula $\langle\langle 1 \rangle\rangle((\Box\Diamond q \vee \Diamond r) \wedge (\Diamond q \vee \Diamond r))$

Full expansions of sets of ATL* formulae. Once we are able to decompose into components every non-primitive ATL* state formulae, it is possible to obtain full expansions of a given set of ATL* state formulae using the following definition:

Definition 2. *Let Γ, Δ be sets of ATL* state formulae and $\Gamma \subseteq \Delta \subseteq cl(\Gamma)$.*

1. *Δ is* patently inconsistent *if it contains $\bot$ or a pair of formulae φ and $\neg\varphi$.*
2. *Δ is a* full expansion *of Γ if it is not patently inconsistent and satisfies the following closure conditions:*
   - *if $\varphi \wedge \psi \in \Delta$ then $\varphi \in \Delta$ and $\psi \in \Delta$;*
   - *if $\varphi \vee \psi \in \Delta$ then $\varphi \in \Delta$ or $\psi \in \Delta$;*
   - *if $\varphi \in \Delta$ is a γ-formula, then at least one γ-component of φ is in Δ and exactly one of these γ-components, say $\gamma_c(\psi, \Psi, S)$, in Δ, denoted $\gamma_l(\varphi, \Delta)$, is designated as the γ-component in Δ linked to the γ-formula φ, as explained below. We also denote by $\gamma_{sl}(\varphi, \Delta)$ the set of path formulae $\gamma_s(\psi, \Psi, S)$, which is linked to the γ-component $\gamma_l(\varphi, \Delta)$*

The set of all full expansions of Γ is denoted by $\mathsf{FE}(\Gamma)$.

Proposition 1. *For any finite set of ATL* state formulae Γ:*

$$\bigwedge \Gamma \equiv \bigvee \left\{ \bigwedge \Delta \mid \Delta \in \mathsf{FE}(\Gamma) \right\}.$$

The proof easily follows from Lemma 1.

The rule **(SR)** adds to the tableau the set of full expansions of a prestate Γ as successor states of Γ.

Rule (SR). Given a prestate Γ, do the following:

1. For each full expansion Δ of Γ add to the pretableau a state with label Δ.
2. For each of the added states Δ, if Δ does not contain any formula of the form $\langle\!\langle A \rangle\!\rangle \bigcirc \varphi$ or $[\![A]\!] \bigcirc \varphi$, add the formula $\langle\!\langle A \rangle\!\rangle \bigcirc \top$ to it;
3. For each state Δ obtained at steps 1 and 2, link Γ to Δ via a $\Longrightarrow$ edge;
4. If, however, the pretableau already contains a state Δ' with label Δ, do not create another copy of it but only link Γ to Δ' via a $\Longrightarrow$ edge.

5 Construction Phase: Dynamic Analysis of Successor Formulae

We recall that the considered agents are those explicitly mentioned in the initial formula η. The rule (**Next**) creates successor prestates to a given state, say Δ, so that the satisfiability of Δ is equivalent to the satisfiability of all the prestates. In our tableau construction procedure, choosing one of the successor formulae contained in Δ is considered as a possible action for every agent. Then each possible action vector is given a set of formulae corresponding to the choice collectively made by every agent. More details about the rationale behind the rule (**Next**) can be found in [3,6]. Moreover, it is worthwhile noticing that the rule (**Next**) is done in such a way so that any created prestate contains at most one formula of the form $[\![A']\!] \bigcirc \psi$, where $A' \neq \mathbb{A}$.

Rule (Next). Given a state Δ, do the following, where σ is a shorthand for $\sigma_{\mathbb{A}}$:

1. List all primitive successor formulae of Δ in such a way that all successor formulae of the form $\langle\!\langle A \rangle\!\rangle \Phi$ precede all formulae of the form $[\![A']\!]\Phi$ where $A' \neq \mathbb{A}$, which themselves precede all formulae of the form $[\![A]\!]\Phi$; let the result be the list

$$\mathbb{L} = \langle\!\langle A_0 \rangle\!\rangle \bigcirc \varphi_0, \ldots, \langle\!\langle A_{m-1} \rangle\!\rangle \bigcirc \varphi_{m-1},$$
$$[\![A'_0]\!] \bigcirc \psi_0, \ldots, [\![A'_{l-1}]\!] \bigcirc \psi_{l-1}, [\![A]\!] \bigcirc \mu_0, \ldots, [\![A]\!] \bigcirc \mu_{n-1}$$

Let $r_\Delta = \max\{m + l, 1\}$; we denote by $D(\Delta)$ the set $\{0, \ldots, r_\Delta - 1\}^{|\mathbb{A}|}$. Then, for every $\sigma \in D(\Delta)$, denote $N(\sigma) := \{i \mid \sigma_i \geqslant m\}$, where σ_i is the ith component of the tuple σ, and let $\mathsf{co}(\sigma) := [\sum_{i \in N(\sigma)}(\sigma_i - m)] \bmod l$.
2. For each $\sigma \in D(\Delta)$ create a prestate:

$$\Gamma_\sigma = \{\varphi_p \mid \langle\!\langle A_p \rangle\!\rangle \bigcirc \varphi_p \in \Delta \text{ and } \sigma_a = p \text{ for all } a \in A_p\}$$
$$\cup \{\psi_q \mid [\![A'_q]\!] \bigcirc \psi_q \in \Delta, \mathsf{co}(\sigma) = q \text{ and } \mathbb{A} - A'_q \subseteq N(\sigma)\}$$
$$\cup \{\mu_r \mid [\![A]\!] \bigcirc \mu_r \in \Delta\}$$

If Γ_σ is empty, add $\top$ to it. Then connect Δ to Γ_σ with $\xrightarrow{\sigma}$.
If, however, $\Gamma_\sigma = \Gamma$ for some prestate Γ that has already been added to the initial tableau, only connect Δ to Γ with $\xrightarrow{\sigma}$.

Example 2. We suppose a state containing the following successor formulae, that we arrange in the following way, where the first line of numbers corresponds to positions among negative successor formulae, and the second line corresponds to positions among successor formulae, with $A \neq \mathbb{A}$.

$$\mathbb{L} = \overset{0}{\langle\langle 1\rangle\rangle} \bigcirc \overset{0}{\langle\langle 1\rangle\rangle} (\square\Diamond q \wedge \Diamond q), \overset{1}{[1]} \bigcirc \overset{1}{[1]}\square\neg q, \overset{1}{[2]} \bigcirc \overset{2}{[2]}\square\Diamond s, [1,2] \bigcirc \neg q$$

The application of the rule **(Next)** on $\mathbb{L}$ gives the following results:

σ	$N(\sigma)$	$\mathsf{co}(\sigma)$	$\Gamma(\sigma)$	σ	$N(\sigma)$	$\mathsf{co}(\sigma)$	$\Gamma(\sigma)$
0,0	$\emptyset$	0	$\langle\langle 1\rangle\rangle(\square\Diamond q \wedge \Diamond q), \neg q$	1,2	$\{1,2\}$	1	$[1]\square\neg q, \neg q$
0,1	$\{2\}$	0	$\langle\langle 1\rangle\rangle(\square\Diamond q \wedge \Diamond q), [2]\square\Diamond s, \neg q$	2,0	$\{1\}$	1	$[1]\square\neg q, \neg q$
0,2	$\{2\}$	1	$\langle\langle 1\rangle\rangle(\square\Diamond q \wedge \Diamond q), \neg q$	2,1	$\{1,2\}$	1	$[1]\square\neg q, \neg q$
1,0	$\{1\}$	0	$\neg q$	2,2	$\{1,2\}$	0	$[2]\square\Diamond s, \neg q$
1,1	$\{1,2\}$	0	$[2]\square\Diamond s, \neg q$				

6   Elimination Phase

The *elimination phase* also works step-by-step. In order to go through one step to another we apply by turns two elimination rules, called **(ER1)** and **(ER2)**, until no more nodes can be eliminated. The rule **(ER1)** detects and deletes nodes with missing successor, while the rule **(ER2)** detects and delete states that do not realize all their eventualities. At each step, we obtain a new intermediate tableau, denoted by T_n^η. We denote by S_n^η the set of nodes (states and prestates) of the intermediate tableau T_n^η.

At the end of the elimination phase, we obtain the *final tableau* for η, denoted by T^η. It is declared *open* if the initial node belongs to S^η, otherwise *closed*. The procedure for deciding satisfiability of η returns "No" if T^η is closed, "Yes" otherwise.

Remark 1. Contrary to the tableau-based decision procedure for ATL^+, we do not eliminate all the prestates at the beginning of the elimination phase. We eliminate them with the rule **(ER1)** only if necessary. This does not have any effect on the result of the procedure, nor any relevant modification in the soundness and completeness proofs, but it makes implementation quicker and easier.

Rule (ER1). Let $\Xi \in S_n^\eta$ be a node (prestate or state).

- In the case where Ξ is a prestate: if all nodes Δ with $\Xi \Longrightarrow \Delta$ have been eliminated at earlier stages, then obtain T_{n+1}^η by eliminating Ξ from T_n^η.
- In the case where Ξ is a state: if, for some $\sigma \in D(\Xi)$, the node Γ with $\Xi \xrightarrow{\sigma} \Gamma$ has been eliminated at earlier stage, then obtain T_{n+1}^η by eliminating Ξ from T_n^η.

In order to define the rule **(ER2)**, we first need to define what is an eventuality in the context of ATL^* and then define how to check whether eventualities are realized or not.

Eventualities. In our context, we consider all γ-formulae as *potential eventualities*. We recall that a γ-formula is of the form $\langle\langle A \rangle\rangle \Phi$ or $[\![A]\!]\Phi$ where $\Phi \neq \bigcirc\varphi$. When constructing a tableau step-by-step as we do in our procedure, it is possible to postpone forever promises encapsulated in operators such as $\mathcal{U}$ as far as we keep promising to satisfy them. We consider that a promise, which is a path formula, is satisfied (or realized) once it is initiated at the current state, which corresponds to an engagement to keep it true if necessary, for any computation starting at that state. So we want to know at a given state and for a given formula whether all promises (or eventualities) are realized. This is the role of the function Realized: $\mathsf{ATL}_p^* \times \mathcal{P}(\mathsf{ATL}_s^*) \times \mathcal{P}(\mathsf{ATL}_p^*) \to \mathbb{B}$, where $\mathbb{B}$ is the set $\{true, false\}$. The first argument of the function Realized is the path formula to study, the second argument is a set of state formulae Θ, and the third argument is a set of path formulae on which one is "engaged". This third argument is exactly what is added with respect to ATL^+ treatment. For our purpose, to know whether a potential eventuality is realized, we use the set Θ to represent the state containing the γ-formula and the set $S = \gamma_{sl}(\Phi, \Theta)$ obtained during the decomposition of Φ and the full expansion of Θ. This last set S is computed in Sect. 4 and corresponds to the set of path formulae initiated in the current state Θ. The definition of Realized is given by recursion on the structure of Φ as follows:

- Realized$(\varphi, \Theta, S) = true$ iff $\varphi \in \Theta$
- Realized$(\Phi_1 \wedge \Phi_2, \Theta, S) = $ Realized$(\Phi_1, \Theta, S) \wedge$ Realized(Φ_2, Θ, S)
- Realized$(\Phi_1 \vee \Phi_2, \Theta, S) = $ Realized$(\Phi_1, \Theta, S) \vee$ Realized(Φ_2, Θ, S)
- Realized$(\bigcirc\Phi_1, \Theta, S) = true$
- Realized$(\Box\Phi_1, \Theta, S) = true$ iff $\Phi_1 \in \Theta \cup S$
- Realized$(\Phi_2 \mathcal{U} \Phi_1, \Theta, S) = true$ iff $\Phi_1 \in \Theta \cup S$

Remark 2. In the two last items, we use the set $\Theta \cup S$ to handle the particular case where Φ_1 is a state formula that is already in the set Θ because of the behaviour of another coalition of agents.

We will see with Definition 4 that if the function Realized declares that an eventuality is not immediately realized at a given state, then we check in the corresponding successor states whether it is realized or not. But, because of the way γ-formulae are decomposed, an eventuality may change its form from one state to another. Therefore, we define the notion of *Descendant potential eventuality* in order to define a parent/child link between potential eventualities and keep track of not yet realized eventualities, and finally check whether the potential eventualities are realized at a given moment.

Definition 3. (Descendant potential eventualities). *Let Δ be a state and let $\xi \in \Delta$ be a potential eventuality of the form $\langle\langle A \rangle\rangle \Phi$ or $[\![A]\!]\Phi$. Suppose the γ-component $\gamma_l(\xi, \Delta)$ in Δ linked to ξ is, respectively, of the form $\psi \wedge \langle\langle A \rangle\rangle\bigcirc\langle\langle A \rangle\rangle\Psi$ or $\psi \wedge [\![A]\!]\bigcirc[\![A]\!]\Psi$. Then the successor potential eventuality of ξ w.r.t. $\gamma_l(\xi, \Delta)$ is the γ-formula $\langle\langle A \rangle\rangle\Psi$ (resp. $[\![A]\!]\Psi$) and it will be denoted by ξ_Δ^1. The notion of descendant potential eventuality of ξ of degree d, for $d > 1$, is defined inductively as follows:*

- *any successor eventuality of ξ (w.r.t. some γ-component of ξ) is a descendant eventuality of ξ of degree 1;*
- *any successor eventuality of a descendant eventuality ξ^n of ξ of degree n is a descendant eventuality of ξ of degree $n + 1$.*

We will also consider ξ to be a descendant eventuality of itself of degree 0.

Realization of Potential Eventualities First, we give some notation:
Notation: Let $\mathbb{L} = \langle\langle A_0 \rangle\rangle \bigcirc \varphi_0, \ldots, \langle\langle A_{m-1} \rangle\rangle \bigcirc \varphi_{m-1}, [\![A_0']\!] \bigcirc \psi_0, \ldots, [\![A_{l-1}']\!] \bigcirc \psi_{l-1}, [\![A]\!] \bigcirc \mu_0, \ldots, [\![A]\!] \bigcirc \mu_{n-1}$ be the list of all primitive successor formulae of $\Delta \in S_0^\eta$, induced as part of application of **(Next)**.

$Succ(\Delta, \langle\langle A_p \rangle\rangle \bigcirc \varphi_p) := \{\Gamma \mid \Delta \xrightarrow{\sigma} \Gamma, \sigma_{\mathsf{a}} = p \text{ for every } \mathsf{a} \in A_p\}$

$Succ(\Delta, [\![A_q']\!] \bigcirc \psi_q) := \{\Gamma \mid \Delta \xrightarrow{\sigma} \Gamma, \mathsf{co}(\sigma) = q \text{ and } \mathbb{A} - A_q' \subseteq N(\sigma)\}$

$Succ(\Delta, [\![A]\!] \bigcirc \mu_r) := \{\Gamma \mid \Delta \xrightarrow{\sigma} \Gamma\}$

Definition 4. (Realization of potential eventualities) *Let $\Delta \in S_n^\eta$ be a state and $\xi \in \Delta$ be a potential eventuality of the form $\langle\langle A \rangle\rangle \Phi$ or $[\![A]\!]\Phi$. Let $S = \gamma_{sl}(\xi, \Delta)$. Then:*

1. *If $\mathsf{Realized}(\Phi, \Delta, S) = true$ then ξ is realized at Δ in $\mathcal{T}_n^\eta$.*
2. *Else, let ξ_Δ^1 be the successor potential eventuality of ξ w.r.t. $\gamma_l(\xi, \Delta)$. If for every $\Gamma \in Succ(\Delta, \langle\langle A \rangle\rangle \bigcirc \xi_\Delta^1)$ (resp. $\Gamma \in Succ(\Delta, [\![A]\!] \bigcirc \xi_\Delta^1)$), there exists $\Delta' \in \mathcal{T}_n^\eta$ with $\Gamma \Longrightarrow \Delta'$ and ξ_Δ^1 is realized at Δ' in $\mathcal{T}_n^\eta$, then ξ is realized at Δ in $\mathcal{T}_n^\eta$.*

Example 3. Let $\Delta = \{\langle\langle 1 \rangle\rangle((\Box\Diamond q \vee \Diamond r) \wedge (\Diamond q \vee \Diamond r)), [\![1]\!]\Box\neg q, [\![2]\!]\Box\Diamond s, [\![1, 2]\!] \bigcirc \neg q, \neg q, \langle\langle 1 \rangle\rangle \bigcirc \langle\langle 1 \rangle\rangle(\Box\Diamond q \wedge \Diamond q), [\![1]\!] \bigcirc [\![1]\!]\Box\neg q, s, [\![2]\!] \bigcirc [\![2]\!]\Box\Diamond s\}$ be a state.
If we consider the potential eventuality $\xi = \langle\langle 1 \rangle\rangle((\Box\Diamond q \vee \Diamond r) \wedge (\Diamond q \vee \Diamond r)) \in \Delta$, $\Phi = (\Box\Diamond q \vee \Diamond r) \wedge (\Diamond q \vee \Diamond r)$ and $S = \gamma_{sl}(\xi, \Delta) = \{\Diamond q\}$, then we obtain the following result:

$$\mathsf{Realized}(\Phi, \Delta, S) = \mathsf{Realized}(\Box\Diamond q \vee \Diamond r, \Delta, S) \wedge \mathsf{Realized}(\Diamond q \vee \Diamond r, \Delta, S)$$
$$= \mathsf{Realized}(\Box\Diamond q, \Delta, S) \vee \mathsf{Realized}(\Diamond r, \Delta, S) \wedge$$
$$\mathsf{Realized}(\Diamond q, \Delta, S) \vee \mathsf{Realized}(\Diamond r, \Delta, S)$$
$$= (true \vee false) \wedge (false \vee false) = false$$

The call of the function $\mathsf{Realized}$ on (Φ, Δ, S) returns $false$, which means that the potential eventuality ξ is not immediately realized. Therefore, we must check in the future if ξ can be realized or not. Concretely, we must check that the descendant potential eventuality $\xi^1 = \langle\langle 1 \rangle\rangle(\Box\Diamond q \wedge \Diamond q)$ is realized at the next states corresponding to the collective choices of all agents to satisfy the successor formula $\langle\langle 1 \rangle\rangle \bigcirc \xi^1$, that is states resulting from the transitions $(0, 0)$, $(0, 1)$ and $(0, 2)$, as seen in Example 2.

Rule (ER2). If $\Delta \in S_n^\eta$ is a state and contains a potential eventuality that is not realized at $\Delta \in \mathcal{T}_n^\eta$, then obtain $\mathcal{T}_{n+1}^\eta$ by removing Δ from S_n^η.

7   Results and Sketches of Proofs

Theorem 1. *The tableau-based procedure for* ATL* *is sound with respect to unsatisfiability, that is if a formula is satisfiable then its final tableau is open.*

To prove soundness, we first prove that from any satisfiable prestate we obtain at least one satisfiable state, and we prove that from any satisfiable state we obtain only satisfiable prestates. Second, we prove that no satisfiable prestate or state can be eliminated via rule **(ER1)** or **(ER2)**, and in particular, if the initial prestate is satisfiable, it cannot be removed, which means that the tableau is open.

Theorem 2. *The tableau-based procedure for* ATL* *is complete with respect to unsatisfiability, that is if a tableau for an input formula is open then this formula is satisfiable.*

To prove completeness, we construct step-by-step a special structure called Hintikka structure from the open tableau and then we prove that a CGM satisfying the initial formula can be obtained from that Hintikka structure.

Theorem 3. *The   tableau-based   procedure   for* ATL* *runs   in   at   most* 3EXPTIME.

We first argue that the number of formulae in the closure of the initial formula η is at most double exponential in the size of η (see Lemma 2). Then we have that the number of states is at most exponential in the size of the closure of η. Therefore the procedure runs in at most 3EXPTIME.

8   Implementation of the Procedure

We propose a prototype implementing our tableau-based decision procedure for ATL*, available on the following web site: http://atila.ibisc.univ-evry.fr/tableau_ATL_star/.

This prototype aims at giving a user-friendly tool to the reader interested in checking satisfiability of ATL* formulae. This is why we provide our prototype as a web application directly ready to be used. The application allows one to enter a formula, or to select one from a predefined list of formulae, and then launch the computation of the corresponding tableau. It returns some statistics about the number of prestates and states generated as well as the initial and final tableaux for the input formula, therefore also an answer on its satisfiability. Explanation on how to use the application is given on the web site.

Our prototype is developed in Ocaml for the computation, and in PHP and JavaScript for the web interface. Binaries of the application can be found on the same web page.

As the main difference between ATL and ATL* comes from path formulae, we mainly focus our test on that point and use the list of tests proposed by Reynolds for CTL* in [7]. This allows us to check that our application gives the same results in term of satisfiability and that our running times for these examples are satisfactory. Moreover, other tests using formulae with non trivial coalitions have been done. Nevertheless a serious benchmark has still to be done, which is a non trivial work, left for the future. Also, we plan to compare theoretically and experimentally our approach with the automata-decision based procedure of [8].

9   Conclusion

In this paper, we propose the first sound, complete and terminating tableau-based decision procedure for ATL*: it is easy to understand and conceptually simple. We also provide the first implementation to decide the satisfiability of ATL* formulae, among which ATL$^+$ formulae. In future works, it would be worthwhile to implement the automata-based decision procedure proposed in [8] and be able to make some practical comparisons. Another perspective is to implement model synthesis with a minimal number of states for satisfiable ATL* formulae.

Acknowledgement. I would like to thank Serenella Cerrito and Valentin Goranko for their advices and proofreading. I also thank the anonymous referees for their helpful criticisms.

References

1. Alur, R., Henzinger, T.A., Kupferman, O.: Alternating-time temporal logic. J. ACM **49**(5), 672–713 (2002)
2. Cerrito, S., David, A., Goranko, V.: Optimal tableaux-based decision procedure for testing satisfiability in the alternating-time temporal logic ATL+. In: Demri, S., Kapur, D., Weidenbach, C. (eds.) IJCAR 2014. LNCS, vol. 8562, pp. 277–291. Springer, Heidelberg (2014). http://arxiv.org/abs/1407.4645
3. Carral, D., Feier, C., Cuenca Grau, B., Hitzler, P., Horrocks, I.: $\mathcal{EL}$-ifying ontologies. In: Demri, S., Kapur, D., Weidenbach, C. (eds.) IJCAR 2014. LNCS, vol. 8562, pp. 464–479. Springer, Heidelberg (2014)
4. Emerson, E.A.: Temporal and modal logics. In: van Leeuwen, J. (ed.) Handbook of Theoretical Computer Science, vol. B, pp. 995–1072. MIT Press (1990)
5. Emerson, E.A., Sistla, A.P.: Deciding full branching time logic. Inf. Control **61**(3), 175–201 (1984)
6. Goranko, V., Shkatov, D.: Tableau-based decision procedures for logics of strategic ability in multiagent systems. ACM Trans. Comput. Log. **11**(1), 1–48 (2009)
7. Reynolds, M.: A faster tableau for CTL. In: Puppis, G., Villa, T. (eds.) Proceedings Fourth International Symposium on Games, Automata, Logics and Formal Verification, GandALF 2013, Borca di Cadore, Dolomites, Italy, 29–31th August 2013. EPTCS, vol. 119, pp. 50–63 (2013). http://dx.doi.org/10.4204/EPTCS.119.7
8. Schewe, S.: ATL* satisfiability Is 2EXPTIME-complete. In: Aceto, L., Damgård, I., Goldberg, L.A., Halldórsson, M.M., Ingólfsdóttir, A., Walukiewicz, I. (eds.) ICALP 2008, Part II. LNCS, vol. 5126, pp. 373–385. Springer, Heidelberg (2008)

Interactive/Automated Theorem Proving
and Applications

A Formalisation of Finite Automata Using Hereditarily Finite Sets

Lawrence C. Paulson[✉]

Computer Laboratory, University of Cambridge, Cambridge, UK
lp15@cam.ac.uk

Abstract. Hereditarily finite (HF) set theory provides a standard universe of sets, but with no infinite sets. Its utility is demonstrated through a formalisation of the theory of regular languages and finite automata, including the Myhill-Nerode theorem and Brzozowski's minimisation algorithm. The states of an automaton are HF sets, possibly constructed by product, sum, powerset and similar operations.

1 Introduction

The theory of finite state machines is fundamental to computer science. It has applications to lexical analysis, hardware design and regular expression pattern matching. A regular language is one accepted by a finite state machine, or equivalently, one generated by a regular expression or a type-3 grammar [6]. Researchers have been formalising this theory for nearly three decades.

A critical question is how to represent the states of a machine. Automata theory is developed using set-theoretic constructions, e.g. the product, disjoint sum or powerset of sets of states. But in a strongly-typed formalism such as higher-order logic (HOL), machines cannot be polymorphic in the type of states: statements such as "every regular language is accepted by a finite state machine" would require existential quantification over types. One might conclude that there is no good way to formalise automata in HOL [5,15].

It turns out that finite automata theory can be formalised within the theory of *hereditarily finite sets*: set theory with the negation of the axiom of infinity. It admits the usual constructions, including lists, functions and integers, but no infinite sets. The type of HF sets can be constructed from the natural numbers within higher-order logic. Using HF sets, we can retain the textbook definitions, without ugly numeric coding. We can expect HF sets to find many other applications when formalising theoretical computer science.

The paper introduces HF set theory and automata (Sect. 2). It presents a formalisation of deterministic finite automata and results such as the Myhill-Nerode theorem (Sect. 3). It also treats nondeterministic finite automata and results such as the powerset construction and closure under regular expression operations (Sect. 4). Next come minimal automata, their uniqueness up to isomorphism, and Brzozowski's algorithm for minimising an automaton [3] (Sect. 5). The paper concludes after discussing related work (Sects. 6–7). The proofs, which are available online [12], also demonstrate the use of Isabelle's *locales* [1].

© Springer International Publishing Switzerland 2015
A.P. Felty and A. Middeldorp (Eds.): CADE-25, LNAI 9195, pp. 231–245, 2015.
DOI: 10.1007/978-3-319-21401-6_15

2   Background

An *hereditarily finite set* can be understood inductively as a finite set of hereditarily finite sets [14]. This definition justifies the recursive definition $f(x) = \sum \{2^{f(y)} \mid y \in x\}$, yielding a bijection $f : \text{HF} \to \mathbb{N}$ between the HF sets and the natural numbers. The linear ordering on HF given by $x < y \iff f(x) < f(y)$ can be shown to extend both the membership and the subset relations.

The HF sets support many standard constructions, even quotients. Equivalence classes are not available in general — they may be infinite — but the linear ordering over HF identifies a unique representative. The integers and rationals can be constructed, with their operations (but not the *set* of integers, obviously). Świerczkowski [14] has used HF as the basis for proving Gödel's incompleteness theorems, and I have formalised his work using Isabelle [13].

Let Σ be a nonempty, finite alphabet of *symbols*. Then Σ^* is the set of *words*: finite sequences of symbols. The empty word is written ϵ, and the concatenation of words u and v is written uv. A *deterministic finite automaton* (DFA) [6,7] is a structure $(K, \Sigma, \delta, q_0, F)$ where K is a finite set of states, $\delta : K \times \Sigma \to K$ is the next-state function, $q_0 \in K$ is the initial state and $F \subseteq K$ is the set of final or accepting states. The next-state function on symbols is extended to one on words, $\delta^* : K \times \Sigma^* \to K$ such that $\delta^*(q, \epsilon) = q$, $\delta^*(q, a) = \delta(q, a)$ for $a \in \Sigma$ and $\delta^*(q, uv) = \delta^*(\delta^*(q, u), v)$. The DFA *accepts* the string w if $\delta^*(q_0, w) \in F$. A set $L \subseteq \Sigma^*$ is a *regular language* if L is the set of strings accepted by some DFA.

A *nondeterministic finite automaton* (NFA) is similar, but admits multiple execution paths and accepts a string if one of them reaches a final state. Formally, an NFA is a structure $(K, \Sigma, \delta, Q_0, F)$ where $\delta : K \times \Sigma \to \mathcal{P}(K)$ is the next-state function, $Q_0 \subseteq K$ a set of initial states, the other components as above. The next-state function is extended to $\delta^* : \mathcal{P}(K) \times \Sigma^* \to \mathcal{P}(K)$ such that $\delta^*(Q, \epsilon) = Q$, $\delta^*(Q, a) = \bigcup_{q \in Q} \delta(q, a)$ for $a \in \Sigma$ and $\delta^*(Q, uv) = \delta^*(\delta^*(Q, u), v)$. An NFA accepts the string w provided $\delta^*(q, w) \in F$ for some $q \in Q_0$.

The notion of NFA can be extended with ϵ-transitions, allowing "silent" transitions between states. Define the transition relation $q \xrightarrow{a} q'$ for $q' \in \delta(q, a)$. Let the ϵ-transition relation $q \xrightarrow{\epsilon} q'$ be given. Then define the transition relation $q \xRightarrow{a} q'$ to allow ϵ-transitions before and after: $(\xrightarrow{\epsilon})^* \circ (\xrightarrow{a}) \circ (\xrightarrow{\epsilon})^*$.

Every NFA can be transformed into a DFA, where the set of states is the powerset of the NFA's states, and the next-state function captures the effect of $q \xRightarrow{a} q'$ on these sets of states. Regular languages are closed under intersection and complement, therefore also under union. They are closed under repetition (Kleene star). Two key results are discussed below:

- The Myhill-Nerode theorem gives necessary and sufficient conditions for a language to be regular. It defines a canonical and minimal DFA for any given regular language. Minimal DFAs are unique up to isomorphism.
- Reorienting the arrows of the transition relation transforms a DFA into an NFA accepting the reverse of the given language. We can regain a DFA using the powerset construction. Repeating this operation yields a minimal DFA for the original language. This is Brzozowski's minimisation algorithm [3].

This work has been done using the proof assistant Isabelle/HOL. Documentation is available online at http://isabelle.in.tum.de/. The work refers to equivalence relations and equivalence classes, following the conventions established in my earlier paper [11]. If R is an equivalence relation on the set A, then $A//R$ is the set of equivalence classes. If $x \in A$, then its equivalence class is $R``\{x\}$. Formally, it is the image of $x$ under $R$: the set of all $y$ such that $(x,y) \in R$. More generally, if $X \subseteq A$ then $R``X$ is the union of the equivalence classes $R``\{x\}$ for $x \in X$.

3   Deterministic Automata; the Myhill-Nerode Theorem

When adopting HF set theory, there is the question of whether to use it for everything, or only where necessary. The set of states is finite, so it could be an HF set, and similarly for the set of final states. The alphabet could also be given by an HF set; then words—lists of symbols—would also be HF sets. Our definitions could be essentially typeless.

The approach adopted here is less radical. It makes a minimal use of HF, allowing stronger type-checking, although this does cause complications elsewhere. Standard HOL sets (which are effectively predicates) are intermixed with HF sets. An HF set has type `hf`, while a (possibly infinite) set of HF sets has type `hf set`. Definitions are polymorphic in the type `'a` of alphabet symbols, while words have type `'a list`.

3.1   Basic Definition of DFAs

The record definition below declares the components of a DFA. The types make it clear that there is indeed a set of states but only a single initial state, etc.

```
record 'a dfa = states :: "hf set"
                init  :: "hf"
                final :: "hf set"
                nxt   :: "hf ⇒ 'a ⇒ hf"
```

Now we package up the axioms of the DFA as a locale [1]:

```
locale dfa =
  fixes M :: "'a dfa"
  assumes init:  "init M ∈ states M"
      and final: "final M ⊆ states M"
      and nxt:   "⋀q x. q ∈ states M ⟹ nxt M q x ∈ states M"
      and finite: "finite (states M)"
```

The last assumption is needed because the `states` field has type `hf set` and not `hf`. The locale bundles the assumptions above into a local context, where they are directly available. It is then easy to define the accepted language.

```
primrec nextl :: "hf ⇒ 'a list ⇒ hf" where
    "nextl q []     = q".
  | "nextl q (x#xs) = nextl (nxt M q x) xs"
definition language :: "('a list) set"  where
  "language ≡ {xs. nextl (init M) xs ∈ final M}"
```

Equivalence relations play a significant role below. The following relation regards two strings as equivalent if they take the machine to the same state [7, p. 90].

definition *eq_nextl* :: "('a list × 'a list) set" **where**
 "eq_nextl ≡ {(u,v). nextl (init M) u = nextl (init M) v}"

Note that *language* and *eq_nextl* take no arguments, but refer to the locale.

3.2 Myhill-Nerode Relations

The Myhill-Nerode theorem asserts the equivalence of three characterisations of regular languages. The first of these is to be the language accepted by some DFA. The other two are connected with certain equivalence relations, called Myhill-Nerode relations, on words of the language.

The definitions below are outside of the locale and are therefore independent of any particular DFA. The predicate *dfa* refers to the locale axioms and expresses that its argument, *M*, is a DFA. The predicate *dfa.language* refers to the constant *language*: outside of the locale, it takes a DFA as an argument.

definition *regular* :: "('a list) set ⇒ bool" **where**
 "regular L ≡ ∃M. dfa M ∧ dfa.language M = L"

The other characterisations of a regular language involve abstract finite state machines derived from the language itself, with certain equivalence classes as the states. A relation is *right invariant* if it satisfies the following closure property.

definition *right_invariant* ::"('a list × 'a list) set ⇒ bool" **where**
 "right_invariant r ≡ (∀u v w. (u,v) ∈ r ⟶ (u@w, v@w) ∈ r)"

The intuition is that if two words u and v are related, then each word brings the "machine" to the same state, and once this has happened, this agreement must continue no matter how the words are extended as u@w and v@w.

A *Myhill-Nerode relation* for a language L is a right invariant equivalence relation of finite index where L is the union of some of the equivalence classes [7, p. 90]. *Finite index* means the set of equivalence classes is finite: finite (UNIV//R).[1] The equivalence classes will be the states of a finite state machine. The equality L = R''A, where A ⊆ L is a set of words of the language, expresses L as the union of a set of equivalence classes, which will be the final states.

definition *MyhillNerode* ::"'a list set ⇒ ('a list * 'a list)set ⇒ bool"
 where MyhillNerode L R ≡ equiv UNIV R ∧ right_invariant R ∧
                    finite (UNIV//R) ∧ (∃A. L = R''A)"

While *eq_nextl* (defined in Sect. 3.1) refers to a machine, the relation *eq_app_right* is defined in terms of a language, L. It relates the words u and v if all extensions of them, u@w and v@w, behave equally with respect to L:

definition *eq_app_right* :: "'a list set ⇒ ('a list * 'a list) set" **where**
 "eq_app_right L ≡ {(u,v). ∀w. u@w ∈ L ⟷ v@w ∈ L}"

---

[1] *UNIV* denotes a typed universal set, here the set of all words.

It is a Myhill-Nerode relation for L provided it is of finite index:

lemma `MN_eq_app_right`:
    `"finite (UNIV // eq_app_right L) ⟹ MyhillNerode L (eq_app_right L)"`

Moreover, every Myhill-Nerode relation R for L refines `eq_app_right L`.

lemma `MN_refines_eq_app_right`: `"MyhillNerode L R ⟹ R ⊆ eq_app_right L"`

This essentially states that `eq_app_right L` is the most abstract Myhill-Nerode relation for L. This will eventually yield a way of defining a minimal machine.

3.3   The Myhill-Nerode Theorem

The Myhill-Nerode theorem says that these three statements are equivalent [6]:

1. The set L is a regular language (is accepted by some DFA).
2. There exists some Myhill-Nerode relation R for L.
3. The relation `eq_app_right L` has finite index.

We have (1) ⇒ (2) because `eq_nextl` is a Myhill-Nerode relation. We have (2) ⇒ (3), by lemma `MN_refines_eq_app_right`, because every equivalence class for `eq_app_right L` is the union of equivalence classes of R, and so `eq_app_right L` has minimal index for all Myhill-Nerode relations. We get (3) ⇒ (1) by constructing a DFA whose states are the (finitely many) equivalence classes of `eq_app_right L`. This construction can be done for every Myhill-Nerode relation.

Until now, all proofs have been routine. But now we face a difficulty: the states of our machine should be equivalence classes of words, but these could be infinite sets. What can be done? The solution adopted here is to map the equivalence classes to the natural numbers, which are easily embedded in HF. Proving that the set of equivalence classes is finite gives us such a map.

Mapping infinite sets to integers seems to call into question the very idea of representing states by HF sets. However, mapping sets to integers turns out to be convenient only occasionally, and it is not necessary: we could formalise DFAs differently, coding symbols (and therefore words) as HF sets. Then we could represent states by representatives (having type `hf`) of equivalence classes. Using Isabelle's type-class system to identify the types (integers, booleans, lists, etc.) that can be embedded into HF, type `'a dfa` could still be polymorphic in the type of symbols. But the approach followed here is simpler.

3.4   Constructing a DFA from a Myhill-Nerode Relation

If R is a Myhill-Nerode relation for a language L, then the set of equivalence classes is finite and yields a DFA for L. The construction is packaged as a locale, which is used once in the proof of the Myhill-Nerode theorem, and again to prove that minimal DFAs are unique. The locale includes not only L and R, but also the set A of accepting states, the cardinality n and the bijection h between the set `UNIV//R` of equivalence classes and the number n as represented in HF. The locale assumes the Myhill-Nerode conditions.

```
locale MyhillNerode_dfa =
  fixes L :: "('a list) set" and R :: "('a list * 'a list) set"
    and A :: "('a list) set" and n :: nat and h :: "('a list) set ⇒ hf"
  assumes eqR: "equiv UNIV R"
      and riR: "right_invariant R"
      and L:   "L = R''A"
      and h:   "bij_betw h (UNIV//R) (hfset (ord_of n))"
```

The DFA is defined within the locale. The states are given by the equivalence classes. The initial state is the equivalence class for the empty word; the set of final states is derived from the set A of words that generate L; the next-state function maps the equivalence class for the word u to that for $u@[x]$. Equivalence classes are not the actual states here, but are mapped to integers via the bijection h. As mentioned above, this use of integers is not essential.

```
definition DFA :: "'a dfa" where
  "DFA = (|states = h ' (UNIV//R),
          init  = h (R '' {[]}),
          final = {h (R '' {u}) | u. u ∈ A},
          nxt   = λq x. h (⋃u ∈ h⁻¹ q. R '' {u@[x]}) |)"
```

This can be proved to be a DFA easily. One proof line, using the right-invariance property and lemmas about quotients [11], proves that the next-state function respects the equivalence relation. Four more lines are needed to verify the properties of a DFA, somewhat more to show that the language of this DFA is indeed L.

The facts proved within the locale are summarised (outside its scope) by the following theorem, stating that every Myhill-Nerode relation yields an equivalent DFA. (The **obtains** form expresses existential and multiple conclusions).

```
theorem MN_imp_dfa:
  assumes "MyhillNerode L R"
  obtains M where "dfa M"  "dfa.language M = L"
                  "card (states M) = card (UNIV//R)"
```

This completes the (3) ⇒ (1) stage, by far the hardest, of the Myhill-Nerode theorem. The three stages are shown below. Lemma *L2_3* includes a result about cardinality: the construction yields a minimal DFA, which will be useful later.

```
lemma L1_2:"regular L ⟹ ∃R. MyhillNerode L R"
lemma L2_3:
  assumes "MyhillNerode L R"
  obtains "finite (UNIV // eq_app_right L)"
          "card (UNIV // eq_app_right L) ≤ card (UNIV // R)"
lemma L3_1: "finite (UNIV // eq_app_right L) ⟹ regular L"
```

4   Nondeterministic Automata and Closure Proofs

As most of the proofs are simple, our focus will be the use of HF sets when defining automata. Our main example is the powerset construction for transforming a nondeterministic automaton into a deterministic one.

4.1   Basic Definition of NFAs

As in the deterministic case, a record holds the necessary components, while a
locale encapsulates the axioms. Component *eps* deals with ϵ-transitions.

```
recordá nfa = states :: "hf set"
              init   :: "hf set"
              final  :: "hf set"
              nxt    :: "hf ⇒ 'a ⇒ hf set"
              eps    :: "(hf * hf) set"
```

The axioms are obvious: the initial, final and next states belong to the set of
states, which is finite. An axiom restricting ϵ-transitions to machine states was
removed, as it did not simplify proofs. Working with ϵ-transitions is messy. It
helps to provide special treatment for NFAs having no ϵ-transitions. Allowing
multiple initial states reduces the need for ϵ-transitions.

```
locale nfa =
  fixes M :: "'a nfa"
  assumes init: "init M ⊆ states M"
      and final: "final M ⊆ states M"
      and nxt:   "⋀q x. q ∈ states M ⟹ nxt M q x ⊆ states M"
      and finite: "finite (states M)"
```

The following function "closes up" a set *Q* of states under ϵ-transitions. Intersection with *states M* confines these transitions to legal states.

```
definition epsclo :: "hf set ⇒ hf set" where
  "epsclo Q ≡ states M ∩ (⋃q∈Q. {q'. (q,q') ∈ (eps M)*})"
```

The remaining definitions are straightforward. Note that *nextl* generalises *nxt*
to take a set of states as well is a list of symbols.

```
primrec nextl :: "hf set ⇒ 'a list ⇒ hf set" where
    "nextl Q []     = epsclo Q"
  | "nextl Q (x#xs) = nextl (⋃q ∈ epsclo Q. nxt M q x) xs"
definition language :: "('a list) set"   where
  "language ≡ {xs. nextl (init M) xs ∩ final M ≠ {}}"
```

4.2   The Powerset Construction

The construction of a DFA to simulate a given NFA is elementary, and is a good
demonstration of the HF sets. The strongly-typed approach used here requires a
pair of coercion functions *hfset* :: "hf ⇒ hf set" and *HF* :: "hf set ⇒ hf"
to convert between HF sets and ordinary sets.

```
lemma HF_hfset: "HF (hfset a) = a"
lemma hfset_HF: "finite A ⟹ hfset (HF A) = A"
```

With this approach, type-checking indicates whether we are dealing with a set of states or a single state. The drawback is that we occasionally have to show that a set of states is finite in the course of reasoning about the coercions, which would never be necessary if we confined our reasoning to the HF world.

Here is the definition of the DFA. The states are ϵ-closed subsets of NFA states, coerced to type hf. The initial and final states are defined similarly, while the next-state function requires both coercions and performs ϵ-closure before and after. We work in locale nfa, with access to the components of the NFA.

definition *Power_dfa* :: *"'a dfa"* **where**
 "Power_dfa = (|dfa.states = HF ' epsclo ' Pow (states M),
          init   = HF(epsclo(init M)),
          final = {HF(epsclo Q) | Q. Q ⊆ states M ∧ Q ∩ final M ≠
{}},
          nxt   = λQ x. HF(⋃q ∈ epsclo (hfset Q). epsclo (nxt M q
x))|)"

Proving that this is a DFA is trivial. The hardest case is to show that the next-state function maps states to states. Proving that the two automata accept the same language is also simple, by reverse induction on lists (the induction step concerns u@[x], putting x at the end). Here, Power.language refers to the language of the powerset DFA, while language refers to that of the NFA.

theorem *Power_language:* *"Power.language = language"*

4.3   Other Closure Properties

The set of languages accepted by some DFA is closed under complement, intersection, concatenation, repetition (Kleene star), etc. [6]. Consider intersection:

theorem *regular_Int:*
  assumes *S:* *"regular S"* **and** *T:* *"regular T"* **shows** *"regular (S ∩ T)"*

The recognising DFA is created by forming the Cartesian product of the sets of states of MS and MT, the DFAs of the two languages. The machines are effectively run in parallel. The decision to represent a set of states by type hf set rather than by type hf means we cannot write dfa.states MS × dfa.states MT, but we can express this concept using set comprehension:

    "(|states = {⟨q1,q2⟩ | q1 q2. q1 ∈ dfa.states MS ∧ q2 ∈ dfa.states
MT},
       init   = ⟨dfa.init MS, dfa.init MT⟩,
       final = {⟨q1,q2⟩ | q1 q2. q1 ∈ dfa.final MS ∧ q2 ∈ dfa.final MT},
       nxt   = λ⟨qs,qt⟩ x. ⟨dfa.nxt MS qs x, dfa.nxt MT qt x⟩|)"

This is trivially shown to be a DFA. Showing that it accepts the intersection of the given languages is again easy by reverse induction.

Closure under concatenation is expressed as follows:

theorem *regular_conc:*
  assumes *S:* *"regular S"* **and** *T:* *"regular T"* **shows** *"regular (S @@ T)"*

The concatenation is recognised by an NFA involving the disjoint sum of the sets of states of *MS* and *MT*, the DFAs of the two languages. The effect is to simulate the first machine until it accepts a string, then to transition to a simulation of the second machine. There are ϵ-transitions linking every final state of *MS* to the initial state of *MT*. We again cannot write `dfa.states MS + dfa.states MT`, but we can express the disjoint sum naturally enough:

```
"(|states = Inl ' (dfa.states MS) ∪ Inr ' (dfa.states MT),
   init  = {Inl (dfa.init MS)},
   final = Inr ' (dfa.final MT),
   nxt   = λq x. sum_case (λqs. {Inl (dfa.nxt MS qs x)})
                          (λqt. {Inr (dfa.nxt MT qt x)}) q,
   eps   = (λq. (Inl q, Inr (dfa.init MT))) ' dfa.final MS|)"
```

Again, it is trivial to show that this is an NFA. But unusually, proving that it recognises the concatenation of the languages is a challenge. We need to show, by induction, that the "left part" of the NFA correctly simulates *MS*.

have `"⋀q. Inl q ∈ ST.nextl {Inl (dfa.init MS)} u ⟷`
`        q = (dfa.nextl MS (dfa.init MS) u)"`

The key property is that any string accepted by the NFA can be split into strings accepted by the two DFAs. The proof involves a fairly messy induction.

have `"⋀q. Inr q ∈ ST.nextl {Inl (dfa.init MS)} u ⟷`
`        (∃ uS uT. uS ∈ dfa.language MS ∧ u = uS@uT ∧`
`         q = dfa.nextl MT (dfa.init MT) uT)"`

Closure under Kleene star is not presented here, as it involves no interesting set operations. The language L^* is recognised by an NFA with an extra state, which serves as the initial state and runs the DFA for L including iteration. The proofs are messy, with many cases. To their credit, Hopcroft and Ullman [6] give some details, while other authors content themselves with diagrams alone.

5   State Minimisation for DFAs

Given a regular language L, the Myhill-Nerode theorem yields a DFA having the minimum number of states. But it does not yield a minimisation algorithm for a given automaton. It turns out that a DFA is minimal if it has no unreachable states and if no two states are *indistinguishable* (in a sense made precise below). This again does not yield an algorithm. *Brzozowski's minimisation algorithm* involves reversing the DFA to create an NFA, converting back to a DFA via powersets, removing unreachable states, then repeating those steps to undo the reversal. Surprisingly, it performs well in practice [3].

5.1   The Left and Right Languages of a State

The following developments are done within the locale `dfa`, and therefore refer to one particular deterministic finite automaton.

The *left language* of a state q is the set of all words w such that $q_0 \overset{w}{\rightarrow}{}^* q$, or informally, such that the machine when started in the initial state and given the word w ends up in q. In a DFA, the left languages of distinct states are disjoint, if they are nonempty.

definition left_lang :: "hf $\Rightarrow$ ('a list) set" **where**
  "left_lang q $\equiv$ {u. nextl (init M) u = q}"

The *right language* of a state q is the set of all words w such that $q \overset{w}{\rightarrow}{}^* q_f$, where q_f is a final state, or informally, such that the machine when started in q will accept the word w. The language of a DFA is the right language of q_0. Two states having the same right language are *indistinguishable*: they both lead to the same words being accepted.

definition right_lang :: "hf $\Rightarrow$ ('a list) set" **where**
  "right_lang q $\equiv$ {u. nextl q u $\in$ final M}"

The *accessible* states are those that can be reached by at least one word.

definition accessible :: "hf set" **where**
  "accessible $\equiv$ {q. left_lang q $\neq$ {}}"

The function path_to returns one specific such word. This function will eventually be used to express an isomorphism between any minimal DFA (one having no inaccessible or indistinguishable states) and the canonical DFA determined by the Myhill-Nerode theorem.

definition path_to :: "hf $\Rightarrow$ 'a list" **where**
  "path_to q $\equiv$ SOME u. u $\in$ left_lang q"
lemma nextl_path_to:
  "q $\in$ accessible $\Longrightarrow$ nextl (dfa.init M) (path_to q) = q"

First, we deal with the problem of inaccessible states. It is easy to restrict any DFA to one having only accessible states.

definition Accessible_dfa :: "'a dfa" **where**
  "Accessible_dfa = (|dfa.states = accessible,
                    init = init M,
                    final = final M $\cap$ accessible,
                    nxt   = nxt M|)"

This construction is readily shown to be a DFA that agrees with the original in most respects. In particular, the two automata agree on left_lang and right_lang, and therefore on the language they accept:

lemma Accessible_language: "Accessible.language = language"

We can now define a DFA to be minimal if all states are accessible and no two states have the same right language. (The formula inj_on right_lang (dfa.states M) expresses that the function right_lang is injective on the set dfa.states M.)

definition `minimal` **where**
    `"minimal ≡ accessible = states M ∧ inj_on right_lang (dfa.states M)"`

Because we are working within the DFA locale, `minimal` is a constant referring to one particular automaton.

5.2   A Collapsing Construction

We can deal with indistinguishable states similarly, defining a DFA in which the indistinguishable states are identified via equivalence classes. This is not part of Brzozowski's minimisation algorithm, but it is interesting in its own right: the equivalence classes themselves are HF sets. We begin by declaring a relation stating that two states are equivalent if they have the same right language.

definition `eq_right_lang :: "(hf × hf) set"` **where**
    `"eq_right_lang ≡ {(u,v). u ∈ states M ∧ v ∈ states M ∧`
                         `right_lang u = right_lang v}"`

Trivially, this is an equivalence relation, and equivalence classes of states are finite (there are only finitely many states). In the corresponding DFA, these equivalence classes form the states, with the initial and final states given by the equivalence classes for the corresponding states of the original DFA. As usual, the function `HF` is used to coerce a set of states to type `hf`.

definition `Collapse_dfa :: "'a dfa"` **where**
    `"Collapse_dfa = ⦇dfa.states = HF ' (states M // eq_right_lang),`
                      `init  = HF (eq_right_lang '' {init M}),`
                      `final = {HF (eq_right_lang '' {q}) | q. q ∈ final M},`
             `nxt = λQ x. HF (⋃q ∈ hfset Q. eq_right_lang '' {nxt M q`
`x})⦈"`

This is easily shown to be a DFA, and the next-state function respects the equivalence relation. Showing that it accepts the same language is straightforward.

lemma `ext_language_Collapse_dfa:`
    `"u ∈ Collapse.language ⟷ u ∈ language"`

5.3   The Uniqueness of Minimal DFAs

The property `minimal` is true for machines having no inaccessible or indistinguishable states. To prove that such a machine actually has a minimal number of states is tricky. It can be shown to be isomorphic to the canonical machine from the Myhill-Nerode theorem, which indeed has a minimal number of states.

Automata M and N are *isomorphic* if there exists a bijection h between their state sets that preserves their initial, final and next states. This conception is nicely captured by a locale, taking the DFAs as parameters:

locale `dfa_isomorphism = M: dfa M + N: dfa N`
                    for `M :: "'a dfa"` **and** `N :: "'a dfa" +`

```
fixes h ::ḧf ⇒ hf"
assumes h: "bij_betw h (states M) (states N)"
    and init : "h (init M) = init N"
    and final: "h ' final M = final N"
    and nxt  : "⋀q x. q ∈ states M ⟹ h(nxt M q x) = nxt N (h q) x"
```

With this concept at our disposal, we resume working within the locale *dfa*, which is concerned with the automaton *M*. If no two states have the same right language, then there is a bijection between the accessible states (of *M*) and the equivalence classes yielded by the relation *eq_app_right language*.

```
lemma inj_right_lang_imp_eq_app_right_index:
  assumes "inj_on right_lang (dfa.states M)"
    shows "bij_betw (λq. eq_app_right language '' {path_to q})
                    accessible  (UNIV // eq_app_right language)"
```

This bijection maps the state *q* to *eq_app_right language '' {path_to q}*. Every element of the quotient *UNIV // eq_app_right language* can be expressed in this form. And therefore, the number of states in a *minimal* machine equals the index of *eq_app_right language*.

```
definition min_states where
  "min_states ≡ card (UNIV // eq_app_right language)"
lemma minimal_imp_index_eq_app_right:
  "minimal ⟹ card(dfa.states M) = min_states"
```

In the proof of the Myhill-Nerode theorem, it emerged that this index was the minimum cardinality for any DFA accepting the given language. Any other automaton, *M'*, accepting the same language cannot have fewer states. This theorem justifies the claim that *minimal* indeed characterises a minimal DFA.

```
theorem minimal_imp_card_states_le:
    "⟦minimal; dfa M'; dfa.language M' = language⟧
    ⟹ card (dfa.states M) ≤ card (dfa.states M')"
```

Note that while the locale *dfa* gives us implicit access to one DFA, namely *M*, it is still possible to refer to other automata, as we see above.

The minimal machine is unique up to isomorphism because every minimal machine is isomorphic to the canonical Myhill-Nerode DFA. The construction of a DFA from a Myhill-Nerode relation was packaged as a locale, and by applying this locale to the given *language* and the relation *eq_app_right language*, we can generate the instance we need.

```
interpretation Canon:
  MyhillNerode_dfa language "eq_app_right language"
                    language min_states index_f
```

Here, *index_f* denotes some bijection between the equivalence classes and their cardinality (as an HF ordinal). It exists (definition omitted) by the definition of cardinality itself. It is the required isomorphism function between *M* and the canonical DFA of Sect. 3.4, which is written *Canon.DFA*.

definition *iso* :: *"hf* ⇒ *hf"* **where**
  "iso ≡ *index_f o (λq. eq_app_right language '' {path_to q})"*

The isomorphism property is stated using locale *dfa_isomorphism*.

theorem *minimal_imp_isomorphic_to_canonical*:
  assumes *minimal*   **shows** *"dfa_isomorphism M Canon.DFA iso"*

Verifying the isomorphism conditions requires delicate reasoning. Hopcroft and Ullman's proof [6, p.29–30] provides just a few clues.

5.4   Brzozowski's Minimisation Algorithm

At the core of this minimisation algorithm is an NFA obtained by reversing all the transitions of a given DFA, and exchanging the initial and final states.

definition *Reverse_nfa* :: *"'a dfa* ⇒ *'a nfa"* **where**
  "Reverse_nfa MS = (|*nfa.states = dfa.states MS,*
                init   = dfa.final MS,
                final = {dfa.init MS},
                nxt   = λq x. {p ∈ *dfa.states MS. q = dfa.nxt MS p x},*
                eps   = {}|)*"*

This is easily shown to be an NFA that accepts the reverse of every word accepted by the original DFA. Applying the powerset construction yields a new DFA that has no indistinguishable states. The point is that the right language of a powerset state is derived from the right languages of the constituent states of the reversal NFA [3]. Those, in turn, are the left languages of the original DFA, and these are disjoint (since the original DFA has no inaccessible states, by assumption).

lemma *inj_on_right_lang_PR*:
  assumes *"dfa.states M = accessible"*
    shows *"inj_on (dfa.right_lang (nfa.Power_dfa (Reverse_nfa M)))*
                (dfa.states (nfa.Power_dfa (Reverse_nfa M)))"

The following definitions abbreviate the steps of Brzozowski's algorithm.

abbreviation *APR* :: *"'x dfa* ⇒ *'x dfa"* **where**
  "APR X ≡ *dfa.Accessible_dfa (nfa.Power_dfa (Reverse_nfa X))"*
definition *Brzozowski* :: *"'a dfa"* **where**
  "Brzozowski ≡ *APR (APR M)"*

  By the lemma proved just above, the *APR* operation yields minimal DFAs.

theorem *minimal_APR*:
  assumes *"dfa.states M = accessible"*
    shows *"dfa.minimal (APR M)"*

Brzozowski's minimisation algorithm is correct. The first APR call reverses the language and eliminates inaccessible states; the second call yields a minimal machine for the original language. The proof uses the theorems just proved.

```
theorem minimal_Brzozowski: "dfa.minimal Brzozowski"
unfolding Brzozowski_def
proof (rule dfa.minimal_APR)
  show "dfa (APR M)"
    by (simp add: dfa.dfa_Accessible nfa.dfa_Power nfa_Reverse_nfa)
next
  show "dfa.states (APR M) = dfa.accessible (APR M)"
    by (simp add: dfa.Accessible_accessible dfa.states_Accessible_dfa
            nfa.dfa_Power nfa_Reverse_nfa)
qed
```

6 Related Work

There is a great body of prior work. One approach involves working constructively, in some sort of type theory. Constable's group has formalised automata [4] in Nuprl, including the Myhill-Nerode theorem. Using type theory in the form of Coq and its Ssreflect library, Doczkal et al. [5] formalise much of the same material as the present paper. They omit ε-transitions and Brzozowski's algorithm and add the pumping lemma and Kleene's algorithm for translating a DFA to a regular expression. Their development is of a similar length, under 1400 lines, and they allow the states of a finite automaton to be given by any finite type. In a substantial development, Braibant and Pous [2] have implemented a tactic for solving equations in Kleene algebras by implementing efficient finite automata algorithms in Coq. They represent states by integers.

An early example of regular expression theory formalised using higher-order logic (Isabelle/HOL) is Nipkow's verified lexical analyser [9]. His automata are polymorphic in the types of state and symbols. NFAs are included, with ε-transitions simulated by an alphabet extended with a dummy symbol.

Recent Isabelle developments explicitly bypass automata theory. Wu et al. [15] prove the Myhill-Nerode theorem using regular expressions. This is a significant feat, especially considering that the theorem's underlying intuitions come from automata. Current work on regular expression equivalence [8,10] continues to focus on regular expressions rather than finite automata.

This paper describes not a project undertaken by a team, but a six-week case study by one person. Its successful outcome obviously reflects Isabelle's powerful automation, but the key factor is the simplicity of the specifications. Finite automata cause complications in the prior work. The HF sets streamline the specifications and allow elementary set-theoretic reasoning.

7 Conclusions

The theory of finite automata can be developed straightforwardly using higher-order logic and HF set theory. We can formalise the textbook proofs: there is no need to shun automata or use constructive type theories. HF set theory can be seen as an abstract universe of computable objects, with many potential applications. One possibility is programming language semantics: using *hf* as the type of values offers open-ended possibilities, including integer, rational and floating point numbers, ASCII characters, and data structures.

Acknowledgements. Christian Urban and Tobias Nipkow offered advice, and suggested Brzozowski's minimisation algorithm as an example. The referees made a variety of useful comments.

References

1. Ballarin, C.: Locales: A module system for mathematical theories. J. Autom. Reasoning **52**(2), 123–153 (2014)
2. Braibant, T., Pous, D.: Deciding Kleene algebras in Coq. Log. Methods Comput. Sci. **8**(1), 1–42 (2012)
3. Champarnaud, J., Khorsi, A., Paranthoën, T.: Split and join for minimizing: Brzozowski's algorithm. In: Balík, M., Simánek, M. (eds.) The Prague Stringology Conference, pp. 96–104. Czech Technical University, Department of Computer Science and Engineering (2002)
4. Constable, R.L., Jackson, P.B., Naumov, P., Uribe, J.C.: Constructively formalizing automata theory. In: Plotkin, G.D., Stirling, C., Tofte, M. (eds.) Proof, Language, and Interaction, pp. 213–238. MIT Press (2000)
5. Doczkal, C., Kaiser, J.-O., Smolka, G.: A constructive theory of regular languages in Coq. In: Gonthier, G., Norrish, M. (eds.) CPP 2013. LNCS, vol. 8307, pp. 82–97. Springer, Heidelberg (2013)
6. Hopcroft, J.E., Ullman, J.D.: Formal Languages and Their Relation to Automata. Addison-Wesley, Boston (1969)
7. Kozen, D.: Automata and computability. Springer, New York (1997)
8. Krauss, A., Nipkow, T.: Proof pearl: regular expression equivalence and relation algebra. J. Autom. Reasoning **49**(1), 95–106 (2012)
9. Nipkow, T.: Verified lexical analysis. In: Grundy, J., Newey, M. (eds.) TPHOLs 1998. LNCS, vol. 1479, pp. 1–15. Springer, Heidelberg (1998)
10. Nipkow, T., Traytel, D.: Unified decision procedures for regular expression equivalence. In: Klein, G., Gamboa, R. (eds.) ITP 2014. LNCS, vol. 8558, pp. 450–466. Springer, Heidelberg (2014)
11. Paulson, L.C.: Defining functions on equivalence classes. ACM Trans. Comput. Logic **7**(4), 658–675 (2006)
12. Paulson, L.C.: Finite automata in hereditarily finite set theory. Archive of Formal Proofs, February 2015. http://afp.sf.net/entries/Finite_Automata_HF.shtml, Formal proof development
13. Paulson, L.C.: A mechanised proof of Gödel's incompleteness theorems using Nominal Isabelle. J. Autom. Reasoning **55**(1), 1–37 (2015). Available online at http://link.springer.com/article/10.1007%2Fs10817-015-9322-8
14. Świerczkowski, S.: Finite sets and Gödel's incompleteness theorems. Dissertationes Mathematicae **422**, 1–58 (2003). http://journals.impan.gov.pl/dm/Inf/422-0-1.html
15. Wu, C., Zhang, X., Urban, C.: A formalisation of the Myhill-Nerode theorem based on regular expressions. J. Autom. Reasoning **52**(4), 451–480 (2014)

SEPIA: Search for Proofs Using Inferred Automata

Thomas Gransden$^{(\boxtimes)}$, Neil Walkinshaw, and Rajeev Raman

Department of Computer Science, University of Leicester, Leicester, UK
tg75@student.le.ac.uk, {nw91,rr29}@leicester.ac.uk

Abstract. This paper describes SEPIA, a tool for automated proof generation in Coq. SEPIA combines model inference with interactive theorem proving. Existing proof corpora are modelled using state-based models inferred from tactic sequences. These can then be traversed automatically to identify proofs. The SEPIA system is described and its performance evaluated on three Coq datasets. Our results show that SEPIA provides a useful complement to existing automated tactics in Coq.

Keywords: Interactive theorem proving · Model inference · Proof automation

1 Introduction

Interactive theorem provers (ITPs) such as Coq [10] and Isabelle [11] are systems that enable the manual development of proofs for a variety of domains. These range from mathematics through to complex software and hardware verification. Thanks to the expressive logics that are used, they provide a very rich programming environment.

Nevertheless, constructing proofs can be a challenging and time-consuming process. A proof development will typically contain many routine lemmas, as well as more complex ones. The ITP system will take care of the bookkeeping and perform simple reasoning steps; however much time is spent manually entering the requisite tactics (even for the most trivial lemmas). In 2008, Wiedijk stated that it takes up to one week to formalize a page of an undergraduate mathematics textbook [14].

To help combat this problem, we present SEPIA (Search for Proofs Using Inferred Automata) – an automated approach designed to assist users of Coq. SEPIA automatically generates proofs by inferring state-based models from previously compiled libraries of successful proofs, and using the inferred models as a basis for automated proof search.

2 Background

This section presents the necessary background required for this paper. We briefly introduce the underlying model inference technique (called MINT), followed by a motivating example.

© Springer International Publishing Switzerland 2015
A.P. Felty and A. Middeldorp (Eds.): CADE-25, LNAI 9195, pp. 246–255, 2015.
DOI: 10.1007/978-3-319-21401-6_16

2.1 Inferring EFSMs with MINT

MINT [13] is an technique designed to infer state machine models from sequences, where the sequencing of events may depend on some underlying data state. Such systems are modelled as extended finite state machines (see Definition 1). EFSMs can be conceptually thought of as conventional finite state machines with an added memory. The transitions in an EFSM not only contain a label, but may also contain guards that must hold with respect to variables contained in the memory.

Definition 1 Extended Finite State Machine. *An Extended Finite State Machine (EFSM) M is a tuple $(S, s_0, F, L, V, \Delta, T)$. S is a set of states, $s_0 \in S$ is the initial state, and $F \subseteq S$ is the set of final states. L is defined as the set of labels. V represents the set of data states, where a single instance v represents a set of concrete variable assignments. $\Delta : V \to \{True, False\}$ is the set of data guards. Transitions $t \in T$ take the form (a, l, δ, b), where $a, b \in S$, $l \in L$, and $\delta \in \Delta$.*

MINT infers EFSMs from sets of *traces*. These can be defined formally as follows:

Definition 2. *A trace $T = \langle e_0, \ldots, e_n \rangle$ is a sequence of n trace elements. Each element e is a tuple (l, v), where l is a label representing the names of function calls or input / output events, and v is a string containing the parameters (this may be empty).*

The inference approach adopted by MINT [13] is an extension of a traditional state-merging approach [9] that has been proven to be successful for conventional (non-extended) finite state machines [12]. Briefly, the model inference starts by arranging the traces into a *prefix-tree*, a tree-shaped state machine that exactly represents the set of given traces. The inference then proceeds by a process of *state-merging*; pairs of states in the tree that are roughly deemed to be equivalent (based on their outgoing sequences) are merged. This merging process yields an EFSM that can accept a broader range of sequences than the initial given set.

The transitions in an EFSM not only imply the sequence in which events can occur, but also place constraints on which parameters are valid. This is done by inferring data-classifiers from the training data – each data guard takes the following form $(l, v, possible)$ where $l \in L$, $v \in V$ and $possible \in \{true, false\}$. When states are merged, the resulting machine is checked to make sure it remains consistent with the data guards.

2.2 Motivating Example

To motivate this work, we consider a typical scenario that arises during interactive proof. Suppose that we are trying to prove the following conjecture: `forall n m p:nat, p + n <= p + m - > n <= m`. The automated Coq tactics [2] have only been able to perform routine reasoning (namely calling the `intros` tactic) to advance the proof to the following:

```
n  :  nat
m  :  nat
p  :  nat
H  :  p + n <= p + m
```
$$n <= m$$

There are 2 theories from the Coq Standard Library called Le.v and Lt.v, that contain proofs about similar properties. The built-in tactics fail to prove the goal. The question we are faced with is this: Given the examples of successful proofs, can we use these to automatically find a proof for the above conjecture?

In previous work [4] we showed how to use MINT to infer EFSM models of Coq proofs. The resulting EFSMs were simply presented and used manually to derive proofs. This work extends our previous approach by automating the search process, allowing proofs to be completed automatically.

3   SEPIA System Description

In this section we describe the SEPIA approach. We present the key stages of the technique. It is available[1] as a ProofGeneral extension that works with Coq. An overview of SEPIA is shown in Figure 1. It contains three main stages:

1. Generate proof traces from a selection of existing Coq theories.
2. Use MINT to infer a model from these proof traces.
3. Systematically search the model, formulating and attempting possible proofs from paths through the model.

Before describing these three steps in more detail, we look at three properties of the approach that are particularly appealing:

Adaptivity. For every iteration, as more valid proofs are discovered they can be incorporated into future cycles to infer more accurate models, forming a 'virtuous loop'. This is a major benefit over the existing built-in automated tactics, which are typically limited to attempting a fixed set of tactics.

Automation. Aside from providing the initial set of theories from which to infer a model, the user is not prompted for any other input. In addition, as will be elaborated later, the overall process typically completes in less than a minute (at least in the context of our experiments).

Ability to identify new proofs. The state-merging process [13] can result in models that accept sequences of tactics which aren't present in the initial set of proofs. These wouldn't necessarily be intuitive, or be spotted from manual scrutiny of the proof library. These can however contain valuable steps that lead to a successful proof.

---

[1] https://bitbucket.org/tomgransden/efsminferencetool.

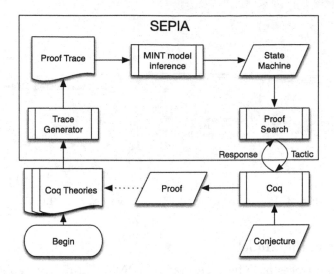

Fig. 1. SEPIA overview

3.1 Generating Traces from Existing Proofs

To begin a proof attempt we must provide one or more Coq theories from which we wish to generate a model. The proofs within the theories must be converted into their corresponding *proof traces* (see Definition 2). This step is identical to the process used in our previous work [4].

Figure 2 shows the proof script from the lemma le_antisym from Le.v and the corresponding proof trace. An important concept in Coq proofs is the semicolon operator. If two (or more) tactics are separated by a semicolon, for example t1;t2, this means apply t1 to the current goal and then apply t2 to *all* generated subgoals. We record the usage of the semicolon in our traces, so that this information can be reused during proof search.

(a) Proof Script

(b) Trace

```
intros n m H;
destruct H as [|m' H];
auto with arith.
intros H1.
absurd (S m' <= m');
auto with arith.
apply le_trans with n;
auto with arith.
```

Event e	Label l	Params v
e_0	intros	"n m H;"
e_1	destruct	"H as [|m' H];"
e_2	auto	"with arith"
e_3	intros	"H1"
e_4	absurd	"(S m' <= m');"
e_5	auto	"with arith"
e_6	apply	"le_trans with n;"
e_7	auto	"with arith"

Fig. 2. Original proof and proof trace for an example lemma

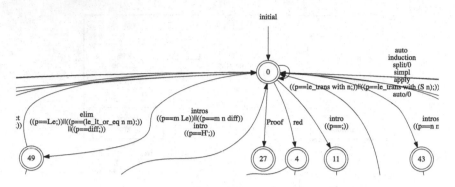

Fig. 3. Portion of inferred EFSM from `Le.v` and `Lt.v`

3.2 Inferring the Model

Once the proof traces have been generated, MINT is invoked to infer a model. There are two main parameters associated with MINT. The inference strategy dictates how states are merged during the inference process. A value called k represents the minimum score before a pair of states can be deemed to be equivalent. An in-depth discussion of these variables is outside the scope of this paper.

A preliminary study (with results online) found that using the state merging strategy `redblue` and $k = 1$ performed reasonably well for the task of interactive proving. These settings are based on the number of proofs discovered, the time taken and the presence of shorter/novel proofs. For the rest of this paper we refer to these as the default settings for MINT. A portion of the EFSM inferred from `Le.v` and `Lt.v` is shown in Fig. 3.

3.3 Searching for a Proof

Once a model has been inferred it can be used to search for candidate proofs. We adopt a breadth-first search as this ensures that if a proof is contained in the model, the shortest one will be returned. An instance of Coq is loaded, and the lemma is stated. The proof search moves through the model and applies the tactics and arguments suggested on each transition.

A timeout or a limit on the number of tactics applied can be provided to control the search. If we reach a point where a proof is found, SEPIA outputs the proof (and some proof search statistics). When running SEPIA on our motivating example we obtain the following result:

```
Proof was: intros m n diff. elim diff; auto with arith.
5611 tactics evaluated.
Inference and search took 0 min, 1 sec
```

The above proof is particularly interesting for two reasons. Firstly, we have managed to prove something completely automatically that Coq's automated tools could not. Secondly, the sequence of tactics (and parameters) was not found anywhere else within `Le.v` or `Lt.v`.

4   Evaluation

In this section we provide an experimental evaluation of our approach. We consider the following research questions:

- **RQ1:** Can proofs be derived automatically using our approach?
  - (a): How many proofs can be found?
  - (b): How long does it take to find a proof?
- **RQ2:** Are there "interesting" characteristics of the proofs?
  - (a): Do the proofs contain new sequences of tactics?
  - (b): Are the proofs shorter?
- **RQ3:** How does our results compare to Coq's built-in automated tactics?

4.1   Methodology

The aim of this evaluation is to assess the practicalities of using our approach in real proof developments. We evaluate SEPIA on three distinct Coq contributions as our datasets. We use a method inspired by k-folds cross-validation [7] in order to study proof attempts made by our approach.

Datasets. The datasets used in this evaluation consist of theories selected from three Coq proof developments. The datasets were chosen mainly for their domain, complexity and size. All theories were selected before the experiments took place. SSreflect[2] contains seven core theories. We select all of these theories as our first dataset. Secondly, MSets[3] is an implementation of finite sets using lists/trees. All eleven theories are selected to form our second dataset. Finally, we use some theories from CompCert[4] Owing to the size of the development, we select a four theories containing both general purpose proofs along with some more specialized ones. Due to the exploratory nature of this evaluation, there are some threats to validity associated with the selection of data. We have only used three Coq datasets, so any results cannot be interpreted to represent performance on all Coq proofs.

Evaluating Proof Attempts. To provide some answers to RQ1, we want to model the following situation: given some existing proofs, can we use these to prove new properties that are not part of the initial collection. To do this, we use an approach inspired by k-folds cross-validation [7].

Each Coq theory file is taken individually and the proofs are randomly partitioned into k non-overlapping sets. We then infer a model from $k-1$ of the sets, and try and prove the lemmas in the remaining set. This process repeats until each set has been used exactly once as the collection of lemmas to be proved.

For each proof attempt, we allow 10,000 tactics to be applied before reporting a failure. The results presented in this paper are from using $k = 10$, a standard

---

[2] http://ssr.msr-inria.inria.fr/doc/ssreflect-1.4/.
[3] https://coq.inria.fr/library/.
[4] http://compcert.inria.fr/doc/index.html.

value for k-folds cross-validation [7]. Other values of k have been investigated and the full set of results are online.

As well as capturing whether a proof attempt was successful or not, when a proof is found we analyse how "interesting" the proof is. First, we check and see whether a proof is shorter than the corresponding hand-curated proof. We also check whether the sequence of tactics was new (i.e. not present in the examples the model was inferred from). These provide us with answers to RQ2.

To investigate RQ3 we also run the Coq automated tools to try and prove each lemma. The following command is issued to Coq: `auto with * || eauto with * || tauto || firstorder || trivial`. This simply attempts to prove a goal by trying all of the automated tactics. The default search depth is used in all cases. Where we can specify lemma databases, we allow any available database to be used during proof search.

4.2   Results

The full results from our experiments are shown in Table 1. The results are presented for each theory, grouped by library. The remainder of this section provides some answers to the research questions defined earlier.

RQ1(a): A significant proportion of the lemmas were proved automatically using our approach. In Table 1, the column headed SEPIA shows the total number of lemmas proved in each theory using our approach. The results suggest that EFSM-based methods are useful at finding proofs automatically. Looking at each dataset as a whole, 32 % (438 out of 1360) of the SSreflect dataset were proved. In MSets, 30 % (211 out of 687) were successfully proved using our approach. In our selection of CompCert theories, there were 25 % (83 out of 335) proved.

RQ1(b): Many proofs were discovered in under 30 seconds. We measured the time required to derive a proof using our approach. These times take into account both the time required to infer the model and the search time. Over 90 % of the proofs were found within 30 s. These results show that when a user invokes the process, a proof will usually be delivered quickly. Overall, a proof can be discovered in a relatively small period of time. Of course, this is encouraging for the user involved in the proof development.

RQ2(a): A quarter of the proofs found were new sequences of tactics. The number of new proofs discovered using our approach are listed under the 'New' column in Table 1. We compare the discovered proof with the ones used to infer the model If the sequence is not contained in an existing proof, then it is considered new and only found as a result of inferring an EFSM. Our results show a significant number of new proofs were discovered, backing up further that

Table 1. Results Summary

Library	Theory	Size	SEPIA			Coq-Tacs
			Total	New	Shorter	
SSreflect	ssrnat	341	135 (39 %)	14	9	59 (17 %)
	ssrbool	240	120 (50 %)	17	10	60 (25 %)
	seq	394	94 (24 %)	14	6	18 (4 %)
	fintype	243	42 (17 %)	15	1	0 (0 %)
	eqtype	82	36 (44 %)	18	2	10 (12 %)
	choice	30	6 (20 %)	0	0	1 (3 %)
	ssrfun	30	5 (16 %)	1	0	7 (23 %)
MSets	avl	26	0 (0 %)	0	0	0 (0 %)
	decide	22	18 (81 %)	0	3	4 (18 %)
	eqproperties	106	43 (40 %)	1	5	47 (44 %)
	facts	65	17 (26 %)	4	8	10 (15 %)
	gentree	61	9 (15 %)	3	3	3 (5 %)
	list	42	8 (19 %)	3	3	3 (7 %)
	positive	67	13 (19 %)	5	4	1 (1 %)
	properties	137	78 (57 %)	9	3	15 (11 %)
	rbt	89	12 (13 %)	10	6	2 (2 %)
	tofiniteset	14	5 (35 %)	2	2	4 (28 %)
	weaklist	27	8 (30 %)	4	5	6 (22 %)
CompCert	cshmgenproof	65	15 (23 %)	14	14	0 (0 %)
	amsgenproof0	57	12 (21 %)	9	9	6 (10 %)
	coqlib	114	36 (31 %)	24	23	16 (14 %)
	values	99	20 (20 %)	17	13	5 (5 %)

EFSMs can be useful for automated proof generation. In SSreflect, a total of 79 proofs were new. In the MSets theories, 41 new proofs were found, and 64 were discovered in CompCert.

RQ2(b): Many proofs discovered were shorter than their original ones.
We have listed the number of shorter proofs found in Table 1 under the Shorter column. When a proof is found, we compare the discovered proof with the original hand-curated one. The length (in terms of tactics used) of both proofs are then compared, to see if we managed to derive a shorter one. In SSreflect, 28 of the proofs found were shorter than their original counterparts. For MSets, 42 of the proofs were shorter, whilst in CompCert 59 of proofs were shorter. The combination of the state merging algorithms and a breadth-first search means we were able to identify these shorter proofs.

RQ3: SEPIA provides an alternative to existing Coq tactics. The column headed Coq-Tacs in Table 1 provides the number of lemmas that were proved using Coq's automated tactics. Despite being relatively limited in the steps that they try, they manage to prove 155 SSreflect lemmas, 95 MSets lemmas and 27 of the CompCert lemmas. On the whole, we see that our approach significantly outperforms the automated tactics in terms of number of lemmas proved. This is to be expected, as they only provide modest automation. Nevertheless, there are occasions where the automated tactics prove more lemmas (in `msetproperties` and `ssrfun` for instance).

5   Related Work

There have been many projects aimed at improving the automation of proofs in ITPs. As we have shown in this work, machine learning can be applied in the context of interactive theorem proving. Specifically, we have shown that the tactics used in proofs can serve as useful features for machine learning algorithms. This is an area that has received moderate attention previously.

Jamnik *et al.* have previously applied an Inductive Logic Programming technique to examples of proofs in the Ωmega system [6]. Given a collection of well chosen proof method sequences, Jamnik *et al.* perform a method of least generalisation to infer what are ultimately regular grammars. The value of even basic models is intuitive. Proofs could be derived automatically using the technique. However, the proof steps learned do not contain any parameters. The parameters required are reconstructed after running the learning technique.

Another approach that concentrated on Isabelle proofs was implemented by Duncan [3]. Duncan's approach was to identify commonly occurring sequences of tactics from a given corpora. After eliciting these tactic sequences, evolutionary algorithms were used to automatically formulate new tactics. The evaluation showed that simple properties could be derived automatically using the technique; however the parameter information was left out of the learning approach.

6   Conclusion and Future Work

This paper has presented SEPIA, an approach to automatically generate proofs in Coq. This has been achieved by applying model inference techniques to interactive proof scripts. We have shown that even learning from tactic sequences, which is admittedly a simplistic view of interactive proofs, can provide effective proof automation. It would be interesting to see what can be achieved by using more sophisticated views such as the proof goal view [5].

The overall process is fully automated our evaluation shows SEPIA performs well on a range of proofs from three varied Coq datasets. It succeeds in proving a number of lemmas that were out of reach for Coq's automated tactics. Additionally, when SEPIA finds a proof it usually does so in seconds.

As well as reusing existing proofs, SEPIA can construct proofs using new tactic sequences. These new sequences might not have been identified if manually analysing proof libraries. In our evaluation, we also identified a number of shorter proofs (by comparing the proofs found using SEPIA to original proofs). This follows the trend of other comparisons of automated and human proofs [1].

We plan to investigate automatic identification of appropriate theories or lemmas that could be used to infer models. Currently, we use whole theories; however it may be the case that only a handful of these proofs are actually useful. By using methods such as ML4PG [8] it may be possible to discover the most useful lemmas from a large collection of theories.

References

1. Alama, J., Kühlwein, D., Urban, J.: Automated and human proofs in general mathematics: an initial comparison. In: Bjørner, N., Voronkov, A. (eds.) LPAR-18 2012. LNCS, vol. 7180, pp. 37–45. Springer, Heidelberg (2012)
2. Bertot, Y., Castéran, P.: Interactive Theorem Proving and Program Development - Coq'Art: The Calculus of Inductive Constructions. Springer, Heidelberg (2004)
3. Duncan, H.: The Use of Data Mining for the Automatic Formation of Tactics. Ph.d. thesis, University of Edinburgh (2007)
4. Gransden, T., Walkinshaw, N., Raman, R.: Mining state-based models from proof corpora. In: Watt, S.M., Davenport, J.H., Sexton, A.P., Sojka, P., Urban, J. (eds.) CICM 2014. LNCS, vol. 8543, pp. 282–297. Springer, Heidelberg (2014)
5. Grov, G., Komendantskata, E., Bundy, A.: A Statistical Relational Learning Challenge Extracting Proof Strategies from Exemplar Proofs. In: ICML-12 Workshop on Statistical Relational Learning (2012)
6. Jamnik, M., Kerber, M., Pollet, M., Benzmüller, C.: Automatic learning of proof methods in proof planning. Logic J. IGPL 11(6), 647–673 (2003)
7. Kohavi, R.: A Study of Cross-validation and Bootstrap for Accuracy Estimation and Model Selection. In: Proceedings of the 14th International Joint Conference on Artificial Intelligence, pp. 1137–1143. Morgan Kaufmann (1995)
8. Komendantskaya, E., Heras, J., Grov, G.: Machine learning in proof general: interfacing interfaces. In: User Interfaces for Theorem Provers, EPTCS, vol. 118, pp. 15–41 (2013)
9. Lang, K.J., Pearlmutter, B.A., Price, R.A.: Results of the abbadingo one DFA learning competition and a new evidence-driven state merging algorithm. In: Honavar, V.G., Slutzki, G. (eds.) ICGI 1998. LNCS (LNAI), vol. 1433, pp. 1–12. Springer, Heidelberg (1998)
10. The Coq Development Team: The Coq Proof Assistant Reference Manual, Version 8.4. LogiCal Project. http://coq.inria.fr/refman
11. Nipkow, T., Paulson, L.C., Wenzel, M.: Isabelle/HOL – A Proof Assistant for Higher-Order Logic, LNCS, vol. 2283. Springer, Heidelberg (2002)
12. Walkinshaw, N., Lambeau, B., Damas, C., Bogdanov, K., Dupont, P.: STAMINA: a competition to encourage the development and assessment of software model inference techniques. Empir. Softw. Eng. 18(4), 791–824 (2013)
13. Walkinshaw, N., Taylor, R., Derrick, J.: Inferring Extended Finite State Machine Models from Software Executions. Empir. Softw. Eng. 1–43 (2015)
14. Wiedijk, F.: Formal proof - getting started. Not. AMS 55(11), 1408–1414 (2008)

Proving Correctness of a KRK Chess Endgame Strategy by Using Isabelle/HOL and Z3

Filip Marić[1](✉), Predrag Janičić[1], and Marko Maliković[2]

[1] Faculty of Mathematics, University of Belgrade, Belgrade, Serbia
{filip,janicic}@matf.bg.ac.rs
[2] Faculty of Humanities and Social Sciences, University of Rijeka, Rijeka, Croatia
marko@ffri.hr

Abstract. We describe an executable specification and a total correctness proof of a King and Rook vs King (KRK) chess endgame strategy within the proof assistant Isabelle/HOL. This work builds upon a previous computer-assisted correctness analysis performed using the constraint solver URSA. The distinctive feature of the present machine verifiable formalization is that all central properties have been automatically proved by the SMT solver Z3 integrated into Isabelle/HOL, after being suitably expressed in linear integer arithmetic. This demonstrates that the synergy between the state-of-the-art automated and interactive theorem proving is mature enough so that very complex conjectures from various AI domains can be proved almost in a "push-button" manner, yet in a rich logical framework offered by the modern ITP systems.

1 Introduction

Chess has always been a target for developing new techniques and approaches of artificial intelligence. One field of chess-related research is concerned with chess endgames where challenges are different from those in openings and midgames. In computer chess playing, endgames are often played based on or analyzed with respect to pre-calculated lookup tables (i.e., endgame databases), containing optimal moves for each legal position. In contrast, *chess endgame strategies* do not necessarily ensure optimal play, but should provide concise, understandable, and intuitive instructions usable both to human and computer players. One of the simplest chess endgames is the *King and Rook vs King (KRK)*. There are several strategies for white for this endgame, generated by humans, semi-automatically, or automatically [10], but only a few of them are really human-understandable. Correctness of a strategy should be ensured – if a player follows the strategy, he should always reach the best possible outcome. Proofs of strategy correctness are typically not given or even not mentioned, although informal proofs are sometimes provided [4]. Proving correctness of chess endgame strategies can be addressed using different approaches [10]. The first approach is a traditional, *"pen-and-paper"* with the drawback of often having missing parts or errors in the arguments. *Computer assisted* proofs can be classified according to two independent dimensions: proofs can be either *indirect* or *direct*, and can be either

A.P. Felty and A. Middeldorp (Eds.): CADE-25, LNAI 9195, pp. 256–271, 2015.
DOI: 10.1007/978-3-319-21401-6_17

informal or *formal*. *Indirect proofs* are based on enumerations and case-analyses. For example, the strategy can be applied to all legal positions and a corresponding endgame-database can be generated, which is then verified using a retrograde procedure (in the style of Thompson's work [13]). *Direct proofs* are high-level, mathematical proofs that explicitly formulate properties of the strategy (preconditions, postconditions, invariants, termination measures), prove them and show that they imply the strategy correctness. *Informal proofs* use unverified programs (either developed in a general-purpose programming language or in some specialized constraint programming system) to check many different positions or to discharge informally stated proof-obligations that somehow contribute to the overall informal correctness arguments. *Formal proofs* are machine-verifiable proofs, checked within a strict logical system of a proof-assistant.

A SAT-based constraint solver URSA [9] has been used for checking key correctness properties for a KRK endgame strategy [10] in a direct, but informal manner. The strategy considered was a slight modification of the one originally formulated by Bratko [4]. Bratko's original paper also contains a very informal correctness proof sketch. The strategy was described within the constraint solving system and several high-level lemmas were formulated and automatically checked by using the power of the constraint solver. The main feature of those proofs is that they required very little human effort and human reasoning. On the other hand, as the authors noted, although the main body of the proof is covered by the checked lemmas, some building blocks were missing to make the proof complete and glued together, mainly due to the lack of expressibility of constraint solving systems (e.g., one cannot express inductive definitions or inductive arguments in a system such as URSA). Also, the proof relies on the definitions specific for the KRK endgame, so there is no link with the rules for the original game of chess. The final conclusion was that, in order to have a full and completely reliable proof, a constraint solver must be replaced by some more expressible reasoning system such as proof-assistants. We believe that modern proof-assistants (e.g., Isabelle/HOL, Coq, HOL-Light), connected to powerful external automated theorem provers and solvers (e.g., Z3, Vampire, Spass, E-prover) are now capable of proving extremely complex combinatorial conjectures, such as those coming from chess. For instance, Isabelle/HOL has been connected to SMT solvers [3], enabling users to employ SMT solvers to discharge complex goals that arise in interactive theorem proving. SMT solvers provide object-level proofs for unsatisfiable formulas and these proofs are then reconstructed within Isabelle/HOL, yielding formal proofs in the above sense.

In this paper, we describe our successful experience with formalizing the KRK endgame correctness proof within Isabelle/HOL. The present formalization is complete and self-contained, and it provides an executable version of the strategy that is proved to be correct (winning for white) with respect to the rules of chess. One of our key goals was that all central lemmas must be proved automatically, if possible — in a "push-button" manner, as it was done within the URSA system. This turned out to be possible, due to the powerful integration of SMT solvers (in particular — Microsoft Research z3 solver) into Isabelle/HOL. In the paper we briefly describe some interesting parts of the formalization. Full formalization is available at http://argo.matf.bg.ac.rs/formalizations/.

2 Chess Rules and Endgame Strategies

In this section we describe the formalization of chess rules and the general theory of chess endgame strategies in Isabelle/HOL.

Chess Rules. The first cornerstone of our formalization are the rules of chess, as given in the FIDE handbook [6]. Hurd has already formalized these rules in HOL [7], and we closely follow his work. Like Hurd, we consider pawnless endgames, and do not consider castling (although our strategy is only for the KRK endgame, we want our basic definitions to be close to general chess rules and to allow later extensions to other endgames, so initial definitions cover other pieces as well). Basic types are defined as follows.

```
side = White | Black
piece = King | Queen | Rook | Bishop | Knight
square = "int × int"
```

We can define many relevant notions using only arithmetic operations and relations over squares coordinates. We show only some examples. The function on_board $(f,r) \longleftrightarrow 0 \leq f \wedge f < F \wedge 0 \leq r \wedge r < R$ checks if the square is *on the board* (global constants $F = 8$ and $R = 8$, for files and ranks, determine the size of the board). We can check if a square sq is *between* two given squares sq_1 and sq_2 either horizontally, vertically, or diagonally (this is denoted by sq_btw sq_1 sq sq_2). We can define the *scope* of each piece (i.e., whether a piece can reach one square from another).

king_scope (f_1,r_1) (f_2,r_2) $\longleftrightarrow$ $|f_1 - f_2| \leq 1 \wedge |r_1 - r_2| \leq 1 \wedge (f_1 \neq f_2 \vee r_1 \neq r_2)$
rook_scope (f_1,r_1) (f_2,r_2) $\longleftrightarrow$ $(f_1 = f_2 \vee r_1 = r_2) \wedge (f_1 \neq f_2 \vee r_1 \neq r_2)$

For two squares of the chessboard, the *Manhattan distance* (mdist (f_1,r_1) $(f_2,r_2)) = |f_1 - f_2| + |r_1 - r_2|$) is the sum of distances along both coordinates, and the *Chebyshev distance* (cdist (f_1,r_1) $(f_2,r_2) = \max |f_1 - f_2||r_1 - r_2|$) is the minimal number of moves a king requires to move between them.

Chess positions can be represented in various ways (e.g., by an 8×8 matrix implicitly mapping positions to pieces, or by a list of piece positions, implicitly mapping pieces to positions). So, instead of fixing a concrete representation, we create an abstraction in a form of an Isabelle/HOL locale [2] and assume that chessboard positions will be represented by some type 'p (usually a record type). Only some values of the type 'p will correspond to valid positions, so we introduce a data-structure invariant pos_inv p that is used to exclude values that are invalid. For example, a type 'p might be a mapping that maps pieces to squares that they are on. In that case, the invariant should require that all pieces map to different squares, since if two pieces are mapped to the same square, the position would clearly be invalid. For each position, we must be able to check whether white or black is on turn (this is done using the function turn p), and for each square to determine if there is a piece on that square and – if yes, what piece it is (this is done using the function on_sq p sq that maps each square sq to either None, or to Some piece and its side).

locale Position =
 fixes pos_inv :: "'p $\Rightarrow$ bool"
 fixes turn :: "'p $\Rightarrow$ side"
 fixes on_sq :: "'p $\Rightarrow$ square $\Rightarrow$ (side $\times$ piece) option"

All chess rules can be defined within this locale, they are parametric, and depend on the type 'p and the above three functions. For example, in a position p, for a square sq we can check if it is *empty* (empty p sq $\longleftrightarrow$ on_sq p sq = None), or *occupied* by a piece of a side sd (occupies p sd sq $\longleftrightarrow$ ($\exists$ pc. on_sq p sq = Some (sd, pc))). In a position p, a square sq_1 attacks sq_2 if the *line between them is clear* (clr_line p sq_1 sq_2 $\longleftrightarrow$ ($\forall$ sq. sq_btw sq_1 sq sq_2 $\longrightarrow$ empty p sq))[1], and if there is a piece on sq_1 such that sq_2 is in its scope.

attacks p sq_1 sq_2 $\longleftrightarrow$ clr_line p sq_1 sq_2 $\wedge$
 (case on_sq p sq_1 of
  None $\Rightarrow$ False
  | Some (_, King) $\Rightarrow$ king_scope sq_1 sq_2
  | Some (_, Rook) $\Rightarrow$ rook_scope sq_1 sq_2
  ...)

A side sd is *in check* in a position p if its king is on a square sq_1, and there is an opponent's piece on some square sq_2 such that it attacks the king on sq_1.

in_chk sd p $\longleftrightarrow$ ($\exists$ sq_1 sq_2. on_sq p sq_1 = Some $(sd,$ King$)$ $\wedge$
         occupies p (opp sd) sq_2 $\wedge$ attacks p sq_2 sq_1)

A *position is legal* if its satisfies the invariant, if all pieces are within the board bounds, and if the opponent of the player on turn is not in check[2].

all_on_board p $\longleftrightarrow$ ($\forall$ sq. $\neg$ empty p sq $\longrightarrow$ on_board sq)

lgl_pos p $\longleftrightarrow$ pos_inv p $\wedge$ all_on_board p $\wedge$ $\neg$ in_chk p (opp (turn p))

Legal moves are defined by the chess rules and from legal positions they lead to legal positions. The function lgl_move p p' checks if the position p' is a result of a legal move from the position p. Finally, we define game outcomes (*checkmate*, *stalemate*, and *draw*).

game_over p $\longleftrightarrow$ lgl_pos p $\wedge$ $\neg$ ($\exists$ p'. lgl_move p p')
checkmate p $\longleftrightarrow$ game_over p $\wedge$ in_chk p (turn p)
stalemate p $\longleftrightarrow$ game_over p $\wedge$ $\neg$ in_chk p (turn p)

A game is drawn if the position is such that neither player can possibly mate. To formalize this, we inductively define the set of positions *reachable* from a given position p_0 by applying only legal moves.

---

[1] Since squares that a knight attacks are not on the same line with the square that it is on, the clear line condition is always satisfied.

[2] This definition is weaker then the one given by FIDE, as it does not take into account reachability from the initial position. Still, this does not threaten the correctness of our results, as we do cover all legal positions in the strong FIDE sense.

$p_0 \in$ **reachable** p_0

$[\![p \in$ **reachable** p_0; **lgl_move** $p\ p']\!] \Longrightarrow p' \in$ **reachable** p_0

draw $p \longleftrightarrow \neg\ (\exists\ p'.\ p' \in$ **reachable** $p \wedge$ **checkmate** $p')$

Endgame Strategies. The *strategy for white* is given by st_wht_move $p\ p'$ — a relation describing all positions p' that can be reached from p by a strategy move. A strategy is deterministic if there is always at most one such position. For each legal position with white on turn, a strategy returns only legal moves. Additionally, a strategy can be characterized by an invariant maintained throughout a play (e.g., in KRK endgame, white rook must not be captured, otherwise the game would be drawn). We define a slot for such invariant (st_inv p) and require that each move of white, and each move of black following a move of white maintains it.

locale Strategy = Position +
  fixes st_wht_move :: "'p $\Rightarrow$'p $\Rightarrow$ bool"
  fixes st_inv :: "'p $\Rightarrow$ bool"
  assumes
  $[\![$lgl_pos p; turn p = White; st_inv p; st_wht_move $p\ p']\!] \Longrightarrow$ lgl_move $p\ p'$
  $[\![$lgl_pos p; turn p = White; st_inv p; st_wht_move $p\ p']\!] \Longrightarrow$ st_inv p'
  $[\![$lgl_pos p; turn p = White; st_inv p; st_wht_move $p\ p'$; lgl_move $p'\ p'']\!]$
    $\Longrightarrow$ st_inv p''

A *strategy play* is a sequence of alternating moves: strategy moves by white, and arbitrary legal moves by black. The set of reachable positions in a play is defined as an inductive set.

st_move $p\ p' \longleftrightarrow$ (turn p = White $\wedge$ st_wht_move $p\ p'$) $\vee$
                    (turn p = Black $\wedge$ lgl_move $p\ p'$)

$p_0 \in$ st_reachable p_0

$[\![p \in$ st_reachable p_0; st_move $p\ p']\!] \Longrightarrow p' \in$ st_reachable p_0

A strategy for white is *winning* if it is terminating and partially correct, i.e., if every strategy play starting from a legal position p_0 with white on turn that satisfies the strategy invariant, terminates in a position where black is mated. If there is no infinite strategy play, there is no set $\mathcal{P}$ containing p_0 such that for each position in $\mathcal{P}$ a strategy move can be made.

st_start $p_0 \longleftrightarrow$ turn p_0 = White $\wedge$ lgl_pos $p_0 \wedge$ st_inv p_0

locale WinningStrategy = Strategy + assumes
  st_start $p_0 \Longrightarrow \neg\ (\exists\ \mathcal{P}.\ p_0 \in \mathcal{P} \wedge (\forall\ p \in \mathcal{P}.\ \exists p' \in \mathcal{P}.\ \text{st\_move } p\ p'))$

  $[\![$st_start p_0; $p \in$ st_reachable p_0; $\neg\ (\exists\ p'.\ \text{st\_move } p\ p')]\!] \Longrightarrow$
  turn p = Black $\wedge$ checkmate p

It can be proved that a strategy is winning for white if there is a well-founded ordering of subsequent white-on-turn positions in each strategy play, if white can always make a strategy move, and it never leads to a stalemate. Therefore, a strategy is winning for white if it meets assumptions of WiningStrategyOrdering (since it a sublocale of the WiningStrategy, i.e., if the assumptions of the former are satisfied, the assumptions of the latter are satisfied too).

```
locale WinningStrategyOrdering = Strategy +
  fixes ordering :: "'p ⇒ ('p ×'p) set"
  assumes
  ⟦st_start p₀⟧ ⟹ wf (ordering p₀)
  ⟦st_start p₀; p ∈ st_reachable p₀; turn p = White;
     st_wht_move p p'; lgl_move p' p''⟧ ⟹ (p'', p) ∈ ordering p₀
  ⟦st_start p₀; p ∈ st_reachable p₀; turn p = White⟧ ⟹ ∃ p'. st_wht_move p p'
  ⟦st_start p₀; p ∈ st_reachable p₀; turn p = White; st_wht_move p p'⟧ ⟹
     ¬ stalemate p'
```

3   KRK Chess Endgame and Bratko-style Strategy

In this section we describe our formalization of KRK chess endgame. We give
a very brief description of the specialization of chess rules for this case and of
Bratko-style strategy for the KRK endgame (we denote it by BTK) [10].

KRK Chess Endgame. Although the KRK endgame follows the general chess
rules introduced in the previous section, due to the specific nature of the game
with just three pieces on the board, most notions can be characterized by much
simpler conditions. Therefore, all general chess definitions are adapted to the
KRK case and are reformulated through alternative definitions. Each such definition is proved to be just a specific instance of its corresponding general chess
definition, and later used to simplify the correctness proofs. Since all following
definitions are based on KRK-specific definitions used in the URSA specification [10], our work shows that the URSA specification follows from general chess
rules.

Since there are only three pieces on the board, each position can be represented by the following simple data-structure.

```
record KRKPosition =
  WK ::"square" (* position of white king *)
  BK ::"square" (* position of black king *)
  WRopt ::"square option" (* position of white rook (None if captured) *)
  WhiteTurn ::"bool" (* Is white on turn? *)
```

Note that the option type is used only for the rook position, as kings must always
be present on the board[3]. The following abbreviations are introduced.

```
BlackTurn p ⟷ ¬ WhiteTurn p,
WR p = the(WRopt p), WRcapt p ⟷ WRopt p = None
```

The KRKPosition record interprets the Position locale, as all required components can be easily implemented.

---

[3] This is only implicitly stated in the FIDE chess rules, as positions are defined to be
legal only if they are reachable from the starting state where both kings are present,
and kings cannot be captured. In our KRK formalization, the condition that both
kings are present is implicitly imposed by the position representation.

KRK.pos_inv p ⟷ WK p ≠ BK p ∧ WRopt p ≠ Some(WK p) ∧ WRopt p ≠ Some(BK p)

KRK.to_move p = (if WhiteTurn p then White else Black)

KRK.on_sq p sq = (if WK p = sq then Some (White, King)

              else if BK p = sq then Some (Black, King)

              else if WRopt p = Some sq then Some (White, Rook)

              else None)"

Once the basic functions are interpreted, instances of all general definitions (e.g., legal positions, legal moves, stalemate, checkmate) for the KRK case are available. However, as we said, most of them are significantly simplified and reformulated, this time without quantifiers, so simpler reasoning methods can be used to reason about their properties. For example, requirement that all pieces are within the board bounds is defined in the following way (compare this with the original definition that uses the universal quantifier).

KRK.all_on_board p ⟷

   on_board (WK p) ∧ on_board (BK p) ∧ (¬ WRcapt p ⟶ on_board (WR p))

It is proved that this simplified KRK.all_on_board p definition is equivalent to the original all_on_board p definition instantiated by the KRKPosition type and its corresponding basic function definitions (lemma "all_on_board p ⟷ KRK.all_on_board p"). Such proofs were not too hard, but Isabelle/HOL could not do them automatically (due to the rich language and the need of reasoning about arbitrarily quantified statements, the record type, tuples, etc.).

The legality of positions can be reduced to requiring that all pieces are on different squares, that kings are not next to each other, and that if white is on turn, then the rook does not attack the black king (in KRK endgames, no diagonal lines but only horizontal and vertical lines need to be considered).

sq_btw_hv (f_1, r_1) (f, r) (f_2, r_2) ⟷

   $(f_1 = f ∧ f = f_2 ∧$ btw r_1 r $r_2) ∨ (r_1 = r ∧ r = r_2 ∧$ btw f_1 f $f_2)$

KRK.WR_attacks_BK p ⟷

   ¬ WRcapt p ∧ rook_scope (WR p) (BK p) ∧ ¬ sq_btw_hv (WR p) (WK p) (BK p)

KRK.kings_separated p ⟷ ¬ king_scope (WK p) (BK p)

KRK.lgl_pos p ⟷ KRK.pos_inv p ∧ KRK.all_on_board p ∧

   KRK.kings_separated p ∧ (WhiteTurn p ⟶ ¬ KRK.WR_attacks_BK p)"

Again, it is formally shown that this simplified definition of KRK.lgl_pos p is equivalent to the original lgl_pos p definition instantiated to the KRK case (lemma "lgl_pos p ⟷ KRK.lgl_pos p").

Moves are defined as functions that modify the record representing the position. Move of the black king is the most complicated (as it can capture a rook).

KRK.moveBK p sq = (let p' = p (| BK := sq, WhiteTurn := True |)

             in if WR p = sq then p' (| WRopt := None |) else p')

With these available, legal moves can be easily characterized. For example, a legal move of the black king can be characterized as follows.

KRK.BK_attacks_sq p sq ⟷ king_scope (BK p) sq

KRK.lgl_move_BK p_1 p_2 ⟷ KRK.lgl_pos p_1 ∧ BlackTurn p_1 ∧ KRK.lgl_pos p_2 ∧

   KRK.BK_attacks_sq p_1 (BK p_2) ∧ p_2 = KRK.moveBK p_1 (WK p_2)

Legal moves of two other pieces are characterized similarly. It is easily proved that all legal moves of black pieces are legal moves of the black king and all legal moves of white pieces are legal moves of either the white king or the white rook.

Bratko-style KRK Strategy Definition. Bratko's strategy can be outlined as follows. Try to *mate* in two moves. If that is not possible, then try to *squeeze* the room — the area to which the black king is confined by the white rook. Otherwise, try to *approach* the black king, to help the rook in squeezing (the approach is towards the *critical square* — a square adjacent to the rook in the direction of the black king). Otherwise, try to maintain the present achievements in the sense of squeeze and approach (i.e. make a waiting move). Otherwise, try to obtain a position such that the rook divides the two kings either vertically or horizontally. The strategy has a number of hidden details (its detailed description consumes more than a full page [10]) and that shows that it is very difficult to have a concise winning strategy (not to mention optimal strategy) even for a simple endgame such as KRK.

One of the central notions in the strategy is *room* (Fig. 1). Following the strategy, white iteratively squeezes the black king and reduces the room, until black can be mated. The room space is always rectangular (e.g., of dimension $f \times r$). Originally, room was measured by its area. However, we noticed that instead of the area, half-perimeter $(f + r)$ can be used, which has equivalent key properties but it does not use multiplication and the arithmetic constraints remain linear. Critical square and room are formalized as follows.

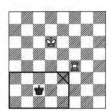

Fig. 1. Room and critical square

```
towards a b = (if a = b then a else if a > b then a - 1 else a + 1)
BTK.critical_sq p = (let (R^f, R^r) = WR p; (k^f, k^r) = BK p
                in (towards R^f k^f, towards R^r k^r))
BTK.room p = (let (R^f, R^r) = WR p; (k^f, k^r) = BK p
              in (if R^f = k^f ∨ R^r = k^r then F + R - 1
                  else let f = if R^f > k^f then R^f else F - 1 - R^f;
                       r = if R^r > k^r then R^r else R - 1 - R^r
                  in f + r))
```

Note that when the black king and the white rook are in line, the black king is not confined, so the room takes the maximal value $(F + R - 1)$.

After some initial moves, an invariant is established that the white rook is not exposed (its king can always approach and protect it, without having to move it) and that it divides two kings (either horizontally or vertically) or in some special cases that they form an L-shaped pattern (kings are in the same row (column), at distance 2, and the rook and the white king are in the same column (row) at distance 1). These notions are formalized as follows.

```
BTK.WR_exposed p ⟷
    (WhiteTurn p ∧ cdist (WK p) (WR p) > cdist (BK p) (WR p) + 1) ∨
    (BlackTurn p ∧ cdist (WK p) (WR p) > cdist (BK p) (WR p))
BTK.WR_divides p ⟷ (let (R^f, R^r) = WR p; (k^f, k^r) = BK p; (K^f, K^r) = WK p
```

    in (btw k^f R^f K^f $\vee$ btw k^r R^r K^r))

BTK.Lpattern p $\longleftrightarrow$ (let (R^f, R^r) = WR p; (k^f, k^r) = BK p; (K^f, K^r) = WK p
    in $(K^r = k^r \wedge |K^f - k^f| = 2 \wedge R^f = K^f \wedge |R^r - K^r| = 1) \vee$
       $(K^f = k^f \wedge |K^r - k^r| = 2 \wedge R^r = K^r \wedge |R^f - K^f| = 1)$

The strategy uses several kinds of moves that are applied in a fixed order (if one kind of move is not applicable, then the next one is tried, and so on). For example, one kind of move is the `ImmediateMate` and it is applicable if white can mate in a single move. The `Squeeze` is applicable if white can reduce the room, while keeping the rook not exposed, dividing the two kings, and avoiding a stalemate position for black. These relations are formalized as follows.

BTK.immediate_mate_cond p $\longleftrightarrow$ KRK.BK_cannot_move p $\wedge$ KRK.WR_attacks_BK p

BTK.squeeze_cond p p' $\longleftrightarrow$
    BTK.room p' < BTK.room p $\wedge$ BTK.WR_divides p' $\wedge$
    $\neg$ BTK.WR_exposed p' $\wedge$ (KRK.BK_cannot_move p' $\longrightarrow$ KRK.WR_attacks_BK p')

In order to apply some rule, its condition must hold but, in addition, no previous moves can be applicable, so their conditions must not hold for any legal move of white. This requires to universally quantify over all possible moves of white pieces. We introduce the function `kings_square` (f, r) k that for values k between 1 and 8, gives coordinates of 8 squares that surround the given central square (f, r). Similarly, the function `rooks_square` (f, r) k for values k between 1 and $F + R$ gives all squares that are in line with the rook (first horizontally, and then vertically). Then we introduce bounded quantification (that is unfolded in the proofs to stay within the quantifier-free fragment) and predicates that encode that a certain kind of move cannot be applied. We show this only for the `ImmediateMate`, as other moves follow a similar pattern.

all_n P n $\longleftrightarrow$ $\forall$ i. $1 \leq i \wedge i \leq n$ $\longrightarrow$ P i

no_mate_WK p $\longleftrightarrow$ all_n 8 (λ k. let sq = kings_square (WK p) k in
    KRK.WK_can_move_to p sq $\longrightarrow$ $\neg$ BTK.immediate_mate_cond (KRK.moveWK p sq))

no_mate_WR p $\longleftrightarrow$ all_n $(F + R)$ (λ k. let sq = rooks_square (WR p) k in
    KRK.WR_can_move_to p sq $\longrightarrow$ $\neg$ BTK.immediate_mate_cond (KRK.moveWR p sq))

no_immediate_mate p $\longleftrightarrow$ no_mate_WK p $\wedge$ no_mate_WR p

Note that a mate cannot occur as a consequence of a white king's move.

    Finally, we introduce the relation BTK.st_wht_move p p' m, encoding that a position p' is reached from a position p after a strategy move of a kind m.

MoveKind = ImmediateMate | ReadyToMate | Squeeze | ApproachDiag |
    ApproachNonDiag | KeepRoomDiag | KeepRoomNonDiag | RookHome | RookSafe

BTK.st_wht_move p p' m $\longleftrightarrow$
(if m = ImmediateMate then
    KRK.lgl_move_WR p p' $\wedge$ BTK.immediate_mate_cond p'
else
    no_immediate_mate p $\wedge$
    if m = ReadyToMate then
        KRK.legal_move_white p p' $\wedge$ BTK.ready_to_mate_cond p'
    else

```
no_ready_to_mate p ∧
  ...
            if m = RookSafe then
                KRK.lgl_move_WR p p' ∧ BTK.rook_safe_cond p p'
            else False)
```

Executable Specification. The relational specification BTK.st_wht_move p p' m is not executable. In many cases, for a given position p there are several possible values of p' and m that satisfy the previous relation. We defined a deterministic, executable function (p', m) = BTK.st_wht_move_fun p that returns a new position and a move type corresponding to the selected strategy move. In most cases this function iterates through all legal moves of white pieces (in some fixed order) until it finds a first move that satisfies the relational specification. Since the iteration order is fixed, the function will be deterministic, but in positions that allow several applicable moves, the choice is made rather arbitrarily (as the iteration order is chosen rather arbitrarily). An interesting exception is the squeeze move. To make the strategy more efficient, the optimal squeeze (the one that confines the black king the most) is always played (if there are several such moves, the first one found in the iterating process is used).

4   Correctness Proofs for Bratko-style Strategy

In this section we describe central correctness arguments for the strategy. All major proof steps were done automatically, by formulating the goals in LIA and applying the SMT solver.

Linear Arithmetic Formulation. The quantifier-free fragment of the theory of linear integer arithmetic (LIA) is very convenient for expressing our goals, so we formulated all our definitions in the language of LIA. This can be seen as an illustration how to prepare a problem (not only chess-related) for solving by automated solvers. Our definitions on this layer usually closely follow previously given definitions for the KRK case and Bratko's (BTK) strategy. However, in our LIA definitions, we never use quantifiers and don't use the record, product, nor the option type that were present on the KRK layer, but only the pure language of LIA. All KRK positions are represented in an unpacked form and functions receive six integers (usually denoted as K^f, K^r for white king file and rank coordinates, k^f, k^r for black king, and R^f, R^r for white rook) instead of a record that collects them. Note that all the following definitions assume that the rook is present on the board, since they are applied only in such situations. Here are some examples.

LIA.on_board sq^f sq^r $\longleftrightarrow$ $0 \leq sq^f \wedge sq^f <$ F $\wedge 0 \leq sq^r \wedge sq^r <$ R

LIA.all_on_board K^f K^r k^f k^r R^f R^r $\longleftrightarrow$
  LIA.on_board K^f K^r $\wedge$ LIA.on_board k^f k^r $\wedge$ LIA.on_board R^f R^r

LIA.king_scope s_1^f s_1^r s_2^f s_2^r $\longleftrightarrow$ $|s_1^f - s_2^f| \leq 1 \wedge |s_1^r - s_2^r| \leq 1 \wedge (s_1^f \neq s_2^f \vee s_1^r \neq s_2^r)$
LIA.pos_inv K^f K^r k^f k^r R^f R^r $\longleftrightarrow$
  $(K^f \neq k^f \vee K^r \neq k^r) \wedge (R^f \neq K^f \vee R^r \neq K^r) \wedge (R^f \neq k^f \vee R^r \neq k^r)$

It is shown that these definitions are equivalent to the KRK ones (under the assumption that the rook is not captured, and that coordinates of pieces are unpacked from the record). For example:

lemma
assumes "WK p = (K^f, K^r)" "BK p = (k^f, k^r)" "WR p = (R^f, R^r)" "$\neg$ WRcapt p"
shows "KRK.pos_inv p $\longleftrightarrow$ LIA.pos_inv K^f K^r k^f k^r R^f R^r"

All KRK definitions and all BTK strategy definitions have their LIA counterparts. The connection between them is quite obvious, so the proofs of equivalence are rather trivial and are proved by Isabelle's native automated tactics. In the following proofs, translation from HOL terms to LIA terms is done manually, but could be automated by implementing a suitable tactic.

Central Theorems. In this section we present our proof that BTK strategy is winning for white i.e., that it interprets the `WinningStrategyOrdering` locale. First, the strategy relation and the invariant are defined.

BTK.st_wht_move p p' $\longleftrightarrow$ ($\exists$ m. BTK.st_wht_move p p' m)

BTK.st_inv p $\longleftrightarrow$ $\neg$ WRcapt p

Before giving proofs, we introduce some auxiliary notions. We are often going to consider full moves (a move of white that follows the strategy, followed by any legal move of the black king)[4].

BTK.st_full_move p_1 p_2 $\longleftrightarrow$ $\exists$ p'_1. BTK.st_wht_move p_1 p'_1 $\wedge$ KRK.lgl_move_BK p'_1 p_2

For positions p_1 and p_2 and a set of strategy move types M, we define the following relations (white is allowed to make a move only if its kind is in M).

BTK.st_wht_move M p p' $\longleftrightarrow$ ($\exists$ $m \in M$. BTK.st_wht_move p p' m)

, BTK.st_full_move M p_1 p_2 $\longleftrightarrow$ $\exists$ p'_1. BTK.st_wht_move M p_1 p'_1 $\wedge$
KRK.lgl_move_BK p'_1 p_2

Next, we show that the BTK.st_wht_move p p' defines a correct strategy, i.e., that it interprets the `Strategy` locale (that all moves are legal, and that the invariant is maintained throughout each strategy play). The following theorem guarantees that every move made by the strategy is legal.

theorem assumes "BTK.st_wht_move p_1 p_2"
          shows   "KRK.lgl_move_WK p_1 p_2 $\vee$ KRK.lgl_move_WR p_1 p_2"

This is trivial to prove, since we explicitly require that a legal move is made in all cases of the definition of BTK.st_wht_move p_1 p_2 m.

It is obvious that a move of white preserves the invariant (white rook remains not captured). The following theorem guarantees that the invariant also remains preserved after any subsequent legal move of black.

theorem assumes "$\neg$ WRcapt p_1" "BTK.st_full_move p_1 p_2"
          shows   "$\neg$ WRcapt p_2"

---

[4] In the chess literature, half-move is sometimes called *ply*, and full-move *move*.

The proof goes as follows. White could not have played the `ImmediateMate` move, since black has made the move. In all other case, except the `ReadyToMate` move, the condition ¬ `BTK.WR_exposed p` is imposed, and it guarantees that the rook cannot be captured. The `ReadyToMate` case is the only non-trivial case and we encode the problem in LIA and employ SMT solvers to discharge the goal.

Next, we prove that the strategy is winning i.e., that it interprets the `Winning-StrategyOrdering` locale (that the strategy is always applicable, that it never leads into stalemate, and that there is a well founded ordering consistent with full strategy moves). The next theorem shows that play can always be continued i.e., that white can always make a strategy move.

theorem assumes "WhiteTurn p_1" "¬ WRcapt p_1" "KRK.lgl_pos p_1"
　　　　　shows　　"∃ p_2. BTK.st_wht_move p_1 p_2"

The proof is based on the following lemma, that guarantees that either `Squeeze`, `RookHome`, or `RookSafe` move are always applicable. The lemma is again proved automatically, by rewriting it into LIA and employing SMT solver.

lemma assumes "WhiteTurn p" "¬ WRcapt p" "KRK.lgl_pos p"
　shows "¬ BTK.no_squeeze p ∨ ¬ BTK.no_rook_safe p ∨ ¬ BTK.no_rook_home p"

The following theorem shows that black is never in a stalemate.

theorem assumes "¬ WRcapt p_1" "BTK.st_wht_move p_1 p_2"
　　　　　shows　　"¬ KRK.stalemate p_2"

This is also proved by analyzing different moves. After the `ImmediateMate`, black is mated and that is not a stalemate. All other moves, except `ReadyToMate`, by their definition require that stalemate did not occur, so they are trivial. The only complicated case is `ReadyToMate`, so we again use SMT solver to discharge it.

Finally, we prove termination. We show that the relation $R = \{(p_2, p_1). \ p_1 \in \texttt{BTK.st\_reachable } p_0 \ \wedge \ \texttt{BTK.st\_full\_move } p_1 \ p_2\}$ is well-founded, for a legal initial position p_0. If it would not be well-founded, then there would be a non-empty set with no minimal element i.e., there would be a non-empty set Q such that a strategy play can always be extended by a strategy move of white, followed by a move of black:

∀ p ∈ Q. p ∈ BTK.st_reachable p_0 ∧ (∃ p' ∈ Q. BTK.st_full_move p p')

Since in such infinite play, white must not make `ImmediateMate` and `ReadyToMate` move, as otherwise, the play would finish in a checkmate position, the following is implied ($\mathcal{M}$ denotes a set of two mate moves, and $\overline{\mathcal{M}}$ denotes its complement).

∀ p ∈ Q. p ∈ BTK.st_reachable p_0 ∧ (∃ p' ∈ Q. BTK.st_full_move $\overline{\mathcal{M}}$ p p')

We show that this is a contradiction. The first observation is that the `RookHome` and `RookSafe` moves can be played only within the first three moves of a strategy play. This is proved by induction, using the following theorem that we proved automatically using LIA and SMT solvers.

theorem assumes "¬ WRcapt p_1"
　　　　　　　　"BTK.st_full_move p_1 p_2" "BTK.st_full_move p_2 p_3"
　　　　　　　　"BTK.st_full_move p_3 p_4" "BTK.st_wht_move p_4 p'_4 m"
　　　　shows　　"$m \neq$ RookHome" "$m \neq$ RookSafe"

Therefore, starting from some position $p'_0 \in Q$ (a position reached after three moves), all moves of white in our infinite strategy play are basic moves ($\mathcal{B}$ denotes the set of basic moves: Squeeze, Approach, or KeepRoom).

$\forall~p \in$ BTK.st_reachable $p'_0 \cap Q.~(\exists~p' \in Q.$ BTK.st_full_move $\mathcal{B}~p~p')$

Next we proved that there is a position p_m that satisfies the following condition.

$p_m \in$ BTK.st_reachable $p'_0 \cap Q \wedge$ BTK.room $p_m \leq 3 \wedge \neg$ BTK.WR_exposed p_m

To prove this, we use the following lemma, stating that when started from a situation where the rook is exposed or the room is grater than 3, after three strategy basic moves, the rook is not exposed anymore and either the room decreased, or it stayed the same, but the Manhattan distance to the critical square decreased. Again, this lemma is proved automatically, by expressing it in language of LIA and employing SMT solver.

theorem assumes "¬ WRcapt p_1" "BTK.WR_exposed p_1 ∨ BTK.room p_1 > 3"
                "BTK.st_full_move $\mathcal{B}~p_1~p_2$" "BTK.st_full_move $\mathcal{B}~p_2~p_3$"
                "BTK.st_full_move $\mathcal{B}~p_3~p_4$"
    shows "(BTK.WR_exposed p_4, BTK.room p_4, BTK.mdist_cs p_4) <
            (BTK.WR_exposed p_1, BTK.room p_1, BTK.mdist_cs p_1)"

Note that the last conclusion is expressed as lexicographic comparison between the ordered triples that contain a bool (BTK.WR_exposed p) and two integers ((BTK.room p and BTK.mdist_cs p). The Boolean value False is considered less than True, and integers are ordered in the standard way. Since these integers are non-negative for all legal positions, this lexicographic ordering is well-founded. Then we consider the following relation.

R' = {(p_2, p_1). ¬ WRcapt p_1 ∧ (BTK.WR_exposed p_1 ∨ BTK.room p_1 > 3) ∧
               BTK.st_full_move $\mathcal{B}~p_1~p_2$}

By the previous theorem, the third power of this relation is a subset of the lexicographic ordering of triples that was well-founded, so the relation R' itself is well-founded. Then, every non-empty set has a minimal element in this relation, so there is an element p_m such that the following holds.

$p_m \in$ BTK.st_reachable $p'_0 \cap Q$
$\forall~p.~(p,~p_m) \in R' \longrightarrow p \notin$ BTK.st_reachable $p'_0 \cap Q$

As we already noted, the play from BTK.st_reachable $p'_0 \cap Q$ can always be extended by a strategy basic move, followed by a move of the black king, i.e., there is a position $p'_m \in Q$ such that BTK.st_full_move $\mathcal{B}~p_m~p'_m$. Then, p'_m would also be in BTK.st_reachable $p'_0 \cap Q$, so $(p'_m, p_m) \notin R'$. Since it holds that ¬ WRcapt p_m and BTK.st_full_move $\mathcal{B}~p_m~p'_m$, it must not hold that BTK.WR_exposed p_1 ∨ BTK.room p_1 > 3, so p_m is the required position, satisfying ¬ BTK.WR_exposed p_m ∧ BTK.room $p_m \leq 3$. From such position, the fifth move by white will be either a ImmediateMate or a ReadyToMate. This holds by the following theorem (again, proved automatically, using LIA and SMT solver).

theorem assumes `"¬ WRcapt `p_0`"` `"BTK.room `p_0` ≤ 3"` `"¬ BTK.WR_exposed `p_0`"`
`"BTK.st_full_move `$\mathcal{B}$` `p_0` `p_1`"` `"BTK.st_full_move `$\mathcal{B}$` `p_1` `p_2`"`
`"BTK.st_full_move `$\mathcal{B}$` `p_2` `p_3`"` `"BTK.st_full_move `$\mathcal{B}$` `p_3` `p_4`"`
`"BTK.st_wht_move `p_4` `p_4'` `m`"`
      shows   `"`$m \in \mathcal{M}$`"`

Since $p_m \in$ `BTK.st_reachable` $p_0' \cap Q$, white is on turn and the play can be infinitely extended by basic moves. However, by the previous theorem, the fifth move of white must be a mating move, giving the final contradiction.

5 Related Work

The presented formalization is, as said, closely related to the one based on constraint solving [10]. Still, the present work is a step forward, since it includes a formal development of relevant chess rules within the proof assistant and all proofs are trustworthy in a stronger sense. Not only that this work glues together conjectures checked earlier by the constraint solver, it also revealed some minor deficiencies (e.g., imprecise definition of legal moves) in the earlier formalization.

Although properties of two-player board games are typically explored using brute-force analyses, other approaches exist, similar to the constraint solving based one. For instance, binary decision diagrams were applied for game-theoretical analysis of a number of games [5]. Again, this approach cannot provide results that can be considered trustworthy in the sense of proof assistants. There is only a limited literature on using interactive theorem proving for analyzing two-player board games. A retrograde chess analysis has been done within Coq, but it does not consider chess strategies [11]. Hurd and Haworth constructed large, formally verified endgame databases, within the HOL system [8]. Their work is focused on endgame tables and it is extremely difficult (if not impossible) to extract some concise strategy descriptions and high-level insights from these tables, so we addressed a different problem and in a different way.

Before using Isabelle/SMT, we formalized the strategy within Coq [12]. However, neither Omega, a built-in solver for quantifier-free formulae in LIA, nor a far more efficient Micromega and corresponding tactics for solving goals over ordered rings (including LIA), were efficient enough. SMTCoq [1] is an external Coq tool that checks proof witnesses coming from external SAT and SMT solvers (zChaff and veriT). Coq implements constructive logic, while veriT reasons classically. SMTCoq was designed to work with type `bool` (which is decidable in Coq) but not with type `Prop` which is natural type for propositions in Coq. Construction of complex theories over type `bool` in Coq can be quite inconvenient and has many pitfalls. There are plans for further improvement of SMTCoq.

6 Conclusions and Further Work

In the presented work, chess — a typical AI domain — has been used as an illustration for showing that the state-of-the-art theorem proving technology has reached the stage when very complex combinatorial conjectures can be proved in a trusted way, with only a small human effort. Our key point is that this is possible thanks to synergy of very expressible interactive provers and very powerful automated provers (SMT solvers in our case). The considered conjectures push the provers up to the limits[5] and while Isabelle/SMT interface can be further improved (e.g., proof checking time could be reduced), our experience with it can be seen as a success story. Our second point is that the presented work can be seen as an exercise not only in automation, but also in suitable formalization of non-trivial combinatorial problems. Namely, computer theorem provers are powerful tools in constructing and checking proofs, but they only work modulo the given definitions. The only way to check definitions is by human inspection, and one must be extremely careful when doing this step. Reducing everything to a small set of basic definitions (as we reduced specific KRK definitions to the basic chess rules) is an important step in ensuring soundness.

For future work, we are planning to analyse different generalizations of the presented central theorem. For example, unlike approaches based on SAT or endgame tables, our approach is not enumerative in its nature and can be used for arbitrary board sizes. We will also use a similar approach for proving other related conjectures in chess and other two-player intellectual games.

Acknowledgments. The authors are grateful to Sascha Böhme and Jasmin Christian Blanchette for their assistance in using SMT solvers from Isabelle/HOL and to Chantal Keller for her assistance in using SMT solvers from Coq. The first and the second author were supported in part by the grant ON174021 of the Ministry of Science of Serbia.

References

1. Armand, M., Faure, G., Grégoire, B., Keller, C., Théry, L., Werner, B.: A modular integration of SAT/SMT solvers to coq through proof witnesses. In: Jouannaud, J.-P., Shao, Z. (eds.) CPP 2011. LNCS, vol. 7086, pp. 135–150. Springer, Heidelberg (2011)
2. Ballarin, C.: Interpretation of locales in isabelle: theories and proof contexts. In: Borwein, J.M., Farmer, W.M. (eds.) MKM 2006. LNCS (LNAI), vol. 4108, pp. 31–43. Springer, Heidelberg (2006)
3. Böhme, S., Weber, T.: Fast LCF-style proof reconstruction for Z3. In: Kaufmann, M., Paulson, L.C. (eds.) ITP 2010. LNCS, vol. 6172, pp. 179–194. Springer, Heidelberg (2010)
4. Bratko, I.: Proving correctness of strategies in the AL1 assertional language. Inform. Process. Lett. **7**(5), 223–230 (1978)

---

[5] The largest SMT formula in the proof has more than 67,000 atoms. Proofs were checked in around 8 CPU minutes on a multiprocessor 1.9GHz machine with 2 GB RAM per CPU when SMT solvers are used in the oracle mode and when SMT proof reconstruction was not performed. SMT proof reconstruction is the slowest part of proof-checking, but it can be done in a quite reasonable time of 29 CPU minutes. The whole formalization has around 12,000 lines of Isabelle/Isar code.

5. Edelkamp, S.: Symbolic exploration in two-player games: preliminary results. In: AIPS 2002 Workshop on Planning via Model-Checking (2002)
6. FIDE. The FIDE Handbook, chapter E.I. The Laws of Chess (2004). Available for download from the FIDE website
7. Hurd, J.: Formal verification of chess endgame databases. In: Theorem Proving in Higher Order Logics: Emerging Trends, Oxford University CLR Report (2005)
8. Hurd, J., Haworth, G.: Data assurance in opaque computations. In: van den Herik, H.J., Spronck, P. (eds.) ACG 2009. LNCS, vol. 6048, pp. 221–231. Springer, Heidelberg (2010)
9. Janičić, P.: URSA: a system for uniform reduction to SAT. Logical Methods Comput. Sci. **8**(3:30) (2012)
10. Maliković, M., Janičić, P.: Proving correctness of a KRK chess endgame strategy by SAT-based constraint solving. ICGA J. **36**(2), 81–99 (2013)
11. Maliković, M., Čubrilo, M.: What were the last moves? Int. Rev. Comput. Softw. **5**(1), 59–70 (2010)
12. Maliković, M., Čubrilo, M., Janičić, P.: Formalization of a strategy for the KRK chess endgame. In: Conference on Information and Intelligent Systems (2012)
13. Thompson, K.: Retrograde analysis of certain endgames. ICCA J. **9**(3), 131–139 (1986)

Inductive Beluga: Programming Proofs

Brigitte Pientka[✉] and Andrew Cave

McGill University, Montreal, QC, Canada
{bpientka,acave1}@cs.mcgill.ca

Abstract. BELUGA is a proof environment which provides a sophisticated infrastructure for implementing formal systems based on the logical framework LF together with a first-order reasoning language for implementing inductive proofs about them following the Curry-Howard isomorphism.

In this paper we describe four significant extensions to BELUGA: (1) we enrich our infrastructure for modelling formal systems with first-class simultaneous substitutions, a key and common concept when reasoning about formal systems (2) we support inductive definitions in our reasoning language which significantly increases BELUGA's expressive power (3) we provide a totality checker which guarantees that recursive programs are well-founded and correspond to inductive proofs (4) we describe an interactive program development environment. Taken together these extensions enable direct and compact mechanizations. To demonstrate BELUGA's strength and illustrate these new features we develop a weak normalization proof using logical relations.

Keywords: Logical frameworks · Dependent types · Proof assistant

1 Introduction

Mechanizing formal systems, given via axioms and inference rules, together with proofs about them plays an important role in establishing trust in formal developments. A key question in this endeavor is how to represent variables, (simultanous) substitution, assumptions, and derivations that depend on assumptions.

BELUGAis a proof environment which provides a sophisticated infrastructure for implementing formal systems based on the logical framework LF [11]. This allows programmers to uniformly specify syntax, inference rules, and derivation trees using higher-order abstract syntax (HOAS) and relieves users from having to build up common infrastructure to mange variable binding, renaming, and (single) substitution. BELUGAprovides in addition support for first-class contexts [16] and simultaneous substitutions [4], two common key concepts that frequently arise in practice. Compared to existing approaches, we consider its infrastructure one of the most advanced for prototyping formal systems [6].

To reason about formal systems, BELUGAprovides a standard first-order proof language with inductive definitions [3] and domain-specific induction principles [18]. It is a natural extension of how we reason inductively about simple

© Springer International Publishing Switzerland 2015
A.P. Felty and A. Middeldorp (Eds.): CADE-25, LNAI 9195, pp. 272–281, 2015.
DOI: 10.1007/978-3-319-21401-6_18

domains such as natural numbers or lists, except that our domain is richer, since it allows us to represent and manipulate derivations trees that may depend on assumptions. Inductive BELUGAsubstantially extends our previous system [19]:

First-class Substitution Variables [4]. We directly support simultaneous substitutions and substitution variables within our LF-infrastructure. Dealing with substitutions manually can lead to a substantial overhead. Being able to abstract over substitutions allows us to tackle challenging examples such as proofs using logical relations concisely without this overhead.

Datatype Definitions [3]. We extend our reasoning language with recursive type definitions. These allow us to express relationships between contexts and derivations (contextual objects). They are crucial in describing semantic properties such as logical relations. They are also important for applications such as type preserving code transformations (e.g. closure conversion and hoisting [1]) and normalization-by-evaluation. To maintain consistency while remaining sufficiently expressive, BELUGAsupports two kinds of recursive type definitions: standard *inductive* definitions, and a form which we call *stratified* types.

Totality Checking [18]. We implement a totality checker which guarantees that a given program is total, i.e. all cases are covering and all recursive calls are well-founded according to a structural subterm ordering. This is an essential step to check that the given recursive program constitutes an inductive proof.

Interactive Proof Development. Similar to other interactive development modes (e.g. Agda [14] or Alf [12]), we support writing holes in programs (i.e. proofs) showing the user the available assumptions and the current goal, and we support automatic case-splitting based on BELUGA's coverage algorithm [5, 18] which users find useful in writing proofs as programs.

The Beluga system, including source code, examples, and an Emacs mode, is available from http://complogic.cs.mcgill.ca/beluga/.

2   Inductive Proofs as Recursive Programs

We describe a weak normalization proof for the simply typed lambda-calculus using logical relations – a proof technique going back to Tait [20] and later refined by Girard [10]. The central idea of logical relations is to define a relation on terms recursively on the syntax of types instead of directly on the syntax of terms themselves; this enables us to reason about logically related terms rather than terms directly. Such proofs are especially challenging to mechanize: first, specifying logical relations themselves typically requires a logic which allows a complex nesting of implications; second, to establish soundness of a logical relation, one must prove a property of well-typed *open* terms under arbitrary instantiations of their free variables. This latter part is typically stated using some notion of simultaneous substitution, and requires various equational properties of these substitutions.

As we will see our mechanization directly mirrors the theoretical development that one would do on paper which we find a remarkably elegant solution.

2.1 Representing Well-Typed Terms and Evaluation in LF

For our example, we consider simply-typed lambda-terms. While we often define the grammar and typing separately, here we work directly with intrinsically typed terms, since it is more succinct. Their definition in the logical framework LF is straightforward. Below, tm defines our family of simply-typed lambda terms indexed by their type. In typical higher-order abstract syntax (HOAS) fashion, lambda abstraction takes a *function* representing the abstraction of a term over a variable. There is no case for variables, as they are treated implicitly. We remind the reader that this is a weak function space – there is no case analysis or recursion, and hence only genuine lambda terms can be represented.

```
LF tp : type =                 LF tm : tp → type =
| b   : tp                     | app : tm (arr T S) → tm T → tm S
| arr : tp → tp → tp;          | lam : (tm T → tm S) → tm (arr T S)
                               | c   : tm b;
```

Our goal is to prove that evaluation of well-typed terms halts. For simplicity, we consider here weak-head reduction that does not evaluate inside abstractions, although our development smoothly extends. We encode the relation step stating that a term steps to another term either by reducing a redex (beta rule) or by finding a redex in the head (stepapp rule). In defining the beta rule, we fall back to LF-application to model substitution. In addition, we define a multi-step relation, called mstep, on top of the single step relation.

```
LF step : tm A → tm A → type =
| beta   : step (app (lam M) N) (M N)
| stepapp: step M M' → step (app M N) (app M' N);

LF mstep : tm A → tm A → type =
| refl   : mstep M M
| onestep: step M N → mstep N M'' → mstep M M';
```

Evaluation of a term halts if there is a value, i.e. either a constant or a lambda-abstraction which it steps to.

```
LF val : tm A → type =          LF halts : tm A → type =
| val/c   : val c               | halts/m : mstep M M' → val M'
| val/lam : val (lam M);                  → halts M;
```

2.2 Representing Reducibility Using Indexed Types

Proving that evaluation of well-typed terms halts cannot be done directly, as the size of our terms may grow when we are using the beta rule. Instead, we define a predicate Reduce on well-typed terms inductively on the syntax of types, often called a reducibility predicate. This enables us to reason about logically related terms rather than terms directly.

– A term M of base type b is reducible if halts M.
– A term M of function type (arr a B) is reducible, if halts M and moreover, for every reducible N of type A, the application app M N is reducible.

Reducibility cannot be directly encoded at the LF layer, since it involves strong implications. We will use an indexed recursive type [3], which allows us to state *properties about well-typed terms* and define the reducibility relation recursively. In our case, it indeed suffices to state reducibility about closed terms; however, in general we may want to state properties about open terms, i.e. terms that may refer to assumptions. In BELUGA, we pair a term M together with the context Ψ in which it is meaningful, written as [Ψ ⊢M]. These are called contextual LF objects [13]. We can then embed contextual objects and types into the reasoning level; in particular, we can state inductive properties about contexts, contextual objects and contextual types.

```
stratified   Reduce : {A:[⊢tp]}{M:[⊢tm A]} ctype =
| I   : [ ⊢halts M] → Reduce [ ⊢b] [ ⊢M]
| Arr: [ ⊢halts M] →
        ({N:[ ⊢tm A]} Reduce [ ⊢A] [ ⊢N] → Reduce [ ⊢B] [ ⊢app M N])
      → Reduce [ ⊢arr A B ] [ ⊢M];
```

Here we state the relation Reduce about the closed term M:[⊢tm A] using the keyword stratified. The constructor I defines that M is reducible at base type, if [⊢halts M]. The constructor Arr defines that a closed term M of type arr A B is reducible if it halts, and moreover for every reducible N of type A, the application app M N is reducible. We write {N:[⊢tm A]} for explicit universal quantification over N, a closed term of type A. To the left of ⊢ in [⊢tm A] is where one writes the context the term is defined in – in this case, it is empty.

In the definition of Reduce, the arrows correspond to usual implications in first-order logic and denote a standard function space, not the weak function space of LF. Contextual LF types and objects are always enclosed with [ ] when they are embedded into recursive data-type definitions in the reasoning language. We note that the definition of Reduce is not (strictly) positive, and hence not inductive, since Reduce appears to the left of an arrow in the Arr case. However, there is a different criterion by which this definition is justified, namely *stratification*. We discuss this point further in Sect. 2.5.

To prove that evaluation of well-typed terms halts, we now prove two lemmas:

1. All closed terms M:[⊢tm A] are reducible, i.e. Reduce [⊢ A] [⊢ M].
2. If Reduce [⊢ A] [⊢ M] then evaluation of M halts, i.e. [⊢ halts M].

The second lemma follows trivially from our definition. The first part is more difficult. It requires a generalization, which says that any well-typed term M under a closing substitution σ is in the relation, i.e. Reduce [⊢ A] [⊢ M[σ]]. To be able to prove this, we need that σ provides reducible instantiations for the free variables in M.

2.3 First-Class Contexts and Simultaneous Substitutions

In BELUGA, we support first-class contexts and simultaneous substitutions. We first define the structure of the context in which a term M is meaningful by defining a context schema: `schema ctx = tm T;`

A context γ of schema `ctx` stands for any context that contains only declarations `x:tm T` for some T. Hence, `x1:tm b, x2:tm (arr b b)` is a valid context, while `a:tp,x:tm a` is not. We can then describe not only closed well-typed terms, but also a term M that is well-typed in a context γ as `[γ ⊢ tm A]` where γ has schema `ctx` [16].

To express that the substitution σ provides reducible instantiations for variables in γ, we again use an indexed recursive type.

```
inductive RedSub : {γ:ctx}{σ:[ ⊢γ]} ctype =
| Nil : RedSub [ ] [ ⊢ ^ ]
| Dot : RedSub [γ] [ ⊢ σ ] → Reduce [ ⊢A] [ ⊢M]
     → RedSub [γ, x:tm A[^]] [ ⊢ σ , M ];
```

In BELUGA, substitution variables are written as σ. Its type is written `[⊢γ]`, meaning that it has domain γ and empty range, i.e. it takes variables in γ to closed terms of the same type. In the base case, the empty substitution, written as `^`, is reducible. In the `Dot` case, we read this as saying: if σ is a reducible substitution (implicitly at type `[⊢γ]`) and M is a reducible term at type A, then σ with M appended is a reducible substitution at type `[⊢γ,x:tm A[^]]` – the domain has been extended with a variable of type A; as the type A is closed, we need to weaken it by applying the empty substitution to ensure it is meaningful in the context γ. For better readability, we subsequently omit the weakening substitution.

2.4 Developing Proofs Interactively

We now have all the definitions in place to prove that any well-typed term M under a closing simultaneous substitution σ is reducible.

Lemma *For all* `M:[γ ⊢ tm A]` *if* `RedSub [γ] [σ]` *then* `Reduce [⊢A] [ ⊢ M[σ]]`.

This statement can be directly translated into a type in BELUGA.

`rec main:{γ:ctx}{M:[γ⊢ tm A]} RedSub [γ] [ ⊢ σ] → Reduce [⊢ A] [ ⊢ M[σ]] = ?;`

Logically, the type corresponds to a first-order logic formula which quantifies over the context γ, the type A, terms M, and substitutions σ. We only quantified over γ and M explicitly and left σ and A free. BELUGA's reconstruction engine [7,17] will infer their types and abstract over them. The type says: for all γ and terms M that have type A in γ, if σ is reducible (i.e. `RedSub [γ] [ ⊢σ]`) then M[σ] is reducible at type A (i.e. `Reduce [⊢A] [⊢M[σ]]`).

We now develop the proof of our main theorem interactively following ideas first developed in the Alf proof editor [12] and later incorporated into Agda [14]. Traditionally, proof assistants such as Coq [2] build a proof by giving commands to a proof engine refining the current proof state. The (partial) proof object corresponding to the proof state is hidden. It is often only checked after the proof has been fully constructed. In BELUGA, as in Alf and Agda, the proof object is the primary focus. We are building (partial) proof objects (i.e. programs) directly. By doing so, we indirectly refine the proof state. Let us illustrate.

Working backwards, we use the introduction rules for universal quantification and implications; mlam-abstraction corresponds to the proof term for universal quantifier introduction and fn-abstraction corresponds to implication introduction. We write ? for the incomplete parts of the proof object.

```
rec main:{γ:ctx}{M:[γ ⊢tm A ]}RedSub [γ] [ ⊢σ] → Reduce [⊢A] [⊢M[σ]] =
  mlam γ, M ⇒ fn rs ⇒ ?;
```

Type checking the above program succeeds, but returns the type of the hole:

```
- Meta-Context:
  {γ : ctx}
  {M : [γ ⊢ tm A]}
-----------------------------------------------------------------------
  - Context:
  main: {γ:ctx}{M:[γ ⊢tm A]} RedSub [γ] [ ⊢σ] → Reduce [ ⊢A] [ ⊢M[σ]]
  rs: RedSub [γ] [ ⊢ σ]
=======================================================================
- Goal Type: Reduce [ ⊢ A] [ ⊢ M[σ]]
```

The meta-context contains assumptions coming from universal quantification, while the context contains assumptions coming from implications. The programmer can refine the current hole by splitting on variables that occur either in the meta-context or in the context using the splitting tactic that reuses our coverage implementation [5,18] to generate all possible cases.

To split on M, our splitting tactic inspects the type of M, namely [γ ⊢ tm A] and automatically generates possible cases using all the constructors that can be used to build a term, i.e. [γ ⊢lam λy.M] and [γ ⊢app M N], and possible variables that match a declaration in the context γ, written here as [γ ⊢#p]. Intuitively writing [γ ⊢ M] stands for a pattern where M stands for a term that may contain variables from the context γ.

```
rec main:{γ:ctx}{M:[γ ⊢tm A]} RedSub [γ] [ ⊢σ] → Reduce [⊢A] [⊢M[σ]] =
  mlam γ, M ⇒ fn rs ⇒ (case [γ ⊢ M] of
  |[γ ⊢ #p] ⇒  ?
  |[γ ⊢ app M N] ⇒  ?
  |[γ ⊢ lam λy. M] ⇒  ?
  |[γ ⊢ c] ⇒  ? );
```

Variable Case. We need to construct the goal Reduce [⊢ T] [⊢ #p[σ]] given a parameter variable #p of type [γ ⊢ tm T]. We use the auxiliary function lookup to retrieve the corresponding reducible term from σ. Note that applying the substitution σ to [γ ⊢ #p] gives us [⊢ #p[σ]].

```
rec lookup:{γ:ctx}{#p:[γ ⊢tm A]}RedSub [γ] [⊢σ] → Reduce [⊢A] [⊢ #p[σ]] = ?;
```

This function is defined inductively on the context γ. The case where γ is empty is impossible, since no variable #p exists. If γ = γ', x:tm T, then there are two cases to consider: either #p stands for x, then we retrieve the last element in the substitution σ together with the proof that it is reducible; if #p stands for another variable in γ', then we recurse. All splits can be done through the splitting tactic.

Application Case. Inspecting the hole tells us that we must construct a proof for Reduce [⊢S] [⊢app M[σ] N[σ]]. BELUGA turned [⊢(app M N)[σ]] silently into [⊢app M[σ] N[σ]] pushing the substitution σ inside. This is one typical example where our equational theory about simultaneous substitution that we support intrinsically in our system comes into play.

Appealing to IH on N, written as the recursive call main [γ] [γ⊢N] rs, returns rN: Reduce [⊢A] [⊢N[σ]]. Appealing to IH on M, written as the recursive call main [γ] [γ ⊢ M] rs, gives us Reduce [⊢arr A B] [⊢M[σ]]. By inversion on the definition of Reduce, we get to the state Reduce [⊢B] [⊢app M[σ] N[σ]] given the assumptions

```
rN: Reduce [ ⊢ A] [ ⊢ N[σ]]
ha: [ ⊢ halts (arr A B) (M[σ])]
f: {N:[ ⊢tm A]} Reduce [ ⊢ A] [ ⊢ N] → Reduce [ ⊢ B] [ ⊢ app (M[σ]) N]
```

Using f and passing to it N together with rN, we can finish this case. Our partial proof object has evolved to:

```
rec main:{γ:ctx}{M:[γ ⊢tm A]} RedSub [γ] [ ⊢σ] → Reduce [⊢A] [⊢M[σ]] =
mlam γ, M ⇒ fn rs ⇒ (case [γ ⊢ M] of
|[γ ⊢ #p] ⇒   lookup [γ] [γ ⊢#p] rs
|[γ ⊢ app M N] ⇒
    let rN      = main [γ] [γ ⊢N ] rs in
    let Arr ha f = main [γ] [γ ⊢M ] rs in f [⊢ _ ] rN
|[γ ⊢ lam λy. M] ⇒  ?
|[γ ⊢ c] ⇒  ? );
```

Abstraction Case. We must find a proof for Reduce [⊢ arr T S] [⊢lam λy.M[σ,y]]. We note again that the substitution σ has been pushed silently inside the abstraction. By definition of Reduce (see the constructor Arr), we need to prove two things: 1) [⊢halts (lam λy.M[σ,y])] and 2) assuming N:[⊢tm T] and rN:Reduce [⊢T] [⊢N] we need to show that Reduce [⊢S] [⊢app (lam λy.M[σ,y]) N]. For part 1), we simply construct the witness [⊢halts/m refl val/lam]. For part 2), we rely on a lemma stating that Reduce is backwards closed under reduction.

```
rec bwd_closed:{S:[ ⊢step M M']} Reduce [⊢A] [⊢M'] → Reduce [⊢A] [⊢M] = ?;
```

Using the fact that N provides a reducible term for x, we appeal to IH on M by recursively calling main [γ,x:tm _] [γ,x⊢M] (Dot rs rN). As a result we obtain rM:Reduce [⊢S] [⊢M[σ,N]]. Now, we argue by the lemma bwd_closed and using the beta rule, that Reduce [⊢S] [⊢app (lam λy.M[σ,y]) N]. While this looks simple, there is in fact some hidden equational reasoning about substitutions. From the beta rule we get [⊢ (λy.M[σ,y]) N] which is not in normal form. To replace y with N, we need to compose the single substitution that replaces y with N with the simultaneous substitution [σ,y]. Again, our equational theory about simultaneous substitutions comes into play.

The complete proof object including the case for constants is given in Fig. 1. Underscores that occur are inferred by BELUGA's type reconstruction.

2.5   Totality Checking

For our programs to be considered proofs, we need to know: 1) Our programs cover all cases 2) They terminate and 3) all datatype definitions are acceptable.

```
rec main:{γ:ctx}{M:[γ⊢tm A]}RedSub [γ] [⊢σ] → Reduce [⊢A] [⊢M[σ]] =
/ total m (main γ a s m) /
mlam γ, M ⇒ fn rs ⇒ case [γ ⊢ M] of
| [γ ⊢ #p] ⇒  lookup [γ] [γ ⊢ #p] rs
| [γ ⊢ lam λx. M] ⇒
  Arr [ ⊢ halts/m refl val/lam]
      (mlam N ⇒ fn rN ⇒
          let rM = main [γ,x:tm _] [γ,x⊢M] (Dot rs rN) in
          bwd_closed [⊢beta] rM)
| [γ ⊢ app M N] ⇒
    let rN     = main [γ] [γ ⊢N ] rs in
    let Arr ha f = main [γ] [γ ⊢M ] rs in   f [ ⊢ _ ] rN
| [γ ⊢  c] ⇒  I [ ⊢ halts/m refl val/c];
```

Fig. 1. Weak Normalization Proof for the Simply-Typed Lambda-Calculus

We verify coverage following [5,18]. If the program was developed interactively it is covering by construction. To verify the program terminates, we verify that the recursive calls are well-founded. We use a totality declaration to specify the argument that is decreasing for a given function. In the given example, the totality declaration tells BELUGAthat main is terminating in the fourth position; the type of main specifies first explicitly the context γ, followed by two implicit arguments for the type A and the substitution σ, that are reconstructed, and then the term M:[γ⊢tm A]. Following [18], we generate valid recursive calls when splitting on M and then subsequently verify that only valid calls are made.

Recursive datatype definitions can be justified in one of two possible ways: by declaring a definition with **inductive**, BELUGAverifies that the definition adheres to a standard strict positivity condition, i.e. there are no recursive occurrences to the left of an arrow. Positive definitions are interpreted inductively, which enables them to be used as a termination argument in recursive functions.

Alternatively, by declaring a definition with **stratified**, BELUGAverifies that there is an index argument which decreases in each recursive occurrence of the definition. This is how our definition of Reduce is justified: it is stratified by its tp index. Such types are not inductive, but rather can be thought of as being constructed in stages, or defined by a special form of large elimination. Consequently, BELUGAdoes not allow stratified types to be used as a termination argument in recursive functions; instead one may use its index.

3   Related Work and Conclusion

There are several approaches to specifying and reasoning about formal systems using higher-order abstract syntax. The Twelf system [15] also provides an implementation of the logical framework LF. However, unlike proofs in BELUGAwhere we implement proofs as recursive functions, proofs in Twelf are implemented as relations. Twelf does not support the ability to reason about contexts, contextual LF objects and first-class simultaneous substitutions. More importantly, it can only encode forall-exists statements and does not support recursive data type definitions about LF objects.

The Abella system [8] provides an interactive theorem prover for reasoning about specifications using higher-order abstract syntax. Its theoretical basis is different and its reasoning logic extends first-order logic with ∇-quantifier [9] which can be used to express properties about variables. Contexts and simultaneous substitutions can be expressed as inductive definitions, but since they are not first-class we must establish properties such as composition of simultaneous substitution, well-formedness of contexts, etc. separately. This is in contrast to our framework where our reasoning logic remains first-order logic, but all reasoning about variables, contexts, simultaneous substitution is encapsulated in our domain, the contextual logical framework. Abella's interactive proof development approach follows the traditional model: we manipulate the proof state by a few tactics such as our splitting tactic and there is no proof object produced that witnesses the proof.

Inductive BELUGA allows programmers to develop proofs interactively by relying on holes. Its expressive power comes on the one hand from indexed recursive datatype definitions on the reasoning logic side and on the other hand from the rich infrastructure contextual LF provides. This allows compact and elegant mechanizations of challenging problems such as proofs by logical relations. In addition to the proof shown here other examples include the mechanization of a binary logical relation for proving completeness of an algorithm for $\beta\eta$-equality and a normalization proof allowing reductions under abstractions, both for the simply-typed lambda calculus.

Acknowledgements. Over the past 6 years several undergraduate and graduate students have contributed to the implementation: A. Marchildon, O. Savary Belanger, M. Boespflug, S. Cooper, F. Ferreira, D. Thibodeau, T. Xue.

References

1. Savary-Belanger, O., Monnier, S., Pientka, B.: Programming type-safe transformations using higher-order abstract syntax. In: Gonthier, G., Norrish, M. (eds.) CPP 2013. LNCS, vol. 8307, pp. 243–258. Springer, Heidelberg (2013)
2. Bertot, Y., Castéran, P.: Interactive Theorem Proving and Program Development. Coq'Art: The Calculus of Inductive Constructions. Springer, Heidelberg (2004)
3. Cave, A., Pientka, B.: Programming with binders and indexed data-types. In: 39th Annual ACM SIGPLAN Symposium on Principles of Programming Languages (POPL 2012), pp. 413–424. ACM Press (2012)
4. Cave, A., Pientka, B.: First-class substitutions in contextual type theory. In: 8th International Workshop on Logical Frameworks and Meta-Languages: Theory and Practice (LFMTP 2013), pp. 15–24. ACM Press (2013)
5. Dunfield, J., Pientka, B.: Case analysis of higher-order data. In: International Workshop on Logical Frameworks and Meta-Languages: Theory and Practice (LFMTP 2008), Electronic Notes in Theoretical Computer Science (ENTCS 228), pp. 69–84. Elsevier (2009)
6. Felty, A.P., Momigliano, A., Pientka, B.: The next 700 Challenge Problems for Reasoning with Higher-order Abstract Syntax Representations: Part 2 - a Survey. Journal of Automated Reasoning (2015)

7. Ferreira, F., Pientka, B.: Bidirectional elaboration of dependently typed languages. In: 16th International Symposium on Principles and Practice of Declarative Programming (PPDP 2014). ACM (2014)

8. Gacek, A.: The abella interactive theorem prover (System Description). In: Armando, A., Baumgartner, P., Dowek, G. (eds.) IJCAR 2008. LNCS (LNAI), vol. 5195, pp. 154–161. Springer, Heidelberg (2008)

9. Gacek, A., Miller, D., Nadathur, G.: Combining generic judgments with recursive definitions. In: 23rd Symposium on Logic in Computer Science. IEEE Computer Society Press (2008)

10. Girard, J.-Y., Lafont, Y., Tayor, P.: Proofs and Types. Cambridge University Press, Cambridge (1990)

11. Harper, R., Honsell, F., Plotkin, G.: A framework for defining logics. J. ACM **40**(1), 143–184 (1993)

12. Magnusson, L., Nordström, B.: The Alf proof editor and its proof engine. In: Barendregt, Henk, Nipkow, Tobias (eds.) TYPES: Types for Proofs and Programs. (LNCS 806), vol. 806, pp. 213–237. Springer, Heidelberg (1994)

13. Nanevski, A., Pfenning, F., Pientka, B.: Contextual modal type theory. ACM Trans. Comput. Logic **9**(3), 1–49 (2008)

14. Norell, U.: Towards a practical programming language based on dependent type theory. Ph.D thesis, Department of Computer Science and Engineering, Chalmers University of Technology, September 2007. Technical Report 33D

15. Pfenning, F., Schürmann, C.: System description: Twelf - a meta-logical framework for deductive systems. In: Ganzinger, H. (ed.) CADE 1999. LNCS (LNAI), vol. 1632, pp. 202–206. Springer, Heidelberg (1999)

16. Pientka, B.: A type-theoretic foundation for programming with higher-order abstract syntax and first-class substitutions. In: 35th Annual ACM SIGPLAN Symposium on Principles of Programming Languages (POPL 2008), pp. 371–382. ACM Press (2008)

17. Pientka, B.: An insider's look at LF type reconstruction: Everything you (n)ever wanted to know. J. Funct. Program. **23**(1), 1–37 (2013)

18. Pientka, B., Abel, A.: Structural recursion over contextual objects, In: 13th Typed Lambda Calculi and Applications (TLCA 2015), LIPIcs-Leibniz International Proceedings in Informatics (2015)

19. Pientka, B., Dunfield, J.: Beluga: a framework for programming and reasoning with deductive systems (System Description). In: Giesl, J., Hähnle, R. (eds.) IJCAR 2010. LNCS, vol. 6173, pp. 15–21. Springer, Heidelberg (2010)

20. Tait, W.: Intensional Interpretations of Functionals of Finite Type I. J. Symb. Log. **32**(2), 198–212 (1967)

New Techniques for Automating and Sharing Proofs

SMTtoTPTP – A Converter for Theorem Proving Formats

Peter Baumgartner[✉]

NICTA and Australian National University, Canberra, Australia
peter.baumgartner@nicta.com.au

Abstract. *SMTtoTPTP* is a converter from proof problems written in the SMT-LIB format into the TPTP TFF format. The SMT-LIB format supports polymorphic sorts and frequently used theories like those of uninterpreted function symbols, arrays, and certain forms of arithmetics. The TPTP TFF format is an extension of the TPTP format widely used by automated theorem provers, adding a sort system and arithmetic theories. *SMTtoTPTP* is useful for, e.g., making SMT-LIB problems available to TPTP system developers, and for making TPTP systems available to users of SMT solvers. This paper describes how the conversion works, its functionality and limitations.

1 Introduction

In the automating reasoning community two major syntax formats have emerged for specifying logical proof problems. They are part of the larger infrastructure initiatives SMT-LIB [1] and TPTP [5], respectively. Both formats are under active development and are widely used for problem libraries and in competitions; both serve as *defacto* standards in the sub-communities of SMT solving and first-order logic theorem proving, respectively.

Over the last years, the theorem provers developed in the mentioned communities have become closer in functionality. SMT solvers increasingly provide support for quantified first-order logic formulas, and first-order logic theorem provers increasingly support reasoning modulo built-in theories, such as integer or rational arithmetic. Likewise, the major respective problem libraries have grown (also) by overlapping problems, i.e., problems that could be fed into both an SMT solver and a first-order theorem prover. This convergence is also reflected in recent CASC competitions. Since 2011, CASC features a competition category comprised of typed first-order logic problems modulo arithmetics (TFA), in which both SMT solvers and first-order logic theorem provers participate.

With these considerations it makes sense to provide a converter between problem formats. In this paper I focus on the more difficult direction, from the SMT-LIB format to the appropriate TPTP format, the *typed first-order TPTP*

NICTA is funded by the Australian Government through the Department of Communications and the Australian Research Council through the ICT Centre of Excellence Program.

A.P. Felty and A. Middeldorp (Eds.): CADE-25, LNAI 9195, pp. 285–294, 2015.
DOI: 10.1007/978-3-319-21401-6_19

format with arithmetics, TFF [6]. This converter, *SMTtoTPTP*, is meant to be useful for, e. g., making the existing large SMT-LIB problem libraries available to (developers of) TPTP systems, and, perhaps more importantly, making TPTP systems available to users used to SMT-LIB. *SMTtoTPTP* may also help embed TPTP systems as sub-systems in other systems that use SMT-LIB as an interface language, e.g., interactive proof assistants. However, the *SMTtoTPTP* support for that needs to remain partial as some SMT-LIB commands, e.g. those related to proof management, are not translatable into TPTP.

On notation: the SMT-LIB documents speak about *sorts* whereas in the TPTP world one has *types*. I use both terms below in their corresponding contexts, which is the only difference for the purpose of this paper.

An *operator* is either a function or a predicate symbol. Unlike the TFF format, SMT-LIB does not formally distinguish between the boolean and other sorts. Hence, all SMT-LIB operators are function symbols.

2   SMT-LIB and TFF

This section provides brief overviews of the SMT-LIB and the TFF formats as far as is needed to make this paper self-contained. See [1,4] and [6,7], respectively, for comprehensive documentation.

SMT-LIB. SMT-LIB provides a language for writing terms and formulas in a sorted (i.e., typed) version of first-order logic, a language for specifying background theories, and a language for specifying logics, i.e., suitably restricted classes of formulas to be checked for satisfiability with respect to a specific background theory. SMT-LIB also provides a command language for interacting with SMT solvers.

All SMT-LIB languages come in a Lisp-like syntax. Assuming a fixed library of logics, an SMT-LIB user writes a *script*, i.e., a sequence of *commands* to specify one or more proof problems for a given logic. A script typically contains commands for sort declarations and definitions, as well as function symbol declarations and definitions. Furthermore it will contain commands for asserting the formulas that make up the proof problem, *assertions* for short.

Sort declarations introduce new sorts by stating their name and arity, e.g., (Pair 2). Sort definitions introduce new sorts in terms of already defined/declared ones. They can be parametric in sort parameters (see Example 2.1 below). Recursive definitions are disallowed.

Function symbol declarations introduce new function symbols together with their argument and result sorts given as expressions over the declared and defined sorts. Again, recursive definitions are disallowed. The semantics of function definitions is given by expansion, i.e., by in lining the definitions everywhere in the script until only declared function symbols remain.

The semantics of (sort parametric) sort definitions is given by expansion, too. Indeed, the SMT-LIB type system is *not* polymorphic. Polymorphism is not a part of the type system, it is a meta-level concept. After expansion of definitions, annotations cannot contain sort parameters, and any (well-sorted) subterm has a sort constructed of declared sorts only.

Example 2.1. The following script demonstrates the SMT-LIB type system:

```
1   (set-logic UFLIA)
2   (declare-sort Pair 2)
3   (define-sort Int-Pair (S) (Pair Int S))
4   (declare-sort Color 0)
5   (declare-fun red () Color)
6   (declare-fun get-int ((Int-Pair Color)) Int)
7   (declare-fun int-color-pair (Int Color) (Pair Int Color))
8   (assert (forall ((i Int) (c Color))
9      (= (get-int (int-color-pair i c)) i)))
10  (check-sat)
```

Here, `Pair` is declared as a 2-ary sort and `Int-Pair` is a defined sort with sort parameter S. In line 4, `Color` is declared as a 0-ary sort. Lines 5–7 declare some function symbols. Notice the use of `Int-Pair` in line 6 in the expression (`Int-Pair Color`), which expands into the sort (`Pair Int Color`). Lines 8–9 contain an assertion, its formula is obvious. The (`check-sat`) command in line 9 instructs the SMT-LIB prover to check the assertions for satisfiability. □

Introducing overloaded function symbols *in scripts* is not supported. However, theories can declare ad-hoc polymorphic function symbols. Indeed, common SMT-LIB theories make heavy use of this. For example, equality (`=`), in the core theory, is of rank $S \times S \mapsto$ `Bool` for any sort S. The theory of arrays declares a binary sort constructor `Array` and select and store operators with ranks `Array`$(S, T) \times S \mapsto T$ and `Array(S,T)` $\times S \times T \mapsto$ `Array`(S, T), respectively, for any sorts S and T.

TFF. The TFF format provides a useful extension of the untyped TPTP first-order logic format by a simple many-sorted type scheme. Types are interpreted by non-empty, pairwise disjoint, domains.

The TFF format described in [6] is just the core TFF0 of a polymorphic typed first-order format TFF1 [2]. The paper [6] also extends TFF by predefined types and operators for integer, rational and real arithmetics, which is the target language of *SMTtoTPTP*.[1]

The TFF format supports declaring (0-ary) types and function and predicate symbols over predefined and these declared types. A TFF file typically contains such declarations along with axioms and conjecture formulas over the input signature given by the declarations. In a refutational setting, conjectures need to be negated before conjoining them with the axioms.

The predefined types are the mentioned arithmetic ones and a type of individuals. Equality and the arithmetic operators are ad-hoc polymorphic over the

---

[1] The correct short name of this language is "TFA", *TFF with arithmetics*. However, most of the features of the translation are arithmetics agnostic, and so I use "TFF".

types. All user-defined, i.e., uninterpreted, operators are monomorphic. In the
input formulas only the variables need explicit typing, which happens in quantifications. Together with the signature information this is enough for checking
well-typedness.

Example 2.2. The following is a TFF specification corresponding to
Example 2.1. It was obtained by the *SMTtoTPTP* program.

```
1    %% Types:
2    tff('Pair', type,'Pair[Int,Color]': $tType).
3    tff('Color', type,'Color': $tType).
4
5    %% Declarations:
6    tff(get_int, type, get_int:'Pair[Int,Color]' > $int).
7    tff(int_color_pair, type, int_color_pair:
8        ($int *'Color') >'Pair[Int,Color]').
9
10   %% Assertions:
11   %% (forall ((i Int) (c Color)) (= (get-int (int-color-pair i c)) i))
12   tff(formula, axiom,
13       ( ! [I:$int, C:'Color'] : (get_int(int_color_pair(I, C)) = I))).
```

A tff-triple consists of a name, a role, and a "formula", in this order. The roles
used by *SMTtoTPTP* are either type, for declaring types and operators with
their ranks, or axiom, for input formulas.

The TFF syntax reserves identifiers starting with capital letters for variables. Non-variable identifiers can always be written between pairs of '-quotes.
The example above makes heavy use of that. Lines 2 and 3 declare the
types 'Pair[Int,Color]' and 'Color', corresponding to the sorts (Pair Int
Color) and Color in Example 2.1. Notice there is no "sort" Int-Pair, as all
occurrences of Int-Pair-expressions have been removed by expansion. Lines 6
and 7 declare the same function symbols as in Example 2.1, however with the
sorts expanded. Finally, the asserted formula in Example 2.1 has its counterpart
in TFF-syntax in lines 11–13 above.   □

3   *SMTtoTPTP* Algorithm

A regular run of *SMTtoTPTP* has four stages.

Parsing. In the first stage the commands in the input file are parsed into
abstract syntax trees (ASTs), one per command. The parser has been conveniently implemented with the Scala Standard Parser Combinator Library. The
ASTs for declarations, definitions and assertions are built over Scala classes
(rather: instances thereof) corresponding to syntactical SMT-LIB entities such
as arithmetic domain elements, constants and functional terms, let-terms, ite-
terms, quantifications, sort expressions, etc.

If the set logic includes the theory of arrays, or the user explicitly asks for
it, the following declarations are added to the ASTs:

```
1   (declare-sort Array 2)
2   (declare-parametric-fun (I E) select ((Array I E) I) E)
3   (declare-parametric-fun (I E) store ((Array I E) I E) (Array I E))
```

The `declare-parametric-fun` command declares parametric function symbols in the obvious way. It is meant to be useful in context with other theories as well. See Sect. 5 below for an example. In particular it provides type-checking for parametric operators for free. The `declare-parametric-fun` command is not in the SMT-LIB. This is not a problem, however, because it is hidden from the user.

Semantic Analysis. In the second stage the ASTs are analyzed semantically. This requires decomposing the commands into their constituents, which can be programmed in a convenient way thanks to Scala's pattern matching facilities. The main result of the analysis are various tables holding signature and other information about declared and defined function symbols and sorts. With these tables, the sort of any subterm in an assertion can be computed by expansion, as explained in Sect. 2. This is important for type checking and in the subsequent stages.

Transformations. In the third stage several transformations on the assertions are carried out, all on the AST level.

Removal of defined functions. All function definitions are transformed into additional assertions. This is done with a universally quantified equation between the function symbol applied to the specified variables and its body. *SMTtoTPTP* thus does not expand function definitions. The rationale is to gain flexibility by letting a theorem prover later decide whether to expand or not.[2]

Let-terms. Both the SMT-LIB and TFF formats feature "let" expressions. Unfortunately they are incompatible. An SMT-LIB let expression works much like a binder for local variables as in functional programming languages. The TFF let construct is used to locally define function or predicate symbols as syntactic macros. *SMTtoTPTP* deals with this problem by transforming SMT-LIB let expressions into existentially quantified formulas over the bound variables. This requires lifting these bindings from the term level to the formula level, thereby avoiding unintended name capturing. More precisely, the transformation works as follows.

Let $\phi[(\texttt{let } ((x\ t\,))\ s\,)]$ be an SMT-LIB `Bool`-sorted term, where t is the term bound to the variable x in s (x must not be free in t). For the purpose of the transformation such a term ϕ must always exist, as let-terms occur in assertions only, which are always `Bool`-sorted. Assume that ϕ is the smallest `Bool`-sorted subformula in an assertion containing a let term, written as above, and that the let-term is an outermost one. Let $\sigma(t)$ denote the sort of term t.

---

[2] Defined functions could also be removed by translation into TFF let-terms, but this is clumsy as it may lead to individual let-terms in every axiom and conjecture. Moreover, let-terms are not supported by many TFF systems.

If $\sigma(t)$ is not `Bool` then the let-term is removed by existential quantification. More precisely, using SMT-LIB syntax, ϕ is replaced by the formula (`exists` $((x\rho\ \sigma(t))$) (`and` (`=` $x\rho$ t) ($\phi[s\rho]$)))), where ρ is a renaming substitution that maps x to a fresh variable. The renaming is needed to avoid unintended variable capturing when lifting x outwards, as usual.

If $\sigma(t)$ is `Bool` then the above transformation is not possible, as TFF does not support quantification over boolean variables. In this case *SMTtoTPTP* removes the let-term by substituting x by t in s.

The above step is repeated until all let-terms are removed. The actual implementation is more efficient and requires one subterm traversal per assertion only.

Alternatively to existential quantification, all let-terms could be handled by substitution. *SMTtoTPTP* does not do that, however, because it may lead to exponentially larger terms.

Ite-terms. Both the SMT-LIB and TFF formats feature "if-then-else" constructs (ite). Fortunately, they are compatible. *SMTtoTPTP* offers the option to either keep ite-expressions in place or to transform them away. The latter is useful because not all TFF systems support ite. For example, the expression (< (+ (ite (< 1 2) 3 4) 5) 6) is transformed into (and (=> (< 1 2) (< (+ 3 5) 6)) (=> (not (< 1 2)) (< (+ 4 5) 6))).

Array axioms. The TFF format has no predefined semantics for arrays. Hence, array axioms need to be generated as needed. This is done by sort-instantiating array axiom templates, for each array-sorted term occurring in the assertions.

Example 3.1. Assume an SMT-LIB specification

```
1   (set-logic AUFLIA)
2   (declare-sort Color 0)
3   (declare-fun red () Color)
4   (declare-fun a () (Array Int Color))
5   (declare-fun b () (Array Int Int))
6   (assert (= (select a 0) red))
```

The sole array-sorted term in assertions here is a, which has the sort (`Array Int Color`). The following axioms are added:

```
1   (forall ((a (Array Int Color)) (i Int) (e Color))
2       (= (select (store a i.e.) i) e))
3   (forall ((a (Array Int Color)) (i Int) (j Int) (e Color))
4       (=> (distinct i j) (= (select (store a i.e.) j) (select a j))))
5   (forall ((a (Array Int Color)) (b (Array Int Color)))
6       (=> (forall ((i Int)) (= (select a i) (select b i))) (= a b)))
```

These are standard axioms for arrays with extensional equality, sorted as required.   □

TFF Generation. In the fourth stage the TFF output is generated. It starts with TFF type declarations tff($name^{\mathrm{TFF}}$, type,σ^{TFF}: \$tType) for every sort σ of every subterm in every assertion. The identifier σ^{TFF} is a TFF identifier for the SMT-LIB sort σ. The TFF type identifier σ^{TFF} is merely a print representation of the sort σ, and $name^{\mathrm{TFF}}$ is a prefix of that. As special cases, the predefined arithmetic SMT-LIB sorts Int and Real are taken as the TFF types \$int and \$real, respectively.

Next, a TFF type declaration consisting of the name and rank is emitted for every operator occurring in the assertions. As explained in Sect. 2, SMT-LIB theories may declare polymorphic function symbols. The equality and arithmetic function symbols pose no problems as these have direct counterparts in TFF. Array expressions, however, involving the polymorphic select and store operators need monomorphization and axioms for each monomorphized operator.

Monomorphization is done by including the operators rank in the name of the TFF operator. More precisely, if ($f\ t_1\ \cdots\ t_n$) is an application of the polymorphic operator f to terms $t_1, \ldots, t_n$, then *SMTtoTPTP* synthesizes an identifier, conveniently a valid TFF one, $f^{\mathrm{TFF}} = \text{`}f\!:\!(\sigma(t_1) * \cdots * \sigma(t_n)\text{>}\sigma_{n+1})\text{'}$. The sort σ_{n+1} is the result sort of the term ($f\ t_1\ \cdots\ t_n$) which is obtained by applying the declaration of f to $\sigma(t_1), \ldots, \sigma(t_n)$. The rank of the operator f^{TFF} to be declared in the generated TFF hence is $\sigma_1^{\mathrm{TFF}} \times \cdots \times^{\mathrm{TFF}} \mapsto \sigma_{n+1}^{\mathrm{TFF}}$. Notice the identifier f^{TFF} contains enough information to distinguish it from other applications of f with different sorts.

A special case occurs when the result sort in the declaration of f contains a free sort parameter. To avoid an error, explicit coercion is needed. For example, the "empty list of integers" could correctly be expressed as the term (as empty (List Int)), cf. Sect. 5. Monomorphization respects such coercions.

Finally, each assertion is written out as a TFF axiom. The axioms are obtained by recursively traversing the assertions' subterms and converting them into TFF terms and formulas. By and large this is straightforward translation from one syntax into another. Some comments:

- The SMT-LIB and TFF syntax of, e.g., operators and variables are rather different. *SMTtoTPTP* tries to re-use the given SMT-LIB identifiers without or only little modifications in the generated TFF. For example, an SMT-LIB variable can often be turned into a TFF variable by capitalizing the first letter.
- Certain SMT-LIB operators are varyadic and carry attributes like chainable, associative or pairwise. These attributes say how to translate n-ary terms over these operators into binary ones. For example, the equality operator is chainable: an expression (= $t_1\ \cdots\ t_n$) is first expanded into the conjunction (and (= $t_1\ t_2$) $\cdots$ (= $t_{n-1}\ t_n$) before converted to TFF.
An exception is the distinct operator, which has a pairwise attribute. An expression (distinct $t_1\ \cdots\ t_n$) is optionally directly translated into the TFF counterpart using the \$distinct predicate symbol. However, \$distinct can be used only as a fact, not under any connective. If not at the top-level of an assertion, the expression is translated into the conjunction of the expressions (not (= $t_i\ t_j$)), for all $i, j = 1, \ldots, n$ with $i \neq j$.

- The array operators select and store are monomorphized.
- SMT-LIB equations between Bool-sorted terms are turned into bi-implications.
- set-option commands carry over their argument into a TFF comment. For example, (set-option :answer 42) translates into %$:answer 42. Some options control the behaviour of *SMTtoTPTP*, e.g., whether to expand ite-terms or not.

Example 3.2. Example 3.1 above is converted into the following TFF. The last two array axioms are omitted for space reasons.

```
1   %% Types:
2   tff('Color', type, 'Color': $tType).
3   tff('Array', type, 'Array[Int,Color]': $tType).
4
5   %% Declarations:
6   tff(red, type, red: 'Color').
7   tff(a, type, a: 'Array[Int,Color]').
8   tff(select, type, 'select:(Array[Int,Color]*Int)>Color':
9       ('Array[Int,Color]' * $int) > 'Color').
10  tff(store, type, 'store:(Array[Int,Color]*Int*Color)>Array[Int,Color]':
11      ('Array[Int,Color]' * $int * 'Color') > 'Array[Int,Color]').
12
13  %% Assertions:
14  %% (= (select a 0) red)
15  tff(formula, axiom,
16     ('select:(Array[Int,Color]*Int)>Color'(a, 0) = red)).
17  %% (forall ((a (Array Int Color)) (i Int) (e Color))
18  %%     (= (select (store a i.e.) i) e))
19  tff(formula, axiom,
20     ( ! [A:'Array[Int,Color]', I:$int, E:'Color'] :
21         ('select:(Array[Int,Color]*Int)>Color'(
22           'store:(Array[Int,Color]*Int*Color)>Array[Int,Color]'(
23             A, I, E), I) = E))).
```

□

4   Limitations

SMTtoTPTP is meant to support a comprehensive subset of the SMT-LIB language and the logics and theories in [4]. Table 1 lists the SMT-LIB language elements for scripts and their status wrt. *SMTtoTPTP*.

Some of the unsupported language elements in Table 1 will be added later, e.g., indexed identifiers such as (_ a 5). Other elements are intrinsicly problematic, in particular the push and pop commands. These commands are used for managing a stack of asserted formulas (typically) for incremental satisfiability checks. This is not supported by the TPTP language, and hence *SMTtoTPTP* throws an error on encountering a **push** or **pop** command. All other commands (e.g., **get-proof**) are untranslatable and can be ignored.

Table 1. Supported SMT-LIB script language elements.

Logics	
Supported: [QF_][A][UF][(L|N)(IA|RA|IRA)]	Unsupported: bitvectors, difference logic
Commands	
Supported: set-logic, declare-sort, define-sort,	Unsupported: push, pop
declare-fun, define-fun, assert, exit	All other commands ignored
Tokens	
Supported: *numeral, decimal, symbol*	Unsupported: *hexadecimal, binary, string*
Other Elements	
Unsupported: indexed identifiers, logic declarations, theory declarations	

SMTtoTPTP supports a fixed set of logics. The regular expression in Table 1 denotes their SMT-LIB names. For example, **QF_AUFLIRA** means "quantifier-free logic over the combined theories of arrays, uninterpreted function symbols, and mixed linear and real arithmetics". Notice that every logic includes the *core* theory, which offers a comprehensive set of boolean-sorted operators, including equality and if-then-else.

SMTtoTPTP does not deal with SMT-LIB theory and logic declarations. As their semantics is described informally, *SMTtoTPTP* can not make much use of them. However, as said, the core theory and the theories of arrays, integer and real arithmetic are built-in.

5   Extensions

Inspired by the Z3 SMT solver [3], *SMTtoTPTP* extends the SMT-LIB standard by datatype definitions. Datatypes can be used to define enumeration types, tuples, records, and recursive data structures like lists, to name a few. The syntax of datatype definitions involves sort parameters and the constructors and destructors for elements of the datatype. Here are some examples:

```
1   (declare-datatypes () ((Color red green blue)))
2   (declare-datatypes (S T) ((Pair (mk-pair (first S) (second T)))))
3   (declare-datatypes (T) ((List nil (insert (head T) (tail (List T))))))
```

Line 1 defines an enumeration datatype with three constructors, as stated. Line 2 defines pairs over the product type S×T, where S and T are type parameters. Line 3 defines the usual polymorphic list datatype, where **nil** and **insert** are constructors, and **head** and **tail** are the destructors for the **insert**-case.

The conversion to TFF of the list datatype with a (List Int) sort instance, for example, is equivalent to the conversion of the following SMT-LIB commands:

```
1    (declare-sort List 1)
2    (declare-parametric-fun (T) nil () (List T))
3    (declare-parametric-fun (T) insert (T (List T)) (List T))
4    (declare-parametric-fun (T) head ((List T)) T)
5    (declare-parametric-fun (T) tail ((List T)) (List T))
6    (assert (forall ((L (List Int)))
7        (or (= L (as nil (List Int))) (= L (insert (head L) (tail L))))))
8    (assert (forall ((N Int) (L (List Int))) (= (head (insert N L)) N)))
9    (assert (forall ((N Int) (L (List Int))) (= (tail (insert N L)) L)))
10   (assert (forall ((N Int) (L (List Int)))
11       (not (= (as nil (List Int)) (insert N L)))))
```

SMTtoTPTP does not do type inference. All occurrences of type-ambiguous constructor terms must be explicitly cast to the proper sort. In the list example, (only) nil terms must be explicitly cast, as in (as nil (List Int)).

The theory of arrays has been extended with constant arrays, i.e., arrays that have the same element everywhere.

6   Other Features

SMTtoTPTP is available at https://bitbucket.org/peba123/smttotptp under a GNU General Public License. The distribution includes the Scala[3] source code and a ready-to-run Java jar-file. *SMTtoTPTP* can also be used as a library for parsing SMT-LIB files into an abstract syntax tree.

References

1. Barrett, C., Stump, A., Tinelli, C.: The SMT-LIB Standard: Version 2.0. In: Gupta, A., Kroening, D. (eds.) SMT Workshop (2010)
2. Blanchette, J.C., Paskevich, A.: TFF1: the TPTP typed first-order form with rank-1 polymorphism. In: Bonacina, M.P. (ed.) CADE 2013. LNCS, vol. 7898, pp. 414–420. Springer, Heidelberg (2013)
3. de Moura, L., Bjørner, N.S.: Z3: an efficient SMT solver. In: Ramakrishnan, C.R., Rehof, J. (eds.) TACAS 2008. LNCS, vol. 4963, pp. 337–340. Springer, Heidelberg (2008)
4. SMT-LIB.: The Satisfiability Modulo Theories Library. http://smt-lib.org/
5. Sutcliffe, G.: The TPTP problem library and associated infrastructure: the FOF and CNF parts, v3.5.0. J. Autom. Reason. **43**(4), 337–362 (2009)
6. Sutcliffe, G., Schulz, S., Claessen, K., Baumgartner, P.: The TPTP typed first-order form with arithmetic. In: Bjørner, N., Voronkov, A. (eds.) LPAR-18 2012. LNCS, vol. 7180, pp. 406–419. Springer, Heidelberg (2012)
7. The TPTP Problem Library for Automated Theorem Proving. http://www.cs.miami.edu/tptp/

---

[3] http://www.scala-lang.org.

CTL Model Checking in Deduction Modulo

Kailiang Ji[✉]

INRIA and Paris Diderot, 23 Avenue d'Italie, CS 81321,
75214 Paris Cedex 13, France
kailiang.ji@inria.fr

Abstract. In this paper we give an overview of proof-search method for CTL model checking based on Deduction Modulo. Deduction Modulo is a reformulation of Predicate Logic where some axioms—possibly all—are replaced by rewrite rules. The focus of this paper is to give an encoding of temporal properties expressed in CTL, by translating the logical equivalence between temporal operators into rewrite rules. This way, the proof-search algorithms designed for Deduction Modulo, such as Resolution Modulo or Tableaux Modulo, can be used in verifying temporal properties of finite transition systems. An experimental evaluation using Resolution Modulo is presented.

Keywords: Model checking · Deduction modulo · Resolution modulo

1 Introduction

In this paper, we express Computation Tree Logic (CTL) [4] for a given finite transition system in Deduction Modulo [6,7]. This way, the proof-search algorithms designed for Deduction Modulo, such as Resolution Modulo [2] or Tableaux Modulo [5], can be used to build proofs in CTL. Deduction Modulo is a reformulation of Predicate Logic where some axioms—possibly all—are replaced by rewrite rules. For example, the axiom $P \Leftrightarrow (Q \lor R)$ can be replaced by the rewrite rule $P \hookrightarrow (Q \lor R)$, meaning that during the proof, P can be replaced by $Q \lor R$ at any time.

The idea of translating CTL to another framework, for instance (quantified) boolean formulae [1,14,16], higher-order logic [12], etc., is not new. But using rewrite rules permits to avoid the explosion of the size of formulae during translation, because rewrite rules can be used on demand to unfold defined symbols. So, one of the advantages of this method is that it can express complicated verification problems succinctly. Gilles Dowek and Ying Jiang had given a way to build an axiomatic theory for a given finite model [9]. In this theory, the formulae are provable if and only if they are valid in the model. In [8], they gave a slight extension of CTL, named SCTL, where the predicates may have arbitrary arities. And they defined a special sequent calculus to write proofs in SCTL. This

K. Ji — This work is supported by the ANR-NSFC project LOCALI (NSFC 61161130530 and ANR 11 IS02 002 01).

A.P. Felty and A. Middeldorp (Eds.): CADE-25, LNAI 9195, pp. 295–310, 2015.
DOI: 10.1007/978-3-319-21401-6_20

sequent calculus is special because it is tailored to each specific finite model M. In this way, a formula is provable in this sequent calculus if and only if it is valid in the model M. In our method, we characterize a finite model in the same way as [9], but instead of building a deduction system, the CTL formulae are taken as terms, and the logical equivalence between different CTL formulae are expressed by rewrite rules. This way, the existing automated theorem modulo provers, for instance iProver Modulo [3], can be used to do model checking directly. The experimental evaluation shows that the resolution based proof-search algorithms is feasible, and sometimes performs better than the existing solving techniques.

The rest of this paper is organized as follows. In Sect. 2 a new variant of Deduction Modulo for one-sided sequents is presented. In Sect. 3, the usual semantics of CTL is presented. Sections 4 and 5 present the new results of this paper: in Sect. 4, an alternative semantics for CTL on finite structures is given; in Sect. 5, the rewrite rules for each CTL operator are given and the soundness and completeness of this presentation of CTL is proved, using the semantics presented in the previous section. Finally in Sect. 6, experimental evaluation for the feasibility of rewrite rules using resolution modulo is presented.

2   Deduction Modulo

One-Sided Sequents. In this work, instead of using usual sequents of the form $A_1, \ldots, A_n \vdash B_1, \ldots, B_p$, we use one-sided sequents [13], where all the propositions are put on the right hand side of the sequent sign $\vdash$ and the sequent above is transformed into $\vdash \neg A_1, \ldots, \neg A_n, B_1, \ldots, B_p$. Moreover, implication is defined from disjunction and negation ($A \Rightarrow B$ is just an abbreviation for $\neg A \lor B$), and negation is pushed inside the propositions using De Morgan's laws. For each atomic proposition P we also have a dual atomic proposition $P^\perp$ corresponding to its negation, and the operator $\perp$ extends to all the propositions. So that the axiom rule can be formulated as

$$\frac{}{\vdash P, Q} \text{ axiom, if } P \text{ and } Q \text{ are dual atomic propositions}$$

Deduction Modulo. A *rewrite system* is a set $\mathcal{R}$ of term and proposition rewrite rules. In this paper, only proposition rewrite rules are considered. A proposition rewrite rule is a pair of propositions $l \hookrightarrow r$, in which l is an atomic proposition and r an arbitrary proposition. For instance, $P \hookrightarrow Q \lor R$. Such a system defines a congruence $\hookrightarrow$ and the relation $\overset{*}{\hookrightarrow}$ is defined, as usual, as the reflexive-transitive closure of $\hookrightarrow$. Deduction Modulo [7] is an extension of first-order logic where axioms are replaced by rewrite rules and in a proof, a proposition can be reduced at any time. This possibility is taken into account in the formulation of *Sequent Calculus Modulo* in Fig. 1. For example, with the axiom $(Q \Rightarrow R) \Rightarrow P$ we can prove the sequent $R \vdash P$. This axiom is replaced by the rules $P \hookrightarrow Q^\perp$ and $P \hookrightarrow R$ and the sequent $R \vdash P$ is expressed as the one-sided sequent $\vdash R^\perp, P$. This sequent has the proof

$$\frac{}{\vdash R^\perp, P} \text{ axiom}$$

as $P \overset{*}{\hookrightarrow} R$.

$$\frac{}{\vdash_{\mathcal{R}} A, B} \ \text{axiom } A \overset{*}{\hookrightarrow} P, B \overset{*}{\hookrightarrow} P^{\perp} \qquad \frac{\vdash_{\mathcal{R}} A, \Delta \quad \vdash_{\mathcal{R}} B, \Delta}{\vdash_{\mathcal{R}} \Delta} \ \text{cut } A \overset{*}{\hookrightarrow} C, B \overset{*}{\hookrightarrow} C^{\perp}$$

$$\frac{\vdash_{\mathcal{R}} \Delta}{\vdash_{\mathcal{R}} A, \Delta} \ \text{weak} \qquad \frac{\vdash_{\mathcal{R}} B, C, \Delta}{\vdash_{\mathcal{R}} A, \Delta} \ \text{contr } A \overset{*}{\hookrightarrow} B, A \overset{*}{\hookrightarrow} C$$

$$\frac{}{\vdash_{\mathcal{R}} A, \Delta} \ \top \ A \overset{*}{\hookrightarrow} \top \qquad \frac{\vdash_{\mathcal{R}} B, \Delta \quad \vdash_{\mathcal{R}} C, \Delta}{\vdash_{\mathcal{R}} A, \Delta} \ \wedge \text{ if } A \overset{*}{\hookrightarrow} B \wedge C$$

$$\frac{\vdash_{\mathcal{R}} B, \Delta}{\vdash_{\mathcal{R}} A, \Delta} \ \vee_1 \ A \overset{*}{\hookrightarrow} B \vee C \qquad \frac{\vdash_{\mathcal{R}} C, \Delta}{\vdash_{\mathcal{R}} A, \Delta} \ \vee_2 \ A \overset{*}{\hookrightarrow} B \vee C$$

$$\frac{\vdash_{\mathcal{R}} C, \Delta}{\vdash_{\mathcal{R}} A, \Delta} \ \exists \ A \overset{*}{\hookrightarrow} \exists x B, (t/x)B \overset{*}{\hookrightarrow} C \qquad \frac{\vdash_{\mathcal{R}} B, \Delta}{\vdash_{\mathcal{R}} A, \Delta} \ \forall \ A \overset{*}{\hookrightarrow} \forall x B, x \notin FV(\Delta)$$

Fig. 1. One-sided Sequent Calculus Modulo

Note that as our system is negation free, all occurrences of atomic propositions are positive. Thus, the rule $P \hookrightarrow A$ does not correspond to an equivalence $P \Leftrightarrow A$ but to an implication $A \Rightarrow P$. In other words, our one-sided presentation of Deduction Modulo is closer to Polarized Deduction Modulo [6] with positive rules only, than to the usual Deduction Modulo. The sequent $\vdash_{\mathcal{R}} \Delta$ has a cut-free proof is represented as $\vdash_{\mathcal{R}}^{cf} \Delta$ has a proof.

3 Computation Tree Logic

Properties of a transition system can be specified by temporal logic propositions. Computation tree logic is a propositional branching-time temporal logic introduced by Emerson and Clarke [4] for finite state systems. Let AP be a set of atomic propositions and p ranges over AP. The set of CTL propositions Φ over AP is defined as follows:

$$\Phi ::= p \mid \neg \Phi \mid \Phi \wedge \Phi \mid \Phi \vee \Phi \mid AX\Phi \mid EX\Phi \mid AF\Phi \mid EF\Phi \mid AG\Phi \mid EG\Phi$$
$$\mid A[\Phi U \Phi] \mid E[\Phi U \Phi] \mid A[\Phi R \Phi] \mid E[\Phi R \Phi]$$

The semantics of CTL can be given using Kripke structure, which is used in model checking to represent the behavior of a system.

Definition 1 (Kripke Structure). *Let AP be a set of atomic propositions. A Kripke structure M over AP is a three tuple $M = (S, \mathsf{next}, \mathsf{L})$ where*

- *S is a finite (non-empty) set of states.*
- *$\mathsf{next} : S \to \mathcal{P}^{+}(S)$ is a function that gives each state a (non-empty) set of successors.*
- *$\mathsf{L} : S \to \mathcal{P}(AP)$ is a function that labels each state with a subset of AP.*

An infinite path is an infinite sequence of states $\pi = \pi_0 \pi_1 \cdots$ s.t. $\forall i \geq 0$, $\pi_{i+1} \in \mathsf{next}(\pi_i)$. Note that the sequence $\pi_i \pi_{i+1} \cdots \pi_j$ is denoted as π_i^j and the path π with $\pi_0 = s$ is denoted as $\pi(s)$.

Definition 2 (Semantics of CTL). *Let p be an atomic proposition. Let φ, φ_1, φ_2 be CTL propositions. The relation $M, s \models \varphi$ is defined as follows.*

$$
\begin{array}{ll}
M, s \models p & \Leftrightarrow p \in \mathsf{L}(s) \\
M, s \models \neg\varphi_1 & \Leftrightarrow M, s \not\models \varphi_1 \\
M, s \models \varphi_1 \wedge \varphi_2 & \Leftrightarrow M, s \models \varphi_1 \text{ and } M, s \models \varphi_2 \\
M, s \models \varphi_1 \vee \varphi_2 & \Leftrightarrow M, s \models \varphi_1 \text{ or } M, s \models \varphi_2 \\
M, s \models AX\varphi_1 & \Leftrightarrow \forall s' \in \mathsf{next}(s), \; M, s' \models \varphi_1 \\
M, s \models EX\varphi_1 & \Leftrightarrow \exists s' \in \mathsf{next}(s), \; M, s' \models \varphi_1 \\
M, s \models AG\varphi_1 & \Leftrightarrow \forall \pi(s), \forall i \geq 0, \; M, \pi_i \models \varphi_1 \\
M, s \models EG\varphi_1 & \Leftrightarrow \exists \pi(s) \text{ s.t. } \forall i \geq 0, \; M, \pi_i \models \varphi_1 \\
M, s \models AF\varphi_1 & \Leftrightarrow \forall \pi(s), \exists i \geq 0 \text{ s.t. } M, \pi_i \models \varphi_1 \\
M, s \models EF\varphi_1 & \Leftrightarrow \exists \pi(s), \exists i \geq 0 \text{ s.t. } M, \pi_i \models \varphi_1 \\
M, s \models A[\varphi_1 U \varphi_2] & \Leftrightarrow \forall \pi(s), \exists j \geq 0 \text{ s.t. } M, \pi_j \models \varphi_2 \text{ and } \forall 0 \leq i < j, \; M, \pi_i \models \varphi_1 \\
M, s \models E[\varphi_1 U \varphi_2] & \Leftrightarrow \exists \pi(s), \exists j \geq 0 \text{ s.t. } M, \pi_j \models \varphi_2 \text{ and } \forall 0 \leq i < j, \; M, \pi_i \models \varphi_1 \\
M, s \models A[\varphi_1 R \varphi_2] & \Leftrightarrow \forall \pi(s), \forall j \geq 0, \text{ either } M, \pi_j \models \varphi_2 \text{ or } \exists 0 \leq i < j \text{ s.t. } M, \pi_i \models \varphi_1 \\
M, s \models E[\varphi_1 R \varphi_2] & \Leftrightarrow \exists \pi(s), \forall j \geq 0, \text{ either } M, \pi_j \models \varphi_2 \text{ or } \exists 0 \leq i < j \text{ s.t. } M, \pi_i \models \varphi_1 \\
\end{array}
$$

4   Alternative Semantics of CTL

In this part we present an alternative semantics of CTL using finite paths only.

Paths with the Last State Repeated (lr-Paths). A finite path is a *lr-path* if and only if the last state on the path occurs twice. For instance s_0, s_1, s_0 is a lr-path. Note that we use $\rho = \rho_0\rho_1 \ldots \rho_j$ to denote a lr-path. A lr-path ρ with $\rho_0 = s$ is denoted as $\rho(s)$, with $\rho_i = \rho_j$ is denoted as $\rho(i,j)$. The length of a path l is expressed by $\mathsf{len}(l)$ and the concatenation of two paths l_1, l_2 is $l_1 \,\hat{}\, l_2$.

Lemma 1. *Let M be a Kripke structure.*

1. *If π is an infinite path of M, then $\exists i \geq 0$ such that π_0^i is a lr-path.*
2. *If $\rho(i,j)$ is a lr-path of M, then $\rho_0^i\,\hat{}\,(\rho_{i+1}^j)^\omega$ is an infinite path.*

Proof. For Case 1, as M is finite, there exists at least one repeating state in π. If π_i is the first state which occurs twice, then π_0^i is a lr-path. Case 2 is trivial. □

Lemma 2. *Let M be a Kripke structure.*

1. *For the path $l = s_0, s_1, \ldots, s_k$, there exists a finite path $l' = s_0', s_1', \ldots, s_i'$ without repeating states s.t. $s_0' = s_0$, $s_i' = s_k$, and $\forall 0 < j < i$, s_j' is on l.*
2. *If there is a path from s to s', then there exists a lr-path $\rho(s)$ s.t. s' is on ρ.*

Proof. For the first case, l' can be built by deleting the cycles from l. The second case is straightforward by the first case and Lemma 1. □

Definition 3 (Alternative Semantics of CTL). *Let p be an atomic proposition and $\varphi, \varphi_1, \varphi_2$ be CTL propositions. The relation $M, s \models_a \varphi$ is defined as follows.*

$$
\begin{aligned}
&M, s \models_a p &&\Leftrightarrow\ p \in L(s) \\
&M, s \models_a \neg\varphi_1 &&\Leftrightarrow\ M, s \not\models_a \varphi_1 \\
&M, s \models_a \varphi_1 \wedge \varphi_2 &&\Leftrightarrow\ M, s \models_a \varphi_1 \text{ and } M, s \models_a \varphi_2 \\
&M, s \models_a \varphi_1 \vee \varphi_2 &&\Leftrightarrow\ M, s \models_a \varphi_1 \text{ or } M, s \models_a \varphi_2 \\
&M, s \models_a AX\varphi_1 &&\Leftrightarrow\ \forall s' \in \text{next}(s),\ M, s' \models_a \varphi_1 \\
&M, s \models_a EX\varphi_1 &&\Leftrightarrow\ \exists s' \in \text{next}(s),\ M, s' \models_a \varphi_1 \\
&M, s \models_a AF\varphi_1 &&\Leftrightarrow\ \forall \rho(s),\ \exists i < \text{len}(\rho) - 1 \text{ s.t. } M, \rho_i \models_a \varphi_1 \\
&M, s \models_a EF\varphi_1 &&\Leftrightarrow\ \exists \rho(s),\ \exists i < \text{len}(\rho) - 1 \text{ s.t. } M, \rho_i \models_a \varphi_1 \\
&M, s \models_a AG\varphi_1 &&\Leftrightarrow\ \forall \rho(s),\ \forall i < \text{len}(\rho) - 1,\ M, \rho_i \models_a \varphi_1 \\
&M, s \models_a EG\varphi_1 &&\Leftrightarrow\ \exists \rho(s),\ \forall i < \text{len}(\rho) - 1,\ M, \rho_i \models_a \varphi_1 \\
&M, s \models_a A[\varphi_1 U\varphi_2] &&\Leftrightarrow\ \forall \rho(s),\ \exists i < \text{len}(\rho) - 1 \text{ s.t. } M, \rho_i \models_a \varphi_2 \text{ and } \forall j < i,\ M, \rho_j \models_a \varphi_1 \\
&M, s \models_a E[\varphi_1 U\varphi_2] &&\Leftrightarrow\ \exists \rho(s),\ \exists i < \text{len}(\rho) - 1 \text{ s.t. } M, \rho_i \models_a \varphi_2 \text{ and } \forall j < i,\ M, \rho_j \models_a \varphi_1 \\
&M, s \models_a A[\varphi_1 R\varphi_2] &&\Leftrightarrow\ \forall \rho(s),\ \forall i < \text{len}(\rho) - 1,\ M, \rho_i \models_a \varphi_2 \text{ or } \exists j < i \text{ s.t. } M, \rho_j \models_a \varphi_1 \\
&M, s \models_a E[\varphi_1 R\varphi_2] &&\Leftrightarrow\ \exists \rho(s),\ \forall i < \text{len}(\rho) - 1,\ M, \rho_i \models_a \varphi_2 \text{ or } \exists j < i \text{ s.t. } M, \rho_j \models_a \varphi_1
\end{aligned}
$$

We now prove the equivalence of the two semantics, that is, $M, s \models \varphi$ iff $M, s \models_a \varphi$. To simplify the proofs, we use a normal form of the CTL propositions, in which all the negations appear only in front of the atomic propositions.

Negation Normal Form. A CTL proposition is in negation normal form (NNF), if the negation $\neg$ is applied only to atomic propositions. Every CTL proposition can be transformed into an equivalent proposition of NNF using the following equivalences.

$$
\begin{aligned}
&\neg\neg\varphi \equiv \varphi \\
&\neg(\varphi \vee \psi) \equiv \neg\varphi \wedge \neg\psi &&\neg AF\varphi \equiv EG\neg\varphi &&\neg A[\varphi U\psi] \equiv E[\neg\varphi R\neg\psi] \\
&\neg(\varphi \wedge \psi) \equiv \neg\varphi \vee \neg\psi &&\neg EF\varphi \equiv AG\neg\varphi &&\neg E[\varphi U\psi] \equiv A[\neg\varphi R\neg\psi] \\
&\neg AX\varphi \equiv EX\neg\varphi &&\neg AG\varphi \equiv EF\neg\varphi &&\neg A[\varphi R\psi] \equiv E[\neg\varphi U\neg\psi] \\
&\neg EX\varphi \equiv AX\neg\varphi &&\neg EG\varphi \equiv AF\neg\varphi &&\neg E[\varphi R\psi] \equiv A[\neg\varphi U\neg\psi]
\end{aligned}
$$

Lemma 3. *Let φ be a CTL proposition of NNF. If $M, s \models \varphi$, then $M, s \models_a \varphi$.*

Proof. By induction on the structure of φ. For brevity, we just prove the two cases where φ is $AF\varphi_1$ and $AG\varphi_1$. The full proof is in [10].

- Let $\varphi = AF\varphi_1$. We prove the contraposition. If there is a lr-path $\rho(s)(j, k)$ s.t. $\forall 0 \le i < k$, $M, \rho_i \not\models \varphi_1$, then by Lemma 1, there exists an infinite path $\rho_0^j{}^\frown(\rho_{j+1}^k)^\omega$, which is a counterexample of $M, s \models AF\varphi_1$. Thus for each lr-path $\rho(s)$, $\exists 0 \le i < \text{len}(\rho) - 1$ s.t. $M, \rho_i \models \varphi_1$ holds. Then by induction hypothesis (IH), for each lr-path $\rho(s)$, $\exists 0 \le i < \text{len}(\rho) - 1$ s.t. $M, \rho_i \models_a \varphi_1$ holds, and thus $M, s \models_a AF\varphi_1$ holds.
- Let $\varphi = AG\varphi_1$. We prove the contraposition. If there is a lr-path $\rho(s)(j, k)$ and $\exists 0 \le i < k$ s.t. $M, \rho_i \not\models \varphi_1$, then by Lemma 1, there exists an infinite path $\rho_0^j{}^\frown(\rho_{j+1}^k)^\omega$, which is a counterexample of $M, s \models AG\varphi_1$. Thus for each lr-path $\rho(s)(j, k)$ and $\forall 0 \le i < k$, $M, \rho_i \models \varphi_1$ holds. Then by IH, for each lr-path $\rho(s)(j, k)$ and $\forall 0 \le i < k$, $M, \rho_i \models_a \varphi_1$ holds, and thus $M, s \models_a AG\varphi_1$ holds. $\qquad\square$

Lemma 4. *Let φ be a CTL proposition of NNF. If $M, s \models_a \varphi$, then $M, s \models \varphi$.*

Proof. By induction on the structure of φ. For brevity, we just prove the two cases where φ is $AF\varphi_1$ and $AG\varphi_1$. The full proof is in [10].

- Let $\varphi = AF\varphi_1$. If there is an infinite path $\pi(s)$ s.t. $\forall j \geq 0$, $M, \pi_j \not\models_a \varphi_1$, then by Lemma 1, there exists $k \geq 0$ s.t. π_0^k is a lr-path, which is a counterexample of $M, s \models_a AF\varphi_1$. Thus for each infinite path $\pi(s)$, $\exists j \geq 0$ s.t. $M, \pi_j \models_a \varphi_1$ holds. Then by IH, for each infinite path $\pi(s)$, $\exists j \geq 0$ s.t. $M, \pi_j \models \varphi_1$ holds and thus $M, s \models AF\varphi_1$ holds.
- Let $\varphi = AG\varphi_1$. Assume that there exists an infinite path $\pi(s)$ and $\exists i \geq 0$, $M, \pi_i \not\models_a \varphi_1$. By Lemma 2, there exists a lr-path $\rho(s)$ s.t. π_i is on ρ, which is a counterexample of $M, s \models_a AG\varphi_1$. Thus for each infinite path $\pi(s)$ and $\forall i \geq 0$, $M, \pi_i \models_a \varphi_1$ holds. Then by IH, for each infinite path $\pi(s)$ and $\forall i \geq 0$, $M, \pi_i \models \varphi_1$ holds and thus $M, s \models AG\varphi_1$ holds. □

Theorem 1. *Let φ be a CTL proposition. $M, s \models \varphi$ iff $M, s \models_a \varphi$.*

5   Rewrite Rules for CTL

The work in this section is to express CTL propositions in Deduction Modulo and prove that for a CTL proposition φ, the translation of $M, s \models_a \varphi$ is provable if and only if $M, s \models_a \varphi$ holds. So we fix such a model $M = (S, \mathsf{next}, \mathsf{L})$. As in [9], we consider a two sorted language $\mathcal{L}$, which contains

- constants $s_1, \ldots, s_n$ for each state of M.
- predicate symbols $\varepsilon_0, \varepsilon_{\sqcap_0}, \varepsilon_{\sqcup_0}, \varepsilon_1, \varepsilon_{\sqcap_1}, \varepsilon_{\sqcup_1}$, in which the binary predicates ε_0, $\varepsilon_{\sqcap_0}$ and $\varepsilon_{\sqcup_0}$ apply to all the CTL propositions, while the ternary predicates $\varepsilon_1, \varepsilon_{\sqcap_1}$ and $\varepsilon_{\sqcup_1}$ only apply to the CTL propositions starting with the temporal connectives AG, EG, AR and ER.
- binary predicate symbols mem for the membership, r for the next-notation.
- a constant nil and a binary function symbol con.

We use x, y, z to denote the variables of the state terms, X, Y, Z to denote the class variables. A class is in fact a set of states, here we use the *class theory*, rather than the *(monadic) second order logic*, is to emphasis that this formalism is a theory and not a logic.

CTL Term. To express CTL in Deduction Modulo, firstly, we translate the CTL proposition φ into a term $|\varphi|$ (*CTL term*). The translation rules are as follows:

$\|p\| = \overline{p},\, p \in AP$	$\|EX\varphi\| = \mathsf{ex}(\|\varphi\|)$	$\|A[\varphi U\psi]\| = \mathsf{au}(\|\varphi\|, \|\psi\|)$
$\|\neg\varphi\| = \mathsf{not}(\|\varphi\|)$	$\|AF\varphi\| = \mathsf{af}(\|\varphi\|)$	$\|E[\varphi U\psi]\| = \mathsf{eu}(\|\varphi\|, \|\psi\|)$
$\|\varphi \wedge \psi\| = \mathsf{and}(\|\varphi\|, \|\psi\|)$	$\|EF\varphi\| = \mathsf{ef}(\|\varphi\|)$	$\|A[\varphi R\psi]\| = \mathsf{ar}(\|\varphi\|, \|\psi\|)$
$\|\varphi \vee \psi\| = \mathsf{or}(\|\varphi\|, \|\psi\|)$	$\|AG\varphi\| = \mathsf{ag}(\|\varphi\|)$	$\|E[\varphi R\psi]\| = \mathsf{er}(\|\varphi\|, \|\psi\|)$
$\|AX\varphi\| = \mathsf{ax}(\|\varphi\|)$	$\|EG\varphi\| = \mathsf{eg}(\|\varphi\|)$	

Note that we use Φ, Ψ to denote the variables of the CTL terms. Both finite sets and finite paths are represented with the symbols con and nil. For the set $S' = \{s_i, \ldots, s_j\}$, we use $[S']$ to denote its term form con$(s_i, \text{con}(\ldots, \text{con}(s_j, \text{nil})\ldots))$. For the path $s_i^j = s_i, \ldots, s_j$, its term form con$(s_j, \text{con}(\ldots, \text{con}(s_i, \text{nil})\ldots))$ is denoted by $[s_i^j]$.

Definition 4 (Semantics). *Semantics of the propositions in $\mathcal{L}$ is as follows.*

$M \models \varepsilon_0(|\varphi|, s)$	$\Leftrightarrow M, s \models_a \varphi$
$M \models r(s, [S'])$	$\Leftrightarrow S' = \text{next}(s)$
$M \models mem(s, [s_0^i])$	$\Leftrightarrow s$ is on the path s_0^i
$M \models \varepsilon_{\sqcap_0}(|\varphi|, [S'])$	$\Leftrightarrow \forall s \in S', M \models \varepsilon_0(|\varphi|, s)$
$M \models \varepsilon_{\sqcup_0}(|\varphi|, [S'])$	$\Leftrightarrow \exists s \in S'$ s.t. $M \models \varepsilon_0(|\varphi|, s)$
$M \models \varepsilon_1(\text{ag}(|\varphi_1|), s, [s_0^i])$	$\Leftrightarrow \forall lr\text{-path } s_0^i {}^\frown s_{i+1}^k(s_{i+1} = s), \forall i < j < k,$ $M \models \varepsilon_0(|\varphi_1|, s_j)$
$M \models \varepsilon_1(\text{eg}(|\varphi_1|), s, [s_0^i])$	$\Leftrightarrow \exists lr\text{-path } s_0^i {}^\frown s_{i+1}^k(s_{i+1} = s), \forall i < j < k,$ $M \models \varepsilon_0(|\varphi_1|, s_j)$
$M \models \varepsilon_1(\text{ar}(|\varphi_1|, |\varphi_2|), s, [s_0^i])$	$\Leftrightarrow \forall lr\text{-path } s_0^k {}^\frown s_{i+1}^k(s_{i+1} = s), \forall i < j < k,$ Either $M \models \varepsilon_0(|\varphi_2|, s_j)$ or $\exists i < m < j$ s.t. $M \models \varepsilon_0(|\varphi_1|, s_m)$
$M \models \varepsilon_1(\text{er}(|\varphi_1|, |\varphi_2|), s, [s_0^i])$	$\Leftrightarrow \exists lr\text{-path } s_0^k {}^\frown s_{i+1}^k(s_{i+1} = s), \forall i < j < k,$ either $M \models \varepsilon_0(|\varphi_2|, s_j)$ or $\exists i < m < j$ s.t. $M \models \varepsilon_0(|\varphi_1|, s_m)$
$M \models \varepsilon_{\sqcap_1}(\text{ag}(|\varphi_1|), [S'], [s_0^i])$	$\Leftrightarrow \forall s \in S', M \models \varepsilon_1(\text{ag}(|\varphi_1|), s, [s_0^i])$
$M \models \varepsilon_{\sqcup_1}(\text{eg}(|\varphi_1|), [S'], [s_0^i])$	$\Leftrightarrow \exists s \in S'$ s.t. $M \models \varepsilon_1(\text{eg}(|\varphi_1|), s, [s_0^i])$
$M \models \varepsilon_{\sqcap_1}(\text{ar}(|\varphi_1|, |\varphi_2|), [S'], [s_0^i])$	$\Leftrightarrow \forall s \in S', M \models \varepsilon_1(\text{ar}(|\varphi_1|, |\varphi_2|), s, [s_0^i])$
$M \models \varepsilon_{\sqcup_1}(\text{er}(|\varphi_1|, |\varphi_2|), [S'], [s_0^i])$	$\Leftrightarrow \exists s \in S'$ s.t. $M \models \varepsilon_1(\text{er}(|\varphi_1|, |\varphi_2|), s, [s_0^i])$

Note that when a proposition $\varepsilon_1(|\varphi|, s, [s_i^j])$ is valid in M, for instance $M \models \varepsilon_1(\text{eg}(|\varphi|), s, [s_i^j])$, $EG\varphi$ may not hold on the state s.

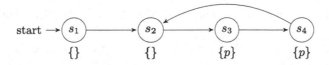

Fig. 2. Example of $\mathcal{L}$

Example 1. For the structure M in Fig. 2, $M \models \varepsilon_1(\text{eg}(\overline{p}), s_3, \text{con}(s_2, \text{con}(s_1, \text{nil})))$ holds because there exists a lr-path, for instance s_1, s_2, s_3, s_4, s_2 such that p holds on s_3 and s_4.

The Rewrite System $\mathcal{R}$. The rewrite system has three components

1. rules for the Kripke structure M (denoted as $\mathcal{R}_M$),
2. rules for the class variables (denoted as $\mathcal{R}_c$),
3. rules for the semantics encoding of the CTL operators (denoted as $\mathcal{R}_{CTL}$).

The Rules of $\mathcal{R}_M$ The rules of $\mathcal{R}_M$ are as follows:

- for each atomic proposition $p \in AP$ and each state $s \in S$, if $p \in \mathsf{L}(s)$, then $\varepsilon_0(\overline{p}, s) \hookrightarrow \top$ is in $\mathcal{R}_M$, otherwise $\varepsilon_0(\mathsf{not}(\overline{p}), s) \hookrightarrow \top$ is in of $\mathcal{R}_M$.
- for each state $s \in S$, take $r(s, [\mathsf{next}(s)]) \hookrightarrow \top$ as a rewrite rule of $\mathcal{R}_M$.

The Rules of $\mathcal{R}_c$ For the class variables, as the domain of the model is finite, there exists two axioms [9] $\forall x(x = x)$, and $\forall x \forall y \forall Z((x = y \vee mem(x, Z)) \Rightarrow mem(x, \mathsf{con}(y, Z)))$. The rewrite rules for these axioms are $x = x \hookrightarrow \top$ and $mem(x, \mathsf{con}(y, Z)) \hookrightarrow x = y \vee mem(x, Z)$. To avoid introducing the predicate "=", the rewrite rules are replaced by the rules ($\mathcal{R}_c$)

$$mem(x, \mathsf{con}(x, Z)) \hookrightarrow \top \text{ and } mem(x, \mathsf{con}(y, Z)) \hookrightarrow mem(x, Z).$$

The Rules of $\mathcal{R}_{CTL}$ The rewrite rules for the predicates carrying the semantic definition of the CTL propositions, are in Fig. 3.

$\varepsilon_0(\mathsf{or}(\varPhi, \varPsi), x) \hookrightarrow \varepsilon_0(\varPhi, x) \vee \varepsilon_0(\varPsi, x) \qquad \varepsilon_0(\mathsf{and}(\varPhi, \varPsi), x) \hookrightarrow \varepsilon_0(\varPhi, x) \wedge \varepsilon_0(\varPsi, x)$

$\varepsilon_0(\mathsf{ax}(\varPhi), x) \hookrightarrow \exists X(r(x, X) \wedge \varepsilon_{\sqcap_0}(\varPhi, X)) \quad \varepsilon_0(\mathsf{ex}(\varPhi), x) \hookrightarrow \exists X(r(x, X) \wedge \varepsilon_{\sqcup_0}(\varPhi, X))$

$\varepsilon_0(\mathsf{af}(\varPhi), x)) \hookrightarrow \varepsilon_0(\varPhi, x) \vee \exists X(r(x, X) \wedge \varepsilon_{\sqcap_0}(\mathsf{af}(\varPhi), X))$

$\varepsilon_0(\mathsf{ef}(\varPhi), x)) \hookrightarrow \varepsilon_0(\varPhi, x) \vee \exists X(r(x, X) \wedge \varepsilon_{\sqcup_0}(\mathsf{ef}(\varPhi), X))$

$\varepsilon_0(\mathsf{ag}(\varPhi), x) \hookrightarrow \varepsilon_1(\mathsf{ag}(\varPhi), x, \mathsf{nil}) \qquad \varepsilon_0(\mathsf{eg}(\varPhi), x) \hookrightarrow \varepsilon_1(\mathsf{eg}(\varPhi), x, \mathsf{nil})$

$\varepsilon_0(\mathsf{au}(\varPhi, \varPsi), x) \hookrightarrow \varepsilon_0(\varPsi, x) \vee (\varepsilon_0(\varPhi, x) \wedge \exists X(r(x, X) \wedge \varepsilon_{\sqcap_0}(\mathsf{au}(\varPhi, \varPsi), X)))$

$\varepsilon_0(\mathsf{eu}(\varPhi, \varPsi), x) \hookrightarrow \varepsilon_0(\varPsi, x) \vee (\varepsilon_0(\varPhi, x) \wedge \exists X(r(x, X) \wedge \varepsilon_{\sqcup_0}(\mathsf{eu}(\varPhi, \varPsi), X)))$

$\varepsilon_0(\mathsf{ar}(\varPhi, \varPsi), x) \hookrightarrow \varepsilon_1(\mathsf{ar}(\varPhi, \varPsi), x, \mathsf{nil}) \qquad \varepsilon_0(\mathsf{er}(\varPhi, \varPsi), x) \hookrightarrow \varepsilon_1(\mathsf{er}(\varPhi, \varPsi), x, \mathsf{nil})$

$\varepsilon_{\sqcap_0}(\varPhi, \mathsf{con}(x, X)) \hookrightarrow \varepsilon_0(\varPhi, x) \wedge \varepsilon_{\sqcap_0}(\varPhi, X) \quad \varepsilon_{\sqcap_0}(\varPhi, \mathsf{nil}) \hookrightarrow \top$

$\varepsilon_{\sqcup_0}(\varPhi, \mathsf{con}(x, X)) \hookrightarrow \varepsilon_0(\varPhi, x) \vee \varepsilon_{\sqcup_0}(\varPhi, X)$

$\varepsilon_1(\mathsf{ag}(\varPhi), x, Y) \hookrightarrow mem(x, Y) \vee (\varepsilon_0(\varPhi, x) \wedge \exists X(r(x, X) \wedge \varepsilon_{\sqcap_1}(\mathsf{ag}(\varPhi), X, \mathsf{con}(x, Y))))$

$\varepsilon_1(\mathsf{eg}(\varPhi), x, Y) \hookrightarrow mem(x, Y) \vee (\varepsilon_0(\varPhi, x) \wedge \exists X(r(x, X) \wedge \varepsilon_{\sqcup_1}(\mathsf{eg}(\varPhi), X, \mathsf{con}(x, Y))))$

$\varepsilon_1(\mathsf{ar}(\varPhi, \varPsi), x, Y) \hookrightarrow$
$mem(x, Y) \vee (\varepsilon_0(\varPsi, x) \wedge (\varepsilon_0(\varPhi, x) \vee \exists X(r(x, X) \wedge \varepsilon_{\sqcap_1}(\mathsf{ar}(\varPhi, \varPsi), X, \mathsf{con}(x, Y)))))$

$\varepsilon_1(\mathsf{er}(\varPhi, \varPsi), x, Y) \hookrightarrow$
$mem(x, Y) \vee (\varepsilon_0(\varPsi, x) \wedge (\varepsilon_0(\varPhi, x) \vee \exists X(r(x, X) \wedge \varepsilon_{\sqcup_1}(\mathsf{er}(\varPhi, \varPsi), X, \mathsf{con}(x, Y)))))$

$\varepsilon_{\sqcap_1}(\varPhi, \mathsf{con}(x, X), Y) \hookrightarrow \varepsilon_1(\varPhi, x, Y) \wedge \varepsilon_{\sqcap_1}(\varPhi, X, Y)$

$\varepsilon_{\sqcap_1}(\varPhi, \mathsf{nil}, Y) \hookrightarrow \top$

$\varepsilon_{\sqcup_1}(\varPhi, \mathsf{con}(x, X), Y) \hookrightarrow \varepsilon_1(\varPhi, x, Y) \vee \varepsilon_{\sqcup_1}(\varPhi, X, Y)$

Fig. 3. Rewrite Rules for CTL Connectives ($\mathcal{R}_{CTL}$)

Now we are ready to prove the main theorem. Our goal is to prove that $M \models \varepsilon_0(|\varphi|, s)$ holds if and only if $\varepsilon_0(|\varphi|, s)$ is provable in Deduction Modulo.

Lemma 5 (Soundness). *For a CTL formula φ of NNF, if the sequent $\vdash_{\mathcal{R}}^{cf}$ $\varepsilon_0(|\varphi|, s)$ has a proof, then $M \models \varepsilon_0(|\varphi|, s)$.*

Proof. More generally, we prove that for any CTL proposition φ of NNF,

- if $\vdash_{\mathcal{R}}^{cf} \varepsilon_0(|\varphi|, s)$ has a proof, then $M \models \varepsilon_0(|\varphi|, s)$.
- if $\vdash_{\mathcal{R}}^{cf} \varepsilon_{\sqcap_0}(|\varphi|, [S'])$ has a proof, then $M \models \varepsilon_{\sqcap_0}(|\varphi|, [S'])$.
- if $\vdash_{\mathcal{R}}^{cf} \varepsilon_{\sqcup_0}(|\varphi|, [S'])$ has a proof, then $M \models \varepsilon_{\sqcup_0}(|\varphi|, [S'])$.
- if $\vdash_{\mathcal{R}}^{cf} \varepsilon_1(|\varphi|, s, [s_i^j])$ has a proof, in which φ is either of the form $AG\varphi_1$, $EG\varphi_1$, $A[\varphi_1 R\varphi_2]$, $E[\varphi_1 R\varphi_2]$, then $M \models \varepsilon_1(|\varphi|, s, [s_i^j])$.
- if $\vdash_{\mathcal{R}}^{cf} \varepsilon_{\sqcap_1}(|\varphi|, [S'], [s_i^j])$ has a proof, in which φ is either of the form $AG\varphi_1$, $A[\varphi_1 R\varphi_2]$, then $M \models \varepsilon_{\sqcap_1}(|\varphi|, [S'], [s_i^j])$.
- if $\vdash_{\mathcal{R}}^{cf} \varepsilon_{\sqcup_1}(|\varphi|, [S'], [s_i^j])$ has a proof, in which φ is either of the form $EG\varphi_1$, $E[\varphi_1 R\varphi_2]$, then $M \models \varepsilon_{\sqcup_1}(|\varphi|, [S'], [s_i^j])$.

By induction on the size of the proof. Consider the different case for φ, we have 18 cases (2 cases for the atomic proposition and its negation, 2 cases for and and or, 10 cases for the temporal connectives ax, ex, af, ef, ag, eg, au, eu, ar, er, 4 cases for the predicate symbols $\varepsilon_{\sqcap_0}$, $\varepsilon_{\sqcup_0}$, $\varepsilon_{\sqcap_1}$, $\varepsilon_{\sqcup_0}$), but each case is easy. For brevity, we just prove two cases. The full proof is in [10].

- Suppose the sequent $\vdash_{\mathcal{R}}^{cf} \varepsilon_0(\mathsf{af}(|\varphi|), s)$ has a proof. As $\varepsilon_0(\mathsf{af}(|\varphi|), s) \hookrightarrow \varepsilon_0(|\varphi|, s) \vee \exists X(r(s, X) \wedge \varepsilon_{\sqcap_0}(\mathsf{af}(|\varphi|), X))$, the last rule in the proof is $\vee_1$ or $\vee_2$. For $\vee_1$, $M \models \varepsilon_0(|\varphi|, s)$ holds by IH, then $M \models \varepsilon_0(\mathsf{af}(|\varphi|), s)$ holds by its semantic definition. For $\vee_2$, $M \models \exists X(r(s, X) \wedge \varepsilon_{\sqcap_0}(\mathsf{af}(|\varphi|), X))$ holds by IH, thus there exists S' s.t. $M \models r(s, [S'])$ and $M \models \varepsilon_{\sqcap_0}(\mathsf{af}(|\varphi|), [S'])$ holds. Then we get $S' = \mathsf{next}(s)$ and for each state s' in S', $M \models \varepsilon_0(\mathsf{af}(|\varphi|), s')$ holds. Now assume $M \not\models \varepsilon_0(\mathsf{af}(|\varphi|), s)$, then there exists a lr-path $\rho(s)(j, k)$ s.t. $\forall 0 \le i < k$, $M \not\models \varepsilon_0(|\varphi|, \rho_i)$. For the path $\rho(s)(j, k)$,
  - if $j \ne 0$, then ρ_1^k is a lr-path, which is a counterexample of $M \models \varepsilon_0(\mathsf{af}(|\varphi|), \rho_1)$.
  - if $j = 0$, then $\rho_1^k {}^\frown \rho_1$ is a lr-path, which is a counterexample of $M \models \varepsilon_0(\mathsf{af}(|\varphi|), \rho_1)$.

  Thus $M \models \varepsilon_0(\mathsf{af}(|\varphi|), s)$ holds by its semantic definition.
- Suppose that $\vdash_{\mathcal{R}}^{cf} \varepsilon_1(\mathsf{ag}(|\varphi|), s, [s_i^j])$ has a proof. As $\varepsilon_1(\mathsf{ag}(|\varphi|), s, [s_i^j]) \hookrightarrow \mathsf{mem}(s, [s_i^j]) \vee (\varepsilon_0(|\varphi|, s) \wedge \exists X(r(s, X) \wedge \varepsilon_{\sqcap_1}(\mathsf{ag}(|\varphi|), X, \mathsf{con}(s, [s_i^j]))))$, the last rule in the proof is $\vee_1$ or $\vee_2$. For $\vee_1$, $M \models \mathsf{mem}(s, [s_i^j])$ holds by IH, thus $s_i^j {}^\frown s$ is a lr-path and $M \models \varepsilon_1(\mathsf{ag}(|\varphi|), s, [s_i^j])$ holds by its semantic definition. For $\vee_2$, $M \models \varepsilon_0(|\varphi|, s)$ and $M \models \exists X(r(s, X) \wedge \varepsilon_{\sqcap_1}(\mathsf{ag}(|\varphi|), X, \mathsf{con}(s, [s_i^j])))$ holds by IH. Thus there exists S' s.t. $M \models r(s, [S']) \wedge \varepsilon_{\sqcap_1}(\mathsf{ag}(|\varphi|), [S'], \mathsf{con}(s, [s_i^j]))$ holds. Then by the semantic definition, $S' = \mathsf{next}(s)$ and for each state $s' \in S'$, $M \models \varepsilon_1(\mathsf{ag}(|\varphi|), s', \mathsf{con}(s, [s_i^j]))$ holds. Thus $M \models \varepsilon_1(\mathsf{ag}(|\varphi|), s, [s_i^j])$ holds by its semantic definition.

Lemma 6 (Completeness). *For a CTL formula φ of NNF, if $M \models \varepsilon_0(|\varphi|, s)$, then the sequent $\vdash_{\mathcal{R}}^{cf} \varepsilon_0(|\varphi|, s)$ has a proof.*

Proof. By induction on the structure of φ. For brevity, here we just prove some of the cases. The full proof is in [10].

- Suppose $M \models \varepsilon_0(\mathsf{af}(|\varphi_1|), s)$ holds. By the semantics of $\mathcal{L}$, there exists a state s' on each lr-path starting from s s.t. $M \models \varepsilon_0(|\varphi_1|, s')$ holds. Thus there exists a finite tree T s.t.
  - T has root s;
  - for each internal node s' in T, the children of s' are labelled by the elements of $\mathsf{next}(s')$;
  - for each leaf s', s' is the first node in the branch starting from s s.t. $M \models \varepsilon_0(|\varphi_1|, s')$ holds.

  By IH, for each leaf s', there exists a proof $\Pi_{(\varphi_1, s')}$ for the sequent $\vdash_{\mathcal{R}}^{cf} \varepsilon_0(|\varphi_1|, s')$. Then, to each subtree T' of T, we associate a proof $|T'|$ of the sequent $\vdash_{\mathcal{R}}^{cf} \varepsilon_0(\mathsf{af}(|\varphi_1|), s')$ where s' is the root of T', by induction, as follows,
  - if T' contains a single node s', then the proof $|T'|$ is as follows:

  $$\dfrac{\Pi_{(\varphi_1, s')}}{\vdash_{\mathcal{R}}^{cf} \varepsilon_0(\mathsf{af}(|\varphi_1|), s')}\ \vee_1$$

  - if $T' = s'(T_1, \ldots, T_n)$, then the proof $|T'|$ is as follows:

  $$\dfrac{\dfrac{}{\vdash_{\mathcal{R}}^{cf} r(s', [\mathsf{next}(s')])}\ \top \qquad \dfrac{|T_1| \quad \cdots \quad |T_n|}{\vdash_{\mathcal{R}}^{cf} \varepsilon_{\sqcap_0}(\mathsf{af}(|\varphi_1|), [\mathsf{next}(s')])}\ \wedge^n}{\dfrac{\dfrac{\vdash_{\mathcal{R}}^{cf} r(s', [\mathsf{next}(s')]) \wedge \varepsilon_{\sqcap_0}(\mathsf{af}(|\varphi_1|), [\mathsf{next}(s')])}{\vdash_{\mathcal{R}}^{cf} \exists X(r(s', X) \wedge \varepsilon_{\sqcap_0}(\mathsf{af}(|\varphi_1|), X))}\ \exists}{\vdash_{\mathcal{R}}^{cf} \varepsilon_0(\mathsf{af}(|\varphi_1|), s')}\ \vee_2}\ \wedge$$

  This way, $|T|$ is a proof of the sequent $\vdash_{\mathcal{R}}^{cf} \varepsilon_0(\mathsf{af}(|\varphi_1|), s)$.

- Suppose $M \models \varepsilon_0(\mathsf{ag}(|\varphi_1|), s)$ holds. By the semantics of $\mathcal{L}$, for each state s' on each lr-path starting from s, $M \models \varepsilon_0(|\varphi_1|, s')$ holds. Thus there exists a finite tree T s.t.
  - T has root s;
  - for each internal node s' in T, the children of s' are labelled by the elements of $\mathsf{next}(s')$;
  - the branch starting from s to each leaf is a lr-path;
  - for each internal node s' in T, $M \models \varepsilon_0(|\varphi_1|, s')$ holds and by IH, there exists a proof $\Pi_{(\varphi_1, s')}$ for the sequent $\vdash_{\mathcal{R}}^{cf} \varepsilon_0(|\varphi_1|, s')$.

  Then, to each subtree T' of T, we associate a proof $|T'|$ of the sequent $\vdash_{\mathcal{R}}^{cf} \varepsilon_1(\mathsf{ag}(|\varphi_1|), s', [s_0'^{k-1}])$ where s' is the root of T' and $s_0'^k (s_k' = s')$ is the branch from s to s', by induction, as follows,
  - if T' contains a single node s', then $s_0'^k$ is a lr-path and the proof is as follows:

  $$\dfrac{\dfrac{}{\vdash_{\mathcal{R}}^{cf} mem(s', [s_0'^{k-1}])}\ \top}{\vdash_{\mathcal{R}}^{cf} \varepsilon_1(\mathsf{ag}(|\varphi_1|), s', [s_0'^{k-1}])}\ \vee_2$$

- if $T' = s'(T_1, \ldots, T_n)$, the proof is as follows:

$$\cfrac{\cfrac{\cfrac{\cfrac{}{\vdash_{\mathcal{R}}^{cf} r(s', [\mathsf{next}(s')])}\; \top \quad \cfrac{|T_1| \quad \cdots \quad |T_n|}{\vdash_{\mathcal{R}}^{cf} \varepsilon_{\sqcap_1}(\mathsf{ag}(|\varphi_1|), [\mathsf{next}(s')], [s_0'^k])}\wedge^n}{\vdash_{\mathcal{R}}^{cf} r(s', [\mathsf{next}(s')]) \wedge \varepsilon_{\sqcap_1}(\mathsf{ag}(|\varphi_1|), [\mathsf{next}(s')], [s_0'^k])}\wedge}{\cfrac{\Pi_{s'}}{\vdash_{\mathcal{R}}^{cf} \varepsilon_0(|\varphi_1|, s') \quad \vdash_{\mathcal{R}}^{cf} \exists X(r(s', X) \wedge \varepsilon_{\sqcap_1}(\mathsf{ag}(|\varphi_1|), X, [s_0'^k]))}}\exists \quad}{\cfrac{\vdash_{\mathcal{R}}^{cf} \varepsilon_0(|\varphi_1|, s') \wedge \exists X(r(s', X) \wedge \varepsilon_{\sqcap_1}(\mathsf{ag}(|\varphi_1|), X, [s_0'^k]))}{\vdash_{\mathcal{R}}^{cf} \varepsilon_1(\mathsf{ag}(|\varphi_1|), s', [s_0'^{k-1}])}\vee_1}$$

This way, as $\varepsilon_0(\mathsf{ag}(|\varphi_1|), s)$ can be rewritten into $\varepsilon_1(\mathsf{ag}(|\varphi_1|), s, \mathsf{nil})$, $|T|$ is a proof for the sequent $\vdash_{\mathcal{R}}^{cf} \varepsilon_0(\mathsf{ag}(|\varphi_1|), s)$. □

Theorem 2 (Soundness and Completeness). *For a CTL proposition φ of NNF, the sequent $\vdash_{\mathcal{R}}^{cf} \varepsilon_0(|\varphi|, s)$ has a proof iff $M \models \varepsilon_0(|\varphi|, s)$ holds.*

6 Applications

6.1 Polarized Resolution Modulo

In Polarized Resolution Modulo, the polarized rewrite rules are taken as one-way clauses [6]. For example, the rewrite rule

$$\varepsilon_1(\mathsf{eg}(\varPhi), x, Y) \hookrightarrow_+ mem(x, Y) \vee (\varepsilon_0(\varPhi, x) \wedge \exists X(r(x, X) \wedge \varepsilon_{\sqcup_1}(\mathsf{eg}(\varPhi), X, \mathsf{con}(x, Y))))$$

is translated into one-way clauses $\underline{\varepsilon_1(\mathsf{eg}(\varPhi), x, Y)} \vee mem(x, Y)^{\perp}$ and $\varepsilon_1(\mathsf{eg}(\varPhi), x, Y)$ $\vee \varepsilon_0(\varPhi, x)^{\perp} \vee r(x, X)^{\perp} \vee \varepsilon_{\sqcup_1}(\mathsf{eg}(\varPhi), X, \mathsf{con}(x, Y))^{\perp}$, in which the underlined literals have the priority to do resolution.

Fig. 4. Resolution Example

Example 2. For the structure M in Fig. 4, we prove that $M, s_1 \models_a EXEGp$. The one-way clauses of M are: $\varepsilon_0(\mathsf{not}(\overline{p}), s_1)$, $\varepsilon_0(\overline{p}, s_2)$, $\varepsilon_0(\overline{p}, s_3)$, $r(s_1, \mathsf{con}(s_2, \mathsf{nil}))$, $r(s_2, \mathsf{con}(s_3, \mathsf{nil}))$, $r(s_3, \mathsf{con}(s_2, \mathsf{nil}))$. The translation of $M, s_1 \models_a EXEGp$ is $\varepsilon_0(\mathsf{ex}(\mathsf{eg}(\overline{p})), s_1)$ and the resolution steps start from

$$\varepsilon_0(\mathsf{ex}(\mathsf{eg}(\overline{p})), s_1)^{\perp}.$$

First apply resolution with $\underline{\varepsilon_0(\mathsf{ex}(\varPhi), x)} \vee r(x, X)^{\perp} \vee \varepsilon_{\sqcup_0}(\varPhi, X)^{\perp}$, with $x = s_1$ and $\varPhi = \mathsf{eg}(\overline{p})$, this yields

$$r(s_1, X)^{\perp} \vee \varepsilon_{\sqcup_0}(\mathsf{eg}(\overline{p}), X)^{\perp}.$$

Then apply resolution with $r(s_1, \mathsf{con}(s_2, \mathsf{nil}))$, with $X = \mathsf{con}(s_2, \mathsf{nil})$, this yields

$$\varepsilon_{\sqcup_0}(\mathsf{eg}(\overline{p}), \mathsf{con}(s_2, \mathsf{nil}))^{\perp}.$$

Then apply resolution with $\varepsilon_{\sqcup_0}(\Phi, \mathsf{con}(x, X)) \vee \varepsilon_0(\Phi, x)^{\perp}$, with $x = s_2$, $X = \mathsf{nil}$ and $\Phi = \mathsf{eg}(\overline{p})$, this yields

$$\varepsilon_0(\mathsf{eg}(\overline{p}), s_2)^{\perp}.$$

Then apply resolution with one-way clause $\varepsilon_0(\mathsf{eg}(\Phi), x) \vee \varepsilon_1(\mathsf{eg}(\Phi), x, \mathsf{nil})^{\perp}$, with $\Phi = \overline{p}$ and $x = s_2$, this yields

$$\varepsilon_1(\mathsf{eg}(\overline{p}), s_2, \mathsf{nil})^{\perp}.$$

Then apply resolution with ($\ddagger$ [1]), with $\Phi = \overline{p}$, $x = s_2$ and $Y = \mathsf{nil}$, this yields

$$\varepsilon_0(\overline{p}, s_2)^{\perp} \vee r(s_2, X)^{\perp} \vee \varepsilon_{\sqcup_1}(\mathsf{eg}(\overline{p}), X, \mathsf{con}(s_2, \mathsf{nil}))^{\perp}.$$

Then apply resolution with $\varepsilon_0(\overline{p}, s_2)$, this yields

$$r(s_2, X)^{\perp} \vee \varepsilon_{\sqcup_1}(\mathsf{eg}(\overline{p}), X, \mathsf{con}(s_2, \mathsf{nil}))^{\perp}.$$

Then apply resolution with $r(s_2, \mathsf{con}(s_3, \mathsf{nil}))$, with $X = \mathsf{con}(s_3, \mathsf{nil})$, this yields

$$\varepsilon_{\sqcup_1}(\mathsf{eg}(\overline{p}), \mathsf{con}(s_3, \mathsf{nil}), \mathsf{con}(s_2, \mathsf{nil}))^{\perp}.$$

Then apply resolution with $\varepsilon_{\sqcup_1}(\Phi, \mathsf{con}(x, X), Y) \vee \varepsilon_1(\Phi, x, Y)^{\perp}$, with $\Phi = \mathsf{eg}(\overline{p})$, $x = s_3$, $X = \mathsf{nil}$ and $Y = \mathsf{con}(s_2, \mathsf{nil})$, this yields

$$\varepsilon_1(\mathsf{eg}(\overline{p}), s_3, \mathsf{con}(s_2, \mathsf{nil}))^{\perp}.$$

Then apply resolution with ($\ddagger$), with $\Phi = \overline{p}$, $x = s_3$, $Y = \mathsf{con}(s_2, \mathsf{nil})$, this yields

$$\varepsilon_0(\overline{p}, s_3)^{\perp} \vee r(s_3, X)^{\perp} \vee \varepsilon_{\sqcup_1}(\mathsf{eg}(\overline{p}), X, \mathsf{con}(s_3, \mathsf{con}(s_2, \mathsf{nil})))^{\perp}.$$

Then apply resolution with $\varepsilon_0(\overline{p}, s_3)$, this yields

$$r(s_3, X)^{\perp} \vee \varepsilon_{\sqcup_1}(\mathsf{eg}(\overline{p}), X, \mathsf{con}(s_3, \mathsf{con}(s_2, \mathsf{nil})))^{\perp}.$$

Then apply resolution with $r(s_3, \mathsf{con}(s_2, \mathsf{nil}))$, with $X = \mathsf{con}(s_2, \mathsf{nil})$, this yields

$$\varepsilon_{\sqcup_1}(\mathsf{eg}(\overline{p}), \mathsf{con}(s_2, \mathsf{nil}), \mathsf{con}(s_3, \mathsf{con}(s_2, \mathsf{nil})))^{\perp}.$$

Then apply resolution with $\varepsilon_{\sqcup_1}(\Phi, \mathsf{con}(x, X), Y) \vee \varepsilon_1(\Phi, x, Y)^{\perp}$, with $\Phi = \mathsf{eg}(\overline{p})$, $x = s_3$, $X = \mathsf{nil}$ and $Y = \mathsf{con}(s_2, \mathsf{nil})$, this yields

$$\varepsilon_1(\mathsf{eg}(\overline{p}), s_2, \mathsf{con}(s_3, \mathsf{con}(s_2, \mathsf{nil})))^{\perp}.$$

Then apply resolution with $\varepsilon_1(\mathsf{eg}(\Phi), x, Y) \vee mem(x, Y)^{\perp}$, with $x = s_2$ and $Y = \mathsf{con}(s_3, \mathsf{con}(s_2, \mathsf{nil}))$, this yields

$$mem(s_2, \mathsf{con}(s_3, \mathsf{con}(s_2, \mathsf{nil})))^{\perp}.$$

Then apply resolution with $mem(x, \mathsf{con}(y, Z)) \vee mem(x, Z)^{\perp}$, with $x = s_2$, $y = s_3$ and $Z = \mathsf{con}(s_2, \mathsf{nil})$, this yields

$$mem(s_2, \mathsf{con}(s_2, \mathsf{nil}))^{\perp}.$$

Then apply resolution with $mem(x, \mathsf{con}(x, Z))$, with $x = s_2$ and $Z = \mathsf{nil}$, this yields the empty clause. Thus $M, s_1 \models_a EXEGp$ holds.

---

[1] $\ddagger$ is $\varepsilon_1(\mathsf{eg}(\Phi), x, Y) \vee \varepsilon_0(\Phi, x)^{\perp} \vee r(x, X)^{\perp} \vee \varepsilon_{\sqcup_1}(\mathsf{eg}(\Phi), X, \mathsf{con}(x, Y))^{\perp}.$

6.2 Experimental Evaluation

In this Section, we give a comparison between Resolution-based and QBF-based verification, that are implemented in iProver Modulo and VERDS [15] respectively. iProver Modulo is a prover by embedding Polarized Resolution Modulo into iProver [11]. The comparison is by proving 24 CTL properties on two kinds of programs: *Programs with Concurrent Processes* and *Programs with Concurrent Sequential Processes*. The programs and CTL properties refer to [16].

Table 1. Experimental Results

iProver/Verds		Con. Processes			Con. Seq. Processes		
Prop	Num	True	False	>20m	True	False	>20m
p_{01}	40	-	40/40	-	23/-	5/4	12/36
p_{02}	40	40/40	-	-	40/40	-	-
p_{03}	40	2/-	37/37	1/3	-	25/15	15/25
p_{04}	40	17/-	-	23/40	-	-	40/40
p_{05}	40	25/34	6/5	9/1	24/24	8/2	8/14
p_{06}	40	31/40	-	9/-	36/31	-	4/9
p_{07}	40	40/40	-	-	40/40	-	-
p_{08}	40	40/40	-	-	40/40	-	-
p_{09}	40	32/32	8/8	-	35/29	5/1	-/10
p_{10}	40	40/40	-	-	40/40	-	-
p_{11}	40	10/10	30/30	-	27/23	8/4	5/13
p_{12}	40	40/40	-	-	40/35	-	-/5
p_{13}	40	-	40/40	-	-	40/40	-
p_{14}	40	3/3	37/37	-	3/3	37/33	-/4
p_{15}	40	5/-	33/33	2/7	-	23/15	17/25
p_{16}	40	19/-	-	21/40	-	-	40/40
p_{17}	40	28/37	3/2	9/1	25/26	5/1	10/13
p_{18}	40	32/40	-	8/-	37/31	-	3/9
p_{19}	40	5/5	35/35	-	6/6	34/34	-
p_{20}	40	15/17	21/21	4/2	12/11	18/22	10/7
p_{21}	40	3/3	37/37	-	3/3	37/37	-
p_{22}	40	3/3	37/37	-	3/3	37/37	-
p_{23}	40	-	40/40	-	-	40/40	-
p_{24}	40	20/25	12/10	8/5	8/8	23/22	9/10
Sum	960	450/449	416/412	94/99	442/393	345/307	173/260

For the *Con. Processes*, each testing case contains 12/24 variables and 3 processes. For the *Con. Seq. Processes*, each testing case contains 12/16 variables and 2 processes.

Table 2. Speed Comparisons

Prop	Num	Con. Processes			Con. Seq. Processes		
		adv/T	adv/F	O(iP/Ver)	adv/T	adv/F	O(iP/Ver)
p_{01}	40	-	0/40	-	-	0/3	25/1
p_{02}	40	40/40	-	-	40/40	-	-
p_{03}	40	-	1/37	2/-	-	11/15	10/-
p_{04}	40	-	-	17/-	-	-	-
p_{05}	40	0/25	3/5	1/9	6/20	2/2	10/4
p_{06}	40	0/31	-	-/9	10/28	-	8/3
p_{07}	40	33/40	-	-	37/40	-	-
p_{08}	40	35/40	-	-	38/40	-	-
p_{09}	40	19/32	0/8	-	22/29	0/1	10/-
p_{10}	40	19/40	-	-	18/40	-	-
p_{11}	40	0/10	0/30	-	9/23	3/4	8/-
p_{12}	40	3/40	-	-	7/35	-	5/-
p_{13}	40	-	38/40	-	-	40/40	-
p_{14}	40	2/3	0/37	-	3/3	23/33	4/-
p_{15}	40	-	0/33	5/-	-	10/14	9/1
p_{16}	40	-	-	19/-	-	-	-
p_{17}	40	0/28	1/2	1/9	8/22	1/1	7/4
p_{18}	40	0/32	-	-/8	11/29	-	8/2
p_{19}	40	2/5	9/35	-	6/6	12/34	-
p_{20}	40	1/15	7/20	1/3	6/11	9/17	2/5
p_{21}	40	2/3	18/37	-	3/3	23/37	-
p_{22}	40	2/3	19/37	-	2/3	22/37	-
p_{23}	40	-	17/40	-	-	25/40	-
p_{24}	40	0/20	1/10	2/5	1/7	4/21	3/2
Sum	960	158/407	114/411	48/43	227/379	185/299	109/22

All the cases are tested on Intel® Core ™ i5-2400 CPU @ 3.10 GHz × 4 with Linux and the testing time of each case is limited to 20 min. The experimental data is presented in Tables 1 and 2[2]. The comparison is based on two aspects: the number of testing cases that can be proved, and the time used if a problem can be proved in both. As can be seen in Table 1, among the 960 testing cases of the *Con. Processes*, 94 of them are timeout in iProver Modulo, while the number in VERDS is 99. For the *Con. Seq. Processes*, among the 960 testing cases, 173 of them are timeout in iProver Modulo, while in VERDS, the number is 260. Table 2 shows that, among the 818 testing cases of the *Con. Processes*, that are both proved in iProver Modulo and VERDS, iProver

---

[2] adv/T(F): has advantage in speed when both return T(F). O(iP/Ver): only solved by iProver/Verds.

Modulo performs better in 272 of them and among the 678 testing cases of the *Con. Seq. Processes*, 412 of them run faster in iProver Modulo.

In summary, for the 1920 testing cases, 1653 (86 %) of them are solved by iProver Modulo, while 1561 (81 %) are solved by VERDS. For all the 1496 testing cases that are both proved, 684 (45.8 %) testing cases run faster in iProver Modulo.

7   Conclusion and Future Work

In this paper, we defined an alternative semantics for CTL, which is bounded to lr-paths. Based on the alternative semantics, a way to embed model checking problems into Deduction Modulo has been presented. Thus this work has given a method to solve model checking problems in automated theorem provers. An experimental evaluation of this approach using resolution modulo has been presented. The comparison with the QBF-based verification showed that automated theorem proving modulo, which performed as well as QBF-based method, can be considered as a new way to quickly determine whether a property is violated in transition system models.

The proof-search method does not work well on proving some temporal propositions, such as the propositions start with *AG*. One of the reasons is during the search steps, it may visit the same state repeatedly. To design new rewrite rules for the encoding of temporal connectives or new elimination rules to avoid this problem remains as future work.

Acknowledgements. I am grateful to Gilles Dowek, for his careful reading and comments.

References

1. Biere, A., Cimatti, A., Clarke, E., Zhu, Y.: Symbolic model checking without BDDs. In: Cleaveland, W.R. (ed.) TACAS 1999. LNCS, vol. 1579, pp. 193–207. Springer, Heidelberg (1999)
2. Burel, G.: Embedding deduction modulo into a prover. In: Dawar, A., Veith, H. (eds.) CSL 2010. LNCS, vol. 6247, pp. 155–169. Springer, Heidelberg (2010)
3. Burel, G.: Experimenting with deduction modulo. In: Bjørner, N., Sofronie-Stokkermans, V. (eds.) CADE 2011. LNCS, vol. 6803, pp. 162–176. Springer, Heidelberg (2011)
4. Clarke Jr., E.M., Grumberg, O., Peled, D.A.: Model Checking. MIT Press, Cambridge, MA, USA (1999)
5. Delahaye, D., Doligez, D., Gilbert, F., Halmagrand, P., Hermant, O.: Zenon modulo: when Achilles outruns the tortoise using deduction modulo. In: McMillan, K., Middeldorp, A., Voronkov, A. (eds.) LPAR-19 2013. LNCS, vol. 8312, pp. 274–290. Springer, Heidelberg (2013)
6. Dowek, G.: Polarized resolution modulo. In: Calude, C.S., Sassone, V. (eds.) TCS 2010. IFIP AICT, vol. 323, pp. 182–196. Springer, Heidelberg (2010)
7. Dowek, G., Hardin, T., Kirchner, C.: Theorem proving modulo. J. Autom. reasoning **31**, 33–72 (2003)
8. Dowek, G., Jiang, Y.: A Logical Approach to CTL (2013). http://hal.inria.fr/docs/00/91/94/67/PDF/ctl.pdf (manuscript)

9. Dowek, G., Jiang, Y.: Axiomatizing Truth in a Finite Model (2013). https://who. rocq.inria.fr/Gilles.Dowek/Publi/classes.pdf (manuscript)

10. Ji, K.: CTL Model Checking in Deduction Modulo. In: Felty, A.P., Middeldorp, A. (eds.) CADE-25, 2015. LNCS, vol. 9195, pp. xx–yy (2015). https://drive.google. com/file/d/0B0CYADxmoWB5UGJsV2UzNnVqVHM/view?usp=sharing (fullpaper)

11. Korovin, K.: iProver – an instantiation-based theorem prover for first-order logic (system description). In: Armando, A., Baumgartner, P., Dowek, G. (eds.) IJCAR 2008. LNCS (LNAI), vol. 5195, pp. 292–298. Springer, Heidelberg (2008)

12. Rajan, S., Shankar, N., Srivas, M.: An Integration of Model Checking with Automated Proof Checking. In: Wolper, P. (ed.) CAV 1995. LNCS, vol. 939, pp. 84–97. Springer, Berlin Heidelberg (1995)

13. Troelstra, A.S., Schwichtenberg, H.: Basic Proof Theory. Cambridge University Press, New York (1996)

14. Zhang, W.: Bounded semantics of CTL and SAT-based verification. In: Breitman, K., Cavalcanti, A. (eds.) ICFEM 2009. LNCS, vol. 5885, pp. 286–305. Springer, Heidelberg (2009)

15. Zhang, W.: VERDS Modeling Language (2012). http://lcs.ios.ac.cn/~zwh/verds/ index.html

16. Zhang, W.: QBF encoding of temporal properties and QBF-based verification. In: Demri, S., Kapur, D., Weidenbach, C. (eds.) IJCAR 2014. LNCS, vol. 8562, pp. 224–239. Springer, Heidelberg (2014)

Quantifier-Free Equational Logic and Prime Implicate Generation

Mnacho Echenim[1,2], Nicolas Peltier[1,4(✉)], and Sophie Tourret[1,3]

[1] Grenoble Informatics Laboratory, Grenoble, France
nicolas.peltier@imag.fr
[2] Grenoble INP - Ensimag, Saint-martin-d'hères, France
[3] Université Grenoble 1, Grenoble, France
[4] CNRS, Toulouse, France

Abstract. An algorithm for generating prime implicates of sets of equational ground clauses is presented. It consists in extending the standard Superposition Calculus with rules that allow attaching hypotheses to clauses to perform additional inferences. The hypotheses that lead to a refutation represent implicates of the original set of clauses. The set of prime implicates of a clausal set can thus be obtained by saturation of this set. Data structures and algorithms are also devised to represent sets of *constrained* clauses in an efficient and concise way.

Our method is proven to be correct and complete. Practical experimentations show the relevance of our method in comparison to existing approaches for propositional or first-order logic.

1 Introduction

We tackle the problem of generating the prime implicates of a quantifier-free equational formula. From a formal point of view, an implicate of a formula S is a clause C such that $S \models C$, and this implicate is prime if for all implicates D such that $D \models C$, we have $C \models D$. In other words, prime implicates are the most general clausal consequences of a formula, and their generation is a more difficult problem than checking satisfiability. Prime implicate generation has many natural applications in artificial intelligence and system verification. It has been extensively investigated in the context of propositional logic [6,12,13,16,17,24,26], but there have been only very few approaches dealing with more expressive logics [14,15,18,19]. The approaches that are capable of handling first-order formulæ are based mainly on unrestricted versions of the resolution calculus (with an explicit encoding of equality axioms) or extensions of the tableau method and do not handle equality efficiently. More recently, algorithms were devised to generate sets of implicants of formulæ interpreted in decidable theories [8], by combining quantifier-elimination (for discarding useless variables) with model building (to construct sufficient conditions for satisfiability). The approach does not apply to equational formulæ with function symbols since this would involve second-order quantifier elimination.

© Springer International Publishing Switzerland 2015
A.P. Felty and A. Middeldorp (Eds.): CADE-25, LNAI 9195, pp. 311–325, 2015.
DOI: 10.1007/978-3-319-21401-6_21

In previous work [9,10] we devised procedures for generating implicates of equational formulæ containing only constant symbols. In this paper, we propose an approach to handle arbitrary uninterpreted function symbols. There are two parts to our contribution. First, in Sect. 3, a calculus is devised to generate implicates. It is based on the standard rules of the Superposition Calculus [20], together with *Assertion* rules allowing the addition of new hypotheses during the proof search. These hypotheses are attached to the clauses as constraints, and once an empty clause is derived, the associated constraint corresponds to the negation of an implicate. This algorithm is completely different from that of [9]: its main advantage is that it remains complete even when applying all the usual restrictions of the Superposition Calculus, and that it allows for a better control of the generated implicates, in case the user is interested only to search for implicates of some particular form. Second, in Sect. 4 we extend the representation mechanism of [9] that uses a trie-based representation of equational clause sets in order to handle function symbols. This extension is not straightforward since, in contrast to [9], we have to encode substitutivity as well as transitivity. We devise data-structures and algorithms to efficiently store equational sets up to redundancy, taking into account the properties of the equality predicate. In Sect. 5 we experimentally compare our approach with existing tools [19,24] for propositional logic and first-order logic respectively. We also compare our method with the approach consisting in encoding equational formulæ as flat clauses by explicitly adding substitutivity axioms and applying the algorithms from [10]. Due to space limitations, the formal proofs are omitted. Proofs are all available at http://membres-lig.imag.fr/tourret/documents/EPT15-long.pdf.

2   Clauses with Uninterpreted Functions in Equational Logic

The theory of equational logic with uninterpreted functions will be denoted by EUF (see [2] for details). Let Σ be a *signature*, and Σ_n the function symbols in Σ of arity n, usually denoted by f, g (and a, b for Σ_0). The notation $\mathfrak{T}(\Sigma)$ stands for the set of well-formed ground terms over *Sigma*, most often denoted by s, t, u, v, w. A well-founded reduction order $\prec$ on $\mathfrak{T}(\Sigma)$ such as Knuth-Bendix Ordering or Recursive Path Ordering [7] is assumed to be given. The subterm of t at position p is denoted by $t|_p$.

A *literal*, usually denoted by l or m, is either an equation (or *atom*, or *positive* literal) $s \simeq t$, or an inequation $s \not\simeq t$ (or *negative* literal). The literal written $s \bowtie t$ can denote either the equation or the inequation between s and t. The literal l^c stands for $s \not\simeq t$ (resp. $s \simeq t$) when l is $s \simeq t$ (resp. $s \not\simeq t$). A literal of the form $s \not\simeq s$ is called a *contradictory* literal (or a contradiction) and a literal of the form $s \simeq s$ is a *tautological* literal (or a tautology). We consider *clauses* as disjunctions (or multisets) of literals and *formulæ* as sets of clauses. If C is a clause and l a literal, $C \backslash l$ denotes the clause C where *all* occurrences of l have been removed (up to commutativity of equality). In Sect. 3, we also consider conjunctions of literals, called *constraints*. For every constraint $\mathcal{X} = \bigwedge_{i=1}^{n} l_i$, $\neg \mathcal{X}$

denotes the clause $\bigvee_{i=1}^{n} l_i^c$. Similarly, if $C = \bigvee_{i=1}^{n} l_i$ then $\neg C \stackrel{\text{def}}{=} \bigwedge_{i=1}^{n} l_i^c$. Empty clauses and constraints are denoted by $\square$ and $\top$ respectively. We often identify sets of clauses with conjunctions.

We define an *equational interpretation* $\mathcal{I}$ as a congruence relation on $\mathfrak{T}(\Sigma)$. A positive literal $l = s \simeq t$ is evaluated to $\top$ (true) in $\mathcal{I}$, written $\mathcal{I} \models l$, if $s =_{\mathcal{I}} t$; otherwise l is evaluated to $\bot$ (false). A negative literal $l = s \not\simeq t$ is evaluated to $\top$ in $\mathcal{I}$ if $s \neq_{\mathcal{I}} t$, and to $\bot$ otherwise. This evaluation is extended to clauses and sets of clauses in the usual way. An interpretation that evaluates C to $\top$ is a *model* of C (often written $\mathcal{M}$ in this paper). A *tautological clause* (or *tautology*) is a clause of which all equational interpretations are models and a *contradiction* is a clause that has no model.

We now associate every clause C with an equivalence relation $\equiv_C$ among terms, defined as the equality relation modulo the constraint $\neg C$, i.e., the smallest congruence containing all pairs (t, s) such that $t \not\simeq s \in C$.

Definition 1. *Let C be a clause, we define for any term s the C-equivalence class of s as $[s]_C = \{t \in \mathfrak{T}(\Sigma) \mid \neg C \models s \simeq t\}$. The corresponding equivalence relation is written $\equiv_C$. The C-representative of a term s, a literal l and a clause D are respectively defined by $s_{|C} \stackrel{\text{def}}{=} \min_{\prec}([s]_C)$, $l_{|C} \stackrel{\text{def}}{=} s_{|C} \bowtie t_{|C}$, for $l = s \bowtie t$, and $D_{|C} \stackrel{\text{def}}{=} \{l_{|C} \mid l \in D\}$*

Example 2. Let $a \prec b \prec c \prec d \prec e \prec g(b) \prec g(c)$ and $C = a \not\simeq g(c) \vee b \not\simeq c \vee d \simeq e$. The clause $D = g(b) \simeq e$ is such that $D_{|C} = a \simeq e$ because $[c]_C = \{b, c\}$, $[g(b)]_C = \{a, g(b), g(c)\}$ and $[e]_C = \{e\}$.

The two following orders on literals are used throughout the paper. Both orders are extended to clauses using the multiset extension and are relaxed (into $\preceq$ and $\leq_\pi$ resp.) by also accepting equal literals or clauses.

1. The total order $\prec$ on terms is extended to literals by considering that a negative literal $t \not\simeq s$ is a set $\{\{t, s\}\}$ and that a positive literal $s \simeq t$ is $\{\{t\}, \{s\}\}$ (see [22]).
2. The total order $<_\pi$ on literals is defined as follows:
   - the equations are all greater than the inequations;
   - for l_1 and l_2 literals with the same polarity, $l_1 <_\pi l_2$ iff $l_1 \prec l_2$.

The order $\prec$ is used, as is usual, to determine which implicates are prime and which are redundant (see Definition 14). The order $<_\pi$ is useful for handling clauses as presented in Sect. 4, but is not used outside of this scope.

Example 3. Let $C = g(a) \not\simeq b \vee c \simeq d$ and $D = a \not\simeq b \vee f(c) \simeq d$, with $a \prec b \prec c \prec d \prec f(c) \prec g(a)$. We have $D \prec C$ and $C <_\pi D$, because on the one hand $a \not\simeq b \prec c \simeq d \prec f(c) \simeq d \prec g(a) \not\simeq b$, and on the other hand $a \not\simeq b <_\pi g(a) \not\simeq b <_\pi c \simeq d <_\pi f(c) \simeq d$.

In propositional logic, testing entailment amounts to a simple inclusion test [6] but things are more complex in EUF because the axioms of transitivity and substitutivity must be taken into account. For example, the clause $e \not\simeq b \vee b \not\simeq c \vee f(a) \simeq f(b)$ is a logical consequence of the clause $e \not\simeq c \vee a \simeq c$ because of these axioms. The following theorem describes the so-called *projection* method for testing entailment in a syntactic way.

Theorem 4. *Let C and D be two non-tautological clauses. The relation $D \models C$ holds iff for every negative literal l in D, the literal $l_{\downarrow C}$ is a contradiction and for every positive literal l in D, there exists a positive literal m in C such that $m_{\downarrow C \vee l^c}$ is tautological.*

Example 5. Given the order $a \prec b \prec c \prec e \prec f(a) \prec f(b)$ on terms, let $D = e \not\simeq c \vee a \simeq c$ and $C = e \not\simeq b \vee b \not\simeq c \vee f(a) \simeq f(b)$, and let $l = e \not\simeq c$ and $m = a \simeq c$ be the literals of D. We have $l_{\downarrow C} = b \not\simeq b$ because $[b]_C = \{b, c, e\}$ and $\min_{\prec}([b]_C) = b$. Moreover the literal $f(a) \simeq f(b) \in C$ is such that $(f(a) \simeq f(b))_{\downarrow C \vee m^c} = f(a) \simeq f(a)$, hence C is redundant w.r.t. D.

In order to avoid having to handle large numbers of equivalent clauses, we define a clausal normal form that is unique up to equivalence.

Definition 6. *A non-tautological clause C is in normal form if:*

1. *every negative literal l in C is such that $l_{\downarrow C \setminus l} = l$;*
2. *every literal $t \simeq s \in C$ is such that $t = t_{\downarrow C}$ and $s = s_{\downarrow C}$;*
3. *there are no two distinct positive literals l, m in C such that $m_{\downarrow l^c \vee C}$ is a tautology;*
4. *C contains no literal of the form $t \not\simeq t$;*
5. *the literals in C occur exactly once in C.*

The normal form equivalent to C is denoted by $C_{\downarrow}$.

In our previous work on prime implicate generation [9], the focus was on strictly flat clauses (i.e., that contain only constant symbols). For the sake of handling non-flat clauses, the clausal normal form (see [10], Definition 4) had to be extended. The differences lie with points 1 and 3 of Definition 6. They respectively strengthen the requirements on negative and positive literals to cover the non-flat ones.

Example 7. Using the same term ordering as in Example 5, the clause $c \not\simeq b \vee e \not\simeq b \vee f(b) \simeq f(a)$ is the normal form of the clauses $c \not\simeq b \vee e \not\simeq b \vee f(c) \simeq f(a)$, $c \not\simeq b \vee e \not\simeq b \vee f(e) \simeq f(a)$, $c \not\simeq e \vee e \not\simeq b \vee f(b) \simeq f(a)$, etc.

Theorem 8. *The normal form of a non-tautological clause C is the $\prec$-smallest clause equivalent to C.*

3   Implicate Generation

Definition 9. *A constrained clause (or c-clause) is a pair $[C \mid \mathcal{X}]$ where C is a clause and $\mathcal{X}$ is a constraint.*

Table 1. Standard Inference Rules

Superposition	$$\dfrac{[r \simeq l \vee C \,|\, \mathcal{X}] \quad [u \bowtie v \vee D \,|\, \mathcal{Y}]}{[u[l] \bowtie v \vee C \vee D \,|\, \mathcal{X} \wedge \mathcal{Y}]}$$	If $u|_p = r$, $r \succ l$, $u \succ v$, and $(r \simeq l)$ and $(u \bowtie v)$ are selected in $(r \simeq l \vee C)$ and $(u \bowtie v \vee D)$ respectively
Factoring	$$\dfrac{[t \simeq u \vee t \simeq v \vee C \,|\, \mathcal{X}]}{[t \simeq v \vee u \not\simeq v \vee C \,|\, \mathcal{X}]}$$	If $t \succ u$, $t \succ v$ and $(t \simeq u)$ is selected in $t \simeq u \vee t \simeq v \vee C$

$[C \mid \top]$ is often written simply as C and referred to as a standard clause. A constraint is *normalized*, or in *normal form*, if the clause $\neg \mathcal{X}$ is in normal form. Note that only non-contradictory constraints can be normalized. Semantically, a constrained clause $[C \mid \mathcal{X}]$ is equivalent to the standard clause $\neg \mathcal{X} \vee C$. For example the c-clause $[c \simeq b \mid f(a) \simeq c \wedge c \not\simeq d]$ is equivalent to $c \simeq b \vee f(a) \not\simeq c \vee c \simeq d$. Intuitively, the intended meaning of a c-clause $[C \mid \mathcal{X}]$ is that the clause C can be inferred provided the literals in $\mathcal{X}$ are added as axioms to the considered clause set. The usual notion of redundancy is extended to c-clauses.

Definition 10. *A c-clause $[C \mid \mathcal{X}]$ is redundant w.r.t. a set of c-clauses S if either $\mathcal{X}$ is unsatisfiable or there exist c-clauses $[D_i \mid \mathcal{Y}_i] \in S$ ($1 \leq i \leq n$) such that $\forall i \in \{1 \ldots n\}\, D_i \preceq C$ and $\mathcal{Y}_i \subseteq \mathcal{X}$, and $\mathcal{X}', D_1, \ldots, D_n \models C$, where $\mathcal{X}'$ denotes the set of literals in $\mathcal{X}$ that are $\prec$-smaller than C.*

We now present an extension of the standard superposition calculus [20] to a constrained superposition calculus referred to as $c\mathcal{SP}$, that is able to generate all prime implicates of a formula up to redundancy. This calculus is composed of the standard superposition rules extended to constrained clauses (Table 1) along with two assertion rules (Table 2). As usual the calculus is parameterized by the ordering $\succ$ on terms and by a selection function sel, where $sel(C)$ contains all maximal literals in C or at least one negative literal. A literal is *selected* in C if it occurs in $sel(C)$. We assume that the clausal part of c-clauses is systematically normalized, which explains the absence of the reflexion rule from Table 1. Note, however, that the constraint part is *not* normalized. Instead, the rules apply only if the constraint of the conclusion is already in normal form, up to the deletion of repeated literals. This strategy greatly prunes the search space, since many inferences can be dismissed. It also preserves deductive-completeness, since intuitively, one can always assume that implicates are in normal form.

Example 11. Consider the following c-clauses (with $f(a) \succ d \succ c \succ b \succ a$): $C : [f(a) \simeq b \mid d \not\simeq c]$, $D : [f(a) \simeq c \mid d \not\simeq c]$, $E : [f(a) \simeq c \mid d \not\simeq a]$. The Superposition rule applies on C and D, yielding: $[b \simeq c \mid d \not\simeq c]$ (the conjunction $d \not\simeq c \wedge d \not\simeq c$ is replaced by $d \not\simeq c$). However, the rule does not apply on C and E because the constraint $d \not\simeq c \wedge d \not\simeq a$ is not in normal form.

The principle of $c\mathcal{SP}$ is to generate the implicates of a formula as constraints of the empty clause. The standard inference rules are used to refute the clausal

Table 2. Assertion Rules

Positive Assertion	$\dfrac{[u \bowtie v \vee C \mid \mathcal{X}]}{[u[s] \bowtie v \vee C \mid \mathcal{X} \wedge t \simeq s]}$	If $u\vert_p = t,\, t \succ s,\, u \succ v$ and $(u \bowtie v)$ is selected in $(u \bowtie v \vee C)$
Negative Assertion	$\dfrac{[t \simeq s \vee C \mid \mathcal{X}]}{[u[s] \bowtie v \vee C \mid \mathcal{X} \wedge u \bowtie v]}$	If $u\vert_p = t,\, t \succ s,\, u \succ v$, and $(t \simeq s)$ is selected in $(t \simeq s \vee C)$

part of c-clauses, while the assertion rules explore the possible implicates by making hypotheses about their literals, and these are stored in the constraint part of the c-clauses. Since only the c-clauses with a refutable clausal part are of interest, the addition of new hypotheses is done only if these hypotheses render a new superposition inference possible (into the clause to which the rule applies for the Pos. Assert. rule and into the asserted literal for the Neg. Assert. rule). In other words, these rules use the fact that $S \models C$ iff $S \wedge \neg C \models \square$ to build implicates literal by literal.

Example 12. The following example shows how to derive the implicate $a \not\simeq d \vee f(c) \simeq f(b)$ from $\{a \simeq b, f(c) \simeq f(d)\}$, given the term ordering $a \prec b \prec c \prec d \prec f(a) \prec f(b) \prec f(c) \prec f(d)$.

$$
\begin{array}{lll}
1 & [f(c) \simeq f(d) \mid \top] & \text{(hyp)} \\
2 & [f(c) \simeq f(a) \mid a \simeq d] & \text{(Pos. AR, 1)} \\
3 & [f(a) \not\simeq f(b) \mid a \simeq d \wedge f(c) \not\simeq f(b)] & \text{(Neg. AR, 2)} \\
4 & [a \simeq b \mid \top] & \text{(hyp)} \\
5 & [f(a) \not\simeq f(a) \mid a \simeq d \wedge f(c) \not\simeq f(b)] & \text{(Sup. 3,4)} \\
6 & [\square \mid a \simeq d \wedge f(c) \not\simeq f(b)] & \text{(Ref. 5)}
\end{array}
$$

The negation of $a \simeq d \wedge f(c) \not\simeq f(b)$ is the desired implicate. Note for instance that the addition of the hypothesis $a \simeq d$ in Clause 1 was possible because it allowed one to replace constant d by a. The Assertion rules merge in a single rule the addition of a new hypothesis followed by a superposition inference from or into this hypothesis.

Theorem 13. *$c\mathcal{SP}$ is sound and deductive-complete, i.e., for any set of clauses S, C is a non-tautological implicate of S iff $c\mathcal{SP}$ generates from S a c-clause $[\square \mid \mathcal{X}]$ such that $\neg \mathcal{X} \models C$.*

Note that S possibly admits infinitely many prime implicates (e.g., $a \simeq b, c \simeq d \models f(a, c, t) \simeq f(b, d, t)$ for every term t). Furthermore, S is not necessarily equivalent to its set of prime implicates, for instance $\{f(a) \simeq a, f(b) \simeq b, a \not\simeq b\} \models f^n(a) \not\simeq f^n(b)$, for every $n \in \mathbb{N}$ and none of the $f^n(a) \not\simeq f^n(b)$ is prime, because $f^{n+1}(a) \not\simeq f^{n+1}(b) \models f^n(a) \not\simeq f^n(b)$ holds for any $n \in \mathbb{N}$.

4   Clause Storage and Redundancy Detection

To store the clauses generated by $c\mathcal{SP}$ and efficiently detect redundancies, a trie-like data structure, the *clausal tree*, is used. It allows one to store efficiently and concisely sets of clauses while taking into account equality axioms. Note that, since our goal is to generate all prime implicates of a formula, we only test one-to-one entailment between clauses (in contrast to the usual practice in automated deduction we cannot discard clauses that are redundant w.r.t. more than one clause since this clause may well be prime). For this reason, we define *e-subsumption*, and we assimilate it to redundancy in the rest of this article.

Definition 14. *Let C and D be two clauses. The clause C e-subsumes the clause D, written $C \leq_e D$, iff $C \models D$ and $C \preceq D$. A c-clause $[C \mid \mathcal{X}]$ e-subsumes a clause $[D \mid \mathcal{Y}]$, written $[C \mid \mathcal{X}] \leq_e [D \mid \mathcal{Y}]$) iff $C \leq_e D$ and $\mathcal{X} \subseteq \mathcal{Y}$.*

Note that both parts of the c-clauses are handled in different ways: the inclusion relation $\subseteq$ used to compare constraints is clearly stronger than the e-subsumption relation $\leq_e$ used for clauses. For instance we have (if $a \succ b \succ c$):

$$[a \not\simeq b \vee f(b) \simeq f(d) \mid \top] \leq_e [a \not\simeq c \vee b \not\simeq c \vee f(c) \simeq f(d) \mid \top], \text{ but}$$
$$[\Box \mid a \simeq b \wedge f(b) \not\simeq f(d)] \not\leq_e [\Box \mid a \simeq c \wedge b \simeq c \wedge f(c) \not\simeq f(d)].$$

Clausal trees are similar to the tries of propositional logic that are trees where the edges are labeled with literals and where some additional ordering constraints ensure the efficiency of the search algorithms. In such a tree, the represented clauses are the branches, that is the disjunction of the literals labeling the edges from root to leaf.

Definition 15. *A clausal tree is inductively defined as either $\Box$, or a set of pairs of the form (l, T') where l is a literal and T' a clausal tree. In addition, a clausal tree T with $(l, T') \in T$ must respect the following conditions:*

– for all l' appearing in T', $l' <_\pi l$,
– there is no clausal tree $T'' \neq T'$ such that $(l, T'') \in T$.

The set of clauses represented by a clausal tree T is defined inductively as follows:

$$\mathcal{C}(T) = \begin{cases} \{\Box\} & \text{if } T = \Box \\ \displaystyle\bigcup_{(l,T')\in T} \left(\bigcup_{D\in\mathcal{C}(T')} \{l \vee D\} \right) & \text{otherwise.} \end{cases}$$

As the definition implies, leaves can be either $\Box$ or $\emptyset$, but in practice if a leaf is labeled with $\emptyset$ (a failure node) then the corresponding branch is irrelevant because a tree of the form $T \cup \{(l, \emptyset)\}$ can be replaced by T without affecting the represented set. The only exception is the empty tree, in which the root is labeled with $\emptyset$. A clausal tree is *normalized* if all the clauses in $\mathcal{C}(T)$ are in normal form. In the following, we assume that all clausal trees are normalized.

Example 16. The structure T below is a clausal tree with the term order $a \prec b \prec c \prec g(c) \prec g(e) \prec f(c) \prec f(d)$.

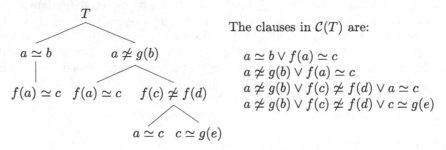

The clauses in $\mathcal{C}(T)$ are:

$$a \simeq b \vee f(a) \simeq c$$
$$a \not\simeq g(b) \vee f(a) \simeq c$$
$$a \not\simeq g(b) \vee f(c) \not\simeq f(d) \vee a \simeq c$$
$$a \not\simeq g(b) \vee f(c) \not\simeq f(d) \vee c \simeq g(e)$$

Notation 17. *Let C be a clause in normal form and T be a clausal tree such that $\forall D \in \mathcal{C}(T)$, $C \vee D$ is in normal form and $\forall l \in D$, $C <_\pi l$. In this case, $C.T$ denotes the clausal tree T' such that $\mathcal{C}(T') = \{C \vee D \mid D \in \mathcal{C}(T)\}$.*

The storage of constrained clauses is similar to that of standard clauses. A main clausal tree is used to store the clausal part of constrained clauses and at each leaf of this tree, a trie is appended to store the different constraints associated to the same clause. Note that, according to Definition 10, constraints are compared using set inclusion instead of logical entailment[1], thus the second tree must be a trie and *not* a clausal tree. In addition, all generated implicates (c-clauses with an empty clausal part) should be stored in a clausal tree in order to remove non-prime implicates.

There are three main operations on clausal trees. The first one consists in checking whether a new clause is redundant w.r.t. an existing one already stored in a clausal tree. The second one removes from a clausal tree all clauses that are redundant w.r.t. a given clause. The last one is the insertion of a new clause into a clausal tree. This last operation is straightforward and thus is not described here. On the contrary, the first two operations are not trivial and are thus carefully detailed in the remaining parts of this section. The algorithms for c-clauses are neither theoretically nor technically challenging compared to the ones for standard clauses. Thus the choice was made to present them only for standard clauses.

The algorithm ISENTAILED (Algorithm 1) tests whether a clause C is redundant w.r.t. a clause in $\mathcal{C}(T)$, where T is a clausal tree. To do so, a call is made to ISENTAILED$(T, \square, C, \square)$ and in the recursive calls to ISENTAILED(T', M, C', N), $M \vee C'$ is equal to C and N represents the path from the root of T to the subtree T'. The principle underlying these calls is to go through the input clause C and tree T while performing the operations necessary to test entailment with the projection method (Theorem 4). Note that it is here that the use of the order $<_\pi$ is crucial. Intuitively, the need for this order stems from the fact that the

---

[1] Using logical entailment makes the calculus incomplete due to the deletion of clauses whose constraint is not in normal form.

negative literals of a clause C are the ones used to project other clauses onto C. In particular for the projection of positive literals, it is necessary to know of all the negative literals that belong to C, while the reverse does not hold.

Algorithm 1. ISENTAILED(T, M, C, N)

Require: T is a clausal tree in normal form, $M \vee C$ and N are clauses in normal form, M is negative and $N \models M \vee C$

Ensure: ISENTAILED(T, M, C, N) = $\top$ iff $\exists D \in \mathcal{C}(T), D \vee N \leq_e M \vee C$

1: **if** $T = \square$ **then return** $N \preceq M \vee C$
2: $T_1 \leftarrow \{(l, T') \in T \mid l_{\downarrow M}$ is a contradiction$\}$
3: **if** $\bigvee_{(l,T') \in T_1}$ ISENTAILED($T', M, C, N \vee l$) **then return** $\top$
4: **if** $C = \square$ **then return** $\bot$
5: $m_1 \leftarrow \min_{<_\pi} \{m \in C\}$
6: **if** m_1 is of the form $u \not\approx v$, with $u \succ v$ **then**
7: $\quad T_2 \leftarrow \{(l, T') \in T \mid l_{\downarrow M} \not\prec_\pi m_1$ and $\nexists w, (l_{\downarrow M} = u \not\approx w,$ with $u \succ w)\}$
8: $\quad$**return** $\bigvee_{(l,T') \in T_2}$ ISENTAILED($l.T', M \vee m_1, C \setminus m_1, N$)
9: **else**
10: $\quad T_3 \leftarrow \{(l, T') \in T \mid C_{\downarrow M \vee l^c}$ contains a tautological literal$\}$
11: $\quad$**return** $\bigvee_{(l,T') \in T_3}$ ISENTAILED($T', M, C, N \vee l$)

Theorem 18. *If T is a clausal tree in normal form, $M \vee C$ and N are clauses in normal form, M is negative and $N \models M \vee C$ then the call* ISENTAILED(T, M, C, N) *terminates and* ISENTAILED(T, M, C, N) = $\top$ *iff $\exists D \in \mathcal{C}(T), D \vee N \leq_e M \vee C$.*

The algorithm PRUNEENTAILED (Algorithm 2) removes from the input tree T all the clauses redundant w.r.t. the input clause C. It proceeds by going through both objects, performing projections and storing the already considered literals in parameters N and M. Once an entailment is established in this way, all that remains is to compare the selected clauses using the order $\prec$ to detect redundancies. This last part is done by the algorithm PRUNEINF (Algorithm 3).

Proposition 19. *Let C and N be clauses in normal form and T be a clausal tree in normal form verifying the preconditions of PRUNEINF. The output tree $T_{out} =$ PRUNEINF(T, C, N) is such that $\mathcal{C}(T_{out}) = \{D_T \in \mathcal{C}(T) \mid C \not\leq_e D_T \vee N\}$.*

Theorem 20. *Let $C \vee M$ and N be clauses in normal form and T be a clausal tree in normal form verifying the preconditions of PRUNEENTAILED. Then the calls PRUNEENTAILED(T, M, C, N) and PRUNEINF(T, C, N) always terminate and $T_{out} =$ PRUNEENTAILED(T, M, C, N) is such that $\mathcal{C}(T_{out}) = \{D \in \mathcal{C}(T) \mid C \vee M \not\leq_e D \vee N\}$.*

Algorithm 2. PRUNEENTAILED(T, M, C, N)

Require: T is a clausal-tree in normal form, $M \vee C$ and N are clauses in normal form,
$M \models N$ and ISENTAILED$(N.T, \Box, C \vee M, \Box) = \bot$.
Ensure: $\mathcal{C}(T_{out}) = \{D \in \mathcal{C}(T) \mid C \vee M \not\preceq_e D \vee N\}$, with $T_{out} =$ PRUNEENTAILED
(T, M, C, N).
1: **if** $C = \Box$ **then return** PRUNEINF(T, M, N)
2: select $m_1 \in C$ such that $m_1{}_{\downarrow N} = \min_{<_\pi} \{m_{\downarrow N} \mid m \in C\}$
3: **if** $m_1{}_{\downarrow N}$ is a contradiction **then**
4:     **return** PRUNEENTAILED$(T, M \vee m_1, C \setminus m_1, N)$
5: **if** $T = \Box$ **then return** T
6: $T_1 \leftarrow \{(l, T') \in T \mid l = u \not\approx v \wedge m_1{}_{\downarrow N} \succeq l\}$
7: $T_{out1} \leftarrow \{(l, \text{PRUNEENTAILED}(T', M, C, N \vee l) \mid$
        $[3]$ $(l, T') \in T_1 \wedge \text{PRUNEENTAILED}(T', M, C, N \vee l) \neq \emptyset\}$
8: **if** m_1 is positive **then**
9:     $T_2 \leftarrow T \setminus T_1$
10:    $T_{out2} \leftarrow \{(l, \text{PRUNEENTAILED}(T', M \vee L_l, C \setminus L_l, N \vee l)) \mid$
        $[4]$ $(l, T') \in T_2 \wedge L_l = \{m \in C \mid l_{\downarrow N \vee m}$ is tautological$\} \wedge$
        $[4]$ $\text{PRUNEENTAILED}(T', M \vee L_l, C \setminus L_l, N \vee l) \neq \emptyset\}$
11:    **return** $T_{out1} \cup T_{out2}$
12: **else**
13:    **return** $T_{out1} \cup T \setminus T_1$

Algorithm 3. PRUNEINF(T, C, N)

Require: T is a clausal-tree in normal form, C in a clause in normal form, N is a
clause in normal form, $C \models N$.
Ensure: $\mathcal{C}(T_{out}) = \{D \in \mathcal{C}(T) \mid C \not\preceq_e D \vee N\}$, with $T_{out} = \text{PRUNEINF}(T, C, N)$.
1: **if** $T = \Box$ and $C \not\preceq N$ **then return** T
2: **if** $C \not\preceq N$ **then**
3:     **return** $\{(l, \text{PRUNEINF}(T', C, N \vee l)) \mid (l, T') \in T \wedge$
        $[4]$ $\text{PRUNEINF}(T', C, N \vee l) \neq \emptyset\}$
4: **return** $\emptyset$

Remark 21. The main difference with the flat version of the algorithms [9] is the handling of clausal tree branches labeled with positive literals, which had to be adapted to the redefined projection method. Also the case in which no redundancy is detected (resp. lines 4&5 of Algorithm 1 & 2) has to be postponed. To ensure the correctness of the algorithms some recursive cases must now be checked first.

5   Experimental Results

We have developed a prototype for generating EUF prime implicates. It uses the Logtk library [5] at its core for term manipulation, and for parsing TPTP inputs [25]. The $c\mathcal{SP}$ rules and the clausal tree operations are built into a Given-Clause loop [23] (in the *Otter* variant). To ensure termination, it is necessary to impose additional conditions on the generated implicates. Indeed, the set of implicates is infinite in general, e.g., $a \not\simeq b, f(a) \simeq a, f(b) \simeq b \models f^n(a) \not\simeq f^n(b)$. In the experiments we only computed implicates built on the set of ground terms occurring in the initial formula. We tested the tool on two sets of problems: a collection of randomly generated formulæ of small size and a set of benchmarks from the SMT-LIB library [3]. All the tests were conducted on a machine equipped with an Intel core i5-3470 CPU and 4×2 GB of RAM.

The first experiment presented is a comparison of different prime implicate generation systems on a set of randomly generated formulæ with a timeout of 5 min. The selected systems are: Zres[2], a prime implicate generation tool for propositional logic [24], SOLAR[3], a prime implicate generation tool for first-order logic [19] which can handle equational formulæ through the use of modification methods [11], cSP, the prime implicate generator prototype for ground equational logic described in this paper, and flat-cSP, the former version of the cSP prototype, that only handles flat clauses. To the best of our knowledge only two other prime implicate generation tools are currently available. One is ritrie [17], a tool that generates propositional prime implicates. This tool was outperformed by Zres in past experiments and we chose not to include it in this set of experiments. The second is the Mistral SMT solver [8] that cannot be compared with the other tools because its prime implicate generation is not complete. More generally the approach in [8] applies to any theory admitting quantifier-elimination but this property does not hold for the logic we consider in the present paper since the elimination of function symbols would require to handle second-order quantification. The input problems were flattened (see e.g. [4] for a definition) for flat-cSP and Zres, and the substitutivity axiom instantiated when necessary. Furthermore, for Zres, these flat equational problems were also converted to propositional ones, by instantiating the transitivity of equality when necessary. In order to perform meaningful comparisons, SOLAR has been parameterized to generate only implicates built on the considered ground terms. Note that Zres generates propositional implicates which can always be translated back into equational clauses built on these terms. The test set consists of randomly generated formulæ of 2 to 4 clauses containing 1 to 3 literals each, with terms of depth between 0 and 2, based on signatures of either 3 or 6 symbols of arity 0 or 1 and 2 constants. Six formulæ are generated in each case, for a total of 144 benchmarks. Although the resulting formulæ are rather small, some of them are complex enough that they timeout on all systems and some produce tens of thousands of implicates, generated after millions of inferences.

The results are summarized in Table 3. Each line corresponds to a system. The column labeled 'successes' indicates the percentage of tests that were completed before the 5 min timeout. The three columns under the label 'SOLAR successes' summarize average results on those tests on which SOLAR terminated before the timeout. The other columns contain results on tests on which Zres terminated but not SOLAR, and on which flat-cSP terminated but not Zres and SOLAR. Finally, the 'timeout' columns expose the mean results on all interrupted

[2] Many thanks to Prof. L. Simon for providing the executable file.

[3] Many thanks to Prof. H. Nabeshima for providing the executable file.

Table 3. Randomly generated formulæ - test results summary

	successes	SOLAR successes			Zres successes			(flat-)cSP			timeouts	
		time(s)	inf	PIs	time(s)	inf	PIs	time(s)	inf	PIs	inf	PIs?
SOLAR	15 %	11.842	663190	506	-	-	-	-	-	-	2452908	28152
Zres	52 %	0.695	X	2986	12.474	X	13804	-	-	-	X	X
flat-cSP	63 %	6.622	5157	74	2.334	3300	158	14.290	11005	348	68959	X
cSP	76 %	0.042	110	21	3.436	1322	47	10.193	1834	79	14714	538

tests. Columns labeled 'time', 'inf.' and 'PIs' respectively give the mean execution time, mean number of inferences and mean number of prime implicates found for each set of tests. The last column is labeled 'PIs?' because due to the timeout, the implicates found are not guaranteed to be prime. Cells labeled with an 'X' indicate that the corresponding data is not accessible.

As shown in the 'successes' column, cSP is the obvious winner in terms of the number of tests handled before timeout. It should also be mentioned that cSP solves all the problems that other systems solve, except for two that are solved only by Zres. The 15 % of problems solved by SOLAR are the simplest of the random formulæ. The results show that SOLAR's approach is very costly both in terms of time and space, although methods to reduce these costs are being investigated[4]. The high number of prime implicates this tool generates compared to those produced by cSP may seem surprising. In fact, SOLAR returns an over-approximation of the result because it does not take into account the equality axioms in its redundancy detection. Thus for example, any literal $t \simeq s$ also appears as $s \simeq t$ and $f(s) \simeq f(t)$ is not detected as redundant w.r.t. $s \simeq t$. Comparatively, the huge number of prime implicates generated by Zres is not surprising at all. It stems directly from the propositional translation of the initial problems and the introduction of new propositional variables. Although Zres is faster than cSP on the problems they both solve, it solves only 52 % of the problems, while cSP solves 76 % of them. The results in the '(flat-)cSP' column are globally higher than those in the 'Zres successes' columns, because the most difficult benchmarks are solved only by cSP and to a lesser extend by flat-cSP. Since cSP solves more problems than flat-cSP and does so faster and with fewer clauses processed, cSP is clearly better adapted to dealing with originally non-flat formulæ. The number of inferences and generated non-redundant implicates when the tool times out illustrate the heavy cost of the cSP inferences and redundancy detection mechanism compared to that of SOLAR. It is a price that seems partly unavoidable to eliminate all redundancies, since this requires complex algorithms.

The second experiment presented uses benchmarks from the QF_AX logic of SMT-LIB [3]. They are synthetic benchmarks that model some properties in the SMT theory of arrays with extensionality, namely: some swappings of elements between cells of an array are commutative (swap benchmarks); and swapping elements between identical cells of equal arrays generate equal arrays (storeinv

---

[4] Personal communication of Prof. Nabeshima.

benchmarks). The benchmarks labeled with `invalid` have been tweaked to falsify the property.

Given that $c\mathcal{SP}$ cannot handle smt-lib inputs or the theory of arrays, we preprocessed the benchmarks by first converting them to TPTP using the `SMTtoTPTP` tool[5] and then applying the method described in [4] to generate equisatisfiable problems free of the axioms of the theory of arrays with extensionality. As shown in [1], these problems can be nontrivial to solve even for state-of-the-art theorem provers like **E** [23] and one cannot expect that the entire set of prime implicates can be generated in reasonable time. We use them mainly to evaluate the impact of our redundancy-pruning technique on the number of superposition inferences carried out by blocking the Assertion rules inferences, allowing the comparison of `cSP` with the **E** theorem prover. The main differences between the methods are the normalization of clauses and the redundancy pruning mechanism. On the one hand, the redundancy pruning algorithm used by `cSP` is weaker because it does not allow for equational simplification or other n-to-one redundancy pruning rules. On the other hand one-to-one redundancy testing is stronger since its uses logical entailment instead of subsumption. The comparison of `cSP` with the **E** theorem prover on these benchmarks shows that the normalization approach can, in some nontrivial cases, reduce the number of processed clauses by an order of magnitude.

Figure 1 presents the most notable results of this experiment, that is the results of the `swap` benchmarks. Among these, only the benchmarks on which both **E** and `cSP` (without Assertion rules) terminate before timeout and without memory overflow were kept, i.e. 76 out of 146. Squares represent the `invalid` benchmarks, i.e. the satisfiable formulæ, while crosses mark the unsatisfiable ones. An interesting observation is that for the largest invalid benchmarks, `cSP` needs to process a smaller number of clauses than **E** before terminating, even 10

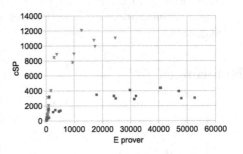

Fig. 1. `swap` benchmarks - comparison of the number of processed clauses for **E** and `cSP`.

times less in the case of the `invalid_swap` benchmarks. The unsatisfiable `swap` benchmarks were run with a timeout of 10 min (the triangles in Fig. 1) and the corresponding results hint that this phenomenon could also be true for larger unsatisfiable problems. This suggests that the redundancy pruning technique based on normalization and clausal trees could be profitably integrated into state-of-the-art superposition-based theorem-provers, at least for ground equational clause sets. However, it might not always be useful, for example the 10 out of 19 `storeinv` benchmarks that do not fail show the opposite tendency.

---

[5] http://users.cecs.anu.edu.au/~baumgart/systems/smttotptp/.

6   Conclusion

In this article, a novel approach for the generation of prime implicates in ground equational logic is presented. It is proved sound and complete and experiments are conducted to compare the approach to state-of-the-art tools. These show that cSP outperforms all other prime implicate generation systems on simple formulæ and can even tackle more involved problems than others, although none of the methods scale well. We also evaluate the impact of the normalization and pruning techniques of cSP compared to the redundancy detection of the E theorem prover. A potential improvement of redundancy detection using these techniques is highlighted. From a practical point of view, the implementation of the cSP prototype leaves rooms for many improvements, for instance a better selection strategy could be used. Note also that tries or clausal trees can be represented as directed acyclic graphs (where identical subtrees are shared) in order to merge suffixes as well as prefixes of clauses. A more drastic evolution would be to integrate cSP to an existing theorem prover to take advantage of its built-in optimizations. On the theoretical side, the $c\mathcal{SP}$ calculus can easily be extended, at least to handle variables, but the extensibility of the redundancy detection method has not been investigated yet. Well-known theoretical limitations may threaten such an extension since the entailment relation in full first-order logic is not decidable [21]. The frontier of what can and cannot be done on the generation of prime implicates in first-order logic is not yet clear and needs further investigations.

References

1. Armando, A., Bonacina, M.P., Ranise, S., Schulz, S.: New results on rewrite-based satisfiability procedures. ACM Trans. Comput. Log. **10**(1), 1–51 (2009)
2. Baader, F., Nipkow, T.: Term Rewriting and All That. Cambridge University Press, Cambridge (1998)
3. Barrett, C., Stump, A., Tinelli, C.: The SMT-LIB standard: Version 2.0. Technical report, Department of Computer Science, The University of Iowa (2010). www.SMT-LIB.org
4. Bonacina, M.P., Echenim, M.: Theory decision by decomposition. J. Symb. Comput. **45**(2), 229–260 (2010)
5. Cruanes, S.:. Logtk: A logic ToolKit for automated reasoning and its implementation. In: 4th Workshop on Practical Aspects of Automated Reasoning (2014)
6. De Kleer, J.: An improved incremental algorithm for generating prime implicates. In: Proceedings of the National Conference on Artificial Intelligence, p. 780. Wiley (1992)
7. Dershowitz, N.: Orderings for term-rewriting systems. In: Proceedings of the 20th Annual Symposium on Foundations of Computer Science, pp. 123–131. IEEE Computer Society, Washington (1979)
8. Dillig, I., Dillig, T., McMillan, K.L., Aiken, A.: Minimum satisfying assignments for SMT. In: Madhusudan, P., Seshia, S.A. (eds.) CAV 2012. LNCS, vol. 7358, pp. 394–409. Springer, Heidelberg (2012)

9. Echenim, M., Peltier, N., Tourret, S.: An approach to abductive reasoning in equational logic. In: Rossi, F. (ed.) IJCAI 2013 - International Joint Conference on Artificial Intelligence, pp. 531–537. AAAI Press, Beijing, August 2013

10. Echenim, M., Peltier, N., Tourret, S.: A rewriting strategy to generate prime implicates in equational logic. In: Demri, S., Kapur, D., Weidenbach, C. (eds.) IJCAR 2014. LNCS, vol. 8562, pp. 137–151. Springer, Heidelberg (2014)

11. Iwanuma, K., Nabeshima, H., Inoue.: Toward an efficient equality computation in connection tableaux: A modification method without symmetry transformation—a preliminary report . First-Order Theorem Proving, p. 19 (2009)

12. Jackson, P.: Computing prime implicates incrementally. In: Kapur, D. (ed.) CADE 1992. LNCS, vol. 607. Springer, Heidelberg (1992)

13. Kean, A., Tsiknis, G.: An incremental method for generating prime implicants/implicates. J. Symb. Comput. **9**(2), 185–206 (1990)

14. Knill, E., Cox, P.T., Pietrzykowski, T.: Equality and abductive residua for horn clauses. Theoret. Comput. Sci. **120**(1), 1–44 (1993)

15. Marquis, P.: Extending abduction from propositional to first-order logic. In: Jorrand, P., Kelemen, J. (eds.) FAIR 1991. LNCS, vol. 535. Springer, Heidelberg (1991)

16. Matusiewicz, A., Murray, N.V., Rosenthal, E.: Prime implicate tries. In: Giese, M., Waaler, A. (eds.) TABLEAUX 2009. LNCS, vol. 5607, pp. 250–264. Springer, Heidelberg (2009)

17. Matusiewicz, A., Murray, N.V., Rosenthal, E.: Tri-based set operations and selective computation of prime implicates. In: Kryszkiewicz, M., Rybinski, H., Skowron, A., Raś, Z.W. (eds.) ISMIS 2011. LNCS, vol. 6804, pp. 203–213. Springer, Heidelberg (2011)

18. Mayer, M.C., Pirri, F.: First order abduction via tableau and sequent calculi. Log. J. IGPL **1**(1), 99–117 (1993)

19. Nabeshima, H., Iwanuma, K., Inoue, K., Ray, O.: SOLAR: an automated deduction system for consequence finding. AI Commun. **23**(2), 183–203 (2010)

20. Nieuwenhuis, R., Rubio, A.: Paramodulation-based theorem proving. In: Robinson, A., Voronkov, A. (eds.) Handbook of Automated Reasoning, pp. 371–443. North Holland, Amsterdam (2001)

21. Schmidt-Schauss, M.: Implication of clauses is undecidable. Theor. Comput. Sci. **59**, 287–296 (1988)

22. Schulz, S.: E - a brainiac theorem prover. AI Commun. **15**(2–3), 111–126 (2002)

23. Schulz, S.: System Description: E 1.8. In: McMillan, K., Middeldorp, A., Voronkov, A. (eds.) LPAR-19 2013. LNCS, vol. 8312, pp. 735–743. Springer, Heidelberg (2013)

24. Simon, L., Del Val A.: Efficient consequence finding. In: International Joint Conference on Artificial Intelligence, vol. 17, pp. 359–370. Lawrence Erlbaum Associates ltd (2001)

25. Sutcliffe, G.: The TPTP problem library and associated infrastructure: The FOF and CNF parts, v3.5.0. J. Autom. Reason. **43**(4), 337–362 (2009)

26. Tison, P.: Generalization of consensus theory and application to the minimization of boolean functions. IEEE Trans. Electron. Comput. **EC–16**(4), 446–456 (1967)

Quantomatic: A Proof Assistant for Diagrammatic Reasoning

Aleks Kissinger and Vladimir Zamdzhiev[✉]

University of Oxford, Oxford, UK
{aleks.kissinger,vladimir.zamdzhiev}@cs.ox.ac.uk

Abstract. Monoidal algebraic structures consist of operations that can have multiple outputs as well as multiple inputs, which have applications in many areas including categorical algebra, programming language semantics, representation theory, algebraic quantum information, and quantum groups. String diagrams provide a convenient graphical syntax for reasoning formally about such structures, while avoiding many of the technical challenges of a term-based approach. Quantomatic is a tool that supports the (semi-)automatic construction of equational proofs using string diagrams. We briefly outline the theoretical basis of Quantomatic's rewriting engine, then give an overview of the core features and architecture and give a simple example project that computes normal forms for commutative bialgebras.

1 Introduction

Quantomatic is a graphical proof assistant. Rather than using terms as the primitive objects in proofs, it uses *string diagrams*. String diagrams provide a simple way of expressing collections of maps or processes that have been plugged together. They consist of boxes representing the processes, and (typed) wires connecting them. Wires are allowed to be open (i.e. not connected to a box) at one or both ends, giving a notion of *input* and *output* for a string diagram (Fig. 1).

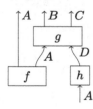

Fig. 1. A string diagram

String diagram notation was first used by Penrose [24] as an alternative to tensor notation for applications in theoretical physics. In 1991, Joyal and Street showed that string diagrams were actually much more general [15], serving to not just represent tensors, but morphisms in any monoidal category. In other words, it is possible to reason about any collection of processes or maps that has well-behaved parallel and sequential composition operations (usually written ⊗ and ∘, respectively) using string diagrams. This includes familiar examples such as functions (where ⊗ := × is just the Cartesian product), and other non-Cartesian examples such as multi-linear maps or matrices over a semi-ring (where ⊗ is a genuine tensor product).

Recently, there has been much interest in diagrammatic theories in a wide variety of areas such as Petri nets [27], programming language semantics [21],

A.P. Felty and A. Middeldorp (Eds.): CADE-25, LNAI 9195, pp. 326–336, 2015.
DOI: 10.1007/978-3-319-21401-6_22

natural language processing [5], systems biology [7], control theory [2,4], program parallelisation [23], and in interactive theorem proving [12]. It has also played a major role in categorical quantum mechanics [1]. In particular, the string diagram-based *ZX-calculus* [6] has found numerous applications within quantum computing (see e.g. [9,13]). String diagrammatic reasoning has also produced results which had been previously unknown [10,11].

The current version of Quantomatic supports the construction of derivations, which are transitive chains of diagram rewrites, as well as simple mechanisms for automated simplification of diagrams and lemma/theorem export and re-use. The theoretical foundations of Quantomatic have been described in previous papers [8,18,22] and this paper is the first system description of the Quantomatic software itself. After introducing the main concepts of diagrammatic reasoning in Sect. 2, we describe how Quantomatic builds derivations and how those derivations can be included in papers or shared in the web in Sect. 3. We show how to implement simplification strategies using a simple combinator language in Sect. 4 and describe an example project involving bialgebras in Sect. 5. Then, we give an overview of the architecture of the system in Sect. 6, and show how it can be extended with new graphical theories. We give details on obtaining Quantomatic and discuss related and future work in Sect. 7.

2   Diagrammatic Reasoning

String diagram rewriting can be seen as a generalisation of (linear) term rewriting.[1] We can see how this works via a simple example. A commutative monoid is a set A, along with a binary operation $(-\cdot-)$ and a constant $e \in A$ such that:

$$(a \cdot b) \cdot c = a \cdot (b \cdot c) \qquad a \cdot e = a = e \cdot a \qquad a \cdot b = b \cdot a \qquad (1)$$

Naturally, we can treat these equations as term rewrite rules, with free variables a, b, c. To apply a rule, we instantiate the free variables, then use it to replace a sub-term. For example, the assignment $\{a := x, b := y \cdot e, c := z\}$ in the first rule yields $(x \cdot (y \cdot e)) \cdot z = x \cdot ((y \cdot e) \cdot z)$, which could be applied in, e.g.:

$$w \cdot ((x \cdot (y \cdot e)) \cdot z) = w \cdot (x \cdot ((y \cdot e) \cdot z)) \qquad (2)$$

We could express the same thing by rewriting string diagrams, which in this case are just trees. Representing $\cdot$ as a node with two inputs and one output and e as a node with just one output, the Equations (1) become:

$$(3)$$

In fact, the variable names on the inputs are no longer necessary. The role of the variables is played by the fact that the LHS and the RHS share a common

---

[1] Non-linear term rewriting can be encoded by introducing special 'copy' and 'delete' nodes which obey certain naturality conditions. However, when $\otimes \neq \times$, these don't exist in general.

boundary. That is, there is a 1-to-1 correspondence between inputs/outputs on the LHS and those on the RHS. The substitution (2) can then be depicted simply as cutting out the LHS of this rule and gluing in the RHS, using the shared boundary:

$$(4)$$

The benefit of this approach is that it treats inputs and outputs symmetrically. For instance, we can define a *cocommutative comonoid* by simply flipping the generators and equations upside-down:

Many interesting and useful structures arise by letting algebraic structures like monoids interact with their 'coalgebraic' counterparts. For example, a *commutative bialgebra* consists of a commutative monoid, a cocommutative comonoid, and three rules governing their interaction:

$$(5)$$

Rewriting with general diagrams proceeds just like the tree rewriting above:

This process of cutting out the LHS and gluing in the RHS along a shared boundary is called *double-pushout (DPO) graph rewriting*. The precise formulation of DPO rewriting for string diagrams is provided in [8].

From hence forth, we will assume all nodes are commutative and cocommutative, or in other words, nodes are invariant to permutations of their inputs and outputs. A current limitation of Quantomatic is that it does not maintain an ordering on inputs/outputs for individual nodes, so this is true by default. A semantics for diagrams with non-commutative nodes is detailed in [19], but is not yet implemented (see Sect. 7).

One of the unique aspects of Quantomatic is that it supports a graphical pattern syntax called *!-box notation* for expressing infinite families of rules, typically involving variable-arity generators. For example, we could alternatively define commutative monoids using n-ary multiplication operations, subject to the rules that adjacent multiplications merge and the '1-ary multiplication' does nothing:

$$(6)$$

One could recursively define these n-ary multiplications as (e.g. left-associated) trees of binary multiplications, where a '0-ary multiplication' is just the unit. Then, by associativity, commutativity, and unit laws, any two trees with the same number of inputs will be equal, from which the two equations above follow.

To represent repetition, we can enclose certain parts of the diagram in !-boxes, which indicate that the marked sub-diagram (along with any wires in or out) can be duplicated any number of times. Replacing the ellipses with !-boxes in (6) yields:

$$\tag{7}$$

An instance of this rule effectively amounts to fixing the number times the contents of b and c are repeated. In order to ensure that all instances are valid string diagram rules (i.e. they share a common boundary), !-box rules must satisfy two well-formedness conditions: (i) the !-boxes on both sides are in bijective correspondance indicated by their labels, and (ii) an input (resp. output) is in a !-box b on the LHS if and only if it is also in b on the RHS, where pairs of inputs or outputs are again indicated by their labels. !-boxes can also be nested in each other, which adds one additional condition, but for simplicity we will ignore this case. More details on !-boxes and their formal semantics can be found in [18].

3 Constructing Proofs in Quantomatic

Quantomatic allows a user to define a set of diagram equations and use them to prove theorems by means of *derivations*. A derivation is simply a transitive chain of rewrite steps, using axioms or other theorems within the project. To begin working in Quantomatic, the user creates a project based on a *graphical theory*, which defines the kinds of admissible nodes in a diagram and how they should be presented to the user (see Sect. 6). At this point, they can define some axioms, i.e. diagram equations (possibly containing !-boxes) subject to the well-formedness conditions listed at the end of Sect. 2.

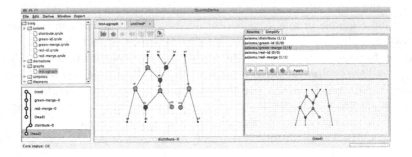

Fig. 2. Derivation editor in Quantomatic

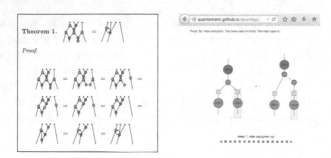

Fig. 3. LaTeX and interactive HTML5 output from Quantomatic

After a set of axioms is defined, they can be used in a derivation. First, the user creates a new graph using the graph editor and chooses to start a new derivation from the menu. The user is then presented with the derivation editor, which is used to explore the derivation history or extend it by applying rewrite rules. The history view on the left shows a chain of proof steps. The history can also be branched off at any step, allowing the user to explore multiple (possibly failed) rewriting paths on the way to producing a proof.

The nodes in this tree are organised into two categories: *proof steps* and *proof heads*. The former represent the application of a rewrite rule. With a proof step selected, the user sees the before and after graphs side-by-side, with the changed portion highlighted. The user can grow the derivation from a proof head. Here, they see the current graph next to a series of controls (as in Fig. 2). If the 'Rewrite' panel is active, Quantomatic will eagerly look for matches of any active rewrite rules on the selected part of the graph on the left. This search is done in parallel, which is especially effective on multi-core machines at providing the desired rule application as soon as possible. Applying a rule will generate a new proof step and advance the proof head. The 'Simplify' panel gives the user access to simplification procedures (see Sect. 4), which will automatically produce proof steps until either the procedure terminates or is interrupted by the user. Once a derivation is complete, it can be exported as a new theorem, which is linked to the derivation and can be used in other derivations.

One of the major advantages of diagrammatic reasoning is that it can produce nice, human-readable proofs. Proofs produced by Quantomatic can be shared in two ways. Graphs, rules, and derivations can be exported as LaTeX and \input directly in to papers (Fig. 3, left). The graphs are rendered using the PGF/TikZ package, and are compatible with graphical editor TikZiT, in case further manual tweaking is required. It is also possible to embed graphs, rules, and derivations from a Quantomatic project in HTML5 using `Quanto.js`. After linking to a Quantomatic project with a `<meta>` tag, this script will substitute specially marked-up `<div>` tags for interactive graphical views of proofs, rendered using `d3.js` (Fig. 3, right).

4   Simplification Procedures

Quantomatic allows for custom simplification procedures (simprocs). These are functions implemented in Poly/ML which send a graph to a lazy sequence of proof steps, which contain the name of the axiom/theorem used, the instantiated rewrite rule, and the rewritten graph. Simprocs are then registered with the Quantomatic GUI by calling `register_simproc`. When a simproc is invoked in the derivation editor, it is passed the current graph, and proof steps are pulled one at a time until either the sequence is exhausted or the user cancels simplification. To construct simprocs, Quantomatic provides a combinator language:

$$
\begin{array}{rcl}
\texttt{++} & \texttt{::} & \texttt{simproc * simproc -> simproc} \\
\texttt{LOOP} & \texttt{::} & \texttt{simproc -> simproc} \\
\texttt{REDUCE} & \texttt{::} & \texttt{rule -> simproc} \\
\texttt{REDUCE\_ALL} & \texttt{::} & \texttt{ruleset -> simproc} \\
\texttt{REDUCE\_WHILE} & \texttt{::} & \texttt{(graph -> bool) -> rule -> simproc} \\
\texttt{type metric} & \texttt{:=} & \texttt{graph -> int} \\
\texttt{REDUCE\_METRIC\_TO} & \texttt{::} & \texttt{int -> metric -> simproc}
\end{array}
$$

The combinator `++` will chain the last graph produced by the first simproc into the second simproc. `LOOP` will repeatedly chain a simproc into itself, until the simproc produces no new proof steps. `REDUCE` will repeatedly apply the first matching of the given rule, and `REDUCE_ALL` does the same, but takes a set of rules. `REDUCE_WHILE` will keep reducing as long as the graph satisfies the given precondition. `REDUCE_METRIC_TO` is useful for

Fig. 4. A simproc in Quantomatic

using non-terminating rules in strategies. It takes an integer k and a function m. It will then repeatedly apply the given rule to a graph g as long as $m(g) > k$ and $m(g)$ is reduced by the rule application.

For terminating, confluent rewrite systems, a single call to `REDUCE_ALL` will usually suffice. However, strategies are very useful for more ill-behaved systems. For example, Fig. 4 shows a simproc that computes pseudo-normal forms for the theory of interacting bialgebras described in [3], which currently has no known convergent completion.

5   Example Project: Bialgebras

As mentioned in Sect. 2, a bialgebra consists of a monoid and a comonoid, subject to three extra equations (5). There is also a more efficient way to define commutative bialgebras, following the n-ary presentation of monoids described in Sect. 2. A commutative bialgebra can be presented in terms of an n-ary multiplication and n-ary comultiplication, subject to the monoid 'tree-merge' rules in (7), as well as the comonoid versions:

$$\text{(8)}$$

and one additional rule. Whenever an n-ary multiplication meets an m-ary comultiplication, the two nodes can be replaced by a complete bipartite graph:

$$\text{(9)}$$

The 5 equations depicted in (7), (8), and (9) can be added to a Quantomatic project. Since they are strongly normalising, the following naïve strategy will compute normal forms:

```
val simps = load_ruleset ["axioms/red-merge","axioms/red-id",
  "axioms/green-merge","axioms/green-id","axioms/distribute"];
register_simproc ("basic_simp", REDUCE_ALL simps);
```

This bialgebra example is a small fragment of the ZX-calculus, which has about 20 basic rules and necessitates non-naïve simplification strategies. The bialgebra example and the ZX-calculus are available on `quantomatic.github.io`.

6 Architecture

Quantomatic consists of two components (Fig. 5): a reasoning engine written in Poly/ML called QuantoCore, and a GUI front-end written in Scala called QuantoDerive. QuantoCore handles matching and rewriting of diagrams, and can be extended via graphical theories. The GUI communicates to the core via a JSON protocol, which spawns independent workers to handle individual matching and rewriting requests. This allows the eager, parallel matching described in Sect. 3. The GUI also communicates directly to Poly/ML using its built-in IDE protocol to register new simprocs written in ML. The core itself can be run in stand-alone mode or within Isabelle/ML. It forms the basis of two other

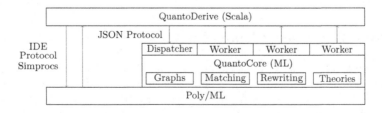

Fig. 5. Architecture of Quantomatic

graph-rewriting projects: QuantoCoSy [17], which generates new graphical theories using conjecture synthesis (cf. [14]), and Tinker [12], which implements a graphical proof strategy language for Isabelle and ProofPower.

Quantomatic is very flexible in terms of the data it can hold on nodes and edges. This can be something as simple as an enumerated type (e.g. a colour), standard types like strings and integers, or more complicated data like linear polynomials, lambda terms, or even full-blown programs. The specification of this data, along with how it should be unified during matching and displayed to the user, is called a *graphical theory*. A graphical theory consists of two parts: a .qtheory file loaded into the GUI and an ML structure loaded into the core. The .qtheory is a JSON file used to register a new theory with the GUI, and provides basic information such as how node/edge data should be displayed to the user.

The ML structure provides four types, which Quantomatic treats as black boxes: nvdata, edata, psubst, and subst. The first two contain node data and edge data, respectively. The third type is for *partial substitutions*, which are used to accumulate state during the course of matching one diagram against another. The fourth type is for *substitutions*, which are partial substitutions that have been completed, or 'solved', after matching is done. Quantomatic accesses these types using several hooks implemented in by theory:

```
     match_nvdata :: nvdata * nvdata -> psubst -> psubst option
      match_edata :: edata * edata -> psubst -> psubst option
     solve_psubst :: psubst -> [subst]
  subst_in_nvdata :: subst -> nvdata -> nvdata
   subst_in_edata :: subst -> edata -> edata
```

The first two hooks are called every time a new node or edge is matched by the graph rewriting engine. The first argument is a pair consisting of the data on the pattern node (resp. edge) and the data on the target node (resp. edge). If the data matches successfully, any updates such as variable instantiations or new constraints are added to the psubst. If it fails (e.g. by introducing unsatisfiable constraints), the function returns NONE. Once matching is done, solve_psubst is invoked to turn the accumulated constraints into an actual instantiation of node/edge data. Since we don't require node/edge data to have most-general unifiers, this is allowed to (lazily) return multiple solutions in general. The final two hooks are used to perform the instantiation of node/edge data on a rewrite rule. Once the theory provides these and a couple of other routine functions (e.g. for (de)serialising data), QuantoCore handles the rest.

7   Availability, Related, and Future Work

Quantomatic is Free and Open Source Software, licensed under GPLv3. The project is hosted by Github, and source code and binaries for GNU/Linux, Mac OSX, and Windows are available from: quantomatic.github.io. Example projects from Sect. 5 are also available from the website. A page showing some of the features of Quanto.js is available at: quantomatic.github.io/quantojs.

There are many tools for graph transformation, Quantomatic is unique in that it is designed specifically for diagrammatic reasoning. In other words, it is a general-purpose proof assistant for string-diagram based theories. Perhaps its closest relatives are general-purpose graph rewriting tools. GROOVE [25] is a tool for graph transformation whose main focus is model checking of object-oriented systems. Like Quantomatic, GROOVE has a mechanism for specifying rules that can match many different concrete graphs, namely *quantified graph transformation rules*. Other graph rewriting tools such as PROGRESS [26] and AGG [28] also have mechanisms that can be used to control application of a single rule to many concrete graphs. All of these mechanisms have quite different semantics from !-box rewriting, owing to the fact that the latter is specifically designed for transforming string diagrams. Its an open question whether any of these mechanisms could encode !-boxes.

There are three major directions in which we hope to extend Quantomatic. The first is in the support of non-commutative vertices and theories. The theoretical foundation for non-commutative graphical theories with !-boxes was given in [19]. A big advantage of this is the ability to define new nodes which could be substituted for entire diagrams. This would allow extension of a theory by arbitrary, possibly recursive definitions. Secondly, we aim to go beyond 'derivation-style' proofs into proper, goal-based backward reasoning. In [16], we introduced the concept of !-box induction, which was subsequently formalised [22]. In conjunction with recursive definitions, this gives a powerful mechanism for introducing new !-box equations. This would also be beneficial even for purely equational proofs, as it is sometimes difficult to coax Quantomatic into performing the correct rewrite step in the presence of too much symmetry. It is also an important stepping stone toward providing QuantoCore with a genuine LCF-style proof kernel. Another, possibly complementary, approach is to integrate Quantomatic with an existing theorem prover, essentially as a 'heavyweight tactic' for the underlying formal semantics of the diagram. In [19], this semantics is presented as a term language with wires as bound pairs of names, and we have had some preliminary success in formalising this language in Nominal Isabelle. Third, it was recently shown in [20] that placing a natural restriction on !-boxes yields a proper subset of context-free graph languages. Another line of future work is to provide support for more general context-free graph languages using vertex replacement grammars. This would allow us to reason about more interesting families of diagrams and borrow proof techniques from the rich literature on context-free graph grammars.

Acknowledgements. In addition to the two authors, Quantomatic has received major contributions from Alex Merry, Lucas Dixon, and Ross Duncan. We would also like to thank David Quick, Benjamin Frot, Fabio Zennaro, Krzysztof Bar, Gudmund Grov, Yuhui Lin, Matvey Soloviev, Song Zhang, and Michael Bradley for their contributions and gratefully acknowledge financial support from EPSRC, the Scatcherd European Scholarship, and the John Templeton Foundation.

References

1. Abramsky, S., Coecke, B.: A categorical semantics of quantum protocols. In: LICS 2004, pp. 415–425. IEEE Computer Society (2004)
2. Baez, J.C., Erbele, J.: Categories in control (2014). arXiv:1405.6881
3. Bonchi, F., Sobociński, P., Zanasi, F.: Interacting bialgebras are frobenius. In: Muscholl, A. (ed.) FOSSACS 2014 (ETAPS). LNCS, vol. 8412, pp. 351–365. Springer, Heidelberg (2014)
4. Bonchi, F., Sobociński, P., Zanasi, F.: Full abstraction for signal flow graphs. In: Principles of Programming Languages, POPL 2015 (2015)
5. Clark, S., Coecke, B., Sadrzadeh, M.: Mathematical foundations for a compositional distributed model of meaning. Linguist. Anal. **36**, 1–4 (2011)
6. Coecke, B., Duncan, R.: Interacting quantum observables: categorical algebra and diagrammatics. New J. Phys. **13**(4), 043016 (2011)
7. Danos, V., Feret, J., Fontana, W., Harmer, R., Krivine, J.: Abstracting the differential semantics of rule-based models: exact and automated model reduction. In: LICS (2010)
8. Dixon, L., Kissinger, A.: Open-graphs and monoidal theories. Math. Struct. Comput. Sci. **23**, 308–359 (2013)
9. Duncan, R., Lucas, M.: Verifying the steane code with quantomatic. In: Quantum Physics and Logic, vol. 2013 (2013)
10. Duncan, R., Perdrix, S.: Rewriting measurement-based quantum computations with generalised flow. In: Abramsky, S., Gavoille, C., Kirchner, C., Meyer auf der Heide, F., Spirakis, P.G. (eds.) ICALP 2010. LNCS, vol. 6199, pp. 285–296. Springer, Heidelberg (2010)
11. Grefenstette, E., Sadrzadeh, M.: Experimental support for a categorical compositional distributional model of meaning. In: Proceedings of the Conference on Empirical Methods in Natural Language Processing (2011)
12. Grov, G., Kissinger, A., Lin, Y.: Tinker, tailor, solver, proof. In: UITP (2014)
13. Hillebrand, A.: Quantum protocols involving multiparticle entanglement and their representations in the ZX-calculus. Master's thesis, Oxford University (2011)
14. Johansson, M., Dixon, L., Bundy, A.: Conjecture synthesis for inductive theories. J. Autom. Reason. **47**(3), 251–289 (2011)
15. Joyal, A., Street, R.: The geometry of tensor calculus I. Adv. Math. **88**(1), 55–112 (1991)
16. Kissinger, A.: Pictures of Processes: Automated Graph Rewriting for Monoidal Categories and Applications to Quantum Computing. Ph.d. thesis, Oxford (2012)
17. Kissinger, A.: Synthesising graphical theories (2012). arXiv:1202.6079
18. Kissinger, A., Merry, A., Soloviev, M.: Pattern graph rewrite systems. In: Proceedings of DCM (2012)
19. Kissinger, A., Quick, D.: Tensors, !-graphs, and non-commutative quantum structures. In: QPL 2014, vol. 172 of EPTCS, pp. 56–67 (2014)
20. Kissinger, A., Zamdzhiev, V.: !-graphs with trivial overlap are context-free. In: Proceedings Graphs as Models, GaM 2015, London, UK, pp. 11–12 , April 2015
21. Melliès, P.-A.: Local states in string diagrams. In: Dowek, G. (ed.) RTA-TLCA 2014. LNCS, vol. 8560, pp. 334–348. Springer, Heidelberg (2014)
22. Merry, A.: Reasoning with !-graphs. Ph.d. thesis, Oxford University (2013)
23. Michaelson, G., Grov, G.: Reasoning about multi-process systems with the box calculus. In: Zsók, V., Horváth, Z., Plasmeijer, R. (eds.) CEFP. LNCS, vol. 7241, pp. 279–338. Springer, Heidelberg (2012)

24. Penrose, R.: Applications of negative dimensional tensors. In: Dowling, T.A., Penrose, R. (eds.) Combinatorial Mathematics and its Applications, pp. 221–244. Academic Press, San Diego (1971)

25. Rensink, A.: The GROOVE simulator: a tool for state space generation. In: Pfaltz, J.L., Nagl, M., Böhlen, B. (eds.) AGTIVE 2003. LNCS, vol. 3062, pp. 479–485. Springer, Heidelberg (2004)

26. Schürr, A.: PROGRESS: a VHL-language based on graph grammars. In: Ehrig, H., Kreoswki, H.-J., Rozenberg, G. (eds.) Graph Grammars and Their Application to Computer Science. LNCS, vol. 532. Springer, Heidelberg (1991)

27. Sobociński, P.: Representations of petri net interactions. In: Gastin, P., Laroussinie, F. (eds.) CONCUR 2010. LNCS, vol. 6269, pp. 554–568. Springer, Heidelberg (2010)

28. Taentzer, G.: AGG: A graph transformation environment for modeling and validation of software. In: Pfaltz, J.L., Nagl, M., Böhlen, B. (eds.) AGTIVE 2003. LNCS, vol. 3062, pp. 446–453. Springer, Heidelberg (2004)

Automating First-Order Logic

Cooperating Proof Attempts

Giles Reger[✉], Dmitry Tishkovsky, and Andrei Voronkov

University of Manchester, Manchester, UK
giles.reger@manchester.ac.uk

Abstract. This paper introduces a pseudo-concurrent architecture for first-order saturation-based theorem provers with the eventual aim of developing it into a truly concurrent architecture. The motivation behind this architecture is two-fold. Firstly, first-order theorem provers have many configuration parameters and commonly utilise multiple strategies to solve problems. It is also common that one of these strategies will solve the problem quickly but it may have to wait for many other strategies to be tried first. The architecture we propose interleaves the execution of these strategies, increasing the likeliness that these 'quick' proofs will be found. Secondly, previous work has established the existence of a synergistic effect when allowing proof attempts to communicate by sharing information about their inferences or clauses. The recently introduced AVATAR approach to splitting uses a SAT solver to explore the clause search space. The new architecture considers sharing this SAT solver between proof attempts, allowing them to share information about pruned areas of the search space, thus preventing them from making unnecessary inferences. Experimental results, using hard problems from the TPTP library, show that interleaving can lead to problems being solved more quickly, and that sharing the SAT solver can lead to new problems being solved by the combined strategies that were never solved individually by any existing theorem prover.

1 Introduction

This paper presents a pseudo-concurrent architecture for first-order saturation-based theorem provers. This work is a first step in a larger attempt to produce a truly concurrent architecture. This architecture allows proof attempts to cooperate and execute in a pseudo-concurrent fashion. This paper considers an instantiation of the architecture for the Vampire prover [8] but the approach is applicable to any saturation-based prover.

Modern first-order provers often use *saturation algorithms* that attempt to saturate the clausified problem with respect to a given inference system. If a contradiction is found in this clause search space then the problem is unsatisfiable. If a contradiction is not found and the search space is saturated, by a complete inference system, it is satisfiable. Even for small problems this search space can grow quickly and provers often employ heuristics to control its exploration.

Andrei Voronkov — Partially supported by the EPSRC grant "Reasoning for Verification and Security".

A.P. Felty and A. Middeldorp (Eds.): CADE-25, LNAI 9195, pp. 339–355, 2015.
DOI: 10.1007/978-3-319-21401-6_23

A major research area is therefore the reduction of this search space using appropriate strategies. One approach is to control how the search space grows by the selection and configuration of different inferences and preprocessing steps. The aim being to find a contradiction more quickly. The notion of *redundancy* allows provers to detect parts of the search space that will not be needed anymore and eliminate redundant inferences; this greatly improves the performance of provers. An additional (incomplete) heuristic for reducing the search space is Vampire's limited resource strategy [12] which discards heavy clauses that are unlikely to be processed given the remaining time and memory resources. Another useful tool in reducing search space explosion is *splitting* [7] where clauses are split so that the search space can be explored in smaller parts. A new, highly successful, approach to splitting is found in the AVATAR architecture [22], which uses a Splitting module with a SAT solver at its core to make splitting decisions.

Although most of theorem provers allow users to choose between various strategies, two issues remain. Firstly, given a problem it is generally not possible to choose a good strategy a priori even if such a strategy is implemented in the theorem prover. Secondly, whilst there are strategies that appear to be good at solving problems on average, there will be problems that can only be solved by strategies which are not good on average.

To deal with these issues, almost all modern automated theorem provers have modes that attempt multiple strategies within a given time, searching for a "fast proof". The work in this paper was motivated by the observation that whilst we are running multiple proof attempts we should allow them to cooperate. This is further supported by work in parallel theorem proving [5,17] that discovered a synergistic effect when clauses were communicated between concurrently running proof attempts.

The pseudo-concurrent architecture presented in this paper introduces two concepts that allow proof attempts to cooperate:

1. Proof attempts are interleaved in a pseudo-concurrent fashion (Sect. 3). This allows "fast proofs" to be found more quickly.
2. The Splitting module from the AVATAR architecture is shared between proof attempts (Sect. 4). This shares information on pruned splitting branches, preventing proof attempts from exploring unnecessary branches.

The performance of automatic theorem provers can be improved in two ways. The first is to solve problems we can already solve more quickly, the second is to solve new problems that we could not solve before. The work presented here targets both ways. Results (Sect. 5) comparing this architecture to a sequential execution of Vampire are very encouraging and demonstrate that this new pseudo-concurrent architecture has great potential to make significant improvements with respect to both of our goals. It is shown that, in general, problems can be solved much faster by this new architecture. This is invaluable for applications such as program analysis or interactive theorem proving, where a large amount of proof attempts can be generated in a short time and the goal is to solve quickly as many of them as possible. Furthermore, by allowing proof attempts to communicate inconsistent splitting decisions we are able to prove new problems that were not solved before by any existing first-order theorem prover.

Further investigation is required to understand how this new architecture can be optimally utilised but it is clear that this approach can improve the general performance of saturation-based provers and solve problems beyond the reach of existing techniques.

2 Vampire and AVATAR

Vampire is a first-order superposition theorem prover. This section reviews its basic structure and components relevant to the rest of this paper. We will use the word *strategy* to refer to a set of configuration parameter values that control proof search and *proof attempt* to refer to a run of the prover using such a strategy.

2.1 Saturation Algorithms

Superposition provers such as Vampire use *saturation algorithms with redundancy elimination*. They work with a search space consisting of a set of clauses and use a collection of generating, simplifying and deleting inferences to explore this space. Generating inferences, such as superposition, extend this search space by adding new clauses obtained by applying inferences to existing clauses. Simplifying inferences, such as demodulation, replace a clause by a simpler one. Deleting inferences, such as subsumption, delete a clause, typically when it becomes redundant (see [1]). Simplifying and deleting inferences must satisfy this condition to preserve completeness.

The goal is to *saturate* the set with respect to the inference system. If the empty clause is derived then the input clauses are unsatisfiable. If no empty clause is derived and the search space is saturated then the input clauses are guaranteed to be satisfiable *only if* a complete strategy is used. A strategy is complete if it is guaranteed that all inferences between non-deleted clauses in the search space will be applied. Vampire includes many incomplete strategies as they can be very efficient at finding unsatisfiability.

All saturation algorithms implemented in Vampire belong to the family of *given clause algorithms*, which achieve completeness via a fair *clause selection* process that prevents the indefinite skipping of old clauses. These algorithms typically divide clauses into three sets, *unprocessed, passive* and *active*, and follow a simple *saturation loop*:

1. Add non-redundant *unprocessed* clauses to *passive*. Redundancy is checked by attempting to *forward simplify* the new clause using processed clauses.
2. Remove processed (passive and active) clauses made redundant by newly processed clauses, i.e. *backward simplify* existing clauses using these clauses.
3. Select a given clause from *passive*, move it to *active* and perform all generating inferences between the given clause and all other active clauses, adding generated clauses to *unprocessed*.

Later we will show how iterations of this saturation loop from different proof attempts can be interleaved. Vampire implements three saturation algorithms:

1. *Otter* uses both passive and active clauses for simplifications.
2. *Limited Resource Strategy (LRS)* [12] extends Otter with a heuristic that discards clauses that are unlikely to be used with the current resources, i.e. time and memory. This strategy is incomplete but also generally the most effective at proving unsatisfiability.
3. DISCOUNT uses only active clauses for simplifications.

There are also other proof strategies that fit into this loop format and can be interleaved with superposition based proof attempts. For example, instance generation [6] saturates the set of clauses with respect to the instance generation rule. As a large focus of this paper is the sharing of AVATAR, which is not compatible with instance generation, we do not consider this saturation algorithm in the following, although it is compatible with the interleaving approach (Sect. 3).

2.2   Strategies in Vampire

Vampire includes more than 50 parameters, including experimental ones. By only varying parameters and values used by Vampire at the last CASC competition, we obtain over 500 million strategies. These parameters control

- Preprocessing steps (24 different parameters)
- The saturation algorithm and related behaviour e.g. clause selection
- Inferences used (16 different kinds with variations)

Even restricting these parameters to a single saturation algorithm and straightforward preprocessing steps, the number of possible strategies is vast. For this reason, Vampire implements a portfolio *CASC mode* [8] that categorises problems based on syntactic features and attempts a sequence of approximately 30 strategies over a five minute period. These strategies are the result of extensive benchmarking and have been shown, experimentally, to work well on unseen problems i.e. those not used for training.

2.3   AVATAR

The search space explored by a saturation-based prover can quickly become full of long and heavy clauses, i.e. those that have many literals that are themselves large. This can dramatically slow down many inferences, which depend on the size of clauses. This is exacerbated by generating inferences, which typically generate long and heavy clauses from long and heavy clauses.

   To deal with this issue we can introduce *splitting*, which is based on the observation that the search space $S \cup (C_1 \vee C_2)$ is unsatisfiable if and only if both $S \cup C_1$ and $S \cup C_2$ are unsatisfiable, for variable disjoint C_1 and C_2. Different approaches to splitting have been proposed [7] and Vampire implements a new approach called AVATAR [22]. The general idea of the AVATAR approach is to allow a SAT solver to make splitting decisions. In the above case the clause $C_1 \vee C_2$ would be represented (propositionally) in the SAT solver, along with other clauses from S, and the SAT solver would decide with component to *assert* in the first-order proof search. Refutations depending on asserted components are given to the SAT solver, restricting future models and therefore the splitting search space. More details can be found elsewhere [11,22]. Later (Sect. 4) we discuss how the AVATAR architecture can be used to communicate between concurrent proof attempts.

3   Interleaved Scheduling

This section introduces the concept of proof attempt interleaving and explains how it is implemented in the pseudo-concurrent architecture, as illustrated in Fig. 1.

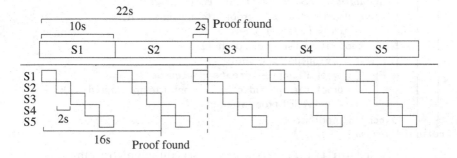

Fig. 1. An illustration of how proof attempt interleaving can prove problems faster.

3.1   Motivation

The intuition behind Vampire's CASC mode is that for many problems there exists a strategy that will solve that problem relatively quickly. Therefore, rather than spending five minutes using a strategy that is 'good on average' it is better to spend a few seconds on each of multiple strategies. However, strategies are attempted sequentially and a problem that can be solved quickly inside an individual strategy, may take a long time to solve within the sequence of strategies.

By interleaving proof attempts (strategies) one can quickly reach these quick solutions. This is illustrated in Fig. 3. There are five strategies and the third strategy solves the problem after 2 s. By interleaving the strategies in blocks of 2 s, the problem is solved in 16 s rather than 22 s. In reality we often have problems solved in a few deciseconds and a sequence of around 30 strategies. So the time savings have the potential to be dramatic.

input : A queue of strategies with local time limits
input : A global time limit and a concurrency limit
output: *Refutation, Satisfiable, Unknown* or *GlobalTimeLimit*

live_list ← []; elapsed ← 0;
for i ← 1 **to** limit **if** size(queue) > 0 **do**
  proof_attempt ← create(pop(queue));
  proof_attempt.budget = 0; proof_attempt.elapsed = 0;
  add(live_list,proof_attempt);
while size(live_list) > 0 **do**
  time_slice ← calculate_time_slice ();
  foreach proof_attempt *in* live_list **do**
    proof_attempt.budget += time_slice ;
    switchIn(proof_attempt);
    while proof_attempt.budget > 0 **do**
      (status, time) ← step(proof_attempt);
      elapsed += time; proof_attempt.elapsed += time ;
      proof_attempt.budget = proof_attempt.budget − time ;
      if elapsed > global **then return** *GlobalTimeLimit* **if** status = *Refutation or Satisfiable* **then return** status **if** proof_attempt.elapsed > proof_attempt.time_limit *or* status = *Unknown* **then**
        remove(live_list,proof_attempt);
        if size(queue) > 0 **then**
          proof_attempt ← create(pop(queue));
          proof_attempt.budget = 0; proof_attempt.elapsed = 0;
          add(live_list,proof_attempt);
    switchOut(proof_attempt);
return *Unknown*

Algorithm 1: Pseudo-concurrent scheduling algorithm

3.2   Interleaving Architecture

Previously we explained how Vampire carries out proof attempts via *saturation algorithms* that iterate a *saturation loop*. The pseudo-concurrent architecture interleaves iterations of this saturation loop from different proof attempts. It is necessary that the granularity of interleaving be at this level as a proof attempt's internal data structures are only guaranteed to be consistent at the end of each iteration.

Each proof attempt is associated with a *context* that contains the structures relevant to that proof attempt. This primarily consists of the saturation algorithm with associated clause sets and indexing structures, as well as configuration parameters local to the proof attempt. In addition to proof attempt's context, there is also a *shared context*, accessible by all proof attempts, that contains a copy of the problem, the problem's signature and sort information, and a global timer. When a proof attempt is *switched in* its local context is loaded and when it is *switched out* this is unloaded.

Algorithm 1 describes the interleaving algorithm. Each strategy is loaded in from a strategy file or the command line and has an individual *local time limit*. These strategies are placed in a queue. There is a *concurrency limit* that controls the number of proof attempts that can run at any one time and a *global time limit* that restricts the run time for the whole algorithm. Global and local time limits are enforced internally by each proof attempt. Initially, the *live list* of proof attempts is populated with this number of proof attempts.

Proof attempt creation is handled by the `create` function. All the contexts and proof attempts are lazily initialised, so that proof attempts that will never run do not take up unnecessary memory.

The proof attempt scheduling algorithm is similar to the standard round-robin scheduling algorithm which is one of the simplest and starvation free scheduling algorithms. Scheduling is performed in circular order and without priority. Its implementation is based on a standard budgeting scheme where each proof attempt is given its own time budget. Ideally each proof attempt would run for a fixed time on each round. However, since it is not possible to know in advance how long a particular proof step will take, time slices for the proof attempts are dynamically calculated.

On each round a proof attempt's budget is increased by a *time slice* computed at the end of the previous round. Each proof attempt will perform multiple proof steps, decreasing its budget by the time it takes to perform each step, until its budget is exhausted. This can make a proof attempt's budget negative.

Initially, the time slice for every proof attempt is one millisecond. At the end of each round the next time slice is computed by `calculate_time_slice`. This calculates the average time it took each proof attempt to make a proof step and selects the smallest of the average times. Such a choice reduces number of scheduling rounds, yet providing a scheduling granularity which is fairly close to the finest. Every 1024 rounds, the time slice is doubled (thus, enlarging the granularity). Increasing the time slice reduces the scheduling overhead and lets proof attempts make reasonable progress on problems requiring a long time to solve. The constant 1024 and the time slice factor are chosen as multiples of two for better efficiency of the scheduling algorithm. Also, 1024 rounds provide approximately one second of running time for each proof attempt with the minimal time slice of one millisecond, that seems a reasonable minimum of time for a proof attempt to make progress.

In contrast to the round-robin scheduling which gives equal time slices for execution of processes, the proof attempt scheduler only tries to give *asymptotically* equal times for running each algorithm in the execution pool. That is, all the proof attempts spend equal execution times whenever every proof attempt runs long enough.

A `step` of a proof attempt results in a (TPTP SZS) status. The most common returned status, which does not trigger any extra action, is *StillRunning*. If the `step` function reports that a proof attempt solved the problem or that all time has been used, the algorithm terminates with that status. If a proof attempt has finished running, either with a local time limit or an *Unknown* result, the proof attempt is replaced by one from the priority queue. An *Unknown* result occurs when a proof attempt is incomplete and saturates its clause set.

4   Proof Attempt Cooperation via the Splitting Module

We are interested in improving performance of a theorem prover by making proof attempts share information. A novel way of doing this, made possible by the AVATAR architecture, is to make them cooperate using the Splitting module. This is done in the following way: clauses generated by different proof attempts are passed to the same Splitting module, and thus processed using the same SAT solver, storing propositional clauses created by all proof attempts.

This idea is based on two observations:

1. SAT solving is cheap compared with first-order theorem proving. Therefore, there should be no noticeable degradation of performance as the SAT solver is dealing with a much larger set of clauses.
2. A SAT solver collecting clauses coming from different proof attempts can make a proof attempt exploit fewer branches.

4.1   Motivation

The clause search space of a problem can be viewed as a (conceptual) splitting tree where each clause is a node and each branch represents splitting that clause into components. The leaves of the splitting tree are the *splitting branches* introduced earlier. Recall that a problem is unsatisfiable if a contradiction is found in each of these splitting branches.

In our pseudo-concurrent architecture the different proof attempts are exploring the same (conceptual) splitting tree. A proof attempt's clause space (the processed clauses, see Sect. 2) covers part of this tree and is determined by the inferences and heuristics it employs. Some proof attempts will have overlapping clause spaces.

By sharing the SAT solver, proof attempts are able to share splitting branches that contain contradictions. Figure 2 illustrates this effect. Proof attempts 1 and 2 have overlapping clause spaces. If one of the proof attempts show that the circled node (representing a clause component) cannot be true then the other proof attempt does not need to consider this branch. Even better, inconsistency of a branch can be established by clauses contributed by more than one proof attempt.

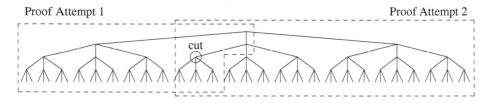

Fig. 2. Illustrating one proof attempt pruning the set of clauses explored by another.

Cooperating in this way reduces the number of splitting branches each proof attempt needs to explore. However, it can also change the behaviour of a strategy as splitting branches may be explored in a different order. Although all splitting branches must be explored to establish unsatisfiability, the order of exploration can impact how quickly a contradiction is found in each splitting branch.

4.2 Organising Cooperation

In reality, splitting requires more structures than simply the SAT solver, as described below. We refer to all structures required to perform splitting as the Splitting module. In this section we describe how the proof attempts interact with the Splitting module. The details of how this is done for a single proof attempt are described elsewhere [22]; we focus on details relevant to multiple proof attempts.

Each proof attempt interacts with the Splitting module independently via a Splitting interface. It passes to the interface clauses to split and splitting branch contradictions it finds. It receives back splitting decisions in the form of clause components to assert (and those to retract). Cooperation is organised so that the assertions it is given by the Splitting module relate only to that proof attempt's clause space.

As illustrated in Fig. 3, the Splitting interface sits between the proof attempts and the SAT solver that makes the splitting decisions. Its role is twofold. Firstly, it must consistently transform clause components into propositional variables to pass to the SAT solver. Secondly, it must transform the interpretation returned by the SAT solver into splitting decisions to pass to the proof attempts. To achieve these two roles, the Splitting interface keeps a variant index and component records.

The variant index ensures that components that are equivalent up to variable renaming and symmetry of equality are translated into the same propositional variable. Importantly, this variant index is common across all proof attempts, meaning that a clause C_1 generated in one proof attempt that is a variant of clause C_2 generated in a different proof attempt will be represented by the same propositional variable in the SAT solver. A (non-trivial) index is used because checking clause equivalence up to variable renaming is a problem equivalent to graph isomorphism [12].

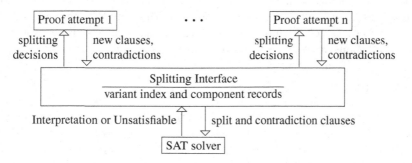

Fig. 3. Multiple proof attempts utilising the same Splitting module

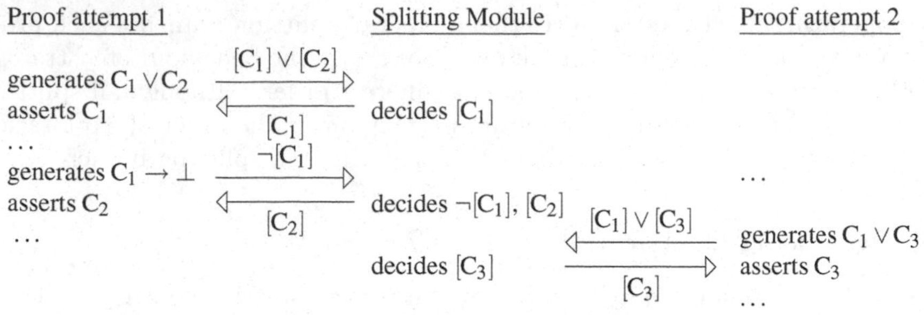

Proof attempt 1   Splitting Module   Proof attempt 2

generates $C_1 \lor C_2$ $\xrightarrow{\;[C_1] \lor [C_2]\;}$

asserts C_1 $\xleftarrow{\;[C_1]\;}$ decides $[C_1]$

...

generates $C_1 \to \bot$ $\xrightarrow{\;\neg[C_1]\;}$

asserts C_2 $\xleftarrow{\;[C_2]\;}$ decides $\neg[C_1], [C_2]$ ...

$\xleftarrow{\;[C_1] \lor [C_3]\;}$ generates $C_1 \lor C_3$

decides $[C_3]$ $\xrightarrow{\;[C_3]\;}$ asserts C_3

...

Fig. 4. Illustrating cooperation via the Splitting Module

Component records store important information about the status of the component. Ordinarily they track information necessary to implement backtracking, i.e. which clauses have been derived from this component and which clauses have been reduced assuming this component. This is important, but not interesting for this work. In our pseudo-concurrent architecture we also track which proof attempt has passed a clause containing this component to the Splitting interface. This allows the Splitting interface to restrict splitting decisions sent to a proof attempt to components generated by that proof attempt.

It is possible to allow more information to flow from the Splitting module to the proof attempts. It would be sound to assert the whole splitting branch in each proof attempt. However, this might pollute the clause space, preventing them from making suitable progress. An approach we have not yet explored is to employ heuristics to select components that should be 'leaked' back to proof attempts.

4.3 Example of Cooperation

Figure 4 illustrates how cooperation can avoid repetition of work. In this scenario the first proof attempt generates a clause consisting of two components C_1 and C_2 and the Splitting module first decides to assert C_1. After some work the first proof attempt shows that C_1 leads to a contradiction and passes this information to the Splitting module. The Splitting module decides that C_2 must hold since C_1 cannot hold. Later, a second proof attempt generates a clause consisting of the components C_1 and C_3. These are passed to the Splitting module and, as C_1 has been contradicted, C_3 is asserted immediately. Here the second proof attempt does not need to spend any effort exploring any branch containing C_1. The component C_1 could have been a large clause and the repetition of the potentially expensive work required to refute it is avoided.

5 Evaluation

We evaluate our pseudo-concurrent architecture by comparing it to the sequential version of Vampire. We are interested in how it currently performs, but also in behaviours that would suggests future avenues for exploration for this new and highly experimental architecture.

5.1   Experimental Setup

We select 1747 problems[1] from the TPTP library [18]. These consist of Unsatisfiable or Theorem problems of 0.8 rating or higher containing non-unit clauses. The rating [19] indicates the hardness of a problem and is the percentage of (eligible) provers that cannot solve a problem. For example, a rating of 0.8 means that only 20 % of (eligible) provers can solve the problem. Therefore, we have selected very hard problems. Note that the rating evaluation does not consider every mode of each prover, so it is possible that a prover can solve a problem of rating 1 using a mode not used in this evaluation. Therefore, we include problems of rating 1 that we know are solvable by some prover, i.e. Vampire. Out of all problems used, 358 are of rating 1.

In this paper we focus on a single set of 30 strategies drawn from the current CASC portfolio mode, however we have explored different sets of strategies and the results presented here are representative. All strategies employ default preprocessing and use a combination of different saturation algorithms, splitting configurations and inferences. We consider a 10 second run of each strategy.

We compare two approaches that use these 30 strategies:

1. The *concurrent* approach utilises the pseudo-concurrent architecture giving a 10 s local time limit to each proof attempt.
2. The *sequential* approach executes the strategies in a predetermined *random* order without any sharing of clauses or resources, i.e. they are run in different provers.

Experiments were run on the StarExec cluster [16], using 160 nodes. The nodes used contain a Intel Xeon 2.4 GHz processor. The default memory limit of 3 GB was used for sequential runs and this was appropriately scaled for concurrent runs, i.e. for 30 strategies this was set to 90 GB.

5.2   Results

Figure 5 plots the number of problems solved by the *concurrent* and *sequential* approaches against the time taken to solve them.

This demonstrates that the *concurrent* approach solved more problems than the *sequential* approach for running times not exceeding 290 s, reaching the maximal difference of 125 problems at 20 s. The lead of the *concurrent* approach is more substantial from the start to about the 85th second. From 207 to 290 s both approaches were roughly on par, but, after 290 s, the *sequential* approach started to gain their lead. So the whole running time can be divided into four intervals: from 0 to 85 s, between 85 and 207 s, from 207 to 290 s, and after 290 s.

The order of strategies used in the *sequential* approach does not effect these results in general. Experiments (not reported here) showed that other random orderings led to a similar curve for the *sequential* approach. However, the reverse ordering does allow the *sequential* approach to take the lead at first as one

---

[1] A list of the selected problems, the executable of our prover, and the results of the experiments are available from http://vprover.org.

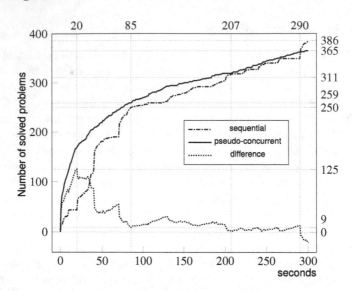

Fig. 5. Number of problems solved by a sequence of proof attempts and the pseudo-concurrent architecture with respect to time.

strategy placed (randomly) at the end of the sequence can solve many problems. This reflects the obvious observation that a single strong strategy can outperform many concurrent strategies at first but the results show that after a short time the combined effect of many strategies leads to more problems being solved.

The *sequential* approach solved 11 problems of rating 1 and the *concurrent* approach solved 14, where 6 were not solved by the *sequential* approach. Of these problems, the *concurrent* approach solved 5 that had not previously been solved by Vampire using any mode. During other experiments, not reported here, the concurrent approach also solved 5 problems not previously solved by any prover, including Vampire. This is significant as Vampire has been run extensively for several years on the TPTP library with many different strategies, while we made very few experiments with the concurrent version.

Overall, the *concurrent* approach solved 63 problems unsolved by the *sequential* approach, which solved 84 unsolved by the *concurrent* approach. Of the problems solved by both approaches, the *concurrent* approach solved problems 1.53 times faster than the *sequential* approach on average. This represents clear advantages for applications where solving problems fast is of key importance. As scheduling is not effected by the global time limit, Fig. 5 can be used to give the number of problems solved for any time limit less than 300s.

Below we discuss how these results can be interpreted in terms of interleaving and sharing the Splitting module.

5.3   The Impact of Interleaving

We contribute much of the success of the *concurrent* approach on the first interval shown in Fig. 5 to the advantage of interleaved proof attempts discussed in Sect. 3.

One can compute the ideal interleaving performance by inspecting the shortest time taken to solve a problem by the sequential proof attempts. The *concurrent* approach roughly follows this ideal in the first interval. For example, after 85 s the *concurrent* approach has solved 259 problems and after 2.83 s (approximately how long each strategy should have run for) the sequential proof attempts had solved 273 problems. However, this deviation suggests that interleaving is not perfect.

After 85 s, the positive effects of interleaving seem to diminish. These effects could be attributed to sharing the Splitting module. However, a separate experiment, shown in Fig. 6, illustrates that there is inherent overhead in the interleaving approach. In this experiment 10 of the strategies were forced not to use splitting and we compare sequential and pseudo-concurrent execution. The interleaving effects are positive to begin with. But we would expect the two approaches to solve the same number of problems overall, and any difference can only be explained by overhead introduced by interleaving, as splitting is not used.

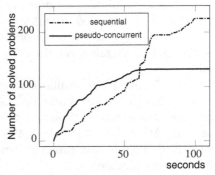

Fig. 6. Comparing pseudo-concurrent and sequential architectures with no splitting.

The cost of switching proof attempts is very small. However, changing proof attempts can have a large impact on memory locality. The average memory consumption for the *concurrent* approach is just under 4 GB, compared with the 80 MB (0.08 GB) used by each sequential strategy on average. Large data structures frequently used by a proof attempt (such as indices) will not be in cache when it is switched in, leading to many cache faults. This is aggravated by the frequent context switching.

This slowdown due to memory overhead can explain the loss of 84 problems. Of these problems, 76 % are solved by the *sequential* approach after 2.85 s (the end of the first interval) and 23 % after 6.9 s (the end of the second interval). The implementation of the new architecture did not focus on minimising memory consumption and future work targeting this issue could alleviate these negative effects.

5.4   The Impact of Sharing AVATAR

We have seen the positive impact strategy interleaving can have, but also the negative impact that running multiple proof attempts in the same memory space can have. Strategy interleaving cannot account for all of the positive results. There were 63 problems solved by the *concurrent* approach and not by the *sequential* approach. This makes up 17 % of problems solved and can be attributed to sharing the Splitting module as non-cooperating interleaved proof-attempts are expected to solve the same problems as the proof attempts running sequentially.

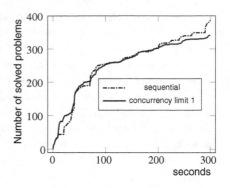

Fig. 7. Running with concurrency limit 1 shows AVATAR sharing effects.

Figure 7 shows the results of running the *concurrent* approach with a concurrency limit of 1 (i.e. running one strategy at a time). This is very similar to the *sequential* approach. However, the *concurrent* approach solves new problems early on. After 20 s it has solved 49 problems that the *sequential* approach has not solved. This is only possible though the additional information shared via the Splitting module.

Furthermore, individual strategies solve many new problems. In one case, a strategy that solves no problems in the *sequential* approach solves 45 problems within the *concurrent* approach with concurrency limit one. This can be explained by the Splitting module letting this proof attempt skip many unnecessary splitting branches to quickly reach a contradiction.

This behaviour is also seen when there is no concurrency limit. One strategy solves 16 problems when sharing the Splitting module, compared with the sequential case; on average local strategies solve 4.8 problems that their sequential counterparts do not solve. This demonstrates that sharing the Splitting module improves the progress of individual strategies generally.

Note that sharing the Splitting module can introduce additional overhead as each call to the SAT solver will be dealing with more propositional variables. This is confirmed by timing results. On average, sequential proof attempts spent roughly 23 % of total running time in the SAT solver whilst proof attempts within the pseudo-concurrent architecture spent roughly 28 %. The represents an increase of approximately 20 %.

5.5   Summary

The experiments in this section show the following:

- Interleaving strategies can find the *quick* proofs, but multiple proof attempts sharing the memory space limits this effect. Further work allowing proof attempts to share certain data structures should alleviate this to some extent.

– Sharing the Splitting module can lead to many new problems being solved that were not solved without this sharing, including very hard problems.

6   Related Work

There are two areas of work related to this new architecture. We first consider the usage of multiple strategies sequentially and then discuss links with parallel theorem proving.

Sequential Usage of Strategies. All modern theorem provers that are competitive in the CASC competition utilise some form of strategy scheduling. The idea was first implemented in the Gandalf prover [21] which dynamically selects a strategy depending on results of performed inferences. Other solvers have similar approaches to Vampire's CASC mode, for example, in the E prover [13] the selection of strategies and generation of schedules for a class of problems is also based on the previous performance of the prover on similar problems. The idea of strategy scheduling has been developed further in the E-MaLeS 1.1 scheduler [9] where schedules for the E prover are produced using kernel-based learning algorithms.

Parallel Theorem Proving. The approach taken in this work has many aspects in common with the area of parallel theorem proving (see [3,20,23] for extensive overviews on problems, architectures, and approaches) including our motivation and methods.

Firstly, previous work with distributed provers (e.g. PARTHEO [14]) and parallel provers (e.g. DISCOUNT [5]) has shown that overall better performance and stability can be achieved by running multiple proof attempts in parallel. Running several theorem provers in parallel without any communication between them also shown to exhibit such a behaviour [2]. As many strategies are explored in parallel, concurrent proof attempts can solve and guarantee answers to more problems. Additionally, it was shown that parallel provers also behave more predictably when run repeatedly on same set of problems. These effects are similar to the consequences of interleaving proof attempts that is described in the paper.

As mentioned in the beginning of this paper, previous work in parallel theorem proving has also observed a synergistic effect [17] for proof attempts that communicate by passing information about their inferences. In this previous work a parallel execution of proof attempts was able to solve a number of hard problems which could not be solved by any individual proof attempt. This effect was also observed in the DISCOUNT parallel theorem prover [5], based on a generic framework for concurrent theorem proving called the *teamwork approach* [4]. Here clause communication is handled by a complex management scheme that passes 'outstanding' clauses between proof attempts. We have observed a similar effect with our sharing of the Splitting module.

Our approach, and the previous parallel theorem provers, run multiple proof attempts concurrently. An alternative approach is to parallelise the internal workings of the prover. In 1990, Slaney and Lusk [15] proposed a parallel architecture for closure algorithms and investigated its application to first-order theorem proving. The work considered how a generalisation of the internal saturation loop in theorem provers can be turned into a multi-threaded algorithm. The resulting ROO theorem prover [10] provides a *fine-grained* splitting of the closure algorithms into concurrent tasks.

7   Conclusion

The presented results show a promising direction in developing concurrent provers. As a future goal we consider developing a truly concurrent thread-safe architecture for saturation-based first-order theorem provers. The main challenge in this respect is to make such a concurrent prover efficient. The advantages of scheduled interleaving of proof attempts demonstrated by the experimental results prove that this is possible. Cooperation of proof attempts via common interfaces like AVATAR is also shown to be fruitful: it was shown that the pseudo-concurrent architecture could utilise this sharing to prove previously unsolvable problems. Also, we have demonstrated that problems can be solved much faster, making the architecture attractive for applications, such as program analysis and interactive theorem proving. Although developing a truly concurrent and efficient prover is an ultimate goal, there are a few other steps can be done to improve the current pseudo-concurrent architecture. One is to develop prover configuration methods which take into account interleaved execution of the strategies. This could utilise machine learning techniques. Another is to allow scheduling modulo priorities assigned to the proof attempts and maximise efficiency of the prover by pre-configuring such parameters. A third step, which is more important and difficult, is to expand cooperation scheme of the proof attempts allowing more useful information to be shared between them with minimal overhead. These directions are interesting, require many experiments, and, hopefully, will lead to new architectures of theorem provers ensuring better stability and efficiency.

References

1. Bachmair, L., Ganzinger, H.: Resolution theorem proving. In: Robinson, A., Voronkov, A. (eds.) Handbook of Automated Reasoning, vol. I, Chap. 2, pp. 19–99. Elsevier Science, Amsterdam (2001)
2. Böhme, S., Nipkow, T.: Sledgehammer: judgement day. In: Giesl, J., Hähnle, R. (eds.) IJCAR 2010. LNCS, vol. 6173, pp. 107–121. Springer, Heidelberg (2010)
3. Bonacina, M.: A taxonomy of parallel strategies for deduction. Ann. Math. Artif. Intell. **29**(1–4), 223–257 (2000)
4. Denzinger, J., Kronenburg., M.: Planning for distributed theorem proving: the teamwork approach. In: Görz, G., Hölldobler, S. (eds.) KI 1996. LNCS, vol. 1137. Springer, Heidelberg (1996)
5. Denzinger, J., Kronenburg, M., Schulz, S.: DISCOUNT – a distributed and learning equational prover. J. Autom. Reasoning **18**(2), 189–198 (1997)

6. Ganzinger, H., Korovin, K.: New directions in instantiation-based theorem proving. In: Proceedings of LICS 2003, pp. 55–64 (2003)
7. Hoder, K., Voronkov, A.: The 481 ways to split a clause and deal with propositional variables. In: Bonacina, M.P. (ed.) CADE 2013. LNCS, vol. 7898, pp. 450–464. Springer, Heidelberg (2013)
8. Kovács, L., Voronkov, A.: First-order theorem proving and VAMPIRE. In: Sharygina, N., Veith, H. (eds.) CAV 2013. LNCS, vol. 8044, pp. 1–35. Springer, Heidelberg (2013)
9. Kühlwein, D., Schulz, S., Urban, J.: E-MaLeS 1.1. In: Bonacina, M.P. (ed.) CADE 2013. LNCS, vol. 7898, pp. 407–413. Springer, Heidelberg (2013)
10. Lusk, E., McCune, W.: Experiments with ROO: a parallel automated deduction system. In: Fronhöfer, B., Wrightson, G. (eds.) Dagstuhl Seminar 1990. LNCS, vol. 590. Springer, Heidelberg (1992)
11. Reger, G., Suda, M., Voronkov, A.: Playing with AVATAR. In: Proceedings of CADE2015 (2015)
12. Riazanov, A., Voronkov, A.: Limited resource strategy in resolution theorem proving. J. Symb. Comp. **36**(1–2), 101–115 (2003)
13. Schulz, S.: System description: E 1.8. In: McMillan, K., Middeldorp, A., Voronkov, A. (eds.) LPAR-19 2013. LNCS, vol. 8312, pp. 735–743. Springer, Heidelberg (2013)
14. Schumann, J., Letz, R.: PARTHEO: a high-performance parallel theorem prover. In: Stickel, M.E. (ed.) CADE 1990. LNCS, vol. 449, pp. 40–56. Springer, Heidelberg (1990)
15. Slaney, J.K., Lusk, E.L.: Parallelizing the closure computation in automated deduction. In: Stickel, M.E. (ed.) CADE 1990. LNCS, vol. 449, pp. 28–39. Springer, Heidelberg (1990)
16. StarExec, https://www.starexec.org
17. Sutcliffe, G.: The design and implementation of a compositional competition-cooperation parallel ATP system. In: Proceedings IWIL-2, number MPI-I-2001-2-006 in MPI für Informatik, Research Report, pp. 92–102 (2001)
18. Sutcliffe, G.: The TPTP problem library and associated infrastructure. J. Autom. Reasoning **43**(4), 337–362 (2009)
19. Sutcliffe, G., Suttner, C.: Evaluating general purpose automated theorem proving systems. Artif. Intell. **131**(1–2), 39–54 (2001)
20. Suttner, C.B., Schumann, J.: Chapter 9 – Parallel automated theorem proving. In: Parallel Processing for Artificial Intelligence, vol. 14 of Machine Intelligence and Pattern Recognition, pp. 209–257. North-Holland (1994)
21. Tammet, T.: Gandalf. J. Autom. Reasoning **18**(2), 199–204 (1997)
22. Voronkov, A.: AVATAR: The architecture for first-order theorem provers. In: Biere, A., Bloem, R. (eds.) CAV 2014. LNCS, vol. 8559, pp. 696–710. Springer, Heidelberg (2014)
23. Wolf, A., Fuchs, M.: Cooperative parallel automated theorem proving. Technical report SFB Bereicht 342/21/97, Technische Universität München (1997)

Towards the Compression of First-Order Resolution Proofs by Lowering Unit Clauses

Jan Gorzny[1]([✉]) and Bruno Woltzenlogel Paleo[2]

[1] University of Victoria, Victoria, Canada
jgorzny@uvic.ca
[2] Vienna University of Technology, Vienna, Austria
bruno@logic.at

Abstract. The recently developed `LowerUnits` algorithm compresses propositional resolution proofs generated by SAT- and SMT-solvers by postponing and lowering resolution inferences involving unit clauses, which have exactly one literal. This paper describes a generalization of this algorithm to the case of first-order resolution proofs generated by automated theorem provers. An empirical evaluation of a simplified version of this algorithm on hundreds of proofs shows promising results.

1 Introduction

Most of the effort in automated reasoning so far has been dedicated to the design and implementation of proof systems and efficient theorem proving procedures. As a result, saturation-based first-order automated theorem provers have achieved a high degree of maturity, with resolution and superposition being among the most common underlying proof calculi. Proof production is an essential feature of modern state-of-the-art provers and proofs are crucial for applications where the user requires certification of the answer provided by the prover. Nevertheless, efficient proof production is non-trivial, and it is to be expected that the best, most efficient, provers do not necessarily generate the best, least redundant, proofs. Therefore, it is a timely moment to develop methods that post-process and simplify proofs. While the foundational problem of simplicity of proofs can be traced back at least to Hilbert's 24th Problem, the maturity of automated deduction has made it particularly relevant today.

For proofs generated by SAT- and SMT-solvers, which use propositional resolution as the basis for the DPLL and CDCL decision procedures, there is now a wide variety of proof compression techniques. Algebraic properties of the resolution operation that might be useful for compression were investigated in [5].

The `Reduce&Reconstruct` algorithm [10] searches for locally redundant subproofs that can be rewritten into subproofs of stronger clauses and with fewer

---

J. Gorzny—Supported by the Google Summer of Code 2014 program.
B. Woltzenlogel Paleo—Stipendiat der Österreichischen Akademie der Wissenschaften (APART).

A.P. Felty and A. Middeldorp (Eds.): CADE-25, LNAI 9195, pp. 356–366, 2015.
DOI: 10.1007/978-3-319-21401-6_24

resolution steps. A linear time proof compression algorithm based on partial regularization was proposed in [2] and improved in [6]. Furthermore, [6] described a linear time algorithm called LowerUnits that delays resolution with unit clauses.

In contrast, for first-order theorem provers, there has been up to now (to the best of our knowledge) no attempt to design and implement an algorithm capable of taking a first-order resolution DAG-proof and efficiently simplifying it, outputting a possibly shorter pure first-order resolution DAG-proof. There are algorithms aimed at simplifying first-order sequent calculus tree-like proofs, based on cut-introduction and Herbrand sequents [7–9]. There is also an algorithm [12] that looks for terms that occur often in any TSTP [11] proof (including first-order resolution DAG-proofs) and introduces abbreviations for these terms. However, as the definitions of the abbreviations are not part of the output proof, it cannot be checked by a pure first-order resolution proof checker.

In this paper, we initiate the process of lifting propositional proof compression techniques to the first-order case, starting with the simplest known algorithm: LowerUnits (described in [6]). As shown in Sect. 3, even for this simple algorithm, the fact that first-order resolution makes use of unification leads to many challenges that simply do not exist in the propositional case. In Sect. 4 we describe an easy to implement algorithm with linear time complexity (with respect to the proof length) which partially overcomes these challenges. In Sect. 5 we present experimental results obtained by applying this algorithm on hundreds of proofs generated with the SPASS theorem prover. The next section introduces the first-order resolution calculus using notations that are more convenient for describing proof transformation operations.

2   The Resolution Calculus

We assume that there are infinitely many variable symbols (e.g. X, Y, Z, X_1, X_2, ...), constant symbols (e.g. a, b, c, a_1, a_2, ...), function symbols of every arity (e.g. f, g, f_1, f_2, ...) and predicate symbols of every arity (e.g. p, q, p_1, p_2,...). A *term* is any variable, constant or the application of an n-ary function symbol to n terms. An *atomic formula* (*atom*) is the application of an n-ary predicate symbol to n terms. A *literal* is an atom or the negation of an atom. The *complement* of a literal ℓ is denoted $\bar{\ell}$ (i.e. for any atom p, $\bar{p} = \neg p$ and $\overline{\neg p} = p$). The set of all literals is denoted $\mathcal{L}$. A *clause* is a multiset of literals. $\bot$ denotes the *empty clause*. A *unit clause* is a clause with a single literal. Sequent notation is used for clauses (i.e. $p_1, \ldots, p_n \vdash q_1, \ldots, q_m$ denotes the clause $\{\neg p_1, \ldots, \neg p_n, q_1, \ldots, q_m\}$). FV (t) (resp. FV (ℓ), FV (Γ)) denotes the set of variables in the term t (resp. in the literal ℓ and in the clause Γ). A *substitution* $\{X_1 \backslash t_1, X_2 \backslash t_2, \ldots\}$ is a mapping from variables $\{X_1, X_2, \ldots\}$ to, respectively, terms $\{t_1, t_2, \ldots\}$. The application of a substitution σ to a term t, a literal ℓ or a clause Γ results in, respectively, the term $t\sigma$, the literal $\ell\sigma$ or the clause $\Gamma\sigma$, obtained from t, ℓ and Γ by replacing all occurrences of the variables in σ by the corresponding terms in σ. The set of all substitutions is denoted $\mathcal{S}$. A *unifier* of a set of literals is a substitution that makes all literals in the set equal. A *resolution proof* is a directed acyclic graph of clauses where the edges correspond to the inference rules of resolution and contraction (as explained in detail in Definition 1). A *resolution refutation* is a resolution proof with root $\bot$.

Definition 1 (First-Order Resolution Proof). *A directed acyclic graph* $\langle V, E, \Gamma \rangle$, *where V is a set of nodes and E is a set of edges labeled by literals and substitutions (i.e. $E \subset V \times 2^{\mathcal{L}} \times \mathcal{S} \times V$ and $v_1 \xrightarrow{\ell}_{\sigma} v_2$ denotes an edge from node v_1 to node v_2 labeled by the literal ℓ and the substitution σ), is a proof of a clause Γ iff it is inductively constructible according to the following cases:*

- **Axiom:** *If Γ is a clause, $\widehat{\Gamma}$ denotes some proof $\langle \{v\}, \varnothing, \Gamma \rangle$, where v is a new (axiom) node.*
- **Resolution:** *If ψ_L is a proof $\langle V_L, E_L, \Gamma_L \rangle$ with $\ell_L \in \Gamma_L$ and ψ_R is a proof $\langle V_R, E_R, \Gamma_R \rangle$ with $\ell_R \in \Gamma_R$, and σ_L and σ_R are substitutions such that $\ell_L \sigma_L = \overline{\ell_R} \sigma_R$ and $\mathrm{FV}\left((\Gamma_L \setminus \{\ell_L\}) \sigma_L \right) \cap \mathrm{FV}\left((\Gamma_R \setminus \{\ell_R\}) \sigma_R \right) = \emptyset$, then $\psi_L \odot_{\ell_L \ell_R}^{\sigma_L \sigma_R} \psi_R$ denotes a proof $\langle V, E, \Gamma \rangle$ s.t.*

$$V = V_L \cup V_R \cup \{v\} \qquad \Gamma = (\Gamma_L \setminus \{\ell_L\}) \sigma_L \cup (\Gamma_R \setminus \{\ell_R\}) \sigma_R$$

$$E = E_L \cup E_R \cup \left\{ \rho(\psi_L) \xrightarrow{\{\ell_L\}}_{\sigma_L} v, \rho(\psi_R) \xrightarrow{\{\ell_R\}}_{\sigma_R} v \right\}$$

where v is a new (resolution) node and $\rho(\varphi)$ denotes the root node of φ. The resolved atom ℓ is such that $\ell = \ell_L \sigma_L = \overline{\ell_R} \sigma_R$ or $\ell = \overline{\ell_L} \sigma_L = \ell_R \sigma_R$.
- **Contraction:** *If ψ' is a proof $\langle V', E', \Gamma' \rangle$ and σ is a unifier of $\{\ell_1, \ldots \ell_n\}$ with $\{\ell_1, \ldots \ell_n\} \subseteq \Gamma'$, then $\lfloor \psi \rfloor_{\{\ell_1, \ldots \ell_n\}}^{\sigma}$ denotes a proof $\langle V, E, \Gamma \rangle$ s.t.*

$$V = V' \cup \{v\} \qquad E = E' \cup \{\rho(\psi') \xrightarrow{\{\ell_1, \ldots \ell_n\}}_{\sigma} v\} \qquad \Gamma = (\Gamma' \setminus \{\ell_1, \ldots \ell_n\}) \sigma \cup \{\ell\}$$

where v is a new (contraction) node, $\ell = \ell_k \sigma$ (for any $k \in \{1, \ldots, n\}$) and $\rho(\varphi)$ denotes the root node of φ. □

The resolution and contraction (factoring) rules described above are the standard rules of the resolution calculus, except for the fact that we do not require resolution to use most general unifiers. The presentation of the resolution rule here uses two substitutions, in order to explicitly handle the necessary renaming of variables, which is often left implicit in other presentations of resolution. When we write $\psi_L \odot_{\ell_L \ell_R} \psi_R$, we assume that the omitted substitutions are such that the resolved atom is most general. We write $\lfloor \psi \rfloor$ for an arbitrary maximal contraction, and $\lfloor \psi \rfloor^{\sigma}$ for a (pseudo-)contraction that does merge no literals but merely applies the substitution σ. When the literals and substitutions are irrelevant or clear from the context, we may write simply $\psi_L \odot \psi_R$ instead of $\psi_L \odot_{\ell_L \ell_R}^{\sigma_L \sigma_R} \psi_R$. The $\odot$ operator is assumed to be left-associative. In the propositional case, we omit contractions (treating clauses as sets instead of multisets) and $\psi_L \odot_{\ell \ell}^{\emptyset \emptyset} \psi_R$ is abbreviated by $\psi_L \odot_{\ell} \psi_R$.

If $\psi = \varphi_L \odot \varphi_R$ or $\psi = \lfloor \varphi \rfloor$, then φ, φ_L and φ_R are *direct subproofs* of ψ and ψ is a *child* of both φ_L and φ_R. The transitive closure of the direct subproof relation is the *subproof* relation. A subproof which has no direct subproof is an *axiom* of the proof.

V_ψ, E_ψ and Γ_ψ denote, respectively, the nodes, edges and proved clause (conclusion) of ψ. If ψ is a proof ending with a resolution node, then ψ_L and ψ_R denote, respectively, the left and right premises of ψ.

3   First-Order Challenges

In this section, we describe challenges that have to be overcome in order to successfully adapt LowerUnits to the first-order case. The first example illustrates the need to take unification into account. The other two examples discuss complex issues that can arise when unification is taken into account in a naively.

Example 1. Consider the following proof ψ, noting that the unit subproof η_2 is used twice. It is resolved once with η_1 (against the literal $p(W)$ and producing the child η_3) and once with η_5 (against the literal $p(X)$ and producing ψ).

$$\frac{\dfrac{\eta_1\colon p(W) \vdash q(Z) \qquad \eta_2\colon\ \vdash p(Y)}{\eta_3\colon\ \vdash q(Z)} \qquad \eta_4\colon p(X), q(Z) \vdash}{\dfrac{\eta_5\colon p(X)\vdash \qquad\qquad\qquad\qquad \eta_2}{\psi\colon \bot}}$$

The result of deleting η_2 from ψ is the proof $\psi \setminus \{\eta_2\}$ shown below:

$$\frac{\eta_1'\colon p(W) \vdash q(Z) \qquad \eta_4'\colon p(X), q(Z) \vdash}{\eta_5'\ (\psi')\colon p(W), p(X) \vdash}$$

Unlike in the propositional case, where the literals that had been resolved against the unit are all syntactically equal, in the first-order case, this is not necessarily the case. As illustrated above, $p(W)$ and $p(X)$ are not syntactically equal. Nevertheless, they are unifiable. Therefore, in order to reintroduce η_2', we may first perform a contraction, as shown below:

$$\frac{\dfrac{\dfrac{\eta_1'\colon p(W) \vdash q(Z) \qquad \eta_4'\colon p(X), q(Z) \vdash}{\eta_5'\colon p(X), p(Y) \vdash}}{\lfloor \eta_5' \rfloor\colon p(U) \vdash} \qquad \eta_2'\colon\ \vdash p(Y)}{\psi^*\colon \bot}$$

Example 2. There are cases, as shown below, when the literals that had been resolved away are not unifiable, and then a contraction is not possible.

$$\frac{\dfrac{\eta_4\colon r(X), p(b) \vdash s(Y) \qquad \dfrac{\eta_1\colon p(a) \vdash q(Y), r(Z) \qquad \eta_2\colon\ \vdash p(X)}{\eta_3\colon\ \vdash q(Y), r(Z)}}{\dfrac{\eta_5\colon p(b) \vdash s(Y), q(Y) \qquad\qquad \eta_6\colon s(Y) \vdash}{\dfrac{\eta_7\colon p(b) \vdash q(Y) \qquad \eta_8\colon q(Y) \vdash}{\eta_9\colon p(b) \vdash}}} \qquad \eta_2}{\psi\colon \bot}$$

If we attempted to postpone the resolution inferences involving the unit η_2 (i.e. by deleting η_2 and reintroducing it with a single resolution inference in the bottom of the proof), a contraction of the literals $p(a)$ and $p(b)$ would be needed. Since these literals are not unifiable, the contraction is not possible. Note that, in principle, we could still lower η_2 if we resolved it not only once but twice when reintroducing it in the bottom of the proof. However, this would lead to no compression of the proof's length.

The observations above lead to the idea of requiring units to satisfy the following property before collecting them to be lowered.

Definition 2. *Let η be a unit with literal ℓ and let $\eta_1, \ldots, \eta_n$ be subproofs that are resolved with η in a proof ψ, respectively, with resolved literals $\ell_1, \ldots, \ell_n$. η is said to satisfy the* pre-deletion unifiability property *in ψ if $\ell_1, \ldots, \ell_n$, and $\overline{\ell}$ are unifiable.*

Example 3. Satisfaction of the pre-deletion unifiability property is not enough. Deletion of the units from a proof ψ may actually change the literals that had been resolved away by the units, because fewer substitutions are applied to them. This is exemplified below:

$$
\frac{
\dfrac{\eta_1\colon r(Y), p(X, q(Y,b)), p(X,Y) \vdash \qquad \eta_2\colon\ \vdash p(U,V)}{\eta_3\colon r(V), p(U, q(V,b)) \vdash} \qquad \eta_4\colon\ \vdash r(W)
}{
\dfrac{\eta_5\colon p(U, q(W,b)) \vdash}{\psi\colon \bot} \qquad\qquad \eta_2
}
$$

If η_2 is collected for lowering and deleted from ψ, we obtain the proof $\psi \setminus \{\eta_2\}$:

$$
\frac{\eta_1'\colon r(Y), p(X, q(Y,b)), p(X,Y) \vdash \qquad \eta_4'\colon\ \vdash r(W)}{\eta_5'(\psi')\colon p(X, q(W,b)), p(X,W) \vdash}
$$

Note that, even though η_2 satisfies the pre-deletion unifiability property (since $p(X, q(Y,b))$ and $p(U, q(W,b))$ are unifiable), η_2 still cannot be lowered and reintroduced by a single resolution inference, because the corresponding modified post-deletion literals $p(X, q(W,b))$ and $p(X,W)$ are actually not unifiable.

The observation above leads to the following stronger property:

Definition 3. *Let η be a unit with literal ℓ_η and let $\eta_1, \ldots, \eta_n$ be subproofs that are resolved with η in a proof ψ, respectively, with resolved literals $\ell_1, \ldots, \ell_m$. η is said to satisfy the* post-deletion unifiability property *in ψ if $\ell_1^{\dagger\downarrow}, \ldots, \ell_m^{\dagger\downarrow}$, and $\overline{\ell_\eta^\dagger}$ are unifiable, where $\ell^\dagger$ is the literal in $\psi \setminus \{\eta\}$ corresponding to ℓ in ψ and $\ell_k^{\dagger\downarrow}$ is the descendant of $\ell_k^\dagger$ in the root of $\psi \setminus \{\eta\}$.*

4   A Linear Greedy Variant of First-Order LowerUnits

The examples shown in the previous section indicate that there are two main challenges that need to be overcome in order to generalize `LowerUnits` to the first-order case:

1. The deletion of a node changes literals. Since substitutions associated with the deleted node are not applied anymore, some literals become more general. Therefore, the reconstruction of the proof during deletion needs to take such changes into account.
2. Whether a unit should be collected for lowering must depend on whether the literals that were resolved with the unit's single literal are unifiable after they are propagated down to the bottom of the proof by the process of unit deletion. Only if this is the case, they can be contracted and the unit can be reintroduced in the bottom of the proof.

The first challenge can be overcome by keeping an additional map from old literals in the input proof to the corresponding more general changed literals in the output proof under construction. The second challenge is harder to overcome. In the propositional case, collecting units and deleting units can be done in two distinct and independent phases (as in `LowerUnits`). In the first-order case, on the other hand, these two phases seem to be so interlaced, that they appear to be in a deadlock: the decision to collect a unit to be lowered depends on what will happen with the proof after deletion, while deletion depends on knowing which units will be lowered. In a naive approach, the deletion algorithm may have to be executed once for every collected unit, and since the number of collected units is in the worst case linear in the length of the proof, the overall runtime complexity is quadratic with respect to the length of the proof.

This section presents `GreedyLinearFirstOrderLowerUnits` (Algorithm 1), a single traversal first-order adaptation of `LowerUnits`, which avoids the quadratic complexity and the implementation difficulties by: (1) ignoring the stricter post-deletion unifiability property and focusing instead on the pre-deletion unifiability property, which is easier to check (lines 13); and (2) employing a greedy contraction approach (lines 19–22) together with substitutions (lines 7–10), in order not to care about bookkeeping. By doing so, compression may not always succeed on all proofs (e.g. Example 3). When compression succeeds, the root clause of the generated proof will be the empty clause (line 24) and the generated proof may be returned. Otherwise, the original proof must be returned (line 25).

5   Experiments

A prototype of a (two-traversal) version of `GreedyLinearFirstOrderLowerUnits` has been implemented in the functional programming language Scala as part of the Skeptik library (https://github.com/Paradoxika/Skeptik) [3]. Before evaluating this algorithm, we first generated several benchmark proofs. This was done by executing the SPASS (http://www.spass-prover.org/) theorem prover on 2280

Input: a proof ψ
Output: a compressed proof ψ^*
Data: a map $.'$, eventually mapping any φ to $\mathtt{delete}(\varphi, \text{Units})$

1   $D \leftarrow \varnothing$;   // set for storing subproofs that need to be deleted
2   Units $\leftarrow \varnothing$;   // stack for storing collected units

3   **for** *every subproof φ, in a top-down traversal of ψ* **do**
4   **if** *φ is an axiom* **then** $\varphi' \leftarrow \varphi$;
5   **else if** $\varphi = \varphi_L \odot_{\ell_L \ell_R}^{\sigma_L \sigma_R} \varphi_R$ **then**
6   **if** $\varphi_L \in D$ *and* $\varphi_R \in D$ **then** add φ to D ;
7   **else if** $\varphi_L \in D$ **then** $\varphi' \leftarrow \lfloor \varphi'_R \rfloor^{\sigma_R}$;
8   **else if** $\varphi_R \in D$ **then** $\varphi' \leftarrow \lfloor \varphi'_L \rfloor^{\sigma_L}$;
9   **else if** $\ell \notin \Gamma_{\varphi'_L}$ **then** $\varphi' \leftarrow \lfloor \varphi'_L \rfloor^{\sigma_L}$;
10   **else if** $\bar{\ell} \notin \Gamma_{\varphi'_R}$ **then** $\varphi' \leftarrow \lfloor \varphi'_R \rfloor^{\sigma_R}$;
11   **else** $\varphi' \leftarrow \varphi'_L \odot_{\ell_L \ell_R}^{\sigma_L \sigma_R} \varphi'_R$;
12   **else if** $\varphi = \lfloor \varphi_c \rfloor_{\{\ell_1, \ldots, \ell_n\}}^{\sigma}$ **then** $\varphi' \leftarrow \lfloor \varphi'_c \rfloor_{\{\ell_1, \ldots, \ell_n\}}^{\sigma}$;

13   **if** *φ is a unit with more than one child satisfying the pre-deletion unifiability property* **then**
14   **push** φ' onto Units ;
15   **add** φ to D ;

   // Reintroduce units
16 $\psi^* \leftarrow \psi'$;
17 **while** Units $\neq \varnothing$ **do**
18     $\varphi' \leftarrow$ **pop** from Units ;
19     $\psi_{\text{next}}^* \leftarrow \lfloor \psi^* \rfloor$;
20   **while** $\Gamma_{\psi_{\text{next}}^*} \neq \psi^*$ **do**
21       $\psi^* \leftarrow \psi_{\text{next}}^*$;
22       $\psi_{\text{next}}^* \leftarrow \lfloor \psi^* \rfloor$;
23   **if** *$\psi^* \odot \varphi'$ is well-defined* **then** $\psi^* \leftarrow \psi^* \odot \varphi'$;
24 **if** $\Gamma_{\psi^*} = \bot$ **then return** ψ^*;
25 **else return** ψ;

Algorithm 1. `GreedyLinearFirstOrderLowerUnits` (single traversal)

real first-order problems without equality of the TPTP Problem Library (among them, 1032 problems are known to be unsatisfiable). In order to generate pure resolution proofs, the advanced inference rules of SPASS were disabled. The Euler Cluster at the University of Victoria was used and the time limit was 300 sec per problem. Under these conditions, SPASS generated 308 proofs.

The evaluation of `GreedyLinearFirstOrderLowerUnits` was performed on a laptop (2.8GHz Intel Core i7 processor with 4 GB of RAM (1333MHz DDR3) available to the Java Virtual Machine). For each benchmark proof ψ, we measured the time needed to compress the proof ($t(\psi)$) and the compression ratio $((|\psi| - |\alpha(\psi)|)/|\psi|)$, where $|\psi|$ is the length of ψ (i.e. the number of axioms, resolution and contractions (ignoring substitutions)) and $\alpha(\psi)$ is the result of applying `GreedyLinearFirstOrderLowerUnits` to ψ. The raw data is available at: http://www.math.uvic.ca/~jgorzny/data/.

The proofs generated by SPASS were small (with lengths from 3 to 49). These proofs are specially small in comparison with the typical proofs generated by SAT- and SMT-solvers, which usually have from a few hundred to a few million nodes. The number of proofs (compressed and uncompressed) per length is shown in Fig. 1 (b). Uncompressed proofs are those which had either no lowerable units to lower or for which `GreedyLinearFirstOrderLowerUnits` failed and returned the original proof. Such failures occurred on only 14 benchmark proofs. Among the smallest of the 308 proofs, very few proofs were compressed. This is to be expected, since the likelihood that a very short proof contain a lowerable unit (or even merely a unit with more than one child) is low. The proportion of compressed proofs among longer proofs is, as expected, larger, since they have more nodes and it is more likely that some of these nodes are lowerable units. 13 out of 18 proofs with length greater than or equal to 30 were compressed.

Figure 1 (a) shows a box-whisker plot of compression ratio with proofs grouped by length and whiskers indicating minimum and maximum compression ratio achieved within the group. Besides the median compression ratio (the horizontal thick black line), the chart also shows the mean compression ratios for all proofs of that length and for all compressed proofs (the red cross and the blue circle). In the longer proofs (length greater than 34), the median and the means are in the range from 5 % to 15 %, which is satisfactory in comparison with the total compression ratio of 7.5 % that has been measured for the propositional `LowerUnits` algorithm on much longer propositional proofs [4].

Figure 1 (c) shows a scatter plot comparing the length of the input proof against the length of the compressed proof. For the longer proofs (circles in the right half of the plot), it is often the case that the length of the compressed proof is significantly lesser than the length of the input proof.

Figure 1 (d) plots the cumulative original and compressed lengths of all benchmark proofs (for an x-axis value of k, the cumulative curves show the sum of the lengths of the shortest k input proofs). The total cumulative length of all original proofs is 4429 while the cumulative length of all proofs after compression is 3929. This results in a total compression ratio of 11.3 %, which is impressive, considering that the inclusion of all the short proofs (in which the presence of lowerable units is a priori unlikely) tends to decrease the total compression ratio. For comparison, the total compression ratio considering only the 100 longest input proofs is 18.4 %.

Figure 1 also indicates an interesting potential trend. The gap between the two cumulative curves seems to grow superlinearly. If this trend is extrapolated, progressively larger compression ratios can be expected for longer proofs. This is compatible with Theorem 10 in [6], which shows that, for proofs generated by eagerly resolving units against all clauses, the propositional `LowerUnits` algorithm can achieve quadratic assymptotic compression. SAT- and SMT-solvers based on CDCL (Conflict-Driven Clause Learning) avoid eagerly resolving unit

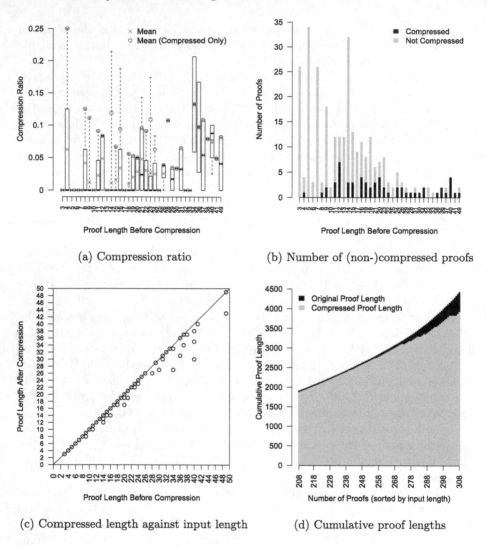

(a) Compression ratio

(b) Number of (non-)compressed proofs

(c) Compressed length against input length

(d) Cumulative proof lengths

Fig. 1. Experimental results

clauses by dealing with unit clauses via boolean propagation on a conflict graph and extracting subproofs from the conflict graph with every unit being used at most once per subproof (even when it was used multiple times in the conflict graph). Saturation-based automated theorem provers, on the other hand, might be susceptible to the eager unit resolution redundancy described in Theorem 10 [6]. This potential trend would need to be confirmed by further experiments with more data (more proofs and longer proofs).

The total time needed by SPASS to solve the 308 problems for which proofs were generated was 2403 s, or approximately 40 min (running on the Euler Cluster and including parsing time and proof generation time for each problem). The total time for `GreedyLinearFirstOrderLowerUnits` to be executed on all 308 proofs was just under 5 sec on a simple laptop (including parsing each proof). Therefore, `GreedyLinearFirstOrderLowerUnits` is a fast algorithm. For a very small overhead in time (in comparison to proving time), it may simplify the proof considerably.

6 Conclusions and Future Work

`GreedyLinearFirstOrderLowerUnits` is our first attempt to lift a propositional proof compression algorithm to the first-order case. We consider it a prototype, useful to evaluate this approach. The results discussed in the previous section are encouraging, especially in comparison with existing results for the propositional case. In the near future, we shall seek improvements of this algorithm as well as other ways to overcome the difficulties related to the post-deletion unifiability property. The difficulties related to unit reintroduction suggest that other propositional proof compression algorithms that do not require reintroduction (e.g. `RecyclePivotsWithIntersection` [6]) might need less sophisticated bookkeeping when lifted to first-order.

The efficiency and versatility of contemporary automated theorem provers depend on inference rules and techniques that go beyond the pure resolution calculus. The generalization of compression algorithms to support such extended calculi will be essential for their usability on a wider range of problems.

References

1. Clarke, E.M., Voronkov, A. (eds.): LPAR-16 2010. LNCS, vol. 6355. Springer, Heidelberg (2010)
2. Bar-Ilan, O., Fuhrmann, O., Hoory, S., Shacham, O., Strichman, O.: Linear-time reductions of resolution proofs. In: Chockler, H., Hu, A.J. (eds.) HVC 2008. LNCS, vol. 5394, pp. 114–128. Springer, Heidelberg (2009)
3. Boudou, J., Fellner, A., Woltzenlogel Paleo, B.: Skeptik: a proof compression system. In: Demri, S., Kapur, D., Weidenbach, C. (eds.) IJCAR 2014. LNCS, vol. 8562, pp. 374–380. Springer, Heidelberg (2014)
4. Boudou, J., Woltzenlogel Paleo, B.: Compression of propositional resolution proofs by lowering subproofs. In: Galmiche, D., Larchey-Wendling, D. (eds.) TABLEAUX 2013. LNCS, vol. 8123, pp. 59–73. Springer, Heidelberg (2013)
5. Fontaine, P., Merz, S., Woltzenlogel Paleo, B.: Exploring and exploiting algebraic and graphical properties of resolution. In: 8th International Workshop on SMT, Edinburgh (2010)
6. Fontaine, P., Merz, S., Woltzenlogel Paleo, B.: Compression of propositional resolution proofs via partial regularization. In: Bjørner, N., Sofronie-Stokkermans, V. (eds.) CADE 2011. LNCS, vol. 6803, pp. 237–251. Springer, Heidelberg (2011)
7. Hetzl, S., Leitsch, A., Reis, G., Weller, D.: Algorithmic introduction of quantified cuts. Theor. Comput. Sci. **549**, 1–16 (2014)
8. Hetzl, S., Leitsch, A., Weller, D., Woltzenlogel Paleo, B.: Herbrand sequent extraction. In: Autexier, S., Campbell, J., Rubio, J., Sorge, V., Suzuki, M., Wiedijk, F. (eds.) AISC 2008, Calculemus 2008, and MKM 2008. LNCS (LNAI), vol. 5144, pp. 462–477. Springer, Heidelberg (2008)

9. Woltzenlogel Paleo, B.: Atomic cut introduction by resolution: proof structuring and compression. In: Clarke, E.M., Voronkov, A. (eds.) LPAR-16 2010. LNCS, vol. 6355, pp. 463–480. Springer, Heidelberg (2010)

10. Rollini, S.F., Bruttomesso, R., Sharygina, N.: An efficient and flexible approach to resolution proof reduction. In: Raz, O. (ed.) HVC 2010. LNCS, vol. 6504, pp. 182–196. Springer, Heidelberg (2010)

11. Sutcliffe, G.: The TPTP problem library and associated infrastructure: The FOF and CNF parts, v3.5.0. J. Autom. Reasoning **43**(4), 337–362 (2009)

12. Vyskočil, J., Stanovský, D., Urban, J.: Automated proof compression by invention of new definitions. In: Clarke, E.M., Voronkov, A. (eds.) LPAR-16 2010. LNCS, vol. 6355, pp. 447–462. Springer, Heidelberg (2010)

Beagle – A Hierarchic Superposition Theorem Prover

Peter Baumgartner[1]($\boxtimes$), Joshua Bax[1], and Uwe Waldmann[2]

[1] NICTA and Australian National University, Canberra, Australia
peter.baumgartner@nicta.com.au
[2] MPI für Informatik, Saarbrücken, Germany

Abstract. *Beagle* is an automated theorem prover for first-order logic modulo built-in theories. It implements a refined version of the hierarchic superposition calculus. This system description focuses on *Beagle*'s proof procedure, background reasoning facilities, implementation, and experimental results.

1 Introduction

This paper describes the automated theorem prover *Beagle*. *Beagle* implements hierarchic superposition [2,7], a calculus for automated reasoning in a hierarchic combination of first-order logic and some background theory. Currently implemented background theories are linear integer and linear rational arithmetics. *Beagle* features new simplification rules for theory reasoning, and well-known ones used for non-theory reasoning. *Beagle* also implements calculus improvements like *weak abstraction* [7] and determining (un)satisfiability w.r.t. quantification over finite integer domains [6].

Beagle is written in Scala, including its implementation of the background reasoners from scratch. Existing SMT solvers can be coupled as background reasoners as well via a textual SMT-LIB interface. *Beagle* accepts problem specifications written in the TFF format (the typed version of the TPTP problem specification language) and in the SMT-LIB format [4,16].

In this paper we describe the above features in more detail and report on *Beagle*'s performance on the TPTP problem library [17] and SMT-LIB benchmarks [16].

2 Hierarchic Theorem Proving

Hierarchic superposition [2,7] is a calculus for automated reasoning in a hierarchic combination of first-order logic and some background theory.[1] We assume

---

NICTA is funded by the Australian Government through the Department of Communications and the Australian Research Council through the ICT Centre of Excellence Program.

[1] Due to a lack of space, we can only give a brief overview of the calculus and of the semantics of hierarchic specifications. We refer to [7] for the details.

A.P. Felty and A. Middeldorp (Eds.): CADE-25, LNAI 9195, pp. 367–377, 2015.
DOI: 10.1007/978-3-319-21401-6_25

that we have a *background ("BG")* prover that accepts as input a set of clauses over a *BG signature* $\Sigma_B = (\Xi_B, \Omega_B)$, where Ξ_B is a set of *BG sorts* and Ω_B is a set of *BG operators*. Terms/clauses over Σ_B and BG-sorted variables are called *BG terms/clauses*. The BG prover decides the satisfiability of Σ_B-clause sets w.r.t. a *BG specification*, that is, a class of term-generated Σ_B-interpretations (called *BG models*) that is closed under isomorphisms. The BG specification is usually some kind of arithmetic, so Ξ_B could for instance be $\{int\}$ and Ω_B could contain the BG operators $0, 1, -1, 2, -2, \ldots, +, -, <, \leq$. We assume that Ω_B also contains infinitely many *parameters* $\alpha, \beta, \ldots$, that is, additional constants that may be interpreted freely by arbitrary elements of the appropriate domain in different models.

The *foreground ("FG")* theorem prover accepts as input a set of clauses over an extended signature $\Sigma = (\Xi, \Omega)$, where $\Xi_B \subseteq \Xi$ and $\Omega_B \subseteq \Omega$. The sorts in $\Xi_F = \Xi \setminus \Xi_B$ and the operator symbols in $\Omega_F = \Omega \setminus \Omega_B$ are called *FG sorts* and *FG operators*. For instance, Ξ_F might be $\{list\}$ and Ω_F could then contain the operators empty $: \rightarrow list$, cons $: int\ list \rightarrow list$, and length $: list \rightarrow int$. We use sans-serif letters to denote FG operators. A Σ-term is an *FG term* if it is not a BG term, that is, if it contains at least one FG operator or FG-sorted variable. We emphasize that for an FG operator f $: \xi_1 \ldots \xi_n \rightarrow \xi_0$ in Ω_F any of the ξ_i may be a BG sort. Consequently, a FG term like length(cons(5, empty)) may have a BG sort. Every FG operator f with a BG range sort $\xi_0 \in \Xi_B$ is called a *free BG-sorted (FG) operator*.

The intended semantics is that of *conservative extensions of the BG specification*, i.e., Σ-interpretations whose restriction to Σ_B is a model of the BG specification. In the concrete example above, that means that we are only interested in models of the FG clause set whose interpretation of the BG sort *int* is the same as in the given BG models; the models may neither identify different elements of the interpretation of *int*, say 5 and 7, nor interpret BG-sorted FG term like length(cons(5, empty)) by some new element that was not present before. We refer to satisfiability in this sense as $\mathcal{B}$-*satisfiability*.

Hierarchic theorem proving requires "abstracting out" terms in preparation for inference rule applications.[2] *Weak abstraction* introduced in [7] abstracts out BG terms other than number constants and variables that occur as subterms of FG terms, so for instance the clause cons(α, cons(x, empty)) $\approx$ cons(3, cons(5 + 2, y)) is converted into

$$z_1 \not\approx \alpha \vee z_2 \not\approx 5 + 2 \vee \text{cons}(z_1, \text{cons}(x, \text{empty})) \approx \text{cons}(3, \text{cons}(z_2, y))$$

whereas length(cons(x, y)) $\approx$ length(y)+1 is left unchanged. See [7] for a discussion of the benefits of weak abstraction.

The FG prover saturates the set of Σ-clauses using the inference rules of hierarchic superposition, such as, e.g.,

$$\text{Negative superposition} \quad \frac{l \approx r \vee C \qquad s[u] \not\approx t \vee D}{\text{abstr}((s[r] \not\approx t \vee C \vee D)\sigma)}$$

---

[2] *Abstracting out* a term t that occurs in a clause $C[t]$ means replacing $C[t]$ by $x \not\approx t \vee C[x]$ for a new variable x.

where σ is an mgu of l and u, These inference rules inherit the ordering and selection restrictions of the standard superposition inference rules [1]; in addition they have the new restriction that only the FG parts of clauses are overlapped. Since the standard inferences can destroy weak abstraction, it is furthermore necessary to apply an explicit weak abstraction to the conclusion. The term ordering $\succ$ needs to satisfy certain properties specific to the hierarchic case, e. g., any concrete number must be smaller than any other ground term. The calculus includes the generic, semantically defined notion of redundancy well-known from standard superposition.

Since the standard superposition inference rules are modified in such a way that only the FG parts of clauses are overlapped, that means in particular that they are never applied to BG clauses derived during the saturation. Such clauses are instead passed to the BG prover. The BG prover implements an inference rule

$$\text{Close} \ \frac{C_1 \quad \cdots \quad C_n}{\square} \qquad \begin{array}{l} \text{if } C_1, \ldots, C_n \text{ are BG clauses and} \\ \{C_1, \ldots, C_n\} \text{ is } \mathcal{B}\text{-unsatisfiable.} \end{array}$$

As soon as one of the two provers derives the empty clause, the input clause set has been shown to be $\mathcal{B}$-unsatisfiable.

The Define *rule.* One of the requirements for the refutational completeness of hierarchic superposition is *sufficient completeness*, i. e., the property that every ground BG-sorted FG term is equal to some BG term. Sufficient completeness of a set of Σ-clauses is a property that is not even recursively enumerable. For certain classes of Σ-clause sets, however, it is possible to establish a variant of sufficient completeness automatically [7,11]: If all BG-sorted FG terms in the input are ground, it suffices to show that each BG-sorted FG term *in the input* is equal to some BG term. This can be achieved by adding a *definition* $\alpha_t \approx t$ for every BG-sorted FG term t occurring in a clause $C[t]$, where α_t is a new parameter (BG constant); afterwards $C[t]$ can be replaced by $C[\alpha_t]$. See [7] for the corresponding *Define* inference rule.

3 Background Reasoning

BG reasoning is represented in *Beagle* as theory specific modules, *"solvers"*, that implement a specific interface. Every solver needs to provide a decision procedure for $\mathcal{B}$-satisfiability of sets of BG clauses. The syntactic fragment of these BG clauses depends on whether free BG-sorted constants are declared as FG constants or as parameters. The former case leads to the A-fragment, the latter case to the EA-fragment. Moreover, if the solver also supports quantifier elimination (QE), the decision procedure receives sets of ground clauses only.

In all examples we use linear integer arithmetic as the background theory.

Quantifier elimination. The solver interface supports specifying a quantifier elimination procedure on BG *formulas*. It is used for eliminating variables that only occur in BG literals. This way, e. g., the clause $\mathsf{P}(x) \vee \neg(x < y) \vee \neg(y < 3)$

becomes $P(x) \vee \neg(x < 2)$ by QE of y from $\neg(x < y) \vee \neg(y < 3)$. Applying QE this way for clause simplification may destroy refutational completeness, since in general the simplification result is possibly larger (under the clause ordering) than the clause being simplified. To avoid this problem, *Beagle* uses QE for clause simplification only during preprocessing. It also stores with each BG clause its ground version, which is sent to the decision procedure.

Splitting. *Beagle* optionally splits (in particular) BG clauses into variable disjoint subclauses. If QE is available and *Beagle* is instructed to, a ground version of each BG clause is added to the current clause set, which is split exhaustively into unit clauses by *Beagle*'s splitting rule. As a consequence, the decision procedure receives sets of unit clauses only, akin to SMT solvers.

Simplification. *Beagle* removes disequations of certain forms from clauses by *unabstraction*. For example, if cautious simplification is chosen, literals of the form $x \not\approx d$ are removed by unabstraction only if d is a concrete number.

Aggressive simplification enables the unabstraction of any term, including FG terms. It can possibly break completeness, since there is no guarantee that the unabstracted clause $C[t]$ is smaller than all possible instances of $C[x] \vee x \not\approx t$. The simplification level of FG clauses is controlled by *Beagle*. Typically only the results of cautious unabstraction are kept; aggressive unabstraction is used to derive unit clauses which may demodulate other clauses, but the unit clauses resulting from unabstraction are not kept.

Beyond that, simplification of arithmetic terms is realized through an internal data structure for simplification rules. The current simplification rules are hard-coded in *Beagle*'s implementation language Scala. Hence they are not user-modifiable, but we might change this in a future version. For each solver there are two sets of simplification rules: *cautious* simplification rules, which are known to preserve both sufficient completeness and refutational completeness, and *aggressive* simplification rules, which in general do not preserve these properties. See below for examples.

Solvers. *Beagle* implements solvers for linear integer arithmetic (LIA) and linear rational arithmetic (LRA). It also accepts linear real arithmetic but the differences are merely syntactic. Alternatively to the built-in LIA solver, existing SMT solvers can be coupled via a textual SMT-LIB interface.

3.1   Linear Integer Arithmetic

Quantifier elimination. The built-in LIA solver is based on Cooper's quantifier elimination algorithm and its improvements as introduced in [8]. It accepts arbitrary BG formulas, in particular conjunctions of clauses. The code structure follows roughly the algorithm described in [10]. The LIA solver is used for both deciding satisfiability of sets of BG clauses and for the elimination of variables as described above.

We have integrated several improvements into Cooper's algorithm to make it more practical. For example, in conjunctions that contain the atomic formulas $\alpha < 5$ and $\alpha < 3$ the former can be removed; a limited form of subsumption. Other simple and cheap techniques include elimination of variables that admit unbounded solutions and elimination of equations $\alpha \approx t$ where α does not occur in t. Furthermore, if a conjunction contains the atomic formulas $s_1 < \alpha, \ldots, s_m < \alpha$ and $\alpha < t_1, \ldots, \alpha < t_n$, given that α does not occur elsewhere, then α can be removed by exhaustive resolution. (Resolution of $s < \alpha$ and $\alpha < t$ yields $s + 1 < t$.) If α does occur somewhere else, then this form of resolution can still be used to prove unsatisfiability when $s + 1 < t$ is false. This is similar to the first step of the Omega test for deciding Presburger arithmetic [14].

The improvements mentioned above often help to solve problems much faster.[3] However, some of them are effective only on conjunctions of literals. In support of this, our algorithm deviates from the standard Cooper algorithm by multiplying out disjunctions that arise from quantifier instantiation. This often avoids deeply structured "or-and" formulas. As a special case, disjunctive normal form is preserved by solving and multiplying out the conjunctions separately.

The final step of Cooper's algorithm involves instantiation over representatives of congruence classes of solutions for the target variable which quite often lead to prohibitively large formulas. Using an improvement suggested in [10] Beagle occasionally defers this instantiation (based on the expected number of instances) until a later round of quantifier elimination. This is done by substituting a fresh variable and terms that describe the solution classes as occasionally a shorter proof of satisfiability/unsatisfiability can be found using a different variable.

When the Close rule applies *Beagle* determines a minimal unsatisfiable subset of the BG clauses passed to the decision procedure. This is advantageous for the main loop's dependency-directed backtracking since cases which only produce BG clauses that are irrelevant for unsatisfiability do not need to be backtracked to. Currently, minimal unsatisfiable subsets are determined by binary search on the whole clause set passed to the (built-in) LIA solver, or by unsatisfiable cores returned by Z3 [12] as a solver.

Simplification and arithmetic terms normalization. The cautious simplification rules for LIA comprise evaluation of arithmetic terms, e.g. $3 \cdot 5$, $3 < 5$, $\alpha + 1 < \alpha + 1$ (equal lhs and rhs terms in inequations), and rules for TPTP-operators, e.g., \$to_rat(5), \$is_int(3.5). For aggressive simplification, integer sorted subterms are brought into a polynomial-like form and are evaluated as much as possible. For example, the term $5 \cdot \alpha + f(3 + 6, \alpha \cdot 4) - \alpha \cdot 3$ becomes $2 \cdot \alpha + f(9, 4 \cdot \alpha)$. These conversions exploit associativity and commutativity of $+$ and $\cdot$. Pure BG formulas always produce proper polynomials, which can be used directly by the QE procedure without further conversions.

Aggressive simplification does not always preserve sufficient completeness. For example, in the clause set $N = \{\mathsf{P}(1 + (2 + \mathsf{f}(x))), \neg\mathsf{P}(1 + (x + \mathsf{f}(x)))\}$ the first clause is aggressively simplified, giving $N' = \{\mathsf{P}(3 + \mathsf{f}(x)), \neg\mathsf{P}(1 + (x + \mathsf{f}(x)))\}$.

---

[3] E.g., the GEG-problems in the TPTP problem library.

Notice that both N and N' are LIA-unsatisfiable, sgi(N) $\cup$ GndTh(LIA) is unsatisfiable, but sgi(N') $\cup$ GndTh(LIA) is satisfiable. Thus, N is (trivially) sufficiently complete while N' is not.

We have also implemented heuristics for normalizing equations and inequations for aggressive simplification. Inequations are normalized by first eliminating the operators $>$, $\geq$ and $\leq$ in terms of $<$. The QE procedure treats $<$ as a primitive, so this is a natural choice. Then, the monomials of the lhs and rhs polynomials are moved around so that only positive signs and only addition of monomials (not subtraction) results. The rationale is to normalize terms by removing unnecessary operators. Similar heuristics apply for equations, which also attempt to arrive at orientable equations. Normalizing (in)equations may remove or install sufficient completeness and destroy refutational completeness. Yet, in our experiments we found that aggressive simplification is far superior to cautious simplification in practice, hence it is enabled by default.

3.2   Other Arithmetic Features

Linear Rational Arithmetics. The solver for LRA comprises a Fourier-Motzkin style quantifier elimination procedure for eliminating BG variables as described in Sect. 3. The decision procedure implements the Simplex algorithm extended to strict inequalities [9]. The cautious simplification rules evaluate arithmetic subterms, and the aggressive simplification rules rewrite sub-terms towards a flat structure by exploiting AC-properties of the operators. Syntactic differences between concrete numbers aside, linear real arithmetics is treated by additional lemmas that are valid in real arithmetics. The LRA solver is not as far developed as the LIA solver.

Nonlinear Arithmetic. *Beagle* features a simplistic treatment of non-linear arithmetics. During preprocessing, every occurrence of a non-linear multiplication subterm $s \cdot t$ is replaced by nlpp(s, t), where nlpp is a dedicated foreground function symbol of the proper arity. As soon as s or t in nlpp(s, t) is replaced by a concrete number, the resulting nlpp is turned into a multiplication term again. In the LIA case, axioms for nlpp are added that express multiplication in terms of repeated addition.

4   Proof Procedure

This section provides a summary of *Beagle*'s proof procedure. The proof procedure follows by and large standard techniques, but treats BG formulas in a specific way on some occasions.

Preprocessing. *Beagle* accepts its input formulas in two alternative syntaxes, TPTP-TFF [19] and SMT-LIB (version 2) [4]. The SMT-LIB language is richer than the TPTP-TFF language due to its support for polymorphic sorts and functions. The SMT-LIB also features predefined theories such as arrays. *Beagle* automatically monomorphizes sorts and function symbols, and it generates array axioms as needed.

Both TPTP-TFF and SMT-LIB provide syntax for full first-order logic (not just clausal logic). *Beagle* has two translators into clause normal form (CNF), a standard one and a Tseitin-style translator which introduces definitions for "complex" subformulas. The default is the standard CNF translator, because it gave better results overall over the problems in the TPTP.

CNF transformation includes Skolemization of existentially quantified variables. *Beagle*'s CNF transformation treats existentially quantified integer variables in a special way, by removing them with QE instead of Skolemization, if possible. For example, the input formula $\forall x : \mathbb{Z}\ P(x) \vee \exists y : \mathbb{Z}\ y \not\approx x + 1$ becomes $\forall x : \mathbb{Z}\ P(x)$, whereas Skolemization would have given $\forall x : \mathbb{Z}\ P(x) \vee f(x) \not\approx x + 1$. In particular, if the input formulas are all BG formulas over the integers, no Skolem functions are introduced, and so *Beagle* is a decision procedure for that class.

Main loop and simplification. Beagle's main loop is the well-known "Discount loop". It maintains two clause sets, *Old* and *New*, where *Old* is initially empty and new is initialized with the input clauses. On each round, a *selected* clause is removed from *New* and simplified by the clauses from *Old* and *New*. The simplified selected clause then is put into *Old* and all inferences between it and the clauses in *Old* are carried out. The resulting clauses are simplified by the *Old* clauses and go into *New* again, this way closing the loop. If a BG clause results, the solver is called with the thus extended current set of all BG clauses.

Implemented simplification techniques include standard ones, like demodulation by unit clauses, proper subsumption deletion, and removing a positive literal L from a clause in presence of a unit clause that instantiates to the complement of L. All clauses in *Old* are mutually simplified. Backward simplification is optional.

By default, a split rule is enabled that breaks clauses into variable-disjoint subclauses and branches out correspondingly. Dependency-directed backtracking is used to avoid exploring irrelevant cases.

The default term ordering is LPO if BG theories are present, otherwise it is KBO. See [7] for properties of the LPO specific to hierarchic superposition.

Fairness. Fairness is achieved by a combination of clause weights and their derivation age. This is controlled by a parameter "weight-age-ratio", a nonnegative number saying how many lightest clauses are selected before an oldest clause is selected. Clause weights are computed in such a way that selection based on weights only would be a fair strategy. In our experiments we used a weight-age-ratio of five.

Auto mode. Beagle includes a simple auto mode. When on, *Beagle* first tries the default flag setting. If there is no conclusive result within half of the given time limit, *Beagle* starts again using a setting where BG variables in the input may be instantiated by BG-sorted FG terms, rather than only by BG terms.

5   Implementation

Beagle implements support for both the TPTP-TFF and SMT-LIB input languages using Scala's parser combinator library. *Beagle*'s internal formula representation follows TFF, so to support the SMT-LIB standard it must perform sort monomorphization and adding axioms for predefined theories like arrays. This is done with the help of the separately developed SMTtoTPTP library [5].

Beagle uses a simple term-indexing scheme which is essentially top symbol hashing. This is used to retrieve term positions eligible for superposition or demodulation within clauses.

Scala specific features. *Beagle* makes heavy use of many built in Scala datastructures, primarily `List`, `Vector` and `Map`. Not only are the implementations well optimised but they also provide powerful abstractions allowing for simple and maintainable code.

Scala's declarative style encourages the use of immutable values, which minimizes data duplication. Scala also provides a lazy evaluation feature, which we have found extremely useful for caching data: e. g. the computation of maximal literals in a clause can be deferred until the clause becomes eligible for an inference and it may never be computed if the clause is simplified first. We found that the Scala REPL interpreter is an invaluable tool for debugging: for example, one could take the (usually large) result of an invalid derivation and programmatically investigate it using functional operators like map or filter.

The simple structure of logic formulas is a good fit for property based testing libraries such as `scalacheck`[4] which use grammars to generate random test data. These data are used as input for properties given as universally quantified predicates.

6   Performance

TPTP. We tried *Beagle* on the first-order problems from the TPTP–v6.1.0 problem library [17] that involve some form of arithmetic, including non-linear, rational and real arithmetics. The experiments were carried out on a MacBook Pro with a 2.3 GHz Intel Core i7 processor and 16 GB of main memory. The CPU time limit was 180 s.

Although *Beagle* detected countersatisfiabilty of some of the (73) non-theorem problems, we discuss in the following the performance on the problems with a "theorem" or "unsatisfiable" status only. Of these 972 problems in total *Beagle* was able to prove 781 using automatic strategy selection. The backup strategy was attempted a total of 21 times and was successful in 15 cases, thereof 13 times in the TPTP DAT category.

Table 1 summarizes the results. Broken down by the TPTP problem category we see that *Beagle*'s best performance was on ARI, DAT and NUM. These are characterized by smaller problem sizes with an arithmetic reasoning component.

---

[4] http://scalacheck.org/.

Table 1. *Beagle* performance on the TPTP "theorem" or "unsatisfiable" problems. The first table breaks down the number of solved problems by category. The second table filters by problem rating. The column ≥ 0.6, for instance, means "all problems with a rating 0.6 or higher".

Category	ARI	DAT	GEG	HWV	MSC	NUM	PUZ	SEV	SWV	SWW	SYN	SYO
Total	539	103	5	88	2	43	1	6	2	177	1	3
Solved	531	98	5	0	2	41	1	2	2	97	0	2

Rating	≥ 0.0	≥ 0.1	≥ 0.2	≥ 0.3	≥ 0.4	≥ 0.5	≥ 0.6	≥ 0.7	≥ 0.8	≥ 0.9	1.0	
Total	972	853	771	527	391	343	253	180	129	97	97	
Solved	781	666	584	340	210	162	85	29	12	2	2	

On the other hand performance was much worse on those problems which involve large problem sizes such as SWW and SWV (translations of model-checking problems). *Beagle* failed to solve any HWV problems (large EPR encodings of bounded model-checking) due to the size of the formulas and emphasis on boolean reasoning. The remaining easy (rated < 0.1) problems that *Beagle* failed to solve were all non-theorems, most involving multiplication operators. The two solvable problems with a rating of 1.0 are ARI536=2.p and DAT086=1.p.

We have also coupled the SMT solver Z3 [12] as an alternative to the built-in LIA solver. In our experiments we also tried a modified split rule that leaves BG subclauses unsplit. In particular, BG clauses are never split then. The rationale is that letting the SMT solver deal with (non-unit) BG clauses might be better than the default FG splitting into sets of unit clauses. As an alternative to the built-in LIA solver and using the modified split rule or not hence gives four base configurations.

We ran *Beagle* in all four base configurations and several additional flag settings. But, surprisingly, Z3 does *not* give better results than the built-in solver. We found that the default split rule is superior to the modified one, both in conjunction with Z3 and the built-in solver. Over all settings, however, almost exactly the same problems are solvable with any of the two solvers and in roughly the same time. This finding might not carry over to problems that require more complex BG reasoning than those in the TPTP.

SMT-LIB. We tested *Beagle* on the 2014 release of SMT-LIB [16] focusing on the logics with an arithmetic component. Specifically these were ALIA, AUFLIA, UFLIA, UF_IDL (integer difference logic) and the corresponding quantifier free problem sets, including QF_LIA. (The LIA category was ignored as it contains only problems from the TPTP). We selected only those problems indicated as unsatisfiable in the problem description and *Beagle* was run with automatic strategy selection (as described above). We found a mix of results: *Beagle* was able to solve a few problems unsolved by SMT solvers[5] yet there were also quite a few problems that were marked as 'trivial' (all SMT solvers in the SMT-Eval 2013 can solve them in under five seconds), which *Beagle* could not solve. Overall

---

[5] For this we used the difficulty ratings given for SMT-Comp 2014.

Beagle solved the following problems by category (QF refers to the quantifier free fragment of the logic to the left):

Logic	ALIA	QF	AUFLIA	QF	UFLIA	QF	UFIDL	QF	QF_IDL	QF_LIA
Total	41	72	4	516	6602	195	62	335	694	2610
Solved	31	40	4	205	1736	155	42	29	24	28

In total *Beagle* solved 89 problems not solved by SMT solvers and these were divided among the following subcategories of 'UFLIA/sledgehammer':

Category	Arrow_Order	FFT	FTA	Hoare	StrongNorm	TwoSquares
Solved	17	2	34	20	2	14

There were many problems which *Beagle* could not parse, as it is not optimized for large problem sets. In total there were 1,391 trivial problems not solved by *Beagle*.

It was not possible to draw broad conclusions about which categories *Beagle* is best suited to. For example, all of the hardest problems *Beagle* solved were among the UFLIA benchmarks, but there were also at least 200 trivial problems from that category that were unsolved (in the 'simplify' and 'simplify2' subcategories). Also it was hypothesised that *Beagle* would perform much worse in the quantifier free fragment, and that was the case for QF_IDL and QF_LIA, but not so for QF_UFLIA and QF_AUFLIA.

CASC-J7. Most recently *Beagle* participated in the CASC-J7 competition [18]. in the TFA division (Typed First-order Arithmetic theorems). For this division the problem set consists of typed first-order problems with an arithmetic component over integers, rationals, or reals, of which roughly half were previously unseen by competitors.

Other solvers entered in the TFA category were *CVC4* [3], *SPASS+T* [13], *Zipperposition* (see [18]), and *Princess* [15]. In terms of overall problems solved *Beagle* placed third equal with 173/200 solutions, only three fewer than the winning solver *CVC4*. *Beagle* performed quite well in terms of mean efficiency (solutions per second multiplied by number of solutions); it was outperformed only by *CVC4* [6].

7 Availability

Beagle is available at https://bitbucket.org/peba123/beagle under a GNU General Public license. The distribution includes the Scala source code and a ready-to-run Java jar-file. A more experimental version of Beagle is maintained at https://bitbucket.org/joshbax189/beagle.

---

[6] For an explanation of how mean efficiency is computed see the CASC-J7 proceedings [18].

References

1. Bachmair, L., Ganzinger, H.: Rewrite-based equational theorem proving with selection and simplification. J. Logic Comput. **4**(3), 217–247 (1994)
2. Bachmair, L., Ganzinger, H., Waldmann, U.: Refutational theorem proving for hierarchic first-order theories. Appl. Algebra Eng. Commun. Comput **5**, 193–212 (1994)
3. Barrett, C., Conway, C.L., Deters, M., Hadarean, L., Jovanović, D., King, T., Reynolds, A., Tinelli, C.: CVC4. In: Gopalakrishnan, G., Qadeer, S. (eds.) CAV 2011. LNCS, vol. 6806, pp. 171–177. Springer, Heidelberg (2011)
4. Barrett, C., Stump, A., Tinelli, C.: The SMT-LIB Standard: Version 2.0. In: Gupta, A., Kroening, D.(eds.) SMT Workshop (2010)
5. Baumgartner, P.: SMTtoTPTP - A converter for theorem proving formats. In: Felty, A., Middeldorp, A. (eds.) CADE-25. LNCS(LNAI), pp. 152–169. Springer, Heidelberg (2015)
6. Baumgartner, P., Bax, J., Waldmann, U.: Finite quantification in hierarchic theorem proving. In: Demri, S., Kapur, D., Weidenbach, C. (eds.) IJCAR 2014. LNCS, vol. 8562, pp. 152–167. Springer, Heidelberg (2014)
7. Baumgartner, P., Waldmann, U.: Hierarchic superposition with weak abstraction. In: Bonacina, M.P. (ed.) CADE 2013. LNCS, vol. 7898, pp. 39–57. Springer, Heidelberg (2013)
8. Cooper, D.C.: Theorem proving in arithmetic without multiplication. In: Machine Intelligence, vol. 7, pp. 91–99. American Elsevier, New York (1972)
9. Dutertre, B., de Moura, L.: A fast linear-arithmetic solver for DPLL(T). In: Ball, T., Jones, R.B. (eds.) CAV 2006. LNCS, vol. 4144, pp. 81–94. Springer, Heidelberg (2006)
10. Harrison, J.: Handbook of Practical Logic and Automated Reasoning. Cambridge University Press, Cambridge (2009)
11. Kruglov, E., Weidenbach, C.: Superposition decides the first-order logic fragment over ground theories. Mathematics in Computer Science, pp. 1–30 (2012)
12. de Moura, L., Bjørner, N.S.: Z3: an efficient SMT solver. In: Ramakrishnan, C.R., Rehof, J. (eds.) TACAS 2008. LNCS, vol. 4963, pp. 337–340. Springer, Heidelberg (2008)
13. Prevosto, V., Waldmann, U.: SPASS+T. In: Sutcliffe, G., Schmidt, R., Schulz, S. (eds.) ESCoR: Empirically Successful Computerized Reasoning. CEUR Workshop Proceedings, pp. 18–33. Seattle, WA, USA (2006)
14. Pugh, W.: The Omega test: a fast and practical integer programming algorithm for dependence analysis. In: ACM/IEEE Conference on Supercomputing, pp. 4–13. ACM (1991)
15. Rümmer, P.: A constraint sequent calculus for first-order logic with linear integer arithmetic. In: Cervesato, I., Veith, H., Voronkov, A. (eds.) LPAR 2008. LNCS (LNAI), vol. 5330, pp. 274–289. Springer, Heidelberg (2008)
16. SMT-LIB, The Satisfiability Modulo Theories Library. http://smt-lib.org/
17. Sutcliffe, G.: The TPTP problem library and associated infrastructure: the FOF and CNF Parts, v3.5.0. J. Autom. Reasoning **43**(4), 337–362 (2009)
18. Sutcliffe, G.: The 7th IJCAR automated theorem proving system competition - CASC-J7. AI Communications, **28** (2015). To appear
19. Sutcliffe, G., Schulz, S., Claessen, K., Baumgartner, P.: The TPTP typed first-order form with arithmetic. In: Bjørner, N., Voronkov, A. (eds.) LPAR-18 2012. LNCS, vol. 7180, pp. 406–419. Springer, Heidelberg (2012)

The Lean Theorem Prover (System Description)

Leonardo de Moura[1]($\boxtimes$), Soonho Kong[2], Jeremy Avigad[2], Floris van Doorn[2], and Jakob von Raumer[2]

[1] Microsoft Research, Redmond, USA
leonardo@microsoft.com
[2] Carnegie Mellon University, Pittsburgh, USA
soonhok@cs.cmu.edu, {avigad,fpv,javra}@andrew.cmu.edu

Abstract. Lean is a new open source theorem prover being developed at Microsoft Research and Carnegie Mellon University, with a small trusted kernel based on dependent type theory. It aims to bridge the gap between interactive and automated theorem proving, by situating automated tools and methods in a framework that supports user interaction and the construction of fully specified axiomatic proofs. Lean is an ongoing and long-term effort, but it already provides many useful components, integrated development environments, and a rich API which can be used to embed it into other systems. It is currently being used to formalize category theory, homotopy type theory, and abstract algebra. We describe the project goals, system architecture, and main features, and we discuss applications and continuing work.

1 Introduction

Formal verification involves the use of logical and computational methods to establish claims that are expressed in precise mathematical terms. These can include ordinary mathematical theorems, as well as claims that pieces of hardware or software, network protocols, and mechanical and hybrid systems meet their specifications. In practice, there is not a sharp distinction between verifying a piece of mathematics and verifying the correctness of a system: formal verification requires describing hardware and software systems in mathematical terms, at which point establishing claims as to their correctness becomes a form of theorem proving. Conversely, the proof of a mathematical theorem may require a lengthy computation, in which case verifying the truth of the theorem requires verifying that the computation does what it is supposed to do.

Automated theorem proving focuses on the "finding" aspect, and strives for power and efficiency, often at the expense of guaranteed soundness. Such systems can have bugs, and typically the correctness relies on extensive testing. In

J. Avigad—Work supported by the AFOSR under MURI Grant Number FA9550-15-1-0053.

J. von Raumer—Visiting student from Karlsruhe Institute of Technology, sponsored by the Baden-Württemberg-Stipendium.

contrast, interactive theorem proving focuses on the *verification* aspect of theorem proving, requiring that every claim is supported by a proof in a suitable axiomatic foundation. This sets a very high standard: every rule of inference and every step of a calculation has to be justified by appealing to prior definitions and theorems, all the way down to basic axioms and rules. In fact, most such systems provide fully elaborated *proof objects* that can be communicated to other systems and checked independently. Constructing such proofs typically requires much more input and interaction from users, but it allows us to obtain deeper and more complex proofs. Finally, we remark that some automated theorem provers do generate proof certificates that can be verified by other systems [8].

The *Lean Theorem Prover*[1] aims to bridge the gap between interactive and automated theorem proving, by situating automated tools and methods in a framework that supports user interaction and the construction of fully specified axiomatic proofs. The goal is to support both mathematical reasoning and reasoning about complex systems, and to verify claims in both domains. Lean is released under the Apache 2.0 license, a permissive open source license that permits others to use and extend the code and mathematical libraries freely. At Carnegie Mellon University, Lean is already being used to formalize category theory, homotopy type theory, and abstract algebra. Lean is an ongoing, long-term effort, and much of the potential for automation will be realized only gradually over time.

Lean's small, trusted kernel is based on dependent type theory, with several configuration options. It can be instantiated with an impredicative sort or propositions, `Prop`, to provide a version of the Calculus of Inductive Constructions (CIC) [5,6]. Moreover, `Prop` can be marked proof-irrelevant if desired. Without an impredicative `Prop`, the kernel implements a version of Martin-Löf type theory [12,23]. In both cases, Lean provides a sequence of non-cumulative type universes, with universe polymorphism.

Lean is meant to be used both as a standalone system and as a software library. SMT solvers can use the Lean API to create proof terms that can be independently checked. The API can be used to export Lean proofs to other systems based on similar foundations (e.g., Coq [3] and Matita [1]). Lean can also be used as an efficient proof checker, and definitions and theorems can be checked in parallel using all available cores on the host machine. When used as a proof assistant, Lean provides a powerful elaborator that can handle higher-order unification, definitional reductions, coercions, overloading, and type classes, in an integrated way. Lean allows users to provide definitions and theorems using a declarative style resembling Mizar [20] and Isabelle/Isar [24]. Lean also provides *tactics* as an alternative (more imperative) approach to constructing (proof) terms as in Coq, HOL-Light [11], Isabelle [17] and PVS [19]. Moreover, the declarative and tactic styles can be freely mixed together.

Lean includes two libraries of formally verified mathematics and basic datastructures. The standard library uses a kernel instantiated with an impredicative and proof-irrelevant `Prop`. This library supports constructive and classical users,

---

[1] http://leanprover.github.io.

and the following axioms can be optionally used: propositional completeness, function extensionality, and strong indefinite description. Lean also contains a library tailored for Homotopy Type Theory (HoTT) [23], using a predicative and proof relevant instantiation of the kernel and higher inductive types (HITs). Future plans to support HoTT include sorts for fibrant type universes.

2   The Kernel

Lean's trusted kernel is implemented in two layers. The first layer contains the type checker and APIs for creating and manipulating terms, declarations, and the environment. This layer consists of 6k lines of C++ code. The second layer provides additional components such as inductive families and quotient types. The inductive family component consists of 700 lines of code. When the kernel is instantiated, one selects which of these components should be used. We have tried to maintain the number of objects manipulated by the kernel to a minimum: the list consists of terms, universe terms, declarations, and environments. Identifiers are encoded as *hierarchical names* [14], i.e. lists of strings/numbers, such as $x.y.1$.

Terms. The term language is a dependent λ-calculus. A term can be a free variable (also called a local constant), a bound variable, a constant (parameterized by universe terms), a function application $f\ t$, a lambda abstraction $\lambda x : A, t$, a function space $\Pi x : A, B$, a sort $\texttt{Type}\ u$ (where u is a universe term), a metavariable, or a macro $m[t_1, \ldots, t_n]$.

Sorts. The sorts $\texttt{Type}\ u$ are used to encode the infinite sequence of universes $\texttt{Type}_0, \texttt{Type}_1, \texttt{Type}_2, \ldots$ An *explicit* universe term is of the form $\mathsf{s}^k\ \mathsf{z}$ (for $k \geq 0$), where z denotes the base universe zero, and s denotes the *successor* universe operator. We use $\texttt{Type}\ \mathsf{z}$ to represent $\texttt{Prop}$ in kernel instantiations that support it. To support universe polymorphism, we also have universe parameters (an identifier), and the operators $\texttt{max}\ u_1\ u_2$ and $\texttt{imax}\ u_1\ u_2$. The universe term $\texttt{max}\ u_1\ u_2$ denotes the universe that is greater than or equal to u_1 and u_2, and is equal to one of them. The universe term $\texttt{imax}\ u_1\ u_2$ denotes the universe zero if u_2 denotes zero, and $\texttt{max}\ u_1\ u_2$ otherwise. The operator $\texttt{imax}$ is only needed for kernel instantiations that have an impredicative $\texttt{Prop}$. In these kernels, given $A :$ $\texttt{Type}\ u_1$ and $B : \texttt{Type}\ u_2$, the type of $\Pi x : A, B$ is $\texttt{Type}\ (\texttt{imax}\ u_1\ u_2)$. The $\texttt{imax}$ operator makes sure that $\Pi x : A, B$ is a proposition when B is a proposition.

Free and bound variables. Free variables have a unique identifier and a type, and bound variables are just a number (a de Bruijn index). By storing the type with each free variable, we do not need to carry around contexts in the type checker and normalizer. As described in [14], this representation simplifies the implementation considerably, and it also minimizes the number of places where calculations with de Bruijn indices must be performed.

Metavariables. In Lean, users may provide *partial constructions*, i.e., constructions containing "holes" that must be filled by the system. These holes (also known as placeholders) are internally represented as metavariables that must be replaced by closed terms that are synthesized by the system. Since only closed terms can be assigned to metavariables, a metavariable that occurs in a context records the parameters it depends on. For example, we encode a hole in the context $(x : nat)$ $(y : bool)$ as $?m\ x\ y$, where $?m$ is a fresh metavariable. As with free variables, every metavariable has a type. We also have universe metavariables to represent "holes" in universe terms.

Macros. Macros, which can be viewed as *procedural attachments*, provide more efficient ways of storing and working with terms. Each macro must provide two procedures, namely, type inference and macro expansion. The type inference procedure `minfer` is responsible for computing the type of a macro application $m[t_1, \ldots, t_n]$, and the macro expansion procedure `mexpand` must expand/eliminate the macro application. The point is that, given a term t of the form $m[t_1, \ldots, t_n]$, `minfer`(t) may be able to infer the type of `mexpand`(t) more efficiently than the kernel type checker, and t may be more compact than `mexpand`(t).

We also use macros to store annotations and hints used by automation such as rewriters and decision procedures. Each macro has a *trust level* represented by a natural number. When the Lean kernel is initialized, the user must provide a trust level ℓ, and the kernel then refuses any term that contains a macro with trust level greater than or equal to ℓ. A kernel initialized with trust level zero does not accept any macro, forcing any macro occurring in declarations to be expanded. The idea is that macros are not part of the trusted code base, but users may choose to trust them "most of the time" when formalizing a system and/or theorem. Note that an independent type checker for Lean does not need to implement support for metavariables or macros.

Environments. An environment stores a sequence of declarations. The kernel currently supports three different kinds of declarations: axioms, definitions and inductive families. Each has a unique identifier, and can be parameterized by a sequence of universe parameters. Every axiom has a type, and every definition has a type and a value. A constant in Lean is just a reference to a declaration. The main task of the kernel is to type check these declarations and refuse type incorrect ones. The kernel does not allow declarations containing metavariables and/or free variables to be added to an environment. Environments are never destructively updated, and are implemented using pure red-black trees, where the keys are hierarchical names.

Inductive families. Inductive families [9] are a form of simultaneously defined collection of algebraic data-structures which can be parameterized over values as well as types. Each inductive family definitions produces introduction rules, elimination rules, and computational rules as described in [9]. As in the CIC, the instances of an inductive family can be in `Prop`, and special rules are used to make sure the eliminator is compatible with proof irrelevance. Finally, when proof irrelevance is enabled in the kernel, axiom K [22] "computes" in Lean (a similar feature is available in Agda [18]). In contrast to Coq, Lean does not have fix-point expressions, `match` expressions, or a termination checker in the kernel. Instead, recursive definitions and pattern matching are compiled into eliminators outside of the kernel.

The type checker. To minimize the amount of code duplication, the type checker plays two roles. First, it is used to validate any declaration sent to the kernel before adding it to an environment. Second, it is used by elaboration procedures that try to synthesize holes in terms provided by the user. Consequently, the type checker is capable of processing terms containing metavariables. When a term contains metavariables, the type checker may produce unification constraints, in which case the resultant type is correct only if the unification constraints can be resolved.

3   Elaboration

The task of the elaborator is to convert a partially specified expression into a fully specified, type-correct term. When typing in a term, users can leave arguments implicit by entering them with an underscore (i.e., a "hole"), leaving it to the elaborator to infer a suitable value. One can also mark arguments implicit by putting them in curly brackets when defining a function, to indicate that they should generally be inferred rather than entered explicitly. For example, the standard library defines the identity function as:

```
definition id {A : Type} (a : A) : A := a
```

As a result, the user can write `id a` rather than `id A a`. It is fairly routine to infer the type `A` given `a : A`. Often the elaborator needs to infer an element of a Π-type, which constitutes a *higher-order* problem. For example, if `e : a = b` is a proof of the equality of two terms of some type `A`, and `H : P` is a proof of some expression involving `a`, the term `subst e H` denotes a proof of the result of replacing some or all the occurrences of `a` in `P` with `b`. Here not just the type `A` is inferred, but also an expression `C : A → Prop` denoting the context for the substitution, that is, the expression with the property that `C a` "reduces" to `P`. Such expressions can be ambiguous. For example, if `H` has type `R (f a a) a`, then with `subst e H` the user may have in mind `R (f b b) b` or `R (f a b) a` among other interpretations, and the elaborator has to rely on context and a backtracking search to find an interpretation that fits. Similar issues arise with proofs by induction, which require the system to infer an induction predicate.

The elaborator should also respect the computational interpretation of terms. It should recognize the equivalence of terms $(\lambda x,\ t)s$ and $t[s/x]$ under beta reduction, as well as $(s,\ t).1$ and s under the reduction rule for pairs. (Terms that are equivalent modulo such reductions are said to be *definitionally equal*.) Unfolding definitions and reducing projections is especially crucial when working with algebraic structures, where many basic expressions cannot even be seen to be type correct without carrying out such reductions.

Lean's elaborator also supports ad-hoc overloading; for example, we can use notation a + b for addition on the natural numbers, integers, and additive groups simultaneously. Each possible interpretation becomes a choice-point in the elaboration process. The elaborator can also detect the need to insert a coercion, say, from nat to int, or from the class of rings to the class of additive groups.

Lean also supports the use of Haskell-style *type classes*. For example, we can define a class has_mul A of types A with an associated multiplication operator, and a class semigroup A of types A with semigroup structure, as follows:

```
structure has_mul [class] (A : Type) :=
(mul : A → A → A)
structure semigroup [class] (A : Type) extends has_mul A :=
(mul_assoc : ∀a b c, mul (mul a b) c = mul a (mul b c))
```

We can then declare appropriate instances of these classes, and instruct the elaborator to synthesize such instances when processing the notation a * b or the generic theorem mul.assoc.

Finally, definitions and proofs can invoke *tactics*, that is, user-defined or built-in procedures that construct various subterms. The elaborator needs to call these procedures at appropriate times during the elaboration process to fill in the corresponding components of a term.

The interactions between these components are subtle, and the main difficulty is that the elaborator has to deal with them all at the same time. A definition or proof may give rise to thousands of constraints requiring a mixture of higher-order unification, disambiguation of overloaded symbols, insertion of coercions, type class inference, and computational reduction. To solve these, the elaborator uses nonchronological backtracking and a carefully tuned algorithm [7].

Recursive equations. Lean provides natural ways of defining recursive functions, performing pattern matching, and writing inductive proofs. Behind the scenes, these are "compiled" down into eliminators and auxiliary definitions automatically generated by Lean whenever we declare an inductive family. This compiler is based on ideas from [4,10,13,21]. The default compilation method supports structural recursion, i.e. recursive applications where one of the arguments is a subterm of the corresponding term on the left-hand-side. Lean can also compile recursive equations using well-founded recursion. The main advantage of the default compilation method is that the recursive equations hold definitionally.

The compiler also supports dependent pattern matching for indexed inductive families. For example, we can define the type vector A n of vectors of type A and length n as follows:

```
inductive vector (A : Type) : nat → Type :=
| nil {} : vector A zero
| cons    : Π {n : nat}, A → vector A n → vector A (succ n)
```

We can then define a function map that applies a binary function f to elements of vectors of type A and B, to produce a vector of elements of type C:

```
definition map {A B C : Type} (f : A → B → C) :
       Π {n : nat}, vector A n → vector B n → vector C n
| map nil      nil     := nil
| map (a::va) (b::vb) := f a b :: map va vb
```

Note that we can omit "unreachable" cases such as map nil (a :: va) because the input vectors have the same length. Behind the scenes, a lot of boilerplate code is needed to reduce these definitions to eliminators for the inductive family.

Type classes. Any family of inductive types can be marked as a *type class*. Then we can declare particular elements of a type class to be *instances*. These provide hints to the elaborator: any time the elaborator is looking for an element of a type class, it can consult a table of declared instances to find a suitable element. What makes type class inference powerful is that one can *chain* instances, that is, an instance declaration can in turn depend on other instances. This causes class inference to chain through instances recursively, backtracking when necessary. The Lean type class resolution procedure can be viewed as a simple λ-Prolog interpreter [15], where the Horn clauses are the user declared instances.

For example, the standard library defines a type class inhabited to enable type class inference to infer a "default" or "arbitrary" element of types that contain at least one element.

```
inductive inhabited [class] (A : Type) : Type :=
mk : A → inhabited A
```

Elements of the class inhabited A are of the form inhabited.mk a, for some element a : A. The following function extracts the corresponding element:

```
definition default (A : Type) [H : inhabited A] : A :=
inhabited.rec (λa, a) H
```

The annotation [H : inhabited A] indicates that H should be synthesized from instance declarations using type class resolution. We can then declare suitable instances for types like nat and Prop. The following declaration shows that if two types A and B are inhabited, then so is their product:

```
definition prod.is_inhabited [instance] {A B : Type}
   (H1 : inhabited A) (H2 : inhabited B) : inhabited (A × B) :=
inhabited.mk (default A, default B)
```

Declarative Proofs. Lean provides a rich notation declaration system [2], and it is used to support human readable proofs similar to the ones found in Mizar and Isabelle/Isar. For example, the have construct introduces an auxiliary subgoal in a longer proof. Internally, the notation have H : p, from s, t produces the term (λ(H : p), t) s. Similarly, show p, from t does nothing more than annotate t with its expected type p. Lean also provides alternative Mizar/Isar-inspired syntax for lambda abstractions: `assume H : p, t` and `take x : A, t`. Calculational proofs, which begin with the keyword `calc`, are a convenient notation for chaining intermediate results that are meant to be composed by basic principles such as the transitivity of equality. The set of binary relation predicates supported in calculational proofs can be freely extended by users. In the following example, we demonstrate some of these features:

```
theorem le.antisymm : ∀ {a b : ℤ}, a ≤ b → b ≤ a → a = b :=
take a b : ℤ, assume (H₁ : a ≤ b) (H₂ : b ≤ a),
obtain (n : ℕ) (Hn : a + n = b), from le.elim H₁,
obtain (m : ℕ) (Hm : b + m = a), from le.elim H₂,
have H₃ : a + of_nat (n + m) = a + 0, from
... -- suppressed rest of the proof due to space limitations
have H₆ : n = 0, from nat.eq_zero_of_add_eq_zero_right H₅,
show a = b, from
  calc
    a = a + 0     : add_zero
    ... = a + n   : H₆
    ... = b       : Hn
```

Namespaces. Lean provides the ability to group definitions, as well as meta-objects such as notation declarations, coercions, rewrite rules and type classes, into nested, hierarchical *namespaces*. The `open` command brings the shorter names and all meta-objects into the current context.

The tactic framework. Tactics provide an alternative approach to constructing terms. We can view a term as a representation of a construction or mathematical proof; tactics are commands, or instructions, that describe how to build such a term. Most automation available in Lean is integrated into the system as tactics. For example, Lean contains a `rewrite` tactic that provides a basic mechanism for performing rewriting. The tactic framework provides a general mechanism for synthesizing metavariables. In this framework, we say a metavariable is a *goal*. A *proof state* contains a sequence of goals; postponed unification constraints; and a substitution which stores already assigned metavariables. A *tactic* is a function that maps a proof state into a stream of proof states, implemented as a lazy list [16]. This is important because some tactics may produce an unbounded stream of proof states. Lean provides all usual combinators (also known as *tacticals*) available in other interactive theorem provers, such as `andthen`, `orelse`, and `try`. Lean also provides the tacticals `par` (for executing tactics concurrently in multiple cores), and `tryfor T n` that fails if tactic `T` does not terminate in `n` milliseconds. Lean also comes equipped with basic tactics such as `apply`, `intro`, `generalize`, `rewrite`, etc. The complete list of tactics is described in [2]. Wherever a term is expected, Lean allows us to insert instead a `begin ... end` block, composed of a sequence of tactics separated by commas. Here is a small example using tactics:

```
theorem test (p q : Prop) : p → q → p ∧ q ∧ p :=
begin
  intro Hp, intro Hq,
  apply and.intro, exact Hp, apply and.intro,
  exact Hq, exact Hp
end
```

4 The User Interface

Lean's standard integrated development environment (IDE) is based on the Emacs editor, and provides continuous elaboration and checking. In the background, the source text is continuously analyzed and annotated with semantic information as it is being edited by the user. The interaction between editor and prover is performed by an asynchronous protocol which exploits parallelism, multi-core hardware, and incremental compilation. The native interface provides all standard features found in advanced IDEs, such as hyperlinks, auto-completion, syntax highlighting, error highlighting, etc. Users can view automatically synthesized terms, implicit coercions, and overloading resolution. If a user makes changes to a file higher in the dependency chain, everything is recompiled in the background, and with caching the changes are propagated almost immediately.

The Javascript bindings for Lean do not contain any native code, and can be used in any modern web browser. They are intended for web applications such as web IDEs[2], "live" tutorial/documentation[3] and online exercises. We have used this infrastructure to develop course material for an interactive theorem proving course[4] being offered in the spring of 2015 at CMU.

5 Conclusion

Lean has been designed with the goal of obtaining a theorem proving system which has all of the following features: an expressive logical foundation for writing mathematical specifications and proofs; an interactive and supportive user interface and environment; a flexible framework for supporting automation; and a rich API that can be used to embed this functionality into other systems. Lean already provides a novel elaboration procedure that can handle higher-order unification, definitional reductions, coercions, overloading, and type classes, in an integrated way. It has a relatively small trusted kernel, making the task of implementing a reference/independent type checker for Lean much simpler. It is also quite fast, with support for multi-core machines and coarse and fine grain parallelism. Lean is an ongoing and long-term effort, and future plans include extensive search procedures, decision procedures, better support for homotopy type theory, and an independent type checker.

---

[2] http://leanprover.github.io/live.

[3] http://leanprover.github.io/tutorial.

[4] http://www.cs.cmu.edu/~emc/15815-s15.

References

1. Asperti, A., Ricciotti, W., Sacerdoti Coen, C., Tassi, E.: The Matita interactive theorem prover. In: Bjørner, N., Sofronie-Stokkermans, V. (eds.) CADE 2011. LNCS, vol. 6803, pp. 64–69. Springer, Heidelberg (2011)
2. Avigad, J., de Moura, L., Kong,S.: Theorem Proving in Lean (2015). http://leanprover.github.io/tutorial/tutorial.pdf
3. Barras, B., Boutin, S., Cornes, C., Courant, J., Filliatre, J.-C., Gimenez, E., Herbelin, H., Huet, G., Munoz, C., Murthy, C. et al.: The Coq proof assistant reference manual: Version 6.1 (1997)
4. Cockx, J., Devriese, D., Piessens, F.: Pattern matching without K. In: Proceedings of the 19th ACM SIGPLAN International Conference on Functional Programming, pp. 257–268. ACM (2014)
5. Coquand, T., Huet, G.: The calculus of constructions. Inf. Comput. **76**(2–3), 95–120 (1988)
6. Coquand, T., Paulin, C.: Inductively defined types. In: COLOG-88 (Tallinn, 1988), pp. 50–66. Springer, Berlin (1990)
7. de Moura, L., Avigad, J., Kong, S., Roux, C.: Elaboration in dependent type theory. Preprint (arXiv)
8. Delahaye, D., Woltzenlogel Paleo, B. (eds.): All about proofs, proofs for all. Mathematical Logic and Foundations, vol. 55 (2015)
9. Dybjer, P.: Inductive families. Formal Aspects Comput. **6**(4), 440–465 (1994)
10. Goguen, H.H., McBride, C., McKinna, J.: Eliminating dependent pattern matching. In: Futatsugi, K., Jouannaud, J.-P., Meseguer, J. (eds.) Algebra, Meaning, and Computation. LNCS, vol. 4060, pp. 521–540. Springer, Heidelberg (2006)
11. Harrison, J.: HOL light: an overview. In: Berghofer, S., Nipkow, T., Urban, C., Wenzel, M. (eds.) TPHOLs 2009. LNCS, vol. 5674, pp. 60–66. Springer, Heidelberg (2009)
12. Martin-Löf, P.: Intuitionistic type theory. Bibliopolis (1984)
13. McBride, C., Goguen, H.H., McKinna, J.: A few constructions on constructors. In: Filliâtre, J.-C., Paulin-Mohring, C., Werner, B. (eds.) TYPES 2004. LNCS, vol. 3839, pp. 186–200. Springer, Heidelberg (2006)
14. McBride, C., McKinna, J.: Functional pearl: I am not a number-I am a free variable. In: Proceedings of the 2004 ACM SIGPLAN Workshop on Haskell, Haskell 2004, pp. 1–9. ACM, New York (2004)
15. Miller, D., Nadathur, G.: Programming with Higher-Order Logic. Cambridge University Press, Cambridge (2012)
16. Nipkow, T., Paulson, L.C.: Isabelle-91. In: Kapur, Deepak (ed.) CADE 1992. LNCS, vol. 607. Springer, Heidelberg (1992)
17. Nipkow, T., Paulson, L.C., Wenzel, M.: Isabelle/HOL: a proof assistant for higher-order logic, vol. 2283. Springer Science and Business Media (2002)
18. Norell, U.: Dependently typed programming in Agda. In: Koopman, P., Plasmeijer, R., Swierstra, D. (eds.) AFP 2008. LNCS, vol. 5832, pp. 230–266. Springer, Heidelberg (2009)
19. Owre, S., Rushby, J., Shankar, N.: PVS: a prototype verification system. In: Kapur, Deepak (ed.) CADE 1992. LNCS, vol. 607. Springer, Heidelberg (1992)
20. Rudnicki, P.: An overview of the Mizar project. In: Proceedings of the 1992 Workshop on Types for Proofs and Programs, pp. 311–330 (1992)
21. Slind, K.: Function definition in higher-order logic. In: von Wright, Joakim, Harrison, J., Grundy, John (eds.) TPHOLs 1996. LNCS, vol. 1125. Springer, Heidelberg (1996)

22. Streicher, T.: Investigations into intensional type theory. Ph.D. thesis, LMU (1993)
23. The Univalent Foundations Program. Homotopy Type Theory: Univalent Foundations of Mathematics. Institute for Advanced Study (2013)
24. Wenzel, M.M.: Isabelle/Isar - a versatile environment for human-readable formal proof documents. Technical report (2002)

System Description: E.T. 0.1

Cezary Kaliszyk[1], Stephan Schulz[2], Josef Urban[3]([⊠]), and Jiří Vyskočil[4]

[1] University of Innsbruck, Innsbruck, Austria
cezary.kaliszyk@uibk.ac
[2] DHBW Stuttgart, Stuttgart, Germany
schulz@eprover.org
[3] Radboud University Nijmegen, Nijmegen, The Netherlands
josef.urban@gmail.com
[4] Czech Technical University in Prague, Prague, Czech Republic
vyskoj1@fel.cvut.cz

Abstract. E.T. 0.1 is a meta-system specialized for theorem proving over large first-order theories containing thousands of axioms. Its design is motivated by the recent theorem proving experiments over the Mizar, Flyspeck and Isabelle data-sets. Unlike other approaches, E.T. does not learn from related proofs, but assumes a situation where previous proofs are not available or hard to get. Instead, E.T. uses several layers of complementary methods and tools with different speed and precision that ultimately select small sets of the most promising axioms for a given conjecture. Such filtered problems are then passed to E, running a large number of suitable automatically invented theorem-proving strategies. On the large-theory Mizar problems, E.T. considerably outperforms E, Vampire, and any other prover that does not learn from related proofs. As a general ATP, E.T. improved over the performance of unmodified E in the combined FOF division of CASC 2014 by 6 %.

1 Introduction

The latest release of the TPTP benchmark library [22], TPTP 6.1.0, contains 20646 problems for theorem provers. More than a third of these problems have more than 100 axioms, more than 10 % (2664) have more than 1000 axioms, and more than 5 % (1231) have more than 10000 axioms.

Traditional (pre-1990) automated theorem proving (ATP) did not focus on such large problems. First experience with larger problems came from Quaife's work in the early 1990s [19]. Quaife identified the selection of relevant axioms as a possible way to handle large specifications, but did not offer detailed solutions.

Currently, large ATP problems are coming from ATP-to-ITP (interactive theorem provers) linkups (*hammers* [3]) such as Sledgehammer [16], HolyHammer [11,12] and MizAR [9], and common-sense reasoning [18] (or reasoning with the world's knowledge [25]) problems. Another interesting recent source of larger

C. Kaliszyk—Supported by the Austrian Science Fund FWF grant P26201.
J. Urban—Supported by NWO grant nr. 612.001.208.

ATP problems is the work in Tarskian geometry by Beeson and Wos [2], containing in some cases over 300 clauses. We strongly believe that today's age of Big Data will lead to more and more large-theory problems, including problems generated from Wikipedia [6], biology textbook encoding [5], and other science domains. Strong and practically usable methods and systems for proving large problems will be crucial for meaningful use of ATP in these new domains.

In the context of ATP/ITP cooperation, a number of methods have been developed recently to attack large problems. In pure ATP systems there has so far been basically just one (highly successful) method for dealing with large problems: the SInE heuristic invented by Hoder, implemented first as a standalone filter [29] and then inside Vampire [7,15] and E [20].

This paper describes E.T., a general large-theory ATP system based on E. It uses a combination of several methods transferred from the recent AITP[1] research. In Sect. 6 we show that the first version of E.T. already performs very well on large problems from the MPTP2078 benchmark, improving over plain E and Vampire by 98 % resp. 22 %. When used as a general ATP, E.T. has improved the performance of plain E in the combined FOF division of CASC 2014 by 6 %. E.T. is available at http://mws.cs.ru.nl/~urban/et10.

2   Overview of E.T

E.T. is intended for solving ATP problems that have one defining feature: they contain a large fraction of axioms that are not necessary for proving the conjecture. A secondary aspect of such large problems is that they often contain lemmas that can be used to construct alternative proofs. Thus it is useful to have a portfolio of complementary strategies that can select different promising axiom subsets and optimize the proof search over them.

We assume a setting in which a sequence of independent problems is presented to the system. In partcular, we do not assume that many related problems are being solved so that one could use consistent symbol and formula names and previous proofs in learning how to prove the next problems. Neither do we assume that common axioms are pre-loaded and expensively pre-processed once. Instead, the reading and preprocessing is done independently for each query.

This setting corresponds to the FOF division of CASC, which uses problems of various origins. It excludes some of the strongest and most obvious learning methods [13] possible in the "Large Theory Batch" scenario, where many problems share a background theory. However, ideas can still be transferred. For example, strategy invention as done by BliStr [28] can be used on sets of problems that do not share symbols and formula names.[2]

Similarly, one can use a number of symbolic and statistical methods successfully used in AITP for extracting useful feature characterizations of the large

---

[1] We will use AITP as an abbreviation for the ATP/ITP cooperation, hinting also at the AI aspects of that topic.

[2] Although never publicly described, similar methods used are one of the main dark sources of Vampire's success.

number of formulas. This has to be done much faster when solving a single large problem, making use of a layered architecture where the layers have different speed/precision trade-offs. Such layered (also called *early-exit*) approaches have been explored in information retrieval and particularly in web-search ranking systems [4], from which large-theory systems like E.T. can draw useful analogies. The extracted features are then used in E.T. as an input to several premise selection algorithms, such as our custom version of the Meng-Paulson filter (MePo) [17] and a non-learning version of the distance-weighted k-nearest neighbor (k-NN) algorithm [10]. The high-level description of E.T.'s processing chain is as follows:

1. The large input problem (potentially containing millions of axioms) is first reduced to several thousands of axioms using three differently parameterized (and reasonably complementary) non-strict versions of E's fast generalized SInE filter (see Sect. 4.1). Further processing is done with the union of these three filtered versions. This first reduction takes about 10 s for a problem with 500000 axioms. This speed is achieved by sharing several SInE preprocessing steps between the differently parameterized SInE selection passes.
2. Very long formulas are removed to prevent blow-ups in the following stages.[3]
3. If the original problem is in FOF, the reduced problem is clausified, removing very long clauses to prevent blow-ups in the feature generation phase.
4. Several tools are used to compute features of the formulas (or clauses) in the reduced problems. The current version can use as features symbols, (variable-normalized) shared terms, and all matching terms. See Sect. 3.
5. Several external premise selectors such as MePo and k-NN (Sect. 4) use these (probabilistically normalized) features to rank the axioms according to their estimated relevance to the conjecture, and problems with varied numbers of the top-ranking axioms are written.
6. Such problems are then (sequentially)[4] passed to E, which then typically applies much more restrictive SInE filtering to them, followed by a pool of the large-theory ATP strategies (Sect. 5).

3   Feature Generators

The feature generators are run on the problems reduced by the first fast SInE layer to several thousand formulas or clauses. Apart from using symbols as features, E.T. also enumerates all terms and subterms in the reduced problems' formulas, using E's fast shared term banks. Different variable normalization schemes can be used to increase or decrease the sharing of such features across formulas, providing different term-based similarity metrics.

A recent addition to the pool of such feature generators is a fast implementation of discrimination trees, enumerating for each formula ϕ all terms in all formulas,

---

[3] The current limit for formula/clause size is 5kb. This filters out only a few of formulas from the large corpora of interest, and in practice does not influence completeness.

[4] E.T. runs its strategies sequentially by default. It is also possible to run the strategies, premise selectors, and feature extractors in parallel when more cores are available.

that are more general than the terms in ϕ. Such features (when suitably probabilistically weighted) provide a better concept of similarity of formulas than any other syntactic features used so far [14]. This feature generator is the reason why the formula and clause sizes need to be kept below certain size (to prevent high quadratic factors), keeping the enumeration of matching terms within seconds for the reduced problem. For the weighting of features, we use the fast IDF scheme that adds practically no overhead while significantly improving the similarity metrics [10].

4   Premise Selectors

Since no previous proofs are available when running E.T., it relies on premise selectors that do not learn from related proofs. In particular, we use a number (currently 28) of differently parameterized E's generalized SInE filters limited to symbolic features in phase #1 and phase #6, and our modified version of the MePo filter using the more expensive features in phase #5. These two methods are briefly described below. Some additional performance is gained (see Table 2) by adding in phase #5 a non-learning version of the distance-weighted k-nearest neighbor (k-NN) algorithm [10], where each formula ϕ only carries the information that it is useful for proving facts with high feature overlap with ϕ.

4.1   Generalized SInE in E

E has native support for axiom selection in large theories. It implements a parametrized and efficient version of Hoder's SInE algorithm [7]. SInE is a fixed point algorithm. It starts with an initial set of formulas deemed necessary for the proof (usually including at least the conjecture), and successively adds formulas *related* to formulas already included, until a fixpoint is reached. Relatedness is based on sharing of at least one function- or predicate symbol between already selected clauses/formulas and new candidates. If all symbols are used, this corresponds to a classical relevance relation. However, this typically selects sets of clauses that are much larger than necessary, and only has limited utility.

Hoder correctly conjectured that rare symbols forge a stronger bond than common symbols, as formulas that share rare symbols are more likely to be part of the same microtheory. Using only the rarest symbols in a formula to find related clauses or formulas turned out to be too strict a relation. Thus, to allow the relaxation of this criterion, Hoder used a *benevolence* parameter that allows not just the symbol with the lowest frequency to be used for the relatedness relation, but also symbols wich occur up to a certain factor more often. E adds the *generosity* concept, which always uses the n least frequent symbols.

E allows the following parametrization of its SInE implementation.

- Frequency can be based on counting formulas/clauses containing a symbol, or on counting individual (sub-)terms.
- The initial set of the fixpoint process can consist of just the conjecture, or it can also include formulas defined as additional hypotheses for a particular proof problem by the user via the TPTP formula role.

– Benevolence and generosity can be set.
– While SInE usually runs to a fixpoint, E can terminate the process after a pre-determined number of iterations
– E also allows hard limits on the axiom set size, either in absolute terms, or as a fraction of the original specification.

This generalized SInE algorithm is implemented in E proper, where it is supported by a meta-level automatic parameterization. It also is available as a stand-alone tool that will efficiently apply several different parameterizations, sharing as much of the work as is possible. This includes parsing, frequency counting, and indexing of clauses and formulas by function symbol.

In E.T., SInE is used in two phases:

1. In phase #1 when the following non-strict (manually adjusted) SInE filters are used to make the later more expensive filters reasonably fast[5]:

    GSinE(CountFormulas, hypos, 3, , , 1500, 1.0)
    GSinE(CountFormulas, hypos, 1.2, , , 1500, 1.0)
    GSinE(CountFormulas, hypos, 30, , , 1500, 1.0)

   This phase takes about 10 s for problems with 500000 axioms, leaving enough time for the next phases when using 60 s time limit.
2. In phase #6, SInE is used in most of the E strategies that are ultimately run on the problems prepared by the previous filters. The parameters for SInE in these strategies are listed in Table 2. They are designed (jointly with other ATP parameters) automatically by the BliStr loop on suitable samples of large-theory problems (in this case Flyspeck and the 1000 Mizar@Turing training problems). The parameters that can be varied are the benevolence, number of iterations, and the absolute maximum axiom size. The rest of the parameters are fixed to the same values as in the non-strict filters above.

4.2   MePo3

MePo3 is an algorithm for assigning predicted relevance based on the Meng-Paulson relevance filter (MePo) [17] modified in several ways.

The algorithm is implemented as a recursive function which is given as input the set of all axioms A a set of weighted features F together with an increment number p. The initial value of F (F_0) are the features of the conjecture C, i.e., $F_0 = F(C)$, where $F(\phi)$ denotes features of a formula ϕ. The weights of the initial features are set to 1. Each recursive call i ($i > 0$) first computes the cosine distance between the remaining (not yet chosen) axioms in A and the given feature vector F_{i-1}. The axioms are then sorted by this distance, and the p axioms with the smallest distance are included in MePo3's answer. For each axiom ϕ included in the answer, its features $F(\phi)$ weighted by a factor of ϕ's distance to F_{i-1} are added to F_{i-1}, resulting in F_i which is then passed to the next recursive call. The algorithm is inspired by the Meng-Paulson filter, however we have introduced several changes:

---

[5] Parameters are used in the order given above. Missing parameters use E's built-in default values.

– MePo3 computes the distance as the cosine distance of the weighted feature vectors, rather than the proportion of relevant features to irrelevant ones.
– MePo includes in the answer the facts that are nearer to the conjecture than a given factor. This factor is modified in the recursive calls. This did not perform well for FOF problems, so we use an included-number in MePo3,
– MePo has a number of special cases that have been built into the algorithm to optimize for Isabelle/HOL, such as bonuses for elimination rules or facts present in the simplifier. MePo3 only has no such optimizations, instead relies on more advanced feature characterizations.

E.T. 0.1 always uses MePo3 with $p = 100$. The two parameters that are varied are the features used, and the number of best premises selected. The same is true for the distance-weighted k-NN, where in the simple scenario without previous proofs the number of best premises selected is always equal to the number of nearest neighbors k. Both MePo3 and k-NN are implemented efficiently in C++. Since the problems passed to them are already reduced to several thousands axioms, running these premise selectors is usually done within seconds, depending on the number of features used. As in E's SInE, a lot of work is shared between the different instances of k-NN and MePo3.

5   E Strategies and Global Optimization

When phase #5 premise selectors have finished, E is run on the filtered problems, using 36 different strategies (see Table 2). Four of these strategies are taken from E's exisiting portfolio, three are various versions of E's auto mode, which itself selects strategies based on problem characteristics.

The remaining 29 strategies have been designed automatically by BliStr, using the Mizar@Turing training problems and a small random sample of the Flyspeck problems. BliStr finds a small set of strategies that solve as many training problems as possible. This is done in an infinite loop which interleaves (i) fast iterative improvement of the strongest strategies on easy problems, (ii) slow evaluation of the newly invented strategies on all problems, and (iii) subsequent update of the candidate set of strong strategies and of the set of easy problems used for the next iterative improvement. The inclusion of the strategies into the final portfolio was done heuristically, based on their joint (greedy) coverage of the Mizar@Turing and Flyspeck problems.

Table 1. ATPs on the large and small MPTP2078 problems, using 60 s time limit.

ATP	E 1.8 (%)	Vampire 2.6 (%)	E.T. 0.1 (%)	Union (%)
Small problems	1213 (58)	1319 (63)	1357 (65)	1416 (68)
Large problems	580 (28)	940 (45)	1148 (55)	1208 (58)
Large/small ratio	0.48	0.71	0.85	0.85

6   Experimental Analysis

For the main evaluation we use the MPTP2078 benchmark [1], used for the large-theory division (Mizar@Turing) of the 2012 CASC@Turing automated reasoning competition [23]. These are 2078 related large-theory problems (conjectures) from the development of the general topological proof of the Bolzano-Weierstrass theorem extracted from the Mizar library. For each conjecture C we assume that all formulas stated earlier in the development can be used to prove C. This results in *large* ATP problems that have 1877 axioms on average. For each conjecture C we also know its ITP (Mizar) proof, from which we can (approximately [1]) determine a much smaller set of axioms that are sufficient for an ATP proof after the translation from Mizar to TPTP [26,27]. This gives rise to *small* ATP problems, where the ATP is significantly advised by the human author of the ITP proof. These small problems contain only 31 axioms on average.

Table 1 compares the performance of E 1.8, Vampire 2.6, and E.T. 0.1 on the MPTP2078 problems. All systems are run with 60 s time limit on a 32-core server with Intel Xeon E5-2670 2.6GHz CPUs, 128 GB RAM, and 20 MB cache per CPU. Each problem is assigned one CPU. On small problems, the three systems do not differ much. Vampire solves 9 % more problems than E, E.T solves 12 % more problems than E and 3 % more than Vampire. All systems together can solve 68 % of the small problems. Differences are larger on large problems, where Vampire solves 62 % more problems than E, E.T. solves 98 % more problems than E, and E.T. solves 22 % more problems than Vampire. An interesting metric is the ratio of the number of large problems solved to the number of small problems solved. For E this ratio is below 0.5, for Vampire it is 0.71, and for E.T. it is 0.85. This suggests that Vampire's large-theory techniques (primarily SInE) are much stronger than those used in the default mode of E, and shows that such techniques in E.T. (i.e., the premise-selection layers) are much more successful than the other systems.

Table 2 sheds more light on how E.T. achieves its performance on large problems. It lists the first 30 strategies (of 49 total) as tried sequentially by E.T., together with their success on the small and large problems. On the small problems, 79 % is solved already by the first two strategies that use only non-strict (or none) SInE filtering. On the large problems, these two strategies solve however only 34 % of the problems, while the next two restrictive strategies solve 40 % of the problems (they are given only the problems unsolved by the first two strategies). The third strategy does only two SInE iterations and takes only 60 best axioms, and the fourth strategy combines MePo (taking 128 best premises) with similarly restrictive SInE.

The second independent evaluation is the FOF division of CASC-J7 [21]. The results of the E-based ATPs and Vampire are shown in Table 3. The overall improvement of E.T. (using E version 1.8) over E (newer version 1.9) is 6 % (18 problems more), and on problems with equality this is 9 %. There is no improvement on the problems without equality. This is likely an artifact of E.T.'s strategy invention being done on Flyspeck and Mizar problems, which almost always use equality. While Vampire solves 11 % more problems than E.T., its

Table 2. The first 30 E.T. strategies run sequentially on the large and small MPTP2078 problems (60 s total time). E.T. exits immediatelly when a strategy finds a proof, therefore the success rates of the strategies are not directly comparable.

nr.	small	large	selector	premises	B_{SInE}	R_{SInE}	L_{SInE}	ATP strategy (name)
1	933	331	ful		2.0		500	b57035dec1c1e73fa888146ae569c7cc8f0
2	140	55	ful					G-E-.208.C18.F1.SE.CS.SP.PS.S0Y
3	28	262	ful		1.1	2	60	eba37f91665fc364eeb63558058658ee9a1
4		198	mepo3-nrm	128	2.0	2	100	88760aa43d575e84b7030b8a6188f74ba5f
5		43	mepo3-nrm	400				G-E-.208.B07.F1.SE.CS.SP.PS.S0Y
6	40	14	knn-nrm	8				G-E-.208.B07.F1.SE.CS.SP.PS.S0Y
7		30	mepo3-std	64	1.5	3	40	1b33b681d9260087e24d422ea286498f4a4
8		24	mepo3-std	512	1.5	3	40	1b33b681d9260087e24d422ea286498f4a4
9		10	mepo3-nrm	128	1.1	1	60	2af8141978cb6a38e97452761cdbd9e1007
10		29	mepo3-std	64				G-E-.208.C18.F1.SE.CS.SP.PS.S0Y
11		18	mepo3-std	512	1.2	2	20000	my8simple.sine13
12	13	6	knn-nrm	20	1.5	4	100	cfee9ff42189552c6557cda7d36f20820c8
13		7	mepo3-std	512	2.0		500	b57035dec1c1e73fa888146ae569c7cc8f0
14		11	mepo3-std	512	1.1	2	60	eba37f91665fc364eeb63558058658ee9a1
15		1	mepo3-std	512				G-E-.208.B07.F1.SE.CS.SP.PS.S0Y
16		2	knn-nrm	96	1.5	4	100	cfee9ff42189552c6557cda7d36f20820c8
17		8	mepo3-nrm	400	6.0	2	20000	92168ebc2ef464a6f2d6a311a4fa90219fd
18		10	knn-nrm	256	2.0	2	100	37be21ea059a2fcb865621e373a97f33a9d
19		5	knn-nrm	64	5.0	2	60	c284f1f10aedfccc65cdb7d9b1210ef814c
20	13	3	knn-nrm	8				G-E-.200.B02.F1.SE.CS.SP.PI.S0S
21	31	5	knn-nrm	20				G-E-.200.B02.F1.SE.CS.SP.PI.S0S
22	46		ful					X-.sauto.schedule
23		8	ful		1.1	1	60	c7bb78cc4c665670e6b866a847165cb4bf9
24			ful		6.0	2	20000	92168ebc2ef464a6f2d6a311a4fa90219fd
25	18	15	cnf		6.0	1	20000	a3154f3180cc47331f1b05c36960c32e480
26	11	3	ful		1.5	4	100	cfee9ff42189552c6557cda7d36f20820c8
27	7	15	cnf		1.2	2	20000	X-.auto.sine03
28	2	1	ful		1.5	4	10	7cec1e0745ab65767d5d930d1f61b255ba3
29	5	6	ful		6.0	1	80	a74b37f2d8b7e35be554fc999f671188cf4
30	32	1	cnf		5.0	6	80	2af8b399ea0b8c22c6fc1b13069ad80214f

nr, small, large: order of the strategy; performance on small/large problems
premises: number of best-ranked premises used by the strategy (for MePo and k-NN)
SInE: B_{SInE} – Benevolence; R_{SInE} – Iteration limit; L_{SInE} – Absolute axiom limit.
Selectors: ful – reduced FOF (after phase #1); cnf – reduced FOF clausified (after phase #3); (mepo3|knn)-std – MePo3 or k-NN using symbols and shared terms with numbered variables (de Bruijn indeces); (mepo3|knn)-nrm – MePo3 or k-NN using symbols and matching terms with all variables renamed to one.
Strategies: The names of BliStr strategies are usually content-based hashes. The names of the original E strategies mirror their main parameters.

Table 3. Vampire and E-based ATPs on the CASC-J7 FOF division.

ATP	Vampire 2.6	E.T. 0.1	E 1.9	VanHElsing 1.0
FOF with Equality	234/250	224/250	205/250	199/250
FOF without Equality	141/150	115/150	116/150	111/150
FOF total	375/400	339/400	321/400	310/400

margin over E.T. on the equational problems is reduced to only 4 %. Quite likely, Vampire's advantage on the problems without equality comes from splitting improvements and integration of SAT-solving [8,30].

7   Conclusion and Future Work

E.T. 0.1 shows very good performance on large problems, while being competitive on the problems from the standard FOF category of CASC. The performance is achieved without relying on slow preprocessing phases and learning from related proofs, however this requires a layered architecture with several filtering phases with different speed/precision trade-offs, and very efficient implementation of the core algorithms, using a lot of sharing and indexing data-structures. The other important aspects of E.T.'s performance are (i) relatively sophisticated features that provide good characterization of formulas, allowing more precise high-level approximation of the search problem, (ii) three non-learning state-of-the-art premise selection methods that complement each other, and (iii) a number of complementary automatically designed large-theory search strategies.

Future work may include addition of further non-learning selection methods such as the model-based selection used in SRASS [24], re-use of the strongest lemmas between the strategies, and, e.g., integration of the more expressive features into E's SInE and into E's clause-evaluation heuristics. Some of the techniques developed for E.T. could be also transferred back to learning systems like MaLARea and the AITP hammers.

References

1. Alama, J., Heskes, T., Kühlwein, D., Tsivtsivadze, E., Urban, J.: Premise selection for mathematics by corpus analysis and kernel methods. J. Autom. Reasoning **52**(2), 191–213 (2014)
2. Beeson, M., Wos, L.: OTTER proofs in Tarskian geometry. In: Demri, S., Kapur, D., Weidenbach, C. (eds.) IJCAR 2014. LNCS, vol. 8562, pp. 495–510. Springer, Heidelberg (2014)
3. Blanchette, J.C., Kaliszyk, C., Paulson, L.C., Urban, J.: Hammering towards QED (2015). http://www4.in.tum.de/~blanchet/h4qed.pdf
4. Cambazoglu, B.B., Zaragoza, H., Chapelle, O., Chen, J., Liao, C., Zheng, Z., Degenhardt, J.: Early exit optimizations for additive machine learned ranking systems. In: Davison, B.D., Suel, T., Craswell, N., Liu, B. (eds.) WSDM, pp. 411–420. ACM, New York (2010)
5. Chaudhri, V.K., Elenius, D., Goldenkranz, A., Gong, A., Martone, M.E., Webb, W., Yorke-Smith, N.: Comparative analysis of knowledge representation and reasoning requirements across a range of life sciences textbooks. J. Biomed. Semant. **5**, 51 (2014)
6. Furbach, U., Glöckner, I., Pelzer, B.: An application of automated reasoning in natural language question answering. AI Commun. **23**(2–3), 241–265 (2010)
7. Hoder, K., Voronkov, A.: Sine Qua Non for large theory reasoning. In: Bjørner, N., Sofronie-Stokkermans, V. (eds.) CADE 2011. LNCS, vol. 6803, pp. 299–314. Springer, Heidelberg (2011)
8. Hoder, K., Voronkov, A.: The 481 ways to split a clause and deal with propositional variables. In: Bonacina, M.P. (ed.) CADE 2013. LNCS, vol. 7898, pp. 450–464. Springer, Heidelberg (2013)
9. Kaliszyk, C., Urban, J.: MizAR 40 for Mizar 40. CoRR, abs/1310.2805 (2013)

10. Kaliszyk, C., Urban, J.: Stronger automation for Flyspeck by feature weighting and strategy evolution. In: Blanchette, J.C., Urban, J. (eds.) PxTP 2013, EPiC Series, vol. 14, pp. 87–95. EasyChair (2013)
11. Kaliszyk, C., Urban, J.: Learning-assisted automated reasoning with Flyspeck. J. Autom. Reasoning **53**(2), 173–213 (2014)
12. Kaliszyk, C., Urban, J.: HOL(y)Hammer: online ATP service for HOL Light. Math. Comput. Sci. **9**(1), 5–22 (2015)
13. Kaliszyk, C., Urban, J., Vyskočil, J.: Machine learner for automated reasoning 0.4 and 0.5. CoRR, abs/1402.2359, PAAR 2014 (2014, to appear)
14. Kaliszyk, C., Urban, J., Vyskočil,J.: Efficient semantic features for automated reasoning over large theories. In: IJCAI (2015, to appear)
15. Kovács, L., Voronkov, A.: First-order theorem proving and VAMPIRE. In: Sharygina, N., Veith, H. (eds.) CAV 2013. LNCS, vol. 8044, pp. 1–35. Springer, Heidelberg (2013)
16. Kühlwein, D., Blanchette, J.C., Kaliszyk, C., Urban, J.: MaSh: machine learning for Sledgehammer. In: Blazy, S., Paulin-Mohring, C., Pichardie, D. (eds.) ITP 2013. LNCS, vol. 7998, pp. 35–50. Springer, Heidelberg (2013)
17. Meng, J., Paulson, L.C.: Lightweight relevance filtering for machine-generated resolution problems. J. Appl. Logic **7**(1), 41–57 (2009)
18. Pease, A., Schulz, S.: Knowledge engineering for large ontologies with sigma KEE 3.0. In: Demri, S., Kapur, D., Weidenbach, C. (eds.) IJCAR 2014. LNCS, vol. 8562, pp. 519–525. Springer, Heidelberg (2014)
19. Quaife, A.: Automated Development of Fundamental Mathematical Theories. Kluwer Academic Publishers, Dordrecht (1992)
20. Schulz, S.: System description: E 1.8. In: McMillan, K., Middeldorp, A., Voronkov, A. (eds.) LPAR-19 2013. LNCS, vol. 8312, pp. 735–743. Springer, Heidelberg (2013)
21. Sutcliffe, G.: Proceedings of the 7th IJCAR ATP system competition. http://www.cs.miami.edu/~tptp/CASC/J7/Proceedings.pdf
22. Sutcliffe, G.: The TPTP problem library and associated infrastructure: the FOF and CNF parts, v3.5.0. J. Autom. Reasoning **43**(4), 337–362 (2009)
23. Sutcliffe, G.: The 6th IJCAR automated theorem proving system competition - CASC-J6. AI Commun. **26**(2), 211–223 (2013)
24. Sutcliffe, G., Puzis, Y.: SRASS - a semantic relevance axiom selection system. In: Pfenning, F. (ed.) CADE 2007. LNCS (LNAI), vol. 4603, pp. 295–310. Springer, Heidelberg (2007)
25. Sutcliffe, G., Suda, M., Teyssandier, A., Dellis, N., de Melo, G.: Progress towards effective automated reasoning with world knowledge. In: Guesgen, H.W., Murray, R.C. (eds.) FLAIRS. AAAI Press, Menlo Park (2010)
26. Urban, J.: MPTP - motivation, implementation, first experiments. J. Autom. Reasoning **33**(3–4), 319–339 (2004)
27. Urban, J.: MPTP 0.2: design, implementation, and initial experiments. J. Autom. Reasoning **37**(1–2), 21–43 (2006)
28. Urban, J.: BliStr: The Blind Strategymaker. CoRR, abs/1301.2683 (2013)
29. Urban, J., Hoder, K., Voronkov, A.: Evaluation of automated theorem proving on the Mizar mathematical library. In: Fukuda, K., Hoeven, J., Joswig, M., Takayama, N. (eds.) ICMS 2010. LNCS, vol. 6327, pp. 155–166. Springer, Heidelberg (2010)
30. Voronkov, A.: AVATAR: the architecture for first-order theorem provers. In: Biere, A., Bloem, R. (eds.) CAV 2014. LNCS, vol. 8559, pp. 696–710. Springer, Heidelberg (2014)

Playing with AVATAR

Giles Reger$^{(\boxtimes)}$, Martin Suda, and Andrei Voronkov

University of Manchester, Manchester, UK
giles.reger@manchester.ac.uk

Abstract. Modern first-order resolution and superposition theorem provers use saturation algorithms to search for a refutation in clauses derivable from the input clauses. On hard problems, this search space often grows rapidly and performance degrades especially fast when long and heavy clauses are generated. One approach that has proved successful in taming the search space is *splitting* where clauses are split into components with disjoint variables and the components are asserted in turn. This reduces the length and weight of clauses in the search space at the cost of keeping track of splitting decisions.

This paper considers the new AVATAR (Advanced Vampire Architecture for Theories And Resolution) approach to splitting which places a SAT (or SMT) solver at the centre of the theorem prover and uses it to direct the exploration of the search space. Using such an approach also allows the propositional part of the search space to be dealt with outside of the first-order prover.

AVATAR has proved very successful, especially for problems coming from applications such as program verification and program analysis as these commonly contain clauses suitable for splitting. However, AVATAR is still a new idea and there is much left to understand. This paper presents an in-depth exploration of this new architecture, introducing new, highly experimental, options that allow us to vary the operation and interaction of the various components. It then extensively evaluates these new options, using the TPTP library, to gain an insight into which of these options are essential and how AVATAR can be optimally applied.

1 Introduction

AVATAR [9] is a new architecture for first-order resolution and superposition theorem provers that places a SAT (or SMT) solver at the centre of the theorem prover to direct exploration of the search space. Certain options control this exploration and this paper describes these options in detail and extensively evaluates how they impact proof-search with the aim of highlighting those parameters that lead to (a) more problems being solved, and (b) problems being solved more efficiently.

Modern first-order resolution and superposition provers use saturation algorithms, i.e., they attempt to construct a saturated set of all clauses derivable

---

A. Voronkov—Partially supported by the EPSRC grant "Reasoning for Verification and Security".

A.P. Felty and A. Middeldorp (Eds.): CADE-25, LNAI 9195, pp. 399–415, 2015.
DOI: 10.1007/978-3-319-21401-6_28

from an initial set. A common issue is a rapidly growing search space containing multi-literal and heavy clauses. A multi-literal clause is one with many literals and a heavy clause is one with many symbol occurrences. Processing such clauses is expensive and typically leads to less of the search space being explored in a given time.

One solution is to throw away such clauses that will probably not be used within the time-limit [6]; however, this destroys completeness as we can no longer saturate the set. Another approach is *splitting*. The idea behind splitting is to take a search space $S \cup \{C_1 \vee C_2\}$ and split it into $S \cup \{C_1\}$ and $S \cup \{C_2\}$, for variable-disjoint C_1 and C_2. The benefit is that in each search space the potentially long and heavy clause $C_1 \vee C_2$ is replaced by one of the shorter and lighter clauses C_1 or C_2. Each search space can be saturated separately. If a refutation is found in both then the original search space is unsatisfiable, but if one is saturated without a refutation then the original search space is satisfiable.

To perform splitting it is necessary to make *splitting decisions*, i.e. assert one component of a clause, and potentially *backtrack* these decisions. Different splitting approaches have been considered in the past. In splitting with backtracking (as seen in SPASS [10]) this is done via a (conceptual) splitting tree where a splitting decision is made and we explore one half of the search space and then backtrack (undo the decision) before exploring the second half. In splitting without backtracking [4], when splitting a clause $C_1 \vee \ldots \vee C_n$ each component C_i is named by a fresh propositional variable p_i and the whole clause is split into clauses $(C_1 \vee \neg p_1), \ldots, (C_{n-1} \vee \neg p_{n-1})$ and $(C_n \vee p_1 \vee \ldots \vee p_{n-1})$. This approach is admittedly easier to implement than splitting with backtracking, but the presence of propositional variables sometimes prevents the prover from performing reductions, which may lead to weaker performance [3].

The AVATAR architecture uses a SAT solver to make splitting decisions. As explained later, the SAT solver is passed information about new clauses and produces a model representing valid branches of the (conceptual) splitting tree. The first-order prover can then assert these components and attempts to find a contradiction, which is then passed back to the SAT solver to prune the search space.

AVATAR proved highly successful in previous work evaluating it against alternative splitting mechanisms [3]. Introducing the architecture helped to solve 421 problems previously unsolvable by Vampire [5] or any other prover. However, its full power, and the best way to use it, is not yet fully understood. Certain architectural choices were based on (informed) intuition and not evaluated experimentally; the aim of this paper is to understand these choices and use this understanding to improve AVATAR.

The paper begins with a description of AVATAR's implementation in Vampire (Sects. 2 to 4). It then introduces and explains new variations to the AVATAR architecture (Sect. 5) that will allow us to better understand the interaction between different parts of the architecture. These variations themselves represent a contribution to the understanding of how AVATAR can be organised.

We finish by presenting an extensive evaluation of these architectural variations (Sect. 6). It is clear that some variations are more useful than others and we discuss the likely cause of these results. This evaluation also considers how our improved understanding of the AVATAR architecture can be used to construct complementary strategies to solve as many problems as possible with as few strategies. We finish having learned much about this new, highly experimental architecture, but also with a number of further questions that will shape the continued improvement of this approach.

2   AVATAR by Example

Whilst the theory of AVATAR was established in [9], and is reviewed later in this paper, the authors feel that the key ideas behind the approach are best demonstrated via an example.

The general architecture of AVATAR consists of a first-order (FO) prover and a SAT solver. The FO prover stores a set of first-order clauses, performs first-order reasoning using a saturation algorithm and passes some clauses to the SAT solver. The SAT solver keeps a set of propositional clauses and produces a model (or an unsat answer) on request from the FO prover.

For our example we consider the following input clauses:

$$q(b) \quad p(x) \vee r(x, z) \quad \neg q(x) \vee \neg s(x) \quad \neg p(x) \vee \neg q(y) \quad s(z) \vee \neg r(x, z) \vee \neg q(w)$$

We check which of these clauses can be split into *components*, i.e. sub-clauses with pairwise disjoint sets of variables. The first three clauses cannot be split and are added directly to the FO prover. The last two clauses can be split into components. Each component is given a unique propositional name. To do this naming in a consistent way we use a *component index*, as seen below. This, for example, ensures that $\neg q(y)$ and $\neg q(w)$ are associated with the same propositional symbol. This results in two propositional *split clauses* representing the first-order clauses.

The theory of splitting tells us we can assert one component and then the other after we find a refutation with the first. We are going to use the SAT solver to make splitting decisions so we pass the representations of the splittable clauses to the SAT solver, but do not yet add any of their respective components to the FO prover.

The state of the FO prover and the SAT solver are shown below, where we write $C \leftarrow A$ to indicate that clause C depends on a (possibly empty) set of assertions (splitting decisions) A:

Component Index	FO	SAT
$0 \mapsto \quad \neg p(x)$	$q(b) \leftarrow \{\}$	$0 \vee 2$
$2 \mapsto \quad \neg q(y)$	$p(x) \vee r(x, z) \leftarrow \{\}$	$2 \vee 4$
$4 \mapsto s(z) \vee \neg r(x, z)$	$\neg q(x) \vee \neg s(x) \leftarrow \{\}$	

The FO prover now requests a model. The SAT solver can assign all variables to true, but let us assume the model is minimized into a *partial* model in which only 2 is true and the values of the remaining variables are undefined. Notice that both split clauses in the SAT solver are satisfied by any total extension of this partial model.

Based on the model we assert $\neg q(y)$, the component corresponding to 2. When adding $\neg q(y)$ the FO prover performs a reduction and $\neg q(x) \vee \neg s(x)$ is subsumed. However, this subsumption is conditional on the assertion 2.

FO	SAT
$q(b) \leftarrow \{\}$	$0 \vee \underline{2}$
$p(x) \vee r(x,z) \leftarrow \{\}$	$\underline{2} \vee 4$
assuming($\{2\}$) $\;\;\cancel{\neg q(x) \vee \neg s(x) \leftarrow \{\}}$	
$\neg q(y) \leftarrow \{2\}$	

The FO prover then performs resolution between $q(b)$ and $\neg q(y) \leftarrow \{2\}$ to get the clause $\bot \leftarrow \{2\}$. A corresponding *contradiction clause* $2 \rightarrow \bot = \neg 2$ is then added to the SAT solver and a new model is computed. This time the $\neg 2$ condition forces the SAT solver to construct a model containing 0, 4 and $\neg 2$. As the assertion under which $\neg q(x) \vee \neg s(x)$ was reduced no longer holds, the deletion of this clause is now undone.

FO	SAT
$q(b) \leftarrow \{\}$	$\underline{0} \vee 2$
$p(x) \vee r(x,z) \leftarrow \{\}$	$2 \vee \underline{4}$
$\neg q(x) \vee \neg s(x) \leftarrow \{\}$	$\underline{\neg 2}$
$\cancel{\neg q(y) \leftarrow \{2\}}$	
$\neg p(x) \leftarrow \{0\}$	
$s(z) \vee \neg r(x,z) \leftarrow \{4\}$	

The FO prover then performs resolution between $s(z) \vee \neg r(x,z) \leftarrow \{4\}$ and $p(x) \vee r(x,z)$ to produce $s(z) \vee p(x) \leftarrow \{4\}$, which is then immediately (conditionally) replaced by $s(z) \leftarrow \{0,4\}$ after performing subsumption resolution with $\neg p(x) \leftarrow \{0\}$. This new clause replaces $s(z) \vee \neg r(x,z) \leftarrow \{4\}$ conditioned on the assertion 0. Finally, the FO prover performs a resolution step between $s(z) \leftarrow \{0,4\}$ and $\neg q(x) \vee \neg s(x)$ to get $\neg q(x) \leftarrow \{0,4\}$.

As $\neg q(x)$ is a known component (up to variable renaming), we can add $0 \wedge 4 \rightarrow 2$ to the SAT solver as the clause $\neg 0 \vee \neg 4 \vee 2$. Now the SAT solver can no longer produce a model and so the input problem is shown unsatisfiable and the prover terminates.

FO	SAT
$q(b) \leftarrow \{\}$	$0 \vee 2$
$p(x) \vee r(x,z) \leftarrow \{\}$	$2 \vee 4$
$\neg q(x) \vee \neg s(x) \leftarrow \{\}$	$\neg 2$
$\neg p(x) \leftarrow \{0\}$	$\neg 0 \vee \neg 4 \vee 2$
assuming($\{0\}$) $\;\;\cancel{s(z) \vee \neg r(x,z) \leftarrow \{4\}}$	
assuming($\{0\}$) $\;\;\cancel{s(z) \vee p(x) \leftarrow \{4\}}$	
$s(z) \leftarrow \{0,4\}$	
$\neg q(x) \leftarrow \{0,4\}$	

3 Proof Attempts in Vampire

In this section we give the relevant background on how proof attempts are carried out in the automated theorem prover Vampire [5]. The next section will show how AVATAR-style splitting fits into this process.

Saturation Algorithms. Superposition provers use *saturation algorithms with redundancy elimination*. These work with a search space consisting of a set of clauses and use a collection of generating, simplifying and deleting inferences to explore this space. The theoretical basis of saturation algorithms is the notion of *redundancy* given, e.g., in [1]. Both simplifying and deletion inferences in saturation algorithms are designed in such a way that they only remove redundant clauses.

All saturation algorithms implemented in Vampire belong to the family of *given clause algorithms*, which achieve completeness via a fair *clause selection* process that prevents the indefinite skipping of old clauses. These algorithms typically divide clauses into three sets, *unprocessed*, *passive* and *active*, and follow a simple *saturation loop*:

1. Add non-redundant *unprocessed* clauses to *passive*. Redundancy is checked by attempting to *forward simplify* the new clause using processed clauses.
2. Remove processed clauses made redundant by new clauses i.e. *backward simplify* existing clauses using the new clauses.
3. Select a given clause from *passive*, move it to *active* and perform all generating inferences between the given clause and all other active clauses, adding generated clauses to *unprocessed*.

Vampire implements three saturation algorithms:

1. Otter uses both passive and active clauses for simplifications.
2. Limited Resource Strategy (LRS) [6] extends Otter with a heuristic that discards clauses that are unlikely to be used with the current resources i.e. time and memory. This strategy is incomplete but also generally the most effective at proving unsatisfiability.
3. Discount uses only active clauses for simplifications.

Inferences. The inferences applied by the saturation algorithm are of three different kinds:

- *Generating inferences* derive new clauses that can be immediately simplified and/or deleted by other kinds of inferences. For example, binary resolution and superposition.
- *Simplifying inferences* replace one clause by another simpler clause. For example, demodulation (rewriting by ordered unit equalities) and subsumption resolution (a variant of binary resolution whose conclusion subsumes one of the premises).
- *Deleting inferences* delete clauses, typically when they become redundant. For example, subsumption and tautology deletion.

CASC Mode. Finally, there is a special competition mode that Vampire can be run in (using `--mode casc`) that attempts a sequence of strategies, chosen based on structural characteristics of the given problem. This is motivated by two observations, firstly that whilst some strategies perform very well on average there is no silver bullet that can solve all problems, and secondly that most solvable problems have a strategy that can solve that problem quickly.

4 Introducing Splitting

As we have previously explained, the search space explored by saturation algorithms can quickly become very large and populated with heavy and long clauses. The technique of splitting, where each component of a clause is asserted in turn, can be used to reduce the search space and improve the prover's performance. This section shows how AVATAR implements this splitting process – a full technical description is given in [9].

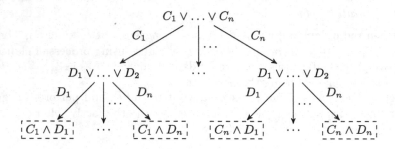

Fig. 1. The conceptual splitting tree

Splitting the Search Space. The general splitting idea can be illustrated by a conceptual splitting tree that is explored during the proof attempt. Every generated clause which can be split is represented by a node and each branch represents a sequence of splitting decisions. When a branch has been found inconsistent *backtracking* occurs and the search moves on to explore a different branch. It can be informative to consider the splitting performed by AVATAR in terms of this splitting tree.

Figure 1 illustrates this splitting tree using clauses $C_1 \vee \ldots \vee C_n$ and $D_1 \vee \ldots \vee D_n$. This tree grows dynamically as further clauses are added to the search space. If every branch contains a contradiction then the problem is unsatisfiable.

Attempting to explore this tree explicitly can be expensive for a number of reasons. Firstly, if clauses share components (i.e. C_i is a variant of D_j) this sharing is not captured by the splitting tree. Secondly, the exploration of the splitting tree is rigid and is difficult to alter based on newly learned information about the components involved. And lastly, information discovered on one branch cannot be easily transferred to a different branch. As we see below, AVATAR implicitly explores this splitting tree by translating the information about splitting components into constraints for a SAT solver and uses the produced model to make component assertions.

The Architecture. Figure 2 illustrates the AVATAR architecture. There are three main parts: the first-order (FO) prover, the SAT solver and the Splitting Interface. The FO prover deals with *clauses with assertions* of the form $D \leftarrow A$ where D is a first-order clause and A is a finite set of propositional variables representing asserted components.

The Splitting Interface manages a mapping between first-order components C and the propositional variable $[C]$ naming that component. The *variant index* ensures that two components C_1 and C_2 are mapped to the same propositional variable if they are equal up to variable renaming, order of literals, and symmetry of equality. This mapping also ensures that the negation of a ground component is translated to the negation of the corresponding propositional variable, i.e. $[\neg C] = \neg[C]$ for every ground component C.[1]

For each component, the Splitting Interface also maintains a *record* which stores:

1. *children* of the component, i.e., clauses that are derived from the component,
2. clauses that were *reduced* by a clause depending on this component.

See below for an explanation of these sets of clauses.

Lastly, to avoid asserting previously asserted components, the Splitting Interface keeps track of the current model previously obtained from the SAT solver. The following sections will explain the communication between the three parts.

Dealing with Assertions in the FO Prover. As we said above, the FO prover is updated to deal with *clauses with assertions*. This affects the way that inferences are carried out in the prover. Firstly, to ensure that assertions are properly propagated, any generating inference of the form

$$\frac{D_1 \quad \cdots \quad D_k}{D}$$

is replaced by the inference

$$\frac{(D_1 \leftarrow A_1) \quad \cdots \quad (D_k \leftarrow A_k)}{(D \leftarrow A_1 \cup \ldots \cup A_k)}$$

and $(D \leftarrow A_1 \cup \ldots \cup A_k)$ is added to the *children* of each component in $A_1 \cup \ldots \cup A_k$ in the component records kept by the Splitting Interface.

Simplifying inferences of the form

$$\frac{D_1 \quad \cdots \quad D_{m-1} \quad \cancel{D_m}}{D} \ .$$

---

[1] This useful optimization is not derictly available for non-ground components. Negating a non-ground component would require skolemization and is not considered in this paper.

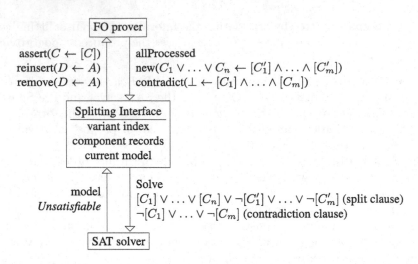

Fig. 2. The AVATAR architecture

previously meant that D is a logical consequence of $D_1, \ldots, D_m$ and D makes D_m redundant. This is replaced by

$$\frac{(D_1 \leftarrow A_1) \quad \cdots \quad (D_{m-1} \leftarrow A_{m-1}) \quad \cancel{(D_m \leftarrow A_m)}}{(D \leftarrow A)},$$

where $A = A_1 \cup \ldots \cup A_m$, distinguishing the following two cases.

1. If $A = A_m$ then $(D_m \leftarrow A_m)$ can be deleted as all the other clauses are based on the same or a weaker set of assertions. It should also be deleted from the *children* of components in A_m.
2. Otherwise, $(D_m \leftarrow A_m)$ can only be *conditionally deleted* as there exists a branch of the splitting tree where the deleted clause is valid but at least one of the side conditions $(D_i \leftarrow A_i)$ is not. To conditionally delete a clause we remove it from the FO prover and add it to the *reduced* set for each component in $A \setminus A_m$.

Exploring the Splitting Tree in AVATAR. When a new splittable clause is selected for processing, the FO prover passes this clause to the Splitting Interface instead of attempting to add it to passive. The Splitting Interface then uses the variant index to translate it into a propositional *split clause* and pass this to the SAT solver.

Once unprocessed is empty, the FO prover sends the `allProcessed` message to the Splitting Interface, which sends the `Solve` message to the SAT solver. The SAT solver then either replies with `Unsatisfiable`, indicating that all splitting branches have been explored, or it returns a new model M.

We allow for partial models which we represent by consistent sets of propositional literals of the form $[C]$ or $\neg[C]$. We require that at least one literal of each propositional clause registered by the SAT solver must be satisfied by the model, but some literals may stay undefined.

Given a new model M and old model $\overline{M}$, the Splitting Interface does the following:

1. For each $[C] \in (\overline{M} \setminus M)$, remove component C and all of its children from the FO prover using remove($D \leftarrow A$). Add any clause $D \leftarrow A$ in the component's reduce set such that $A \subseteq M$, using reinsert($D \leftarrow A$).
2. For each $[C] \in (M \setminus \overline{M})$, add component C to the FO prover using assert($C \leftarrow [C]$) and add each of the component's children $D \leftarrow A$ such that $A \subseteq M$, using reinsert($D \leftarrow A$).

Removing the children of removed components is necessary as they rely on assertions that are no longer true. Reinserting a clause that has been reduced with the help of a removed component retracts the now no longer supported deletion of the clause. Reinserting the children of a component means that clauses generated from this component on previous branches are brought into this branch. In both cases, we only reinsert those clauses that have all their assertions true in the new model M.

Split clauses introduce new branches into the conceptual splitting tree, although note that due to the use of the variant index some of these branches may be shared. To prune the splitting tree we need *contradiction clauses*. When the FO prover produces a contradiction with assertions, this contradiction is passed to the Splitting Interface, which performs the translation into a propositional *contradiction clause* and sends this to the SAT solver. This contradiction forces the model to change. Notice that a contradiction clause can cut off many branches of the splitting tree.

5   Varying the Architecture

We now consider some of the choices made in the architecture of AVATAR, how we may change these, and what effects these changes may have.

Adding Components. When providing information about new clauses to the SAT solver (in the form of component clauses) we need to decide what to do with *nonsplittable* clauses, i.e., those that cannot be split into multiple components. We consider two values for this option (named nonsplittable_components):

1. none: do not add any non-splittable component,
2. known: add such a clause if it has previously been introduced as a component.

The example in Sect. 2 uses this second option value when it adds the component clause $\neg 0 \lor \neg 4 \lor 2$ as the clause $\neg q(x)$ is nonsplittable. With the option set to none, the FO prover would have performed an additional resolution step to produce the contradiction clause $\neg 0 \lor \neg 4$. By using the known component, we constrained the split tree explored by the SAT solver and thus avoided performing the additional inference.

Constructing a Model. Previously, we referred to the SAT solver as just a SAT solver, but different SAT solvers take a different amount of time to construct a model and potentially also construct different models. With the option `sat_solver`, we can vary which SAT solver we use. In this paper we consider our own SAT solver and Minisat (version 2.2) [2] using the default options. As a SAT solver, Minisat is better than Vampire's native solver. Our aim was to understand whether a better SAT solver results in a better overall performance of AVATAR.

Partial Models and Minimization. While models produced by a standard SAT solver are total, AVATAR may work with partial models in which some literals are undefined, provided each such model makes true at least one literal in every clause. Total models may result in adding unnecessary assertions to the FO prover, for instance, when they set to true multiple literals from a single split clause. This corresponds to the exploration of multiple splitting branches at once, an effort which is often wasted as each of the branches usually needs to be considered separately later on as well.

We can attempt to minimize the total model produced by the SAT solver by dropping literals that are not needed for satisfying any clause and thus to restrict the exploration to a single branch. We use a simple greedy procedure for the minimization, whose result is a partial model, a sub-model of the original one. Minimization is controlled by the option `minimize_model`. Again, we can vary how we choose to do this.

1. `off`: We use the total model.
2. `all`: We minimize with respect to all clauses.
3. `sco`: We minimize with respect to split clauses only.

Note that the `sco` option value is sound, because we always start minimizing from a total model which satisfies all the clauses.

Asserting Complements. Another factor to consider is the possibility of treating ground components specially as we are able to assert these negatively to the FO prover. That is, when the SAT solver sets the value of a ground component C to false in the model we can assert $\neg C \leftarrow [\neg C]$ to the FO prover, even if this is not needed to satisfy any split clause. This is controlled by the option `add_complementary`. We are prevented from asserting both C and $\neg C$ at the same time as the mapping from components to propositional variables ensures that $[\neg C] = \neg[C]$ for ground components.

When to do Splitting. Previously we described the splitting process occurring at clause introduction, i.e., when we attempt to move it into passive. Alternatively we can consider splitting a clause at *activation*. This is controlled by the option `split_at_activation`. This delays the expense of splitting but also delays the benefits of additional information being passed to the SAT solver. For example, a subset of passive clauses may already be propositionally unsatisfiable, but we will not discover this until all clauses in this subset become activated.

To Delete or to Deactivate. In the previous presentation, clauses that are deactivated due to switching the splitting branch are reasserted when they become valid again. Remembering these clauses may cost us a lot of memory. Moreover, some of these clauses may never need to be reasserted, if they depend on a partial branch which will not be visited anymore. With the option `delete_deactivated`, we delete these clauses instead and later only reassert the respective component clause $C \leftarrow [C]$, which is sufficient for completeness. The downside is that we may need to recompute some of these delete clauses if a particular partial branch is revisited.

Currently even deactivated clauses are removed from the term indexing structures used for efficient inferences. Providing an option to preserve deactivated clauses in these structures remains further work, and may prove beneficial as deleting and inserting clauses into indexing structures can become very costly.

Clearing the Assertions. SAT solvers typically perform DPLL splitting and may, at some point, derive that a propositional literal must be true in any possible model. These are called zero-implied literals as their truth value is decided at the zeroth level. This information can be used to remove the corresponding assertions from clauses in the FO prover as these are redundant. This can reduce the number of conditional clause deletions as any deletions conditional only on zero-implied components can be considered unconditional. This option is controlled by `handle_zero_implied`.

Summary. Table 1 describes the Vampire options we consider in this work, i.e., those we will vary in experiments later. All other options will be fixed at their default value. Note that some options are experimental and may not be included in future releases of Vampire.

Table 1. The Vampire options of interest.

Option	Short name	Considered values (<u>default</u>)
`saturation_algorithm`	sa	<u>lrs</u>, discount
`sat_solver`	sas	<u>vampire</u>, minisat
`nonsplittable_components`	ssnc	<u>known</u>, none
`minimize_model`	smm	<u>sco</u>, off, all
`add_complementary`	ssac	<u>ground</u>, none
`split_at_activation`	sac	<u>off</u>, on
`delete_deactivated`	sdd	<u>on</u>, off
`handle_zero_implied`	shzi	<u>off</u>, on

6 Experiments

In this section we experimentally evaluate the impact of the different variations of the AVATAR architecture on the performance of theorem prover Vampire.

Designing Experiments. The aim of these experiments is to evaluate how effective the different architectural variations are. To do that we need to understand what we mean by effective. The existence of the CASC portfolio mode is a testament to the fact that there is no best strategy. In fact, the value of a strategy is difficult to understand. Some strategies perform very well on average but cannot solve problems solvable by other strategies. The motivation behind CASC mode is that a collection of strategies, each of which may be bad on average, can easily outperform a collection of strategies, each of which is good on average. However, within a collection of strategies we need those that can solve many problems as time limits do not usually allow for running too many strategies. Therefore, the aim of these experiments is to identify those options that allow us to solve previously unsolved problems as well as those options that help us solve the most problems.

Experimental Setup. For our benchmarks we use TPTP [7] problems containing non-unit clauses with a rating of 0.5 or higher. The TPTP rating [8] is the percentage of (eligible) provers that cannot solve a problem, thus, for example, a rating of 0.5 means that half of (eligible) provers can solve the problem and a rating of 1 means that the problem cannot be solved by any of these provers. However, the rating evaluation does not use every mode of each prover, so it is possible that a prover used to generate ratings can solve a problem of rating 1 using a different mode. We only include problems in our experiment that we know are solvable by some prover, e.g., Vampire. This led to the selection of 3823 problems.[2]

For the experiments, we took all combinations of options discussed in Sect. 5. This cartesian product (cube) gives us 384 strategies and represents almost 1.5 million experiments in total. We ran experiments with a time limit of 10 seconds[3], meaning that our results reasonably reflect the utility of the strategy when placed within a cocktail of other strategies. We used the default values for all options not explicitly stated.

Experiments were run on the StarExec[4] cluster, using 160 nodes. The nodes used contain a Intel Xeon 2.4 GHz processor. Experiments used Vampire's default memory limit of 3 GB; this memory limit was only reached in rare cases (<0.05 %).

Time Spent in the SAT Solver. The experiments show that time spent in the SAT solver does not generally dominate. On average, 9.6 % of the time was spent in the SAT solver. In 8.8 % of the experiments, calls to the SAT solver took more than 50 % of the time and in 0.5 % of the experiments calls to the SAT solver took more than 90 % of the time[5].

Best and Worst Strategies. In total, 1444 problems (38 % of all problems) were solved by at least one of the considered strategies, of these 328 were of rating 0.8 or higher. Table 2 shows the performance of the worst and best strategy with respect to the number of solved problems and the values of options that define them. We can see that the best strategy only solves 1103 problems which amounts

---

[2] A list of the selected problems, the executable of our prover as well as the results of the experiment are available from http://vprover.org.

[3] Note that previous experiments [3] used longer time limits.

[4] https://www.starexec.org.

[5] Only runs which took at least one second to complete are considered here.

Table 2. Best and worst strategies with respect to the number of problems solved, option values that define them, the number of problems solved by the 10 % worst and best strategies in union, respectively, and the respective proportional representation of the option values in these strategies.

	worst	worst 10 %	best	best 10 %
problems solved	796	1149	1103	1223
saturation_algorithm	lrs	61 %	discount	100 %
sat_solver	vampire	100 %	minisat	63 %
nonsplittable_components	none	79 %	known	47 %
minimize_model	off	63 %	all	42 %
add_complementary	ground	53 %	ground	100 %
split_at_activation	off	100 %	on	100 %
delete_deactivated	off	55 %	on	53 %
handle_zero_implied	on	50 %	off	50 %

to about 76 % of all the problems solved. The table also shows the performance of two meta-strategies, one consisting of the union of the 10 % worst and the other of the 10 % best strategies, and, in the lower part, the percentage of the 10 % worst and best strategies which use the same value for a particular option as the ultimate worst and best strategy, respectively.[6]

Perhaps the most surprising observation is that lrs does not appear at all amongst the 10 % of the best strategies. We suspect that LRS, which was not adapted to AVATAR, misinterprets the remaining amount of resources available for proving, because it does not take into account the part of the split tree that still needs to explored. Attempting to confirm this hypothesis is one possible direction for future work.

Another interesting fact is that both the worst and the best strategy employ the value ground for the add_complementary option. This option value is definitely useful (all the best strategies use it), but may have some shortcomings, because it is also used by the majority of the worst strategies.

When interpreting the results for minimize_model, one should keep in mind that this option has three possible values and so the result of 42 % for all with the best strategies is significant. On the other hand, Table 2 indicates that the effect of delete_deactivated and, especially, of handle_zero_implied is close to random.

Importance of Particular Options. To better determine the importance of individual options, we put the number of problems solved with a particular value of an option into Table 3. On a per option basis, the table also shows (in parenthesis) the number of problems solved only by a strategy using a particular

---

[6] A different statistic, not shown in the table, is the performance of strategies at the 10 % mark from each end of the sorted order (quantiles), which were 865 and 1072, respectively.

Table 3. Number of problems (uniquely) solved with a particular option value.

saturation_algorithm		add_complementary	
lrs	1287 (142)	none	1372 (20)
discount	1302 (157)	ground	1424 (72)

sat_solver		split_at_activation	
vampire	1375 (38)	off	1345 (83)
minisat	1406 (69)	on	1361 (99)

nonsplittable_components		delete_deactivated	
none	1416 (27)	off	1427 (31)
known	1417 (28)	on	1413 (17)

minimize_model		handle_zero_implied	
off	1402 (15)	off	1428 (16)
all	1412 (17)	on	1428 (16)
sco	1401 (11)		

value of the option and not by any strategy using any of the other values. This means the value is necessary for solving these problems.

An option is important if it has at least two values each necessary for solving many problems. This perspective implies that saturation_algorithm is the most important option in our experiment and split_at_activation the most important one for AVATAR per se. When focusing on individual values, we can see that minisat helps to solve more problems than vampire, that the value ground should be preferred over none for add_complementary, and that it perhaps does not pay off to keep the value sco for minimize_model.

Conditional Projections. Having collected the data for all the possible combinations of option values one can also ask questions such as what would Table 3 look like if we focused only on strategies where a particular option is fixed to a certain value. This allows us to distinguish generally good values of options from those that are only good under certain conditions.

For example, we observed that while with discount we could solve 39 more problems when split_at_activation was turned on, this did not happen for lrs, where we could solve 1208 problems with split_at_activation off, but only 1202 problems with the option on. This is most like related to the fact that LRS uses clauses in passive for simplifications and therefore benefits from these clauses being already split.

Also, both the lrs and the vampire perspective significantly favour the value known over none for nonsplittable_components, while in Table 3 these two values seem to behave similarly. In the former conditional projection, none solves only 1242 while known 1266, in the latter, none solves 1333 and known 1347. This phenomenon seems to be quite difficult to explain and should be further explored.

Table 4. Sequence of strategies to greedily cover all solved problems. For space reasons, short names are used for options (see Table 1).

	1	2	3	4	5
Contribution	1103	114	45	31	21
Solves	1103	943	905	948	1081
Nominal order	1	155	283	141	23
sa	discount	lrs	discount	lrs	discount
sas	minisat	minisat	vampire	minisat	vampire
ssnc	known	known	known	known	none
smm	all	sco	sco	off	sco
ssac	ground	none	ground	ground	ground
sac	on	off	off	on	on
sdd	on	off	on	off	on
shzi	off	on	off	on	on

Greedy Problem Coverage. Next we consider how the strategies could be greedily ordered to cover all problems solved, i.e., we attempt to produce a (greedy) CASC portfolio mode. We require 61 strategies in total to cover all problems, with the last 32 strategies only contributing one additional problem each.

Table 4 gives the first five strategies in this greedily produced portfolio sequence along with the number of problems each strategy contributes to the portfolio, how many problems that strategy normally solves, and the nominal order in all strategies (with respect to number of problems solved).

We first note that we require both strategies that are good on average and also those that solve problems uniquely. In the sequence of strategies, 72 % come from the bottom half of strategies in terms of number of problems solved. It is also interesting to note that some option values, such as sco for minimize_model, that were previously seen to contribute little, are needed here.

Further Lessons Learned. One of the interesting lessons learned with these experiments is that the choice of a SAT solver significantly influences the performance of a strategy. This suggests that the queries passed to the SAT solver are by no means easy (as we originally assumed) and that on many problems the solver takes over a considerable part of the required reasoning.

Moreover, efficiently dealing with the incremental nature of the presented queries becomes a relevant factor in AVATAR. When restricting solution times to a maximum of 1 second, vampire became the solver of choice for the best strategy with respect number of solved problems. The vampire solver was designed with the AVATAR application in mind, and therefore deals well with the incremental usage required. However, as it is not as highly tuned as minisat, its performance tails off quickly as the size of the problem increases. This may explain the observed behaviour.

Another aspect influenced by the choice of a SAT solver is the inherent "quality" (from the perspective of AVATAR) of the models it produces. It is clear that the produced model affect how the splitting tree is explored, but not yet clear why one solver may produce 'better' models in general. Further investigations will consider the SAT solver options themselves and how varying these affects the models produced.

7 Conclusion

AVATAR is a new and highly successful architecture. While previously used saturation algorithms, their variations and options have been studied for decades, almost nothing is known about options that can improve AVATAR even further. Likewise, almost nothing is known about the behaviour of various existing options in presence of AVATAR. This is the first paper that both introduces AVATAR specific options and investigates their behaviour. We believe this is the first in many studies by us, and others, exploring this novel architecture.

The usage of a SAT solver to perform splitting operations is a novel idea, which has the potential to change how modern first-order theorem provers explore the clause search space. The architectural variations explored in this paper help us better understand the optimal configuration for this new form of splitting.

We found that the importance of the individual options for solving additional problems varies, and while the important ones should be kept and further explored, removing the ones that seem to have negligible influence on the performance of AVATAR could simplify the implementation and improve its maintainability.

We also discovered that the efficiency of the SAT solver is very important for the overall performance of AVATAR. This is not only in terms of proving time, but also their ability to handle incrementality. We observed cases where performance suffered as a result of insufficient support for incremental usage; this suggests that improving SAT solvers in this respect can improve AVATAR. We have also identified new questions to be answered, for example how the model produced by the SAT solver interacts with the exploration of the splitting tree.

An additional discovery is that the limited resource strategy, thought to be the best strategy within Vampire for showing unsatisfiability, does not interact well with the way in which AVATAR explores the clause space. This suggests that further investigation is required to establish how best to adapt the LRS approach to AVATAR.

Whilst the results from the current architecture are impressive, there is more that can be squeezed from this idea. One major area of interest is replacing the SAT solver with an SMT solver, allowing it reason on the theory level.

References

1. Bachmair, L., Ganzinger, H.: Resolution theorem proving. In: Robinson, A., Voronkov, A. (eds.) Handbook of Automated Reasoning Chap. 2, pp. 19–99. Elsevier Science, North-Holland (2001)
2. Eén, N., Sörensson, N.: An extensible SAT-solver. In: Giunchiglia, E., Tacchella, A. (eds.) SAT 2003. LNCS, vol. 2919, pp. 502–518. Springer, Heidelberg (2004)
3. Hoder, K., Voronkov, A.: The 481 ways to split a clause and deal with propositional variables. In: Bonacina, M.P. (ed.) CADE 2013. LNCS, vol. 7898, pp. 450–464. Springer, Heidelberg (2013)
4. Riazanov, A., Voronkov, A.: Splitting without backtracking. In: Nebel, B. (ed.) 17th International Joint Conference on Artificial Intelligence, IJCAI 2001, vol. 1, pp. 611–617 (2001)
5. Riazanov, A., Voronkov, A.: The design and implementation of Vampire. AI Commun. **15**(2–3), 91–110 (2002)
6. Riazanov, A., Voronkov, A.: Limited resource strategy in resolution theorem proving. J. Symbolic Comput. **36**(1–2), 101–115 (2003)
7. Sutcliffe, G.: The TPTP problem library and associated infrastructure. J. Autom. Reasoning **43**(4), 337–362 (2009)
8. Sutcliffe, G., Suttner, C.: Evaluating general purpose automated theorem proving systems. Artif. Intell. **131**(1–2), 39–54 (2001)
9. Voronkov, A.: AVATAR: the architecture for first-order theorem provers. In: Biere, A., Bloem, R. (eds.) CAV 2014. LNCS, vol. 8559, pp. 696–710. Springer, Heidelberg (2014)
10. Weidenbach, C.: Combining superposition, sorts and splitting. In: Robinson, A., Voronkov, A. (eds.) Handbook of Automated Reasoning Chap. 27, pp. 1965–2013. Elsevier Science, North-Holland (2001)

Combinations

A Polite Non-Disjoint Combination Method: Theories with Bridging Functions Revisited

Paula Chocron[1], Pascal Fontaine[2], and Christophe Ringeissen[2]($\boxtimes$)

[1] IIIA-CSIC, Bellaterra, Catalonia, Spain
[2] INRIA, Université de Lorraine and LORIA, Nancy, France
Christophe.Ringeissen@loria.fr

Abstract. The Nelson-Oppen combination method is ubiquitous in Satisfiability Modulo Theories solvers. However, one of its major drawbacks is to be restricted to disjoint unions of theories. We investigate the problem of extending this combination method to particular non-disjoint unions of theories connected via bridging functions. The motivation is, e.g., to solve verification problems expressed in a combination of data structures connected to arithmetic with bridging functions such as the length of lists and the size of trees. We present a sound and complete combination procedure à la Nelson-Oppen for the theory of absolutely free data structures, including lists and trees. This combination procedure is then refined for standard interpretations. The resulting theory has a nice politeness property, enabling combinations with arbitrary decidable theories of elements.

1 Introduction

Solving the satisfiability problem modulo a theory given as a union of decidable sub-theories naturally calls for combination methods. The Nelson-Oppen combination method [9] is now ubiquitous in SMT (Satisfiability Modulo Theories) solvers. However, this technique imposes strong assumptions on the theories in the combination; in the classical scheme [9,17], the theories notably have to be signature-disjoint and stably infinite. Many recent advances aim to go beyond these two limitations.

The design of a combination method for non-disjoint unions of theories is clearly a hard task [7,18]. To stay within the frontiers of decidability, it is necessary to impose restrictions on the theories in the combination; and at the same time, those restrictions should not be such that there is no hope of concrete applications for the combination scheme. For this reason, it is worth exploring specific classes of non-disjoint combinations of theories that appear frequently in software specification, and for which it would be useful to have a simple combination procedure. An example is the case of shared sets, where sets are represented

This work has been partially supported by the project ANR-13-IS02-0001 of the Agence Nationale de la Recherche, by the European Union Seventh Framework Programme under grant agreement no. 295261 (MEALS), and by the STIC AmSud MISMT.

© Springer International Publishing Switzerland 2015
A.P. Felty and A. Middeldorp (Eds.): CADE-25, LNAI 9195, pp. 419–433, 2015.
DOI: 10.1007/978-3-319-21401-6_29

by unary predicates [4,19]. In this context, the cardinality operator can also be considered; notice that this operator is a bridging function from sets to natural numbers [22]. In this paper, we investigate the case of bridging functions between data structures and a target theory, for instance, the length of lists, in which case the target theory is a fragment of arithmetic. Here, non-disjointness arises from connecting two disjoint theories via a third theory defining the bridging function. This problem is of prime interest for software verification [6,14,15,23], in particular for the verification of recursive (functional) programs with functions defined by pattern-matching. For instance, a satisfiability procedure for data structures combined with bridging functions is the core reasoning engine of the verification tool Leon targeting Scala programs [16]. To solve instances of this problem, dedicated techniques have been developed [15,20], and general frameworks, based on locality [14] or superposition [1,3,10], are also applicable. In particular, the contributions by Zarba [20], Sofronie-Stokkermans [14], and Suter et al. [15] have given rise to the straight combination approach highlighted in this paper. In [20], Zarba presents a procedure for checking satisfiability of lists with length by using a reduction to arithmetic, and a similar reduction applies to multisets with multiplicity [21]. The motivation was to relax the stably-infiniteness assumption in Nelson-Oppen's procedure, e.g., to be able to consider multisets over a finite domain of elements. In that line of work, Zarba focuses on *standard* interpretations. For instance, the standard interpretation for lists corresponds to the case where lists are interpreted as finite lists of elements. Sofronie-Stokkermans [14] relies on locality properties to show that the definition of the function connecting the theories can be eliminated (using instantiations by ground terms). The subtle problem of restricting interpretations to standard ones is also discussed but, in contrast to our approach, only the case of an infinite domain of elements is considered. In [15], Suter et al. present a dedicated procedure for standard interpretations that is sound and complete for *sufficiently surjective* abstraction functions.

We investigate here an approach by reduction from non-disjoint to disjoint combination. It is an alternative to a non-disjoint combination approach à la Ghilardi [2,7], for which some assumptions on the shared (target) theory are required. Ghilardi's approach has been applied to combine data structures with fragments of arithmetic, like Integer Offsets [11] and Abelian groups [10]; it is however difficult to go beyond Abelian groups and consider for instance any decidable fragment of arithmetic as a shared theory. The approach by reduction does not impose such limitations, and any (decidable) fragment of arithmetic is suitable for the target (shared) theory. The resulting combination procedure is correct for (*arbitrary* interpretations of) absolutely free data structures. Our correctness proof is not based on locality principles [14], but relies on the construction of a combined model in the line of the Nelson-Oppen procedure. Eventually, the outcome of our approach bears similarities with the locality-based procedure.

Then we focus on the problem of adapting this combination procedure to get a satisfiability procedure for the restricted class of *standard* interpretations of absolutely free data structures. The correctness of the combined satisfiability procedure for standard interpretations is based on a *politeness* property, previously introduced to consider disjoint combinations of some data structure theories with any theory of elements [8,13]. This paper is a first application of politeness to non-disjoint combinations. The interest of applying politeness is twofold. First, it provides a way to relate satisfiability in standard interpretations to satisfiability in the class of all interpretations. Second, it permits to solve in a modular way the satisfiability problem in the combination of (1) standard interpretations of a data structure theory extended with a bridging function and (2) any arbitrary theory of elements. The resulting combined satisfiability procedure has some similarities with the one studied in [12,15,16].

Our combination procedures for arbitrary/standard interpretations are first illustrated on the prominent case of lists with length [6]: this is a simple but meaningful case to grasp the concepts and techniques developed in the paper. But our study is not limited to that particular case, and we show that our combination procedures apply to the general case of trees with bridging functions.

The rest of the paper is organized as follows. Section 2 recalls basic concepts and notations. The combination problem is presented in Sect. 3 and the related combination procedure in Sect. 4. In Sect. 5, we focus on the restriction to standard interpretations for the cases of lists (Sects. 5.1–5.2) and trees (Sect. 5.3), by considering appropriate bridging functions and the combination problem with an arbitrary theory of elements. Omitted proofs can be found in [5].

2 Preliminaries

We assume an enumerable set of variables V and a first-order many-sorted signature Σ given by a set of sorts and sets of function and predicate symbols (equipped with an arity). Nullary function symbols are called constant symbols. We assume that, for each sort σ, the equality "$=_\sigma$" is a logical symbol that does not occur in Σ and that is always interpreted as the identity relation over (the interpretation of) σ; moreover, as a notational convention, we omit the subscript for sorts and we simply use the symbol $=$. The notions of Σ-terms, atomic Σ-formulas and first-order Σ-formulas are defined in the usual way. In particular an atomic formula is either an equality, or a predicate symbol applied to the right number of well-sorted terms. Formulas are built from atomic formulas, Boolean connectives ($\neg$, $\wedge$, $\vee$, $\Rightarrow$, $\equiv$), and quantifiers ($\forall$, $\exists$). A literal is an atomic formula or the negation of an atomic formula. A flat equality is either of the form $t_0 = t_1$ or $t_0 = f(t_1, \ldots, t_n)$ where each term $t_0, \ldots, t_n$ is a variable or a constant. A disequality $t_0 \neq t_1$ is flat when each term t_0, t_1 is a variable or a constant. A flat literal is either a flat equality or a flat disequality. An *arrangement* over a finite set of variables V is a maximal satisfiable set of well-sorted equalities and disequalities $x = y$ or $x \neq y$, with $x, y \in V$. Given a quantifier-free formula φ, an *arranged form* of φ is any conjunction of φ with an arrangement over the variables in φ. For n distinct variables $x_1, \ldots, x_n$, the set of literals $\{x_i \neq x_j \mid i \neq j, \ i, j = 1, \ldots, n\}$ is denoted by $\{x_1 \neq \cdots \neq x_n\}$.

Free variables are defined in the usual way, and the set of free variables of a formula φ is denoted by $Var(\varphi)$. Given a sort σ, $Var_\sigma(\varphi)$ denotes the set of variables of sort σ in $Var(\varphi)$. A formula with no free variables is closed, and a formula without variables is ground. A universal formula is a closed formula $\forall x_1 \ldots \forall x_n. \varphi$ where φ is quantifier-free. A (finite) Σ-theory is a (finite) set of closed Σ-formulas. Two theories are disjoint if no predicate symbol or function symbol appears in both respective signatures.

From the semantic side, a Σ-*interpretation* $\mathcal{I}$ comprises a non-empty pairwise disjoint domains D_σ for every sort σ, a sort- and arity-matching total function $\mathcal{I}[f]$ for every function symbol f, a sort- and arity-matching predicate $\mathcal{I}[p]$ for every predicate symbol p, and an element $\mathcal{I}[x] \in D_\sigma$ for every variable x of sort σ. By extension, an interpretation defines a value in D_σ for every term of sort σ, and a truth value for every formula. We may write $\mathcal{I} \models \varphi$ whenever $\mathcal{I}[\varphi] = \top$. A Σ-structure is a Σ-interpretation over an empty set of variables.

A model of a formula (theory) is an interpretation that evaluates the formula (resp. all formulas in the theory) to true. A formula or theory is satisfiable (or consistent) if it has a model; it is unsatisfiable otherwise. A formula G is T-satisfiable if it is satisfiable in the theory T, that is, if $T \cup \{G\}$ is satisfiable. A T-model of G is a model of $T \cup \{G\}$. A formula G is T-unsatisfiable if it has no T-models. A theory T is *stably infinite* if any T-satisfiable set of literals is satisfiable in a model of T whose domain is infinite. A Σ-theory T can be equivalently defined as a pair $T = (\Sigma, \mathbf{A})$, where $\mathbf{A}$ is a class of Σ-structures. We may write $\mathcal{A} \in T$ when $T = (\Sigma, \mathbf{A})$ and $\mathcal{A} \in \mathbf{A}$. Given theories $T_i = (\Sigma_i, \mathbf{A}_i)$ for $i = 1, 2$, the *combination* of T_1 and T_2 is the theory $T_1 \oplus T_2 = (\Sigma_1 \cup \Sigma_2, \mathbf{A})$ where $\mathbf{A}$ is the set of $\Sigma_1 \cup \Sigma_2$-structures $\mathcal{A}$ such that the Σ_i-structure $\mathcal{A}^{\Sigma_i}$ (defined by restricting $\mathcal{A}$ to interpret only symbols in Σ_i) is in $\mathbf{A}_i$ for $i = 1, 2$.

3   The Combination Problem

Consider a many-sorted Σ_s-theory T_s and a many-sorted Σ_t-theory T_t (s and t stand for source and target respectively) such that T_s and T_t have disjoint function symbols (but sorts can be shared by Σ_s and Σ_t). We consider a function f mapping elements from T_s to elements in T_t. This function is defined by some axioms expressed in the signature $\Sigma_s \cup \Sigma_t \cup \{f\}$. The set of axioms defining f is called T_f.

In the following, the theory T_s is the theory of Absolutely Free Data Structures [14] (AFDS, for short) and T_f is a bridging theory connecting AFDS to an arbitrary (target) theory T_t. For simplicity, we only consider Absolutely Free Data Structures without selectors. In Sect. 5, we will argue that selectors are not mandatory when standard interpretations are considered.

Definition 1. *Consider a set of sorts* Elem, *and a sort* struct $\notin$ Elem. *Let* Σ *be a signature whose set of sorts is* $\{$struct$\} \cup$ Elem *and whose function symbols* $c \in \Sigma$ *(called* constructors*) have arities of the form:*

$$c : \mathsf{e}_1 \times \cdots \times \mathsf{e}_m \times \mathtt{struct} \times \cdots \times \mathtt{struct} \to \mathtt{struct}$$

where $e_1, \ldots, e_m \in \texttt{Elem}$. *Consider the following axioms (where upper case letters denote implicitly universally quantified variables)*

$$
\begin{array}{ll}
(Inj_c) & c(X_1, \ldots, X_n) = c(Y_1, \ldots, Y_n) \Rightarrow \bigwedge_{i=1}^{n} X_i = Y_i \\
(Dis_{c,d}) & c(X_1, \ldots, X_n) \neq d(Y_1, \ldots, Y_m) \\
(Acyc_\Sigma) & X \neq t[X] \ \textit{if } t \textit{ is a non-variable } \Sigma\textit{-term}
\end{array}
$$

The theory of Absolutely Free Data Structures over Σ is

$$
AFDS_\Sigma = \Big(\bigcup_{c \in \Sigma} Inj_c \Big) \cup \Big(\bigcup_{c,d \in \Sigma, c \neq d} Dis_{c,d} \Big) \cup Acyc_\Sigma
$$

From now on, T_s is $AFDS_{\Sigma_s}$ (see [5] for a T_s-satisfiability procedure).

Example 1. The theory of lists is an example of AFDS where the constructors are *cons* : $\texttt{elem} \times \texttt{list} \rightarrow \texttt{list}$ and *nil* : $\texttt{list}$. Similarly (binary) trees are also a classical AFDS example, where the constructor operator is, e.g., *cons* : $\texttt{elem} \times \texttt{tree} \times \texttt{tree} \rightarrow \texttt{tree}$. The theory of pairs (of numbers) is another example of AFDS where the constructor is *cons* : $\texttt{num} \times \texttt{num} \rightarrow \texttt{struct}$.

Given a tuple e of terms of sorts in $\texttt{Elem}$ and a tuple t of terms of sort $\texttt{struct}$, the tuple e, t may be written $e; t$ to distinguish terms of sort $\texttt{struct}$ from the other ones. A bridging theory is a set of equational axioms defining a bridging function by structural induction over a set of constructors.

Definition 2. *Let Σ be a signature as given in Definition 1 and let Σ_t be a signature such that Σ and Σ_t have distinct function symbols, and may share sorts, except $\texttt{struct}$. A bridging function $f \notin \Sigma \cup \Sigma_t$ has arity $\texttt{struct} \rightarrow \texttt{t}$ where $\texttt{t}$ is a sort in Σ_t. A bridging theory T_f associated with a bridging function f has the form:*

$$
T_f = \bigcup_{c \in \Sigma} \Big\{ \ \forall e \forall t_1, \ldots, t_n. \ f(c(e; t_1, \ldots, t_n)) = f_c(e; f(t_1), \ldots, f(t_n)) \ \Big\}
$$

where $f_c(x; y)$ denotes a Σ_t-term. When x does not occur in $f_c(x; y)$ for any $c \in \Sigma$, we say that T_f is $\texttt{Elem}$-independent.

Remark that the notation $f_c(x; y)$ does not enforce all elements of $x; y$ to occur in the term $f_c(x; y)$: only elements in x of sort in Σ_t can occur in $f_c(x; y)$, and there is no occurrence of x in $f_c(x; y)$ in the case of an $\texttt{Elem}$-independent bridging theory. Throughout the paper, we assume that for any constant c in Σ, f_c denotes a constant in Σ_t, and the equality $f(c) = f_c$ occurs in T_f. For instance, in the case of length of lists, $\ell(nil) = \ell_{nil} = 0$.

Example 2 (Example 1 continued). Many useful bridging theories fall into the above definition such as:

- Length of lists: $\ell(cons(e, y)) = 1 + \ell(y)$, $\ell(nil) = 0$
- Sum of lists of numbers: $lsum(cons(e, y)) = e + lsum(y)$, $lsum(nil) = 0$
- Sum of pairs of numbers: $psum(cons(e, e')) = e + e'$

Among the above bridging theories, only the length of lists is $\texttt{Elem}$-independent.

4   A Combination Procedure for Bridging Functions

We introduce a combination method for a non-disjoint union of theories $T = T_s \cup T_f \cup T_t$ as stated in Sect. 3, where the source theory T_s is $AFDS_{\Sigma_s}$, T_t is an arbitrary target Σ_t-theory, and the bridging theory T_f follows Definition 2. It is worth noticing that T_t is not required to be stably infinite. We describe below a decision procedure for checking the T-satisfiability of sets of literals. As usual, the input set of literals is first purified to get a separate form.

Definition 3. *A set of literals φ is in* separate form *if $\varphi = \varphi_{struct} \cup \varphi_{elem} \cup \varphi_t \cup \varphi_f$ where:*

- *φ_{struct} contains only flat literals of forms $x = y$, $x \neq y$ or $x = c(e; x_1, \ldots, x_n)$ where x, $x_1, \ldots, x_n$ and y are variables of sort $\mathtt{struct}$ and c is a constructor*
- *φ_{elem} contains only literals of sorts in $\Sigma_s \backslash (\Sigma_t \cup \{\mathtt{struct}\})$*
- *φ_t contains only Σ_t-literals*
- *φ_f contains only flat equalities of the form $f_x = f(x)$, where f_x denotes a variable associated with $f(x)$, such that f_x and $f(x)$ occur once in φ_f and each variable of sort $\mathtt{struct}$ in φ_{struct} occurs in φ_f.*

It is easy to convert any set of literals into an equisatisfiable separate form by introducing fresh variables to denote impure terms.

Unlike classical disjoint combination methods, guessing only one arrangement on the shared variables is not sufficient to get a modular decision procedure.

Definition 4. *Given a set of literals $\varphi = \varphi_{struct} \cup \varphi_{elem} \cup \varphi_t \cup \varphi_f$ in separate form and two arrangements*

- *Γ over the variables of sorts in $\Sigma_s \cap \Sigma_t$ occurring in both φ_{struct} and $\varphi_t \cup \varphi_f$;*
- *Γ' over the variables of sort $\mathtt{struct}$ in $\varphi_{struct} \cup \varphi_f$;*

the combinable separate form *of φ corresponding to Γ, Γ' is $(\varphi_{struct} \cup \Gamma_{struct}) \cup \varphi_{elem} \cup (\varphi_t \cup \Gamma_t) \cup \varphi_f$ where*

$$\Gamma_{struct} = \Gamma \cup \Gamma'$$
$$\Gamma_t = \Gamma \cup \{f_x = f_y \mid x = y \in \Gamma'\}$$
$$\cup \{f_x = f_c(e; f_{x_1}, \ldots, f_{x_n}) \mid x = c(e; x_1, \ldots, x_n) \in \varphi_{struct}\}$$

Any separate form extends to finitely many combinable separate forms. Also the separate form is T-equivalent to the disjunction of those combinable separate forms. From now on, we will only consider combinable separate forms and assume that a combinable separate form $\varphi_{struct} \cup \varphi_{elem} \cup \varphi_t \cup \varphi_f$ includes Γ_{struct} and Γ_t respectively in φ_{struct} and φ_t. The T-satisfiability of combinable separate forms can be checked in a modular way (see [5] for the correctness proof):

Theorem 1. *A **combinable** separate form $\varphi_{struct} \cup \varphi_{elem} \cup \varphi_t \cup \varphi_f$ is T-satisfiable if and only if $\varphi_{struct} \cup \varphi_{elem}$ is T_s-satisfiable and φ_t is T_t-satisfiable.*

Notice that φ_f is not used when checking satisfiability: these constraints are indeed encoded within φ_t, according to Definition 4.

Example 3. Consider the theory of (acyclic) lists with a length function ℓ and the set of literals $\varphi = \{x = cons(a, cons(b, z)), \ell(x) + 1 = \ell(z)\}$, where the sort for elements is not the sort of integers.

1. **Variable Abstraction and Partition.** Formula φ is separated into
   - $\varphi_{list} : \{y = cons(b, z), x = cons(a, y)\}$
   - $\varphi_{elem} : \emptyset$
   - $\varphi_{int} : \{\ell_x + 1 = \ell_z\}$
   - $\varphi_\ell : \{\ell_x = \ell(x), \ell_y = \ell(y), \ell_z = \ell(z)\}$
2. **Decomposition.** To build the combinable separate form, let Γ_{list} be the only arrangement over the list variables satisfiable together with φ_{list}, i.e. $\{x \neq y \neq z\}$. By Definition 4, Γ_{int} is $\{\ell_y = \ell_z + 1, \ell_x = \ell_y + 1\}$.
3. **Check.** The set $\varphi_{list} \cup \varphi_{elem} \cup \Gamma_{list}$ is satisfiable in the theory of lists. However $\varphi_{int} \cup \Gamma_{int}$ is unsatisfiable in the theory of linear arithmetic (over the integers). The original set of literals φ is thus unsatisfiable.

The next satisfiable formula is used as a running example in Sect. 5.

Example 4. Consider the following set of literals

$$\varphi = \{x_1 = cons(d, y_1), x_2 = cons(d, y_2), x_1 \neq x_2 \neq y_1 \neq y_2 \neq y_3, \ell(y_2) = \ell(y_3)\}$$

1. **Variable Abstraction and Partition.** Formula φ is separated into
   - $\varphi_{list} : \{x_1 = cons(d, y_1), x_2 = cons(d, y_2), x_1 \neq x_2 \neq y_1 \neq y_2 \neq y_3\}$
   - $\varphi_{elem} : \emptyset$
   - $\varphi_{int} : \{\ell_{y_2} = \ell_{y_3}\}$
   - $\varphi_\ell : \{\ell_{x_1} = \ell(x_1), \ell_{x_2} = \ell(x_2), \ell_{y_1} = \ell(y_1), \ell_{y_2} = \ell(y_2), \ell_{y_3} = \ell(y_3)\}$
2. **Decomposition.** Formula φ_{list} already includes the arrangement $\Gamma_{list} = \{x_1 \neq x_2 \neq y_1 \neq y_2 \neq y_3\}$, and $\Gamma_{int} = \{\ell_{x_1} = \ell_{y_1} + 1, \ell_{x_2} = \ell_{y_2} + 1\}$.
3. **Check.** The set $\varphi_{list} \cup \varphi_{elem} \cup \Gamma_{list}$ is satisfiable in the theory of lists. The set $\varphi_{int} \cup \Gamma_{int}$ is also satisfiable in the theory of linear arithmetic (over the integers), e.g. $\ell_{x_1} = 4, \ell_{x_2} = 3, \ell_{y_1} = 3, \ell_{y_2} = 2, \ell_{y_3} = 2$. Thus φ is satisfiable.

5 Standard Interpretations

Now consider the satisfiability problem modulo data structure theories defined as classes of *standard* structures, where each interpretation domain of struct contains only the finite terms generated by the constructors and the elements in the interpretation domains of Elem. Standard structures are specific models of the (axiomatized) theories considered in previous sections. We investigate the possibility to get a satisfiability procedure for standard interpretations by applying the combination method of Sect. 4. We first study the particular case of lists, and then the general case of trees corresponding to the standard interpretations of absolutely free data structures.

5.1   Lists with Length

Definition 5. *Consider the signature* $\Sigma_{list} = \Sigma \cup \{\ell : \texttt{list} \rightarrow \texttt{int}\} \cup \Sigma_{int}$ *where* $\Sigma = \{cons : \texttt{elem} \times \texttt{list} \rightarrow \texttt{list}, nil : \texttt{list}\}$, $\Sigma_{int} = \{0 : \texttt{int}, 1 : \texttt{int}, + : \texttt{int} \times \texttt{int} \rightarrow \texttt{int}, \leq : \texttt{int} \times \texttt{int}\}$, *and* $\texttt{elem} \neq \texttt{int}$. *A standard* list*-interpretation* $\mathcal{A}$ *is a* Σ_{list}*-interpretation satisfying the following conditions:*

- $|A_{\texttt{elem}}| > 1$;
- $A_{\texttt{list}} = (A_{\texttt{elem}})^*$ *where* $(A_{\texttt{elem}})^*$ *is the set of finite sequences* $\langle e_1, \ldots, e_n \rangle$ *for* $n \geq 0$ *and* $e_1, \ldots, e_n \in A_{\texttt{elem}}$;
- $A_{\texttt{int}} = \mathbb{Z}$ *and* $0, 1, +, \leq$ *are interpreted according to their standard interpretation in* $\mathbb{Z}$;
- $\mathcal{A}[nil] = \langle \rangle$;
- $\mathcal{A}[cons](e, \langle e_1, \ldots, e_n \rangle) = \langle e, e_1, \ldots, e_n \rangle$, *for* $n \geq 0$ *and* $e, e_1, \ldots, e_n \in A_{\texttt{elem}}$;
- $\mathcal{A}[\ell](\langle \rangle) = 0$;
- $\mathcal{A}[\ell](\langle e_1, \ldots, e_n \rangle) = n$.

The theory of (standard interpretations) of lists with length *is the pair* $T_{list}^{si} = (\Sigma_{list}, \mathbf{A})$, *where* $\mathbf{A}$ *is the class of all standard list-structures.*

Remark 1. Definition 5 excludes the case of lists built over only one element. In that singular case, the length function ℓ is bijective, which means that any disequality $x \neq y$ between list-variables can be equivalently translated into an int-disequality $\ell_x \neq \ell_y$. It thus suffices to extend the combination procedure in Sect. 4 with this additional translation expressing the bijectivity of ℓ. The satisfiability of the int-part of the resulting separate form gives a model for the theory of lists on only one element. In the case of lists, and for simplicity, we thus impose the restriction of at least two elements to standard interpretations.

In this paper, we have chosen to define standard interpretations without selectors. Indeed, selectors would be partial functions defined only on non-empty lists, and could be seen as syntactic sugar: any equality $e = car(x)$ (resp. $y = cdr(x)$) can be equivalently expressed as an equality $x = cons(e, x')$ (resp. $x = cons(d, y)$) where x' (resp. d) is a fresh variable. Thus we define T_{list}^{si} without selectors. As shown below, we can relate T_{list}^{si}-satisfiability to satisfiability modulo the **combined** theory of lists with length T_{list} defined as (the class of all the models of) the union of theories $AFDS_\Sigma \cup T_\ell \cup T_\mathbb{Z}$ where $\Sigma = \{cons : \texttt{elem} \times \texttt{list} \rightarrow \texttt{list}, nil : \texttt{list}\}$, $T_\ell = \{\forall e \forall y. \ \ell(cons(e, y)) = 1 + \ell(y), \ \ell(nil) = 0\}$, and $T_\mathbb{Z}$ denotes the theory of linear arithmetic over the integers. Since $T_{list}^{si} \models T_{list}$, a T_{list}^{si}-satisfiable formula is also T_{list}-satisfiable. For the converse implication, we need to preprocess the formulas. To build a standard interpretation from the model construction of Theorem 1, we need additional arithmetic constraints to state that each value of a length variable corresponds to the length of some finite list. These constraints are used to get witness formulas as defined in [8] for the combination of polite theories [13].

Definition 6 (Finite witnessability). *Let* Σ *be a signature,* S *be a set of sorts in* Σ, *T a* Σ*-theory, and* φ *a quantifier-free* Σ*-formula. A quantifier-free* Σ*-formula* ψ *is a* finite witness *of* φ *in* T *with respect to* S *if:*

1. φ and $(\exists \bar{v})\psi$ are T-equivalent, where $\bar{v} = Var(\psi) \setminus Var(\varphi)$;
2. for any arranged form ψ' of ψ, if ψ' is T-satisfiable then there exists a T-interpretation $\mathcal{A}$ satisfying ψ' such that $A_\sigma = \bigcup_{v \in Var_\sigma(\psi')} \mathcal{A}[v]$, for each $\sigma \in S$.

T is finitely witnessable *with respect to* S if there exists a computable function witness such that, for every quantifier-free Σ-formula φ, witness(φ) is a finite witness of φ in T with respect to S.

For the theory T_{list}^{si}, witnesses w.r.t. {elem} are derived from *range constraints*: given a set of literals φ in separate form and a natural number n, a *range constraint for φ bounded by* n is a set of literals $\mathfrak{c} = \{\mathfrak{c}(f_x) \mid f_x \in Var_{int}(\varphi_f)\}$ where $\mathfrak{c}(f_x)$ is either $f_x = i$ $(0 \le i < n)$ or $f_x \ge n$. A range constraint $\mathfrak{c}$ for φ is *satisfiable* if $\varphi_{int} \cup \mathfrak{c}$ is satisfiable in $\mathbb{Z}$. In the case of T_{list}^{si}, the role of range constraints is to perform a guessing of values for length variables. Beyond a limit value n, depending on the input formula, there are enough different lists to satisfy the disequalities between lists, and then to build a standard interpretation.

Proposition 1. *For any set of literals φ in combinable separate form, there exists a finite set of satisfiable range constraints $\mathfrak{C}$ such that*

- *φ is T_{list}^{si}-equivalent to $\bigvee_{\mathfrak{c} \in \mathfrak{C}}(\varphi \wedge \mathfrak{c})$*
- *For any $\mathfrak{c} \in \mathfrak{C}$, $\varphi \wedge \mathfrak{c}$ admits a witness denoted witness$(\varphi \wedge \mathfrak{c})$ such that any arranged form of witness$(\varphi \wedge \mathfrak{c})$ is T_{list}^{si}-satisfiable iff it is T_{list}-satisfiable.*

Proof. Since φ is a combinable separate form, it implies a unique arrangement over list-variables. Let m be the number of the corresponding equivalence classes over list-variables. We define the bound n used in range constraints as $n = \lceil \log_2(m) \rceil$ to have, for any $i \ge n$, m different lists of length i built over two elements. The set $\mathfrak{C}$ is defined as the set of all satisfiable range constraints bounded by n. Let us now define the witness of a range constraint $\mathfrak{c}$:

- $witness_{rc}(\{\ell_x = 0\} \cup \mathfrak{c}) = \{x = nil\} \cup witness_{rc}(\mathfrak{c})$
- $witness_{rc}(\{\ell_x = i\} \cup \mathfrak{c}) = \{x = cons(e_1, \ldots cons(e_i, nil) \ldots)\} \cup witness_{rc}(\mathfrak{c})$
  if $0 < i < n$, where $e_1, \ldots, e_i$ are fresh elem-variables
- $witness_{rc}(\{\ell_x \ge n\} \cup \mathfrak{c}) = witness_{rc}(\mathfrak{c})$

Then, $witness(\varphi \wedge \mathfrak{c}) = (e \ne e') \wedge \varphi \wedge \mathfrak{c} \wedge witness_{rc}(\mathfrak{c})$, where e, e' are two distinct fresh elem-variables.

Consider an arbitrary arrangement arr over variables in $witness(\varphi \wedge \mathfrak{c})$. If $witness(\varphi \wedge \mathfrak{c}) \wedge arr$ is T_{list}-satisfiable, then it is possible to construct (by using syntactic unification, see [5]) a T_{list}-equivalent set of literals φ' whose list-part contains only flat disequalities and equalities of the following forms:

(1) flat equalities $v = x$ such that v occurs once in φ',
(2) equalities $x = t$, where t is a nil-terminated list and x occurs once in the equalities of φ',
(3) equalities $x = cons(d, y)$, where x and y cannot be equal to nil-terminated lists (by applying the variable replacement of syntactic unification).

Let us now define a T_{list}^{si}-interpretation. First, the equalities in (1) can be discarded since v occurs once in φ'. The interpretation of variables occurring in (2) directly follows from φ'. It remains to show how to interpret variables occurring in (3). Note that each of these variables has a length greater or equal than n, otherwise it would occur in (2). The solved form φ' defines a (partial) ordering $>$ on these variables: $x > y$ if $x = cons(d, y)$ occurs in φ'. Each minimal variable y with respect to $>$ is interpreted by a fresh nil-terminated list not occurring in φ' whose elements are (the representatives of) e, e', and whose length is the interpretation of ℓ_y (this is possible by definition of n and the fact that $\ell_y \geq n$). Then, the interpretation of non-minimal variables follows from the equalities (3) in φ'. By construction, distinct variables are interpreted by distinct lists. In other words, the list-disequalities introduced by arr are satisfied by this interpretation. Furthermore, any equality $\ell_x = \ell(x)$ in φ_ℓ is satisfied by this interpretation since φ is a combinable separate form. Therefore, all literals of φ' are true in this interpretation, and so we have built a T_{list}^{si}-model of $witness(\varphi \wedge \mathfrak{c}) \wedge arr$.

Moreover the above construction is a way to build a T_{list}^{si}-model such that the **elem** sort is interpreted as the set of interpreted **elem**-variables in the witness. So the *witness* function satisfies the requirements of Definition 6.   □

Example 5. Consider the T_{list}^{si}-satisfiability of the combinable separate form built in Example 4: $\varphi = \varphi_\ell \cup \{x_1 = cons(d, y_1), x_2 = cons(d, y_2), x_1 \neq x_2 \neq y_1 \neq y_2 \neq y_3, \ell_{x_1} = \ell_{y_1} + 1, \ell_{x_2} = \ell_{y_2} + 1, \ell_{y_2} = \ell_{y_3}\}$. The five list-variables imply that range constraints are bounded by $n = 3$. There are 4^5 possible range constraints (each variable can be equal to 0, 1, 2 or greater than or equal to 3). We now focus on few satisfiable range constraints and their related witnesses (the remaining ones are handled similarly).

1. $\mathfrak{c} = \{\ell_{x_1} = \ell_{x_2} = 1, \ell_{y_1} = \ell_{y_2} = \ell_{y_3} = 0\}$. To obtain a witness of φ and $\mathfrak{c}$, we add $y_1 = y_2 = y_3 = nil$, $x_1 = cons(e_{x_1}, nil)$, and $x_2 = cons(e_{x_2}, nil)$. It follows that $e_{x_1} = e_{x_2} = d$ and $x_1 = x_2$ which contradicts φ.
2. $\mathfrak{c} = \{\ell_{x_1} \geq 3, \ell_{y_1} = \ell_{x_2} = 2, \ell_{y_2} = \ell_{y_3} = 1\}$. The witness leads to
   - $y_1 = cons(e'_{y_1}, cons(e_{y_1}, nil))$, $y_2 = cons(e_{y_2}, nil)$, $y_3 = cons(e_{y_3}, nil)$
   - $x_1 = cons(d, y_1) = cons(d, cons(e'_{y_1}, cons(e_{y_1}, nil)))$
   - $x_2 = cons(d, y_2) = cons(d, cons(e_{y_2}, nil))$

   All the list-variables are instantiated by distinct lists, provided the arrangement over **elem**-variables is such that $e_{y_2} \neq e_{y_3}$ and ($e_{y_1} \neq e_{y_2}$ or $e'_{y_1} \neq d$).
3. $\mathfrak{c} = \{\ell_{x_1} = 1, \ell_{y_1} = 0, \ell_{x_2} \geq 3, \ell_{y_2} \geq 3, \ell_{y_3} \geq 3\}$. The related witness is equisatisfiable to $\varphi \cup \mathfrak{c} \cup \{y_1 = nil, e \neq e'\}$, which is satisfiable by considering:
   - $y_2 = cons(e, cons(e, cons(e, nil)))$, $y_3 = cons(e, cons(e, cons(e', nil)))$
   - $y_1 = nil, x_1 = cons(d, nil), x_2 = cons(d, cons(e, cons(e, cons(e, nil))))$

The following sections will demonstrate that witnesses are not only interesting for T_{list}^{si}-satisfiability but also for the combination of T_{list}^{si} with an arbitrary theory for elements. However, when T_{list}^{si} is considered alone, there is a much simpler T_{list}^{si}-satisfiability procedure (see [5] for the proof).

Theorem 2. *Let φ be a set of literals in combinable separate form, and let $\mathfrak{C}$ be the finite set of satisfiable range constraints of φ **bounded by** 1. The formula φ is T_{list}^{si}-satisfiable iff there exists a satisfiable range constraint $\mathfrak{c} \in \mathfrak{C}$ such that $witness(\varphi \wedge \mathfrak{c})$ is T_{list}-satisfiable.*

5.2   Combining Lists with an Arbitrary Theory of Elements

As shown below, T_{list}^{si} is actually a polite theory, and so it can be combined with an arbitrary disjoint theory of elements, using the combination method designed for polite theories [8,13]. By definition, a polite theory is both finite witnessable and smooth.

Definition 7 (Smoothness and Politeness).   *Consider a set $S = \{\sigma_1, \ldots, \sigma_n\}$ of sorts in a signature Σ. A Σ-theory T is smooth with respect to S if:*

- *for every T-satisfiable quantifier-free Σ-formula φ,*
- *for every T-interpretation $\mathcal{A}$ satisfying φ,*
- *for every cardinal number $\kappa_1, \ldots, \kappa_n$ such that $\kappa_i \geq |A_{\sigma_i}|$, for $i = 1, \ldots, n$,*

there exists a T-model $\mathcal{B}$ of φ such that $|B_{\sigma_i}| = \kappa_i$ for $i = 1, \ldots, n$. A Σ-theory T is polite with respect to S if it is both smooth and finitely witnessable with respect to S.

The smoothness of the theory of standard interpretations of lists has been shown in [13], and this is preserved while considering the length function. By definition of T_{list}^{si}, any set of elements can be used to build the lists (since ℓ is $\{\texttt{elem}\}$-independent). Hence a T_{list}^{si}-satisfiable formula remains T_{list}^{si}-satisfiable when adding elements (of sort $\texttt{elem}$), and so T_{list}^{si} is smooth. The finite witnessability of T_{list}^{si} is a consequence of Proposition 1.

Proposition 2. *T_{list}^{si} is polite with respect to $\{\texttt{elem}\}$.*

Consider the satisfiability problem in the disjoint combination $T_{list}^{si} \oplus T_{elem}$ where T_{elem} is a mono-sorted Σ_{elem}-theory over the shared sort $\texttt{elem}$. Due to the politeness of T_{list}^{si}, we can directly use the combination method initiated in [13] for polite theories, and this leads to the following result.

Theorem 3. *Let φ be a set of literals in combinable separate form, and let $\mathfrak{C}$ be the finite set of satisfiable range constraints introduced in Proposition 1. The formula φ is $T_{list}^{si} \oplus T_{elem}$-satisfiable iff there exists a satisfiable range constraint $\mathfrak{c} \in \mathfrak{C}$ and an arrangement arr such that (1) $witness(\varphi \wedge \mathfrak{c}) \wedge arr$ is T_{list}-satisfiable and (2) $\varphi_{elem} \wedge arr$ is T_{elem}-satisfiable, where arr is an arrangement over the variables of sort $\texttt{elem}$ in $witness(\varphi \wedge \mathfrak{c})$.*

Example 6. Recall the formula from Example 4 in its combinable separate form and suppose we add a new literal stating that the sum of the lengths of y_1, y_2 and y_3 is three: $\varphi = \varphi_\ell \cup \{x_1 = cons(d, y_1), x_2 = cons(d, y_2), x_1 \neq x_2 \neq y_1 \neq y_2 \neq y_3, \ell_{x_1} = \ell_{y_1} + 1, \ell_{x_2} = \ell_{y_2} + 1, \ell_{y_2} = \ell_{y_3}, \ell_{y_1} + \ell_{y_2} + \ell_{y_3} = 3\}$ and consider the theory of elements $T_{elem} = \{a \neq b, (\forall x : \texttt{elem}.\ x = a \vee x = b)\}$. There are now only two satisfiable range constraints:

1. $\ell_{x_1} \geq 3, \ell_{y_1} \geq 3, \ell_{x_2} = 1, \ell_{y_2} = 0, \ell_{y_3} = 0$, which leads to $\ell_{x_1} = 4$ and $\ell_{y_1} = 3$. But this is T_{list}^{si}-unsatisfiable, as it requires $y_2 = nil$ and $y_3 = nil$, which makes $y_2 \neq y_3$ false.
2. $\ell_{x_1} = 2, \ell_{y_1} = 1, \ell_{x_2} = 2, \ell_{y_2} = 1, \ell_{y_3} = 1$, which implies
   - $y_1 = cons(e_{y_1}, nil), y_2 = cons(e_{y_2}, nil), y_3 = cons(e_{y_3}, nil)$
   - $x_1 = cons(d, cons(e_{y_1}, nil)), x_2 = cons(d, cons(e_{y_2}, nil))$
   But this requires $e_{y_1} \neq e_{y_2} \neq e_{y_3}$, which is T_{elem}-unsatisfiable.

Hence φ is $T_{list}^{si} \oplus T_{elem}$-unsatisfiable.

Let us now assume T_{elem} is stably infinite. Since T_{list}^{si} is stably infinite too, the classical Nelson-Oppen combination method applies to $T_{list}^{si} \oplus T_{elem}$ by using the T_{list}^{si}-satisfiability procedure stated in Theorem 2. This leads to a result similar to Theorem 3, where it is sufficient to guess only few particular range constraints.

Proposition 3. *Assume T_{elem} is stably infinite. Let φ be a set of literals in combinable separate form, and let $\mathfrak{C}$ be the finite set of satisfiable range constraints of φ bounded by 1. The formula φ is $T_{list}^{si} \oplus T_{elem}$-satisfiable iff there exists a satisfiable range constraint $\mathfrak{c} \in \mathfrak{C}$ and an arrangement arr such that (1) witness($\varphi \wedge \mathfrak{c}$) $\wedge$ arr is T_{list}-satisfiable and (2) $\varphi_{elem} \wedge$ arr is T_{elem}-satisfiable, where arr is an arrangement over the variables of sort elem in witness($\varphi \wedge \mathfrak{c}$).*

In the above proposition, arr is an arrangement over the variables of sort elem in φ since $Var(\varphi) = Var(witness(\varphi \wedge \mathfrak{c}))$. Indeed $witness(\varphi \wedge \mathfrak{c})$ only extends $\varphi \cup \mathfrak{c}$ with an equality $x = nil$ for each $l_x = 0$ in $\mathfrak{c}$.

5.3   Trees with Bridging Functions over the Integers

The combination method presented for standard interpretations of lists can be extended to standard interpretations of any AFDS theory, as discussed below.

Definition 8. *Consider the signature $\Sigma_{tree} = \Sigma \cup \{f : \text{struct} \to \text{int}\} \cup \Sigma_{int}$ where Σ and Σ_{int} are signatures respectively as in Definitions 1 and 5 such that $\text{int} \notin \text{Elem}$, and let T_f be an Elem-independent bridging theory as in Definition 2. A standard tree-interpretation $\mathcal{A}$ is a Σ_{tree}-interpretation satisfying the following conditions:*

- *$\mathcal{A}_{\text{struct}}$ is the set of Σ-terms of sort struct built with elements in $(\mathcal{A}_e)_{e \in \text{Elem}}$;*
- *$\mathcal{A}_{\text{int}} = \mathbb{Z}$ and $0, 1, +, \leq$ are interpreted according to their standard interpretation in $\mathbb{Z}$;*
- *$\mathcal{A}[c] = c$ for each constant constructor $c \in \Sigma$;*
- *$\mathcal{A}[c](e, t_1, \ldots, t_n) = c(e, t_1, \ldots, t_n)$ for each non-constant constructor $c \in \Sigma$, tuple e of elements in $(\mathcal{A}_e)_{e \in \text{Elem}}$, and $t_1, \ldots, t_n \in \mathcal{A}_{\text{struct}}$;*
- *$\mathcal{A}[f](c) = f_c$ for each constant constructor $c \in \Sigma$;*
- *$\mathcal{A}[f](c(e, t_1, \ldots, t_n)) = f_c(e, \mathcal{A}[f](t_1), \ldots, \mathcal{A}[f](t_n))$ for each non-constant constructor $c \in \Sigma$, tuple e of elements in $(\mathcal{A}_e)_{e \in \text{Elem}}$, and $t_1, \ldots, t_n \in \mathcal{A}_{\text{struct}}$.*

The theory of (standard interpretations) of trees with bridging function f is the pair $T_{tree}^{si} = (\Sigma_{tree}, \mathbf{A})$, where $\mathbf{A}$ is the class of all standard tree-structures.

Let T_{tree} be the **combined** theory of trees with the bridging function f defined as (the class of all the models of) the union of theories $AFDS_\Sigma \cup T_f \cup T_\mathbb{Z}$. If a formula is T_{tree}^{si}-satisfiable, then it is also T_{tree}-satisfiable. For the converse implication, we proceed like for lists by introducing witnesses. Witnesses can easily be computed when f is the height or the size of trees. Hence, in a way analogous to what has been done for lists (cf. Proposition 1), there is a method to reduce T_{tree}^{si}-satisfiability to T_{tree}-satisfiability. As shown below, T_{tree}^{si} is polite, and so we can obtain a $T_{tree}^{si} \oplus T_{elem}$-satisfiability procedure by combining the satisfiability procedures for T_{tree} and T_{elem} (analogous to Theorem 3). The following assumptions enable us to extend the proofs developed for lists to the case of trees.

Definition 9. *Let T be a theory defined as a class of standard tree-structures. For any $\mathcal{A} \in T$, let $F_\mathcal{A}^{-1}(n) = \{t \mid \mathcal{A}[f](t) = n\}$. The bridging function f is gently growing in T if*

1. *for any $n \in \mathbb{Z}$ and any $\mathcal{A} \in T$, $F_\mathcal{A}^{-1}(n) \neq \emptyset \iff n \geq 0$;*
2. *for any $n \geq 0$ and any $\mathcal{A} \in T$, $|F_\mathcal{A}^{-1}(n)| < |F_\mathcal{A}^{-1}(n+1)|$;*
3. *there exists a computable function $b : \mathbb{N} \to \mathbb{N}$ such that for any $n > 1$ and any $\mathcal{A} \in T$, $|F_\mathcal{A}^{-1}(b(n))| \geq n$;*
4. *for any $n \geq 0$, one can compute a finite non-empty set $F^{-1}(n)$ of terms with variables of sorts in* Elem *such that*

$$T \models f(x) = n \iff (\exists \bar{v} . \bigvee_{t \in F^{-1}(n)} x = t) \quad \text{where } \bar{v} = Var(F^{-1}(n))$$

Proposition 4. *Let $\Sigma = \{cons : \text{elem} \times \text{struct} \times \cdots \times \text{struct} \to \text{struct},\ nil : \text{struct}\}$. Assume that cons is of arity strictly greater than 2 and consider the following bridging theories:*

- *Size of trees: $sz(cons(e, y_1, \ldots, y_n)) = 1 + \sum_{i=1}^{n} sz(y_i)$, $sz(nil) = 0$*
- *Height of trees: $ht(cons(e, y_1, \ldots, y_n)) = 1 + \max_{i \in [1,n]} ht(y_i)$, $ht(nil) = 0$*

If $f = sz$ or $f = ht$, then f is gently growing in T_{tree}^{si}.

To prove the above proposition, the function b of Definition 9 can be defined as the identity over $\mathbb{N}$, but it is possible to get a better bound, e.g., thanks to Catalan numbers [23] for the size of trees. When *cons* is of arity 2, sz and ht coincide with the length ℓ. In that case, ℓ is gently growing in T_{list}^{si} that corresponds to $T_{tree}^{si} \cup \{\exists v, v' : \text{elem} . v \neq v'\}$.

Proposition 5. *If f is gently growing in T_{tree}^{si}, then T_{tree}^{si} is polite w.r.t.* Elem.

Theorem 3 (for lists) can be rephrased for trees and gives:

Corollary 1. *Assume f is gently growing in T_{tree}^{si}. Let $T_{tree}^{si} \oplus T_{elem}$ be a disjoint combination where T_{elem} is a many-sorted Σ_{elem}-theory over the sorts in* Elem. *$T_{tree}^{si} \oplus T_{elem}$-satisfiability is decidable if T_{elem}-satisfiability is decidable.*

Consider a theory T_{tree}^{si} as in Proposition 4 where $f = sz$ or $f = ht$. Similarly to Theorem 2, T_{tree}^{si}-satisfiability reduces to T_{tree}-satisfiability by guessing only range constraints bounded by 1. If T_{elem} is stably infinite, then we get a combination method for $T_{tree}^{si} \oplus T_{elem}$-satisfiability as in Proposition 3.

6 Conclusion

This paper describes (Sect. 4) a non-deterministic combination method à la Nelson-Oppen for unions of theories including absolutely free data structures connected to target theories via bridging functions. Similarly to the classical Nelson-Oppen method, implementations of this non-deterministic combination method should be based not on guessings but on more practical refinements. But this lightweight approach is in the line with disjoint combination procedures embedded in SMT solvers, and is thus amenable to integration in those tools.

We reuse the notions of witness and politeness (Sect. 5), already introduced for non-stably infinite disjoint combinations, to adapt satisfiability procedures to standard interpretations. Hence, the combination method for polite theories is applicable to combine the theory of standard interpretations of lists (trees) with an arbitrary disjoint theory for elements. The case of standard interpretations of lists (trees) over integer elements has not been tackled but can be solved using the approach discussed for a stably infinite theory of elements.

To complete this work, we are currently investigating more (data structure) theories with bridging functions for which the combination method of Sect. 4 is sound and complete. Another natural continuation consists in considering standard interpretations modulo non-absolutely free constructors [14], e.g., associative-commutative operators to specify multisets.

Acknowledgments. We are grateful to Jasmin Blanchette and to the anonymous reviewers for many constructive remarks.

References

1. Armando, A., Bonacina, M.P., Ranise, S., Schulz, S.: New results on rewrite-based satisfiability procedures. ACM Trans. Comput. Log. **10**(1), 4 (2009)
2. Baader, F., Ghilardi, S.: Connecting many-sorted theories. J. Symb. Log. **72**(2), 535–583 (2007)
3. Baumgartner, P., Waldmann, U.: Hierarchic superposition with weak abstraction. In: Bonacina, M.P. (ed.) CADE 2013. LNCS, vol. 7898, pp. 39–57. Springer, Heidelberg (2013)
4. Chocron, P., Fontaine, P., Ringeissen, C.: A gentle non-disjoint combination of satisfiability procedures. In: Demri, S., Kapur, D., Weidenbach, C. (eds.) IJCAR 2014. LNCS, vol. 8562, pp. 122–136. Springer, Heidelberg (2014). http://hal.inria.fr/hal-00985135
5. Chocron, P., Fontaine, P., Ringeissen, C.: A Polite Non-Disjoint Combination Method: Theories with Bridging Functions Revisited (Extended Version) (2015). http://hal.inria.fr
6. Fontaine, P., Ranise, S., Zarba, C.G.: Combining lists with non-stably infinite theories. In: Baader, F., Voronkov, A. (eds.) LPAR 2004. LNCS (LNAI), vol. 3452, pp. 51–66. Springer, Heidelberg (2005)
7. Ghilardi, S.: Model-theoretic methods in combined constraint satisfiability. J. Autom. Reasoning **33**(3–4), 221–249 (2004)

8. Jovanović, D., Barrett, C.: Polite theories revisited. In: Fermüller, C.G., Voronkov, A. (eds.) LPAR-17. LNCS, vol. 6397, pp. 402–416. Springer, Heidelberg (2010)
9. Nelson, G., Oppen, D.C.: Simplification by cooperating decision procedures. ACM Trans. Program. Lang. Syst. 1(2), 245–257 (1979)
10. Nicolini, E., Ringeissen, C., Rusinowitch, M.: Combinable extensions of abelian groups. In: Schmidt, R.A. (ed.) CADE-22. LNCS, vol. 5663, pp. 51–66. Springer, Heidelberg (2009)
11. Nicolini, E., Ringeissen, C., Rusinowitch, M.: Combining satisfiability procedures for unions of theories with a shared counting operator. Fundam. Inf. 105(1–2), 163–187 (2010)
12. Pham, T.-H., Whalen, M.W.: An improved unrolling-based decision procedure for algebraic data types. In: Cohen, E., Rybalchenko, A. (eds.) VSTTE 2013. LNCS, vol. 8164, pp. 129–148. Springer, Heidelberg (2014)
13. Ranise, S., Ringeissen, C., Zarba, C.G.: Combining data structures with nonstably infinite theories using many-sorted logic. In: Gramlich, B. (ed.) FroCoS 2005. LNCS (LNAI), vol. 3717, pp. 48–64. Springer, Heidelberg (2005)
14. Sofronie-Stokkermans, V.: Locality results for certain extensions of theories with bridging functions. In: Schmidt, R.A. (ed.) CADE-22. LNCS, vol. 5663, pp. 67–83. Springer, Heidelberg (2009)
15. Suter, P., Dotta, M., Kuncak, V.: Decision procedures for algebraic data types with abstractions. In: Hermenegildo, M.V., Palsberg, J. (eds.) Principles of Programming Languages (POPL), pp. 199–210. ACM, New York (2010)
16. Suter, P., Köksal, A.S., Kuncak, V.: Satisfiability modulo recursive programs. In: Yahav, E. (ed.) Static Analysis. LNCS, vol. 6887, pp. 298–315. Springer, Heidelberg (2011)
17. Tinelli, C., Harandi, M.T.: A new correctness proof of the Nelson-Oppen combination procedure. In: Baader, F., Schulz, K.U. (eds.) Frontiers of Combining Systems (FroCoS), Applied Logic, pp. 103–120. Kluwer Academic Publishers (1996)
18. Tinelli, C., Ringeissen, C.: Unions of non-disjoint theories and combinations of satisfiability procedures. Theoret. Comput. Sci. 290(1), 291–353 (2003)
19. Wies, T., Piskac, R., Kuncak, V.: Combining theories with shared set operations. In: Ghilardi, S., Sebastiani, R. (eds.) FroCoS 2009. LNCS, vol. 5749, pp. 366–382. Springer, Heidelberg (2009)
20. Zarba, C.G.: Combining lists with integers. In: International Joint Conference on Automated Reasoning (Short Papers), Technical report DII 11/01, pp. 170–179. University of Siena (2001)
21. Zarba, C.G.: Combining multisets with integers. In: Voronkov, A. (ed.) CADE 2002. LNCS (LNAI), vol. 2392, pp. 363–376. Springer, Heidelberg (2002)
22. Zarba, C.G.: Combining sets with cardinals. J. Autom. Reasoning 34(1), 1–29 (2005)
23. Zhang, T., Sipma, H.B., Manna, Z.: Decision procedures for term algebras with integer constraints. Inf. Comput. 204(10), 1526–1574 (2006)

Exploring Theories with a Model-Finding Assistant

Salman Saghafi$^{(\boxtimes)}$, Ryan Danas, and Daniel J. Dougherty

Worcester Polytechnic Institute, Worcester, MA, USA
salmans@wpi.edu

Abstract. We present an approach to understanding first-order theories by exploring their models. A typical use case is the analysis of artifacts such as policies, protocols, configurations, and software designs. For the analyses we offer, users are not required to frame formal properties or construct derivations. Rather, they can explore examples of their designs, confirming the expected instances and perhaps recognizing bugs inherent in surprising instances.

Key foundational ideas include: the information preorder on models given by homomorphism, an inductively-defined refinement of the Herbrand base of a theory, and a notion of provenance for elements and facts in models. The implementation makes use of SMT-solving and an algorithm for minimization with respect to the information preorder on models.

Our approach is embodied in a tool, Razor, that is complete for finite satisfiability and provides a read-eval-print loop used to navigate the set of finite models of a theory and to display provenance.

1 Introduction

Suppose $\mathcal{T}$ is a first-order theory. If $\mathcal{T}$ specifies a software artifact written by a user, such as an access-control policy, a description of a protocol, or a software design, our user will want to understand whether or not the logical consequences of $\mathcal{T}$ match her expectations. A standard approach using automated deduction tools offers the following workflow: (i) the user specifies, as a sentence σ, some typical property she hopes will hold about the system, then (ii) checks whether σ is provable from $\mathcal{T}$, using a theorem-prover or a proof-assistant.

An alternative approach is to explore the *models* of $\mathcal{T}$. This is of course logically at least as rich as the deductive approach, since σ will hold iff $\mathcal{T} \cup \neg\sigma$ has no models. But the model-exploring approach offers a wider range of affordances to the user than does deduction. For one thing, if property σ fails of $\mathcal{T}$, it can be instructive to see *example situations,* that is, to see concrete models of $\mathcal{T} \cup \neg\sigma$. This will be especially useful if we can offer tools to help our user *understand* these examples ("what is that element doing there? why is that fact true?").

More radically, our user might use a model-building tool to explore models of $\mathcal{T}$ *without having to articulate logical consequences.* For example, if $\mathcal{T}$ describes

This material is based upon work supported by the National Science Foundation under Grant No. CNS-1116557.

A.P. Felty and A. Middeldorp (Eds.): CADE-25, LNAI 9195, pp. 434–449, 2015.
DOI: 10.1007/978-3-319-21401-6_30

a policy for accessing a building, our user can explore the question, "who can enter after 5 pm?" by expressing "someone enters after 5 pm" as a sentence σ and asking for models of $(\mathcal{T} \cup \{\sigma\})$. The resulting models may capture situations that confirm the user's expectation, but there also may be models with unanticipated settings, allowing surprising accesses, which uncover gaps in the policy.

Model-finding is an active area of investigation [1–8]. But—with some exceptions noted below in Related Work—existing model-finders compute an essentially random set of models, present them to the user in arbitrary order, provide no facility for exploring the space of models in a systematic way, and offer little help to users in understanding a given model. We will clarify below what we mean by "understanding" a model, but the notion has clear intuitive force. For example when a sysadmin is debugging a firewall policy, a typical question at hand is: what rule blocked (or allowed) this packet?

As our main contribution, we initiate a theory of *exploration* of finite models, with two main components: (i) a notion of *provenance* as a way to explain why elements are in the model and why properties are true of them, and (ii) strategies for traversing the models of an input theory by *augmentation*. Our approach is realized in a model-finding assistant, Razor.[1] We call Razor a model-finding *assistant* because users interact with it to build and examine models.

Minimality and the Chase. At the core of our approach is the notion of a homomorphism between models (Sect. 2) and the preorder $\preceq$ determined by homomorphism. A homomorphism preserves information, so that if $\mathbb{A}$ and $\mathbb{B}$ are each models of some phenomenon and $\mathbb{A} \preceq \mathbb{B}$, with $\mathbb{B} \not\preceq \mathbb{A}$, then we prefer to show $\mathbb{A}$ to the user, at least initially, since it has less "extraneous" information than $\mathbb{B}$. The theoretical foundations of our tool derive from the classical Chase algorithm from database theory (Sect. 3), and our core algorithm (Sect. 4) builds models that are minimal in the homomorphism ordering.

Provenance. As a direct consequence of the fact that Razor ultimately computes Chase-models, Razor can display provenance information for elements and facts. Any element in a Chase-model is there in response to a sentence in the user's input (Sect. 3.2), indeed as a witness for a particular existential quantifier in the input theory. Similarly, any atomic fact of a Chase-model is there because of the requirement that a particular input sentence hold. Razor keeps track of these justifications—we call them "naming" and "blaming," respectively—and can answer provenance queries from the user.

Augmentation. Focusing on Chase models promotes conceptual clarity by allowing the user to focus only on models with no inessential aspects. But the user can access other models of the theory by augmenting models by new facts. When a user asks to augment a given model $\mathbb{M}$ of a theory $\mathcal{T}$ by some fact F, other consequences may be entailed by $\mathcal{T}$, perhaps "disjunctive" consequences. Razor thus computes a stream consisting of all the *minimal* extensions of $\mathbb{M}$ by the augmenting fact F (Sect. 3.1). There may be none: F may be inconsistent with the state of affairs $\mathbb{M}$, which may be of real significance. The (relative)

---

[1] http://salmans.github.io/Razor/.

minimality of the resulting models ensures that provenance information can be computed over them as well. Most important of all, this augmentation will be under the control of the user.

Implementation. We have found it more efficient to implement a variation on the Chase, which leverages an SMT-solver to handle the difficulties arising form disjunctions and equations (Sect. 4). A key ingredient of this approach is the use of a refinement of the notion of the Herbrand base of a theory, the *possible facts* set defined in Algorithm 3.

The REPL. Since the original input theory need not be a Horn theory, we will not expect unique minimal models. Razor provides a read-eval-print loop in which users can (i) ask for the next model in the current stream, (ii) play "what-if?" by augmenting the currently-displayed model with a new fact or (iii) ask for the provenance of elements or facts in the current model.

1.1 Related Work

Model-Finding. The development of algorithms for the generation of finite models is an active area of research. The prominent method is "MACE-style" [2], embodied in tools such as Paradox [3], Kodkod [6], which reduce the problem to be solved into propositional logic and employ a SAT-solver. "Instance based" methods for proof search can be adapted to compute finite models [5,7,9]. Our approach is related to the bottom-up model generation [4] method and the refutationally complete solution presented in [10]. Our techniques for bounding the search are related to those presented in [11]. Closer in spirit to our goals are lightweight formal methods tools such as Alloy [12] and Margrave [13,14]. The goals of these works differ from ours in that their main concern is usually not the *exploration* of the space of all models of a theory.

Minimality. Logic programming languages produce single, *least* models as a consequence of their semantics. In more specialized settings, generation of minimal models usually relies on dedicated techniques, often based on tableaux [15] or hyperresolution [16]. Aluminum [17] supports exploration by returning minimal models: it instruments the model-finding engine of Alloy. It thus inherits the limitation that it requires user-supplied bounds, and it cannot generate provenance information. The Cryptographic Protocol Shapes Analyzer [18] also generates minimal models. However, its application domain and especially algorithms are quite different from ours. The Network Optimized Datalog tool [19], which has been released as a part of Z3 [20], presents limited minimization and provenance construction for reasoning about beliefs in the context of network reachability policies.

Geometric Logic. The case for geometric logic as a logic of observable properties was made clearly by Abramsky [21] and has been explored as a notion of specification by several authors [22,23] Geometric logic for theorem-proving was introduced in [24] and generalized in [5]. The crucial difference with the current work is of course the fact that we focus on model-finding and exploration.

Chase. Our model-finding is founded on the Chase, an algorithm well-known in the database community [25–27]. Challenges arise for us in managing the complexity that arises due to disjunction, and in treating equality. Our strategy for addressing these challenges comprises Sect. 4.

2   Preliminaries

We work over a first-order signature with relation symbols (including equality). As syntactic sugar for users, we allow function symbols in the concrete syntax. It turns out to be convenient and flexible to interpret such function symbols as partial functions. What this means formally is that when relation symbols are translated to function symbols in the usual way, theories are augmented with axioms ensuring singled-valuedness but not necessarily with totality axioms. *Skolemization* will play an important role in the following, especially as regards provenance of elements; we assume familiarity with the basic notions.

Models, and the notion of satisfaction of a formula in a model, are defined in the usual way. If $\mathcal{T}$ is a theory we write $Mod(\mathcal{T})$ for the class of models of $\mathcal{T}$. If $\Sigma \subseteq \Sigma^+$ are signatures and $\mathbb{M}$ is a Σ^+-model then the *reduct* of $\mathbb{M}$ to Σ is obtained in the obvious way by ignoring the relations of Σ^+ not in Σ. A *homomorphism* from $\mathbb{M}$ to $\mathbb{N}$ is a map from the domain of $\mathbb{M}$ to the domain of $\mathbb{N}$, $h : |\mathbb{M}| \to |\mathbb{N}|$, such that for every relational symbol R and tuple $\langle e_1, \ldots e_n \rangle$ of elements of $|\mathbb{M}|$, if $\mathbb{M} \models R[e_1, \ldots, e_n]$ then $\mathbb{N} \models R[h(e_1), \ldots, h(e_n)]$. We write $\mathbb{M} \preccurlyeq \mathbb{N}$ for the preorder defined by the existence of a homomorphism from $\mathbb{M}$ to $\mathbb{N}$. We say that $\mathbb{M}$ is a *minimal model* in a class $\mathcal{C}$ of models if $\mathbb{N} \in \mathcal{C}$ and $\mathbb{N} \preccurlyeq \mathbb{M}$ implies $\mathbb{M} \preccurlyeq \mathbb{N}$. A set $\mathcal{M}$ of models is a *set-of-support* for a class $\mathcal{C}$ if for each $\mathbb{N} \in \mathcal{C}$ there is some $\mathbb{M} \in \mathcal{M}$ with $\mathbb{M} \preccurlyeq \mathbb{N}$.

2.1   Logic in Geometric Form

A *positive-existential* formula (PEF) is one built from atomic formulas (including $\top$ and $\bot$) using $\wedge$, $\vee$, and $\exists$. If $\alpha(\vec{x})$ is a PEF true of a tuple $\vec{e}$ in a model $\mathbb{M}$ then the truth of this fact is supported by a finite fragment of $\mathbb{M}$. Thus if $\mathbb{M}$ satisfies α with $\vec{e}$ and $\mathbb{M}$ is expanded, by adding new elements and/or new facts, $\alpha(\vec{x})$ still holds of $\vec{e}$ in the resulting model. For this reason, properties defined by PEF are sometimes called *observable* properties [21].

It is a classical result that PEFs are precisely the formulas preserved under homomorphisms; Rossman [28] has shown that this holds even if we restrict attention to finite models only. Thus the homomorphism preorder captures the observable properties of models: this is the sense in which we view this preorder as an "information-preserving" one.

A sentence is *geometric* if it is of the form $\forall \vec{x} (\varphi \Rightarrow \psi)$, where φ and ψ are PEFs. It is often convenient to suppress writing the universal quantification explicitly. We sometimes refer to φ and ψ respectively as the *body* and *head* of $\varphi \Rightarrow \psi$. Note that an empty conjunction may be regarded as truth ($\top$) and an empty disjunction as falsehood ($\bot$). So we may view a universally quantified

PEF, or a universally quantified negated PEF, as a geometric sentence. A theory is in *geometric form* if it consists of a set of geometric sentences.[2] Thus logic in geometric form is the logic of implications between observable properties.

By routine logical manipulations we may assume that every geometric sentence $\varphi \Rightarrow \psi$ is in *standard* form

$$\alpha(\vec{x}) \Rightarrow \bigvee_i (\exists y_{i1} \ldots \exists y_{ip}.\beta_i(\vec{x}, y_{i1}, \ldots, y_{ip})),$$

where α and each β_i is a conjunction of atoms.

Transformation to Geometric Form. The sense in which geometric form is—and is not—a restriction is delicate, but interesting. As is well-known, any theory is equisatisfiable with one in conjunctive normal form, by introducing Skolem functions. And modulo trivial equivalences, such a sentence is a geometric one. But Skolemization has consequences for *user-centered* model-finding. For example, traditionally, Skolem functions are *total*, and of course it is easy to achieve this abstractly by making arbitrary choices if necessary. But for reasons connected with computing provenance and keeping models finite, it is much more convenient to work with partial functions, or in other words, at-most-single valued relations.

It is easy to check that $\mathcal{T}$ can be put in geometric form—without Skolemization —whenever each axiom is an $\forall\exists$ sentence, with the caveat that no existential quantifier has within its scope both an atom with negative polarity and one of positive polarity. This circumstance arises infrequently in practice. We prefer to avoid Skolemization if possible. Any Skolemization necessary for putting a theory in this form is considered to happen "off stage".

3   Model-Finding via the Chase

In this section we outline the essential features of the Chase, since it is the most natural setting for understanding the way that minimality and provenance drive a general model-finding framework based on geometric form. It turns out that a straightforward implementation of the Chase algorithm is too inefficient. In Sect. 4 we describe the strategy we use in Razor to build the same models the Chase would construct but using SMT-solving technology for efficiency.

It is easiest to present the standard Chase as a non-deterministic procedure. We assume given an infinite set K of symbols used to construct elements of the model: at any stage of the process we will have identified a finite subset K' of K and (if the theory $\mathcal{T}$ involves equality) a congruence relation over K'. The congruence classes are the elements of the model.

Assume that the input theory $\mathcal{T}$ is presented in standard geometric form. At a given stage, if the current model $\mathbb{M}$ is not yet a model of $\mathcal{T}$ then there is some

---

[2] The term "geometric" arises from the original study of this class of formulas in the nexus between algebraic geometry and logic [29].

sentence σ of $\mathcal{T}$ false in $\mathbb{M}$:

$$\sigma \equiv \alpha(x_1, \ldots, x_k) \Rightarrow \bigvee_i (\exists y_{i1} \ldots \exists y_{ip}.\beta_i(x_1, \ldots, x_k, y_{i1}, \ldots, y_{ip})). \qquad (1)$$

That is, there is an environment (a mapping from variables to elements of $\mathbb{M}$) $\eta \equiv \{x_1 \mapsto e_1, \ldots, x_k \mapsto e_k\}$ such that $\alpha[\vec{e}]$ holds in $\mathbb{M}$ yet for no i does $\beta_i[\vec{e}]$ hold. The data (σ, η) determines a *chase-step*. We may execute such a chase-step and return a new model $\mathbb{N}$; this process proceeds as follows:

1. Choose some disjunct $\beta_i(x_1, \ldots, x_k, y_{i1}, \ldots, y_{ip})$
2. Choose new elements $k_1, \ldots, k_{ip}$ from K
   and add them to the domain,
3. Add facts, that is, enrich the relations of $\mathbb{M}$, to ensure the truth of
   $\beta_i(e_1, \ldots, e_k, k_1, \ldots, k_{ip})$.
   Here we have slightly abused notation, since some atom in β_i may be an equality, in which case we must enrich the congruence relation to identify the appropriate elements of $|\mathbb{M}| \cup \{k_1, \ldots k_{ip}\}$.

A chase-step can be viewed as a database repair of the failure of the current model to satisfy the dependency expressed by σ.

There are three possible outcomes of a run of the Chase. (i) It may halt with success if we reach model $\mathbb{M}$ where we cannot apply a step, *i.e.* when $\mathbb{M} \models \mathcal{T}$. (ii) It may halt with failure, if there is a sentence $\alpha \Rightarrow \perp$ of $\mathcal{T}$ and we reach a model $\mathbb{M}$ in which some instance of α holds, (iii) It may fail to terminate.

Properties of the Chase. Theorem 1 records the basic properties of the Chase. These results are adaptations of well-known [27,30] results in database theory. A run of the Chase is said to be *fair* if—in the notation above—every pair of possible choices for σ and η will be eventually evaluated.

Theorem 1. *Let $\mathcal{T}$ be a geometric theory. Then $\mathcal{T}$ is satisfiable if and only if there is a fair run of the Chase, starting with the empty model, that does not fail. Let $\mathcal{U}$ be the set (possibly infinite) of models obtained by fair runs of the Chase. Then $\mathcal{U}$ is a set-of-support for $Mod(\mathcal{T})$: for any model $\mathbb{M}$ of $\mathcal{T}$, there is a $\mathbb{U} \in \mathcal{U}$ and a homomorphism from $\mathbb{U}$ to $\mathbb{M}$.*

Note that the Theorem implies that the fair Chase is refutationally complete. If $\mathcal{T}$ has no models, then in any fair run of the Chase, each set of non-deterministic choices will eventually yield failure. By König's Lemma, then, the Chase process will halt.

Termination and Decidability. In general, termination of the Chase for an arbitrary theory is undecidable [31]. However, Fagin *et al.* [30] define a syntactic condition on theories, known as *weak acyclicity*, by which the Chase is guaranteed to terminate. Briefly, one constructs a directed graph whose nodes are positions in relations and whose edges capture possible "information flow" between positions; a theory is weakly acyclic if there are no cycles of a certain form in this graph. (The notion of weakly acyclicity in [30] is defined for theories without disjunction, but the obvious extension of the definition to the general case supports the argument for termination in the general case.)

Observe that if $\mathcal{T}$ is such that all runs of the Chase terminate, then—by König's Lemma—there is a finite set of models returned by the Chase. Thus we can compute a finite set that jointly provides a set-of-support for all models of $\mathcal{T}$ relative to the homomorphism order $\preccurlyeq$.

Since weak acyclicity implies termination of the Chase we may conclude that weakly acyclic theories have the finite model property. Furthermore, entailment of positive-existential sentences from a weakly acyclic theory is decidable, as follows. Suppose $\mathcal{T}$ is weakly acyclic and α is a positive-existential sentence. Let $\mathbb{A}_1, \ldots, \mathbb{A}_n$ be the models of $\mathcal{T}$. To check that α holds in all models of $\mathcal{T}$ it suffices to test α in each of the (finite) models $\mathbb{A}_i$, since if $\mathbb{B}$ were a counter-model for α, and $\mathbb{A}_i$ the chase-model such that $\mathbb{A}_i \preccurlyeq \mathbb{B}$, then $\mathbb{A}_i$ would be a counter-model for α, recalling that positive-existential sentences are preserved by homomorphisms. This proof technique was used recently [32] to show decidability for the theory of a class of Diffie-Hellman key-establishment protocols.

The Bounded Chase. For theories that do not enjoy termination of the Chase, we must resort to bounding our search. A traditional way to do so, used by tools such as Alloy, Margrave, and Aluminum, is to use user-supplied upper bounds on the domain of the model. Razor uses a somewhat more subtle device, which is outlined in [33], but which we cannot detail here for lack of space.

3.1   Augmentation: Exploring the Set of Models

Let $\mathcal{T}$ be a geometric theory and $\mathbb{M}$ be a model of $\mathcal{T}$. Razor allows the user to *augment* $\mathbb{M}$ with an additional positive-existential formula α resulting in an extension model $\mathbb{N}$ of $\mathcal{T}$ such that α is true in $\mathbb{N}$.

$\mathbb{N}$ can be computed by a run of the Chase starting with a model $\mathbb{M}' \equiv \mathbb{M} \cup \{\alpha\}$. A key point is that if α entails other observations given $\mathcal{T}$ and the facts already in $\mathbb{M}$, those observations will be added to the resulting model. And the augmentation may *fail* if adding α to $\mathbb{M}$ is inconsistent with $\mathcal{T}$.

Theorem 2. *Let $\mathbb{N}$ be a finite model of the theory $\mathcal{T}$. Suppose that $\mathbb{M}$ is a finite model returned by the Chase with $\mathbb{M} \preccurlyeq \mathbb{N}$. Then there is a finite sequence of augmentations on $\mathbb{M}$ resulting in a model isomorphic to $\mathbb{N}$.*

In particular, if $\mathcal{T}$ is weakly acyclic, then for every $\mathbb{N}$ there is a Chase model $\mathbb{M}$ and a finite sequence of augments of $\mathbb{M}$ yielding $\mathbb{N}$.

3.2   Provenance and the Witnessing Signature

A crucial aspect of our approach to constructing and reasoning about models is a notation for *witnessing* an existential quantifier.

Notation. Given a sentence $\alpha \Rightarrow \bigvee_i (\exists y_{i1} \ldots \exists y_{ip}.\beta_i(\vec{x}, y_{i1}, \ldots, y_{ip}))$, we assign a unique, fresh, witnessing (partial) function symbol f_{ik}^{σ} to each quantifier $\exists y_{ik}$. This determines an associated sentence $\alpha \Rightarrow \bigvee_i \beta_i(\vec{x}, f_{i1}^{\sigma}(\vec{x}), \ldots, f_{ip}^{\sigma}(\vec{x}))$ in an expanded signature, the witnessing signature.

This is closely related to Skolemization of course, but with the important difference that our witnessing functions are partial functions, and this witnessing is not a source-transformation of the input theory. This alternate representation of geometric sentences allows us to define a refined version of the Chase, that maintains bookkeeping information about elements and facts. Specifically: *each element of a model built using the Chase will have a closed term of the witnessing signature associated with it: this is that element's "provenance".*

To illustrate, consider a chase-step as presented earlier, using a formula such as Formula 1, whose associated sentence over the witnessing signature is

$$\sigma^{\mathsf{w}} \equiv \alpha(x_1, \ldots, x_k) \Rightarrow \bigvee_i \beta_i(x_1, \ldots, x_k, f_{i1}^{\sigma}(\vec{x}), \ldots, f_{ip}^{\sigma}(\vec{x})) \tag{2}$$

In the chase-step, when $\vec{x}$ is instantiated by $\vec{e}$, the elements $k_1, \ldots, k_{ip}$ added in line 2 are naturally "named" by $f_{i1}^{\sigma}(\vec{e}), \ldots, f_{ip}^{\sigma}(\vec{e})$. Proceeding inductively, each of the e_j will be named by a closed term t_j, so that the elements $k_1, \ldots, k_{ip}$ added in line 2 have, respectively, the provenance $f_{i1}^{\sigma}(\vec{t}), \ldots, f_{ip}^{\sigma}(\vec{t})$.

It is possible that an element enjoys more than one provenance, in the case when a chase-step equates two elements.

Also observe that every fact added (in line 3) to the model being constructed can be "blamed" on the pair (σ, η), that is, the sentence and binding that fired the rule. *This is that fact's provenance.*

4 Implementation

A naive implementation of the Chase in our setting can be computationally prohibitive, due to the need to fork different branches of the model-construction in the presence of disjunctions. Instead, we take advantage of SAT-solving technology to navigate the disjunctions. The use of SAT-solving is of course the essence of MACE-style model-finding, but the difference here is that we do not simply work with the ground instances of the input theory, $\mathcal{T}$, over a fixed set of constants. Rather we compute a ground theory $\mathcal{T}^*$ consisting of a sufficiently large set of instantiations of $\mathcal{T}$ by closed terms of the witness signature for $\mathcal{T}$.

Since we want to handle theories with equality, we want to construct models of $\mathcal{T}^*$ modulo equality reasoning and the theory of uninterpreted functions, so we use an *SMT-solver*. We utilize Z3 (QF_UFBV) as the backend SMT-solver.

A straightforward use of SMT-solving would result in losing control over the model-building: even though the elements of a model of $\mathcal{T}^*$ returned would have appropriate provenance, the solver may make unnecessary relations hold between elements, and may collapse elements unnecessarily. So we follow the SMT-solving with a *minimization* phase, in which we eliminate relational facts that are not necessary, and even "un-collapse" elements when possible.

BUILDMODEL (Algorithm 1) presents the overall process by which models of an input theory $\mathcal{T}$ are generated. The GROUND procedure (line 2, given as Algorithm 3) consists of construction of ground instances of $\mathcal{T}$ by a run of a variation of the Chase, where every disjunction in the head of geometric sentences is replaced by a conjunction. In this way we represent all repair-branches that could be taken in a Chase step over the original $\mathcal{T}$. Such a computation creates a refined Skolem-Herbrand base, containing a set of *possible facts* $\mathbb{P}^{\mathcal{T}}$ that could be true in any Chase model of $\mathcal{T}$. (Some care is required to handle contingent equalities; space does not permit a detailed explanation here.)

The *anonymization* procedure (line 3, not detailed here) constructs a flat theory $\mathcal{T}^{\mathsf{K}}$ by replacing every term in $\mathcal{T}^*$ over the witness signature with constants from a signature Σ^{K}. The theory $\mathcal{T}^{\mathsf{K}}$ is in a form that can be fed to the underlying model-finding and minimization algorithms by a call to NEXT (Algorithm 2). Finally, Razor returns the set models $\mathcal{U}$ produced by model-finding and minimization, reduced to the signature of the original input $\mathcal{T}$.

Algorithm 1. Razor

1: **function** BUILDMODEL($\mathcal{T}$)   ▷ $\mathcal{T}$ over signature Σ
2:    $(\mathcal{T}^*, \mathbb{P}^{\mathcal{T}}) \leftarrow$ GROUND($\mathcal{T}$)   ▷ $\mathcal{T}^*$ over the witness signature Σ^{w}
3:    $\mathcal{T}^{\mathsf{K}} \leftarrow$ ANONYMIZE($\mathcal{T}^*$)   ▷ $\mathcal{T}^{\mathsf{K}}$ over the anonymized signature Σ^{K}
4:    $\mathcal{U} \leftarrow \emptyset$
5:    $\mathbb{M} \leftarrow$ NEXT($\mathcal{T}^{\mathsf{K}}, \mathcal{U}$)
6:    **while** $\mathbb{M} \neq$ **unsat do**
7:        $\mathcal{U} \leftarrow \mathcal{U} \cup \{\mathbb{M}\}$
8:        $\mathbb{M} \leftarrow$ NEXT($\mathcal{T}^{\mathsf{K}}, \mathcal{U}$)
9:    **return** REDUCT($\mathcal{U}$)   ▷ Reduct of models in $\mathcal{U}$ to Σ

Algorithm 2. Next Model

Require:
    $\mathcal{T}$ is ground and flat
    for all $\mathbb{U} \in \mathcal{U}$, $\mathbb{U} \models \mathcal{T}$ and $\mathbb{U}$ is homomorphically minimal

1: **function** NEXT($\mathcal{T}, \mathcal{U}$)
2:    $\Phi \leftarrow \bigcup_i \{$FLIP($\mathbb{U}_i$)$\}$ for all $\mathbb{U}_i \in \mathcal{U}$ ▷ Flip axioms about existing models.
3:    **if** exists $\mathbb{M}$ such that $\mathbb{M} \models (\mathcal{T} \cup \Phi)$ **then**   ▷ Ask the SMT-solver for $\mathbb{M}$.
4:        $\mathbb{N} \leftarrow$ MINIMIZE($\mathcal{T}, \mathbb{M}$)
5:        **return** $\mathbb{N}$
6:    **else**
7:        **return unsat**   ▷ No more models.

Algorithm 3. Grounding

1: **function** GROUND($\mathcal{G}$)
2:     $\mathbb{P}^{\mathcal{G}} \leftarrow \emptyset$ $\triangleright$ $\mathbb{P}^{\mathcal{G}}$ is initially the empty model
3:     $\mathcal{G}^* \leftarrow \emptyset$ $\triangleright$ $\mathcal{G}^*$ is initially an empty theory
4:     **repeat**
5:         choose $\sigma \equiv \varphi \Rightarrow \psi \in \mathcal{G}$
6:         **for each** λ where $\lambda\varphi \in \mathbb{P}^{\mathcal{G}}$ **do**
7:             $\mathbb{P}^{\mathcal{G}} \leftarrow$ EXTEND($\mathbb{P}^{\mathcal{G}}, \sigma, \lambda$)
8:             $\mathcal{G}^* \leftarrow \mathcal{G}^* \cup \{$INSTANTIATE($\mathbb{P}^{\mathcal{G}}, \sigma, \lambda$)$\}$
9:     **until** $\mathcal{G}^*$ and $\mathbb{P}^{\mathcal{G}}$ are changing
10:     **return** ($\mathcal{G}^*, \mathbb{P}^{\mathcal{G}}$)

11: **function** EXTEND($\mathbb{M}, \varphi \Rightarrow_{\vec{x}} \psi, \eta$)
12:     **if** $\psi = \bot$ **then fail**
13:     $\mathbb{N} \leftarrow \mathbb{M}$
14:     **for each** disjunct $\exists^{f_1} y_1, \ldots, \exists^{f_m} y_m. \bigwedge_{j=1}^n P_j$ in ψ **do**
15:         $|\mathbb{N}| \leftarrow |\mathbb{N}| \cup \{[\![f_i(\vec{x})]\!]_\eta^{\mathbb{M}} \mid 1 \le i \le m\}$
16:         $\mu \leftarrow \eta[y_1 \mapsto [\![f_1(\vec{x})]\!]_\eta^{\mathbb{M}}, \ldots, y_m \mapsto [\![f_m(\vec{x})]\!]_\eta^{\mathbb{M}}]$
17:         $\mathbb{N} \leftarrow \mathbb{N} \cup \{P_1[\mu(\vec{x}, \vec{y})], \ldots, P_n[\mu(\vec{x}, \vec{y})]\}$
18:     **return** $\mathbb{N}$

19: **function** INSTANTIATE($\mathbb{P}^{\mathcal{G}}, \varphi \Rightarrow_{\vec{x}} \bigvee_i \exists^{f_{i1}} y_{i1} \ldots \exists^{f_{im}} y_{im}.\psi_i, \eta$)
20:     $\mu \leftarrow \eta[y_{ij} \mapsto [\![f_{ij}(\vec{x})]\!]_\eta^{\mathbb{P}^{\mathcal{G}}}]$ ($1 \le j \le m$)
21:     **return** $\mu\sigma$

Algorithm 4. Minimize

Require: $\mathbb{M} \models \mathcal{T}$

1: **function** MINIMIZE($\mathcal{T}, \mathbb{M}$)
2:     **repeat**
3:         $\mathbb{N} \leftarrow \mathbb{M}$
4:         $\mathbb{M} \leftarrow$ REDUCE($\mathbb{M}$)
5:     **until** $\mathbb{M} = $ **unsat** $\triangleright$ Cannot reduce
6:     **return** $\mathbb{N}$ $\triangleright$ $\mathbb{N}$ is a minimal model for $\mathcal{T}$

The NEXT algorithm (Algorithm 2) accepts a set $\mathcal{U}$ of minimal models under the homomorphism ordering and returns a minimal model $\mathbb{M}$ for $\mathcal{T}$ that is not reachable from any of the models in $\mathcal{U}$ via homomorphism. The FLIP procedure (line 2, not detailed here) on an existing model $\mathbb{U} \in \mathcal{U}$ records the disjunction of the negation of all facts (including equational facts) true in $\mathbb{U}$: this guarantees that the next model returned by the solver will not be reachable from any of the models in $\mathcal{U}$ via homomorphism. The call to MINIMIZE (Algorithm 4) on line 4

Algorithm 5. Reduce

Require: $\mathbb{M} \models \mathcal{T}$

1: **function** REDUCE($\mathcal{T}$, $\mathbb{M}$)
2:    $\nu \leftarrow$ NEGPRESERVE($\mathcal{T}$,$\mathbb{M}$)
3:    $\varphi \leftarrow$ FLIP($\mathcal{T}$,$\mathbb{M}$)
4:    **if** exists $\mathbb{N}$ such that $\mathbb{N} \models \mathcal{T} \cup \{\nu \wedge \varphi\}$ **then** ▷ Ask the SMT solver for $\mathbb{N}$
5:       **return** $\mathbb{N}$
6:    **else**
7:       **return unsat**   ▷ $\mathbb{M}$ is minimal.

---

reduces the next model returned by the solver to a homomorphically minimal one by repeated invocations of the solver. In every reduction step i, the solver is asked for a model $\mathbb{M}_i$ that satisfies

- the input theory $\mathcal{T}$.
- the *negation preserving* axiom of $\mathbb{M}_{i-1}$, which is the conjunction of all facts (including equational facts) that are false in $\mathbb{M}_{i-1}$.
- the *flip axioms* about the model $\mathbb{M}_{i-1}$ from the previous step.

It can be shown that for every model $\mathbb{M}$, REDUCE($\mathcal{T}, \mathbb{M}$) $\prec \mathbb{M}$. The reduction process continues until the solver returns "unsatisfiable".

Theorem 3. *Fix a relational theory in geometric form $\mathcal{T}$. Let $\mathbb{P}^{\mathcal{T}}$ and $\mathcal{T}^*$ be a set of possible facts and its corresponding ground theory for $\mathcal{T}$, constructed by the Chase-based grounding algorithm. Let $\mathbb{M}$ be a model in the witness signature for $\mathcal{T}^*$ and $\mathbb{M}^-$ the reduct of $\mathbb{M}$ to the signature of $\mathcal{T}$.*

1. *(Soundness.) If $\mathbb{M} \models \mathcal{T}^*$ and $\mathbb{M}$ is homomorphically minimal, then $\mathbb{M}^-$ is a model of $\mathcal{T}$.*
2. *(Completeness.) If $\mathbb{M}$ is constructed by the Chase and $\mathbb{M}^- \models \mathcal{T}$ then $\mathbb{M}$ is a model of $\mathcal{T}^*$.*

Proof (Sketch). For (1): Let $\sigma \equiv \varphi \Rightarrow (\bigvee_i \exists^{f_{i1}} y_{i1} \ldots \exists^{f_{ip}} y_{ip} . \psi_i)$ be a sentence in $\mathcal{T}$. Let $\vec{x}$ be the free variables of σ. We show that if $\mathbb{M}^- \models_\eta \varphi$ for environment η, then $\mathbb{M}^- \models_\eta (\bigvee_i \exists^{f_{i1}} y_{i1} \ldots \exists^{f_{ip}} y_{ip} . \psi_i)$: because $\mathbb{M}$ is minimal, the facts in $\mathbb{M}$ are *contained* in $\mathbb{P}^{\mathcal{T}}$, and since φ is positive, $\eta\varphi \in \mathbb{P}^{\mathcal{T}}$. Therefore, by the construction of $\mathcal{T}^*$, a sentence $\varphi[\vec{t}] \Rightarrow \bigvee_i \psi_i[\vec{t}, \vec{u}_i]$ exists in $\mathcal{T}^*$ where $\vec{t} = \eta\vec{x}$, and for each u_{ij} in $\vec{u}_i$, $u_{ij} = f_{ij}(\vec{t})$ ($1 \leq j \leq p$). Observe that because $\mathbb{M}^- \models_\eta \varphi$ then $\mathbb{M} \models \varphi[\vec{t}]$ as $\vec{t}$ are witnesses for the elements that are images of $\vec{x}$ in η. Finally, since $\mathbb{M}$ is a model of $\mathcal{T}^*$, then $\mathbb{M} \models \psi_i[\vec{t}, \vec{u}_i]$ for some i. Therefore, it follows that $\mathbb{M}^- \models_\eta (\bigvee_i \exists y_{i1}, \ldots, \exists y_{ip} . \psi_i)$.

For (2): Let $\sigma^* \equiv \varphi[\vec{t}] \Rightarrow \bigvee_i \psi_i[\vec{t}, \vec{u}_i]$ be a sentence in $\mathcal{T}^*$. By definition, σ^* is an instance of a sentence $\sigma \equiv \bigvee_i \varphi \Rightarrow (\exists^{f_{i1}} y_{i1} \ldots \exists^{f_{ip}} y_{ip} . \psi_i)$ by a substitution that sends the free variables $\vec{x}$ of σ to $\vec{t}$ and $\vec{y}_i$ to $\vec{u}_i$. Moreover, for each u_{ij} in $\vec{u}_i$ ($1 \leq i \leq p$), $u_{ij} = f_{ij}(\vec{t})$.

Assume $\mathbb{M} \models \varphi[\vec{t}]$. Then, $\mathbb{M}^- \models_\eta \varphi$ where the environment η sends the variables in $\vec{x}$ to the elements $\vec{e}$ in $\mathbb{M}^-$ that are denoted by $\vec{t}$ in $\mathbb{M}$. Because $\mathbb{M}^-$ is a chase-model for $\mathcal{T}$, then for some i, $\mathbb{M}^- \models_\lambda \exists y_{i1} \ldots \exists y_{ip} \ . \ \psi_i$ where $\lambda = \eta[y_{ij} \mapsto \mathbf{d_j}]$ $(1 \leq j \leq p)$. Let u_{ij} denote $\mathbf{d_j}$ in $\mathbb{M}$ under λ. Therefore, $\mathbb{M} \models \psi_i[\vec{t}, \vec{u}_i]$ follows.

It remains to show that a set-of-support computed by the minimization algorithm for $\mathcal{T}^*$ (modulo anonymization) is in fact a set-of-support for $\mathcal{T}$.

Theorem 4. *Fix a theory $\mathcal{T}$ in geometric form over a signature Σ. Let $\mathcal{T}^*$ be computed by a run of the grounding algorithm on $\mathcal{T}$.*

1. *The set $\mathcal{U}$ of models computed during Algorithm 1 is a set-of-support for $\mathcal{T}^*$.*
2. *The reducts $\mathcal{U}^-$ of $\mathcal{U}$ to Σ, returned by Algorithm 1, is a set-of-support for $\mathcal{T}$.*

Proof (Sketch) For (1): In every call of NEXT for a set of models $\mathcal{U}_i$, the flip axioms about the models in $\mathcal{U}_i$

ensure that every model $\mathbb{M}$ returned satisfies $\mathbb{M} \not\preccurlyeq \mathbb{U}$ for each $\mathbb{U} \in \mathcal{U}$. If an $\mathbb{M}$ is returned by the solver, a model $\mathbb{U}$ with $\mathbb{U} \preccurlyeq \mathbb{M}$ will be added to $\mathcal{U}_i$.

For (2): Let $\mathbb{A}$ be a model of $\mathcal{T}$. By Theorem 1 a chase-model $\mathbb{M}$ over the witness signature exists such $\mathbb{M}^- \preccurlyeq \mathbb{A}$. By Theorem 3, part (2) $M \models \mathcal{T}^*$. By part (1) of this theorem there exists a model $\mathbb{U} \in \mathcal{U}$ such that $\mathbb{U} \preccurlyeq \mathbb{M}$. Therefore, for the reduct $\mathbb{U}^-$ of $\mathbb{U}$ to Σ, $\mathbb{U}^- \preccurlyeq \mathbb{A}$.

5 Examples

From the Alloy Repository. Our main focus is on theories developed by hand. A natural source of such theories is the Alloy repository. We ran Razor on 11 theories from the Alloy book, suitably translated to Razor's input language. The following summarizes the experience; space precludes a detailed report.

For 6 theories Razor returned a complete set-of-support in unbounded mode; the time to return the first model was less than a second. For the remaining 5, we had to run in bounded mode in order to return models within 5 min. For 2 of these, iterative deepening succeed in finding a bound that was sufficient for finding a complete set of support for all finite models: the times-to-first-model were 375 ms and 17.2 s, respectively. For the other 3, with respective bounds 1, 2, and 3, we computed models quickly at the respective bounds but incrementing the bound led to a 5-min timeout. In all cases, once the first model was found, subsequent models were completed in negligible time.

From TPTP. We performed several experiments running Razor on the satisfiable problems in the TPTP problem repository [34] Razor's current performance on these problems is not satisfactory: it frequently fails to terminate within a five-minute bound. Razor tends to perform better on problems that are developed by hand, have a limited number of predicates, and don't include relations with high arity. Future developments in Razor's implementation will improve performance; a long-term research question is exploring the tradeoffs between efficiency and the kind of enhanced expressivity we offer.

Extended Example: Lab Door Security. Here is an introductory example, demonstrating a specification in Razor of a simple policy for access to a our local lab, and typical queries about the policy. The sentences below capture the following policy specification.

Logic and Systems are research groups in lab (1–2). Research group members must be able to enter the lab (3). Key or card access allows a person to enter (4–5). To enter a lab, a member must have a key or card (6). Only members have cards (7). Employees grant keys to people (8–9). Systems members are not allowed to have keys (10).

```
 1.  LabOf('Logic,'TheLab);
 2.  LabOf('Systems, 'TheLab);
 3.  MemberOf(p,r) & LabOf(r,l) => Enter(p,l);
 4.  HasKey(p,k) & KOpens(k,l) => Enter(p,l);
 5.  COpens(cardOf(p),l) => Enter(p,l);
 6.  Enter(p,l) => COpens(cardOf(p),l)
              |  exists k. HasKey(p,k) & KOpens(k,l);
 7.  COpens(cardOf(p), l) => exists r. MemberOf(p,r) & LabOf(r,l);
 8.  HasKey(p,k) => exists e. Grant(e,p,k) & Employee(e);
 9.  Grant(e,p,k)  => HasKey(p,k)
10.  MemberOf(p,'Systems) & HasKey(p,k)
              & KOpens(k,'TheLab) => Falsehood;
```

The user can ask if a thief can access the lab without being Logic or Systems member:

```
Enter('Thief,'TheLab);
```

First Model: Granted a Key The first model exhibits that there is no policy restriction on employee key-granting capabilities:

```
Enter    = {(p1, 11)}       'TheLab  = 11
Employee = {(e1)}           'Systems = r2
Grant    = {(e1,p1,k1)}     'Logic   = r1
HasKey   = {(p1,k1)}        'Thief   = p1
KOpens   = {(k1,11)}
LabOf    = {(r1,11), (r2,11)}
```

The user can investigate if the thief p1 and the employee e1 can be the same person in this example? This may be done by augmenting the model with aug p1 = e1. The augmentation results in one model (not shown).

Second Model: Third research group The second example is more curious:

```
Enter    = {(p1, 11)}       'TheLab  = 11
COpens   = {(c1, 11)}       'Systems = r2
MemberOf = {(e1, r3)}       'Logic   = r1
cardOf   = {(p1, c1)}       'Thief   = p1
LabOf    = {(r1, 11), (r2, 11), (r3, 11)}
```

The user may ask "where did this third research group come from?", then user look at the provenance information about r3 by running origin r3. Razor pinpoints an instance of the causal sentence (sentence 7):

```
7: COpens(c1, l1) => MemberOf(e1, r3) & LabOf(r3, l1)
```

This policy rule does not restrict which research groups live in the lab. Such a restriction would force the mystery group r3 to be the Systems or Logic group. The user confirms this policy fix by applying aug r3 = r2. The augmentation produces no counter examples; the fix is valid. The research group r3 exists because the thief has a card. By asking blame COpens(c1, l1), the user sees why:

```
6: Enter(p1, l1) => COpens(c1, l1)
|  HasKey(p1, k1) & KOpens(k1, l1)
```

The thief has a card because the user's query said he could enter the lab. He could also have a key, which is evident in the first model. Why does the thief belong to a research group in this scenario, but not in the previous? Being a research group member is a consequence of having a card; not for having a key. Belonging to a research group when having a key is extraneous information. Razor does not include this scenario in the minimal model returned.

Software-Defined Networks. At Razor's web page http://salmans.github.io/Razor/ one can find a more advanced extended example, showing how Razor can reason about controller programs for Software-Defined Networks. For a program P in the declarative networking program Flowlog [35], we show how to define a theory $\mathcal{T}_P$ such that a model $\mathbb{M}$ of $\mathcal{T}_P$ is a snapshot of the state of the system at a moment in time. The user can augment $\mathbb{M}$ by a fact capturing a network event, and the resulting models correspondingly capture the next state of the system. In this way, *augmentation acts as a stepper in a debugger.*

6   Future Work

Highlights of the ongoing work on this project include (i) work on efficiency of the model-building, (ii) taking real advantage of the fact that we incorporate an SMT-solver, to work more effectively when part of a user's input theory has a known decision procedure, and (iii) an improved user interface for the tool, including a more sophisticated GUI for presenting models, and parsers to allow input in native formats such as Description Logic, firewall specifications, XAMCL, and cryptographic protocols.

Acknowledgements. We benefitted from discussions with Henning Günther, Joshua Guttman, Daniel Jackson, Shriram Krishnaturthi, Tim Nelson, and John Ramsdell. The name of our tool is homage to Ockham's Razor (William of Occam 1285–1349): "Pluralitas non est ponenda sine neccesitate".

References

1. Zhang, J., Zhang, H.: SEM: a system for enumerating models. In: International Joint Conference On Artificial Intelligence (1995)
2. McCune, W.: MACE 2.0 Reference Manual and Guide. CoRR (2001)
3. Claessen, K., Sörensson, N.: New techniques that improve MACE-Style finite model finding. In: CADE Workshop on Model Computation-Principles, Algorithms, Applications (2003)
4. Baumgartner, P., Schmidt, R.A.: Blocking and other enhancements for bottom-up model generation methods. In: Furbach, U., Shankar, N. (eds.) IJCAR 2006. LNCS (LNAI), vol. 4130, pp. 125–139. Springer, Heidelberg (2006)
5. de Nivelle, H., Meng, J.: Geometric resolution: a proof procedure based on finite model search. In: Furbach, U., Shankar, N. (eds.) IJCAR 2006. LNCS (LNAI), vol. 4130, pp. 303–317. Springer, Heidelberg (2006)
6. Torlak, E., Jackson, D.: Kodkod: a relational model finder. In: Grumberg, O., Huth, M. (eds.) TACAS 2007. LNCS, vol. 4424, pp. 632–647. Springer, Heidelberg (2007)
7. Baumgartner, P., Fuchs, A., De Nivelle, H., Tinelli, C.: Computing finite models by reduction to function-free clause logic. J. Appl. Logic 7(1), 58–74 (2009)
8. Reynolds, A., Tinelli, C., Goel, A., Krstić, S.: Finite model finding in SMT. In: Sharygina, N., Veith, H. (eds.) CAV 2013. LNCS, vol. 8044, pp. 640–655. Springer, Heidelberg (2013)
9. Korovin, K., Sticksel, C.: iProver-Eq: an instantiation-based theorem prover with equality. In: Giesl, J., Hähnle, R. (eds.) IJCAR 2010. LNCS, vol. 6173, pp. 196–202. Springer, Heidelberg (2010)
10. Bry, F., Torge, S.: A deduction method complete for refutation and finite satisfiability. In: Dix, J., Fariñas del Cerro, L., Furbach, U. (eds.) JELIA 1998. LNCS (LNAI), vol. 1489, pp. 122–138. Springer, Heidelberg (1998)
11. Baumgartner, P., Suchanek, F.M.: Automated reasoning support for first-order ontologies. In: Alferes, J.J., Bailey, J., May, W., Schwertel, U. (eds.) PPSWR 2006. LNCS, vol. 4187, pp. 18–32. Springer, Heidelberg (2006)
12. Jackson, D.: Software Abstractions, 2nd edn. MIT Press, London (2012)
13. Fisler, K., Krishnamurthi, S., Meyerovich, L.A., Tschantz, M.C.: Verification and change-impact analysis of access-control policies. In: International Conference on Software Engineering (2005)
14. Nelson, T., Barratt, C., Dougherty, D.J., Fisler, K., Krishnamurthi, S.: The margrave tool for firewall analysis. In: USENIX Large Installation System Administration Conference (2010)
15. Niemelä, I.: A tableau calculus for minimal model reasoning. In: Workshop on Theorem Proving with Analytic Tableaux and Related Methods (1996)
16. Bry, F., Yahya, A.: Positive unit hyperresolution tableaux and their application to minimal model generation. J. Autom. Reasoning 25(1), 35–82 (2000)
17. Nelson, T., Saghafi, S., Dougherty, D.J., Fisler, K., Krishnamurthi, S.: Aluminum: principled scenario exploration through minimality. In: International Conference on Software Engineering (2013)
18. Doghmi, S.F., Guttman, J.D., Thayer, F.J.: Searching for shapes in cryptographic protocols. In: Grumberg, O., Huth, M. (eds.) TACAS 2007. LNCS, vol. 4424, pp. 523–537. Springer, Heidelberg (2007)
19. Lopes, N., Bjorner, N., Godefroid, P., Jayaraman, K., Varghese, G.: Checking beliefs in dynamic networks. Technical report, Microsoft Research (2014)

20. de Moura, L., Bjørner, N.S.: Z3: An efficient SMT solver. In: Ramakrishnan, C.R., Rehof, J. (eds.) TACAS 2008. LNCS, vol. 4963, pp. 337–340. Springer, Heidelberg (2008)
21. Abramsky, S.: Domain theory in logical form. Ann. Pure Appl. Logic **51**, 1–77 (1991)
22. Vickers, S.: Geometric logic as a specification language. In: Imperial College Department of Computing Workshop on Theory and Formal Methods (1995)
23. Sofronie-Stokkermans, V.: Sheaves and Geometric Logic and Applications to Modular Verification of Complex Systems. Electronic Notes on Theoretical Computer Science **230**, 161–187 (2009)
24. Bezem, M., Coquand, T.: Automating coherent logic. In: Sutcliffe, G., Voronkov, A. (eds.) LPAR 2005. LNCS (LNAI), vol. 3835, pp. 246–260. Springer, Heidelberg (2005)
25. Maier, D., Mendelzon, A.O., Sagiv, Y.: Testing implications of data dependencies. ACM Trans. Database Syst. **4**, 445–469 (1979)
26. Beeri, C., Vardi, M.Y.: A proof procedure for data dependencies. J. ACM **31**(4), 718–741 (1984)
27. Deutsch, A., Tannen, V.: XML queries and constraints, containment and reformulation. ACM Symposium on Theory Computer Science (2005)
28. Rossman, B.: Existential positive types and preservation under homomorphisms. In: IEEE Logic in Computer Science. IEEE (2005)
29. Makkai, M., Reyes, G.E.: First Order Categorical Logic. Springer, Heidelberg (1977)
30. Fagin, R., Kolaitis, P.G., Miller, R.J., Popa, L.: Data exchange: semantics and query answering. In: Calvanese, D., Lenzerini, M., Motwani, R. (eds.) ICDT 2003. LNCS, vol. 2572, pp. 207–224. Springer, Heidelberg (2002)
31. Deutsch, A., Nash, A., Remmel, J.: The chase revisited. In: ACM Symposium on Principles of Database Systems (2008)
32. Dougherty, D.J., Guttman, J.D.: Decidability for lightweight Diffie-Hellman protocols. In: IEEE Symposium on Computer Security Foundations, pp. 217–231 (2014)
33. Saghafi, S., Dougherty, D.J.: Razor: provenance and exploration in model-finding. In: 4th Workshop on Practical Aspects of Automated Reasoning (PAAR) (2014)
34. Sutcliffe, G.: The TPTP problem library and associated infrastructure: The FOF and CNF parts, v3.5.0. J. Autom. Reasoning **43**(4), 337–362 (2009)
35. Nelson, T., Ferguson, A.D., Scheer, M., Krishnamurthi, S.: Tierless programming and reasoning for software-defined networks. NSDI, April (2014)

Abstract Interpretation as Automated Deduction

Vijay D'Silva[1] and Caterina Urban[2]([⊠])

[1] Google Inc., San Francisco, USA
[2] École Normale Supérieure, Paris, France
urban@di.ens.fr

Abstract. Algorithmic deduction and abstract interpretation are two widely used and successful approaches to implementing program verifiers. A major impediment to combining these approaches is that their mathematical foundations and implementation approaches are fundamentally different. This paper presents a new, logical perspective on abstract interpreters that perform reachability analysis using non-relational domains. We encode reachability of a location in a control-flow graph as satisfiability in a monadic, second-order logic parameterized by a first-order theory. We show that three components of an abstract interpreter, the lattice, transformers and iteration algorithm, represent a first-order, substructural theory, parametric deduction and abduction in that theory, and second-order constraint propagation.

1 Introduction

Two major approaches to automated reasoning about programs are those based on SAT and SMT solvers and those based on abstract interpretation. In the solver-based approaches, a property of a program is encoded by formulae in a logic or theory and a solver is used to check if the property holds. In abstract interpretation, a property of a program is expressed in terms of fixed points and fixed point approximation techniques are used to calculate and reason about fixed points [6]. The complementary strengths of these approaches has led to a decade of theoretical and practical effort to combine them.

The strengths of SMT solvers include efficient Boolean reasoning, complete reasoning in certain theories, theory combination, proof generation and interpolation. Recent research has demonstrated that deduction algorithms have applications in program analysis besides solving formulae. DPLL(T) and CDCL have been lifted to implement property-guided, path-sensitive analyses [9,15]. Stålmarck's method has been used to refine abstract transformers [26], interpolants have been used to refine widening operators [12] and unification has been used to obtain complete reasoning about restricted families of programs [28]. The Nelson-Oppen procedure, though less general than reduced product [7,8], works as an algorithmic domain combinator [13].

Conversely, the strengths of abstract interpreters are the use of approximation to overcome the theoretical undecidability and practical scalability issues in

© Springer International Publishing Switzerland 2015
A.P. Felty and A. Middeldorp (Eds.): CADE-25, LNAI 9195, pp. 450–464, 2015.
DOI: 10.1007/978-3-319-21401-6_31

program verification and the use of widening operators to derive invariants. The large number of abstract domains enables application-specific reasoning and the flexibility to choose the trade-off between precision and efficiency. Ideas from abstract interpretation have been incorporated in SMT and constraint solvers by using abstract domains for theory propagation [19,29], joins for space-efficient representation [3], and widening for generalization [18]. Algorithms based on fixed points have been used to implement alternatives to DPLL(T) [4,27].

Nonetheless, there remain obstacles to combining these two approaches. Conceptually, SMT algorithms are expressed in terms of models and proofs while abstract interpretation is presented in terms of lattices, transformers and fixed points. These mathematical differences translate into practical differences in the interfaces implemented by solvers and abstract interpreters and type of results they produce, leading to further impediments to combining the two approaches.

This paper presents a logical account of a family of reachability analyses based on abstract interpretation. We encode reachability as satisfiability in a weak, monadic, second-order logic. A classic result of Büchi shows that a formula in the weak monadic second-order theory of one successor (WS1S) is satisfiable exactly if the models of that formula form a regular language [5,30,31]. If an automaton is viewed as a finite-state program, Büchi's theorem encodes reachability as satisfiability in WS1S. We extend a part of this result to a logic WS1S(T) interpreted over finite sequences of first-order structures.

Much of this paper is concerned with logical characterizations of the components of an abstract interpreter. The lattice in an analyzer represents a substructural, first-order theory, with the proof system for the theory generating the partial order of the lattice. Transformers for conditionals implement deduction and abduction modulo the theory. The invariant map constructed by abstract interpreters is a strict generalization of partial assignments in SAT and SMT solvers and fixed point iteration is second-order constraint propagation. Due to space restrictions, we defer proofs of statements to the full version of the paper.

2 Reachability as Second-Order Satisfiability

The contribution of this section is the logic WS1S(T), which is an extension of Büchi's WS1S with a theory. To simplify reasoning about programs in this logic, we restrict the class of models that are usually considered for WS1S.

Notation. We use $\hat{=}$ for definition. Let $\mathscr{P}(S)$ denote the set of all subsets of S, called the powerset of S, and $\mathscr{F}(S)$ denote the finite subsets of S. Given a function $f : A \to B$, $f[a \mapsto b]$ denotes the function that maps a to b and maps c distinct from a to $f(c)$.

2.1 Weak Monadic Second Order Theories of One Successor

Our syntax contains first-order variables *Vars*, functions *Fun* and predicates *Pred*. The symbols x, y, z range over *Vars*, f, g, h range over *Fun* and P, Q, R range over *Pred*. We also use a set *Pos* of first-order *position variables* whose elements are i, j, k and a set $SVar$ of *monadic second-order variables* denoted X, Y, Z. Second-order variables are uninterpreted, unary predicates. We also use a unary successor function suc and a binary, successor predicate Suc.

Our logic consists of three families of formulae called state, transition and trace formulae, which are interpreted over first-order structures, pairs of first-order structures and finite sequences of first-order structures respectively. The formulae are named after how they are interpreted over programs.

$$t ::= x \mid f(t_0, \ldots, t_n) \qquad\qquad\qquad \text{Term}$$
$$\varphi ::= P(t_0, \ldots, t_n) \mid \varphi \wedge \varphi \mid \neg\varphi \qquad\qquad \text{State Formula}$$
$$\psi ::= suc(x) = t \mid \psi \wedge \psi \mid \neg\psi \qquad\qquad \text{Transition Formula}$$
$$\Phi ::= X(i) \mid Suc(i, j) \mid \varphi(i) \mid \psi(i)$$
$$\mid \Phi \wedge \Phi \mid \neg\Phi \mid \exists i : Pos.\Phi \qquad\qquad \text{Trace formula}$$

State formulae are interpreted with respect to a *theory* $\mathcal{T}$ given by a first-order interpretation (Val, I), which defines functions $I(f)$, relations $I(P)$, and equality $=_{\mathcal{T}}$ over values in Val. A *state* maps variables to values and $State \triangleq Vars \rightarrow Val$ is the set of states. The value $[\![t]\!]_s$ of a term t in a state s is defined as usual.

$$[\![x]\!]_s \triangleq s(x) \qquad\qquad [\![f(t_1, \ldots, t_k)]\!]_s \triangleq I(f)([\![t_0]\!]_s, \ldots, [\![t_n]\!]_s)$$

As is standard, $s \models_{\mathcal{T}} \varphi$ denotes that s is a model of φ in the theory $\mathcal{T}$.

$$s \models_{\mathcal{T}} P(t_0, \ldots, t_n) \text{ if } ([\![t_0]\!]_s, \ldots, [\![t_n]\!]_s) \in I(P)$$
$$s \models_{\mathcal{T}} \varphi \wedge \psi \text{ if } s \models_{\mathcal{T}} \varphi \text{ and } s \models_{\mathcal{T}} \psi \qquad\qquad s \models_{\mathcal{T}} \neg\varphi \text{ if } s \not\models_{\mathcal{T}} \varphi$$

The semantics of Boolean operators is defined analogously for transition and trace formulae, so we omit them in what follows. A *transition* is a pair of states (r, s) and a transition formula is interpreted at a transition.

$$(r, s) \models P(t_0, \ldots, t_n) \text{ if } ([\![t_0]\!]_r, \ldots, [\![t_n]\!]_r) \in I(P)$$
$$(r, s) \models suc(x) = t \text{ if } [\![x]\!]_s =_{\mathcal{T}} [\![t]\!]_r$$

A *trace of length* k is a sequence $\tau : [0, k-1] \rightarrow State$. We call $\tau(m)$ the *state at position* m, with the implicit qualifier $m < k$. A *k-assignment* $\sigma : (Pos \rightarrow \mathbb{N}) \cup (SVar \rightarrow \mathscr{F}(\mathbb{N}))$ maps position variables to $[0, k-1]$ and second-order variables to *finite subsets* of $[0, k-1]$. A k-assignment satisfies that $\{\sigma(X) \mid X \in SVar\}$ partitions the interval $[0, k-1]$. We explain the partition condition shortly. A ws1s(t) *structure* (τ, σ) consists of a trace τ of length k and a k-assignment σ. A trace formula is interpreted with respect to a ws1s(t) structure.

$$(\tau, o) \models X(i) \text{ if } \sigma(i) \text{ is in } \sigma(X)$$
$$(\tau, \sigma) \models \varphi(i) \text{ if } \tau(\sigma(i)) \models_{\mathcal{T}} \varphi$$
$$(\tau, \sigma) \models \psi(i) \text{ if } \sigma(i) < k-1 \text{ and } (\tau(\sigma(i)), \tau(\sigma(i)+1)) \models \psi$$
$$(\tau, \sigma) \models Suc(i, j) \text{ if } \sigma(i) + 1 = \sigma(j)$$
$$(\tau, \sigma) \models \exists i : Pos.\Phi \text{ if } (\tau, \sigma[i \mapsto n]) \models \Phi \text{ for some } n \text{ in } \mathbb{N}$$

Note that $\varphi(i)$ is interpreted at the state at position $\sigma(i)$ and $\psi(i)$ at the transition from $\sigma(i)$. The semantics of $\psi(i)$ is only defined if $\sigma(i)$ is not the last position on τ. A trace formula Φ is *satisfiable* if there exists a trace τ and assignment σ such that $(\tau, \sigma) \models \Phi$. We assume standard shorthands for $\lor$ and $\Rightarrow$ and write $\Phi \models \Psi$ for $\models \Phi \Rightarrow \Psi$.

Example 1. The WS1S formula $First(i) \triangleq \forall j. \neg Suc(j, i)$ is true at the first position on a trace and $Last(i) \triangleq \forall j. \neg Suc(i, j)$ is true at the last position. See [30,31] for more examples. WS1S(T) has no second-order quantification so the encoding of transitive closure in WS1S does not carry over. Transitive closure may be encoded if the underlying theory is powerful enough.   ◁

2.2 Encoding Reachability in WS1S(T)

Büchi showed that the models of a WS1S formula form a regular language and vice-versa. The modern proof of this statement [30,31] encodes the structure of a finite automaton using second-order variables. We now extend this construction to encode a control-flow graph (CFG) by a WS1S(T) formula.

A *command* is an assignment $x := t$ of a term t to a first-order variable x, or is a condition $[\varphi]$, where φ is a state formula. A CFG $G = (Loc, E, \mathbf{in}, Ex, stmt)$ consists of a finite set of locations Loc including an initial location $\mathbf{in}$, a set of exit locations Ex, edges $E \subseteq Loc \times Loc$, and a labelling $stmt : E \to Cmd$ of edges with commands. To assist the presentation, we require that every location is reachable from $\mathbf{in}$, and that exit locations have no successors. This requirement is not fundamental to our results.

We define an execution semantics for CFGs. We assume that terms in commands are interpreted over the same first-order structure as state formulae. The formula $Same_V \triangleq \bigwedge_{x \in V} succ(x) = x$ expresses that variables in the set V are not modified in a transition and $Trans_c$ is the *transition formula* for a command.

$$Trans_c \triangleq \begin{cases} b \implies Same_{Vars} & \text{if } c = [b] \\ suc(x) = t \land Same_{Vars \setminus \{x\}} & \text{if } c = x := t \end{cases}$$

A *transition relation* for a command c is the set of models Rel_c of $Trans_c$. We write $Trans_e$ and Rel_e for the transition formula and relation of the command $stmt(e)$. An *execution of length* k is a sequence $\rho = (m_0, s_0), \ldots, (m_{k-1}, s_{k-1})$ of location and state pairs in which each $e = (m_i, m_{i+1})$ is an edge in E and the pair of states (s_i, s_{i+1}) is in the transition relation Rel_e. A location m is *reachable* if there is an execution ρ of some length k such that $\rho(k-1) = (m, s)$ for some state s.

The safety properties checked by abstract interpreters are usually encoded as reachability of locations in a CFG. The formula $Reach_{G,L}$ below encodes reachability of a set of locations L in a CFG G as satisfiability in WS1S(T). The first line below is an *initial constraint*, the second is a set of *transition constraints* indexed by locations, and the third line encodes *final constraints*.

$$Reach_{G,L} \doteq \forall i.First(i) \implies X_{\mathtt{in}}(i)$$
$$\wedge \bigwedge_{v \in Loc} \forall i.\forall j.X_v(j) \wedge Suc(i,j) \implies \bigvee_{(u,v) \in E} Trans_{(u,v)}(i) \wedge X_u(i)$$
$$\wedge \forall j.Last(j) \implies \bigvee_{u \in L} X_u(j)$$

We explain this formula in terms of a structure (τ, σ). The trace τ contains valuations of variables but has no information about locations. A second-order variable X_v represents the location v and $\sigma(X_v)$ represents the points in τ when control is at v. The initial constraint ensures that execution begins in $\mathtt{in}$. The final constraint ensures that execution ends in one of the locations in L. In a transition constraint, $X_v(j) \wedge Suc(i,j)$ expresses that the state $\tau(j)$ is visited at location v and its consequent expresses that the state $\tau(i)$ must have been visited at a location u that precedes v in the CFG and that $(\tau(i), \tau(j))$ must be in the transition relation (u,v).

Theorem 1. *Some location in a set L in a CFG G is reachable if and only if the formula $Reach_{G,L}$ is satisfiable.*

Proof. [$\Leftarrow$] If a location $w \in L$ is reachable, there is an execution $\rho \doteq (u_0, s_0), \ldots,$ (u_{k-1}, s_{k-1}) with $u_0 = \mathtt{in}$ and $u_{k-1} = w$. Define the structure (τ, σ) with $\tau \doteq s_0, \ldots, s_{k-1}$ and $\sigma \doteq \{X_u \mapsto \{i \mid \rho(i) = (u,s), s \in State\} \mid u \in Loc\}$. We show that (τ, σ) is a model of $Reach_{G,L}$. Since $u_0 = \mathtt{in}$ and $u_{k-1} = w$, the initial and final constraints are satisfied. In the transition constraint, if $X_v(j)$ holds, there is some $(u_i, s_i), (u_{i+1}, s_{i+1})$ in ρ with $u_{i+1} = v$. Thus, the transition (s_i, s_{i+1}) satisfies the transition formula $Trans_{(u_i, v)}$.

[$\Rightarrow$] Assume (τ, σ) is a model of $Reach_{G,L}$. Define a sequence ρ with $\rho(i) \doteq (u, \tau(i))$ where $i \in \sigma(X_u)$. As σ induces a partition, there is a unique u with i in $\sigma(X_u)$. We show that ρ is an execution reaching L. The initial constraint guarantees that $\rho(0)$ is at $\mathtt{in}$ and the final constraints guarantee that ρ ends in L. The transition constraints ensure that every step in the execution traverses an edge in G and respects the transition relation of the edge. $\square$

We believe this is a simple yet novel encoding of reachability, a property widely checked by abstract interpreters, in a minor extension of a well-known logic. The translation from second-order logics is at the heart of the automata-based verification, so we believe this work connects abstract interpretation to the automata-theoretic approach to program verification in a fundamental, yet novel way. In other second-order characterizations of correctness [2,11], it is invariants and not executions that are encoded by satisfying assignments. Moreover, those encodings do not connect to the automata-theoretic approach.

Example 2. A CFG G and the formula $Reach_{G,Ex}$ for a program with a variable x of type $\mathbb{Z}$ are shown in Fig. 1. Executions that start with a strictly negative value of x neither terminate nor reach $\mathtt{ex}$. For brevity, we write a state as the value of x. The execution $(\mathtt{in}, 1), (a, 1), (\mathtt{in}, 0), (\mathtt{ex}, 0)$ reaches $\mathtt{ex}$. It is encoded by the model (τ, σ), with $\sigma \triangleq \{X_{\mathtt{in}} \mapsto \{0, 2\}, X_a \mapsto \{1\}, X_{\mathtt{ex}} \triangleq \{3\}\}$ and $\tau = 1, 1, 0, 0$. Note that σ partitions $SVar$ because each position on the trace corresponds to a unique location. No structure (τ, σ) in which x is strictly negative in $\tau(0)$ satisfies $Reach_{G, Ex}$.   ◁

$$(\forall i. First(i) \Rightarrow X_{\mathtt{in}}(i)) \wedge (\forall i. Last(i) \Rightarrow X_{\mathtt{ex}}(i))$$
$$\wedge \; \forall i. \forall j. X_{\mathtt{in}}(j) \wedge Suc(i, j) \Rightarrow (suc(x) = x - 1)(i) \wedge X_a(i)$$
$$\wedge \; \forall i. \forall j. X_a(j) \wedge Suc(i, j) \; \Rightarrow ((x \neq 0 \Rightarrow suc(x) = x)(i) \wedge X_{\mathtt{in}}(i)$$
$$\wedge \; \forall i. \forall j. X_{\mathtt{ex}}(j) \wedge Suc(i, j) \Rightarrow (x = 0 \Rightarrow suc(x) = x)(i) \wedge X_{\mathtt{in}}(i))$$

Fig. 1. A CFG for a program with non-terminating executions and a WS1S(T) formula over the theory of integer arithmetic encoding the reachability of $\mathtt{ex}$.

Note that a program invariant would include all reachable states, but a model of $Reach_{G, L}$ only involves states that occur on a single execution. We emphasise that we are not considering arbitrary formulae in WS1S(T). The formula $Reach_{G, L}$ is a conjunction of constraints in which the initial, final and transition constraints have a fixed structure. The second-order variables and first-order program variables are free, but the first-order position variables are bound.

3 Lattices and Substructural First-Order Theories

The contribution of this section is to relate first-order substructural theories with the lattices in abstract domains. We show that certain lattices used in practice are Lindenbaum-Tarski algebras of theories that we introduce here.

3.1 First-Order Substructural Theories

For this section, assume a set of variables $Vars$ and a first-order theory of integer arithmetic with the standard functions and relations and let $\models_{\mathbb{Z}}$ define the semantics of quantifier-free first-order formulae. A *logical language* $(\mathcal{L}, \vdash_{\mathcal{L}})$ consists of a set of formulae and a proof system. The grammar below defines a set of formulae in terms of atomic formulae, logical constants and connectives.

We introduce formulae and calculi for a *sign logic*, a *constant logic* and an *interval logic*, with the names deriving from the abstract domains being modelled. The formulae in our logics are closed under conjunction but not under disjunction or negation. There are only three atomic formulae in $\mathcal{S}$. The infinitely many atomic formulae in $\mathcal{C}$ are equalities between a variable and an integer, and atomic formulae in $\mathcal{I}$-formulae involve upper bounds and lower bounds on variables. The three logics contain the logical constant tt, denoting true, but instead

Table 1. Proof rules for the core calculus and its extensions. The core calculus $\vdash_{\text{CORE}}$ contains rules for introduction (I), cut (CUT), weakening (WL), contraction (CL) and permutation (PL) on the left, conjunction ($\wedge$L$_1$, $\wedge$L$_2$, $\wedge$R), false (ffL), in which $\varphi(x)$ has only one free variable x, and true (ttR).

$$
\boxed{
\begin{array}{c}
\text{The core calculus } \vdash_{\text{CORE}} \\[2mm]
\dfrac{}{\varphi \vdash \varphi}\,\text{I} \qquad
\dfrac{}{\text{ff}_x \vdash \varphi(x)}\,\text{ffL} \qquad
\dfrac{\Gamma \vdash \varphi \qquad \varphi, \Sigma \vdash \psi}{\Gamma, \Sigma \vdash \psi}\,\text{CUT} \qquad
\dfrac{}{\Gamma \vdash \text{tt}}\,\text{ttR} \\[4mm]
\dfrac{\Gamma \vdash \psi}{\Gamma, \varphi \vdash \psi}\,\text{WL} \qquad
\dfrac{\Gamma, \varphi, \varphi \vdash \psi}{\Gamma, \varphi \vdash \psi}\,\text{CL} \qquad
\dfrac{\Gamma, \varphi, \psi \vdash \theta}{\Gamma, \psi, \varphi \vdash \theta}\,\text{PL} \\[4mm]
\dfrac{\Gamma, \varphi \vdash \theta}{\Gamma, \varphi \wedge \psi \vdash \theta}\,\wedge\text{L}_1 \qquad
\dfrac{\Gamma, \psi \vdash \theta}{\Gamma, \varphi \wedge \psi \vdash \theta}\,\wedge\text{L}_2 \qquad
\dfrac{\Gamma \vdash \varphi \qquad \Sigma \vdash \psi}{\Gamma, \Sigma \vdash \varphi \wedge \psi}\,\wedge\text{R}
\end{array}
}
$$

of a constant for false, we have a family ff_x parameterized by variables.

$$
\begin{array}{ll}
\varphi ::= x < 0 \mid x = 0 \mid x > 0 \mid \text{ff}_x \mid \text{tt} \mid \varphi \wedge \varphi & \mathcal{S} \\
\varphi ::= x = k \mid \text{ff}_x \mid \text{tt} \mid \varphi \wedge \varphi & \mathcal{C} \\
\varphi ::= x \leq k \mid x \geq k \mid \text{ff}_x \mid \text{tt} \mid \varphi \wedge \varphi & \mathcal{I}
\end{array}
$$

A calculus $\vdash_{\text{CORE}}$ for the logical core of these logics is shown in Table 1. We use sequents of the form $\Gamma, \Sigma \vdash_{\mathcal{L}} \varphi$, where the *antecedents* Γ and Σ are sequences of formulae, and the *consequent* φ is a single, first-order formula. We write $\bigwedge \Gamma$ for the conjunction of the sequence elements in Γ. A calculus $\vdash_{\mathcal{L}}$ is *sound* if every derivable sequent $\Gamma \vdash_{\mathcal{L}} \psi$ satisfies that $\models_{\mathbb{Z}} \bigwedge \Gamma \Rightarrow \psi$. The semantics of ff_x in $\models_{\mathbb{Z}}$ is ff. Two formulae are *inter-derivable* if the sequents $\varphi \vdash_{\mathcal{L}} \psi$ and $\psi \vdash_{\mathcal{L}} \varphi$ are both derivable.

Sequent calculi usually contain structural, logical and cut rules, and in the case of theories also theory rules. Our logics are *substructural* because the sequents have a restricted structure, lack right structural rules, and lack rules for disjunction, negation and implication. Our non-standard treatment of false is influenced by the way abstract domains reason about contradictions.

We review the theory rules for our logics. The reader should be warned that these logics have a restricted syntax and weak proof systems so the set of derivations is limited. We claim no responsibility for any disappointment arising from how uninteresting the derivable theorems are. The calculus $\vdash_{\mathcal{S}}$, in Fig. 2, extends $\vdash_{\text{CORE}}$ with rules for deriving ff_x from conjunctions of atomic formulae. The calculus for $\mathcal{C}$ is similar to that for $\mathcal{S}$ with the theory rule below instead. The calculus $\mathcal{I}$ in Fig. 4 contains rules for modifying upper and lower bounds.

$$
[m \neq_{\mathbb{Z}} n] \dfrac{}{\Gamma, x = m \wedge x = n \vdash \text{ff}_x}\,\text{ffR}_4
$$

Example 3. Figure 3 contains a derivation of $x < 0 \vdash_{\mathcal{S}} x < 0 \wedge \text{tt}$. The converse $x < 0 \wedge \text{tt} \vdash_{\mathcal{S}} x < 0$ is derivable with I and $\wedge$L$_1$, showing that $x < 0$ and $x < 0 \wedge \text{tt}$ are inter-derivable.

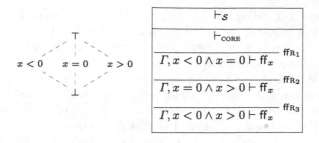

Fig. 2. A lattice of signs and a calculus that generates it.

$$\dfrac{\overline{\rule{0pt}{8pt}\,x<0\vdash_{\mathcal{S}_1} x<0\,}\ \text{I} \qquad \overline{\rule{0pt}{8pt}\,x<0\vdash_{\mathcal{S}_1}\text{tt}\,}\ \text{ttR}}{\dfrac{x<0,x<0\vdash_{\mathcal{S}_1} x<0\wedge\text{tt}}{x<0\vdash_{\mathcal{S}_1} x<0\wedge\text{tt}}\ \text{CL}}\ \wedge\text{R}$$

Fig. 3. A derivation in the sign calculus $\vdash_{\mathcal{S}}$.

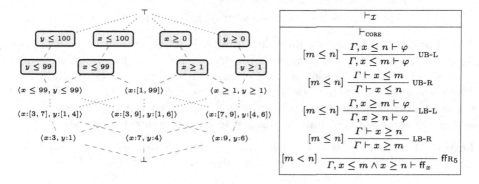

Fig. 4. The domain of intervals over two variables and a calculus for interval logic.

An abstract interpreter reasoning about variable values can derive a sequent $y\le 0, x\le 5\wedge x\ge 7\vdash_{\mathcal{I}}\text{ff}_x\wedge y\le 3$ showing that the inconsistency arises from x or $x\le 2, y\le 0\wedge y\ge 3\vdash_{\mathcal{I}}\text{ff}_y\wedge x\le 3$ showing an inconsistency from y.   ◁

Theorem 2. *The calculi $\vdash_{\mathcal{S}}$, $\vdash_{\mathcal{C}}$, and $\vdash_{\mathcal{I}}$ are sound.*

The proof is by induction on the structure of a derivation. This soundness theorem is used to show an isomorphism between the lattices generated by these calculi and the lattices they model.

3.2   Lattices from Substructural Theories

We recall elementary lattice theory. A lattice $(A,\sqsubseteq,\sqcap,\sqcup)$ is a partially ordered set (poset) with a binary greatest lower bound $\sqcap$, called the *meet*, and a binary least upper bound $\sqcup$, called the *join*. A poset with only a meet is called a meet-semi-lattice. A lattice is *bounded* if it has a greatest element $\top$, called *top*, and a least element $\bot$ called *bottom*. The notion of isomorphism for lattices is standard.

Pointwise lifting is an operation that lifts the order and operations of a lattice to functions on the lattice. Consider the functions $f, g : S \to A$, where S is a set and A a lattice as above. The pointwise order $f \sqsubseteq g$ holds if $f(x) \sqsubseteq g(x)$ for all x, while the pointwise meet $f \sqcap g$ maps x in S to $f(x) \sqcap g(x)$. The pointwise lift of other relations and operations on A is similarly defined.

Tarski related logic and lattices by extending a construction of Lindenbaum to generate Boolean algebras from propositional calculi and first-order sentences. We use a generalization of this construction to formulae with free variables [20]. We write $[\varphi]_{\mathcal{L}}$ for the equivalence class of φ with respect to an equivalence relation $\equiv_{\mathcal{L}}$. A logic $(\mathcal{L}, \vdash_{\mathcal{L}})$ that is closed under conjunction generates the *Lindenbaum-Tarski algebra* $(\mathcal{L}/\equiv_{\mathcal{L}}, \preccurlyeq, \curlywedge)$ below.

$$\varphi \equiv_{\mathcal{L}} \psi \text{ if } \varphi \vdash_{\mathcal{L}} \psi \text{ and } \psi \vdash_{\mathcal{L}} \varphi.$$

$$[\varphi]_{\mathcal{L}} \preccurlyeq [\psi]_{\mathcal{L}} \text{ if } \theta_1 \vdash_{\mathcal{L}} \theta_2 \text{ for some } \theta_1 \in [\varphi]_{\mathcal{L}}, \text{ and } \theta_2 \in [\psi]_{\mathcal{L}}.$$

$$[\varphi]_{\mathcal{L}} \curlywedge [\psi]_{\mathcal{L}} \hat{=} [\theta_1 \wedge \theta_2]_{\mathcal{L}} \text{ where } \theta_1 \in [\varphi]_{\mathcal{L}}, \text{ and } \theta_2 \in [\psi]_{\mathcal{L}}.$$

The relation $\equiv_{\mathcal{L}}$, defined by inter-derivability, is an equivalence whose classes form the carrier set of the algebra. Logical connectives generate operators. Though Lindenbaum-Tarski algebras of standard logics have been studied in depth, the algebras for the substructural theories we consider have not. To prove the lemma below, we show that derivability induces a partial order on the equivalence classes of $\equiv_{\mathcal{L}}$ and that conjunction induces a greatest lower bound.

Lemma 1. *Let $(\mathcal{L}, \vdash)$ be a quantifer-free first-order language closed under conjunction and $\vdash$ be a sound calculus that extends $\vdash_{\mathrm{CORE}}$. The Lindenbaum-Tarski algebra of $\mathcal{L}$ is a meet-semi-lattice.*

We now recall certain lattices used in abstract interpretation and show that they are isomorphic to the Lindenbaum-Tarski algebras of the logics we introduced. The lattice of signs $(Sign, \sqsubseteq)$ is shown in Fig. 2. The lattice of integer constants $(Const, \sqsubseteq)$ consists of the elements $\mathbb{Z} \cup \{\bot, \top\}$, with $\bot$ and $\top$ as bottom and top, and all other elements being incomparable. The lattice of integer intervals $(Itv, \sqsubseteq)$, consists of the set $\{[a, b] \mid a \leq b, a \in \mathbb{Z} \cup \{-\infty\}, b \in \mathbb{Z} \cup \{\infty\}\}$ and a special element $\bot$ denoting the empty interval. The partial order is standard and $[-\infty, \infty]$ is the top element.

An *abstract environment* is a function $\varepsilon : Vars \to D$, from program variables to a lattice D that represents approximations of variable values. A lattice of abstract environments is derived from a lattice D by pointwise lifting.

Theorem 3. *The Lindenbaum-Tarski algebra of each of the logics $\mathcal{S}, \mathcal{C}, \mathcal{I}$ over a set of variables Vars, is isomorphic to the pointwise lift of each of the lattices Sign, Const, and Itv to abstract environments over Vars.*

To provide intuition for the proof, we detail here the case for the logic $\mathcal{S}$ and the lattice *Sign* over one variable. To prove that the Lindenbaum-Tarski algebra of $\mathcal{S}$ over a variable x is isomorphic to *Sign* we have to show that there are five equivalence classes, and that $\preccurlyeq$ and $\curlywedge$ are as in Fig. 2. The five candidate equivalence classes are $\{[\mathsf{ff}_x]_{\mathcal{S}}, [\mathsf{tt}]_{\mathcal{S}}, [x < 0]_{\mathcal{S}}, [x = 0]_{\mathcal{S}}, [x > 0]_{\mathcal{S}}\}$. The proof that there are at most five equivalence classes is by induction on the structure of formulae. The proof argument is that every conjunct in $\mathcal{S}$ is inter-derivable from a formula in one of these classes. The proof that there are at least five equivalence classes relies on the soundness of $\vdash_{\mathcal{S}}$. If there are fewer than five equivalence classes, there are consequences derivable in $\vdash_{\mathcal{S}}$ that do not hold semantically. Observe that this proof argument holds only because every lattice element represents a different set of structures. In abstract interpretation parlance, this argument only applies to abstractions in which the concretization function is injective.

Next, we define a function $h : \mathcal{S}/\equiv_{\mathcal{S}} \to \textit{Sign}$ that maps equivalence classes to corresponding lattice elements. To show that h is an isomorphism we argue by induction on formula structure for comparable equivalence classes and by appeal to soundness for incomparable equivalence classes. This argument generalizes to a finite number of variables because all the logics we have considered only involve one-place predicates. The shaded elements in Fig. 4 are the images of the formulae shown under the isomorphism.

4 Abstract Transformers, Deduction and Abduction

The constant and interval domains are used in practice even though, as shown in the previous section, they have weak proof systems. In this section, we adapt Tarski's notion of consequence operators to logically model abstract transformers for conditionals. Consequence operators provided an approach to algebraically modelling deduction. These transformers can be viewed as enriching a weak proof system $\vdash_{\mathcal{L}}$ with the ability to reason about formulae that are not definable in $\mathcal{L}$.

We consider again a quantifier-free first-order theory $\mathcal{T}$ with semantics $\models_{\mathcal{T}}$ and a logical language $(\mathcal{L}, \vdash_{\mathcal{L}})$, where $\mathcal{L} \subseteq \mathcal{T}$. Deduction in $\mathcal{L}$ *with respect to a formula* $\varphi \in \mathcal{T}$ is formalized by a *deduction function* $ded_{\varphi} : \mathscr{F}(\mathcal{L}) \to \mathscr{F}(\mathcal{L})$ between finite sets of formulae in $\mathcal{L}$. A deduction function is sound if for finite $\Theta \subseteq \mathcal{L}$, and $\theta \in ded_{\varphi}(\Theta)$, $\varphi \wedge \bigwedge \Theta \models_{\mathcal{T}} \theta$. That is, the deduced formulae are consequences of the arguments and, crucially, the parameter φ. The formula φ acts as an external hint to boost the capabilities of a weak deductive system. The parameter φ may not exist in $\mathcal{L}$, so there may not be a rule of the form $\varphi, \Gamma \vdash_{\mathcal{L}} \theta$ corresponding to an application of the deduction function.

Similarly, we model abduction by a function that generates antecedents given consequents. An abduction function $abd_{\varphi} : \mathscr{F}(\mathcal{L}) \to \mathscr{F}(\mathcal{L})$ derives antecedents in $\mathcal{L}$ with respect to a parameter φ. An abduction function is sound if for all Θ, and $\theta \in abd_{\varphi}(\Theta)$, $\varphi \wedge \theta \models_{\mathcal{T}} \bigwedge \Theta$.

Example 4. This examples illustrates how deduction with respect to a formula enables reasoning that is not possible in the lattice itself. Let $\varphi \triangleq 3y - 1 > 0 \wedge x = -y$ be a formula in a theory. We define one possible sound deduction function ded_φ for consequences in $\mathcal{S}$.

$$ded_\varphi(\{\mathtt{tt}\}) = ded_\varphi(\{y > 0\}) = ded_\varphi(\{x < 0\}) = \{y > 0, x < 0\}$$
$$ded_\varphi(\{\mathtt{ff}_x\}) = ded_\varphi(\{x = 0\}) = ded_\varphi(\{x > 0\}) = ded_\varphi(\{y = 0\}) = \{\mathtt{ff}_x, \mathtt{ff}_y\}$$

The results of applying ded_φ shown above are the only two possibilities, even for sets of formulae not shown above.

The difference between ded_φ and classical consequence operators is that we make fewer assumptions on properties of ded_φ in the same way our lattices make fewer structural assumptions than classical logics. Recall that a set of formulae C is *consequence-closed* with respect to $\vdash_\mathcal{L}$ if for all φ in C, if $\varphi \vdash_\mathcal{L} \theta$, then θ is in C. The *consequence closure* of C is the smallest consequence-closed set containing C. If $\Gamma \vdash_\mathcal{L} \theta$, the consequence closure of Γ contains the consequence closure of θ. A deduction function inverts this relationship, because it strengthens its argument using φ. That is, the consequence closure of $ded_\varphi(\Theta)$ is a superset of the consequence closure of Θ because it contains formulae derived using φ.

Deduction functions, when factored through the Lindenbaum-Tarski equivalence relation, give rise to sound transformers for conditionals. To make this precise, we require the notion of a concretization function from abstract interpretation. Let $(A, \sqsubseteq, \sqcap)$ be a bounded lattice and $(\mathcal{P}(State), \subseteq, \cap)$ be the powerset of states with the subset order. We say that A is an *abstraction of* $\mathcal{P}(State)$ if there is a monotone function $\gamma : A \to \mathcal{P}(State)$ satisfying that $\gamma(\top) = State$ and $\gamma(\bot) = \emptyset$. Requiring that $\bot$ has an empty concretization is non-standard but is required for a logical treatment of false.

Recall that $Rel_{[\varphi]}$ is the transition relation for a conditional. A function $post_{[\varphi]} : A \to A$ is a *sound successor transformer* for the conditional $[\varphi]$ if the set of structures obtained by applying $post_{[\varphi]}$ overapproximates the structures obtained by applying the transition relation: $Rel_{[\varphi]}(\gamma(a)) \subseteq \gamma(post_{[\varphi]}(a))$. Dually, a function $\widetilde{pre}_{[\varphi]} : A \to A$ is a *sound predecessor transformer* for the conditional $[\varphi]$ if the set of structures obtained by applying $\widetilde{pre}_{[\varphi]}$ underapproximates the structures obtained by applying the transition relation backwards: $\gamma(\widetilde{pre}_{[\varphi]}(a)) \subseteq \{s \mid Rel_{[\varphi]}(\{s\}) \subseteq \gamma(a)\}$.

To relate these transformers to deduction and abduction functions, we lift the functions above to operate on the Lindenbaum-Tarski algebra. We write $\equiv$ instead of $\equiv_\mathcal{L}$ for brevity.

$$ded_\varphi^\equiv : \mathcal{L}/\equiv \;\to\; \mathcal{L}/\equiv \qquad ded_\varphi^\equiv([\theta]_\equiv) \mathrel{\hat{=}} \bigwedge \{[\psi]_\equiv \mid \psi \in ded_\varphi([\theta]_\equiv)\}$$
$$abd_\varphi^\equiv : \mathcal{L}/\equiv \;\to\; \mathcal{L}/\equiv \qquad abd_\varphi^\equiv([\theta]_\equiv) \mathrel{\hat{=}} [\psi]_\equiv \text{ for some } \psi \in abd_\varphi([\theta]_\equiv)$$

The result of deduction on the Lindenbaum-Tarski algebra is a meet of equivalence classes of formulae in order to obtain the strongest consequence possible. Assuming that an equivalence class consists of only finitely many formulae, this result is well-defined. If an equivalence class is not finite, a finite number of representatives can be used instead. The lift of abduction is not the dual of deduction. Instead, the result of lifting abduction is the equivalence class of one of the formulae that result from abduction. This is because we want the weakest possible abduction but our logics lack disjunction. Using the lattice-theoretic join in algebras where the join exists may lead to unsound abduction.

Theorem 4. *Let ded_φ and abd_φ be sound deduction and abduction transformers and $(\mathcal{L}, \vdash_\mathcal{L})$ be a logical language closed under conjunction with a calculus that extends $\vdash_{\text{CORE}}$. Then, the lifted functions $ded_\varphi^{\overline{\equiv}}$ and $abd_\varphi^{\overline{\equiv}}$ are sound successor and predecessor transformers for the conditional $[\varphi]$.*

We have not modelled transformers for assignment because we have not identified a satisfying treatment of substitution and quantification that factors through the Lindenbaum-Tarski construction.

5 Abstract Interpreters as Second-Order Solvers

An abstract interpreter for reachability analysis combines a lattice with transformers to derive program invariants. We have shown that lattices approximate state formulae, and that deduction and abduction functions approximate transition formulae. We now show that the steps in fixed point iteration can be understood as second-order propagation. Logically, a fixed point iterator can be viewed as an SMT solver for trace formulae.

We introduce *abstract assignments* to model approximations of trace formulae. We have chosen this term to emphasise the similarity to partial assignments in SAT solvers. Let $(A, \sqsubseteq, \sqcap)$ be a lattice that is an abstraction of the lattice of states $(\mathscr{P}(State), \subseteq, \cap)$. Recall that $SVar$ is the set of second-order variables. The *lattice of abstract assignments* is $(Asg_A, \sqsubseteq, \sqcap)$, where $Asg_A \triangleq SVar \to A$ is the set of abstract assignments and the order and meet are defined pointwise.

Let $Struct$ be the set of pairs (τ, σ) of WS1S(T) structures. We show that the lattice of abstract assignments is an abstraction of $(\mathscr{P}(Struct), \subseteq, \cup)$. An abstract assignment represents sets of WS1S(T) structures analogous to the way a partial assignment in a DPLL-based SAT solver represents all assignments that extend to undefined variables. The set of WS1S(T) structures represented by an abstract assignment is given by the concretization $conc : Asg_A \to \mathscr{P}(Struct)$ below.

$$conc(asg) \triangleq \{(\tau, \sigma) \mid \text{for all } X \in SVar.\, \{\tau(i) \mid i \in \sigma(X)\} \subseteq \gamma(asg(X))\}$$

Explained in terms of states, an abstract assignment represents structures by the set of states at each program location but forgets the order between states.

We present the run of an abstract interpreter as a solver for $Reach_{G,L}$. An abstract interpreter begins with the variable map $\lambda Y.\top$ indicating that nothing is known about the satisfiability of $Reach_{G,L}$, so every structure is potentially a model of $Reach_{G,L}$. An abstract assignment is updated using a propagation rule. If a location is not reachable, the formula is unsatisfiable, as deduced by the conflict rule.

$$asg \rightsquigarrow asg[X_v \mapsto d], \quad \text{where } d = \bigsqcup_{(u,v) \in E} \left\{ post_{(u,v)}(asg(X_v)) \right\} \quad \text{Propagate}$$

$$asg \rightsquigarrow \text{unsat} \qquad \text{if } asg(X_v) = \bot, \text{ for some } v \in L \qquad \text{Conflict}$$

Propagation modifies an abstract assignment similar to the way Boolean constraint propagation (BCP) updates a partial assignment with two key differences. One is that rather than values, the assignment is updated with elements of a lattice. The second is that in BCP, before decisions are made, every value that is undefined becomes tt or ff, becoming strictly more precise. With abstract assignments, the assignments to X_v within an SCC with more than one node, will, in general, get weaker. We have not modelled termination concerns, which are addressed with widening and narrowing operators. The theorem below expresses the soundness of fixed point iteration without widening and narrowing in terms of satisfiability.

Theorem 5. *If the repeated application of the propagation and conflict rules leads to* unsat, *the formula* $Reach_{G,L}$ *is unsatisfiable.*

6 Related Work, Discussion and Conclusion

The development of novel combinations of automated deduction and abstract interpretation is a driving force behind much current research, which we surveyed in the introduction. Consult the dissertations [14,22] and Dagstuhl seminar notes [17] for a detailed treatment of this research. Such work has been applied to design new SMT solvers [4], program analyzers [10,23], and has helped automate the construction of program analyzers [24,25,27].

However, our experience has been that crucial aspects of solvers such as branching and conflict analysis heuristics are difficult to characterize lattice-theoretically due to their combinatorial nature. In this work, we have initiated a complementary research programme by giving logical characterizations of instances of abstract interpretation. To relate logics to lattices, we have combined ideas from substructural logic with the Lindenbaum-Tarski construction [20] and Tarski's algebraic treatment of deduction.

A more abstract approach would be to use the framework of Stone duality, which uses category theory to relate lattices, topological spaces and logics. Stone duality was extended to programs by Abramsky [1] who related domains in semantics to intuitionistic, modal, fixed point logic. Jensen [16] applied Abramsky's work to extract a logic from a specific abstract interpretation: strictness analysis.

In this paper, we have modelled logics that lack disjunction and have weaker proof systems than those considered in approaches based on Stone duality. The closest study to ours is by Schmidt [21], who articulated the idea that the partial order of an abstract domain defines its proof theory. In terms of algebraic logic, Schmidt's work can be understood as identifying logical characterization of families of lattices in abstract interpretation as free algebras of the Lindenbaum-Tarski construction. In comparison, our work has focused on characterizing specific lattices as theories.

Conclusion. This work is a first step towards a logical description of the internals of an abstract interpreter in the mathematical and algorithmic vocabulary of SAT and SMT solvers. The results in this paper make precise widespread folk intuition about the logical basis of certain abstract interpreters. Though our results are unsurprising, we believe the techniques we have used are novel and connect ideas from substructural logic, algebraic logic and satisfiability research. In using Büchi's construction, we have also connected abstract interpretation with the automata-theoretic approach to logic and verification. Folk knowledge asserts that transformers for assignments provide a form of quantifier elimination. We have not modelled these transformers here because we are missing a rigorous treatment that integrates with the Lindenbaum-Tarski construction.

In terms of solver architecture, the simple abstract interpreter we have considered can be viewed as a second-order theory solver that only updates assignments. This view provides a direct route to integrating branching heuristics, conflict analysis, and variable selection. We have begun these investigations and hope that this exposition enables the automated deduction community to participate in the same.

References

1. Abramsky, S.: Domain theory and the logic of observable properties. Ph.D. thesis, University of London (1987)
2. Aiken, A.: Introduction to set constraint-based program analysis. Sci. Comput. Program. **35**, 79–111 (1999)
3. Bjørner, N., Duterte, B., de Moura, L.: Accelerating lemma learning using joins - DPLL(⊔). In: Cervesato, I., Veith, H., Voronkov, A. (eds.) LPAR (2008)
4. Brain, M., D'silva, V., Griggio, A., Haller, L., Kroening, D.: Deciding floating-point logic with abstract conflict driven clause learning. Formal Methods Syst. Des. **45**(2), 213–245 (2014)
5. Büchi, J.R.: On a decision method in restricted second order arithmetic. In: Logic, Methodology and Philosophy of Science, pp. 1–11. Stanford University Press (1960)
6. Cousot, P., Cousot, R.: Abstract interpretation: a unified lattice model for static analysis of programs by construction or approximation of fixpoints. In: POPL, pp. 238–252. ACM Press (1977)
7. Cousot, P., Cousot, R.: Systematic design of program analysis frameworks. In: POPL, pp. 269–282. ACM Press (1979)
8. Cousot, P., Cousot, R., Mauborgne, L.: Theories, solvers and static analysis by abstract interpretation. J. ACM **59**(6), 31:1–31:56 (2013)
9. D'Silva, V., Haller, L., Kroening, D.: Abstract conflict driven learning. In: Giacobazzi, R., Cousot, R. (eds.) POPL, pp. 143–154. ACM Press (2013)
10. D'Silva, V., Haller, L., Kroening, D., Tautschnig, M.: Numeric bounds analysis with conflict-driven learning. In: Flanagan, C., König, B. (eds.) TACAS 2012. LNCS, vol. 7214, pp. 48–63. Springer, Heidelberg (2012)
11. Grebenshchikov, S., Lopes, N.P., Popeea, C., Rybalchenko, A.: Synthesizing software verifiers from proof rules. In: Vitek, J., Lin, H., Tip, F. (eds.) PLDI, pp. 405–416. ACM Press (2012)
12. Gulavani, B.S., Chakraborty, S., Nori, A.V., Rajamani, S.K.: Automatically refining abstract interpretations. In: Ramakrishnan, C.R., Rehof, J. (eds.) TACAS 2008. LNCS, vol. 4963, pp. 443–458. Springer, Heidelberg (2008)

13. Gulwani, S., Tiwari, A.: Combining abstract interpreters. In: Schwartzbach, M.I., Ball, T. (eds.) PLDI, pp. 376–386. ACM Press (2006)
14. Haller, L.C.R.: Abstract satisfaction. Ph.D. thesis, University of Oxford (2014)
15. Harris, W.R., Sankaranarayanan, S., Ivančić, F., Gupta, A.: Program analysis via satisfiability modulo path programs. In: Hermenegildo, M., Palsberg, J. (eds.) POPL, pp. 71–82 (2010)
16. Jensen, T.P.: Strictness analysis in logical form. In: Hughes, J. (ed.) Functional Programming Languages and Computer Architecture. LNCS, vol. 523, pp. 352–366. Springer, Heidelberg (1991)
17. Kroening, D., Reps, T.W., Seshia, S.A., Thakur, A.V.: Decision procedures and abstract interpretation (Dagstuhl seminar 14351). Dagstuhl Rep. 4(8), 89–106 (2014)
18. Leino, K.R.M., Logozzo, F.: Using widenings to infer loop invariants inside an SMT solver, or: a theorem prover as abstract domain. In: Workshop on Invariant Generation, pp. 70–84. RISC Report 07–07 (2007)
19. Pelleau, M., Truchet, C., Benhamou, F.: Octagonal domains for continuous constraints. In: Lee, J. (ed.) CP 2011. LNCS, vol. 6876, pp. 706–720. Springer, Heidelberg (2011)
20. Rasiowa, H., Sikorski, R.: The Mathematics of Metamathematics. Polish Academy of Science, Warsaw (1963)
21. Schmidt, D.A.: Internal and external logics of abstract interpretations. In: Logozzo, F., Peled, D.A., Zuck, L.D. (eds.) VMCAI 2008. LNCS, vol. 4905, pp. 263–278. Springer, Heidelberg (2008)
22. Thakur, A.V.: Symbolic abstraction: algorithms and applications. Ph.D. thesis, The University of Wisconsin - Madison (2014)
23. Thakur, A.V., Breck, J., Reps, T.W.: Satisfiability modulo abstraction for separation logic with linked lists. In: Rungta, N., Tkachuk, O. (eds.) SPIN, pp. 58–67 (2014)
24. Thakur, A., Elder, M., Reps, T.: Bilateral algorithms for symbolic abstraction. In: Miné, A., Schmidt, D. (eds.) SAS 2012. LNCS, vol. 7460, pp. 111–128. Springer, Heidelberg (2012)
25. Thakur, A.V., Lal, A., Lim, J., Reps, T.W.: Posthat and all that: automating abstract interpretation. Electr. Notes Theor. Comput. Sci. 311, 15–32 (2015)
26. Thakur, A., Reps, T.: A Generalization of Stålmarck's method. In: Miné, A., Schmidt, D. (eds.) SAS 2012. LNCS, vol. 7460, pp. 334–351. Springer, Heidelberg (2012)
27. Thakur, A., Reps, T.: A method for symbolic computation of abstract operations. In: Madhusudan, P., Seshia, S.A. (eds.) CAV 2012. LNCS, vol. 7358, pp. 174–192. Springer, Heidelberg (2012)
28. Tiwari, A., Gulwani, S.: Logical interpretation: static program analysis using theorem proving. In: Pfenning, F. (ed.) CADE 2007. LNCS (LNAI), vol. 4603, pp. 147–166. Springer, Heidelberg (2007)
29. Truchet, C., Pelleau, M., Benhamou, F.: Abstract domains for constraint programming, with the example of octagons. In: Symbolic and Numeric Algorithms for Scientific, Computing, pp. 72–79 (2010)
30. van den Elsen, S.: Weak monadic second-order theory of one successor. Seminar: Decision Procedures (2012)
31. Vardi, M.Y., Wilke, T.: Automata: from logics to algorithms. In: Logic and Automata: History and Perspectives [in Honor of Wolfgang Thomas], pp. 629–736 (2008)

Hybrid Sytems and Program Synthesis

A Uniform Substitution Calculus
for Differential Dynamic Logic

André Platzer$^{(\boxtimes)}$

Computer Science Department, Carnegie Mellon University, Pittsburgh, USA
aplatzer@cs.cmu.edu

Abstract. This paper introduces a new proof calculus for *differential dynamic logic* (d$\mathcal{L}$) that is entirely based on *uniform substitution*, a proof rule that substitutes a formula for a predicate symbol everywhere. Uniform substitutions make it possible to rely on *axioms* rather than axiom schemata, substantially simplifying implementations. Instead of subtle schema variables and soundness-critical side conditions on the occurrence patterns of variables, the resulting calculus adopts only a finite number of ordinary d$\mathcal{L}$ formulas as axioms. The static semantics of differential dynamic logic is captured exclusively in uniform substitutions and bound variable renamings as opposed to being spread in delicate ways across the prover implementation. In addition to sound uniform substitutions, this paper introduces *differential forms* for differential dynamic logic that make it possible to internalize differential invariants, differential substitutions, and derivations as first-class axioms in d$\mathcal{L}$.

1 Introduction

Differential dynamic logic (d$\mathcal{L}$) [4,6] is a logic for proving correctness properties of hybrid systems. It has a sound and complete proof calculus relative to differential equations [4,6] and a sound and complete proof calculus relative to discrete systems [6]. Both sequent calculi [4] and Hilbert-type axiomatizations [6] have been presented for d$\mathcal{L}$ but only the former has been implemented. The implementation of d$\mathcal{L}$'s sequent calculus in KeYmaera makes it straightforward for users to prove properties of hybrid systems, because it provides rules performing natural decompositions for each operator. The downside is that the implementation of the rule schemata and their side conditions on occurrence constraints and relations of reading and writing of variables as well as rule applications in context is nontrivial and inflexible in KeYmaera.

The goal of this paper is to identify how to make it straightforward to implement the axioms and proof rules of differential dynamic logic by writing down a finite list of *axioms* (concrete formulas, not axiom schemata that represent an infinite list of axioms subject to sophisticated soundness-critical schema variable matching implementations). They require multiple axioms to be combined with

---

All proofs are in a companion report [9]. This material is based upon work supported by the National Science Foundation by NSF CAREER Award CNS-1054246.

© Springer International Publishing Switzerland 2015
A.P. Felty and A. Middeldorp (Eds.): CADE-25, LNAI 9195, pp. 467–481, 2015.
DOI: 10.1007/978-3-319-21401-6_32

one another to obtain the effect that a user would want for proving a hybrid system conjecture. This paper argues that this is still a net win for hybrid systems, because a substantially simpler prover core is easier to implement correctly, and the need to combine multiple axioms to obtain user-level proof steps can be achieved equally well by appropriate tactics, which are not soundness-critical.

To achieve this goal, this paper follows observations for differential game logic [8] that highlight the significance and elegance of *uniform substitutions*, a classical proof rule for first-order logic [2, §35,40]. Uniform substitutions uniformly instantiate predicate and function symbols with formulas and terms, respectively, as functions of their arguments. In the presence of the nontrivial binding structure that nondeterminism and differential equations of hybrid programs induce for the dynamic modalities of differential dynamic logic, flexible but sound uniform substitutions become more complex for d$\mathcal{L}$, but can still be read off elegantly from its static semantics. In fact, d$\mathcal{L}$'s static semantics is solely captured[1] in the implementation of uniform substitution (and bound variable renaming), thereby leading to a completely modular proof calculus.

This paper introduces a static and dynamic semantics for *differential-form* d$\mathcal{L}$, proves coincidence lemmas and uniform substitution lemmas, culminating in a soundness proof for uniform substitutions (Sect. 3). It exploits the new *differential forms* that this paper adds to d$\mathcal{L}$ for internalizing differential invariants [5], differential cuts [5,7], differential ghosts [7], differential substitutions, total differentials and Lie-derivations [5,7] as first-class citizens in d$\mathcal{L}$, culminating in entirely modular axioms for differential equations and a superbly modular soundness proof (Sect. 4). This approach is to be contrasted with earlier approaches for differential invariants that were based on complex built-in rules [5,7]. The relationship to related work from previous presentations of differential dynamic logic [4,6] continues to apply except that d$\mathcal{L}$ now internalizes differential equation reasoning axiomatically via differential forms.

2 Differential-Form Differential Dynamic Logic

2.1 Syntax

Formulas and hybrid programs (HPs) of d$\mathcal{L}$ are defined by simultaneous induction based on the following definition of terms. Similar simultaneous inductions are used throughout the proofs for d$\mathcal{L}$. The set of all *variables* is $\mathcal{V}$. For any $V \subseteq \mathcal{V}$ is $V' \stackrel{\text{def}}{=} \{x' : x \in V\}$ the set of *differential symbols* x' for the variables in V. Function symbols are written f, g, h, predicate symbols p, q, r, and variables $x, y, z \in \mathcal{V}$ with differential symbols $x', y', z' \in \mathcal{V}'$. Program constants are a, b, c.

Definition 1 (Terms). Terms *are defined by this grammar (with* $\theta, \eta, \theta_1, \ldots, \theta_k$ *as terms,* $x \in \mathcal{V}$ *as variable,* $x' \in \mathcal{V}'$ *differential symbol, and* f *function symbol):*

$$\theta, \eta ::= x \mid x' \mid f(\theta_1, \ldots, \theta_k) \mid \theta + \eta \mid \theta \cdot \eta \mid (\theta)'$$

---

[1] This approach is dual to other successful ways of solving the intricacies and subtleties of substitutions [1,3] by imposing occurrence side conditions on axiom schemata and proof rules, which is what uniform substitutions can get rid of.

Number literals such as 0,1 are allowed as function symbols without arguments that are always interpreted as the numbers they denote. Beyond differential symbols x', *differential-form* dℒ allows *differentials* $(\theta)'$ of terms θ as terms for the purpose of axiomatically internalizing reasoning about differential equations.

Definition 2 (Hybrid program). Hybrid programs *(HPs) are defined by the following grammar (with α, β as HPs, program constant a, variable x, term θ possibly containing x, and formula ψ of first-order logic of real arithmetic):*

$$\alpha, \beta ::= a \mid x := \theta \mid x' := \theta \mid ?\psi \mid x' = \theta \,\&\, \psi \mid \alpha \cup \beta \mid \alpha; \beta \mid \alpha^*$$

Assignments $x := \theta$ of θ to variable x, tests $?\psi$ of the formula ψ in the current state, differential equations $x' = \theta \,\&\, \psi$ restricted to the evolution domain constraint ψ, nondeterministic choices $\alpha \cup \beta$, sequential compositions $\alpha; \beta$, and nondeterministic repetition α^* are as usual in dℒ [4,6]. The effect of the *differential assignment $x' := \theta$* to differential symbol x' is similar to the effect of the assignment $x := \theta$ to variable x, except that it changes the value of the differential symbol x' around instead of the value of x. It is not to be confused with the differential equation $x' = \theta$, which will follow said differential equation continuously for an arbitrary amount of time. The differential assignment $x' := \theta$, instead, only assigns the value of θ to the differential symbol x' discretely once at an instant of time. Program constants a are uninterpreted, i.e. their behavior depends on the interpretation in the same way that the values of function symbols f and predicate symbols p depends on their interpretation.

Definition 3 (dℒ formula). *The* formulas of (differential-form) differential dynamic logic *(dℒ) are defined by the grammar (with* dℒ *formulas ϕ, ψ, terms $\theta, \eta, \theta_1, \ldots, \theta_k$, predicate symbol p, quantifier symbol C, variable x, HP α):*

$$\phi, \psi ::= \theta \geq \eta \mid p(\theta_1, \ldots, \theta_k) \mid C(\phi) \mid \neg\phi \mid \phi \wedge \psi \mid \forall x\, \phi \mid \exists x\, \phi \mid [\alpha]\phi \mid \langle\alpha\rangle\phi$$

Operators $>, \leq, <, \vee, \rightarrow, \leftrightarrow$ are definable, e.g., $\phi \rightarrow \psi$ as $\neg(\phi \wedge \neg\psi)$. Likewise $[\alpha]\phi$ is equivalent to $\neg\langle\alpha\rangle\neg\phi$ and $\forall x\, \phi$ equivalent to $\neg\exists x\, \neg\phi$. The modal formula $[\alpha]\phi$ expresses that ϕ holds after all runs of α, while the dual $\langle\alpha\rangle\phi$ expresses that there is a run of α after which ϕ holds. *Quantifier symbols C (with formula ϕ as argument)*, i.e. higher-order predicate symbols that bind all variables of ϕ, are unnecessary but internalize contextual congruence reasoning efficiently.

2.2   Dynamic Semantics

A state is a mapping from variables $\mathcal{V}$ and differential symbols $\mathcal{V}'$ to $\mathbb{R}$. The set of states is denoted $\mathcal{S}$. Let ν_x^r denote the state that agrees with state ν except for the value of variable x, which is changed to $r \in \mathbb{R}$, and accordingly for $\nu_{x'}^r$. The interpretation of a function symbol f with arity n (i.e. with n arguments) is a smooth function $I(f) : \mathbb{R}^n \rightarrow \mathbb{R}$ of n arguments.

Definition 4 (Semantics of terms). *For each interpretation I, the* semantics of a term θ in a state $\nu \in \mathcal{S}$ is its value in $\mathbb{R}$. It is defined inductively as follows

1. $[\![x]\!]^I \nu = \nu(x)$ *for variable* $x \in \mathcal{V}$
2. $[\![x']\!]^I \nu = \nu(x')$ *for differential symbol* $x' \in \mathcal{V}'$
3. $[\![f(\theta_1, \ldots, \theta_k)]\!]^I \nu = I(f)([\![\theta_1]\!]^I \nu, \ldots, [\![\theta_k]\!]^I \nu)$ *for function symbol* f
4. $[\![\theta + \eta]\!]^I \nu = [\![\theta]\!]^I \nu + [\![\eta]\!]^I \nu$
5. $[\![\theta \cdot \eta]\!]^I \nu = [\![\theta]\!]^I \nu \cdot [\![\eta]\!]^I \nu$
6. $[\![(\theta)']\!]^I \nu = \sum_x \nu(x') \dfrac{\partial [\![\theta]\!]^I}{\partial x}(\nu) = \sum_x \nu(x') \dfrac{\partial [\![\theta]\!]^I \nu_x^X}{\partial X}$

Time-derivatives are undefined in an isolated state ν. The clou is that differentials can still be given a local semantics: $[\![(\theta)']\!]^I \nu$ is the sum of all (analytic) spatial partial derivatives of the value of θ by all variables x (or rather their values X) multiplied by the corresponding tangent described by the value $\nu(x')$ of differential symbol x'. That sum over all variables $x \in \mathcal{V}$ has finite support, because θ only mentions finitely many variables x and the partial derivative by variables x that do not occur in θ is 0. The spatial derivatives exist since $[\![\theta]\!]^I \nu$ is a composition of smooth functions, so smooth. Thus, the semantics of $[\![(\theta)']\!]^I \nu$ is the *differential*[2] of (the value of) θ, hence a differential one-form giving a real value for each tangent vector (i.e. vector field) described by the values $\nu(x')$. The values $\nu(x')$ of the differential symbols x' describe an arbitrary tangent vector or vector field. Along the flow of (the vector field of a) differential equation, though, the value of the differential $(\theta)'$ coincides with the analytic time-derivative of θ (Lemma 8). The interpretation of predicate symbol p with arity n is an n-ary relation $I(p) \subseteq \mathbb{R}^n$. The interpretation of quantifier symbol C is a functional $I(C)$ mapping subsets $M \subseteq \mathcal{S}$ to subsets $I(C)(M) \subseteq \mathcal{S}$.

Definition 5 (d$\mathcal{L}$ semantics). *The semantics of a* d$\mathcal{L}$ *formula* ϕ, *for each interpretation* I *with a corresponding set of states* $\mathcal{S}$, *is the subset* $[\![\phi]\!]^I \subseteq \mathcal{S}$ *of states in which* ϕ *is true. It is defined inductively as follows*

1. $[\![\theta \geq \eta]\!]^I = \{\nu \in \mathcal{S} \; : \; [\![\theta]\!]^I \nu \geq [\![\eta]\!]^I \nu\}$
2. $[\![p(\theta_1, \ldots, \theta_k)]\!]^I = \{\nu \in \mathcal{S} \; : \; ([\![\theta_1]\!]^I \nu, \ldots, [\![\theta_k]\!]^I \nu) \in I(p)\}$
3. $[\![C(\phi)]\!]^I = I(C)([\![\phi]\!]^I)$ *for quantifier symbol* C
4. $[\![\neg\phi]\!]^I = ([\![\phi]\!]^I)^C = \mathcal{S} \setminus [\![\phi]\!]^I$
5. $[\![\phi \wedge \psi]\!]^I = [\![\phi]\!]^I \cap [\![\psi]\!]^I$
6. $[\![\exists x\, \phi]\!]^I = \{\nu \in \mathcal{S} \; : \; \nu_x^r \in [\![\phi]\!]^I \text{ for some } r \in \mathbb{R}\}$
7. $[\![\langle\alpha\rangle\phi]\!]^I = [\![\alpha]\!]^I \circ [\![\phi]\!]^I = \{\nu \; : \; \omega \in [\![\phi]\!]^I \text{ for some } \omega \text{ such that } (\nu, \omega) \in [\![\alpha]\!]^I\}$
8. $[\![[\alpha]\phi]\!]^I = [\![\neg\langle\alpha\rangle\neg\phi]\!]^I = \{\nu \; : \; \omega \in [\![\phi]\!]^I \text{ for all } \omega \text{ such that } (\nu, \omega) \in [\![\alpha]\!]^I\}$

A d$\mathcal{L}$ *formula* ϕ *is valid in* I, *written* $I \models \phi$, *iff* $[\![\phi]\!]^I = \mathcal{S}$, *i.e.* $\nu \in [\![\phi]\!]^I$ *for all* ν. *Formula* ϕ *is valid, written* $\models \phi$, *iff* $I \models \phi$ *for all interpretations* I.

The interpretation of a program constant a is a state-transition relation $I(a) \subseteq \mathcal{S} \times \mathcal{S}$, where $(\nu, \omega) \in I(a)$ iff a can run from initial state ν to final state ω.

---

[2] A slight abuse of notation rewrites the differential as $[\![(\theta)']\!]^I = d[\![\theta]\!]^I = \sum_{i=1}^n \frac{\partial [\![\theta]\!]^I}{\partial x^i} dx^i$ when $x^1, \ldots, x^n$ are the variables in θ and their differentials dx^i form the basis of the cotangent space, which, when evaluated at a point ν whose values $\nu(x')$ determine the tangent vector alias vector field, coincides with Definition 4.

Definition 6 (Transition semantics of HPs). *For each interpretation I, each HP α is interpreted semantically as a binary transition relation $[\![\alpha]\!]^I \subseteq \mathcal{S} \times \mathcal{S}$ on states, defined inductively by*

1. $[\![a]\!]^I = I(a)$ *for program constants a*
2. $[\![x := \theta]\!]^I = \{(\nu, \nu_x^r) \;:\; r = [\![\theta]\!]^I \nu\} = \{(\nu, \omega) \;:\; \omega = \nu \text{ except } [\![x]\!]^I \omega = [\![\theta]\!]^I \nu\}$
3. $[\![x' := \theta]\!]^I = \{(\nu, \nu_{x'}^r) \;:\; r = [\![\theta]\!]^I \nu\} = \{(\nu, \omega) \;:\; \omega = \nu \text{ except } [\![x']\!]^I \omega = [\![\theta]\!]^I \nu\}$
4. $[\![?\psi]\!]^I = \{(\nu, \nu) \;:\; \nu \in [\![\psi]\!]^I\}$
5. $[\![x' = \theta \,\&\, \psi]\!]^I = \{(\nu, \omega) \;:\; I, \varphi, \models x' = \theta \wedge \psi, \text{ i.e. } \varphi(\zeta) \in [\![x' = \theta \wedge \psi]\!]^I$ *for all $0 \leq \zeta \leq r$, for some function $\varphi : [0, r] \to \mathcal{S}$ of some duration r for which all $\varphi(\zeta)(x') = \frac{d\varphi(t)(x)}{dt}(\zeta)$ exist and $\nu = \varphi(0)$ on $\{x'\}^{\complement}$ and $\omega = \varphi(r)\}$; i.e., φ solves the differential equation and satisfies ψ at all times. In case $r = 0$, the only condition is that $\nu = \omega$ on $\{x'\}^{\complement}$ and $\omega(x') = [\![\theta]\!]^I \omega$ and $\omega \in [\![\psi]\!]^I$.*
6. $[\![\alpha \cup \beta]\!]^I = [\![\alpha]\!]^I \cup [\![\beta]\!]^I$
7. $[\![\alpha; \beta]\!]^I = [\![\alpha]\!]^I \circ [\![\beta]\!]^I = \{(\nu, \omega) : (\nu, \mu) \in [\![\alpha]\!]^I, (\mu, \omega) \in [\![\beta]\!]^I\}$
8. $[\![\alpha^*]\!]^I = ([\![\alpha]\!]^I)^* = \bigcup_{n \in \mathbb{N}} [\![\alpha^n]\!]^I$ *with $\alpha^{n+1} \equiv \alpha^n; \alpha$ and $\alpha^0 \equiv ?true$*

where ρ^* denotes the reflexive transitive closure of relation ρ.

The initial values $\nu(x')$ of differential symbols x' do *not* influence the behavior of $(\nu, \omega) \in [\![x' = \theta \,\&\, \psi]\!]^I$, because they may not be compatible with the time-derivatives for the differential equation, e.g. in $x' := 1; x' = 2$, with a x' mismatch.

Functions and predicates are interpreted by I and are only influenced indirectly by ν through the values of their arguments. So $p(e) \to [x := x + 1]p(e)$ is valid if x is not in e since the change in x does not change whether $p(e)$ is true (Lemma 2). By contrast $p(x) \to [x := x + 1]p(x)$ is invalid, since it is false when $I(p) = \{d \;:\; d \leq 5\}$ and $\nu(x) = 4.5$. If the semantics of p were to depend on the state ν, then there would be no discernible relationship between the truth-values of p in different states, so not even $p \to [x := x + 1]p$ would be valid.

2.3   Static Semantics

The static semantics of d$\mathcal{L}$ and HPs defines some aspects of their behavior that can be read off directly from their syntactic structure without running their programs or evaluating their dynamical effects. The most important aspects of the static semantics concern free or bound occurrences of variables. Bound variables x are those that are bound by $\forall x$ or $\exists x$, but also those that are bound by modalities such as $[x := 5y]$ or $\langle x' = 1 \rangle$ or $[x := 1 \cup x' = 1]$ or $[x := 1 \cup ?true]$.

The notions of free and bound variables are defined by simultaneous induction in the subsequent definitions: free variables for terms $(\mathrm{FV}(\theta))$, formulas $(\mathrm{FV}(\phi))$, and HPs $(\mathrm{FV}(\alpha))$, as well as bound variables for formulas $(\mathrm{BV}(\phi))$ and for HPs $(\mathrm{BV}(\alpha))$. For HPs, there will be a need to distinguish must-bound variables $(\mathrm{MBV}(\alpha))$ that are bound/written to on all executions of α from (may-)bound variables $(\mathrm{BV}(\alpha))$ which are bound on some (not necessarily all) execution paths of α, such as in $[x := 1 \cup (x := 0; y := x + 1)]$, which has bound variables $\{x, y\}$ but must-bound variables only $\{x\}$, because y is not written to in the first choice.

Definition 7 (Bound variable). *The set $BV(\phi) \subseteq \mathcal{V} \cup \mathcal{V}'$ of bound variables of dL formula ϕ is defined inductively as*

$$BV(\theta \geq \eta) = BV(p(\theta_1, \ldots, \theta_k)) = \emptyset$$
$$BV(C(\phi)) = \mathcal{V} \cup \mathcal{V}'$$
$$BV(\neg\phi) = BV(\phi)$$
$$BV(\phi \wedge \psi) = BV(\phi) \cup BV(\psi)$$
$$BV(\forall x\, \phi) = BV(\exists x\, \phi) = \{x\} \cup BV(\phi)$$
$$BV([\alpha]\phi) = BV(\langle\alpha\rangle\phi) = BV(\alpha) \cup BV(\phi)$$

Definition 8 (Free variable). *The set $FV(\theta) \subseteq \mathcal{V} \cup \mathcal{V}'$ of free variables of term θ, i.e. those that occur in θ, is defined inductively as*

$$FV(x) = \{x\}$$
$$FV(x') = \{x'\}$$
$$FV(f(\theta_1, \ldots, \theta_k)) = FV(\theta_1) \cup \cdots \cup FV(\theta_k)$$
$$FV(\theta + \eta) = FV(\theta \cdot \eta) = FV(\theta) \cup FV(\eta)$$
$$FV((\theta)') = FV(\theta) \cup FV(\theta)'$$

The set $FV(\phi)$ of free variables of dL formula ϕ, i.e. all those that occur in ϕ outside the scope of quantifiers or modalities binding it, is defined inductively as

$$FV(\theta \geq \eta) = FV(\theta) \cup FV(\eta)$$
$$FV(p(\theta_1, \ldots, \theta_k)) = FV(\theta_1) \cup \cdots \cup FV(\theta_k)$$
$$FV(C(\phi)) = \mathcal{V} \cup \mathcal{V}'$$
$$FV(\neg\phi) = FV(\phi)$$
$$FV(\phi \wedge \psi) = FV(\phi) \cup FV(\psi)$$
$$FV(\forall x\, \phi) = FV(\exists x\, \phi) = FV(\phi) \setminus \{x\}$$
$$FV([\alpha]\phi) = FV(\langle\alpha\rangle\phi) = FV(\alpha) \cup (FV(\phi) \setminus MBV(\alpha))$$

Soundness requires that $FV([\alpha]\phi)$ is not defined as $FV(\alpha) \cup (FV(\phi) \setminus BV(\alpha))$, otherwise $[x := 1 \cup y := 2]x \geq 1$ would have no free variables, but its truth-value depends on the initial value of x, demanding $FV([x := 1 \cup y := 2]x \geq 1) = \{x\}$.

The static semantics defines which variables are free so may be read ($FV(\alpha)$), which are bound ($BV(\alpha)$) so may be written to somewhere in α, and which are must-bound ($MBV(\alpha)$) so must be written to on all execution paths of α.

Definition 9 (Bound variable). *The set $BV(\alpha) \subseteq \mathcal{V} \cup \mathcal{V}'$ of bound variables of HP α, i.e. all those that may potentially be written to, is defined inductively:*

$$BV(a) = \mathcal{V} \cup \mathcal{V}' \qquad\qquad \textit{for program constant } a$$
$$BV(x := \theta) = \{x\}$$
$$BV(x' := \theta) = \{x'\}$$

$$BV(?\psi) = \emptyset$$
$$BV(x' = \theta \,\&\, \psi) = \{x, x'\}$$
$$BV(\alpha \cup \beta) = BV(\alpha; \beta) = BV(\alpha) \cup BV(\beta)$$
$$BV(\alpha^*) = BV(\alpha)$$

Definition 10 (Must-bound variable). *The set $MBV(\alpha) \subseteq BV(\alpha) \subseteq \mathcal{V} \cup \mathcal{V}'$ of* must-bound variables *of HP α, i.e. all those that must be written to on all paths of α, is defined inductively as*

$$MBV(a) = \emptyset \qquad\qquad \text{for program constant } a$$
$$MBV(\alpha) = BV(\alpha) \qquad\qquad \text{for other atomic HPs } \alpha$$
$$MBV(\alpha \cup \beta) = MBV(\alpha) \cap MBV(\beta)$$
$$MBV(\alpha; \beta) = MBV(\alpha) \cup MBV(\beta)$$
$$MBV(\alpha^*) = \emptyset$$

Definition 11 (Free variable). *The set $FV(\alpha) \subseteq \mathcal{V} \cup \mathcal{V}'$ of* free variables *of HP α, i.e. all those that may potentially be read, is defined inductively as*

$$FV(a) = \mathcal{V} \cup \mathcal{V}' \qquad\qquad \text{for program constant } a$$
$$FV(x := \theta) = FV(x' := \theta) = FV(\theta)$$
$$FV(?\psi) = FV(\psi)$$
$$FV(x' = \theta \,\&\, \psi) = \{x\} \cup FV(\theta) \cup FV(\psi)$$
$$FV(\alpha \cup \beta) = FV(\alpha) \cup FV(\beta)$$
$$FV(\alpha; \beta) = FV(\alpha) \cup (FV(\beta) \setminus MBV(\alpha))$$
$$FV(\alpha^*) = FV(\alpha)$$

Unlike x, the left-hand side x' of differential equations is not added to the free variables of $FV(x' = \theta \,\&\, \psi)$, because its behavior does not depend on the initial value of differential symbols x', only the initial value of x. Free and bound variables are the set of all variables $\mathcal{V}$ and differential symbols $\mathcal{V}'$ for program constants a, because their effect depends on the interpretation I, so may read and write all $FV(a) = BV(a) = \mathcal{V} \cup \mathcal{V}'$ but not on all paths $MBV(a) = \emptyset$. Subsequent results about free and bound variables are, thus, vacuously true when program constants occur. Corresponding observations hold for quantifier symbols.

The static semantics defines which variables are readable or writable. There may not be any run of α in which a variable is read or written to. If $x \notin FV(\alpha)$, though, then α cannot read the value of x. If $x \notin BV(\alpha)$, it cannot write to x.

The *signature*, i.e. set of function, predicate, quantifier symbols, and program constants in ϕ is denoted by $\Sigma(\phi)$ (accordingly for terms and programs). It is defined like $FV(\phi)$ except that all occurrences are free. Variables in $\mathcal{V} \cup \mathcal{V}'$ are interpreted by state ν. The symbols in $\Sigma(\phi)$ are interpreted by interpretation I.

2.4   Correctness of Static Semantics

The following result reflects that HPs have bounded effect: for a variable x to be modified during a run of α, x needs the be a bound variable in *HP* α, i.e.

$x \in BV(\alpha)$, so that α can write to x. The converse is not true, because α may bind a variable x, e.g. by having an assignment to x, that never actually changes the value of x, such as $x := x$ or because the assignment can never be executed.

Lemma 1 (Bound effect lemma). *If $(\nu, \omega) \in [\![\alpha]\!]^I$, then $\nu = \omega$ on $BV(\alpha)^{\complement}$.*

Similarly, only $BV(\phi)$ change their value during the evaluation of formulas.

The value of a term only depends on the values of its free variables. When evaluating a term θ in two states $\nu, \tilde{\nu}$ that differ widely but agree on the free variables $FV(\theta)$ of θ, the values of θ in both states coincide. Accordingly for different interpretations I, J that agree on the symbols $\Sigma(\theta)$ that occur in θ. Recall that all proofs and additional examples are in a companion report [9].

Lemma 2 (Coincidence lemma). *If $\nu = \tilde{\nu}$ on $FV(\theta)$ and $I = J$ on $\Sigma(\theta)$, then $[\![\theta]\!]^I \nu = [\![\theta]\!]^J \tilde{\nu}$.*

By a more subtle argument, the values of $d\mathcal{L}$ formulas also only depend on the values of their free variables. When evaluating $d\mathcal{L}$ formula ϕ in two states ν, $\tilde{\nu}$ that differ but agree on the free variables $FV(\phi)$ of ϕ, the (truth) values of ϕ in both states coincide. Lemmas 3 and 4 are proved by simultaneous induction.

Lemma 3 (Coincidence lemma). *If $\nu = \tilde{\nu}$ on $FV(\phi)$ and $I = J$ on $\Sigma(\phi)$, then $\nu \in [\![\phi]\!]^I$ iff $\tilde{\nu} \in [\![\phi]\!]^J$.*

In a sense, the runs of an HP α also only depend on the values of its free variables, because its behavior cannot depend on the values of variables that it never reads. That is, if $\nu = \tilde{\nu}$ on $FV(\alpha)$ and $(\nu, \omega) \in [\![\alpha]\!]^I$, then there is an $\tilde{\omega}$ such that $(\tilde{\nu}, \tilde{\omega}) \in [\![\alpha]\!]^J$ and ω and $\tilde{\omega}$ agree in some sense. There is a subtlety, though. The resulting states ω and $\tilde{\omega}$ will only continue to agree on $FV(\alpha)$ and the variables that are bound on the particular path that α took for the transition $(\nu, \omega) \in [\![\alpha]\!]^I$. On variables z that are neither free (so the initial states ν and $\tilde{\nu}$ have not been assumed to coincide) nor bound on the particular path that α took, ω and $\tilde{\omega}$ may continue to disagree, because z has not been written to. Yet, ω and $\tilde{\omega}$ agree on the variables that are bound on *all* paths of α, rather than somewhere in α. That is on the must-bound variables of α. If initial states agree on (at least) all free variables $FV(\alpha)$ that HP α may read, then the final states agree on those as well as on all variables that α must write, i.e. on $MBV(\alpha)$.

Lemma 4 (Coincidence lemma). *If $\nu = \tilde{\nu}$ on $V \supseteq FV(\alpha)$ and $I = J$ on $\Sigma(\alpha)$ and $(\nu, \omega) \in [\![\alpha]\!]^I$, then there is an $\tilde{\omega}$ such that $(\tilde{\nu}, \tilde{\omega}) \in [\![\alpha]\!]^J$ and $\omega = \tilde{\omega}$ on $V \cup MBV(\alpha)$.*

3   Uniform Substitutions

The uniform substitution rule US_1 from first-order logic [2, §35,40] substitutes *all* occurrences of predicate $p(\cdot)$ by a formula $\psi(\cdot)$, i.e. it replaces all occurrences of $p(\theta)$, for any (vectorial) term θ, by the corresponding $\psi(\theta)$ simultaneously:

$$(\text{US}_1) \quad \frac{\phi}{\phi_{p(\cdot)}^{\psi(\cdot)}} \qquad\qquad (\text{US}) \quad \frac{\phi}{\sigma(\phi)}$$

Rule US$_1$ [8] requires all relevant substitutions of $\psi(\theta)$ for $p(\theta)$ to be admissible and requires that no $p(\theta)$ occurs in the scope of a quantifier or modality binding a variable of $\psi(\theta)$ other than the occurrences in θ; see [2, §35,40].

This section considers a constructive definition of this proof rule that is more general: US. The d$\mathcal{L}$ calculus uses uniform substitutions that affect terms, formulas, and programs. A *uniform substitution* σ is a mapping from expressions of the form $f(\cdot)$ to terms $\sigma f(\cdot)$, from $p(\cdot)$ to formulas $\sigma p(\cdot)$, from $C(\_)$ to formulas $\sigma C(\_)$, and from program constants a to HPs σa. Vectorial extensions are accordingly for uniform substitutions of other arities $k \geq 0$. Here $\cdot$ is a reserved function symbol of arity zero and $\_$ a reserved quantifier symbol of arity zero. Figure 1 defines the result $\sigma(\phi)$ of applying to a d$\mathcal{L}$ formula ϕ the *uniform substitution* σ that uniformly replaces all occurrences of function f by a (instantiated) term and all occurrences of predicate p or quantifier C by a (instantiated) formula as well as of program constant a by a program. The notation $\sigma f(\cdot)$ denotes the replacement for $f(\cdot)$ according to σ, i.e. the value $\sigma f(\cdot)$ of function σ at $f(\cdot)$. By contrast, $\sigma(\phi)$ denotes the result of applying σ to ϕ according to Fig. 1 (likewise for $\sigma(\theta)$ and $\sigma(\alpha)$). The notation $f \in \sigma$ signifies that σ replaces f, i.e. $\sigma f(\cdot) \neq f(\cdot)$. Finally, σ is a total function when augmented with $\sigma g(\cdot) = g(\cdot)$ for all $g \notin \sigma$. Accordingly for predicate symbols, quantifiers, and program constants.

Definition 12 (Admissible uniform substitution). *The uniform substitution σ is U-admissible for ϕ (or θ or α, respectively) with respect to the set $U \subseteq \mathcal{V} \cup \mathcal{V}'$ iff $FV(\sigma|_{\Sigma(\phi)}) \cap U = \emptyset$, where $\sigma|_{\Sigma(\phi)}$ is the restriction of σ that only replaces symbols that occur in ϕ and $FV(\sigma) = \bigcup_{f \in \sigma} FV(\sigma f(\cdot)) \cup \bigcup_{p \in \sigma} FV(\sigma p(\cdot))$ are the* free variables *that σ introduces. The uniform substitution σ is admissible for ϕ (or θ or α, respectively) iff all admissibility conditions during its application according to Fig. 1 hold, which check that the bound variables U of each operator are not free in the substitution on its arguments, i.e. σ is U-admissible. Otherwise the substitution clashes so its result $\sigma(\phi)$ ($\sigma(\theta)$ or $\sigma(\alpha)$) is not defined.*

US is only applicable if σ is admissible for ϕ. In all subsequent results, all applications of uniform substitutions are required to be defined (no clash).

3.1   Correctness of Uniform Substitutions

Let I_p^R denote the interpretation that agrees with interpretation I except for the interpretation of predicate symbol p, which is changed to $R \subseteq \mathbb{R}$. Accordingly for predicate symbols of other arities, for function symbols f, and quantifiers C.

Corollary 1 (Substitution adjoints). *The adjoint interpretation $\sigma_\nu^* I$ to substitution σ for I, ν is the interpretation that agrees with I except that for each function symbol $f \in \sigma$, predicate symbol $p \in \sigma$, quantifier $C \in \sigma$, and program constant $a \in \sigma$:*

$$\sigma(x) = x \qquad \text{for variable } x \in \mathcal{V}$$
$$\sigma(x') = x' \qquad \text{for differential symbol } x' \in \mathcal{V}'$$
$$\sigma(f(\theta)) = (\sigma(f))(\sigma(\theta)) \stackrel{\text{def}}{=} \{\cdot \mapsto \sigma(\theta)\}(\sigma f(\cdot)) \quad \text{for function symbol } f \in \sigma$$
$$\sigma(g(\theta)) = g(\sigma(\theta)) \qquad \text{for function symbol } g \notin \sigma$$
$$\sigma(\theta + \eta) = \sigma(\theta) + \sigma(\eta)$$
$$\sigma(\theta \cdot \eta) = \sigma(\theta) \cdot \sigma(\eta)$$
$$\sigma((\theta)') = (\sigma(\theta))' \qquad \text{if } \sigma \ \mathcal{V} \cup \mathcal{V}'\text{-admissible for } \theta$$

$$\sigma(\theta \geq \eta) \equiv \sigma(\theta) \geq \sigma(\eta)$$
$$\sigma(p(\theta)) \equiv (\sigma(p))(\sigma(\theta)) \stackrel{\text{def}}{=} \{\cdot \mapsto \sigma(\theta)\}(\sigma p(\cdot)) \quad \text{for predicate symbol } p \in \sigma$$
$$\sigma(q(\theta)) \equiv q(\sigma(\theta)) \qquad \text{for predicate symbol } q \notin \sigma$$
$$\sigma(C(\phi)) \equiv \sigma(C)(\sigma(\phi)) \stackrel{\text{def}}{=} \{_\_ \mapsto \sigma(\phi)\}(\sigma C(\_)) \ \text{if } \sigma \ \mathcal{V} \cup \mathcal{V}'\text{-admissible for } \phi, \ C \in \sigma$$
$$\sigma(C(\phi)) \equiv C(\sigma(\phi)) \qquad \text{if } \sigma \ \mathcal{V} \cup \mathcal{V}'\text{-admissible for } \phi, \ C \notin \sigma$$
$$\sigma(\neg\phi) \equiv \neg\sigma(\phi)$$
$$\sigma(\phi \wedge \psi) \equiv \sigma(\phi) \wedge \sigma(\psi)$$
$$\sigma(\forall x\, \phi) = \forall x\, \sigma(\phi) \qquad \text{if } \sigma \ \{x\}\text{-admissible for } \phi$$
$$\sigma(\exists x\, \phi) = \exists x\, \sigma(\phi) \qquad \text{if } \sigma \ \{x\}\text{-admissible for } \phi$$
$$\sigma([\alpha]\phi) = [\sigma(\alpha)]\sigma(\phi) \qquad \text{if } \sigma \ \mathrm{BV}(\sigma(\alpha))\text{-admissible for } \phi$$
$$\sigma(\langle\alpha\rangle\phi) = \langle\sigma(\alpha)\rangle\sigma(\phi) \qquad \text{if } \sigma \ \mathrm{BV}(\sigma(\alpha))\text{-admissible for } \phi$$

$$\sigma(a) \equiv \sigma a \qquad \text{for program constant } a \in \sigma$$
$$\sigma(b) \equiv b \qquad \text{for program constant } b \notin \sigma$$
$$\sigma(x := \theta) \equiv x := \sigma(\theta)$$
$$\sigma(x' := \theta) \equiv x' := \sigma(\theta)$$
$$\sigma(x' = \theta \,\&\, \psi) \equiv x' = \sigma(\theta) \,\&\, \sigma(\psi) \qquad \text{if } \sigma \ \{x, x'\}\text{-admissible for } \theta, \psi$$
$$\sigma(?\psi) \equiv ?\sigma(\psi)$$
$$\sigma(\alpha \cup \beta) \equiv \sigma(\alpha) \cup \sigma(\beta)$$
$$\sigma(\alpha; \beta) \equiv \sigma(\alpha); \sigma(\beta) \qquad \text{if } \sigma \ \mathrm{BV}(\sigma(\alpha))\text{-admissible for } \beta$$
$$\sigma(\alpha^*) \equiv (\sigma(\alpha))^* \qquad \text{if } \sigma \ \mathrm{BV}(\sigma(\alpha))\text{-admissible for } \alpha$$

Fig. 1. Recursive application of uniform substitution σ

$$\sigma_\nu^* I(f) : \mathbb{R} \to \mathbb{R}; \ d \mapsto \llbracket \sigma f(\cdot) \rrbracket^{I_d^d} \nu$$

$$\sigma_\nu^* I(p) = \{d \in \mathbb{R} \ : \ \nu \in \llbracket \sigma p(\cdot) \rrbracket^{I_d^d}\}$$

$$\sigma_\nu^* I(C) : \wp(\mathbb{R}) \to \wp(\mathbb{R}); \ R \mapsto \llbracket \sigma C(\_) \rrbracket^{I_R^R}$$

$$\sigma_\nu^* I(a) = \llbracket \sigma a \rrbracket^I$$

If $\nu = \omega$ on $FV(\sigma)$, then $\sigma_\nu^* I = \sigma_\omega^* I$. If σ is U-admissible for ϕ (or θ or α, respectively) and $\nu = \omega$ on $U^\complement$, then

$$\llbracket \theta \rrbracket^{\sigma_\nu^* I} = \llbracket \theta \rrbracket^{\sigma_\omega^* I} \ \textit{i.e.} \ \llbracket \theta \rrbracket^{\sigma_\nu^* I} \mu - \llbracket \theta \rrbracket^{\sigma_\omega^* I} \mu \ \textit{for all } \mu$$
$$\llbracket \phi \rrbracket^{\sigma_\nu^* I} = \llbracket \phi \rrbracket^{\sigma_\omega^* I}$$
$$\llbracket \alpha \rrbracket^{\sigma_\nu^* I} = \llbracket \alpha \rrbracket^{\sigma_\omega^* I}$$

Substituting equals for equals is sound by the compositional semantics of d$\mathcal{L}$. The more general uniform substitutions are still sound, because interpretations of uniform substitutes correspond to interpretations of their adjoints. The semantic modification of adjoint interpretations has the same effect as the syntactic uniform substitution, proved by simultaneous induction.

Lemma 5 (Uniform substitution lemma). *The uniform substitution σ and its adjoint interpretation $\sigma_\nu^* I, \nu$ to σ for I, ν have the same term semantics:*

$$[\![\sigma(\theta)]\!]^I \nu = [\![\theta]\!]^{\sigma_\nu^* I} \nu$$

Lemma 6 (Uniform substitution lemma). *The uniform substitution σ and its adjoint interpretation $\sigma_\nu^* I, \nu$ to σ for I, ν have the same formula semantics:*

$$\nu \in [\![\sigma(\phi)]\!]^I \text{ iff } \nu \in [\![\phi]\!]^{\sigma_\nu^* I}$$

Lemma 7 (Uniform substitution lemma). *The uniform substitution σ and its adjoint interpretation $\sigma_\nu^* I, \nu$ to σ for I, ν have the same program semantics:*

$$(\nu, \omega) \in [\![\sigma(\alpha)]\!]^I \text{ iff } (\nu, \omega) \in [\![\alpha]\!]^{\sigma_\nu^* I}$$

3.2   Soundness

The uniform substitution lemmas are the key insights for the soundness of US. US is only applicable if the uniform substitution is defined (does not clash).

Theorem 1 (Soundness of uniform substitution). *US is sound and so is its special case US$_1$. That is, if their premise is valid, then so is their conclusion.*

4   Differential Dynamic Logic Axioms

Proof rules and axioms for a Hilbert-type axiomatization of d$\mathcal{L}$ from prior work [6] are shown in Fig. 2, except that, thanks to rule US, axioms and rules now comprise the finite list of d$\mathcal{L}$ formulas in Fig. 2 as opposed to an infinite collection of axioms from a finite list of axiom schemata along with schema variables, side conditions, and implicit instantiation rules. Soundness of the axioms in Fig. 2 follows from the soundness of corresponding axiom schemata [6], but would be easier to prove standalone, because it is a finite list of formulas without the need to prove soundness for all their instantiations. The rules in Fig. 2 are *axiomatic rules*, i.e. pairs of concrete formulas instantiated by US. Further, $\bar{x}$ is the vector of all relevant variables, which is finite-dimensional, or, in practice, considered as a built-in vectorial term. Proofs in the uniform substitution d$\mathcal{L}$ calculus use US (and bound renaming such as $\forall x\, p(x) \leftrightarrow \forall y\, p(y)$) to instantiate the axioms from Fig. 2 to the required form. CT,CQ,CE are congruence rules, which are included for efficiency to use axioms in any context even if not needed for completeness.

Real Quantifiers. Besides (decidable) real arithmetic (whose use is denoted ℝ), complete axioms for first-order logic can be adopted to express universal instantiation $\forall$i, distributivity $\forall \rightarrow$, and vacuous quantification V$_\forall$.

$(\forall \text{i}) \; (\forall x\, p(x)) \rightarrow p(f)$
$(\forall \rightarrow) \; \forall x\, (p(x) \rightarrow q(x)) \rightarrow (\forall x\, p(x) \rightarrow \forall x\, q(x))$
$(\text{V}_\forall) \; p \rightarrow \forall x\, p$

$$\langle \cdot \rangle \ \langle a \rangle p(\bar{x}) \leftrightarrow \neg[a]\neg p(\bar{x})$$

$$[:=] \ [x := f]p(x) \leftrightarrow p(f)$$

$$[?] \ [?q]p \leftrightarrow (q \to p)$$

$$[\cup] \ [a \cup b]p(\bar{x}) \leftrightarrow [a]p(\bar{x}) \wedge [b]p(\bar{x})$$

$$[;] \ [a;b]p(\bar{x}) \leftrightarrow [a][b]p(\bar{x})$$

$$[*] \ [a^*]p(\bar{x}) \leftrightarrow p(\bar{x}) \wedge [a][a^*]p(\bar{x})$$

$$K \ [a](p(\bar{x}) \to q(\bar{x})) \to ([a]p(\bar{x}) \to [a]q(\bar{x}))$$

$$I \ [a^*](p(\bar{x}) \to [a]p(\bar{x})) \to (p(\bar{x}) \to [a^*]p(\bar{x}))$$

$$V \ p \to [a]p$$

$$G \ \frac{p(\bar{x})}{[a]p(\bar{x})}$$

$$\forall \ \frac{p(x)}{\forall x \, p(x)}$$

$$MP \ \frac{p \to q \quad p}{q}$$

$$CT \ \frac{f(\bar{x}) = g(\bar{x})}{c(f(\bar{x})) = c(g(\bar{x}))}$$

$$CQ \ \frac{f(\bar{x}) = g(\bar{x})}{p(f(\bar{x})) \leftrightarrow p(g(\bar{x}))}$$

$$CE \ \frac{p(\bar{x}) \leftrightarrow q(\bar{x})}{C(p(\bar{x})) \leftrightarrow C(q(\bar{x}))}$$

Fig. 2. Differential dynamic logic axioms and proof rules

The Significance of Clashes. US clashes for substitutions that introduce a free variable into a bound context. Even an occurrence of $p(x)$ in a context where x is bound does not allow mentioning x in the replacement except in the $\cdot$ places:

$$\text{clash} \not{\frac{[x := f]p(x) \leftrightarrow p(f)}{[x := x+1]x \neq x \leftrightarrow x+1 \neq x}} \qquad \sigma = \{f \mapsto x+1, p(\cdot) \mapsto (\cdot \neq x)\}$$

US can directly handle even nontrivial binding structures, though, e.g. from $[:=]$ with the substitution $\sigma = \{f \mapsto x^2, p(\cdot) \mapsto [(z := \cdot + z)^*; z := \cdot + yz]y \geq \cdot\}$:

$$\text{US} \frac{[x := f]p(x) \leftrightarrow p(f)}{[x := x^2][(z := x+z)^*; z := x+yz]y \geq x \leftrightarrow [(z := x^2+z)^*; z := x^2+yz]y \geq x^2}$$

5 Differential Equations and Differential Axioms

Section 4 leverages the first-order features of dℒ and US to obtain a finite list of axioms without side-conditions. They lack axioms for differential equations, though. Classical calculi for dℒ have axioms for replacing differential equations with a quantifier for time $t \geq 0$ and an assignment for their solutions $\bar{x}(t)$ [4,6]. Besides being limited to simple differential equations, such axioms have the inherent side-condition "if $\bar{x}(t)$ is a solution of the differential equation $x' = \theta$ with symbolic initial value x". Such a side-condition is more difficult than occurrence and read/write conditions, but equally soundness-critical. This section leverages US and the new differential forms in dℒ to obtain a logically internalized version of differential invariants and related proof rules for differential equations [5,7] as axioms (without schema variables and free of side-conditions). These axioms can prove properties of more general "unsolvable" differential equations. They can also prove all properties of differential equations that can be proved with solutions [7] while guaranteeing correctness of the solution as part of the proof.

5.1   Differentials: Invariants, Cuts, Effects, and Ghosts

Figure 3 shows axioms for proving properties of differential equations (DW–DS) as well as axioms for differential substitutions ($[':=]$), and differential axioms for differentials ($+'$, $\cdot'$, $\circ'$). Axioms identifying $(x)' = x'$ for variables $x \in \mathcal{V}$ and $(f)' = 0$ for functions f and number literals of arity 0 are used implicitly. Some axioms use reverse implications ($\phi \leftarrow \psi$) $\equiv$ ($\psi \rightarrow \phi$) for emphasis.

DW $[x' = f(x) \,\&\, q(x)]q(x)$

DC $([x' = f(x) \,\&\, q(x)]p(x) \leftrightarrow [x' = f(x) \,\&\, q(x) \wedge r(x)]p(x)) \leftarrow [x' = f(x) \,\&\, q(x)]r(x)$

DE $[x' = f(x) \,\&\, q(x)]p(x, x') \leftrightarrow [x' = f(x) \,\&\, q(x)][x' := f(x)]p(x, x')$

DI $[x' = f(x) \,\&\, q(x)]p(x) \leftarrow (q(x) \rightarrow p(x) \wedge [x' = f(x) \,\&\, q(x)](p(x))')$

DG $[x' = f(x) \,\&\, q(x)]p(x) \leftrightarrow \exists y\, [x' = f(x), y' = a(x)y + b(x) \,\&\, q(x)]p(x)$

DS $[x' = f \,\&\, q(x)]p(x) \leftrightarrow \forall t{\geq}0 \left((\forall 0{\leq}s{\leq}t\, q(x + fs)) \rightarrow [x := x + ft]p(x)\right)$

$[':=]$ $[x' := f]p(x') \leftrightarrow p(f)$

$+'$ $(f(\bar{x}) + g(\bar{x}))' = (f(\bar{x}))' + (g(\bar{x}))'$

$\cdot'$ $(f(\bar{x}) \cdot g(\bar{x}))' = (f(\bar{x}))' \cdot g(\bar{x}) + f(\bar{x}) \cdot (g(\bar{x}))'$

$\circ'$ $[y := g(x)][y' := 1]\left((f(g(x)))' = (f(y))' \cdot (g(x))'\right)$

Fig. 3. Differential equation axioms and differential axioms

Differential weakening axiom DW internalizes that differential equations can never leave their evolution domain $q(x)$. DW implies[3] $[x' = f(x) \,\&\, q(x)]p(x) \leftrightarrow [x' = f(x) \,\&\, q(x)](q(x) \rightarrow p(x))$ also called DW, whose (right) assumption is best proved by G. The *differential cut* axiom DC is a cut for differential equations. It internalizes that differential equations staying in $r(x)$ stay in $p(x)$ iff $p(x)$ always holds after the differential equation that is restricted to the smaller evolution domain $\&\, q(x) \wedge r(x)$. DC is a differential variant of modal modus ponens K.

Differential effect axiom DE internalizes that the effect on differential symbols along a differential equation is a differential assignment assigning the right-hand side $f(x)$ to the left-hand side x'. Axiom DI internalizes *differential invariants*, i.e. that a differential equation stays in $p(x)$ if it starts in $p(x)$ and if its differential $(p(x))'$ always holds after the differential equation $x' = f(x) \,\&\, q(x)$. The differential equation also vacuously stays in $p(x)$ if it starts outside $q(x)$, since it is stuck then. The (right) assumption of DI is best proved by DE to select the appropriate vector field $x' = f(x)$ for the differential $(p(x))'$ and a subsequent DW,G to make the evolution domain constraint $q(x)$ available as an assumption. For simplicity, this paper focuses on atomic postconditions for which $(\theta \geq \eta)' \equiv (\theta > \eta)' \equiv (\theta)' \geq (\eta)'$ and $(\theta = \eta)' \equiv (\theta \neq \eta)' \equiv (\theta)' = (\eta)'$, etc. Axiom DG internalizes *differential ghosts*, i.e. that additional differential equations can be added if

---

[3] $[x' = f(x) \,\&\, q(x)](q(x) \rightarrow p(x)) \rightarrow [x' = f(x) \,\&\, q(x)]p(x)$ derives by K from DW. The converse $[x' = f(x) \,\&\, q(x)]p(x) \rightarrow [x' = f(x) \,\&\, q(x)](q(x) \rightarrow p(x))$ derives by K since G derives $[x' = f(x) \,\&\, q(x)](p(x) \rightarrow (q(x) \rightarrow p(x)))$.

their solution exists long enough. Axiom DS solves differential equations with the help of DG,DC. Vectorial generalizations to systems of differential equations are possible for the axioms in Fig. 3.

The following proof proves a property of a differential equation using differential invariants without having to solve that differential equation. One use of US is shown explicitly, other uses of US are similar for DI,DE,$[':=]$ instances.

$$
\text{DI} \cfrac{\text{DE} \cfrac{\text{CE} \cfrac{\text{G} \cfrac{[':=] \cfrac{\text{R} \cfrac{*}{x^3 \cdot x + x \cdot x^3 \geq 0}}{[x' := x^3] x' \cdot x + x \cdot x' \geq 0}}{[x' = x^3][x' := x^3] x' \cdot x + x \cdot x' \geq 0}}{\text{CQ} \cfrac{\text{US} \cfrac{*}{\cfrac{(f(\bar{x}) \cdot g(\bar{x}))' = (f(\bar{x}))' \cdot g(\bar{x}) + f(\bar{x}) \cdot (g(\bar{x}))'}{\cfrac{(x \cdot x)' = (x)' \cdot x + x \cdot (x)'}{(x \cdot x)' = x' \cdot x + x \cdot x'}}}}{\cfrac{(x \cdot x)' \geq 0 \leftrightarrow x' \cdot x + x \cdot x' \geq 0}{(x \cdot x \geq 1)' \leftrightarrow x' \cdot x + x \cdot x' \geq 0}}}{[x' = x^3][x' := x^3](x \cdot x \geq 1)'}}{[x' = x^3](x \cdot x \geq 1)'}}{x \cdot x \geq 1 \rightarrow [x' = x^3] x \cdot x \geq 1}
$$

Previous calculi [5, 7] collapse this proof into a single proof step with complicated built-in operator implementations that silently perform the same reasoning in an opaque way. The approach presented here combines separate axioms to achieve the same effect in a modular way, because they have individual responsibilities internalizing separate logical reasoning principles in *differential-form* d$\mathcal{L}$.

5.2   Differential Substitution Lemmas

The key insight for the soundness of DI is that the analytic time-derivative of the value of a term η along a differential equation $x' = \theta \,\&\, \psi$ agrees with the values of its differential $(\eta)'$ along the vector field of that differential equation.

Lemma 8 (Differential lemma). *If $I, \varphi \models x' = \theta \wedge \psi$ holds for some flow $\varphi : [0, r] \to \mathcal{S}$ of any duration $r > 0$, then for all $0 \leq \zeta \leq r$:*

$$
[\![(\eta)']\!]^I \varphi(\zeta) = \frac{\mathrm{d}[\![\eta]\!]^I \varphi(t)}{\mathrm{d}t}(\zeta)
$$

The key insight for the soundness of differential effects DE is that differential assignments mimicking the differential equation are vacuous along that differential equation. The differential substitution resulting from a subsequent use of $[':=]$ is crucial to relay the values of the time-derivatives of the state variables x along a differential equation by way of their corresponding differential symbol x'. In combination, this makes it possible to soundly substitute the right-hand side of a differential equation for its left-hand side in a proof.

Lemma 9 (Differential assignment). *If $I, \varphi \models x' = \theta \wedge \psi$ holds for some flow $\varphi : [0, r] \to \mathcal{S}$ of any duration $r \geq 0$, then*

$$
I, \varphi \models \phi \leftrightarrow [x' := \theta] \phi
$$

The final insights for differential invariant reasoning for differential equations are syntactic ways of computing differentials, which can be internalized as axioms $(+', \cdot', \circ')$, since differentials are syntactically represented in differential-form d$\mathcal{L}$.

Lemma 10 (Derivations). *The following equations of differentials are valid:*

$$(f)' = 0 \qquad\qquad \textit{for arity 0 functions/numbers } f \quad (1)$$
$$(x)' = x' \qquad\qquad \textit{for variables } x \in \mathcal{V} \quad (2)$$
$$(\theta + \eta)' = (\theta)' + (\eta)' \quad (3)$$
$$(\theta \cdot \eta)' = (\theta)' \cdot \eta + \theta \cdot (\eta)' \quad (4)$$
$$[y := \theta]\,[y' := 1]\big((f(\theta))' = (f(y))' \cdot (\theta)'\big) \quad \textit{for } y, y' \notin \theta \quad (5)$$

5.3 Soundness

Theorem 2 (Soundness). *The d$\mathcal{L}$ axioms and proof rules in Figs. 2 and 3 are sound, i.e. the axioms are valid formulas and the conclusion of a rule is valid if its premises are. All US instances of the proof rules (with $FV(\sigma) = \emptyset$) are sound.*

6 Conclusions

With differential forms for local reasoning about differential equations, uniform substitutions lead to a simple and modular proof calculus for differential dynamic logic that is entirely based on axioms and axiomatic rules, instead of soundness-critical schema variables with side-conditions in axiom schemata. The US calculus is straightforward to implement and enables flexible reasoning with axioms by contextual equivalence. Efficiency can be regained by tactics that combine multiple axioms and rebalance the proof to obtain short proof search branches. Contextual equivalence rewriting for implications is possible when adding monotone quantifiers C whose substitution instances limit _ to positive polarity.

Acknowledgment. I thank the anonymous reviewers for their helpful feedback.

References

1. Church, A.: A formulation of the simple theory of types. J. Symb. Log. **5**(2), 56–68 (1940)
2. Church, A.: Introduction to Mathematical Logic, vol. I. Princeton University Press, Princeton (1956)
3. Henkin, L.: Banishing the rule of substitution for functional variables. J. Symb. Log. **18**(3), 201–208 (1953)
4. Platzer, A.: Differential dynamic logic for hybrid systems. J. Autom. Reas. **41**(2), 143–189 (2008)
5. Platzer, A.: Differential-algebraic dynamic logic for differential-algebraic programs. J. Log. Comput. **20**(1), 309–352 (2010)
6. Platzer, A.: The complete proof theory of hybrid systems. In: LICS, pp. 541–550. IEEE (2012)
7. Platzer, A.: The structure of differential invariants and differential cut elimination. Log. Meth. Comput. Sci. **8**(4), 1–38 (2012)
8. Platzer, A.: Differential game logic. CoRR abs/1408.1980 (2014)
9. Platzer, A.: A uniform substitution calculus for differential dynamic logic. CoRR abs/1503.01981 (2015)

Program Synthesis Using Dual Interpretation

Ashish Tiwari[✉], Adrià Gascón, and Bruno Dutertre

SRI International, Menlo Park, CA, USA
tiwari@csl.sri.com

Abstract. We present an approach for component-based program synthesis that uses two distinct interpretations for the symbols in the program. The first interpretation defines the semantics of the program. It is used to specify functional requirements. The second interpretation is used to capture nonfunctional requirements that may vary by application. We present a language for program synthesis from components that uses dual interpretation. We reduce the synthesis problem to an exists-forall problem, which is solved using the exists-forall extension of the SMT-solver Yices. We use our approach to synthesize bitvector manipulation programs, padding-based encryption schemes, and block cipher modes of operations.

Keywords: Program synthesis · Syntax-guided synthesis · Abstract interpretation · Encryption · SMT solving · Exists-forall solving

1 Introduction

A program is often given a concrete semantics that forms the basis of all reasoning and analysis. This semantics is typically defined over a *concrete domain* or an *abstraction* of this concrete domain, as in type checking and abstract interpretation [4]. In first-order logic, semantics is specified by a collection of structures, but there is often a single canonical structure, such as a Herbrand model minimal in some ordering, which forms the basis of reasoning. Are there any benefits in using two or more different and incomparable structures as bases for reasoning?

Type systems in programming languages can be viewed as providing second interpretations. However, they are mostly abstractions of the concrete semantics. Examples of second interpretations unrelated to the concrete semantics can be found in language-based security where ideas such as security-type systems and, more generally, semantic-based security are explored [15]. Many security properties, for example noninterference, are not concerned with the functionality of a program, but *how* it implements such functionality in the presence of a malicious

This work was sponsored, in part, by ONR under subaward 60106452-107484-C under prime grant N00014-12-1-0914, and the National Science Foundation under grant CCF-1423296. The views, opinions, and/or findings contained in this report are those of the authors and should not be interpreted as representing the official views or policies, either expressed or implied, of the funding agencies.

© Springer International Publishing Switzerland 2015
A.P. Felty and A. Middeldorp (Eds.): CADE-25, LNAI 9195, pp. 482–497, 2015.
DOI: 10.1007/978-3-319-21401-6_33

adversary. The analysis of such *nonfunctional properties* usually benefits from having a second semantics modeling the attacker's view of the program.

We use dual interpretations for performing program synthesis. We illustrate our approach by synthesizing cryptographic schemes. A correct cryptographic scheme must satisfy two different properties. First, every encryption scheme should have a corresponding decryption scheme. This functional correctness property can be decided using the concrete semantics of the program. Second, we must guarantee that the encryption scheme is secure (in some attacker model). This property is not functional and it is difficult to specify using the concrete semantics. Instead, it is sometimes possible to reason (conservatively) about security properties using a second, completely different, meaning of the program. This observation motivates the dual interpretation approach of this paper.

Ostensibly, the prospect of having two different semantics for programs seems to be a potentially troublesome idea. However, in theory, it is not much different from having just one concrete semantics since one could merge the two semantics by considering the product of the two domains. For reasoning though, it is still beneficial to consider the two semantics separately.

Our goal is to automatically generate correct programs using components or functions from a library. The synthesized program must satisfy both functional and nonfunctional requirements. We use a primary concrete semantics to specify the functional requirements and a second alternate semantics to specify the nonfunctional requirements. The second interpretation is also used to restrict the set of candidate programs. We present a language for writing program sketches and specifying both types of requirements. We solve the synthesis problem by compiling it to an exists-forall formula, which our tool currently solves using the exists-forall solver of Yices [5]. We provide experimental evidence of the value of the language and our synthesis approach by presenting a collection of examples that were automatically synthesized using our tool.

Related Work. Our work is inspired by recent progress in the area of program synthesis. Synthesizing a program from an abstract specification is not achievable in practice but *template-based* synthesis has shown a lot of promise [1,17,18]. In this approach, the designer provides a *template* that captures the shape of the intended solution(s) together with the specification. A synthesis algorithm fills in the details. This general idea has been successfully applied to several domains. For example, imperative programs can be obtained from a given sketch, as long as their intended behavior is also provided [16]; efficient bitvector manipulations can be synthesized from naïve implementations [10]; agent behavior in distributed algorithms can be synthesized from a description of a global goal [8]; and deobfuscated code can be obtained using similar ideas [11]. Although all these applications rely on template-based synthesis, different synthesis algorithms are used in different domains. Logically, most of the synthesis algorithms are solving an *exists-forall* problem.

Recently, we have implemented an exists-forall solver as part of the SMT-solver Yices [5,7]. In this paper, we present a language for specifying sketches, which are partially specified programs, but unlike any previous work on synthesis, we use two different interpretations for the program symbols. We perform synthesis by explicitly generating an exists-forall formula in Yices syntax and then use Yices to solve it. Unrelated to synthesis, ideas similar to dual interpretation have appeared in the form of "derived programs" [19] and "shadow variables" [14].

The component-based program synthesis problem was formulated in [10], but the interest in [10] was only on functional requirements. Here, we also consider nonfunctional requirements, which forces us to reason with two different semantics of the same program. We use benchmarks from [10] in this paper.

2   Component-Based Program Synthesis

We assume that programs are constructed from a library of components. We are interested in constructing straight-line programs using the library. A straight-line program can be viewed as a term over the signature of the library.

Let Σ be a signature consisting of constant and function symbols. Let Vars denote a set of (input) variables. Let $\mathrm{Terms}(\Sigma, \mathrm{Vars})$ denote the set of all terms defined over the signature Σ and variables Vars. A term t in $\mathrm{Terms}(\Sigma, \mathrm{Vars})$ naturally corresponds to a straight-line program whose inputs are the variables occurring in t. For example, the term $f(g(x), x)$ corresponds to the following program:

$$\texttt{input } x; \;\; y_1 := g(x); \;\; y_2 := f(y_1, x); \;\; \texttt{output } y_2$$

We give meaning to programs by using a *structure* (Dom, Int) where the domain Dom is a nonempty set, and the interpretation Int maps every constant $c \in \Sigma$ to an element $c^{\mathrm{Int}} \in \mathrm{Dom}$ and every function symbol $f \in \Sigma$ with arity n to a concrete function $f^{\mathrm{Int}} : \mathrm{Dom}^n \mapsto \mathrm{Dom}$. The mapping Int is extended to all ground terms $\mathrm{Terms}(\Sigma \cup \mathrm{Dom}, \emptyset)$ is the usual way: if $f \in \Sigma$ then $f(t_1, \ldots, t_n)^{\mathrm{Int}}$ is defined as $f^{\mathrm{Int}}(t_1^{\mathrm{Int}}, \ldots, t_n^{\mathrm{Int}})$ recursively; and if $e \in \mathrm{Dom}$ then $e^{\mathrm{Int}} = e$.

In the program-synthesis terminology, the symbols Σ and their interpretation Int form the *library* of *components*.

A substitution $\sigma : \mathrm{Vars} \mapsto \mathrm{Dom}$ maps the (input) variables Vars to elements in Dom. The meaning of a (program) term $t \in \mathrm{Terms}(\Sigma, \mathrm{Vars})$ is defined as follows: given an assignment σ to input variables, the program t computes the output $(t\sigma)^{\mathrm{Int}}$, where $t\sigma \in \mathrm{Terms}(\Sigma \cup \mathrm{Dom}, \emptyset)$ denotes the result of replacing every x by $\sigma(x)$ in t.

Example 1. Let $\Sigma = \{(f : 2), (g : 2), (h : 1), (c : 3), (d : 2), (e : 0)\}$ be a signature. Let Dom be the set of bitvectors of an arbitrary but constant length k, and let Int be the function

$$\begin{array}{lll} \mathrm{Int}(f) = \texttt{bv-xor} & \mathrm{Int}(g) = \texttt{bv-and} & \mathrm{Int}(h) = \texttt{bv-neg} \\ \mathrm{Int}(c) = \texttt{ite} & \mathrm{Int}(d) = \texttt{bv-sgt} & \mathrm{Int}(e) = 0\ldots01 \end{array}$$

where bv-xor is bitwise xor, bv-and is bitwise and, bv-neg is the negative in 2s complement, ite is if-then-else, bv-sgt is signed greater-than, and $0\ldots01$ is the bitvector representing 1. Consider the term $s = c(d(x,y),x,y)$. Under the meaning defined by the structure (Dom, Int), s corresponds to a program computing the maximum of two binary integers of length k. Note that the term s is *equivalent* to the term $t = f(g(f(x,y),h(d(x,y))),y)$ under the semantics defined by (Dom, Int), but t does not use c.

We want to synthesize programs that satisfy functional, as well as, non-functional, requirements.

2.1 Functional Requirements

A functional requirement states that the output value computed by the program satisfies some property. A functional requirement is specified by a subset $\phi_{\texttt{fspec}} \subseteq \texttt{Dom}^{n+1}$. A term $t \in \texttt{Terms}(\Sigma, \{i_1, \ldots, i_n\})$ satisfies the functional requirement $\phi_{\texttt{fspec}}$ if, for all $e_1, \ldots, e_n \in \texttt{Dom}$, it is the case that

$$(e_1, \ldots, e_n, (t\sigma)^{\texttt{Int}}) \in \phi_{\texttt{fspec}} \quad \text{where } \sigma := \{i_1 \mapsto e_1, \ldots, i_n \mapsto e_n\} \qquad (1)$$

2.2 Nonfunctional Requirements

Nonfunctional requirements concern properties of programs that are unrelated to the function being computed by the program (in the primary interpretation). Our key idea is that nonfunctional requirements can be captured as functional requirements of a second alternate interpretation of the program. This alternate interpretation is given by a structure (DomB, IntB), where DomB is a domain and IntB is an interpretation such that

- IntB maps a constant $c \in \Sigma$ to a subset $c^{\texttt{IntB}} \in 2^{\texttt{DomB}}$, and
- IntB maps a function $f \in \Sigma$ of arity n to a (non-deterministic) function $f^{\texttt{IntB}} : \texttt{DomB}^n \mapsto 2^{\texttt{DomB}}$

The interpretation IntB is extended to $\texttt{Terms}(\Sigma \cup \texttt{DomB}, \emptyset)$ using the recursive rule:

(1) if $t_i^{\texttt{IntB}} \subseteq \texttt{DomB}, i = 1, \ldots, n$ are the interpretations of terms $t_1, \ldots, t_n$ and $f \in \Sigma$, then $f(t_1, \ldots, t_n)^{\texttt{IntB}}$ is the set

$$\{y \in \texttt{DomB} \mid y \in f^{\texttt{IntB}}(y_1, \ldots, y_n), y_i \in t_i^{\texttt{IntB}} \text{ for } i = 1, \ldots, n\}$$

(2) if $e \in \texttt{DomB}$, then $e^{\texttt{IntB}} = \{e\}$.

For a program that takes n inputs, a nonfunctional requirement is given by $\phi_{\texttt{nfspec}} \subseteq \texttt{DomB}^{n+1}$. Formally, a program $t \in \texttt{Terms}(\Sigma, \{i_1, \ldots, i_n\})$ satisfies the nonfunctional requirement $\phi_{\texttt{nfspec}}$ if *there exists* a tuple $(e_1, \ldots, e_n, e) \in \phi_{\texttt{nfspec}}$ such that

$$e \in (t\theta)^{\texttt{IntB}} \quad \text{where } \theta := \{i_1 \mapsto e_1, \ldots, i_n \mapsto e_n\} \qquad (2)$$

The requirement ϕ_{nfspec} constrains the values (from DomB) that can be assigned to the inputs and outputs of the program, whereas IntB implicitly constrains the values (from DomB) that can be assigned to intermediate program variables.

Remark 1. The second interpretation structure (DomB, IntB) can be used to force *type correctness*. The values in DomB could be possible *types*, and for each $f \in \Sigma$, f^{IntB} would specify possible types for the output of f. The requirement ϕ_{nfspec} would capture well-typedness of programs, and it says that the inputs, outputs and all intermediate program variables can be assigned types such that the program is well-typed. More generally, (DomB, IntB) can be used to carry a predicate abstraction of the first semantics, or perform abstract interpretation [4].

In general, the second interpretation structure need not be an abstraction of the primary interpretation. The second interpretation serves two purposes. First, it can be used to state nonfunctional requirements. Second, it can be used to prune the synthesis search space, since a program (term) that can not be "typed" can be pruned early.

Example 2. Consider the signature Σ from Example 1. We can define a second interpretation (DomB, IntB), where DomB := {true, false} and for all symbols F in Σ of arity n, let $\text{IntB}(F)(b_1, \ldots, b_n) = \{\bigvee_i b_i\}$. If the input variables x, y are interpreted as {true}, then a term t will be interpreted as {true} iff it contains an input variable. A ground term, such as $f(e, e)$, will get an interpretation {false}. Thus, if we pick the requirement $\phi_{\text{nfspec}} := \{(\text{true}, \ldots, \text{true})\}$, then a program will satisfy this requirement only if it uses (at least one of) its inputs to compute its output.

Remark 2. Even when the formula $\phi_{\text{nfspec}} = \text{DomB}^{n+1}$ places no constraint on the values (from DomB) assigned to the inputs and output, the second interpretation may still constrain the set of valid programs. As an extreme case, if f^{IntB} maps everything to $\emptyset$, for some $f \in \Sigma$, then a program t can not use f, because if it did, we would be unable to assign a value from DomB to all of its intermediate variables (subterms of t).

2.3   Problem Definition

We now define the component-based program synthesis problem with functional and nonfunctional requirements as follows. We also add a size requirement on the synthesized program to enable solvability of the problem.

Definition 1 (Program Synthesis with Dual Requirements). *Given two structures* (Dom, Int) *and* (DomB, IntB) *that provide two different interpretations for the symbols in* Σ, *a size requirement* N, *a functional requirement* $\phi_{\text{fspec}} \subseteq \text{Dom}^{n+1}$ *and a nonfunctional requirement* $\phi_{\text{nfspec}} \subseteq \text{DomB}^{n+1}$, *the component-based program synthesis problem seeks to find a term* $t \in \text{Terms}(\Sigma, \{i_1, \ldots, i_n\})$ *of size* N *such that for all* $e_1, \ldots, e_n \in \text{Dom}$ *the condition in Eq. 1 holds, and for some* $(e_1, \ldots, e_n, e) \in \phi_{\text{nfspec}}$, *the condition in Eq. 2 holds.*

3   Synthesis Approach

The program-synthesis problem formulated in Definition 1 can be reduced to an exists-forall formula, which is then solved using an off-the-shelf solver.

Let $\texttt{subterms}(t)$ denote the set of all subterms of the term t. Henceforth, fix $\texttt{Vars} = \{i_1, \ldots, i_n\}$.

Consider the program synthesis with dual requirements problem in Definition 1. The problem can be rewritten in logical notation as follows:

$$\exists t \in \texttt{Terms}(\Sigma, \texttt{Vars}) : \texttt{size}(t) = N \wedge$$
$$(\forall \sigma : \texttt{Vars} \mapsto \texttt{Dom} : (\sigma(i_1), \ldots, \sigma(i_n), (t\sigma)^{\texttt{Int}}) \in \phi_{\texttt{fspec}}) \wedge$$
$$(\exists \tau : \texttt{subterms}(t) \mapsto \texttt{DomB} : (\tau(i_1), \ldots, \tau(i_n), \tau(t)) \in \phi_{\texttt{nfspec}} \wedge$$
$$(\forall \underbrace{f(s_1, \ldots, s_m)}_{s} \in \texttt{subterms}(t) : \tau(s) \in f^{\texttt{IntB}}(\tau(s_1), \ldots, \tau(s_m)))) \quad (3)$$

Clearly, a witness for t in this formula is a solution to the synthesis problem.

We define the size $\texttt{size}(t)$ of a term t to be the cardinality of $\texttt{subterms}(t)$; that is, the size of a minimal DAG representing t. Since we assume that Σ is finite, there are only finitely many terms of size N, and hence the first existential corresponds to a finite search. Since the cardinality of $\texttt{subterms}(t)$ is N, the second existential ($\exists \tau$) reduces to existence of N elements of $\texttt{DomB}$. The next $\forall$ quantifier (over subterms of t) is over a finite set and hence it is just a shorthand for a large conjunction. Finally, the remaining $\forall$ quantifier ($\forall \sigma$) is over n elements of $\texttt{Dom}$, and thus, we map our synthesis problem to an exists-forall problem in the theory of $\texttt{Dom}$ and $\texttt{DomB}$.

To increase expressiveness and improve scalability, we need an approach that allows a user to prune the search space for t as much as possible. We have designed a language that not only allows users to specify the program synthesis problem with dual requirements (Definition 1), but also allows users to constrain the search space. We briefly describe this language next.

3.1   Synudic: A Language for Synthesis Using Dual Interpretations on Components

Synudic (Synthesis using dual interpretation on components) is a language for specifying program synthesis problems with dual requirements (Definition 1). It also allows users to provide additional restriction on the structure of the program to be synthesized.

We call a well-formed Synudic program a *sketch* since it need not be a complete executable program, but only an incomplete program with a specification. A synudic sketch consists of a description of the signature Σ, the first interpretation ($\texttt{Dom}, \texttt{Int}$), the second interpretation ($\texttt{DomB}, \texttt{IntB}$) and the specifications $\phi_{\texttt{fspec}}$ and $\phi_{\texttt{nfspec}}$. Along with a size bound N, this completely specifies the synthesis problem from Definition 1. However, a Synudic sketch contains one more important piece of information: it specifies a regular tree grammar that syntactically restricts the number of possible terms (from $\mathtt{Terms}(\Sigma, \mathtt{Vars})$) that need to be considered (as possibe values for t in Formula 3).

For concrete syntax and examples of Synudic language, we refer to the Synudic webpage [9]. Logically, a Synudic sketch specifies an instance of the "program synthesis with dual requirements" problem, along with an additional regular tree grammar that restricts the class of straight-line programs of interest.

We present an instance of the program synthesis problem in a form that is logically equivalent to its Synudic description. Given a bitvector x, consider the function $\mathtt{rightmost1}$ that returns a bitvector that has 1 only at the position of the rightmost 1 in x. For example, $\mathtt{rightmost1}(10110) = 00010$. We want to synthesize a two-line program for computing $\mathtt{rightmost1}$. An example of a concrete Synudic sketch for solving this problem could contain following information:

Σ: The library $\Sigma = \{\mathtt{bvand}, \mathtt{bvneg}, \mathtt{bvone}\}$.
Dom: The domain Dom consists of bitvectors of length 5.
Int: The interpretation Int maps $\mathtt{bvand}$ to bitwise "and", $\mathtt{bvneg}$ to unary minus (in 2s complement notation), and $\mathtt{bvone}$ to the constant 00001 in Dom.
DomB: The second domain is $\mathtt{DomB} = \{\mathtt{true}, \mathtt{false}\}$.
IntB: The second interpretation IntB maps $\mathtt{bvand}$ to $\{(x, y, z) \mid z = x \vee y\}$, $\mathtt{bvneg}$ to $\{(x, y) \mid y = x\}$ and $\mathtt{bvone}$ to $\{\mathtt{false}\}$.
Requirements: The predicate $\phi_{\mathtt{fspec}} \subseteq \mathtt{Dom}^2$ is $\{(x, y) \mid y = \mathtt{rightmost1}(x)\}$. The predicate $\phi_{\mathtt{nfspec}} \subseteq \mathtt{DomB}^2$ is $\{(\mathtt{true}, \mathtt{true})\}$.
Regular Tree Grammar: The following grammar constrains the space of programs.

$$
\begin{aligned}
\mathtt{In} \quad &\overset{1}{:=} \quad i_1 \\
\mathtt{Prgm} \quad &\overset{na}{:=} \quad \mathtt{bvand(\ Prgm|In,\ Prgm|In\ )} \\
&\quad | \quad \mathtt{bvneg(\ Prgm|In\ )} \\
&\quad | \quad \mathtt{bvone}
\end{aligned}
$$

Each production in the tree grammar can be seen as a program block. The first production (for In) corresponds to a program block that has one line that outputs the value of the input variable i_1. The second production (for Prgm) corresponds to a program block that has na lines, where na is a parameter (that we will set to 2 since we are interested in synthesizing a two line program) and each line can use any function from Σ. The arguments of the functions can come from block In or from previous lines of this block.

One possible program generated by the above grammar would be $y_1 := i_1; y_2 := \mathtt{bvand}(y_1, y_1); y_3 := \mathtt{bvone}$. This program does not satisfy the two requirements. In particular, it does not satisfy the nonfunctional requirement because the output (y_3) does not depend on the input (i_1). Note that the Boolean "type" attached to each value just denotes whether the input was syntactically used to compute that value.

The description of the function $\mathtt{rightmost1}$ can use nested "if-then-else" statements. However, by fixing Σ and the grammar as above, we are forced to find an implementation for `rightmost1` that does not use "if-then-else". A program that is a solution for the Synudic sketch above is $y_1 := i_1; y_2 := \text{bvneg}(y_1); y_3 := \text{bvand}(y_2, y_1)$.

Remark 3. The regular tree grammar in Synudic can be used to specify one fixed concrete program, as well as, a completely unknown program. A concrete straight-line program can be written using blocks of length 1 in which there is just one option for the right-hand side expression. On the other extreme, an arbitrary straight-line program of length n over a library containing functions $f_1, \ldots, f_m$ can be written as

$$\text{Prgm} \overset{n}{:=} f_1(\text{ Prgm}|\text{In}, \ldots, \text{Prgm}|\text{In }) \mid \quad \cdots \quad \mid f_n(\text{ Prgm}|\text{In}, \ldots, \text{Prgm}|\text{In })$$

where `In` is the block generating the inputs. When performing synthesis, finding one program from the set of *all* n line programs can be difficult. Our language allows users to specify program search space that falls anywhere in between these two extremes. The use of grammars to restrict the space of programs is also used in syntax-guided synthesis [1].

3.2   From Synudic Sketches to Yices ∃∀ Formulas

Given a Synudic sketch, we developed a tool that generates the corresponding exists-forall formula in Yices syntax: logically, this formula is equivalent to the Formula 3 except that it also includes the restriction coming from the regular tree grammar. In our current version of the Synudic syntax, `Dom` and `DomB` are defined as Yices types, and the interpretations `Int` and `IntB` are defined using Yices functions to ease translation into ∃∀ Yices.

The ∃∀ Yices formula is obtained as follows. Let *lineType* be an enumeration type $\{line_1, \ldots, line_N\}$ representing the N lines in the program. Let $\overline{N} = \{1, \ldots, N\}$. Let Σ be the enumeration type representing the functions in the library. Assume maximum arity of any function in Σ is 2. The set of existential variables X consists of (a) variables $fSymbol_i$, $i \in \overline{N}$ of type Σ, where the value $fSymbol_i$ denotes the function used in i-th line, (b) variables arg_{i1}, arg_{i2}, $i \in \overline{N}$, of type *lineType*, where the value arg_{ij} denotes the line number that generates the j-th argument of $fSymbol_i$ on line i, and (c) variables typ_i, $i \in \overline{N}$ of type `DomB`, where typ_i is the value from `DomB` that is "computed" on line i. These $4N$ existential variables are constrained by the following formulas: (a) a formula ϕ_{gr} that forces variables $fSymbol_i$ and arg_{i1}, arg_{i2} to take values consistent with the given tree grammar, and (b) a formula ϕ_{IntB} that forces variables typ_i to take values consistent with the second interpretation `IntB` and the given nonfunctional requirement ϕ_{nfspec}. We generate $\phi_{\text{gr}} \wedge \phi_{\text{IntB}}$ as just one formula and not as (a conjunction of) two separate formulas.

The set of universal variables Y consists of the variables $varg_{i1}, varg_{i2}, vout_i$, $i \in \overline{N}$, of type `Dom`, where $varg_{ij}$ is the *value* of the j-th argument of $fSymbol_i$ on line i, and $vout_i$ is the value in `Dom` computed on line i. We generate a formula ϕ_{Int} that is true only if the Y variables take values consistent with the first interpretation Int and the program structure captured in the X variables. The final $\exists\forall$ Yices formula is:

$$\exists X : (\phi_{\text{gr}} \wedge \phi_{\text{IntB}} \wedge \forall Y, y : ((\phi_{\text{Int}} \wedge \phi_{\text{fspec}}(i_1, \ldots, i_n, y)) \Rightarrow \phi_{\text{fspec}}(i_1, \ldots, i_n, vout_N)))$$

The translation borrows ideas from the translation proposed in [10], but extends those ideas to handle dual interpretations and tree grammar restrictions, which were both absent in [10].

Our tool calls the exists-forall solver of Yices on the generated $\exists\forall$ formula. If there is a solution, the tool outputs the model for the existential variables, which can be used to obtain the concrete program. By giving an appropriate command-line argument, the tool can also search for alternate (more than one) solutions for the same sketch.

We next describe case studies from two domains - synthesis of bitvector manipulation tricks and synthesis of cryptographic schemes.

4   Bitvector Manipulation Programs

As a baseline, we evaluate our approach on bitvector manipulation benchmarks from [10,20]. The goal of these experiments is to show that (a) synthesis benchmarks that have been used before can be specified in the Synudic language, and (b) features supported by Synudic can be used to speed-up the synthesis process.

A simplified version of one our benchmark examples was described in Sect. 3.1. (The version used in our experiments had a larger library.) We note a few salient features of all the bitvector synthesis benchmarks.

(1) First, we use bitvectors of length 5 as Dom. It turns out that the algorithms that are synthesized to work on bitvectors of length 5 also work on bitvectors of arbitrary length. This observation was already made in [10]. We just note here that our language allows the user to set Dom to any type (supported by Yices).

(2) We use the usual bitvector operations, such as bitwise or, and, xor, as well as arithmetic functions on bitvectors, such as add and subtract, in the library. Certain examples also need functions that perform bitvector comparison, shift right, and division. We included them in the library whenever they were needed.

(3) Subtracting 1 is a common operation. We have two options: either we can include a subtract 1 operation as a library primitive, or we can include the subtraction operation and a function that generates the constant 1 in the library. Our language can support both choices. Using the former option usually speeds up the synthesis process.

(4) We used the Booleans as DomB. The Boolean value associated to a program variable keeps track of whether "the input was used to compute the value of that program variable", as shown in Example 2. For the bitvector examples, the second interpretation was not strictly required (since there was no nonfunctional requirement).

We present the results from bitvector benchmarks in Table 1. Synthesizing longer programs takes longer, and increasing the library size usually increases the time taken for synthesis (Columns 5 and 7), but in some cases, the rise is

Table 1. Bitvector benchmarks: Column #lines is the number of lines in the synthesized program, #lib is the number of functions in the library used for synthesis, time denotes the time (in seconds) taken for the tool to synthesize the program, and timet denotes the time taken when using a second interpretation to prune search space.

Name	Function $x(,y) \mapsto z$	#lines	#lib	time	#lib	time	timet
Rightmost 1 off	$u10^* \mapsto u00^*$	3	6	0.24	8	0.5	0.5
Isolate rightmost 1	$u10^* \mapsto 0^*10^*$	2	7	0.18	9	0.2	0.2
Average	$z = \frac{x+y}{2}$	4	4	2.9	7	27	5.4
Mask for 10*$	$u10^* \mapsto 011^*$	3	7	0.2	9	0.2	0.5
Maximum	$z = \max(x,y)$	4	4	77	7	238	86
Turnoff 1^+0^*	$u1^+0^* \mapsto u0^+0^*$	5	6	21	8	102	2
next# same#1s	$\min z$ s.t. $z > x$, $z|_1 = x|_1$	8	5	154	6	$\frac{500}{TO}$	54

steep (third example computing "average"). To evaluate the benefit of pruning using the second interpretation, we added a second interpretation to enforce that certain library components are used (at most) once, and the running times with the second interpretation added are shown in the last column in Table 1. As expected, our running times in Column 7 are comparable to those reported in [10]. In some cases, our tool synthesized "new" procedures that were semantically equivalent variants of the known procedures, see [9] for such examples.

5   Cryptographic Constructions

We now provide examples of how dual interpretations are useful for the synthesis of cryptographic constructions. We first provide an example from public key cryptography inspired by the work in [2] that consist on synthesizing padding schemes. Our second example is related to symmetric key encryption, and builds upon the work presented in [13].

5.1   Synthesis of Padding-Based Encryption Schemes

In public key cryptography, padding is the process of preparing a message for encryption. A modern form of padding is OAEP, which is often paired with RSA public key encryption. Padding schemes, and in particular OAEP, satisfy the goals of (1) converting a deterministic encryption scheme, e.g. RSA, into a probabilistic one, and (2) ensuring that a portion of the encrypted message cannot be decrypted without being able to invert the full encryption.

Inspired by the success of the tool Zoocrypt in synthesizing padding-based encryption schemes [2] (and their corresponding security proofs), we used our synthesis tool for exploring the same space.

The library Σ of components consists of two unary hash functions, G and H, a binary $\oplus$ function, and the identity function. Padding with 0 is not modeled explicitly since it can be added as a post-processing step to make the hash functions applicable on its arguments.

There are two inputs: the message m and a random number r. The goal is to construct a pair of values – intuitively, an encrypted message m' and a key k – that can be concatenated and encrypted (by RSA, say f) and the result $f(m'||k)$ can be sent on the network. The functional requirement stems from the fact that the receiver should be able to get back m from $f(m'||k)$ assuming it has access to f^-. Thus, we synthesize two blocks of code – an encryption block for generating m' and k from m and r, and a decryption block for generating m from m' and k. The functional requirement states that the output of the decryption block should be equal to the input m. The nonfunctional requirement states that the two values m' and k that are transmitted should essentially be random.

To capture the nonfunctional requirement, we give a second interpretation to the program (sketch). As DomB we used bitvectors of length 5, since that was enough to carry the information that a certain value was "essentially random":
(a) The first bit keeps information about the size of the computed value. This information is necessary to produce type correct programs, since we have hash functions mapping bitvectors of one size to another.
(b) The second bit is set if the data value is essentially the same as a random value in its domain. It is difficult to carry forward this information precisely, so we use conservative typing rules to update the value of the second type-bit during each operation.
(c) The third and fourth bits are set if the top function application is the hash function G and H, respectively. This information is used to update the second bit of the type.
(d) The fifth bit is set if the top function application is the xor function. This information is used for the same purpose as the previous two type-bits.

The complete Synudic sketch for the example can be found at [9]. We note two things. First, we used fixed length bitvectors as Dom. The length choice is arbitrary: larger bitlengths would mean more computational resources would be required to solve the synthesis problem, but smaller bitlengths could lead to synthesis of schemes that do not work for arbitrary sizes. Second, the interpretations of H and G had to be concretized to bitvector (Yices) functions, but they had to be picked carefully so that they satisfy (exactly) the algebraic relations the actual functions satisfy. This may not be possible always, in which case one should choose interpretations that are likely to lead to correct solutions. In such cases, we rely on a *post-processing security verification tool* to formally verify the correctness and security properties of the synthesized schemes.

We used our tool to synthesize different padding schemes using different program lenths for the encryption block and the decryption block. Some example synthesized schemes are shown in Fig. 1. Again, we do not show the padding with 0 that is required to make arguments reach the required bitvector length. Note that the OAEP scheme [3] was also generated: it is the last schemes in Fig. 1. But smaller padding-based schemes were also found by the tool. Similar schemes have also been reported in [2].

$$f(G(r)\|(G(r) \oplus m))$$
$$f(r\|(G(r) \oplus m))$$
$$f(G(r \oplus H(m))\|(G(r \oplus H(m)) \oplus m))$$
$$f((r \oplus H(m))\|(G(r \oplus H(m)) \oplus m))$$
$$f((G(r) \oplus m)\|(H(G(r) \oplus m) \oplus r))$$

Fig. 1. Some automatically synthesized padding-based encryption schemes.

5.2 Synthesis of Block Ciphers Modes of Operation

A *block cipher* consists of (deterministic) algorithms for encrypting and decrypting fixed-length *blocks* of data. It has one algorithm for computing the encryption function $F : \{0,1\}^l \times \{0,1\}^{l_k} \to \{0,1\}^l$ and one for computing the decryption function $F^- : \{0,1\}^l \times \{0,1\}^{l_k} \to \{0,1\}^l$ such that for any *block* $B \in \{0,1\}^l$ of exactly l bits and for any *key* $k \in \{0,1\}^{l_k}$ of l_k bits, $F^-(F(B,k),k) = B$. We denote $F(.,k)$ by F_k and $F^-(.,k)$ by F_k^-. An example of block cipher is the standardized AES for which $l = 128$.

Roughly speaking (see [12] for a formal definition), a block cipher (F, F^-) is secure against the so-called chosen plaintext attacks (in the standard model) if, for a fixed random key k, an attacker allowed to query F_k has negligible probability of distinguishing F_k from a random permutation, given certain limitations on the computational power of the attacker and the number of times F_k can be queried.

A *mode of operation* is a pair of algorithms that use a symmetric block cipher algorithm, e.g. AES, to encrypt/decrypt amounts of data *larger* than a block. A secure mode of operation must provide the same level of security as its associated block cipher. For example, the encryption algorithm of the popular Cipher Block Chaining (CBC) mode is depicted in Fig. 2. CBC, when equipped with a secure block cipher, provides IND$-CPA security, i.e. an attacker cannot distinguish its output from an uniformly random string with significant probability (under certain constraints on the computational power of the attacker). Note that CBC encryption consists of an initialization algorithm, where a random initialization vector IV is produced, followed by n copies of a block processing algorithm, where exactly one value is fed from one copy to the next one. This structure is common to many popular modes of operation.

Most of the previous approaches to the formal verification and synthesis of block cipher modes of operation (and certainly the ones considered in this paper) build upon the observation that these kind of programs can be constructed using a limited set of operations such as xor, concatenation, generation of random values, and evaluation of the block cipher.

Recent effort in the automation of the analysis of block cipher modes include [6,13]. In contrast to [6], which suffers from the limitation that the analyzed mode must operate on a *fixed* number of blocks, the work in [13] models the operation that is carried out when encrypting a *single* block, exploiting the common structure of block cipher modes of operation mentioned above. In [13],

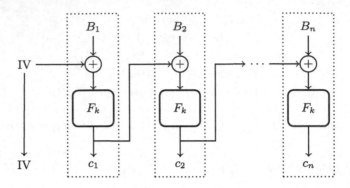

Fig. 2. The CBC mode of operation for the encryption of an n-block message. The dotted boxes correspond to the multiple copies of the block processing procedure.

the encryption algorithm of a mode of operation is described as a pair of straight-line programs (`Init`, `Block`). `Init` models the initialization phase of the mode of operation. In the case of the CBC mode of Fig. 2, `Init` would correspond to the generation of the random value IV. On the other hand, `Block` corresponds to the algorithm that, given a value coming from the previous iteration (or the initialization phase) and a certain message block m, produces the ciphertext for m and the value to be fed to the next iteration of the mode of operation. For the CBC mode, the different instances of `Block` correspond to the subalgorithms in dotted boxes in Fig. 2. While `Init` is very simple in that, roughly speaking, it may contain only a random number generation operation, `Block` might contain an arbitrary number of xor operations and evaluations of F_k for a fixed value of k. A further relevant structural restriction in the straight-line programs `Init` and `Block` is that the output of every operation in the program must be used exactly once in the rest of the program, with the exception of an additional operation called dup implementing the identity function and whose output must be used twice.

As main contribution in [13], the authors present a type system T that guarantees that type correct modes (`Init`, `Block`) encode secure modes of operation. Then, synthesis of secure modes is performed by enumerating straight-line programs satisfying the constraints above and filtering out the ones that are not type correct w.r.t. T. This check is implemented by means of an SMT solver. An ad hoc procedure is used to further guarantee that the resulting mode of operation admits a decryption algorithm.

In the example presented in this section, we encoded the synthesis approach from [13] as a *program synthesis with dual requirements* problem (Definition 1) Instead of separately filtering modes of operation that are not decryptable as done in [13], we encoded the existence of a decoding algorithm as a *functional requirement*. That has the advantage that encryption algorithms are synthesized together with their corresponding decryption procedure. Moreover, the constraint that Init and Block must be type correct w.r.t. T can be naturally encoded as a *nonfunctional requirement* in our language.

While reducing the synthesis of block ciphers modes of operation to the *program synthesis with dual requirements* problem has many advantages, our approach also suffers from some limitations when compared to [13]. The main limitation is that, whereas the approach of [13] is completely symbolic, we need to provide a domain Dom and interpretation Int for every operation (including the permutation F_k) that can be handled by ∃∀-Yices. For Dom we chose bitvectors of length 5. While the operations xor, dup1, and dup2 have natural operations in the domain of bitvectors, the interpretation of F should be picked carefully so that it satisfies (exactly) the algebraic relations the actual functions satisfy. We picked left-rotation by two for the interpretation of F. Although in principle a poor choice might cause invalid schemes to be accepted as decryptable, this can be easily avoided in many cases, or one could use a verification step a-posteriori. The full sketch used for this example can be found in [9].

Using our tool we could synthesize the well-known modes ECB, OFB, CFB, CBC, and PCBC, also automatically found in [13], as well as some variants of those. The parameters na, nb, nc in Fig. 3 denote the number of lines of code in the Init, Block, and decryption blocks. The reported times corresponds to a complete exploration of the search space for those parameters. For example, the second row of the first table means that, with parameters $na = 2$, $nb = 6$, $nc = 3$, it took our tool 6.07 seconds to conclude that *exactly* two instances of the sketch are secure and decryptable modes of operation. The modes marked with an asterisk (*) correspond to redundant variants of the corresponding mode.

Parameters			Modes	Time(s)
na	nb	nc		
2	4	3	CBC	3.25
2	6	3	CBC* OFB*	6.07
2	6	4	CBC*	22.9
2	6	5	CBC OFB* CFB	5.77

Parameters			Modes	Time(s)
na	nb	nc		
2	7	6	OFB variant CBC variant	6.34
2	8	5	CBC* OFB* CFB*	39.47
2	9	5	PCBC OFB variant	109.82

Fig. 3. Results of the synthesis of block cipher modes of operation using Synudic.

6 Conclusion

We presented an approach for program synthesis that relies on using two different interpretations for the program variables. The dual interpretation approach enables specification of both functional and nonfunctional requirements, and pruning the synthesis search space. We solve the synthesis problem by converting it to an $\exists\forall$ Yices formula, and using Yices to solve it. We applied our approach to synthesize bitvector manipulation tricks, padding-based encryption schemes and block cipher modes of operation.

Our current implementation is limited in several ways. First, the Yices existsforall solver handles only bitvectors, Booleans, and linear arithmetic expressions. Hence, only these types can be used to define the two interpretations. Our synthesis language allows synthesis of only straight-line programs, and does not allow, for example, synthesis of functions that are used within other functions. Such extensions are left for future work.

Acknowledgments. We thank the anonymous reviewers and N. Shankar for helpful comments.

References

1. Alur, R., Bodík, R., Juniwal, G., Martin, M.M.K., Raghothaman, M., Seshia, S.A., Singh, R., Solar-Lezama, A., Torlak, E., Udupa, A.: Syntax-guided synthesis. In: Formal Methods in Computer-Aided Design, FMCAD, pp. 1–17 (2013)
2. Barthe, G., Crespo, J.M., Kunz, C., Schmidt, B., Gregoire, B., Lakhnech, Y., Zanella-Beguelin, S.: Fully automated analysis of padding-based encryption in the computational model (2013). http://www.easycrypt.info/zoocrypt/
3. Bellare, M., Rogaway, P.: Optimal asymmetric encryption. In: De Santis, A. (ed.) EUROCRYPT 1994. LNCS, vol. 950, pp. 92–111. Springer, Heidelberg (1995)
4. Cousot, P., Cousot, R.: Abstract interpretation: a unified lattice model for static analysis of programs by construction or approximation of fixpoints. In: 4th ACM Symposium on Principles of Programming Languages, POPL 1977, pages 238–252 (1977)
5. Dutertre, B.: Yices 2.2. In: Biere, A., Bloem, R. (eds.) CAV 2014. LNCS, vol. 8559, pp. 737–744. Springer, Heidelberg (2014)
6. Gagné, M., Lafourcade, P., Lakhnech, Y., Safavi-Naini, R.: Automated verification of block cipher modes of operation, an improved method. In: Garcia-Alfaro, J., Lafourcade, P. (eds.) FPS 2011. LNCS, vol. 6888, pp. 23–31. Springer, Heidelberg (2012)
7. Gascón, A., Subramanyan, P., Dutertre, B., Tiwari, A., Jovanovic, D., Malik, S.: Template-based circuit understanding. In: Formal Methods in Computer-Aided Design, FMCAD, pp. 83–90. IEEE (2014)
8. Gascón, A., Tiwari, A.: A synthesized algorithm for interactive consistency. In: Badger, J.M., Rozier, K.Y. (eds.) NFM 2014. LNCS, vol. 8430, pp. 270–284. Springer, Heidelberg (2014)
9. Gascón, A., Tiwari, A.: Synudic: synthesis using dual interpretation on components (2015). http://www.csl.sri.com/users/tiwari/softwares/auto-crypto/
10. Gulwani, S., Jha, S., Tiwari, A., Venkatesan, R.: Synthesis of loop-free programs. In: Proceedings of ACM Conference on Programming Language Design and Implementation, PLDI, pp. 62–73 (2011)
11. Jha, S., Gulwani, S., Seshia, S.A., Tiwari, A.: Oracle-guided component-based program synthesis. In: Proceedings of ICSE, vol. 1, pp. 215–224. ACM (2010)

12. Katz, J., Lindell, Y.: Introduction to Modern Cryptography. Chapman and Hall/CRC Press, Boca Raton (2007)
13. Malozemoff, A.J., Katz, J., Green, M.D.: Automated analysis and synthesis of block-cipher modes of operation. In: IEEE 27th Computer Security Foundations Symposium, CSF, pp. 140–152. IEEE (2014)
14. Morgan, C.: The shadow knows: refinement and security in sequential programs. Sci. Comput. Program. **74**(8), 629–653 (2009)
15. Sabelfeld, A., Myers, A.C.: Language-based information-flow security. IEEE J. Sel. Areas Commun. **21**(1), 5–19 (2003)
16. Solar-Lezama, A.: Program sketching. STTT **15**(5–6), 475–495 (2013)
17. Solar-Lezama, A., Rabbah, R.M., Bodík, R., Ebcioglu, K.: Programming by sketching for bit-streaming programs. In: PLDI (2005)
18. Solar-Lezama, A., Tancau, L., Bodík, R., Saraswat, V., Seshia, S.: Combinatorial sketching for finite programs. In: ASPLOS (2006)
19. Talcott, C.: A theory for program and data type specification. TCS **104**(1), 129–159 (1992)
20. Warren, H.S.: Hacker's Delight. Addison-Wesley Longman Publishing Co. Inc., Boston (2002)

Logics and Systems for Program Verification

Automated Theorem Proving for Assertions in Separation Logic with All Connectives

Zhé Hóu[1](✉), Rajeev Goré[1], and Alwen Tiu[2]

[1] Research School of Computer Science,
The Australian National University, Canberra, Australia
zhe.hou@anu.edu.au
[2] School of Computer Engineering, Nanyang Technological University,
Singapore, Singapore

Abstract. This paper considers Reynolds's separation logic with all logical connectives but without arbitrary predicates. This logic is not recursively enumerable but is very useful in practice. We give a sound labelled sequent calculus for this logic. Using numerous examples, we illustrate the subtle deficiencies of several existing proof calculi for separation logic, and show that our rules repair these deficiencies. We extend the calculus with rules for linked lists and binary trees, giving a sound, complete and terminating proof system for a popular fragment called symbolic heaps. Our prover has comparable performance to Smallfoot, a prover dedicated to symbolic heaps, on valid formulae extracted from program verification examples; but our prover is not competitive on invalid formulae. We also show the ability of our prover beyond symbolic heaps, our prover handles the largest fragment of logical connectives in separation logic.

1 Introduction

Separation logic (SL) was invented to verify the correctness of programs that mutate possibly shared data structures [30]. SL is an extension of Hoare logic with logical connectives $\top^*, *, -\!\!*$ from the logic of bunched implications (BI) [29] to capture the empty heap, heap composition, and heap extension respectively, and a predicate $\mapsto$ to describe singleton heaps. Reynolds [34] coupled the semantics for the above extensions with classical connectives, making Boolean BI (BBI) the basis of separation logic, although there are earlier versions that consider intuitionistic additive connectives [33]. Using BI logics to enable local reasoning has proven very successful, and many variants of separation logic have been developed. For instance, separation logic for higher-order store [32], bunched typing [3], concurrency [6], owned variables [4,31], rely/guarantee reasoning [37], abstract data types [22], amongst many.

These separation logics require proof methods to reason about their assertion languages, and since most separation logic variants are based on the original SL, automated tools usually respect Reynolds's semantics [34]. However, most existing tools for Reynolds-like semantics SL, such as Smallfoot [1], jStar [13], VeriStar [35], SLP [27], and Asterix [28], are all restricted to small fragments,

© Springer International Publishing Switzerland 2015
A.P. Felty and A. Middeldorp (Eds.): CADE-25, LNAI 9195, pp. 501–516, 2015.
DOI: 10.1007/978-3-319-21401-6_34

most notably, the symbolic heaps fragment of Berdine et al. [2]. On the other hand, there are also existing tools that handle larger fragments than symbolic heaps, but for non-Reynolds semantics, e.g., Lee and Park's theorem prover [24], and Thakur et al.'s unsatisfiability checker [36], cf. Sects. 4 and 6.

There is a growing demand from the program verification community to move beyond symbolic heaps and to deal with $-\!*$, which is ignored in most SL fragments. Having $-\!*$ is a desirable feature, since many algorithms/programs are verified using this connective, especially when expressing tail-recursive operations [26], iterators [23], septraction in rely/guarantee [37] etc. Moreover, $-\!*$ is useful in the weakest precondition calculus for SL, which introduces $-\!*$ "in each statement in the program being analysed" [25]. See the introduction of [24] and [36] for other examples requiring $-\!*$. In addition to $-\!*$, allowing arbitrary combinations of logical connectives is also useful when describing overlaid data structures [16], properties such as cross-split can be useful in proof search in this setting [14]. Nevertheless, existing tools for SL with Reynolds's semantics do not support the reasoning for all logical connectives. Thus, an important area of research is to obtain a practical proof system for SL with all connectives.

SL is not recursively enumerable in general [5,10], neither is the fragment we consider here, so there is no finite, sound and complete proof system for this logic, and computability results are not our focus. Interested readers are referred to [11,12] for other fragments of SL and their decidability and complexity results. Building upon the labelled sequent calculi for propositional abstract separation logics (PASL, cf. [19]), we give a sound, w.r.t. Reynolds's semantics, proof method that is useful in program verification. Since we focus on SL with heap model semantics here, although [19] is complete for PASL, it is not comparable to this work in terms of provability. We extend PASL with inference rules for quantifiers, equality and the $\mapsto$ predicate. The latter involves heaps and stores in the semantics in a very subtle way, making this study error-prone. Some subtle mistakes in the literature are discussed in Sect. 4.

Capturing data structures is important since they are frequently used in program verification. We extend our proof system with treatments for singly linked lists based on similar rules in Smallfoot [2]. Binary trees can be handled similarly; see [18]. We also move beyond symbolic heaps to consider arbitrary combinations of logical connectives. We show that our proof method is complete w.r.t. symbolic heaps. We give a sound, complete, and terminating proof search procedure for symbolic heaps. Our implementation is competitive with Smallfoot on valid formulae, but not on invalid formulae, taken from benchmarks extracted from program verification problems. In addition, we demonstrate that our prover can handle a wider range of formulae than existing tools, thus it handles the largest fragment of SL in terms of logical connectives, and paves the way to more sophisticated program verification using Reynolds's SL.

2   Separation Logic

Separation logic generally refers to a combination of a programming language, an assertion logic and a specification logic for Hoare triples [21]. Here, we focus on the assertion logic that is compliant with Reynolds's semantics [34].

Following Reynolds [34], we consider all *values* as *integers*, an infinite number of which are *addresses*. *Atoms*, containing *nil*, form a subset of values that is disjoint from addresses. *Heaps* are finite partial functions from addresses to values, and *stores* are total functions from finite sets of *program variables* to values. These are formalised as below:

$$Val = Int \qquad\qquad Atoms \cup Addr \subseteq Int \qquad\qquad nil \in Atoms$$
$$Atoms \cap Addr = \emptyset \qquad\qquad H = Addr \rightharpoonup_{fin} Val \qquad\qquad S = Var \to Val$$

We assume a set of program variables, ranged over by x, y, z, and a constant *nil*. An *expression* is either a constant or a program variable. Expressions are denoted by e. We ignore arithmetic expressions such as those allowed by Reynolds [34].

The syntax for formulae is given by:

$$F ::= e = e' \mid e \mapsto e' \mid e \mapsto e', e'' \mid \bot \mid F \to F \mid \top^* \mid F * F \mid F \twoheadrightarrow F \mid \exists x.F$$

The only atomic formulae are $\bot$, $\top^*$, $(e = e')$, $(e \mapsto e')$, and $(e \mapsto e', e'')$. The latter two are called the "points-to" predicates. The domain of the quantifier is the set of values. We assume the usual notion of free and bound variables in formulae. We prefer to write $\top^*$ for the empty heap constant *emp* to be consistent with the prior work for BBI and PASL [19,20]. The points-to predicate $e \mapsto e'$ denotes a singleton heap sending the value of e to the value of e'. The connectives $*$ and $\twoheadrightarrow$ denote heap composition and heap extension respectively. These two connectives are interpreted with the binary operator $\circ$ defined as $h_1 \circ h_2 = h_1 \cup h_2$ when h_1, h_2 have disjoint domains, and undefined otherwise. A *state* is a pair (s, h) of a store and a heap.

A *separation logic model* is a pair (S, H) of stores and heaps, both are nonempty as defined previously. The forcing relation between a state and the formulae is formally defined in Table 1. We write $[\![e]\!]_s$ to denote the valuation of an expression e by looking up the value of variables in e in the store s. We fix that $[\![nil]\!]_s = nil$. We write $s[x \mapsto v]$ to denote a stack that is identical to s, except possibly on the valuation of x, i.e., $s[x \mapsto v](x) = v$ and $s[x \mapsto v](y) = s(y)$ for $y \neq x$. A formula F is true at the state (s, h) if $(s, h) \Vdash F$, and it is *valid* if $(s, h) \Vdash F$ for every $s \in S, h \in H$.

The literature contains the following useful abbreviations:

$$e \mapsto \_ \equiv \exists x. e \mapsto x \qquad e \mapsto e_1, \cdots, e_n \equiv (e \mapsto e_1) * \cdots * (e + n - 1 \mapsto e_n)$$

The multi-field points-to predicate $e \mapsto e_1, \cdots, e_n$ has different interpretations in the literature. In Reynolds's notation, the formula $e \mapsto e_1, e_2$ is equivalent to $(e \mapsto e_1) * (e + 1 \mapsto e_2)$, thus it is a heap of size two. However, in other versions of SL, the set of heaps may be defined as finite partial functions from addresses to pairs of values [7,10], as shown below left:

$$H = Addr \rightharpoonup_{fin} Val \times Val \qquad H = Addr \rightharpoonup_{fin} (Fields \to Val)$$

In this setting the formula $e \mapsto e_1, e_2$ is a singleton heap. A more general case can be found in the definition of symbolic heaps [2] with heaps defined as shown

Table 1. The semantics of the assertion logic of separation logic.

$s, h \Vdash \bot$ iff never	$s, h \Vdash \top^*$ iff $h = \emptyset$
$s, h \Vdash e = e'$ iff $[\![e]\!]_s = [\![e']\!]_s$	$s, h \Vdash A \rightarrow B$ iff $s, h \Vdash A$ implies $s, h \Vdash B$
$s, h \Vdash e \mapsto e'$ iff $dom(h) = \{[\![e]\!]_s\}$ and $h([\![e]\!]_s) = [\![e']\!]_s$	
$s, h \Vdash \exists x.A$ iff $\exists v \in Val$ such that $s[x \mapsto v], h \Vdash A$	
$s, h \Vdash A * B$ iff $\exists h_1, h_2.(h_1 \circ h_2 = h$ and $s, h_1 \Vdash A$ and $s, h_2 \Vdash B)$	
$s, h \Vdash A \!-\!\!* B$ iff $\forall h_1, h_2.(h_1 \circ h = h_2$ and $s, h_1 \Vdash A)$ implies $s, h_2 \Vdash B)$	

above right with a slight modification to make addresses a subset of values. *Fields* are simply the names for the data being pointed to.

The syntax of SL in this paper is more expressive than the popular *symbolic heaps* fragment of SL [2], which is restricted to the following syntax:

$$P ::= e = e' \mid \neg P \qquad\qquad \Pi ::= \top \mid P \mid \Pi \wedge \Pi$$
$$S ::= e \mapsto [f : e] \qquad\qquad \Sigma ::= \top^* \mid S \mid \Sigma * \Sigma$$

The $\mapsto$ predicate in symbolic heaps allows a list $[f : e]$ of fields, where f is the name of a field, and e is the content. Symbolic heaps are pairs $\Pi \wedge \Sigma$. The entailment of symbolic heaps is written as $\Pi \wedge \Sigma \vdash \Pi' \wedge \Sigma'$. Symbolic heaps also allow formulae of the form $e \mapsto \_$ which does not specify the content of the heap.

3 LS_{SL}: A Labelled Sequent Calculus for SL

Let LVar be an infinite set of *label variables*, the set $\mathcal{L}$ of *labels* is LVar$\cup\{\epsilon\}$, where $\epsilon \notin$ LVar is a label constant. Labels are ranged over by h. We may sometimes use "heap" to mean a label h or a $\mapsto$ atomic formula. A *labelled formula* has the form $h : F$, where h is a label and F is a formula. We use ternary relational atoms $(h_1, h_2 \rhd h_3)$ to indicate that the composition of the heaps represented by h_1, h_2 gives the heap represented by h_3. A *sequent* takes the form $\mathcal{G}; \Gamma \vdash \Delta$ where $\mathcal{G}$ is a set of ternary relational atoms, Γ, Δ are sets of labelled formulae, and ; denotes set union. Thus $\Gamma; h : A$ is the union of Γ and $\{h : A\}$. The left hand side of a sequent is the *antecedent* and the right hand side is the *succedent*.

The labelled sequent calculus LS_{SL} consists of inference rules taken from $LS_{PASL} + D + CS$ [19] with the addition of some special *id* rules, a cut rule for $=$, and the general rules for $\exists$ and $=$, as shown in Figs. 1 and 2, and the rules for the $\mapsto$ predicate, as shown in Fig. 3, which are new to this paper. In these figures we write A, B for formulae. Although our proof system is incomplete for SL with heap model semantics and it may not be complete even for the quantifier-free fragment, the underlying system $LS_{PASL} + D + CS$ is complete for PASL with disjointness and cross-split [19]. The inference rules for the $\mapsto$ predicate with two fields are analogous to the rules in Fig. 3.

A *label substitution* is a mapping from label variables to labels, which is an identity map except for a finite subset of $LVar$. We write $[h'_1/h_1, \ldots, h'_n/h_n]$ for a label substitution which maps h_i to h'_i. Label substitutions are extended to

$$\frac{}{\mathcal{G}; \Gamma; h : e_1 \mapsto e_2 \vdash h : e_1 \mapsto e_2; \Delta} \; id \qquad \frac{}{\mathcal{G}; \Gamma; h : e_1 \mapsto e_2, e_3 \vdash h : e_1 \mapsto e_2, e_3; \Delta} \; id_2$$

$$\frac{\mathcal{G}; \Gamma[e_1/e_2] \vdash \Delta[e_1/e_2] \qquad \mathcal{G}; \Gamma \vdash h : e_1 = e_2; \Delta}{\mathcal{G}; \Gamma \vdash \Delta} \; cut_=$$

$$\frac{}{\mathcal{G}; \Gamma; h : \bot \vdash \Delta} \; \bot L \qquad \frac{\mathcal{G}[\epsilon/h]; \Gamma[\epsilon/h] \vdash \Delta[\epsilon/h]}{\mathcal{G}; \Gamma; h : \top^* \vdash \Delta} \; \top^* L \qquad \frac{}{\mathcal{G}; \Gamma \vdash \epsilon : \top^*; \Delta} \; \top^* R$$

$$\frac{\mathcal{G}; \Gamma \vdash h : A; \Delta \qquad \mathcal{G}; \Gamma; h : B \vdash \Delta}{\mathcal{G}; \Gamma; h : A \to B \vdash \Delta} \; \to L \qquad \frac{\mathcal{G}; \Gamma; h : A \vdash h : B; \Delta}{\mathcal{G}; \Gamma \vdash h : A \to B; \Delta} \; \to R$$

$$\frac{(h_1, h_2 \rhd h_0); \mathcal{G}; \Gamma; h_1 : A; h_2 : B \vdash \Delta}{\mathcal{G}; \Gamma; h_0 : A * B \vdash \Delta} \; * L \qquad \frac{(h_1, h_0 \rhd h_2); \mathcal{G}; \Gamma; h_1 : A \vdash h_2 : B; \Delta}{\mathcal{G}; \Gamma \vdash h_0 : A \mathbin{-\!\!*} B; \Delta} \; \mathbin{-\!\!*} R$$

$$\frac{(h_1, h_2 \rhd h_0); \mathcal{G}; \Gamma \vdash h_1 : A; h_0 : A * B; \Delta \qquad (h_1, h_2 \rhd h_0); \mathcal{G}; \Gamma \vdash h_2 : B; h_0 : A * B; \Delta}{(h_1, h_2 \rhd h_0); \mathcal{G}; \Gamma \vdash h_0 : A * B; \Delta} \; * R$$

$$\frac{(h_1, h_0 \rhd h_2); \mathcal{G}; \Gamma; h_0 : A \mathbin{-\!\!*} B \vdash h_1 : A; \Delta \qquad (h_1, h_0 \rhd h_2); \mathcal{G}; \Gamma; h_0 : A \mathbin{-\!\!*} B; h_2 : B \vdash \Delta}{(h_1, h_0 \rhd h_2); \mathcal{G}; \Gamma; h_0 : A \mathbin{-\!\!*} B \vdash \Delta} \; \mathbin{-\!\!*} L$$

$$\frac{\mathcal{G}; \Gamma; h : A[y/x] \vdash \Delta}{\mathcal{G}; \Gamma; h : \exists x.A \vdash \Delta} \; \exists L \qquad \frac{\mathcal{G}; \Gamma \vdash h : A[e/x]; h : \exists x.A; \Delta}{\mathcal{G}; \Gamma \vdash h : \exists x.A; \Delta} \; \exists R$$

$$\frac{\mathcal{G}; \Gamma\theta \vdash \Delta\theta}{\mathcal{G}; \Gamma; h : e_1 = e_2 \vdash \Delta} \; = L \qquad \frac{}{\mathcal{G}; \Gamma \vdash h : e = e; \Delta} \; = R$$

Side conditions:
Each label being substituted cannot be ϵ, each expression being substituted cannot be *nil*.
In $= L$, $\theta = mgu(\{(e_1, e_2)\})$.
In $* L$ and $\mathbin{-\!\!*} R$, the labels h_1 and h_2 do not occur in the conclusion.
In $\exists L$, y is not free in the conclusion.

Fig. 1. Logical rules in LS_{SL}.

mappings between labelled formulae and labelled sequents in the obvious way. An *expression substitution* is defined similarly, where the domain is the set of program variables and the codomain is the set of expressions. We use θ (possibly with subscripts) to range over expression substitutions, and write $e\theta$ for the result of applying θ to e. Given a set of pairs of expressions $E = \{(e_1, e_1'), \ldots, (e_n, e_n')\}$, a *unifier* for E is an expression substitution θ such that $\forall i,\ e_i\theta = e_i'\theta$. We assume the usual notion of the *most general unifier* (mgu) from logic programming. We denote with $mgu(E)$ the most general unifier of E when it exists. The formulae (resp. relational atoms) shown explicitly in the conclusion of a rule are called *principal formulae* (resp. *principal relational atoms*). A formula F is *provable* or *derivable* if there is a derivation of the sequent $\vdash h : F$ for an arbitrary $h \in$ LVar.

A *label mapping* ρ is a function $\mathcal{L} \to H$ such that $\rho(\epsilon) = \emptyset$. We define an *extended separation logic model* (S, H, s, ρ) as a separation logic model plus a stack and a label mapping.

Theorem 1 (Soundness). *For any formula F, and for any $h \in$ LVar, if the sequent $\vdash h : F$ is derivable in LS_{SL}, then F is valid in Reynolds's semantics.*

$$\frac{(h, \epsilon \triangleright h); \mathcal{G}; \Gamma \vdash \Delta}{\mathcal{G}; \Gamma \vdash \Delta} \; U \qquad \frac{(h_3, h_5 \triangleright h_0); (h_2, h_4 \triangleright h_5); (h_1, h_2 \triangleright h_0); (h_3, h_4 \triangleright h_1); \mathcal{G}; \Gamma \vdash \Delta}{(h_1, h_2 \triangleright h_0); (h_3, h_4 \triangleright h_1); \mathcal{G}; \Gamma \vdash \Delta} \; A$$

$$\frac{(h_2, h_1 \triangleright h_0); (h_1, h_2 \triangleright h_0); \mathcal{G}; \Gamma \vdash \Delta}{(h_1, h_2 \triangleright h_0); \mathcal{G}; \Gamma \vdash \Delta} \; E \qquad \frac{(\epsilon, \epsilon \triangleright h_2); \mathcal{G}[\epsilon/h_1]; \Gamma[\epsilon/h_1] \vdash \Delta[\epsilon/h_1]}{(h_1, h_1 \triangleright h_2); \mathcal{G}; \Gamma \vdash \Delta} \; D$$

$$\frac{(\epsilon, h_2 \triangleright h_2); \mathcal{G}[h_2/h_1]; \Gamma[h_2/h_1] \vdash \Delta[h_2/h_1]}{(\epsilon, h_1 \triangleright h_2); \mathcal{G}; \Gamma \vdash \Delta} \; Eq_1 \qquad \frac{(\epsilon, h_1 \triangleright h_1); \mathcal{G}[h_1/h_2]; \Gamma[h_1/h_2] \vdash \Delta[h_1/h_2]}{(\epsilon, h_1 \triangleright h_2); \mathcal{G}; \Gamma \vdash \Delta} \; Eq_2$$

$$\frac{(h_1, h_2 \triangleright h_0); \mathcal{G}[h_0/h_3]; \Gamma[h_0/h_3] \vdash \Delta[h_0/h_3]}{(h_1, h_2 \triangleright h_0); (h_1, h_2 \triangleright h_3); \mathcal{G}; \Gamma \vdash \Delta} \; P \qquad \frac{(h_1, h_2 \triangleright h_0); \mathcal{G}[h_2/h_3]; \Gamma[h_2/h_3] \vdash \Delta[h_2/h_3]}{(h_1, h_2 \triangleright h_0); (h_1, h_3 \triangleright h_0); \mathcal{G}; \Gamma \vdash \Delta} \; C$$

$$\frac{(h_5, h_6 \triangleright h_1); (h_7, h_8 \triangleright h_2); (h_5, h_7 \triangleright h_3); (h_6, h_8 \triangleright h_4); (h_1, h_2 \triangleright h_0); (h_3, h_4 \triangleright h_0); \mathcal{G}; \Gamma \vdash \Delta}{(h_1, h_2 \triangleright h_0); (h_3, h_4 \triangleright h_0); \mathcal{G}; \Gamma \vdash \Delta} \; CS$$

Fig. 2. Structural rules in LS_{SL}.

$$\frac{}{\mathcal{G}; \Gamma; \epsilon : e_1 \mapsto e_2 \vdash \Delta} \mapsto L_1 \qquad \frac{(h_1, h_0 \triangleright h_2); \mathcal{G}; \Gamma; h_1 : e_1 \mapsto e_2 \vdash \Delta}{\mathcal{G}; \Gamma \vdash \Delta} \; HE$$

$$\frac{\begin{array}{l}(\epsilon, h_0 \triangleright h_0); \mathcal{G}[\epsilon/h_1, h_0/h_2]; \Gamma[\epsilon/h_1, h_0/h_2]; h_0 : e_1 \mapsto e_2 \vdash \Delta[\epsilon/h_1, h_0/h_2] \\ (h_0, \epsilon \triangleright h_0); \mathcal{G}[\epsilon/h_2, h_0/h_1]; \Gamma[\epsilon/h_2, h_0/h_1]; h_0 : e_1 \mapsto e_2 \vdash \Delta[\epsilon/h_2, h_0/h_1]\end{array}}{(h_1, h_2 \triangleright h_0); \mathcal{G}; \Gamma; h_0 : e_1 \mapsto e_2 \vdash \Delta} \mapsto L_2$$

$$\frac{}{(h_1, h_2 \triangleright h_0); \mathcal{G}; \Gamma; h_1 : e \mapsto e_1; h_2 : e \mapsto e_2 \vdash \Delta} \mapsto L_3 \qquad \frac{\mathcal{G}; \Gamma\theta; h : e_1\theta \mapsto e_2\theta \vdash \Delta\theta}{\mathcal{G}; \Gamma; h : e_1 \mapsto e_2; h : e_3 \mapsto e_4 \vdash \Delta} \mapsto L_4$$

$$\frac{\mathcal{G}[h_1/h_2]; \Gamma[h_1/h_2]; h_1 : e_1 \mapsto e_2 \vdash \Delta[h_1/h_2]}{\mathcal{G}; \Gamma; h_1 : e_1 \mapsto e_2; h_2 : e_1 \mapsto e_2 \vdash \Delta} \mapsto L_5 \qquad \frac{}{\mathcal{G}; \Gamma; h : nil \mapsto e \vdash \Delta} \; NIL$$

$$\frac{(h_3, h_4 \triangleright h_1); (h_5, h_6 \triangleright h_2); \mathcal{G}; \Gamma; h_3 : e_1 \mapsto e_2; h_5 : e_1 \mapsto e_3 \vdash \Delta \qquad (h_1, h_2 \triangleright h_0); \mathcal{G}; \Gamma \vdash \Delta}{\mathcal{G}; \Gamma \vdash \Delta} \; HC$$

Side conditions:
Each label being substituted cannot be ϵ, each expression being substituted cannot be nil.
In $\mapsto L_4$, $\theta = mgu(\{(e_1, e_3), (e_2, e_4)\})$.
In HE, h_0 occurs in conclusion, h_1, h_2, e_1 are fresh.
In HC, h_1, h_2 occur in the conclusion, $h_0, h_3, h_4, h_5, h_6, e_1, e_2, e_3$ are fresh in the premise.

Fig. 3. Pointer rules in LS_{SL}.

The rules $\mapsto L_1, L_2$ specify that $e_1 \mapsto e_2$ is a singleton heap, so it cannot be empty, nor a composite heap. These were all anticipated in [19]. However, the $\mapsto L_3$ rule proposed in [19], which says that any two singleton heaps with the same address are identical, is unsound for Reynolds's semantics. The corresponding $\mapsto L_3$ rule in Fig. 3 is correct, which states that it is fine to have two singleton heaps with the same address, but they cannot be combined to form another heap. The rules $\mapsto L_5$ and $\mapsto L_4$ state that singleton heaps are uniquely determined by the $\mapsto$ relation. The rule NIL states that nil is not a valid address.

Since the set of addresses is infinite, we can extend any heap with fresh addresses, giving rise to the rule HE. The rule HC captures heap composition:

given any two heaps h_1, h_2, either they can be combined, giving the right premise; or they cannot be combined, hence their domains intersect, i.e., there is some e_1 whose value is in this intersection, yielding the left premise. To our knowledge, proof systems for SL in the literature do not have rules similar to HE and HC, which enable us to prove many formulae that no other systems can prove.

4 Comparison with Existing Proof Calculi

This section compares and contrasts our calculus with existing proof calculi for "separation logics" and points out some subtleties in the literature.

Formula 1 says that any heap can be combined with a composite heap:

$$\neg(((\neg T^*) * (\neg T^*)) {-\!*} \perp) \tag{1}$$

The key to proving this formula is to show that any heap can be extended with a heap that contains at least two singleton mappings. This can be done using the rule HE. We show here the key part of the derivation for the above formula.

$$\frac{\dfrac{\dfrac{(h_2, h_3 \rhd h_4); (h_0, h_1 \rhd h_2); h_0 : ((\neg T^*) * (\neg T^*)) {-\!*} \perp; h_1 : e_1 \mapsto e_2; h_3 : e_3 \mapsto e_4 \vdash}{(h_0, h_1 \rhd h_2); h_0 : ((\neg T^*) * (\neg T^*)) {-\!*} \perp; h_1 : e_1 \mapsto e_2 \vdash} \; {}_{HE}}{; h_0 : ((\neg T^*) * (\neg T^*)) {-\!*} \perp \vdash} \; {}_{HE}}{; \vdash h_0 : \neg(((\neg T^*) * (\neg T^*)) {-\!*} \perp)} \; {}_{\neg R}$$

To our knowledge, current proof systems for separation logic lack this kind of mechanism. It is possible to prove this formula by changing or adding some rules in resource graph tableaux [15], but their proof relies on the restriction that every l in $(l \mapsto e)$ is an address. Thus their method cannot be used in a more general situation like ours.

Formula 2 is another interesting example, it is not valid in Reynolds's separation logic and not provable in LS_{SL}, but is provable in Lee and Park's system [24].

$$(((e_1 \mapsto e_2) * T) {-\!*} \perp) \vee (((e_1 \mapsto e_3) * T) {-\!*} \neg((e_1 \mapsto e_2) {-\!*} \perp)) \vee (e_2 = e_3) \tag{2}$$

The meaning of Formula 2 is not straightforward, but we can construct a counter-model for it in Reynolds's semantics by trying to derive it in LS_{SL}. The following sequent will occur in the backward proof search for Formula 2:

$$(h_5, h_6 \rhd h_1); (h_7, h_8 \rhd h_3); (h_1, h_0 \rhd h_2); (h_3, h_0 \rhd h_4);$$
$$h_5 : (e_1 \mapsto e_2); h_7 : (e_1 \mapsto e_3); h_4 : (e_1 \mapsto e_2) {-\!*} \perp \vdash h_0 : (e_2 = e_3)$$

It is easy to see that the above is a counter-model, and h_5 cannot be combined with h_4. By using Park et al.'s rule $Disj{-\!*}$, we obtain

$$(v_1, v_2 \rhd h_1); (v_2, v_3 \rhd h_3); (v_1, h_4 \rhd w); (v_3, h_2 \rhd w)$$

where v_1, v_2, v_3, w are fresh. The common subheap v_2 of h_1 and h_3 cannot contain h_5, so h_5 must be in v_1. However, if v_1 can be combined with h_4, so can h_5, contradiction. Thus their rule $Disj{-\!*}$ is unsound in Reynolds's semantics.

As mentioned before, our proof system is not complete. For a valid example that cannot be proved by LS_{SL}, consider Formula 3:

$$\top^* \vee (\exists e1, e2.(e1 \mapsto e2)) \vee ((\neg \top^*) * (\neg \top^*)) \tag{3}$$

This formula is valid because any heap can only be either (1) an empty heap, or (2) a singleton heap, or (3) a composite heap. To prove Formula 3, we can add a $\mapsto R$ rule with four premises:

$$\frac{\begin{array}{ll}(1) & (h_1, h_2 \rhd h); (h_1 \neq \epsilon); (h_2 \neq \epsilon); \mathcal{G}; \Gamma \vdash h : e_1 \mapsto e_2; \Delta \\ (2) & \mathcal{G}; \Gamma; h : e_1 \mapsto e_3 \vdash h : e_2 = e_3; h : e_1 \mapsto e_2; \Delta \\ (3) & \mathcal{G}; \Gamma; h : e_3 \mapsto e_4 \vdash h : e_1 = e_3; h : e_1 \mapsto e_2; \Delta \\ (4) & \mathcal{G}[\epsilon/h]; \Gamma[\epsilon/h] \vdash \epsilon : e_1 \mapsto e_2; \Delta[\epsilon/h]\end{array}}{\mathcal{G}; \Gamma \vdash h : e_1 \mapsto e_2; \Delta} \mapsto R$$

The rule $\mapsto R$ essentially negates the semantics for $\mapsto$, giving four possibilities when $e_1 \mapsto e_2$ is false at a heap h: (1) h is a composite heap, so it is possible to split it into two non-empty heaps; (2) h is a singleton heap, its address is the value of e_1, but it does not map this address to the value of e_2; (3) h is a singleton heap, but its address is not the value of e_1; (4) h is the empty heap. We will also need a new type of expression, namely inequality of labels as shown in the first premise. The rule $\mapsto R$ is not included in our proof system for efficiency. For more interesting formula and their derivations, see [18].

5 Inference Rules for Data Structures

Many data structures can be defined inductively by using separation logic's assertion language [7]. We focus here on the widely used singly linked lists. The treatment for binary trees is similar [18]. We adopt several rules from Berdine et al.'s method for symbolic heaps entailment [2] and extend these rules with new ones for formulae outside the symbolic heaps fragment.

We use the definition of linked lists for provers for symbolic heaps [2,7], i.e.,

$$ls(e_1, e_2) \iff (e_1 = e_2 \wedge \top^*) \vee (e_1 \neq e_2 \wedge \exists x.(e_1 \mapsto x * ls(x, e_2)))$$

to facilitate comparison between our prover and the other provers. The inference rules for singly linked lists are given in Fig. 4. The rules LS_6 and LS_7 are for non-symbolic heaps, they handle cases where two lists overlap. There $ds(e, e')$ stands for a data structure that starts from the address e, and ends with e'. We use $ad(e)$ for a data structure that *may* contain the address of value of e, and use $G(ad(e))$ in the succedent to ensure that $ad(e)$ is non-empty.

For LS_8, suppose the heap h_1 is a data structure from e_1 to e_2, and h_3 is a data structure that mentions e_3. By $G(ad(e_3))$ in the succedent, we know that h_3 is non-empty and indeed contains the address of e_3. Since $(h_1, h_3 \rhd h_4)$ holds, the address e_3 is not in the domain of h_1. The labelled formula $h_0 : ls(e_1, e_3)$ in the succedent indicates that h_0 should also make $ds(e_1, e_2) * ls(e_2, e_3)$ false,

$$\dfrac{\mathcal{G};\Gamma[e_1/e_2] \vdash \Delta[e_1/e_2]}{\mathcal{G};\Gamma;\epsilon : ls(e_1,e_2) \vdash \Delta}\ LS_1 \qquad \dfrac{}{\mathcal{G};\Gamma \vdash \epsilon : ls(e,e);\Delta}\ LS_2 \qquad \dfrac{\mathcal{G};\Gamma;h : \top^{*} \vdash \Delta}{\mathcal{G};\Gamma;h : ls(e,e) \vdash \Delta}\ LS_3$$

$$\dfrac{\mathcal{G};\Gamma[nil/e];h : \top^{*} \vdash \Delta[nil/e]}{\mathcal{G};\Gamma;h : ls(nil,e) \vdash \Delta}\ LS_4 \qquad \dfrac{}{\mathcal{G};\Gamma;h : A \vdash h : A;\Delta}\ id_a$$

$$\dfrac{\mathcal{G};\Gamma\theta_1;h : \top^{*} \vdash \Delta\theta_1 \qquad \mathcal{G};\Gamma\theta_2;h : ls(e_1\theta_2,e_2\theta_2) \vdash \Delta\theta_2}{\mathcal{G};\Gamma;h : ls(e_1,e_2);h : ls(e_3,e_4) \vdash \Delta}\ LS_5$$

$$\dfrac{(h_1,h_2 \rhd h_0);\mathcal{G};\Gamma;h_1 : ds(e_1,e_2);h_0 : ls(e_1,e_3);h_2 : ls(e_2,e_3) \vdash \Delta}{(h_1,h_2 \rhd h_0);\mathcal{G};\Gamma;h_1 : ds(e_1,e_2);h_0 : ls(e_1,e_3) \vdash \Delta}\ LS_6$$

$$\dfrac{(h_1,h_2 \rhd h_0);\mathcal{G};\Gamma;h_1 : ds(e_2,e_3);h_0 : ls(e_1,e_3);h_2 : ls(e_1,e_2) \vdash \Delta}{(h_1,h_2 \rhd h_0);\mathcal{G};\Gamma;h_1 : ds(e_2,e_3);h_0 : ls(e_1,e_3) \vdash \Delta}\ LS_7$$

$$\dfrac{\begin{array}{c}(h_1,h_2 \rhd h_0);(h_1,h_3 \rhd h_4);\mathcal{G}; \\ \Gamma;h_1 : ds(e_1,e_2);h_3 : ad(e_3)\end{array} \vdash h_2 : ls(e_2,e_3);h_0 : ls(e_1,e_3);h : G(ad(e_3));\Delta}{(h_1,h_2 \rhd h_0);(h_1,h_3 \rhd h_4);\mathcal{G};\Gamma;h_1 : ds(e_1,e_2);h_3 : ad(e_3) \vdash h_0 : ls(e_1,e_3);h : G(ad(e_3));\Delta}\ LS_8$$

$$\dfrac{}{(h_1,h_2 \rhd h_0);\mathcal{G};\Gamma;h_1 : ad(e_1);h_2 : ad(e_1)' \vdash h_3 : G(ad(e_1));h_4 : G(ad(e_1)');\Delta}\ IC$$

Abbreviations and side conditions:
$ds(e,e')$ is either $(e \mapsto e')$ or $ls(e,e')$.
$ad(e)$ stands for one of $(e \mapsto e')$, $(e \mapsto e',e'')$, $ls(e,e')$, or $tr(e)$, for some e', e''. Similarly for $ad(e)'$.
$G(ad(e))$ is defined as $G(e \mapsto e') \equiv G(e \mapsto e',e'') \equiv \bot$, $G(ls(e,e')) \equiv (e = e')$, $G(tr(e)) \equiv (e = nil)$.
In LS_5, $\theta_1 = mgu(\{(e_1,e_2),(e_3,e_4)\})$ and $\theta_2 = mgu(\{(e_1,e_3),(e_2,e_4)\})$.
In LS_8, if e_3 is nil, then $(h_1,h_3 \rhd h_4)$, $h_3 : ad(e_3)$ and $h : G(ad(e_3))$ in the conclusion are optional.
In LS_8, if $ds(e_1,e_2)$ is $(e_1 \mapsto e_2)$, then $(h_1,h_3 \rhd h_4)$, $h_3 : ad(e_3)$ and $h : G(ad(e_3))$ in the conclusion are optional, on the condition that $h' : (e_1 = e_3)$ occurs in the RHS of the conclusion, for some h'.

Fig. 4. Inference rules for data structures.

thus by an $*R$ application on this formula using $(h_1,h_2 \rhd h_0)$, the branch with $h_1 : ds(e_1,e_2)$ in the succedent can be closed, and we only have the other branch with $h_2 : ls(e_2,e_3)$ in the succedent. There are two special cases as indicated by the side conditions. First, if e_3 is nil, then e_3 can never be an address. Thus we do not need $(h_1,h_3 \rhd h_4)$, $h_3 : ad(e_3)$ and $h : G(ad(e_3))$ in the conclusion. Second, if $ds(e_1,e_2)$ is a singleton heap $(e_1 \mapsto e_2)$, then we only require that e_3 does not have the same value as e_1, thus $(h_1,h_3 \rhd h_4)$, $h_3 : ad(e_3)$ and $h : G(ad(e_3))$ can be neglected as long as $(e_1 = e_3)$ occurs in the succedent.

The rules IC and id_a respectively generalise $\mapsto L_3$ and id. Thus IC captures that two data structures that contain the same address cannot be composed by $*$, and id_a simply forbids a heap to make a formula both true and false.

We refer to the labelled system LS_{SL} plus the rules introduced in Fig. 4 and the rules for binary trees (not shown here, but can be found in [18]) as $LS_{SL} + DS$. The soundness of $LS_{SL} + DS$ for Reynolds's semantics can be proved in the same way as previously showed for Theorem 1.

Recall that symbolic heaps employ slightly different semantics for the multi-field $\mapsto$ predicate, and treat it as a singleton heap. This reading would not make

$$\frac{\mathcal{G}; \Gamma; h : e \mapsto \alpha, \alpha' \vdash h : e \mapsto \alpha; \Delta}{\mathcal{G}; \Gamma; \epsilon : e \mapsto \alpha \vdash \Delta}$$

$$\frac{}{\mathcal{G}; \Gamma; h : nil \mapsto \alpha \vdash \Delta} \qquad \frac{\mathcal{G}; \Gamma\theta; h : e_1 \mapsto \alpha \vdash \Delta\theta}{\mathcal{G}; \Gamma; h : e_1 \mapsto \alpha; h : e_2 \mapsto \alpha' \vdash \Delta} \quad \theta = mgu(\{(e_1, e_2), (\alpha, \alpha')\})$$

$$\frac{(\epsilon, h_0 \triangleright h_0); \mathcal{G}[\epsilon/h_1][h_0/h_2]; \Gamma[\epsilon/h_1][h_0/h_2]; h_0 : e_1 \mapsto \alpha \vdash \Delta[\epsilon/h_1][h_0/h_2]}{(h_0, \epsilon \triangleright h_0); \mathcal{G}[\epsilon/h_2][h_0/h_1]; \Gamma[\epsilon/h_2][h_0/h_1]; h_0 : e_1 \mapsto \alpha \vdash \Delta[\epsilon/h_2][h_0/h_1]}{(h_1, h_2 \triangleright h_0); \mathcal{G}; \Gamma; h_0 : e_1 \mapsto \alpha \vdash \Delta}$$

$$\frac{\mathcal{G}[h_1/h_2]; \Gamma[h_1/h_2]; h_1 : e_1 \mapsto \alpha, \alpha'; h_1 : e_1 \mapsto \alpha \vdash \Delta[h_1/h_2]}{\mathcal{G}; \Gamma; h_1 : e_1 \mapsto \alpha, \alpha'; h_2 : e_1 \mapsto \alpha \vdash \Delta}$$

Fig. 5. Generalised rules for $\mapsto$ with arbitrary fields in non-Reynolds's semantics.

sense in our setting because our logic is based on Reynolds's semantics. Here we develop a branch of our system by compromising both kinds of semantics and viewing $(e_1 \mapsto e_2, e_3)$ as a singleton heap that maps the value of e_1 to the value of e_2, and the next address contains the value of e_3. We give the generalised $\mapsto$ rules for non-Reynolds's semantics in Fig. 5 where α, α' denote any number of fields. For the non-Reynolds's semantics, the rules in Fig. 4 need to be adjusted so that $ds(e1, e2)$ now considers $(e_1 \mapsto e_2, \alpha)$ and $ad(e)$ considers $(e \mapsto \alpha)$. We refer to the variant of $LS_{SL} + DS$ with these changes and the addition of rules in Fig. 5 as $LS'_{SL} + DS$, which is complete for the symbolic heaps fragment; see [18] for the proof.

Theorem 2. *Any symbolic heap formula provable in $LS'_{SL} + DS$ is valid, and any valid symbolic heap formula is provable in $LS'_{SL} + DS$.*

6   Proof Search and Experiment

This section describes proof search and automated reasoning based on the system $LS'_{SL} + DS$, these tactics can also be used on the variant $LS_{SL} + DS$.

We have implemented our labelled calculus $LS'_{SL} + DS$ as a prover called Separata+, in which several restrictions for the logical and structural rules are incorporated without sacrificing provability. See Figs. 1 and 2 for the related inference rules in LS'_{SL}. Some of these restrictions are also used in the prover for PASL [19]. The rule U only creates identity relations for existing labels. The rule A is only applicable when the following holds: if the principal relational atoms are $(h_1, h_2 \triangleright h_0)$ and $(h_3, h_4 \triangleright h_1)$, then the conclusion does not contain $(h_3, h \triangleright h_0)$ and $(h_2, h_4 \triangleright h)$, or any commutative variants of them, for any h.

In applying the cross-split rule CS, we choose the principal relational atoms such that the parent label has the least number of children. Other strategies to apply cross-split are possible; see e.g., [24]. Calcagno et al. [10] showed how to deal with $-\!\!*$ formulae in the quantifier-free fragment, but we do not know whether their results hold for our SL. Nevertheless, inspired by their result, the rules HE, HC in our prover are driven by $-\!\!*$ formulae in the antecedent. We first define a notion of the *size* of a formula as below:

$$|e \mapsto e'| = |e \mapsto e', e''| = 1 \qquad |e = e'| = 0 \qquad |\bot| = 0 \qquad |A * B| = |A| + |B|$$
$$|A \to B| = max(|A|, |B|) \qquad |\exists x.A| = |A| \qquad |\top^*| = 1 \qquad |A \twoheadrightarrow B| = |B|$$

Given a labelled formula $h : A \twoheadrightarrow B$ in the antecedent of a sequent, we allow to use the HE rule to extend h for at most $max(|A|, |B|)/2 + 1$ times instead of $max(|A|, |B|)$ as indicated in [10], because we do not worry about completeness w.r.t. SL here. The HC rule is restricted to only combine three types of heaps: any singleton heaps that occur as subformulae of $A \twoheadrightarrow B$; any heaps created by HE for $A \twoheadrightarrow B$; and any compositions of the previous two. The restrictions on HE and HC are parameters which can be fine-tuned for specific applications.

The atomic formula $e \mapsto \_$, translated to $\exists x.(e \mapsto x)$, is the only type of formula in symbolic heaps that involves quantifiers. Since nested quantifiers are forbidden in symbolic heaps, the $\exists R$ rule can be restricted so that it only instantiates the quantified variable with an existing expression or nil. We call this restricted version $\exists R'$. Although not explicitly allowed in the symbolic heaps fragment nor in our assertion logic, some symbolic heaps provers can recognise numbers. To match them, we check when a rule wants to globally replace a number (expression) by another number, and close the branch immediately because two distinct numbers should not be made equal. The rule $cut_=$ is restricted to apply only on existing expressions and the constant nil.

Our proof search procedure for $LS'_{SL} + DS$ builds in the above tactics, and applies the first applicable rule in the following order:

1. Any zero-premise rule.
2. Any unary rule that involves global substitutions.
3. Any other unary non-structural rule except $\exists R$.
4. Any binary rule that involves global substitutions except $cut_=$.
5. $\to L$.   6. $*R$, $\twoheadrightarrow L$ and $\exists R'$.   7. U, E, A, CS.   8. $cut_=$.

Theorem 3 (Termination for symbolic heaps). *The proof search procedure for $LS'_{SL} + DS$ is complete and terminating for the symbolic heaps fragment.*

The experiments were run on a machine with a Core i7 2600 3.4 GHz processor and 8 GB memory, in Ubuntu 14.04. The code is written in OCaml. Our prover and test suites can be found at [17]. The proof of theorems is in [18].

Our first experiment compares our prover with state-of-the-art provers for symbolic heaps using the Clones benchmark from Navarro and Rybalchenko [27], which is generated from "real life" list manipulating programs and specifications involved in verification. We filter out problems that contain a data structure that we do not consider in this paper, the remaining set consists of 164 valid formulae and 39 invalid formulae. Each Clones test set has the same type of formulae, but the length (number of copies of subformulae) of formulae increases from Clones 1 to Clones 10. We compare our prover with Asterix, Smallfoot, and Cyclist$_{SL}$ [8], the last of which is designed for a $\forall\exists$ DNF-like fragment of separation logic. Cyclist$_{SL}$ cannot recognise numbers, and there are 17 formulae in each Clones test set that cannot be parsed by it (counted as not proved).

Table 2. Experiment 1: the Clones benchmark. Times are in seconds.

Test suite	Test suite with 164 valid formulae						Test suite with 39 invalid formulae					
	Separata+		Cyclist$_{SL}$		Smallfoot		Separata+		Cyclist$_{SL}$		Smallfoot	
	proved	avg. time	proved	avg. time	proved	avg. time	dis-proved	avg. time	dis-proved	avg. time	dis-proved	avg. time
Clones 1	164	0.01	147	0.04	164	0.00	39	0.09	0	-	39	0.00
Clones 2	160	0.02	137	0.17	164	0.00	23	3.37	0	-	39	0.00
Clones 3	159	0.07	126	0.48	164	0.01	9	1.78	0	-	39	0.01
Clones 4	159	0.30	117	0.11	164	0.03	6	7.89	0	-	39	0.02
Clones 5	158	0.03	115	0.13	164	0.15	2	0.52	0	-	39	0.10
Clones 6	158	0.08	114	0.29	164	0.65	2	20.10	0	-	39	0.40
Clones 7	158	0.18	106	0.01	162	0.75	0	-	0	-	39	0.00
Clones 8	158	0.42	106	0.01	160	0.83	0	-	0	-	38	2.10
Clones 9	158	0.89	106	0.01	157	0.36	0	-	0	-	38	5.37
Clones 10	157	1.19	106	0.01	157	0.83	0	-	0	-	32	3.54

Asterix proved every test set with an average of 0.01 s and 100 % successful rate.

Table 2 shows the results of the first experiment. Time out is 50 s. The proved column for each prover shows the number of formulae the prover proves or disproves within the time out, the avg. time column shows the average time used when successfully proving a formula. Unsuccessful attempts counted in average time. Asterix outperformed all the compared provers. Cyclist$_{SL}$ is not complete, so it might terminate without giving a proof. It also cannot determine if a formula is invalid. However, the advantage of Cyclist$_{SL}$ is not in its performance, but in its generality. For example, Cyclist$_{SL}$ can be easily extended to handle other inductive definitions, this is not the case for the other provers in comparison. Separata+ and Smallfoot have similar performance on valid formulae, but Separata+ is not efficient on invalid formulae.

The second experiment features some formulae outside the symbolic heaps fragment, thus we cannot find other provers to compare with, except for a recent work by Thakur, Breck, and Reps [36]. However, their semantics assume acyclic heaps. For example, $(e_1 \mapsto e_2) * (e_2 \mapsto e_1)$ is a satisfiable formula in Reynolds's semantics, but is unsatisfiable in Thakur et al.'s semantics. The fragment of separation logic they consider has "septraction" $A \twoheadrightarrow \circledast B$, defined as $\neg(A \twoheadrightarrow \neg B)$, and only allows classical negation on atomic formulae. Table 3 shows some formulae derived from [36, Table 3], using the definition of septraction as given above. The other formulae from [36, Table 3] are not included, as they are unsatisfiable in Reynolds's semantics. Formula T3.3 to T3.13 in Table 3 are identified as "beyond the scope of existing tools" [36]. More specifically, Formula T3.1, T3.3 and T3.4 describe overlapping data structures; the other formulae in Table 3 demonstrate the use of list and septraction. For example, Formula T3.6 is an instance of an elimination rule for $\twoheadrightarrow \circledast$ and linked list segment in [9].

Maeda, Sato, and Yonezawa [26] provide more examples that use $\twoheadrightarrow$ in program verification. Many of their inferences, e.g. [26, Section 3.1], are easily proved by Separata+ if their syntax is carefully translated into ours, such as Formula 4, which captures a property described in their original type system.

Table 3. Selected formulae from [36, Table3] translated via $A \twoheadrightarrow\!\circledast B \equiv \neg(A \twoheadrightarrow\!\ast \neg B)$.

	Formula
T3.1	$ls(e_1, e_2) \wedge \top^* \wedge \neg(e_1 = e_2)$
T3.2	$\neg((e_1 \mapsto e_2) \twoheadrightarrow\!\ast \neg \top) \wedge ((e_1 \mapsto e_2) \ast \top)$
T3.3	$(ls(e_1, e_2) \ast \neg ls(e_2, e_3)) \wedge ls(e_1, e_3)$
T3.4	$ls(e_1, e_2) \wedge ls(e_1, e_3) \wedge \neg \top^* \wedge \neg(e_2 = e_3)$
T3.5	$\neg(ls(e_1, e_2) \twoheadrightarrow\!\ast \neg ls(e_1, e_2)) \wedge \neg \top^*$
T3.6	$\neg((e_3 \mapsto e_4) \twoheadrightarrow\!\ast \neg ls(e_1, e_4)) \wedge ((e_3 = e_4) \vee \neg ls(e_1, e_3))$
T3.7	$\neg(\neg((e_2 \mapsto e_3) \twoheadrightarrow\!\ast \neg ls(e_2, e_4)) \twoheadrightarrow\!\ast \neg ls(e_1, e_4)) \wedge \neg ls(e_1, e_3)$
T3.8	$\neg(\neg((e_2 \mapsto e_3) \twoheadrightarrow\!\ast \neg ls(e_2, e_4)) \twoheadrightarrow\!\ast \neg ls(e_3, e_1)) \wedge (e_2 = e_4)$
T3.9	$\neg((e_1 \mapsto e_2) \twoheadrightarrow\!\ast \neg ls(e_1, e_3)) \wedge (\neg ls(e_2, e_3) \vee ((\top \wedge ((e_1 \mapsto e_4) \ast \top)) \vee (e_1 = e_3)))$
T3.10	$\neg((ls(e_1, e_2) \wedge \neg(e_1 = e_2)) \twoheadrightarrow\!\ast \neg ls(e_3, e_4)) \wedge \neg(e_3 = e_1) \wedge (e_4 = e_2) \wedge \neg ls(e_3, e_1)$
T3.11	$\neg(e_3 = e_4) \wedge \neg(ls(e_3, e_4) \twoheadrightarrow\!\ast \neg ls(e_1, e_2)) \wedge (e_4 = e_2) \wedge \neg ls(e_1, e_3)$
T3.12	$\neg((ls(e_1, e_2) \wedge \neg(e_1 = e_2)) \twoheadrightarrow\!\ast \neg ls(e_3, e_4)) \wedge \neg(e_3 = e_2) \wedge (e_3 = e_1) \wedge \neg ls(e_2, e_4)$
T3.13	$\neg(\neg((e_2 \mapsto e_3) \twoheadrightarrow\!\ast \neg ls(e_2, e_4)) \twoheadrightarrow\!\ast \neg ls(e_3, e_1)) \wedge (\neg ls(e_4, e_1) \vee (e_2 = e_4))$

Separata+ proved the negation of each listed formula within 0.01 s.

Table 4. Mutated clones benchmark for formulae in Table 3.

Test suite	Separata+		Test suite	Separata+	
	Proved	avg. time		Proved	avg. time
MClones 1	26/26	2.96s	MClones 6	18/26	16.44s
MClones 2	23/26	8.76s	MClones 7	17/26	3.97s
MClones 3	20/26	7.00s	MClones 8	15/26	2.93s
MClones 4	20/26	0.62s	MClones 9	16/26	8.43s
MClones 5	20/26	22.35s	MClones 10	14/26	10.71s

$$(ls(e_0, nil) \twoheadrightarrow\!\ast (ls(e_0, nil) \ast (ls(e_0, nil) \twoheadrightarrow\!\ast ((ls(e_1, nil) \twoheadrightarrow\!\ast ls(e_2, nil)) \ast (e_1 \mapsto e_3) \ast$$
$$ls(e_0, nil))) \ast (ls(e_0, nil) \twoheadrightarrow\!\ast ls(e_3, nil)))) \rightarrow (ls(e_0, nil) \twoheadrightarrow\!\ast (((ls(e_1, nil)$$
$$\twoheadrightarrow\!\ast ls(e_2, nil)) \ast (e_1 \mapsto e_3) \ast ls(e_0, nil)) \ast (ls(e_0, nil) \twoheadrightarrow\!\ast ls(e_3, nil))))$$
$$(4)$$

To challenge our prover further, we build larger formulae generated from Table 3, Formulae 1, 4 (and some formulae similar to 4) and some formulae in [18], totalling 26 formulae inexpressible in symbolic heaps. We use the "clone" method [1] to generate larger formulae, but we make the formulae "harder" by randomly switching the order of starred subformulae. We call these test suites "MClones". The test results are shown in Table 4. The MClones 1 set contains 26 original formulae. The experiment method is the same as before, except that the timeout is set to 500 s. The successful rate drops as the number of cloned subformulae increases. The average time used to prove a formulae, however, fluctuates, because we do not count the timed out attempts. In both experiments, the first test suite (Clones 1 and MClones 1) contains the original formulae in program verification. These formulae can be easily proved by Separata+.

7   Conclusion

We have presented a labelled sequent calculus LS_{SL} for Reynolds's SL. The syntax allows all the logical connectives in SL including $*$, $-\!*$ and quantifiers, the predicate $\mapsto$ and equality. It is impossible to obtain a finite, sound and complete sequent system for this logic, so we focused on soundness, usefulness, and efficiency. With the extension to handle data structures, our proof method is sound, complete, and terminating for the widely used symbolic heaps fragment. Our prover Separata+ showed comparable results as that for Smallfoot on proving valid formulae, although Separata+ does not perform well when the formula is invalid, which may be due to our inference rules having to cover a larger fragment. However, Separata+ can deal with many formulae that, to our knowledge, no other provers for Reynolds's SL can prove. Some of these formulae are taken from existing (manual) proofs to verify algorithms/programs. These indicate that our method would be useful, at least as a part of the tool chain, for program verification with more sophisticated use of separation logic.

Acknowledgment. The third author is partly supported by NTU start-up grant M4081190.020.

References

1. Berdine, J., Calcagno, C., O'Hearn, P.W.: Smallfoot: modular automatic assertion checking with separation logic. In: de Boer, F.S., Bonsangue, M.M., Graf, S., de Roever, W.-P. (eds.) FMCO 2005. LNCS, vol. 4111, pp. 115–137. Springer, Heidelberg (2006)
2. Berdine, J., Calcagno, C., O'Hearn, P.W.: Symbolic execution with separation logic. In: Yi, K. (ed.) APLAS 2005. LNCS, vol. 3780, pp. 52–68. Springer, Heidelberg (2005)
3. Berdine, J., O'Hearn, P.W.: Strong update, disposal, and encapsulation in bunched typing. Electron. Notes Theor. Comput. Sci. **158**, 81–98 (2006)
4. Bornat, R., Calcagno, C., Yang, H.: Variables as resource in separation logic. In: MFPS, vol. 155 of ENTCS, pp. 247–276 (2006)
5. Brochenin, R., Demri, S., Lozes, E.: On the almighty wand. Inform. Comput. **211**, 106–137 (2012)
6. Brookes, S.: A semantics for concurrent separation logic. Theor. Comput. Sci. **375**(1–3), 227–270 (2007)
7. Brotherston, J., Distefano, D., Petersen, R.L.: Automated cyclic entailment proofs in separation logic. In: Björner, N., Sofronie-Stokkermans, V. (eds.) CADE 2011. LNCS, vol. 6803, pp. 131–146. Springer, Heidelberg (2011)
8. Brotherston, J., Gorogiannis, N., Petersen, R.L.: A generic cyclic theorem prover. In: Jhala, R., Igarashi, A. (eds.) APLAS 2012. LNCS, vol. 7705, pp. 350–367. Springer, Heidelberg (2012)

9. Calcagno, C., Parkinson, M., Vafeiadis, V.: Modular safety checking for fine-grained concurrency. In: Riis Nielson, H., Filé, G. (eds.) SAS 2007. LNCS, vol. 4634, pp. 233–248. Springer, Heidelberg (2007)

10. Calcagno, C., Yang, H., O'Hearn, P.W.: Computability and complexity results for a spatial assertion language for data structures. In: Hariharan, R., Mukund, M., Vinay, V. (eds.) FSTTCS 2001. LNCS, vol. 2245, pp. 108–119. Springer, Heidelberg (2001)

11. Demri, S., Galmiche, D., Larchey-Wendling, D., Méry, D.: Separation logic with one quantified variable. In: Hirsch, E.A., Kuznetsov, S.O., Pin, J.É., Vereshchagin, N.K. (eds.) CSR 2014. LNCS, vol. 8476, pp. 125–138. Springer, Heidelberg (2014)

12. Demri, S., Deters, M.: Expressive completeness of separation logic with two variables and no separating conjunction. In: CSL/LICS, Vienna (2014)

13. Distefano, D., Matthew, P.: jStar: towards practical verification for java. ACM Sigplan Not. **43**, 213–226 (2008)

14. Dockins, R., Hobor, A., Appel, A.W.: A fresh look at separation algebras and share accounting. In: Hu, Z. (ed.) APLAS 2009. LNCS, vol. 5904, pp. 161–177. Springer, Heidelberg (2009)

15. Galmiche, D., Méry, D.: Tableaux and resource graphs for separation logic. J. Logic Comput. **20**(1), 189–231 (2007)

16. Hobor, A., Villard, J.: The ramifications of sharing in data structures. In: POPL 2013, pp. 523–536. ACM, New York, NY, USA (2013)

17. Hóu, Z.: Separata+. http://users.cecs.anu.edu.au/zhehou/

18. Hóu, Z.: Labelled Sequent Calculi and Automated Reasoning for Assertions in Separation Logic. Ph.D. thesis, The Australian National University (2015). Submitted

19. Hóu, Z., Clouston, R., Goré, R., Tiu, A.: Proof search for propositional abstract separation logics via labelled sequents. In: POPL, pp. 465–476. ACM (2014)

20. Hóu, Z., Tiu, A., Goré, R.: A labelled sequent calculus for BBI: proof theory and proof search. In: Galmiche, D., Larchey-Wendling, D. (eds.) TABLEAUX 2013. LNCS, vol. 8123, pp. 172–187. Springer, Heidelberg (2013)

21. Jensen, J.: Techniques for model construction in separation logic. Report (2013)

22. Jensen, J.B., Birkedal, L.: Fictional separation logic. In: Seidl, H. (ed.) Programming Languages and Systems. LNCS, vol. 7211, pp. 377–396. Springer, Heidelberg (2012)

23. Krishnaswami, N.R.: Reasoning about iterators with separation logic. In: SAVCBS, pp. 83–86. ACM (2006)

24. Lee, W., Park, S.: A proof system for separation logic with magic wand. In: POPL 2014, pp. 477–490. ACM, New York, NY, USA (2014)

25. Maclean, E., Ireland, A., Grov, G.: Proof automation for functional correctness in separation log. J. Logic Comput. (2014)

26. Maeda, T., Sato, H., Yonezawa, A.: Extended alias type system using separating implication. In: TLDI 2011, pp. 29–42. ACM, New York, NY, USA (2011)

27. Navarro Pérez, J.A., Rybalchenko, A.: Separation logic + superposition calculus = heap theorem prover. In: PLDI 2011, pp. 556–566. ACM, USA (2011)

28. Navarro Pérez, J.A., Rybalchenko, A.: Separation logic modulo theories. In: Shan, C. (ed.) APLAS 2013. LNCS, vol. 8301, pp. 90–106. Springer, Heidelberg (2013)

29. O'Hearn, P.W., Pym, D.J.: The logic of bunched implications. Bull. Symbolic Logic **5**(2), 215–244 (1999)

30. O'Hearn, P.W., Reynolds, J.C., Yang, H.: Local reasoning about programs that alter data structures. In: Fribourg, L. (ed.) CSL 2001 and EACSL 2001. LNCS, vol. 2142, pp. 1–19. Springer, Heidelberg (2001)

31. Parkinson, M., Bornat, R., Calcagno, C.: Variables as resource in hoare logics. In: 21st LICS (2006)
32. Reus, B., Schwinghammer, J.: Separation logic for higher-order store. In: Ésik, Z. (ed.) CSL 2006. LNCS, vol. 4207, pp. 575–590. Springer, Heidelberg (2006)
33. Reynolds, J.C.: Intuitionistic reasoning about shared mutable data structure. In: Millennial Perspectives in Computer Science, pp. 303–321. Palgrave (2000)
34. Reynolds, J.C.: Separation logic: a logic for shared mutable data structures. In: LICS, pp. 55–74. IEEE (2002)
35. Stewart, G., Beringer, L., Appel, A.W.: Verified heap theorem prover by paramodulation. In: ICFP, pp. 3–14. ACM (2012)
36. Thakur, A., Breck, J., Reps, T.: Satisfiability modulo abstraction for separation logic with linked lists. Technical report. University of Wisconsin (2014)
37. Vafeiadis, V., Parkinson, M.: A marriage of rely/guarantee and separation logic. In: Caires, L., Vasconcelos, V.T. (eds.) CONCUR 2007. LNCS, vol. 4703, pp. 256–271. Springer, Heidelberg (2007)

KeY-ABS: A Deductive Verification Tool for the Concurrent Modelling Language ABS

Crystal Chang Din[(✉)], Richard Bubel, and Reiner Hähnle

Department of Computer Science, Technische Universität Darmstadt,
Darmstadt, Germany
{crystald,bubel,haehnle}@cs.tu-darmstadt.de

Abstract. We present KeY-ABS, a tool for deductive verification of concurrent and distributed programs written in ABS. KeY-ABS allows to verify data dependent and history-based functional properties of ABS models. In this paper we give a glimpse of system workflow, tool architecture, and the usage of KeY-ABS. In addition, we briefly present the syntax, semantics and calculus of KeY-ABS Dynamic Logic (ABSDL). The system is available for download.

1 Introduction

KeY-ABS is a deductive verification system for the concurrent modelling language ABS [1,8,11]. It is based on the KeY theorem prover [2]. KeY-ABS provides an interactive theorem proving environment and allows one to prove properties of object-oriented and concurrent ABS models. The concurrency model of ABS has been carefully engineered to admit a proof system that is modular and permits to reduce correctness of concurrent programs to reasoning about sequential ones [3,5]. The deductive component of KeY-ABS is an axiomatization of the operational semantics of ABS in the form of a sequent calculus for first-order dynamic logic for ABS (ABSDL). The rules of the calculus that axiomatize program formulae define a symbolic execution engine for ABS. The system provides heuristics and proof strategies that automate ≥90 % of proof construction. For example, first-order reasoning, arithmetic simplification, symbolic state simplification, and symbolic execution of loop- and recursion-free programs are performed mostly automatically. The remaining user input typically consists of universal and existential quantifier instantiations.

ABS is a rich language with Haskell-like (first-order) datatypes, Java-like objects and thread-based as well as actor-based concurrency. In contrast to model checking, KeY-ABS allows to verify complex functional properties of systems with unbounded size [6]. In this paper we concentrate on the design of the KeY-ABS prover and its usage. KeY-ABS consists of around 11,000 lines of Java code (KeY-ABS + reused parts of KeY: ca. 100,000 lines). The rule base consists of ca. 10,000 lines written in KeY's *taclets* rule description language [2].

This work was done in the context of the EU project FP7-610582 *Envisage: Engineering Virtualized Services* (http://www.envisage-project.eu).

A.P. Felty and A. Middeldorp (Eds.): CADE-25, LNAI 9195, pp. 517–526, 2015.
DOI: 10.1007/978-3-319-21401-6_35

At http://www.envisage-project.eu/?page_id=1558 the KeY-ABS tool and a screencast showing how to use it can be downloaded.

2   The Design of KeY-ABS

2.1   System Workflow

The input files to KeY-ABS comprise (i) an *.abs* file containing ABS programs and (ii) a *.key* file containing the class invariants, functions, predicates and specific proof rules required for this particular verification case. Given these input files, KeY-ABS opens a proof obligation selection dialogue that lets one choose a target method implementation. From the selection the proof obligation generator creates an ABSDL formula. By clicking on the **Start** button the verifier will try to automatically prove the generated formula. A positive outcome shows that the target method preserves the specified class invariants. In the case that a subgoal cannot be proved automatically, the user is able to interact with the verifier to choose proof strategies and proof rules manually. The reason for a formula to be unprovable might be that the target method implementation does not preserve one of the class invariants, that the specified invariants are too weak/too strong or that additional proof rules are required. The workflow of KeY-ABS is illustrated in Fig. 1.

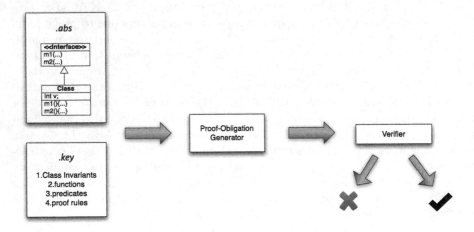

Fig. 1. Verification workflow of KeY-ABS

2.2   The Concurrency Model of ABS

In ABS [1,8,11] two different kinds of concurrency are supported depending on whether two objects belong to the same or to different *concurrent object groups*

(COGs). The affinity of an object to a COG is determined at creation time. The creator decides whether the object should be assigned to a new COG or to the COG of its creator. Within a COG several threads might exist, but only one of these threads (and hence one object) can be active at any time. Another thread can only take over when the current active thread explicitly releases control. In other words, ABS realizes *cooperative scheduling* within a COG. All interleaving points occur syntactically in an ABS program in the form of an *await* or *suspend* statement by which the current thread releases control.

While one COG represents a single processor with task switching and shared memory, two different COGs run actually in parallel and are separated by a network. As a consequence, objects within the same COG may communicate either by asynchronous or by synchronous method invocation, while objects living on different COGs *must* communicate with asynchronous method invocation and message passing. Any asynchronous method invocation creates a so called *future* as its immediate result. Futures are a handle for the result value once it becomes available. Attempting to access an unavailable result blocks the current thread and its COG until the result value is available. To avoid this, retrieval of futures is usually guarded with an *await* statement that, instead of blocking, releases control in case of an unavailable result. Futures are first-class citizens and can be assigned to local variables, object fields, and passed as method arguments.

In contrast to current industrial programming languages such as C++ or Java which support multithreaded concurrency, ABS has a fully formalized concurrency model and natively supports distributed computation.

2.3 Verification Approach for ABS Programs

An asynchronous method call in ABS does not transfer the execution control from the caller to the callee, but leads to a new process on the called object. Remote field access is not supported by the language, so there is no shared variable communication between different objects. This is a stronger notion of privacy than in Java where other instances of the same class can access private fields. Consequently, different concurrent objects in ABS do not have aliases to the same object. Hence, state changes made by one object do not affect other objects. The concurrency model of ABS is compositional by design.

To make verification of ABS programs modular, KeY-ABS follows the monitor [9] approach. We define invariants for each ABS class and reason locally about each class by KeY-ABS. Each class invariant is required to hold after initialization in all class instances, before any process release point, and upon termination of each method call. Consequently, whenever a process is released, either by termination of a method execution or by a release point, the thread that gains execution control can rely on the class invariant to hold. The proof rule for compositional reasoning about ABS programs is given and proved sound in [5], by which we obtain system invariants from the proof results by KeY-ABS.

To write meaningful invariants of concurrent systems, it must be possible to refer to previous communication events. The observable behavior of a system can be described by *communication histories* over observable events [10]. Since

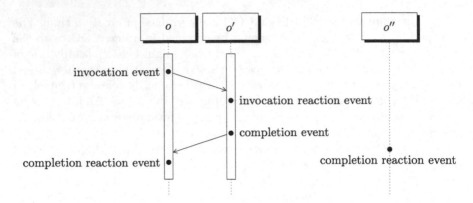

Fig. 2. History events and when they occur on objects o, o' and o''

message passing in ABS does not transfer the execution control from the caller to the callee, in KeY-ABS we consider separate events for method invocation, for reacting upon a method call, for resolving a future, and for fetching the value of a future. Each event can only be observed by one object, namely the object that generates it. Assume an object o calls a method on object o' and generates a future identity fr associated to the method call. An invocation message is sent from o to o' when the method is invoked. This is reflected by the *invocation event* generated by o and illustrated by the sequence diagram in Fig. 2. An *invocation reaction event* is generated by o' once the method starts execution. When the method terminates, the object o' generates the *completion event*. This event reflects that the associated future is resolved, i.e., it contains the called method's result. The *completion reaction event* is generated by the caller o when fetching the value of the resolved future. Since future identities may be passed to a third object o'', that object may also fetch the future value, reflected by another *completion reaction event*, generated by o'' in Fig. 2. All events generated by one object forms the local history of the object. When composing objects, the local histories of the composed objects are merged to a common history, containing all the events of the composed objects [14].

2.4   Syntax and Semantics of the KeY-ABS Logic

Specification and verification of ABS models is done in KeY-ABS dynamic logic (ABSDL). ABSDL is a typed first-order logic plus a box modality: For a sequence of executable ABS statements S and ABSDL formulae P and Q, the formula $P \rightarrow [S]Q$ expresses: If the execution of S starts in a state where the assertion P holds and the program terminates normally, then the assertion Q holds in the final state. Verification of an ABSDL formula proceeds by symbolic execution of S, where state modifications are handled by the *update* mechanism [2]. An *elementary update* has the form $U = \{loc := val\}$, where loc is a *location* expression and val is its new value term. Updates can only be applied to formulae or terms, i.e. $U\phi$. Semantically, the validity of $U\phi$ in state s is defined as the validity of ϕ in state s', which is state s with the modification of *loc* according to the update U. There are operations for sequential as well as parallel composition of updates. Typically, loop- and recursion-free sequences of program statements can be turned into updates fully automatically. Given an ABS method m with body mb and a class invariant I, the ABSDL formula $I \rightarrow [mb]I$ expresses that the method m preserves the class invariant.

KeY-ABS natively supports concurrency in its program logic. In ABSDL we express properties of a system in terms of histories. This is realized by a dedicated, global program variable `history`, which contains the object local histories as a sequence of events. The history events themselves are elements of datatype `HistoryLabel`, which defines for each event type a constructor function. For instance, a completion event is represented as $compEv(o, fr, m, e)$ where o is the callee, fr the corresponding future, m the method name, and e the return result of the method execution. In addition to the history formalisation as a sequence of events, there are a number of built-in functions and predicates that allow to express common properties concerning histories. For example, function $getFuture(e)$ returns the future identity contained in the event e, and predicate $isInvocationEv(e)$ returns true if event e is an invocation event.

The type system of KeY-ABS reflects the ABS type system. Besides the type `HistoryLabel`, the type system of ABSDL contains, for example, the sequence type `Seq`, the root reference type `any`, the super type `ABSAnyInterface` of all ABS objects, the future type `Future`, and the type `null`, which is a subtype of all reference types. Users can define their own functions, predicates and types, which are used to represent the interfaces and abstract data types of a given ABS program.

2.5 Rule Formalisation

The user can interleave the automated proof search implemented in KeY-ABS with interactive rule application. For the latter, the KeY-ABS prover has a graphical user interface that is built upon the idea of direct manipulation. To apply a rule, the user first selects a focus of application by highlighting a (sub-)formula or a (sub-)term in the goal sequent. The prover then offers a choice of rules applicable at this focus. Rule schema variable instantiations are mostly inferred by matching. Figure 3 shows an example of proof rule selection in KeY-ABS. The user is about to apply the *awaitExp* rule that executes an await statement.

Another way to apply rules and provide instantiations is by drag and drop. The user simply drags an equation onto a term, and the system will try to rewrite the term with the equation. If the user drags a term onto a quantifier the system will try to instantiate the quantifier with this term.

The interaction style is closely related to the way rules are formalised in KeY-ABS. All rules are defined as *taclets* [2]. Here is a (slightly simplified) example:

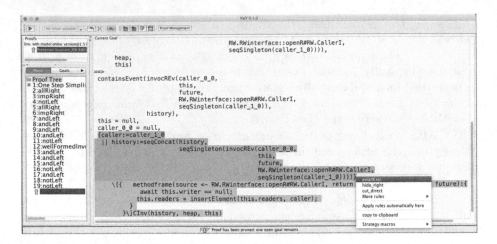

Fig. 3. Proof rule selection

$\texttt{\textbackslash find}\ ([\{method(source \leftarrow m, return \leftarrow (var : r,\ fut : u)) : \{\texttt{return}\ exp; \}\}]\phi)$
$\texttt{\textbackslash replacewith}\ (\{\texttt{history} := seqConcat(\texttt{history}, compEv(\texttt{this}, u, m, exp)))\}\phi)$
$\texttt{\textbackslash heuristics}\ (simplify\_prog)$

The rule symbolically executes a **return** statement inside a method invocation.
It applies the *update* mechanism to the variable **history**, which is extended with
a *completion event* capturing the termination and return value of the method
execution. The **find** clause specifies the potential application focus. The taclet
will be offered to the user on selecting a matching focus. The action clause
replacewith modifies the formula in focus. The **heuristics** clause provides
priority information to the parameterized automated proof search strategy. The
taclet language is quickly mastered and makes the rule base easy to maintain
and extend. A full account of the taclet language is given in [2].

2.6 KeY-ABS Architecture

Figure 4 depicts the principal architecture of the KeY-ABS   system.
KeY-ABS is based on the KeY 2.0
platform—a verification system for
Java. To be able to reuse most parts of
the system, we had to generalize various subsystems and to abstract away
from their Java specifics. For instance,
the rule application logic of KeY made
several assumptions which are valid for
Java but not for other programming
languages. Likewise, the specification

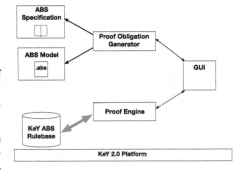

Fig. 4. The architecture of KeY-ABS

framework of KeY, even though it provided general interfaces for contracts and invariants, made implicit assumptions that were insufficient for our communication histories and needed to be factored out. After refactoring the KeY system provides core subsystems (rule engine, proof construction, search strategies, specification language, proof management etc.) that are independent of the specific program logic or target language. These are then extended and adapted by the ABS and Java backends.

The proof obligation generator needs to parse the source code of the ABS model and the specification. For the source code we use the parser as provided by the ABS toolkit [1,15] with no changes. The resulting abstract syntax tree is then converted into KeY's internal representation. The specification parser for ABSDL formulas is an adapted version of the parser for JavaDL [2]. The rule base for ABSDL reuses the language-independent theories of the KeY tool, such as arithmetic, sequences and first-order logic. The rules for symbolic execution have been written from scratch for ABS as well as the formalisation of the history datatype.

3   The Usage of KeY-ABS

The ABS language was designed around a concurrency model whose analysis stays manageable. The restriction of the ABS concurrency model, specifically the fact that scheduling points are syntactically in the code, makes it possible to define a compositional specification and verification method. This is essential for being able to scale verification to non-trivial programs, because it is possible to specify and verify each ABS method separately, without the need for a global invariant. KeY-ABS follows the Design-by-Contract paradigm with an emphasis on specification of class invariants for concurrent and distributed programs in ABS. In the following we will show some examples of how and what we can specify in a class invariant.

A history-based class invariant in ABSDL can relate the state of an object to the local history of the system. A simple banking system is verified in [3] by KeY-ABS , where an invariant ensures that the value of the account balance (a class attribute) always coincides with the value returned by the most recent call to a deposit or withdraw method (captured in the history). Here we use a more ambitious case study to illustrate this style of class invariant. In Fig. 5 an ABS implementation of the classic reader-writer problem [4] is shown. The RWController class provides read and write operations to clients and four methods to synchronize reading and writing activities: openR, closeR, openW and closeW.

The class attribute *readers* contains a set of clients currently with read access and *writer* contains the client with write access. The set of *readers* is extended by execution of openR or openW, and is reduced by closeR or closeW. The *writer* is added by execution of openW and removed by closeW. Two class invariants of the reader-writer example are (slightly simplified) shown in Fig. 6, in which the invariants *isReader* and *isWriter* express that the value of class attributes *readers* and *writer* are always equal to the set of relevant callers extracted from

```
class RWController implements RWinterface {
  Set<CallerI> readers = EmptySet; CallerI writer = null;

  Unit openR(CallerI caller){
    await writer == null;
    readers = insertElement(readers, caller);}

  Unit closeR(CallerI caller){
    readers = remove(readers, caller);}

  Unit openW(CallerI caller){
    await writer == null;
    writer = caller; readers = insertElement(readers, caller);}

  Unit closeW(CallerI caller){
    await writer == caller;
    writer = null; readers = remove(readers, caller);}

  String read(CallerI caller, Int key){...}

  Unit write(CallerI caller, Int key, String value){...}
```

Fig. 5. The controller class of the RW example in ABS

```
\invariants(Seq historySV, Heap heapSV, ABSAnyInterface self) {
  isReader : RW.RWController {
    RW.RWController::self.readers
    = currentReaders(historySV)
  };

  isWriter : RW.RWController {
    insertElement(EmptySet, RW.CallerI::self.writer)
    = currentWriter(historySV)
  };
}
```

Fig. 6. Class invariants of the RW example

the current history. The keyword **invariants** opens a section where invariants can be specified. Its parameters declare program variables that can be used to refer to the history (historySV), the heap (heapSV, implicit by attribute access), and the current object (self, similar as Java's *this*).

The functions *currentReaders*(h) and *currentWriter*(h) are defined inductively over the history h to capture a set of existing callers to the corresponding methods. The statistics of verifying these two invariants are in Fig. 7. For each of the six methods of the RWController class we show it satisfies *isReader* and *isWriter*. For instance, the proof tree for verifying the invariant *isReader* for method openR contains 3884 nodes and 12 branches. Verification of this case study was automatic except for a few instantiations of quantifiers and the rule application on inductive functions.[1]

---

[1] The complete model of the reader-writer example with all formal specifications and proofs is available at
https://www.se.tu-darmstadt.de/se/group-members/crystal-chang-din/rw.

invariants\methods	openR	closeR	openW	closeW	read	write
isReader	3884 − 12	1147 − 7	2836 − 9	3904 − 12	5459 − 26	3572 − 35
isWriter	2735 − 9	739 − 5	3891 − 12	3818 − 12	5056 − 29	4058 − 32

Fig. 7. Verification Result of RW example: # nodes − # branches

A history-based class invariant in ABSDL can also express structural properties of the history, for example, that an event ev_1 occurs in the history before an event ev_2 is generated. To formalize this kind of class invariant, quantifiers and indices of sequences are used to locate the events at certain positions of the history. Recently, we applied the KeY-ABS system to a case study of an ABS model of a Network-on-Chip (NoC) packet switching platform [12], called ASPIN (Asynchronous Scalable Packet Switching Integrated Network) [13]. It is currently the largest ABS program we have proved by KeY-ABS . We proved that ASPIN drops no packets and a packet is never sent in a circle by compositional reasoning. The ABS model, the specifications and the proof rules can be found in [6]. Both styles of class invariants mentioned above were used. The KeY-ABS verification approach to the NoC case study deals with an *unbounded* number of objects and is valid for generic NoC models for any m × n mesh in the ASPIN chip as well as any number of sent packets.

The global history of the whole system is formed by the composition of the local history of each instance of the class in the system. A *global invariant* can be obtained as a conjunction of instances of the class invariants verified by KeY-ABS for all objects in the system, adding wellformedness of the global history [7]. This allows to prove *global* safety properties of the system using *local* rules and symbolic execution, such as absence of packet loss and no circular packet sending. In contrast to model checking this allows us to deal effectively with unbounded target systems without suffering from state explosion.

4   Conclusion

We presented the KeY-ABS formal verification tool for the concurrent modelling language ABS. ABS is a rich, fully executable language with unbounded data structures and Java-like control structures as well as objects. It offers thread-based as well as actor-based concurrency with the main restriction being that scheduling points are made syntactically in the code ("cooperative scheduling"). KeY-ABS implements a compositional proof system [4,5] for ABS. Its architecture is based on the state-of-art Java verification tool KeY and KeY-ABS reuses some of KeY's infrastructure.

KeY-ABS is able to verify global, functional properties of considerable complexity for unbounded systems. At the same time, the degree of automation is high. Therefore, KeY-ABS is a good alternative for the verification of unbounded, concurrent systems where model checking is not expressive or scalable enough.

References

1. The ABS tool suite. https://github.com/abstools/abstools. Accessed 17 May 2015
2. Beckert, B., Hähnle, R., Schmitt, P.H. (eds.): Verification of object-oriented software: the KeY approach. LNCS (LNAI), vol. 4334. Springer, Heidelberg (2007)
3. Bubel, R., Montoya, A.F., Hähnle, R.: Analysis of executable software models. In: Bernardo, M., Damiani, F., Hähnle, R., Johnsen, E.B., Schaefer, I. (eds.) SFM 2014. LNCS, vol. 8483, pp. 1–25. Springer, Heidelberg (2014)
4. Din, C.C., Dovland, J., Johnsen, E.B., Owe, O.: Observable behavior of distributed systems: component reasoning for concurrent objects. J. Logic Algebraic Program. **81**(3), 227–256 (2012)
5. Din, C.C., Owe, O.: Compositional reasoning about active objects with shared futures. Formal Aspects Comput. **27**(3), 551–572 (2015)
6. Din, C.C., Tarifa, S.L.T., Hähnle, R., Johnsen, E.B.: The NoC verification case study with KeY-ABS. Technical report, Department of Computer Science, Technische Universität Darmstadt, Germany, February 2015
7. Dovland, J., Johnsen, E.B., Owe, O.: Verification of concurrent objects with asynchronous method calls. In: Proceedings of the IEEE International Conference on Software Science, Technology & Engineering (SwSTE 2005), pp. 141–150. IEEE Computer Society Press, February 2005
8. Hähnle, R.: The abstract behavioral specification language: a tutorial introduction. In: Giachino, E., Hähnle, R., de Boer, F.S., Bonsangue, M.M. (eds.) FMCO 2012. LNCS, vol. 7866, pp. 1–37. Springer, Heidelberg (2013)
9. Hoare, C.A.R.: Monitors: an operating system structuring concept. Commun. ACM **17**(10), 549–557 (1974)
10. Hoare, C.A.R.: Communicating Sequential Processes. Prentice-Hall International Series in Computer Science, Upper Saddle River (1985)
11. Johnsen, E.B., Hähnle, R., Schäfer, J., Schlatte, R., Steffen, M.: ABS: a core language for abstract behavioral specification. In: Aichernig, B.K., de Boer, F.S., Bonsangue, M.M. (eds.) FMCO 2010. LNCS, vol. 6957, pp. 142–164. Springer, Heidelberg (2011)
12. Kumar, S., Jantsch, A., Millberg, M., Öberg, J., Soininen, J., Forsell, M., Tiensyrjä, K., Hemani, A.: A network on chip architecture and design methodology. In: 2002 IEEE Computer Society Annual Symposium on VLSI (ISVLSI 2002), Pittsburgh, PA, USA, 25–26 April 2002, pp. 117–124 (2002)
13. Sheibanyrad, A., Greiner, A., Panades, I.M.: Multisynchronous and fully asynchronous NoCs for GALS architectures. IEEE Des. Test Comput. **25**(6), 572–580 (2008)
14. Soundararajan, N.: A proof technique for parallel programs. Theoret. Comput. Sci. **31**(1–2), 13–29 (1984)
15. Wong, P.Y.H., Albert, E., Muschevici, R., Proença, J., Schäfer, J., Schlatte, R.: The ABS tool suite: modelling, executing and analysing distributed adaptable object-oriented systems. STTT **14**(5), 567–588 (2012)

KeYmaera X: An Axiomatic Tactical Theorem Prover for Hybrid Systems

Nathan Fulton[1]([⊠]), Stefan Mitsch[1], Jan-David Quesel[1], Marcus Völp[1,2], and André Platzer[1]

[1] Computer Science Department, Carnegie Mellon University,
Pittsburgh, PA 15213, USA
{nathanfu,smitsch,jquesel,aplatzer}@cs.cmu.edu
[2] Technische Universität Dresden, 01157 Dresden, Germany
marcus.voelp@tu-dresden.de

Abstract. KeYmaera X is a theorem prover for *differential dynamic logic* (d$\mathcal{L}$), a logic for specifying and verifying properties of hybrid systems. Reasoning about complicated hybrid systems models requires support for sophisticated proof techniques, efficient computation, and a user interface that crystallizes salient properties of the system. KeYmaera X allows users to specify custom proof search techniques as *tactics*, execute these tactics in parallel, and interface with partial proofs via an extensible user interface.

Advanced proof search features—and user-defined tactics in particular—are difficult to check for soundness. To admit extension and experimentation in proof search without reducing trust in the prover, KeYmaera X is built up from a *small trusted kernel*. The prover kernel contains a list of sound d$\mathcal{L}$ axioms that are instantiated using a uniform substitution proof rule. Isolating all soundness-critical reasoning to this prover kernel obviates the intractable task of ensuring that each new proof search algorithm is implemented correctly. Preliminary experiments suggest that a single layer of tactics on top of the prover kernel provides a rich language for implementing novel and sophisticated proof search techniques.

1 Introduction

Computational control of physical processes such as cyber-physical systems introduces complex interactions between discrete and continuous dynamics. Developing techniques for reasoning about this interaction is important to prevent software bugs from causing harm in the real world. For this reason, formal verification of safety-critical software is upheld as best practice [4].

This material is based upon work supported by the National Science Foundation under NSF CAREER Award CNS-1054246, NSF CNS-1035800, and CNS-0931985, and by ERC under PIOF-GA-2012-328378 (Mitsch on leave from Johannes Kepler University Linz).
KeYmaera X is available for download from http://keymaerax.org/.

A.P. Felty and A. Middeldorp (Eds.): CADE-25, LNAI 9195, pp. 527–538, 2015.
DOI: 10.1007/978-3-319-21401-6_36

Verifying correctness properties about cyber-physical systems requires analyzing the system's discrete and continuous dynamics together in a hybrid system [2]. For example, establishing the correctness of an adaptive cruise control system in a car requires reasoning about the computations of the controller together with the resulting physical motion of the car. Theorem proving is a useful technique for proving correctness properties of hybrid systems [11]. Theorem proving complements model checking and reachability analysis, which are successful at finding bugs in discrete systems.

A theorem prover for hybrid systems must be sound to ensure trustworthy proofs, and should be flexible to enable efficient proof search. This paper presents KeYmaera X, a hybrid system theorem prover that meets these conflicting goals. Its design emphasizes a clear separation between a small soundness-critical prover kernel and the rest of the theorem prover. This separation ensures trust in the prover kernel and allows extension of the prover with user-defined proof strategies and custom user interfaces.

We build on experience with KeYmaera [15], an extension of the KeY theorem prover [1]. The success of KeYmaera in cyber-physical systems is due, in part, to its support for reasoning about programs with differential equations and its integration of real arithmetic decision procedures. Case studies include adaptive cruise control and autonomous automobile control, the European Train Control System, aircraft collision avoidance maneuvers, autonomous robots, and surgical robots. Despite the prior successes of KeYmaera, however, its monolithic architecture makes it increasingly difficult to scale to large systems. Aside from soundness concerns, a monolithic architecture precludes extensions necessary for proofs of larger systems, parallel proof search, or proof strategies for specific analyses such as model refinement or monitor synthesis.

KeYmaera X is a clean-slate reimplementation to replace KeYmaera. KeYmaera X focuses on a small trusted prover kernel, extensive tactic support for steering proof search, and a user interface intended to support a mixture of interactive and automatic theorem proving. KeYmaera X improves on automation when compared to KeYmaera for our ModelPlex case study: it automates the otherwise ≈60 % manual steps in [8].

2   KeYmaera X Feature Overview

Hybrid Systems. Hybrid dynamical systems [2,12] are mathematical models for analyzing the interaction between discrete and continuous dynamics.

Hybrid automata [2] are a machine model of hybrid systems. A hybrid automaton is a finite automaton over an alphabet of real variables. Variables may instantaneously take on new values upon state transitions. Unlike classical finite automata, each state is associated with a continuous dynamical system (modeled using ordinary differential equations) defined over an evolution domain. Whenever the system enters a new state, the variables of the system evolve according to the continuous dynamics and within the evolution domain associated with that state. Hybrid automata are not conducive to *compositional*

Table 1. Hybrid Programs

Program statement	Meaning
$\alpha; \beta$	Sequential composition of α and β.
$\alpha \cup \beta$	Nondeterministic choice ($\cup$) executes either α or β.
α^*	Repeats α zero or more times.
$x := \theta$	Evaluate the expression θ and assign its result to x.
$x := *$	Assigns some arbitrary real value to x.
$\{x_1' = \theta_1, ..., x_n' = \theta_n \& F\}$	Continuous evolution along the differential equation system $x_i' = \theta_i$ for an arbitrary duration within the region described by formula F.
$?F$	Tests if formula F is true at current state, aborts otherwise.

reasoning; to establish a property about a hybrid automaton, it does not suffice to establish that property about each component of a decomposed system.

Hybrid programs [10–12], in contrast, are a compositional programming language model of hybrid dynamics. They extend regular programs with differential equations. A syntax and informal semantics of hybrid programs is given in Table 1.

Differential Dynamic Logic. Differential dynamic logic ($d\mathcal{L}$) [10–12] is a first-order multimodal logic for specifying and proving properties of hybrid programs. Each hybrid program α is associated with modal operators $[\alpha]$ and $\langle\alpha\rangle$, which express state reachability properties of the parametrizing program. For example, $[\alpha]\phi$ states that the formula ϕ is true in any state reachable by the hybrid program α. Similarly, $\langle\alpha\rangle\phi$ expresses that the property ϕ is true after some execution of α. The $d\mathcal{L}$ formulas are generated by the EBNF grammar

$$\phi ::= \theta_1 \backsim \theta_2 \mid \neg\phi \mid \phi \wedge \psi \mid \phi \vee \psi \mid \phi \rightarrow \psi \mid \phi \leftrightarrow \psi \mid \forall x\,\phi \mid \exists x\,\phi \mid [\alpha]\phi \mid \langle\alpha\rangle\phi$$

where θ_i are arithmetic expressions over the reals, ϕ and ψ are formulas, α ranges over hybrid programs, and $\backsim$ is a comparison operator $=, \neq, \geq, >, \leq, <$.

Example 1. The following $d\mathcal{L}$ formula describes a safety property for a car model.

$$\underbrace{v \geq 0 \wedge A > 0}_{\text{initial condition}} \rightarrow [(\underbrace{(a := A \cup a := 0)}_{ctrl} ; \underbrace{\{p' = v, \, v' = a\}}_{plant})^*]\ \underbrace{v \geq 0}_{\text{postcondition}} \tag{1}$$

Formula (1) expresses that a car, when started with non-negative velocity $v \geq 0$ and positive acceleration $A > 0$ (left-hand side of the implication), will always drive forward ($v \geq 0$) after executing $(ctrl; plant)^*$, i.e. running *ctrl* followed by the differential equation *plant* arbitrarily often. Since there are no evolution domain constraints in *plant* that limit the duration, each continuous evolution has an arbitrary duration $r \in R_{\geq 0}$. As its decisions, *ctrl* lists that the car can either accelerate $a := A$ or coast $a := 0$, while *plant* describes the motion of the car (position p changes according to velocity v, velocity v according to the

Table 2. Dynamics of tactic combinators

Tactic combinator	Meaning
t ::= b	b Basic tactics.
$\mid$ t & u	Executes t and, if successful, then executes u.
$\mid$ $t\mid u$	Executes t only if t is applicable. If t is not applicable, then u is executed.
$\mid$ t^*	Repeats t until t is no longer applicable.
$\mid$ $<(u_1, \ldots, u_k)$	Applied to a goal with k subgoals, each u_i is executed on the ith subgoal.
$\mid$ label(ℓ)	Labels the current goal with label ℓ.
$\mid$ onLabel(ℓ, t)	Executes tactic t only if the goal is labeled ℓ.
$\mid$ ifT$(c)(u, v)$	Executes u if c is true, and executes v otherwise.

chosen acceleration a). Details on dℒ are in the literature [10–12], including a tutorial on modeling and proving in KeYmaera [16].

Proofs in KeYmaera X. Proofs in KeYmaera X are built up from three components (kernel primitives): a small set of dℒ axioms (not axiom schemata) [14] from its axiomatization [12], bound variable renaming and uniform substitution [13,14], and the propositional fragment of the dℒ sequent calculus [10]. Even if unnecessary in theory [12,14], the propositional fragment of the dℒ sequent calculus is included in the prover kernel because the implementation is easy to check for soundness and significantly improves the efficiency of the prover during proof search.

The KeYmaera X prover kernel implements a Hilbert system for dℒ [12] as a uniform substitution calculus with bound variable renaming and uniform substitution [14]. A typical proof in KeYmaera X involves a succession of cuts of axioms, followed by uniform substitution and variable renaming to align the current goal with the cut-in axiom, and use the instantiated axiom by fast contextual equivalence rewriting [14].

Kernel Primitives and the dℒ Sequent Calculus. Although the Hilbert-style prover kernel is helpful for ensuring soundness, manually constructing proofs from kernel primitives is prohibitively tedious. To automate proof construction, KeYmaera X provides a library of basic tactics and a set of tactic combinators.

Basic tactics implement the dℒ sequent calculus [10,11] in terms of kernel primitives. Some dℒ proof rules are trivial to implement in terms of kernel primitives; for example, ImplyRight is a tactic that just applies the corresponding proof rule in the kernel's propositional sequent calculus implementation. Other dℒ sequent rules compose multiple prover kernel primitives (e.g., the Differential Invariant proof rule [14] for proving properties of differential equations without solving them).

Tactical Proving. The tactic combinator language (see Table 2) provides a mechanism for combining basic and other pre-existing tactics to build proof search strategies. All tactics—whether built-in or constructed using combinators—are applied to a sequent or a set of sequents called a goal. Tactics have an applicability condition and a dynamic semantics, both of which may depend upon the goal to which the tactic is applied.

The applicability condition associated with each tactic defines a set of sequents at which the tactic may *possibly* succeed. Applicability for built-in tactics is defined by their author, and these applicability conditions extend automatically to terms of the combinator language. The dynamic semantics of a tactic is ultimately a sequence of kernel primitives that are applied to the current goal. All tactics may either succeed or fail on error, and errors are propagated through combinator terms.

The sequential composition combinator $(t \ \& \ u)$ is similar to the semi-colon in a C-like programming language, and is used in a similar way. The tactic $t \ \& \ u$ is applicable when the first tactic (t) is applicable. The tactic results in an error under three conditions: if t results in an error, if u is not applicable at the result of t, or if u results in an error.

The either combinator $(t|u)$ is useful when writing tactics that apply at many possible syntactic forms (e.g., a tactic that symbolically executes any hybrid program). It is applicable when either t or u is applicable. The applicable tactic is executed and the other is ignored; if both are applicable, then t is executed and u is ignored. The tactic $t|u$ results in an error if the executed tactic results in an error.

The Kleene star (t^*) saturates the tactic t by applying t as often as possible, which is useful when writing general-purpose tactics. The tactic t^* is always applicable and results in an error if any iteration of t results in an error.

Branching composition $(< (u_1, ..., u_k))$ is useful for handling branching proofs (e.g., any proof that uses invariants or involves disjunctive assumptions). The tactic is always applicable, and errors when applied to a goal with a non-k number of subgoals or if any u_i is inapplicable or results in an error. Branching $(< (u_1, ..., u_k))$ has a sequential semantics given by applying each u_i sequentially. The parallel semantics of branching depends upon scheduling and synchronization, which are defined in terms of a proof tree with And/Or-branching as in Fig. 1. KeYmaera X's proof search engine is discussed in Sect. 3.

Finally, labels are useful for structuring branching proofs. Many built-in tactics that generate multiple subgoals provide labels for each subgoal, which can be matched against using the `onLabel` combinator. The tactic $\text{onLabel}((\ell_1, t_1), ..., (\ell_k, t_k))$ is applicable if any of the labels ℓ_i exists in the current goal and executes the corresponding constituent tactic t_i, resulting in an error if t_i results in an error.

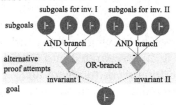

Fig. 1. Proof tree data structure

Proof search strategies are expressed using combinators. While generic proof search strategies exist (e.g., Master), KeYmaera X allows user-defined custom proof search strategies expressed as tactics. The full Scala language is available when implementing proof search strategies, but KeYmaera X also exposes an interface for running pure combinator tactics. Where automated tactics fail, users can interact with the prover by manually applying proof rules or by selecting the appropriate tactic and any necessary input (e.g., loop invariants). The following tactic example illustrates the tactic language by providing a detailed strategy for proving the safety property of Example 1 (note, that the tactic Master with invariant $v \geq 0$ would prove the example fully automatically as well but it is instructive to see the shape of the proof in a detailed proof tactic).

```
ImplyRight & Loop("v>=0") & onLabel(
  ("base case", Master),
  ("induction step", ImplyRight & Seq & Choice & AndRight & <
    (Assign & ODESolve & Master ,
     Assign & ODESolve & Master) ),
  ("use case", Master) )
```

At every execution step the strategy applies to the topmost operator, starting with the implication in (1) followed by induction with invariant $v \geq 0$ to handle the loop in the box modality. The loop induction tactic generates three labeled subgoals.

The subgoals labeled "base case" and "use case" are handled by the Master tactic, a general-purpose tactic for proving d$\mathcal{L}$ formulas. Master tries non-branching propositional tactics and hybrid program tactics, then applies any branching in propositional tactics, then searches for invariants, and finally resorts to quantifier elimination.

The tactic for the induction step follows the structure of the program. Seq handles the sequential composition between *ctrl* and *plant*, then Choice & AndRight split the non-deterministic choice $a := A \cup a := 0$. On the resulting two sub-branches, the assignments $a := A$ and $a := 0$ are handled, followed by ODESolve, which solves the differential equations of *plant*. The remaining nonmodal goals are proved by Master.

3   KeYmaera X Tool Architecture

KeYmaera X was designed to achieve powerful automation of hybrid systems theorem proving while ensuring soundness. The architecture of KeYmaera X (Fig. 2) is separated into a small, soundness-critical kernel and an extensive tactic framework to regain and exceed the convenience of powerful proof rules. A scheduler multiplexes tactics to worker threads to utilize available CPU cores. It also manages calls to external tools, such as real quantifier elimination and differential equation solving. On top of proof tactics and scheduling, the HyDRA server provides components for proof tree simplification, tactic search and custom tactic scheduling policies, as well as for storing and accessing proofs. These components can be accessed remotely through a REST-API. The KeYmaera X

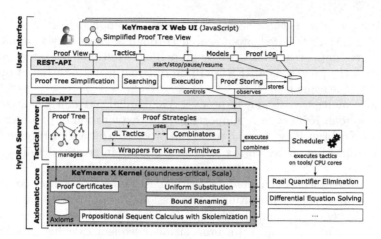

Fig. 2. KeYmaera X architecture: soundness-critical kernel is shown in dark with a dashed border

web user interface, implemented in JavaScript, uses this REST-API to communicate with the server. The remaining subsections are organized around Fig. 2.

HyDRA: Hybrid Distributed Reasoning Architecture. KeYmaera X has an isolated prover kernel, which offers a restricted interface to the remaining system components. The prover kernel operates in terms of *proof certificates*, which capture certified provability in the kernel. A proof certificate means that from certain premises the prover can soundly show a particular conclusion (e.g., a rule AndRight would have two premises, one for each conjunct, whereas an axiom has no premises). KeYmaera X ensures soundness by construction; it disallows construction of proof certificates that do not correspond to a correct derivation. That way, the prover kernel does not need to care about how proof certificates relate to each other, as long as it ensures that proof certificates only originate from within the kernel. To achieve this, components outside the soundness-critical kernel, such as tactics, the user interface and the framework for parallel execution, receive at most read-only access to proof certificates. All mechanisms for creating new proof certificates—rewrites corresponding to the axioms of d$\mathcal{L}$, uniform substitution, bound variable renaming, Skolemization and the rules of the propositional sequent calculus—are contained in the kernel. Proof certificates are managed in an And/Or proof tree outside the prover kernel, so that tactics and users have access to the proof history (Fig. 1 denotes And-branches with solid lines between nodes in the proof tree, whereas Or-branches are depicted using dashed lines).

Correctness of the prover depends on the soundness of Scala's pattern matching capabilities in a similar way that Isabelle [9] depends upon the correctness of Standard ML. Our selection of Scala is motivated by our need to interact with Mathematica and a web server. The Scala ecosystem is also attractive from the perspective of supporting parallel proof search and other advanced proof search features.

Collaboration and Distributed Search. KeYmaera X supports collaborative proving and parallel, distributed proof search through a client-server architecture and proof tree data structures with Or-branching. Multiple user interfaces may interact with the prover via a REST-API on different goals, or attempt different strategies on the same goal.

Similarly, multiple goals may be processed in parallel and multiple tactics tried on the same goal. KeYmaera X supports parallel exploration of proof strategies by means of Or-branching alternatives in the proof-tree data structure and by its continuation-passing tactics library, which we explain in greater detail below.

KeYmaera X Kernel. The soundness-critical KeYmaera X kernel consists of: (i) algebraic data types representing d$\mathcal{L}$ expressions and proof certificates; (ii) the axioms of differential dynamic logic [14]; (iii) bound variable renaming and uniform substitution rules [14]; (iv) a propositional sequent calculus with Skolemization [10]. To a lesser extent, the kernel also features expression parsing and printing. KeYmaera X bans them from the soundness-critical kernel by dynamically checking whether pretty-printing reparses to the original expressions and by declaring the pretty-printed property to be proved rather than the textual representation in input files.

The entire prover kernel has a size of about 1700 lines of Scala code (LOC). Parsing and printing weighs in at another 1700 LOC. Proofs are certified by an LCF-style design in which only the small list of certified proof rules can create proof certificates. All this puts verifying the kernel in feasible range: The axiomatic portion of the kernel uses primarily algebraic data types and recursive functions defined over these types, so mechanizing the theory of KeYmaera X in a higher-order proof assistant and possibly performing code extraction appears feasible.

KeYmaera X implements rules from the propositional sequent calculus, bound variable renaming, and, most importantly, uniform substitution. These rules are the basis for constructing all proofs. Tactics are constructed from axioms by aligning them with the current goal using bound variable renaming and uniform substitution. The axiom base from which proofs are constructed is kept small (49 axioms and 17 additional derived axioms) and syntactically close to the way it is presented in papers and books. Since the axioms cannot be proven within the system itself, this design is crucial to allow manual inspection to ensure that the system's foundation is sound and well chosen.

KeYmaera X relies on external tools as real arithmetic decision procedures. Arithmetic facts are stored as lemmas that are verified by the decision procedures. These lemmas are collected together with the resulting proof and, thus, can be fed into different decision procedures to increase trust in their correctness or retained as arithmetic assumptions. The dependence on external tools is minimized compared to KeYmaera [15]. Differential equation solvers are removed from the trusted kernel and arithmetic is used exclusively at the leaves of the proof tree.

Runtime and Scheduler for Executing Tactics. Tactics and kernel primitives (through their wrappers) as well as external tools are not invoked directly from the user interface but passed to a scheduler. The scheduler multiplexes tactics

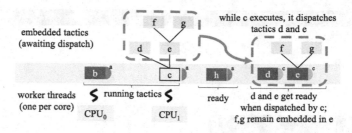

Fig. 3. Tactic scheduling using continuations

to worker threads for parallel execution and manages limited parallelism and blocking on external tools.

To achieve this, the scheduler instantiates one worker thread per CPU core and in addition one worker thread for each blocking link to external tools. By blocking we mean a link that requires the worker thread to wait for a result after it has passed the request to a tool. In addition, KeYmaera X tactics are schedulable objects comprised of a main body and a continuation, which can be passed to other tactics to regain control after completion, in particular if they have been executed on a different CPU core.

Figure 3 illustrates the dispatching of tactics and the role of continuations. A tactic a (not shown) has dispatched the tactics b, c and h for parallel execution by inserting them into the global priority-sorted ready list from which the worker threads on CPU_1 selected c, which it currently executes. Worker threads always pick the highest-prioritized ready tactic from the ready list and execute them non-preemptively (i.e., they first complete a started tactic before they look for the next one). Tactic c represents any tactic that would add multiple independent tactics to the queue, such as the $<(d, e)$ tactic. The tactics d and e are associated with subgoals of the goal at which c is applied. Once a tactic has been associated to a proof node and a continuation, the tactic is ready for dispatch into the scheduler's ready list. The result of dispatching of d and e is shown on the right of Fig. 3 when following the arrow. Tactic e is a combinator (e.g., $e = f$ & g) with embedded tactics f, g. Because e did not yet execute and because g will execute on the subgoal yet-to-be produced by f, these tactics are not ready yet.

To regain control after d and e complete, c has passed a continuation to both tactics (c is the parent of the continuation). Continuations are invoked once the body of a tactic completes. A continuation can inspect the result and the completion status (success or failure) of the completed tactic, as well as its parent to make decisions about the next proof step based on whether or not the proof changed.

User Interface. The KeYmaera X system features multiple interfaces: (i) a Scala-API for accessing the axiomatic core and tactical prover programmatically from (standalone) Scala and Java applications; (ii) a REST-API intended for remote access to the HyDRA server; and (iii) a graphical web-based user interface for point-and-click interaction. The Scala-API is designed for tight integration of KeYmaera X into other programs. It is the basis for the HyDRA server and used

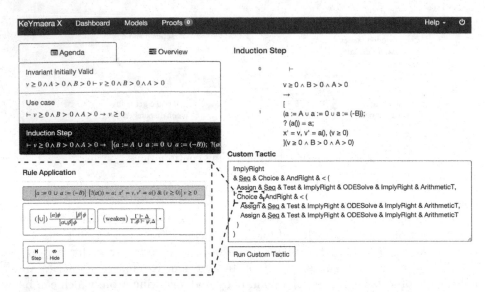

Fig. 4. A tactic for closing the induction step of a simple hybrid car model. The dotted selection illustrates what the Apply Rule dialog would look like just before executing the second `Choice` in the custom tactic.

in the development process for unit testing. The REST-API wraps the Scala-API in a web application and gives access to server functionality: it identifies the "resources" at the HyDRA server (such as goals in a proof tree, formulas in a sequent, and tactics) using hierarchical URLs and uses standard HTTP requests to manipulate these resources. On top of that, KeYmaera X provides a native web interface for managing proofs and lemma databases, as well as for interactive and tactical proving sessions. Figure 4 shows the web interface during an interactive proving session. In the web interface, proof trees are collapsed for presentation into simplified views, which highlight proof steps at the granularity of $d\mathcal{L}$ sequent rules but shortcut through the axiom-application steps that we introduced to improve confidence in soundness. Custom tactics can be specified using the combinator language of Sect. 2. Alternatively, proof rules such as ODESolve can be selected directly by clicking on the formula, as illustrated in Fig. 4.

4   Related Work

KeYmaera X is the first theorem prover to unify Hilbert systems and Gentzen-style sequent calculi by combining uniform substitution with a flexible tactics mechanism. Hilbert systems simplify reasoning about soundness, which reduces the complexity and risk associated with extending the theorem prover with new proof search techniques or new logic fragments. This distinction separates KeYmaera X from other deductive verification systems such as KeY [3] and KeYmaera [15].

LCF-style theorem provers, including Isabelle [9], feature both a minimal trusted kernel as well as support for tactics. These tools influenced the design of KeYmaera X. Most major theorem provers, including Coq [7] and Isabelle [9], also provide user interfaces. In [5], similar to KIV [6], a tactical theorem prover for verifying software is presented. Unlike these, KeYmaera X is particularly well-suited to the analysis of hybrid dynamical systems with their differential equations.

Other successful tools exist for hybrid systems; however, apart from KeYmaera, none based on the rigor of a sound logic let alone a small kernel. A comparison of d$\mathcal{L}$ with other approaches to analysis of hybrid systems is provided in the literature [11].

Acknowledgments. The authors thank the anonymous reviewers for their helpful feedback, and Ran Ji for help with testing and extending KeYmaera X.

References

1. Ahrendt, W., Baar, T., Beckert, B., Bubel, R., Giese, M., Hähnle, R., Menzel, W., Mostowski, W., Roth, A., Schlager, S., Schmitt, P.H.: The KeY tool. Softw. Syst. Model. 4(1), 32–54 (2005)
2. Alur, R., Courcoubetis, C., Henzinger, T.A., Ho, P.-H.: Hybrid automata: an algorithmic approach to the specification and verification of hybrid systems. In: Grossman, R.L., Ravn, A.P., Rischel, H., Nerode, A. (eds.) HS 1991 and HS 1992. LNCS, vol. 736, pp. 209–229. Springer, Heidelberg (1993)
3. Beckert, B., Hähnle, R., Schmitt, P.H. (eds.): Verification of Object-Oriented Software. LNCS (LNAI), vol. 4334, pp. 453–479. Springer, Heidelberg (2007)
4. Bowen, J., Stavridou, V.: Safety-critical systems, formal methods and standards. Softw. Eng. J. 8(4), 189–209 (1993)
5. Felty, A., Howe, D.: Tactic theorem proving with refinement-tree proofs and metavariables. In: Bundy, A. (ed.) CADE 1994. LNCS, vol. 814, pp. 605–619. Springer, Heidelberg (1994)
6. Heisel, M., Reif, W., Stephan, W.: Tactical theorem proving in program verification. In: Stickel, M.E. (ed.) CADE 1990. LNCS, vol. 449, pp. 117–131. Springer, Heidelberg (1990)
7. The Coq development team: The Coq proof assistant reference manual. LogiCal project, version 8.0 (2004). http://coq.inria.fr
8. Mitsch, S., Platzer, A.: ModelPlex: verified runtime validation of verified cyber-physical system models. In: Bonakdarpour, B., Smolka, S.A. (eds.) RV 2014. LNCS, vol. 8734, pp. 199–214. Springer, Heidelberg (2014)
9. Nipkow, T., Paulson, L.C., Wenzel, M. (eds.): Isabelle/HOL: A Proof Assistant for Higher-Order Logic. LNCS, vol. 2283. Springer, Heidelberg (2002)
10. Platzer, A.: Differential dynamic logic for hybrid systems. J. Autom. Reas. 41(2), 143–189 (2008)
11. Platzer, A.: Logical Analysis of Hybrid Systems: Proving Theorems for Complex Dynamics. Springer, Heidelberg (2010)
12. Platzer, A.: Logics of Dynamical Systems. In: LICS, pp. 13–24. IEEE (2012)
13. Platzer, A.: Differential Game Logic. CoRR abs/1408.1980 (2014)

14. Platzer, A.: A uniform substitution calculus for differential dynamic logic. In: Felty, A.P., Middeldorp, A. (eds.) CADE-25. LNCS, vol. 9195, pp. xx–yy. Springer, Heidelberg (2015)
15. Platzer, A., Quesel, J.-D.: KeYmaera: a hybrid theorem prover for hybrid systems (system description). In: Armando, A., Baumgartner, P., Dowek, G. (eds.) IJCAR 2008. LNCS (LNAI), vol. 5195, pp. 171–178. Springer, Heidelberg (2008)
16. Quesel, J.D., Mitsch, S., Loos, S., Aréchiga, N., Platzer, A.: How to model and prove hybrid systems with KeYmaera: a tutorial on safety. STTT (2015)

Tableaux Methods for Propositional Dynamic Logics with Separating Parallel Composition

Philippe Balbiani and Joseph Boudou[(✉)]

Institut de Recherche en Informatique de Toulouse,
CNRS — Toulouse University, Toulouse, France
joseph.boudou@irit.fr

Abstract. PRSPDL is a propositional dynamic logic with an operator for parallel compositions of programs. We first give a complexity upper bound for this logic. Then we focus on the class of ◁-deterministic frames and give tableaux methods for two fragments of PRSPDL over this class of frames.

1 Introduction

Propositional dynamic logic (PDL) is a multi-modal logic designed to reason about the behaviors of programs [11]. With each program α is associated a modal operator $[\alpha]$, formulas $[\alpha]\varphi$ being read "all executions of α from the current state lead to a state where φ holds". The set of programs is structured by some operators: the sequential composition $\alpha \,;\, \beta$ of programs α and β corresponds to the composition of the accessibility relations $R(\alpha)$ and $R(\beta)$; test $\varphi?$ on formula φ corresponds to the identity relation restricted to the states at which φ holds; the iteration α^* corresponds to the reflexive and transitive closure of $R(\alpha)$. PDL has been extensively studied and a great deal is known about its complexity and proof theory [9,11,12,15,16,18]. Moreover, since PDL's programs are abstract, variants of PDL has been devised for different fields, like knowledge representation and linguistic.

A limitation of PDL is the lack of constructs to reason about concurrency. Different extensions of PDL have been devised to overcome this limitation, for instance interleaving PDL [1], PDL with intersection (IPDL) [8] and the concurrent dynamic logic [17]. PDL with storing, recovering and parallel composition (PRSPDL) [4] is another extension of PDL with a construct for parallel composition of programs. The key difference of PRSPDL is that, for the execution of the program $\alpha \parallel \beta$, α and β are executed in parallel *on two different substates* of the initial state. Hence, $\alpha \parallel \beta$ being executable at some state does not imply that α or β is executable at that state. Since states can be separated in substates and merged back, PRSPDL is related to the Boolean logic of bunched implication (BBI) [19]. Indeed, a multiplicative conjunction semantically similar

---

J. Boudou—Our research is supported by the "French National Research Agency" (DynRes contract ANR-11-BS02-011).

A.P. Felty and A. Middeldorp (Eds.): CADE-25, LNAI 9195, pp. 539–554, 2015.
DOI: 10.1007/978-3-319-21401-6_37

to the one found in BBI can be defined in PRSPDL. Hence PRSPDL is a modal logics of separation like logics in [5–7], and is closely related to the process logic MBIc [6]. The differences between PRSPDL and MBIc are the lack of sequential compositions in MBIc making it strictly less expressive [2] and the associativity of the separation relation making the satisfiability problem harder [14]. The combination of separation and concurrency provided by PRSPDL suggests interesting applications. For instance, in the field of program verification, a dynamic and concurrent logic on heaps akin to separation logics [10,20] may be envisioned.

PRSPDL has many similarities with IPDL. Like IPDL, PRSPDL lacks the tree model property. Moreover, due to formulas of the form $[\alpha \parallel \beta]\,\varphi$, there is no sets of formulas of the language comparable to the Fischer-Ladner closure for PDL [11] on which the filtration method could be applied. Hence, studying PRSPDL's computability is hard and the only result currently known about the computability of PRSPDL is its undecidability when interpreted over the class of separated frames [3]. This difficulties, added to the usual complications due to the iteration construct, make the conception of a tableaux method for PRSPDL a real challenge. We overcome all this difficulties by adapting techniques from [16]: compound programs are allowed as edge's labels of the constructed model and new atomic formulas are used to identify states reachable by some programs. The added value of our paper consists in the presence of the iteration construct and in a new extended definition of the Fischer-Ladner closure.

In this paper, three variants of PRSPDL are studied. First, PRSPDL, formally defined in Sect. 2, is proved in Sect. 3 to be faithfully translatable into IPDL with converse. This result conveys a 2EXPTIME complexity upper bound. Then the fragment of PRSPDL without the special programs of storing and recovering, interpreted over the class of $\lhd$-deterministic frames, is considered. A Fischer-Ladner closure for an extension of this fragment is defined in Sect. 4 and a sound and complete tableaux method is exhibited in Sect. 5. Finally, an optimal decision procedure for the fragment of PRSPDL without storing, recovering and iteration, interpreted over $\lhd$-deterministic frames, is given in Sect. 6, proving this fragment to be PSPACE-complete.

2   PRSPDL

Let Π_0 a countable set of atomic programs (denoted $a, b \ldots \ldots$) and Φ_0 be a countable set of propositional variables (denoted $p, q \ldots \ldots$). The sets Π and Φ of programs and formulas are defined as follows:

$$\alpha, \beta := a \mid (\alpha \,;\, \beta) \mid \varphi? \mid \alpha^* \mid (\alpha \parallel \beta) \mid s_1 \mid s_2 \mid r_1 \mid r_2$$
$$\varphi := p \mid \bot \mid \neg\varphi \mid [\alpha]\,\varphi$$

We define the abbreviations $\top \doteq \neg\bot$ and $\langle\alpha\rangle\varphi \doteq \neg\,[\alpha]\,\neg\varphi$. The Boolean operators can be defined too, for instance $\varphi \to \psi \doteq [\varphi?]\,\psi$. Moreover, a multiplicative conjunction related to BBI may be defined as $\varphi * \psi \doteq \neg\,[\varphi? \parallel \psi?]\,\bot$. Parentheses may be omitted for clarity, but they are taken into account when counting occurrences of symbols. Double negations are implicitly eliminated. We write $|\alpha|$ and $|\varphi|$ for the number of occurrences of symbols in any program α and any formula φ respectively. We define two fragments of PRSPDL's language:

- $\mathcal{L}_{;?*\|}$ is the set of PRSPDL's formulas and programs with no occurrences of the symbols s_1, s_2, r_1 and r_2;
- $\mathcal{L}_{;?\|}$ is the set of PRSPDL's formulas and programs with no occurrences of the symbols s_1, s_2, r_1, r_2 and $*$.

A frame is a tuple $(W, R, \lhd)$ where W is a non-empty set of states, R is a function associating a binary relation over W to each atomic program and $\lhd$ is a ternary relation over W. Intuitively, $x\,R\,(a)\,y$ means that the program a can be executed in state x, reaching state y. Similarly, $x \lhd (y, z)$ means that x can be split into the substates y and z or equivalently that y and z can be merged to obtain x. When the merging of states is functional, the frame is said to be $\lhd$-deterministic. This is a common restriction, for instance in separation logics [10,20]. Formally, a frame is $\lhd$-deterministic iff for all $x, y, w_1, w_2 \in W$, if $x \lhd (w_1, w_2)$ and $y \lhd (w_1, w_2)$ then $x = y$. The class of all frames is denoted by $\mathcal{C}_{\text{all}}$ and the class of $\lhd$-deterministic frames by $\mathcal{C}_{\lhd\text{-det}}$. A model is a tuple $(W, R, \lhd, V)$ where $(W, R, \lhd)$ is a frame and V is a function associating a subset of W to each propositional variable. A model is $\lhd$-deterministic iff its frame is $\lhd$-deterministic. The forcing relation $\vDash$ is defined by parallel induction along with the extension of R to all programs:

$$\mathcal{M}, x \vDash p \quad \text{iff } x \in V(p)$$
$$\mathcal{M}, x \vDash \bot \quad \text{never}$$
$$\mathcal{M}, x \vDash \neg\varphi \quad \text{iff } \mathcal{M}, x \nvDash \varphi$$
$$\mathcal{M}, x \vDash [\alpha]\varphi \quad \text{iff } \forall z \in W,\ xR\,(\alpha)z \text{ implies } \mathcal{M}, z \vDash \varphi$$
$$xR\,(\alpha\,;\beta)y \quad \text{iff } \exists z \in W,\ xR\,(\alpha)z \text{ and } zR\,(\beta)y$$
$$xR\,(\varphi?)y \quad \text{iff } x = y \text{ and } \mathcal{M}, x \vDash \varphi$$
$$xR\,(\alpha^*)y \quad \text{iff } xR\,(\alpha)^*y$$
$$\text{where} R\,(\alpha)^* \text{ is the reflexive transitive closure of} R\,(\alpha)$$
$$xR\,(\alpha \parallel \beta)y \quad \text{iff } \exists w_1, w_2, w_3, w_4 \in W,$$
$$x \lhd (w_1, w_2)\,,\, w_1 R\,(\alpha)w_3, w_2 R\,(\beta)w_4 \text{ and } y \lhd (w_3, w_4)$$
$$xR\,(s_i)y \quad \text{iff } \exists z_1, z_2 \in W,\ y \lhd (z_1, z_2) \text{ and } x = z_i$$
$$xR\,(r_i)y \quad \text{iff } \exists z_1, z_2 \in W,\ x \lhd (z_1, z_2) \text{ and } y = z_i$$

A formula φ is *satisfiable in a class $\mathcal{C}$ of frames* iff there exists a model $\mathcal{M} = (W, R, \lhd, V)$ and a state $w \in W$ such that $(W, R, \lhd) \in \mathcal{C}$ and $\mathcal{M}, w \vDash \varphi$. The satisfiability problem for a fragment $\mathcal{L}$ of PRSPDL over a class $\mathcal{C}$ of frames is the decision problem answering whether a formula in $\mathcal{L}$ is satisfiable in $\mathcal{C}$.

3   Complexity Upper Bound for PRSPDL

In order to illustrate the close relationship between PRSPDL and IPDL, we provide a faithful translation from PRSPDL to PDL with intersection and converse (ICPDL) [15]. This translation conveys an upper bound for the complexity of the satisfiability problem of PRSPDL with respect to $\mathcal{C}_{all}$. Given the same sets Π_0 and Φ_0 as for PRSPDL and three new atomic programs b_0, b_1, b_2, the language of ICPDL is defined by:

$$\alpha, \beta := a \mid (\alpha\,;\beta) \mid (\alpha \cup \beta) \mid \varphi? \mid \alpha^* \mid \alpha^- \mid (\alpha \cap \beta) \mid b_0 \mid b_1 \mid b_2$$
$$\varphi := p \mid \perp \mid \neg\varphi \mid [\alpha]\varphi$$

A model for ICPDL is a tuple $\mathcal{M} = (W, R, V)$ where W is a non-empty set of states, $R : \Pi_0 \cup \{b_0, b_1, b_2\} \longrightarrow \mathcal{P}(W^2)$ and $V : \Phi_0 \longrightarrow \mathcal{P}(W)$. See [15] for the definition of the forcing relation $\vDash_{IC}$. The translation function τ from PRSPDL to ICPDL is defined by inductively replacing all subprograms of the form $\alpha \parallel \beta$ by $b_0\,;\left((b_1\,;\tau(\alpha)\,;b_1^-) \cap (b_2\,;\tau(\beta)\,;b_2^-)\right)\,;b_0^-$, all subprograms of the form s_i for $i \in \{1,2\}$ by $b_0^-\,;b_i^-$ and all subprograms of the form r_i for $i \in \{1,2\}$ by $b_0\,;b_i$. Given a PRSPDL's formula φ, let $\{a_1, \ldots\ldots, a_n\}$ be the set of atomic programs occurring in φ. The ICPDL program $\pi(\varphi)$ is defined by $\pi(\varphi) = \left(a_1 \cup \ldots\ldots \cup a_n \cup b_0 \cup b_0^- \cup b_1 \cup b_1^- \cup b_2 \cup b_2^-\right)^*$.

Proposition 1. *A PRSPDL formulas $\varphi \in \Phi$ is satisfiable if and only if the ICPDL formula $\tau(\varphi) \wedge [\pi(\varphi)](\langle b_1\rangle\top \leftrightarrow \langle b_2\rangle\top)$ is satisfiable.*

As a corollary, by [15]:

Corollary 1. *The satisfiability problem of PRSPDL with respect to the class of all frames is in 2EXPTIME.*

4 Fischer-Ladner Closure over $\mathcal{L}_{;?*\parallel}$

In this section, we consider the fragment $\mathcal{L}_{;?*\parallel}$ of PRSPDL. We define the sets $\Pi_{;?*\parallel} = \Pi \cap \mathcal{L}_{;?*\parallel}$ and $\Phi_{;?*\parallel} = \Phi \cap \mathcal{L}_{;?*\parallel}$ of programs and formulas of $\mathcal{L}_{;?*\parallel}$. In traditional tableaux methods for PDL-like logics, the formulas appearing in a tableau all belong to a Fischer-Ladner closure [11]. In the case of PRSPDL, due to the parallel composition construct, the Fischer-Ladner closure must be defined in an extension of the language.

4.1 Placeholders and Marking Functions

In order to decompose formulas of the form $[\alpha \parallel \beta]\varphi$ into subformulas, parallel compositions are distinguished using indices and new atomic formulas called placeholders are added. The sets Π_{PH}, Φ_{pure} and Φ_{PH} of *annotated programs*, *pure formulas* and *annotated formulas* respectively, are defined by parallel induction as follows:

$$\alpha, \beta := a \mid (\alpha\,;\beta) \mid \varphi? \mid \alpha^* \mid (\alpha \parallel_i \beta)$$
$$\varphi := p \mid \perp \mid \neg\varphi \mid [\alpha]\varphi$$
$$\psi := \varphi \mid (i, j) \mid \neg\psi \mid [\alpha]\psi$$

where i ranges over $\mathbb{N}$ and j over $\{1, 2\}$. Moreover, for any $i \in \mathbb{N}$, there must be at most one occurrence of $\|_i$ in any pure formula. The integers below the parallel composition symbols are called *indices*. Formulas of the form (i, j) are called *placeholders*.

To interpret annotated formulas, if placeholders were simply considered as new propositional variables, it would be impossible to ensure that whenever $w \lhd (x, y)$ and $\mathcal{M}, x \vDash [\alpha \|_i \beta] \varphi$ then $\mathcal{M}, x \vDash [\alpha] (i, 1)$ and $\mathcal{M}, y \vDash [\beta] (i, 2)$. Therefore we interpret placeholders using *marking functions* which assign subsets of W to placeholders. The set of all such functions is denoted by B_W. The *empty marking function* $m_W^\varnothing \in B_W$ binds the empty set to all placeholders. The 4-ary forcing relation $\vDash_F$ is defined on all models $\mathcal{M} = (W, R, \lhd, V)$, all $w \in W$, all $m \in B_W$ and all $\varphi \in \varPhi_{PH}$ by parallel induction along with the extension of R to all annotated programs, in a similar way than for PRSPDL except:

$$\mathcal{M}, x, m \vDash_F (i, j) \text{ iff } x \in m(i, j)$$

$$xR(\varphi?)y \qquad \text{iff } x = y \text{ and } \mathcal{M}, x, m_W^\varnothing \vDash_F \varphi$$

$$xR(\alpha \|_i \beta)y \qquad \text{iff } \exists w_1, w_2, w_3, w_4 \in W,$$
$$x \lhd (w_1, w_2), w_1 R(\alpha)w_3, w_2 R(\beta)w_4 \text{ and } y \lhd (w_3, w_4)$$

There exists a forgetful epimorphism $^\top : \varPhi_{\text{pure}} \longrightarrow \varPhi_{;?*\|}$ associating to each pure formula φ the formula $\overline{\varphi}$ obtained by removing all indices in φ. Thanks to the following lemma, which can be easily proved by induction on $|\varphi|$, we will consider satisfiability of pure formulas instead of satisfiability of $\mathcal{L}_{;?*\|}$ formulas.

Lemma 1. *For all $\varphi_0 \in \varPhi_{pure}$, $\mathcal{M}, w, m_W^\varnothing \vDash_F \varphi_0$ iff $\mathcal{M}, w \vDash \overline{\varphi_0}$. Moreover, for all $m \in B_W$, $\mathcal{M}, w, m_W^\varnothing \vDash_F \varphi_0$ iff $\mathcal{M}, w, m \vDash_F \varphi_0$*

4.2   Fischer-Ladner Closure

Following [11], given an annotated formulas φ over $\varPi_0$ and $\varPhi_0$, we will define the *closure* $\mathrm{FL}(\varphi_0)$ of φ by applying the rules in Fig. 1.

Lemma 2. *The cardinality of $\mathrm{FL}(\varphi_0)$ is linear in $|\varphi|$.*

We will be mainly interested in closures of pure formulas. We define the set $\mathrm{SP}(\varphi_0) = \{\alpha \mid \exists \varphi, [\alpha] \varphi \in \mathrm{FL}(\varphi_0)\}$ of subprograms of any pure formula φ_0.

$$\frac{[a]\,\varphi}{\varphi} \qquad\qquad\qquad \frac{\varphi}{\neg\varphi}$$

$$\frac{[\varphi?]\,\psi}{\varphi \quad \psi} \qquad\qquad\qquad \frac{[\alpha\,;\,\beta]\,\varphi}{[\alpha]\,[\beta]\,\varphi}$$

$$\frac{[\alpha^*]\,\varphi}{[\alpha]\,[\alpha^*]\,\varphi \quad \varphi} \qquad\qquad \frac{[\alpha \|_i \beta]\,\varphi}{[\alpha](i, 1) \quad [\beta](i, 2) \quad \varphi}$$

Fig. 1. Fischer-Ladner closure calculus

Lemma 3. *For any pure formula φ_0 and any $i \in \mathbb{N}$, there is at most one formula of the form $[\alpha \parallel_i \beta] \varphi$ in $\mathrm{FL}(\varphi_0)$.*

The function G_{φ_0} is defined such that for all $i \in \mathbb{N}$, if there exists $\alpha, \beta \in \Pi_{PH}$ verifying $[\alpha \parallel_i \beta] \varphi \in \mathrm{FL}(\varphi_0)$ then $G_{\varphi_0}(i) = \varphi$, otherwise $G_{\varphi_0}(i) = \top$. When the index φ_0 is obvious from the context, we write G instead of G_{φ_0}.

5   Tableaux Method for $\mathcal{L}_{;?*\parallel}$ over $\mathcal{C}_{\lhd\text{-det}}$

In this section we introduce a tableaux method for pure formulas interpreted over $\mathcal{C}_{\lhd\text{-det}}$. To deal with the merging of states at the end of parallel compositions, we borrow some ideas from [16]. Firstly, non-atomic programs are allowed as label of edges in the built structure. Secondly, placeholders are used in order to ensure that formulas of the form $[\alpha \parallel_i \beta] \varphi$ are propagated.

5.1   Rules of the Tableaux Method

Given a set W of states, a judgment about W is either:

- a judgment $x \colon \varphi$ stating that x must satisfy φ;
- a judgment $(x, y) \colon \alpha$ stating that y can be reached from x by α;
- a judgment $(x, y, z) : \Delta$ with $\Delta \in \{F, B\}$, stating that x can be decomposed forwardly (if $\Delta = F$) or backwardly (if $\Delta = B$) into y and z.

A judgment j *involves* a state x iff x appears on the left of j. A structure is a tuple $\mathcal{S} = (W, J, K)$ where W is a set of states, J a set of judgments about W and $K \subseteq J$ a subset of inactive judgments. A tableau $\mathcal{T}$ for a pure formula φ_0 is an ordered, possibly infinite tree whose nodes are labeled with structures, the root being labeled with the initial structure $(\{w_0\}, \{w_0 \colon \varphi_0\}, \emptyset)$ for some w_0. Successor nodes are constructed in accordance with the rules from Figs. 2, 3, 4, 5 and 6. The rules have the general form

$$\frac{X_0}{X_1 \mid \ \ldots\ldots\ \mid X_\ell} C$$

where X_0 is the set of premises, $(X_k)_{k \in 1 \ldots \ell}$ are the sets of conclusions, C is the set of side conditions and $\ell > 0$. The rules $(\square)$, $(\square\parallel 1F)$, $(\square\parallel 1B)$, $(\square\parallel 0\top)$ and $(\square\parallel 0\bot)$ are called *universal*. States denoted by n, n_1, n_2, n_3 and n_4 in the conclusions must be fresh. A rule instantiation is applicable to a node η_0 labeled with $\mathcal{S}_0 = (W_0, J_0, K_0)$ if all the following conditions are met:

- the instantiation $\overline{X_0}$ of the set of premises is a subset of $J_0 \setminus K_0$,
- all side conditions' instantiations are satisfied,
- if the rule is universal then for all $k \in 1 \ldots \ell$, there is a judgment j_k in X'_ks instantiation such that $j_k \notin J_0$.

$$\square \ \frac{x\colon [a]\varphi \quad (x,y)\colon a}{y\colon \varphi}$$

$$\Diamond 1 \ \frac{x\colon \langle\alpha\rangle\varphi}{(x,n)\colon \alpha \quad n\colon \varphi} \ \mathrm{size}(\alpha)=1 \qquad\qquad \Diamond 0 \ \frac{x\colon \langle\alpha\rangle\varphi}{(x,x)\colon \alpha \quad x\colon \varphi} \ \mathrm{size}(\alpha)=0$$

$$\Diamond* \ \frac{x\colon \langle\alpha\rangle\varphi}{(x,n)\colon \alpha \quad n\colon \varphi \ \mid \ (x,x)\colon \alpha \quad x\colon \varphi} \ \mathrm{size}(\alpha)=*$$

Fig. 2. Basic rules of $\mathcal{L}_{;?*\|}$'s tableaux calculus

$$\square? \ \frac{x\colon [\varphi?]\psi}{x\colon \neg\varphi \ \mid \ x\colon \psi} \qquad\qquad\qquad \Diamond? \ \frac{(x,x)\colon \varphi?}{x\colon \varphi}$$

Fig. 3. Test rules of $\mathcal{L}_{;?*\|}$'s tableaux calculus

$$\square; \ \frac{x\colon [\alpha\,;\beta]\varphi}{x\colon [\alpha][\beta]\varphi} \qquad\qquad\qquad \Diamond;00 \ \frac{(x,x)\colon \alpha\,;\beta}{(x,x)\colon \alpha \quad (x,x)\colon \beta}$$

$$\Diamond;0\bullet \ \frac{(x,y)\colon \alpha\,;\beta}{Y_{0\bullet}} \ \mathrm{size}(\alpha)=0, x\neq y \qquad \Diamond;\bullet 0 \ \frac{(x,y)\colon \alpha\,;\beta}{Y_{\bullet 0}} \ \mathrm{size}(\beta)=0, x\neq y$$

$$\Diamond;11 \ \frac{(x,y)\colon \alpha\,;\beta}{Y_{11}} \ \mathrm{size}(\alpha)=\mathrm{size}(\beta)=1$$

$$\Diamond;1* \ \frac{(x,y)\colon \alpha\,;\beta}{Y_{\bullet 0} \ \mid \ Y_{11}} \ \mathrm{size}(\alpha)=1, \mathrm{size}(\beta)=*$$

$$\Diamond;*1 \ \frac{(x,y)\colon \alpha\,;\beta}{Y_{0\bullet} \ \mid \ Y_{11}} \ \mathrm{size}(\alpha)=*, \mathrm{size}(\beta)=1$$

$$\Diamond;** \ \frac{(x,y)\colon \alpha\,;\beta}{Y_{0\bullet} \ \mid \ Y_{\bullet 0} \ \mid \ Y_{11}} \ \mathrm{size}(\alpha)=\mathrm{size}(\beta)=*, x\neq y$$

$$Y_{0\bullet} = \{(x,x)\colon \alpha, \ (x,y)\colon \beta\}$$
$$Y_{\bullet 0} = \{(x,y)\colon \alpha, \ (y,y)\colon \beta\}$$
$$Y_{11} = \{(x,n)\colon \alpha, \ (n,y)\colon \beta\}$$

Fig. 4. Sequence rules of $\mathcal{L}_{;?*\|}$'s tableaux calculus

$$\square* \ \frac{x\colon [\alpha^*]\varphi}{x\colon \varphi \quad x\colon [\alpha][\alpha^*]\varphi}$$

$$\Diamond*\neq \ \frac{(x,y)\colon \alpha^*}{(x,y)\colon \alpha \ \mid \ (x,n)\colon \alpha \quad (n,y)\colon \alpha^*} \ x\neq y$$

Fig. 5. Iteration rules of $\mathcal{L}_{;?*\|}$'s tableaux calculus

$$\Box\|1F \quad \frac{x\colon [\alpha \|_i \beta]\varphi \quad\quad (x,y,z)\colon F}{y\colon [\alpha](i,1) \quad\quad z\colon [\beta](i,2)} \quad \text{size}\,(\alpha \|_i \beta) \neq 0$$

$$\Box\|1B \quad \frac{y\colon (i,1) \quad\quad z\colon (i,2) \quad\quad (x,y,z)\colon B}{x\colon G(i)}$$

$$\Box\|0T \quad \frac{x\colon [\alpha \|_i \beta]\varphi}{x\colon \varphi \quad | \quad x\colon [\text{desiter}\,(\alpha \|_i \beta)]\perp} \quad \text{size}\,(\alpha \|_i \beta) \neq 1$$

$$\Box\|0\perp \quad \frac{x\colon [\alpha \|_i \beta]\perp \quad\quad (x,y,z)\colon \Delta}{y\colon [\alpha]\perp \quad | \quad z\colon [\beta]\perp} \quad \text{size}\,(\alpha \|_i \beta) = 0$$

$$\Diamond\|00 \quad \frac{(x,x)\colon \alpha \|_i \beta}{(x,n_1,n_2)\colon F \quad (n_1,n_1)\colon \alpha \quad (n_2,n_2)\colon \beta}$$

$$\Diamond\|0. \quad \frac{(x,y)\colon \alpha \|_i \beta}{Z_{0.}} \quad \text{size}(\alpha) = 0, x \neq y \quad\quad \Diamond\|.0 \quad \frac{(x,y)\colon \alpha \|_i \beta}{Z_{.0}} \quad \text{size}(\beta) = 0, x \neq y$$

$$\Diamond\|11 \quad \frac{(x,y)\colon \alpha \|_i \beta}{Z_{11}} \quad \text{size}(\alpha) = 1, \text{size}(\beta) = 1$$

$$\Diamond\|1* \quad \frac{(x,y)\colon \alpha \|_i \beta}{Z_{.0} \quad | \quad Z_{11}} \quad \text{size}(\alpha) = 1, \text{size}(\beta) = *$$

$$\Diamond\|*1 \quad \frac{(x,y)\colon \alpha \|_i \beta}{Z_{0.} \quad | \quad Z_{11}} \quad \text{size}(\alpha) = *, \text{size}(\beta) = 1$$

$$\Diamond\|** \quad \frac{(x,y)\colon \alpha \|_i \beta}{Z_{0.} \quad | \quad Z_{.0} \quad | \quad Z_{11}} \quad \text{size}(\alpha) = \text{size}(\beta) = *, x \neq y$$

$$Z_{0.} = \{(x,n_1,n_2)\colon F, \ (y,n_1,n_4)\colon B, \ (n_1,n_1)\colon \alpha, \ (n_2,n_4)\colon \beta\}$$
$$Z_{.0} = \{(x,n_1,n_2)\colon F, \ (y,n_3,n_2)\colon B, \ (n_1,n_3)\colon \alpha, \ (n_2,n_2)\colon \beta\}$$
$$Z_{11} = \{(x,n_1,n_2)\colon F, \ (y,n_3,n_4)\colon B, \ (n_1,n_3)\colon \alpha, \ (n_2,n_4)\colon \beta\}$$

Fig. 6. Parallel composition rules of $\mathcal{L}_{;?*\|}$'s tableaux calculus

When applying a rule instantiation, the ℓ child nodes $\eta_1, \ldots\ldots, \eta_\ell$ of η_0, labeled with $\mathcal{S}_1, \ldots\ldots, \mathcal{S}_\ell$, are created such that $\mathcal{S}_k = (W_0 \cup F_k, J_0 \cup X_k, K_0 \cup Q)$ where F_k is the set of fresh states corresponding to n, n_1, n_2, n_3 or n_4 in X_k, $\overline{X_k}$ is the instantiation of X_k and $Q = \overline{X_0}$ except for the universal rules for which $Q = \emptyset$. The size function in the side conditions is defined by:

$$\text{size}(\varphi?) = 0 \quad\quad \text{size}(\alpha\,;\beta) = \text{size}(\alpha \| \beta) = \begin{cases} 0 & \text{if size}(\alpha) = \text{size}(\beta) = 0 \\ 1 & \text{if size}(\alpha) = 1 \text{ or size}(\beta) = 1 \\ * & \text{otherwise} \end{cases}$$

$$\text{size}(a) = 1$$

$$\text{size}(\alpha*) = \begin{cases} 0 & \text{if size}(\alpha) = 0 \\ * & \text{otherwise} \end{cases}$$

The rules ensure that for any judgment $(x, y)\colon \alpha \in J$, if $\mathrm{size}(\alpha) = 0$ then $x = y$ and if $\mathrm{size}(\alpha) = 1$ then $x \neq y$. When $\mathrm{size}(\alpha) = *$, both cases must be considered. For instance rule $(\Diamond *)$ may be seen as the disjunction of the rules $(\Diamond 1)$ and $(\Diamond 0)$. When a program α of size $*$ is considered as having size 0, it is implicitly replaced by $\mathrm{desiter}(\alpha)$. The function $\mathrm{desiter} \colon \Pi_{PH} \longrightarrow \Pi_{PH}$ substitutes each occurrence of subprograms of the form α^* with $\top$? The replacement is made explicit in the right-hand side conclusion of rule $(\square \| 0 \top)$ in order to enable the application of rule $(\square \| 0 \bot)$ afterward. Obviously, if $\mathrm{size}(\alpha) \neq 1$ then $\mathrm{size}\,(\mathrm{desiter}\,(\alpha)) = 0$.

For judgments of the form $(x, y, z)\colon \Delta$, we distinguish forward $(\Delta = F)$ and backward $(\Delta = B)$ decompositions. The rules ensure that if $(x, y, z)\colon \Delta \in J$, $(x', y', z')\colon \Delta' \in J$ and y' and z' are reachable from y and z respectively, then either $(y, z) = (y', z')$ or $\Delta = F$ and $\Delta' = B$. This property is used in rules $(\square \| 1F)$ and $(\square \| 1B)$ to ensure that no new judgments about a state is added after all successors of that state have been added (see Lemma 7 on page 12). Rules $(\square \| 1F)$ and $(\square \| 1B)$ ensure that if $x\colon [\alpha \|_i \beta]\, \varphi \in J$ then for any state $y \neq x$ reachable from x by $\alpha \|_i \beta$, $y\colon \varphi \in J$. They make use of placeholders and function G from Sect. 4. Similarly, rules $(\square \| 0 \top)$ and $(\square \| 0 \bot)$ ensure that if $x\colon [\alpha \|_i \beta]\, \varphi \in J$ then either $x\colon \varphi \in J$ or x is not reachable from x by $\alpha \|_i \beta$. When $\mathrm{size}(\alpha \|_i \beta) = *$, since the rules $(\square \| 1F)$ and $(\square \| 0 \top)$ are both universal, they could be both applied on the same judgment $x\colon [\alpha \|_i \beta]\, \varphi$.

In a tableau, a maximal path from the root is called a *branch*. For any branch $\mathcal{B}$, we write $W_{\mathcal{B}}$ (resp. $J_{\mathcal{B}}$) for the union of the W (resp. J) such that there exists a node in $\mathcal{B}$ labeled with (W, J, K) for some J (resp. W) and K. A structure $\mathcal{S} = (W, J, K)$ is *inconsistent* if there exists $x \in W$ such that $x\colon \bot \in J$ or both $x\colon \varphi \in J$ and $x\colon \neg\varphi \in J$ for some $\varphi \in \Phi_{PH}$. A branch is *open* if its nodes are all labeled with a consistent structure. A branch $\mathcal{B}$ is *saturated* iff for any node $\eta \in \mathcal{B}$ labeled with $\mathcal{S} = (W, J, K)$ and any rule's instantiation π applicable on $\mathcal{S}$, there exists a node η' in $\mathcal{B}$ labeled with $\mathcal{S}' = (W', J', K')$ and such that one of π's conclusions sets is a subset of J'. A branch $\mathcal{B}$ is *demand-satisfied* iff for any node $\eta \in \mathcal{B}$ labeled with $\mathcal{S} = (W, J, K)$ and any judgment in J of the form $(x, y)\colon \alpha^*$ there is a node $\eta' \in \mathcal{B}$ labeled with $\mathcal{S}' = (W', J', K')$ and a list $x_0, \ldots\ldots, x_m \in W'$ such that $x_0 = x$, $x_m = y$ and for all $i < m$, $(x_i, x_{i+1})\colon \alpha \in J'$. A tableau is *satisfying* if it has an open saturated demand-satisfied branch. We will prove that for any pure formula φ_0, there exists a satisfying tableau for φ_0 if and only if φ_0 is satisfiable.

5.2 Soundness

We prove the soundness of the tableaux method by interpreting branches into a satisfying model. The use of placeholders necessitates the selection of marking functions to interpret judgments. We introduce the notion of *twines* to select those marking functions. Intuitively, a twine corresponds to an equivalence class of states by the transitive and symmetric closure of the relation obtained as the union of the accessibility relation (by any program) and the relation linking two states iff they are mergeable (by $\lhd$).

Formally, Let $\mathcal{B}$ be a branch from a tableau for φ_0. The set Θ of twines of $\mathcal{B}$ is defined as $\Theta = W_{\mathcal{B}}^2 \cup \{\theta_0\}$ with θ_0 not being a member of $W_{\mathcal{B}}^2$. A twine function t assigns a twine to each state in $W_{\mathcal{B}}$. The function t is constructed from the root of $\mathcal{B}$ as follows:

1. If x is the unique state in the label of the root, then $t(x) = \theta_0$.
2. If x has been added by an application of a rule which did not add a judgment of the form (z, w_1, w_2): F (rules $(\Diamond 1)$, $(\Diamond *)$, $(\Diamond ; 11)$, $(\Diamond ; 1*)$, $(\Diamond ; *1)$, $(\Diamond ; **)$ and $(\Diamond * \neq)$) then $t(x) = t(y)$, y being any state involved in the premises of the rule instantiation. A careful analysis of the rules shows that the choice of y does not matter, because whenever (y_1, y_2): $\alpha \in J_{\mathcal{B}}$ then $t(y_1) = t(y_2)$.
3. If x has been added by an application of a rule which did add a judgment of the form (z, w_1, w_2): F (rules $(\Diamond \| 00)$, $(\Diamond \| 0\bullet)$, $(\Diamond \| \bullet 0)$, $(\Diamond \| 11)$, $(\Diamond \| 1*)$, $(\Diamond \| *1)$ and $(\Diamond \| * *)$), then $t(x) = (w_1, w_2)$.

The set Θ^+ of *active twines of* $\mathcal{B}$ is the image of the twine function. For any twine $\theta \in \Theta^+ \setminus \{\theta_0\}$ there exists a unique tuple $(x, w_1, w_2) \in W_{\mathcal{B}}^3$ such that $\theta = (w_1, w_2)$ and (x, w_1, w_2): $F \in J_{\mathcal{B}}$. In that case, we write $\triangleleft \theta$ for $t(x)$.

Given a branch $\mathcal{B}$ with twine function t, a structure $\mathcal{S} = (W, J, K)$ labeling a node in $\mathcal{B}$ and a model $\mathcal{M}' = (W', R', \triangleleft', V')$, a pair (f, g) is an *interpretation of* $\mathcal{S}$ *into* $\mathcal{M}'$ *with respect to* $\mathcal{B}$ if f is a function from W to W' and g a function from Θ^+ to $B_{W'}$ such that for all $x, y, z \in W$, $x', y', z' \in W'$, $\varphi \in \Phi_{PH}$, $\alpha \in \Pi_{PH}$, $\Delta \in \{F, B\}$, $\theta \in \Theta^+ \setminus \{\theta_0\}$ and $i \in \mathbb{N}$:

$$x: \varphi \in J \Rightarrow \mathcal{M}', f(x), g\,(t(x)) \vDash_F \varphi \tag{1}$$

$$(x, y): \alpha \in J \Rightarrow f(x) R'\,(\alpha) f(y) \tag{2}$$

$$(x, x): \alpha \in J \Rightarrow f(x) R'\,(\text{desiter}(\alpha)) f(x) \tag{3}$$

$$(x, y): \alpha \in J, x \neq y \text{ and size}(\alpha) = * \Rightarrow (f(x), f(y)) \notin R'(\text{desiter}(\alpha)) \tag{4}$$

$$(x, y, z): \Delta \in J \Rightarrow f(x) \triangleleft'\,(f(y), f(z)) \tag{5}$$

$$x' \triangleleft'\,(y', z') \wedge y' \in g(\theta)(i, 1) \wedge z' \in g(\theta)(i, 2) \Rightarrow \mathcal{M}', x', g(\triangleleft \theta) \vDash_F G(i) \tag{6}$$

If there is such an interpretation, $\mathcal{S}$ is said to be interpretable in $\mathcal{M}'$ with respect to $\mathcal{B}$. If the label of each node in $\mathcal{B}$ is interpretable in $\mathcal{M}'$ with respect to $\mathcal{B}$, then $\mathcal{B}$ is interpretable in $\mathcal{M}'$.

Obviously, interpretable branches are open and the rules preserve the interpretability. By ordering the applicable rule instantiations in a queue, a strategy for rule applications can be easily defined such that a saturated tableau is obtained for all pure formulas. Then, to prove Proposition 2 below, it suffices to prove the following lemma, which is done by selecting the interpretable branch where the leftmost child of nodes on which rule $(\Diamond * \neq)$ is applied is chosen whenever possible.

Lemma 4. *If φ_0 is satisfiable and $\mathcal{T}$ is a tableau for φ_0 in which all open branches are saturated, then $\mathcal{T}$ has an open saturated demand-satisfied branch.*

Proposition 2. *If $\varphi_0 \in \Phi_{pure}$ is satisfiable, there exists a tableaux for φ_0 with an open saturated demand-satisfied branch.*

5.3   Completeness

We now consider a satisfying tableau $\mathcal{T}$ for φ_0. We will construct a model satisfying φ_0. Since $\mathcal{T}$ is satisfying, it has an open saturated demand-satisfied branch $\mathcal{B}$. The model $\mathcal{M} = (W, R, \lhd, V)$ and the marking function m are defined as follows:

$$W = W_\mathcal{B}$$
$$R(a) = \left\{ (x,y) \in W^2 \mid (x,y) : a \in J_\mathcal{B} \right\}, \; \forall a \in \Pi_0$$
$$\lhd = \left\{ (x,y,z) \in W^3 \mid \exists \Delta \in \{F, B\}, \; (x,y,z) : \Delta \in J_\mathcal{B} \right\}$$
$$V(p) = \left\{ x \in W \mid x : p \in J_\mathcal{B} \right\}, \; \forall p \in \Phi_0$$
$$m(i,j) = \left\{ x \in W \mid x : (i,j) \in J_\mathcal{B} \right\}, \; \forall (i,j) \in \mathbb{N} \times \{1,2\}$$

By construction of $\mathcal{T}$, $\mathcal{M}$ is $\lhd$-deterministic. By induction on $|\varphi|$ and $|\alpha|$, the following truth lemma can be proved.

Lemma 5. *For all $x, y \in W$, $\varphi \in \Phi_{PH}$ and $\alpha \in \Pi_{PH}$,*

$$x : \varphi \in J_\mathcal{B} \Rightarrow \mathcal{M}, x, m \vDash_F \varphi \tag{7}$$
$$(x,y) : \alpha \in J_\mathcal{B} \Rightarrow x R(\alpha) y \tag{8}$$

The proof of Lemma 5 necessitates various properties of function R:

- If $x R(\alpha \parallel_i \beta) y$ and $x \neq y$, then there exists $w_1, w_2, w_3, w_4 \in W$ such that $(x, w_1, w_2) : F \in J_\mathcal{B}$, $(y, w_3, w_4) : B \in J_\mathcal{B}$, $w_1 R(\alpha) w_3$ and $w_2 R(\beta) w_4$.
- For all $x \in W$, $\alpha, \beta \in \Pi_{PH}$ and $i \in \mathbb{N}$, if $x R(\alpha \parallel_i \beta) x$ then there exists $w_1, w_2 \in W$ such that $x \lhd (w_1, w_2)$, $w_1 R(\alpha) w_1$ and $w_2 R(\beta) w_2$.
- For all $x \in W$ and $\alpha \in \Pi_{PH}$, if $x R(\alpha) x$ then size$(\alpha) \neq 1$.
- For all $x \in W$ and $\alpha \in \Pi_{PH}$, if $x R(\alpha) x$ then $x R(\text{desiter}(\alpha)) x$.

Our completeness result immediately follows from Lemma 5.

Proposition 3. *For any pure formula φ_0, if there exists a satisfying tableau for φ_0, then φ_0 is satisfiable.*

6   Optimal Decision Procedure for $\mathcal{L}_{;?\parallel}$ Over $\mathcal{C}_{\lhd\text{-det}}$

In this section we establish the complexity of the satisfiability problem of $\mathcal{L}_{;?\parallel}$ over $\mathcal{C}_{\lhd\text{-det}}$. The fragment $\mathcal{L}_{;?\parallel}$ is the iteration-free fragment of $\mathcal{L}_{;?*\parallel}$. Therefore, we reuse the constructions from the previous sections. We write $\Pi_{0,PH}$ for the set of iteration-free annotated programs, $\Phi_{0,PH}$ for the set of iteration-free annotated formulas and $\Phi_{0,\text{pure}}$ for the set of iteration-free pure formulas. It can be easily checked that for all $\varphi_0 \in \Phi_{0,\text{pure}}$, $\text{FL}(\varphi_0) \subseteq \Phi_{0,PH}$.

6.1 Semantic Tableaux Method

The rules of $\mathcal{L}_{;?\|}$'s tableaux calculus are the rules $(\Box)$, $(\Diamond 1)$, $(\Diamond 0)$, $(\Box?)$, $(\Diamond?)$, $(\Box;)$, $(\Diamond;00)$, $(\Diamond;11)$, $(\Diamond;0_\bullet)$, $(\Diamond;_\bullet0)$, $(\Box\|1F)$, $(\Box\|1B)$, $(\Box\|0\top)$, $(\Box\|0\bot)$, $(\Diamond\|00)$, $(\Diamond\|0_\bullet)$, $(\Diamond\|_\bullet0)$ and $(\Diamond\|11)$ from Figs. 2, 3, 4 and 6 along with the rule $(\Diamond;D)$ from Fig. 7. The rule $(\Diamond;D)$ is needed to ensure local saturation in the strategy of Sect. 6.2. Using the same techniques as in Sect. 5, we can prove that for any pure formula $\varphi_0 \in \Phi_{0,\text{pure}}$, there exists a tableau for φ_0 with an open saturated branch if and only if φ_0 is satisfiable.

$$\Diamond;D \quad \frac{x\colon \langle \alpha\,;\,\beta\rangle\varphi}{x\colon \langle\alpha\rangle\langle\beta\rangle\varphi}$$

Fig. 7. Additional rule for $\mathcal{L}_{;?\|}$'s tableaux calculus

Let $\text{FL}^+(\varphi_0) = \text{FL}(\varphi_0) \cup \{[\alpha]\bot \mid \alpha \in \text{SP}(\varphi_0)\}$. The following lemma can be proved by induction on the length of the path from the root of the tableau to the node η.

Lemma 6. *Given any structure $\mathcal{S} = (W, J, K)$ labeling a node in a tableau for φ_0, for any judgment $x\colon \varphi \in J$, $\varphi \in \text{FL}(\varphi_0)[\varphi_0][^+]$ and for any judgment $(x, y)\colon \alpha \in J$, $\alpha \in \text{SP}(\varphi_0)$.*

6.2 Optimal Decision Procedure

We will prove that the nondeterministic procedure DECISION defined on the next page solves the satisfiability problem of $\mathcal{L}_{;?\|}$ over $\mathcal{C}_{\lhd\text{-det}}$ in polynomial space. Called with a pure formula φ_0, this procedure constructs a branch of a tableau for φ_0 and returns SAT if this branch is open and saturated. In order to reduce memory usage, the procedure ensures that after any application of an instantiation π of the rules $(\Diamond 1)$ and $(\Diamond\|00)$, no new judgments can be added which involve only the states in π's premises, see Lemma 7 below. A *local rule* is a rule which is neither $(\Diamond 1)$ nor $(\Diamond\|00)$. A structure is *locally saturated* iff no local rule instantiations can be applied to it. A rule's instantiation π is *appropriate to $\mathcal{S}$ and x* iff π is applicable to $\mathcal{S}$ and either π is an instantiation of a local rule or the instantiations of the premises involve only x.

Lemma 7. *Let $\mathcal{S} = (W, J, K)$ be a locally saturated structure labeling a node η in a branch of a tableau. For any descendent node η' of η with label $\mathcal{S}' = (W', J', K')$ and any judgment $\mathsf{j} \in J'$ involving only one state $x \in W'$, if $x \in W$ then $\mathsf{j} \in J$.*

DECISION first creates the structure for the root node (line 1). Then it locally saturates this structure without adding any new state (line 2–3). Finally it calls the recursive EXTEND procedure and check whether it returns the empty structure. The empty structure is used by EXTEND as a marker for a closed branch.

Procedure 1. DECISION

Input: A pure iteration-free formula $\varphi_0 \in \Phi_{0,\text{pure}}$.
Output: SAT or UNKNOWN.
Data: A structure $\mathcal{S} = (W, J, K)$.

1 $\mathcal{S} \leftarrow (\{w_0\}, \{w_0 : \varphi_0\}, \emptyset)$
2 **while** *there is a local rule's instantiation π applicable to $\mathcal{S}$* **do**
3      $\mathcal{S} \leftarrow$ a nondeterministically chosen successor of $\mathcal{S}$ by π
4 $\mathcal{S} \leftarrow$ EXTEND$(\mathcal{S}, w_0)$
5 **if** $W \neq \emptyset$ **then** return SAT
6 **else** return UNKNOWN

Procedure 2. EXTEND

Input: A locally saturated structure $\mathcal{S} = (W, J, K)$ and a state $x \in W$.
Output: A (possibly empty) structure $\mathcal{S}_f = (W_f, J_f, K_f)$.
Data: A set J_0 of judgments and a structure $\mathcal{S}' = (W', J', K')$.

7   $J_0 \leftarrow \{j \in J \mid j \text{ involves only } x\}$
8   $\mathcal{S}' \leftarrow (W, J_0, K \cap J_0)$

9   **while** *there is a rule's instantiation π appropriate to $\mathcal{S}'$ and x* **do**
10      $\mathcal{S}' \leftarrow$ a nondeterministically chosen successor of $\mathcal{S}'$ by π

11 **if** $\mathcal{S}'$ *is inconsistent* **then**
12      $\mathcal{S}_f \leftarrow (\emptyset, \emptyset, \emptyset)$
13 **else**
14      $\mathcal{S}_f \leftarrow (W', J \cup J', K \cup K')$
15   **foreach** $y \in W' \setminus W$ **do**
16          $\mathcal{S}_f \leftarrow$ EXTEND $(\mathcal{S}_f, y)$
17 **return** $\mathcal{S}_f$

EXTEND operates in two steps. Firstly, in the *existential loop* (lines 9–10), successors of x are added and the structure is locally saturated. Secondly, in the *universal loop* (lines 15–16), EXTEND is recursively called for each state created by the existential loop. The following properties can be proved:

Lemma 8. *1. At each run of* EXTEND, *the existential loop adds a number of new states bounded by a polynomial in* $|\varphi_0|$.
2. During a call to DECISION(φ_0), *the recursion depth of the calls to* EXTEND *is bounded by a polynomial in* $|\varphi_0|$.
3. DECISION(φ_0) *returns SAT only if a saturated branch for* φ_0 *has been constructed.*

By Lemmas 2 and 6, the number of judgments added by the loop at lines 2–3 is polynomial in $|\varphi_0|$. Hence this loop terminates. By Lemma 8, the number of new states added by the existential loop (lines 9–10) is polynomial in $|\varphi_0|$. Therefore, by Lemmas 2 and 6, the cardinality of J' is always bounded by a polynomial in $|\varphi_0|$. Hence the existential loop terminates. By Lemmas 8 and

Köning's lemma, the execution tree of EXTEND is finite, hence the whole procedure terminates. Moreover, each call to EXTEND($\mathcal{S}, x$) needs to keep track of the judgments involving only states in $\{x\} \cup (W' \setminus W)$ and the number of this judgments is polynomial in $|\varphi_0|$. Finally, by Lemma 8, only a polynomial number of such configurations have to be stored. Consequently,

Proposition 4. *All executions of* DECISION(φ_0) *terminate and* DECISION *can be implemented using polynomial space.*

Given a pure formula $\varphi_0 \in \Phi_{0,\text{pure}}$, the set of executions of DECISION(φ_0) corresponds to a collection Γ of tableaux for φ_0 where each execution corresponds to a branch of a tree. If φ_0 is satisfiable, by soundness of the tableaux method, there is an open branch in each tree of Γ. Since DECISION returns UNKOWN only when the corresponding branch is close, there is an execution of DECISION(φ_0) returning SAT. Conversely, by Lemma 8, if an execution of DECISION(φ_0) returns SAT, the corresponding branch $\mathcal{B}$ is saturated. Since $\mathcal{B}$ is open too, by completeness of the tableaux method, φ_0 is satisfiable. As a result:

Proposition 5. *The nondeterministic procedure* DECISION *is a decision procedure for the satisfiability problem of* $\mathcal{L}_{;?\|}$ *with respect to* $\mathcal{C}_{\lhd\text{-det}}$.

By Propositions 4 and 5 and Savitch's Theorem, the satisfiability problem of the fragment $\mathcal{L}_{;?\|}$ with respect to $\mathcal{C}_{\lhd\text{-det}}$ is in PSPACE. PSPACE-hardness is given by the obvious embedding of the modal logic K. Hence:

Proposition 6. *The satisfiability problem of the fragment* $\mathcal{L}_{;?\|}$ *over the class* $\mathcal{C}_{\lhd\text{-det}}$ *is PSPACE-complete.*

7   Conclusion and Future Works

We have given a complexity upper bound for the satisfiability of PRSPDL over the class of all frames, a sound and complete tableaux method for the fragment of PRSPDL without special programs interpreted over $\lhd$-deterministic frames and an optimal decision procedure for an iteration-free fragment of PRSPDL over $\lhd$-deterministic frames. Both our complexity results answer questions left open in [2–4]. Moreover, we proved the addition of $\lhd$-deterministic parallel composition to PDL without choice and iteration does not increase its complexity. We leave for future works to prove this interesting property for the full language.

Because of the characteristics PRSPDL shares with IPDL (for which there is no tableaux method to date), our tableaux method has some peculiarities borrowed from [16] (mainly compound programs as edge's labels) which are difficult to combine with the iteration construct. Hence, because of the rules ($\Box *$) and ($\Diamond * \neq$), our tableaux method for $\mathcal{L}_{;?*\|}$ over $\mathcal{C}_{\lhd\text{-det}}$ does not terminate. But we believe this tableaux method can be modified to become a decision procedure. For instance, by adding dynamic blocking as in [13], it might be possible to represent infinite saturated tableaux by finite unsaturated ones. A finite model property would be helpful. The notion of twines and the Fischer-Ladner closure introduced in this paper could be useful to prove this property.

In PDL, nondeterministic choice adds technical difficulties without changing neither the expressive power nor the complexity of the logic. It has been omitted in this paper for the sake of simplicity, leaving the study of PRSPDL with nondeterministic choice for future works.

References

1. Abrahamson, K.: Modal logic of concurrent nondeterministic programs. In: Kahn, G. (ed.) Semantics of Concurrent Computation. LNCS, vol. 70, pp. 21–33. Springer, Heidelberg (1979)
2. Balbiani, P., Boudou, J.: Iteration-free PDL with storing, recovering and parallel composition: a complete axiomatization, J. Logic Comput. (2015)
3. Balbiani, P., Tinchev, T.: Definability and computability for PRSPDL. In: Advances in Modal Logic, pp. 16–33. College Publications (2014)
4. Benevides, M.R.F., de Freitas, R.P., Viana, J.P.: Propositional dynamic logic with storing, recovering and parallel composition. ENTCS **269**, 95–107 (2011)
5. Brochenin, R., Demri, S., Lozes, É.: Reasoning about sequences of memory states. Ann. Pure Appl. Logic **161**(3), 305–323 (2009)
6. Collinson, M., Pym, D.J.: Algebra and logic for resource-based systems modelling. Math. Struct. Comput. Sci. **19**(5), 959–1027 (2009)
7. Courtault, J.R., Galmiche, D.: A Modal BI logic for dynamic resource properties. In: Artemov, S., Nerode, A. (eds.) LFCS 2013. LNCS, vol. 7734, pp. 134–148. Springer, Heidelberg (2013)
8. Danecki, R.: Nondeterministic propositional dynamic logic with intersection is decidable. In: Skowron, A. (ed.) Computation Theory. LNCS, vol. 208, pp. 34–53. Springer, Heidelberg (1984)
9. De Giacomo, G., Massacci, F.: Combining deduction and model checking into tableaux and algorithms for converse-PDL. Inf. Comput. **162**(1–2), 117–137 (2000)
10. Demri, S., Deters, M.: Separation logics and modalities: a survey. J. Appl. Non-Class. Logics **25**(1), 50–99 (2015)
11. Fischer, M.J., Ladner, R.E.: Propositional dynamic logic of regular programs. J. Comput. Syst. Sci. **18**(2), 194–211 (1979)
12. Goré, R., Widmann, F.: An optimal on-the-fly tableau-based decision procedure for PDL-satisfiability. In: Schmidt, R.A. (ed.) CADE-22. LNCS, vol. 5663, pp. 437–452. Springer, Heidelberg (2009)
13. Horrocks, I., Sattler, U.: A description logic with transitive and inverse roles and role hierarchies. J. Log. Comput. **9**(3), 385–410 (1999)
14. Kurucz, Á., Németi, I., Sain, I., Simon, A.: Decidable and undecidable logics with a binary modality. J. Logic Lang. Inf. **4**(3), 191–206 (1995)
15. Lutz, C.: PDL with intersection and converse is decidable. In: Ong, L. (ed.) CSL 2005. LNCS, vol. 3634, pp. 413–427. Springer, Heidelberg (2005)
16. Massacci, F.: Decision procedures for expressive description logics with intersection, composition, converse of roles and role identity. In: IJCAI, pp. 193–198. Morgan Kaufmann (2001)
17. Peleg, D.: Concurrent dynamic logic. J. ACM **34**(2), 450–479 (1987)

18. Pratt, V.R.: A near-optimal method for reasoning about action. J. Comput. Syst. Sci. **20**(2), 231–254 (1980)
19. Pym, D.J.: The semantics and Proof Theory of the Logic of Bunched Implications. Applied Logic Series, vol. 26. Kluwer Academic Publishers, Dordrecht (2002)
20. Reynolds, J.C.: Separation logic: a logic for shared mutable data structures. In: LICS, pp. 55–74. IEEE Computer Society (2002)

Unification

Regular Patterns in Second-Order Unification

Tomer Libal[(✉)]

INRIA, Paris, France
tomer.libal@inria.fr

Abstract. The second-order unification problem is undecidable. While unification procedures, like Huet's pre-unification, terminate with success on unifiable problems, they might not terminate on non-unifiable ones. There are several decidability results for infinitary unification, such as for monadic second-order problems. These results are based on the regular structure of the solutions of these problems and by computing minimal unifiers. In this paper we describe a refinement to Huet's pre-unification procedure for arbitrary second-order signatures which, in some cases, terminates on problems on which the original pre-unification procedure fails to terminate. We show that the refinement has, asymptotically, the same complexity as the original procedure. Another contribution of the paper is the identification of a new decidable class of second-order unification problems.

1 Introduction

The unification principle has many uses in Computer Science. Due to the undecidability of the higher-order unification problem, many applications find it necessary to restrict the use of unification to decidable classes only. This can either be achieved by applying unification on fragments of higher-order logic problems, whose unifiability is known to be decidable or by restricting unification procedures to search for an incomplete set of unifiers. Among the fragments of the first kind we can find Miller's higher-order pattern unification [15,17] and decidable sub-classes of context unification [6,12,18,20]. When we need to consider arbitrary higher-order unification problems, we must search for an incomplete set of unifiers.

Most higher-order theorem provers, such as Isabelle [16], TPS [2] and LEO II [4] and III [24], rely on Huet's pre-unification procedure [9] for the unification of higher-order terms. Since the procedure does not terminate, these theorem provers must search for incomplete finite sets of unifiers only. The most common way to obtain such a set is by bounding the depth of the terms in the co-domain of the unifiers. When one is interested in complete sets of unifiers, one must accept non-termination.

Tomer Libal—Funded by the ERC Advanced Grant ProofCert.

A.P. Felty and A. Middeldorp (Eds.): CADE-25, LNAI 9195, pp. 557–571, 2015.
DOI: 10.1007/978-3-319-21401-6_38

The common practice of establishing the decidability of a new class of infinitary unification problems is by proving that their complete sets of unifiers can be described by a finite regular expression. One can then use the exponent of periodicity theorem [10,19], in order to prove the existence of minimal unifiers. Among the classes of unification problems decided by this technique are not only those over monadic signatures [7,14,25] but also their extensions to problems over arbitrary signatures for which the unifiers are restricted to have a limited number of occurrences of the bound variables [12,13,18,20,21]. The common property of all these classes is that complete sets of unifiers can be described by regular terms. By regularity we mean the ability to describe infinite sets using finite descriptions.

Unfortunately, unrestricted unification over arbitrary signatures does not enjoy this property, even when restricted to very simple second-order languages, as was shown by Farmer [8].

Many interesting problems, among them unification problems generated in the search of theorems of second-order arithmetic, do not fall within these classes. For these problems, non-termination of unification seems inevitable.

In this paper we present a procedure for second-order pre-unification which terminates on more classes of unification problems than Huet's pre-unification procedure while keeping to the same complexity class. This is achieved by a new technique of extending the non-regular complete sets of unifiers of these problems into regular complete supersets of unifiers. We then prove the existence of minimal members in these supersets. Empty complete supersets of unifiers will imply the emptiness of the respective complete sets of unifiers and those, will prove the non-unifiability of the respective problems. We prove the soundness and completeness of this procedure.

As a second contribution, we use the structures developed in this paper in order to recognize a new class of second-order unification problems, whose complete sets of unifiers are regular and which can be decided by the new pre-unification procedure. We believe that this approach can lead to more decidable classes of second-order unification problems.

Other similar works includes the work of Abdulrab et al. for the finite representation of all unifiers for sub-classes of the string unification problem by using graphs and regular expressions [1], that of Zaionc for the regular expression description of complete sets of unifiers for monadic second-order unification [25] and the work of Le Chenadec for the description of first-order cycles using finite automata [11]. These works differ from the current one in that they cover problems whose complete sets of unifiers are regular.

The paper is organized as follows. In the next section we give some definitions and notations which will be used throughout the paper. The main section is dedicated to the construction of complete supersets of unifiers, the establishment of some of their properties and the presentation of the pre-unification procedure. We then prove the correctness of the procedure and its improved termination over classes of problems when compared to Huet's procedure. We conclude by proving the asymptotic equivalence of the complexity of the two procedures.

Due to space considerations, we omit the proofs of all the theorems and lemmas appearing in the paper. The interested reader can find these proofs on the author's website[1].

2 Preliminaries

2.1 Typed Lambda Calculus

In this section we will present the logical language that will be used throughout the paper. The language is a version of Church's simple theory of types [5] with an η-conversion rule as presented in [3,22] and with implicit α-conversions. Most of the definitions in this section are adapted from [22].

Let $\mathfrak{T}_0$ be a set of basic types, then the set of types $\mathfrak{T}$ is generated by $\mathfrak{T} := \mathfrak{T}_0 | \mathfrak{T} \to \mathfrak{T}$. Let Σ be a signature of function symbols and let $\mathfrak{V}$ be a countably infinite set of variable symbols. The function ar denotes the *arity* of each function symbol and variable according to its type in the usual way. *Variables* are normally denoted by the letters x, y, z and *function symbols* by the letters f, g, h. We sometimes use subscripts and superscript as well. We sometimes add a superscript to symbols in order to specify their type. The set Term^α of terms of type α is generated by $\mathsf{Term}^\alpha := f^\alpha | x^\alpha | \lambda x^\beta . \mathsf{Term}^\gamma | \mathsf{Term}^{\beta \to \alpha}(\mathsf{Term}^\beta)$ where $f \in \Sigma, x \in \mathfrak{V}$ and $\alpha \in \mathfrak{T}$ (in the abstraction, $\alpha = \beta \to \gamma$). Applications throughout the paper will be associated to the right. We will sometimes omit brackets in applications when the meaning is clear. The set Term denotes the set of all terms. *Subterms*, *positions* and *position prefixes* are defined as usual. *Sizes of positions* denote the length of the path to the position. We denote the subterm of t at position p by $t|_p$. *Bound* and *free variables* are defined as usual. Given a term t, we denote by $\mathsf{hd}(t)$ its *head symbol* and distinguish between *flex* terms, whose head is a free variable and *rigid* terms, whose head is a function symbol or a bound variable. *Rigid positions* are positions such that no flex subterm is in a prefix position. The *depth* of a term t, denoted by $\mathsf{d}(t)$, is the size of the maximal rigid position in t. The *order* of types are denoted by order and are defined as usual. The order of a term, denoted using the same symbol, is the order of its type.

Substitutions and their *composition* ($\circ$) are defined as usual. We denote by $\sigma|_W$ the substitution obtained from substitution σ by restricting its domain to variables in W. We extend the application of substitutions to terms in the usual way and denote it by postfix notation. Variable capture is avoided by implicitly renaming variables to fresh names upon binding. A substitution σ is *more general* than a substitution θ, denoted $\sigma \leq_s \theta$, if there is a substitution δ such that $\sigma \circ \delta = \theta$.

We assume that all the terms considered in this paper are in β-normal and η-expanded forms [22]. We further assume that all substitutions are idempotent [23] and contain only terms in β-normal and η-expanded forms in their codomain. This allows us to deal with normal forms implicitly (see [22] for more

---

[1] http://logic.at/staff/shaolin/papers/holunif_proofs.pdf.

information). Equality between terms is always assumed to be α-equality. Each application of a λ-term to another is always converted implicitly into β−normal form.

We introduce also a vector notation $\overline{t_n}$ for the sequence of terms $t_1, \ldots, t_n$. Currying and uncurrying is applied implicitly as well.

We will sometimes refer to the position $0 < i \leq n$ of a term s in the sequence by $t_1, \ldots, s_{@i}, \ldots, t_n$.

2.2   Contexts and Pre-unification

The majority of the definitions in this section are taken from [21,22].

Terms of the form $\lambda z^\alpha . s^\alpha$ where z occurs in s exactly once are called *contexts* and are denoted by $s([.])$ where $[.]$ is considered as the "hole" of the term. We denote by $\mathrm{mpath}(C)$ the *main path* of the context C which is the position of the hole in the context C.

Unification problems (or systems) are sets of terms $t \doteq s$, called *equations*, where t and s are of the same type. Based on whether t and s are flex or rigid, we make a distinction between *flex-flex*, *flex-rigid* and *rigid-rigid* equations. Systems are considered closed under symmetry of $\doteq$.

A substitution σ *unifies* an equation $t \doteq s$ if $t\sigma = s\sigma$. It unifies a system if it unifies all its equations. We denote the *set of all unifiers* of a system S by $\mathrm{Unifiers}(S)$. Let $\cong$ be the least congruence relation on **Term** which contains $\{(t, s) \mid \mathrm{hd}(t), \mathrm{hd}(s) \in \mathfrak{V}\}$. A substitution σ *pre-unifies* an equation $t \doteq s$ if $t\sigma \cong s\sigma$. It pre-unifies a system if it pre-unifies all its equations. The completing substitution ξ_S for a system S maps every two variables in S of the same type to the same fresh variable. It is simple to prove that if σ pre-unifies a system S, then $\sigma \circ \xi$ unifies S [22]. A *complete set of pre-unifiers* for a system S, denoted by $\mathrm{PreUnifiers}(S)$, is a set of substitutions such that $\{\sigma \circ \xi_S \mid \sigma \in \mathrm{PreUnifiers}(S)\} \subseteq \mathrm{Unifiers}(S)$ and for every $\theta \in \mathrm{Unifiers}(S)$ there exists $\sigma \in \mathrm{PreUnifiers}(S)$ such that $\sigma|_{dom(\theta)} \leq \theta$.

An equation $x \doteq t$ in η-normal form is called *solved* in system S if x does not occur elsewhere in S. We call x a solved variable in S. An equation is *pre-solved* in a system S if it is either solved in S or flex-flex. A system is solved (pre-solved) if all its equations are solved (pre-solved). We denote by σ_S the substitution obtained from mapping x to t in all solved equations $x \doteq t$ in S.

Imitation partial bindings and *projection partial bindings* are defined in [22] and are denoted, respectively, by $\mathrm{PB}(f, \alpha)$ and $\mathrm{PB}(i, \alpha)$ where $\alpha \in \mathfrak{T}$, $f \in \Sigma$ and $0 < i$. Briefly, partial bindings are substitutions which are used in order to approximate the (possibly infinite) number of final mappings for variables occurring in flex-rigid equations. By either imitating the head symbol of the rigid equation or by projecting one of the bound variables of the mapping for the variable, the set of partial bindings is always finite.

Huet's pre-unification procedure PUA, as presented by Snyder and Gallier [22], is given in Fig. 1.

$$\frac{S}{S \cup \{A \doteq A\}} \; \text{(Delete)} \qquad \frac{S \cup \{\lambda\overline{z_k}.s_1 \doteq \lambda\overline{z_k}.t_1, \ldots, \lambda\overline{z_k}.s_n \doteq \lambda\overline{z_k}.t_n\}}{S \cup \{\lambda\overline{z_k}.f(\overline{s_n}) \doteq \lambda\overline{z_k}.f(\overline{t_n})\}} \; \text{(Decomp)}$$

$$\frac{S\sigma \cup \{x \doteq \lambda\overline{z_k}.t\} \qquad x \notin \mathrm{FV}(t) \wedge \sigma = [\lambda\overline{z_k}.t/x]}{S \cup \{\lambda\overline{z_k}.x(\overline{z_k}) \doteq \lambda\overline{z_k}.t\}} \; \text{(Bind)}$$

$$\frac{S \cup \{x \doteq u, \lambda\overline{z_k}.x^\alpha(\overline{s_n}) \doteq \lambda\overline{z_k}.f(\overline{t_m})\} \qquad u \in \mathrm{PB}(f,\alpha)}{S \cup \{\lambda\overline{z_k}.x^\alpha(\overline{s_n}) \doteq \lambda\overline{z_k}.f(\overline{t_m})\}} \; \text{(Imitate)}^1$$

$$\frac{S \cup \{x \doteq u, \lambda\overline{z_k}.x^\alpha(\overline{s_n}) \doteq \lambda\overline{z_k}.a(\overline{t_m})\} \qquad 0 < i \le k, u = \mathrm{PB}(i,\alpha)}{S \cup \{\lambda\overline{z_k}.x^\alpha(\overline{s_n}) \doteq \lambda\overline{z_k}.a(\overline{t_m})\}} \; \text{(Project)}^2$$

$$\frac{\bot}{S \cup \{\lambda\overline{z_k}.f(\overline{s_n}) \doteq \lambda\overline{z_k}.g(\overline{t_n})\}} \; \text{(Symbol-Clash)}^{3,4}$$

1. where $f \in \Sigma$.
2. where either $a \in \Sigma$ or $a = z_i$ for some $0 < j \le k$.
3. where $f, g \in \Sigma$ and $f \ne g$.
4. this rule is redundant with regard to soundness and completeness but has implications with regard to termination and appears in [9].

Fig. 1. PUA- Huet's pre-unification procedure

Theorem 1 (Soundness of PUA [9]). *If S' is obtained from a unification system S using PUA and is in pre-solved form, then $\sigma_{S'}|_{FV(S)} \in PreUnifiers(S)$.*

Theorem 2 (Completeness of PUA [9]). *If $\theta \in PreUnifiers(S)$ for a unification system S, then there exists a pre-solved system S', which is obtainable from S using PUA such that $\sigma_{S'}|_{FV(S)} \le_s \theta$.*

Remark 3. The procedure PUA contains two kinds of non-determinism. On the one hand, we need to choose an equation at each step and on the other, we need to choose which rule to apply to it. In [22] it is argued that completeness is only affected by the second kind of non-determinism and more precisely, by the choice between the (Imitate) and (Project) rules. The first case is a "don't-care" non-determinism while the second is a "don't-know" non-determinism. We will use this fact in the rest of the paper and allow ourselves to choose specific equations to process without harming completeness.

3   The Refinement Procedure

In order to simplify definitions and proofs, we consider only first-order functions symbols of arbitrary arity and second-order unary variables. Note that even very simple second-order unification classes are undecidable [8]. An extension to the general second-order case is straightforward. We discuss the possibility to extend the method to the general higher-order case in the conclusion.

In this section we will be interested in trying to obtain failure information from cyclic equations.

3.1   Cyclic Equations and Their Properties

Let the relations $x < y$ and $x = y$ be defined for equations $C(x\overline{t_n}) \doteq D(y\overline{s_m})$ where C and D are contexts and where $\text{mpath}(C) < \text{mpath}(D)$ and $\text{mpath}(C) = \text{mpath}(D)$ respectively. Define a partial order over variables by the transitive closure of the union of the two relations (under further restrictions on the symmetry of $=$, see [13,21] for a full definition). A set of equations is *cyclic* if the partial order generated over the set contains the relation $x < x$ for some variable x occurring in the set. An example of a cycle is the set $\{xa \doteq f(yc, b), g(yd) \doteq g(ze), zb \doteq xb\}$. Cycles capture the idea that using PUA, one can, in some cases, obtain again (a variation of) the original set of equations.

The next result exemplifies the role of cycles in the non-termination of second-order unification.

Theorem 4 (Levy [13]). *It is decidable whether a second-order unification problem not containing cycles has a unifier.*

In this paper we will focus on a certain kind of second-order cycles of the following form.

Definition 5 (Cyclic equations). *Let e be an equation of the form $\lambda \overline{z_n}.x_0 t \doteq \lambda \overline{z_n}.C(x_0 s)$. e is called a cyclic equation where C is a context. t, s and C may contain the variables $\overline{z_n}$ but not the variable x_0. We denote the fact that e is cyclic by the predicate $\mathtt{cyclic}(e)$.*

We next prove that the restriction on C not to contain x_0 can be avoided.

Lemma 6. *Let e be an equation $\lambda \overline{z_n}.x_0 t \doteq \lambda \overline{z_n}.C(x_0 s)$ and assume further, without loss of generality, that for all occurrences of x_0 in C, the sizes of their positions are not smaller than the size of the position of the hole in C (otherwise, define the hole to be the position of the minimal such occurrence). Then we can obtain, using the rules of PUA, an equation $\lambda \overline{z_n}.w_0 t \doteq \lambda \overline{z_n}.C'(w_0 s)$ where w_0 does not occur in C' for some context C'.*

Definition 7 (Progressive context). *Given a cyclic equation e, where $C = C_1 \ldots C_m$ such that for all $0 < i \leq m$, $C_i = f_i(r_i^1, \ldots, [.], \ldots, r_i^{n_i})$ where $n_i = \mathtt{ar}(f_i) - 1$. Define also, for all $m < i$, $C_i = f_k(y_{i-m}^1 s, \ldots, [.], \ldots, y_{i-m}^{n_k} s)$ where $k = ((i-1) \mod m) + 1$ and y_{i-m}^j for $0 < j \leq n_k$ are new variables. We define the progressive context D_i^e for all $0 \leq i$ as follows:*

– for all $0 \leq i$, $D_i^e = C_{i+1} \ldots C_{i+m}$.

We will use the cycle $x_0 t \doteq f(r_1, g(x_0 s, r_2))$ as a running example. This cycle is interesting as it has instances which are unifiable and instances which are not unifiable. For the unifiable ones, both Huet's procedure and the one presented here will compute a complete set of pre-unifiers. For the non-unifiable ones, Huet's will fail to terminate while our procedure, as proved in Theorem 37 and under some additional restrictions as defined in Definition 35, will terminate with failure. An example for a unifiable instance is for $t = f(a, g(f(f(a, a), a)), r_1 = a, r_2 = a$ and $s = f(a, a)$. For obtaining a non-unifiable instance which corresponds to Definition 35, just replace s from the previous instance with $f(a, b)$.

Example 8. Given the cycle $x_0 t \doteq f(r_1, g(x_0 s, r_2))$ (having $m = 2$), its progressive contexts for $0 \leq i \leq 2$ are

- $D_0 = C_1 C_2$, $D_1 = C_2 C_3$ and $D_2 = C_3 C_4$

where

- $C_1 = f(r_1, [.])$.
- $C_2 = g([.], r_2)$.
- $C_3 = f(y_1, [.])$.
- $C_4 = g([.], y_2)$.

In the rest of this paper, e will refer to equations of this form and $t, s, C, m, k, n_i, r_i^j$ and y_i^j will refer to the corresponding values in e.

As mentioned in Remark 3, the "don't-know" non-determinism affects the completeness of PUA and it is not hard to see that it is also the cause of its non-termination. The way to improve termination and to define additional decidable classes will depend, therefore, on refining the possible "don't-know" choices allowed in the search.

We will first define the result of applying (Imitate) and (Project) (plus some additional deterministic rules) on cyclic equations.

Definition 9 ($\mathfrak{I}$ and $\mathfrak{P}$). *Given a cyclic equation e, for all $0 \leq i$, we define $\mathfrak{I}(i)$, $\mathfrak{I}^*(i)$ and $\mathfrak{P}(i)$ inductively as follows:*

- $\mathfrak{P}(0) = \mathfrak{I}(0) = \mathfrak{I}^*(0) = \emptyset$.
- *if $0 < i \leq m$ then $\mathfrak{I}^*(i) = \mathfrak{I}^*(i-1) \cup \{\lambda \overline{z_n}.y_i^j t \doteq \lambda \overline{z_n}.r_i^j \mid 1 \leq j \leq n_i\}$.*
- *if $m < i$ then $\mathfrak{I}^*(i) = \mathfrak{I}^*(i-1) \cup \{\lambda \overline{z_n}.y_i^j t \doteq \lambda \overline{z_n}.y_{i-m}^j s \mid 1 \leq j \leq n_i\}$.*
- *for all $0 < i$, $\mathfrak{I}(i) = \mathfrak{I}^*(i) \cup \{\lambda \overline{z_n}.x_i t \doteq \lambda \overline{z_n}.D_i^e(x_i s)\}$.*
- *for all $0 < i$, $\mathfrak{P}(i) = \mathfrak{I}^*(i-1) \cup \{\lambda \overline{z_n}.t \doteq \lambda \overline{z_n}.D_{i-1}^e(s)\}$.*

Using these definitions, one can now describe the search conducted by PUA graphically as can be seen in Fig. 2.

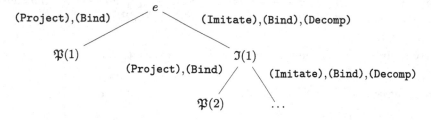

Fig. 2. The "don't-know" non-determinism in PUA

Example 10. Extending Example 8, we get the following values for $0 < i \leq 3$:

- $\mathfrak{P}(1)$ is $t \doteq f(r_1, g(s, r_2))$ which is equivalent to $t \doteq D_0(s)$.
- $\mathfrak{I}(1)$ is $\{x_1 t \doteq g(f(y_1 s, x_1 s), r_2), y_1 t \doteq r_1\}$ which is equivalent to $\{x_1 t \doteq D_1(x_1 s), y_1 t \doteq r_1\}$.

- $\mathfrak{P}(2)$ is $\{t \doteq g(f(y_1s, s), r_2), y_1t \doteq r_1\}$ which is equivalent to $\{t \doteq D_1(s), y_1t \doteq r_1\}$.
- $\mathfrak{I}(2)$ is $\{x_2t \doteq f(y_1s, g(x_2s, y_2s)), y_2t \doteq r_2, y_1t \doteq r_1\}$ which is equivalent to $\{x_2t \doteq D_2(x_2s), y_2t \doteq r_2, y_1t \doteq r_1\}$.
- $\mathfrak{P}(3)$ is $\{t \doteq D_2(s), y_2t \doteq r_2, y_1t \doteq r_1\}$
- $\mathfrak{I}(3)$ is $\{x_3t \doteq D_3(x_3s), y_3t \doteq y_1s, y_2t \doteq r_2, y_1t \doteq r_1\}$

The correctness of this description is proved next.

Lemma 11. *Let e be a cyclic equation, then, up to the renaming of the free variables and for all $0 \leq i$, the application of (Imitate), (Bind) and (Decomp) on $\lambda\overline{z_n}.x_it \doteq \lambda\overline{z_n}.D_i^e(x_is)$ results in a set of equations containing $\lambda\overline{z_n}.x_{i+1}t \doteq \lambda\overline{z_n}.D_{i+1}^e(x_{i+1}s)$.*

We call this cycle the *principle cycle* of the application of (Imitate).

Lemma 12. *Let $S \cup \{e\}$ be a unification problem where e is a cyclic equation. Then, there is a substitution τ such that $FV(S \cup \{e\}) \cap dom(\tau) = \{x_0\}$ and such that the following holds, up to the renaming of the free variables:*

- *assume we repeatedly apply i times (Imitate), (Bind) and (Decomp) on e and the generated principal cycles, then the obtained unification problem is $(S \cup \mathfrak{I}(i))\tau$.*
- *assume we apply a (Project) and (Bind) after $i-1$ applications of (Imitate), (Bind) and (Decomp) on e and the generated principal cycles, then the obtained problem is $(S \cup \mathfrak{P}(i))\tau$.*

A simple but crucial fact that will enable us to enlarge the non-regular sets of solutions of PUA into regular supersets is the following.

Proposition 13. *Let S be a unification problem and let $S' \subset S$, then:*

- *S is unified by a substitution σ only if S' is unified by σ.*
- *$PreUnifiers(S) \subseteq PreUnifiers(S')$.*

We can now prove that each derivation of e must, at some point, use the above sequence of rules.

Lemma 14. *For any S, $\sigma \in PreUnifiers(S \cup \{e\})$ iff there is $0 < i$ and substitutions θ and τ such that $\theta \in PreUnifiers((S \cup \{e\} \cup \mathfrak{P}(i))\tau)$, $\theta|_{FV(S\cup\{e\})} \leq_s \sigma$ and $\tau|_{FV(S\cup\{e\})} \leq_s \sigma$.*

By taking $\theta \circ \tau$, the next corollary follows immediately.

Corollary 15. *For any S, $\sigma \in PreUnifiers(S \cup \{e\})$ only if there is $0 < i$ and a substitution θ such that $\theta \in PreUnifiers(S \cup \{e\} \cup \mathfrak{P}(i))$, and $\theta|_{FV(S\cup\{e\})} \leq_s \sigma$.*

The definitions of the generated sets $\mathfrak{P}(i)$ for all $0 < i$, are given inductively. We notice that the sets, for $i > m$, are made of two components:

- the inductive part which includes all equations $\{\lambda\overline{z_n}.y_l^j t \doteq \lambda\overline{z_n}.y_{l-m}^j s \mid 1 \leq j \leq n_k\}$, for all $m < l \leq i$.
- the base part which includes the equations $\{\lambda\overline{z_n}.t \doteq \lambda\overline{z_n}.D_{i-1}^e(s)\}$ and $\{\lambda\overline{z_n}.y_l^j t \doteq \lambda\overline{z_n}.r_l^j \mid 1 \leq j \leq n_k\}$ for just $0 < l \leq m$.

We will use this distinction in the next section.

3.2   The Refinement Procedure

In this section we will show how to obtain a superset of all unifiers of a cycle such that this superset will not be defined inductively. This will allow us to give a finite representation of this set which will be used in order to improve termination.

The next sets are constructed without the inductive part mentioned earlier.

Definition 16 ($\mathfrak{P}^-$). *Given a cyclic equation e, we define $\mathfrak{P}^-$ for all $0 < i$ as follows:*

- *if $0 < i \leq m + 1$ then $\mathfrak{P}^-(i) = \mathfrak{P}(i)$.*
- *if $m + 1 < i$ then $\mathfrak{P}^-(i) = \mathfrak{I}^*(m) \cup \{\lambda \overline{z_n}.t \doteq \lambda \overline{z_n}.D_{i-1}^e(s)\}$.*

The equation $\lambda \overline{z_n}.t \doteq \lambda \overline{z_n}.D_{i-1}^e(s)$ is called the projected equation of $\mathfrak{P}^-(i)$.

As can be seen from the definition, for $i > m$, $\mathfrak{P}^-$ is defined in a non-inductive way as it depends on a fixed set $\mathfrak{I}^*(m)$.

Example 17. The values for the equation from Example 8 for $i = 3, 4, 5, 6, 7$ are:

- $\mathfrak{P}^-(3) = \{t \doteq D_2(s), y_2 s \doteq r_2, y_1 s \doteq r_1\} = \mathfrak{P}(3)$.
- $\mathfrak{P}^-(4) = \{t \doteq D_3(s), y_2 s \doteq r_2, y_1 s \doteq r_1\} \subseteq \mathfrak{P}(4)$.
- $\mathfrak{P}^-(5) = \{t \doteq D_4(s), y_2 s \doteq r_2, y_1 s \doteq r_1\} \subseteq \mathfrak{P}(5)$.
- $\mathfrak{P}^-(6) = \{t \doteq D_5(s), y_2 s \doteq r_2, y_1 s \doteq r_1\} \subseteq \mathfrak{P}(6)$.
- $\mathfrak{P}^-(7) = \{t \doteq D_6(s), y_2 s \doteq r_2, y_1 s \doteq r_1\} \subseteq \mathfrak{P}(7)$.

Together with Proposition 13 and Corollary 15, we can now prove two lemmas asserting that these new sets are indeed complete supersets of unifiers.

Lemma 18. *prjm   For all $0 < i$, $\mathfrak{P}^-(i) \subseteq \mathfrak{P}(i)$.*

Lemma 19. *For all S and for all $0 < i$, $\mathtt{PreUnifiers}(S \cup \mathfrak{P}(i)) \subseteq \mathtt{PreUnifiers}(S \cup \mathfrak{P}^-(i))$.*

The fact that the sets $\mathfrak{P}^-(i)$, for $i > m$, are not defined in an inductive way, will enable us to simplify the description of their pre-unifiers. In the next lemma we will prove that iterating the ⟨Imitate⟩ rule beyond the first $3m$ iterations gives no further information about the unifiability of the set.

Lemma 20. *For all S and for all $3m < i$, $S \cup \mathfrak{P}^-(i)$ is unifiable iff $S \cup \mathfrak{P}^-(i-m)$ is unifiable. Moreover, if σ is a pre-unifier of $S \cup \mathfrak{P}^-(i - m)$, then σ' is a pre-unifier of $S \cup \mathfrak{P}^-(i)$ where $dom(\sigma') = dom(\sigma) \setminus \{y_l^j \mid 0 < j \leq n_k, i - 2m \leq l < i - m\} \cup \{y_l^j \mid 0 < j \leq n_k, i - m \leq l < i\}$ and $\sigma'(y_l^j) = \sigma(y_{l-m}^j)$ for all $i - m \leq l < i$ and $0 < j \leq n_k$ where $k = ((i - 1) \mod m) + 1$.*

The intuition behind this lemma is demonstrated in the following example.

Example 21. Take $\mathfrak{P}^-(5)$ and $\mathfrak{P}^-(7)$ (remember that $m = 2$) from Example 17:

- $\mathfrak{P}^-(5) = \{t \doteq f(y_3 s, g(s, y_4 s)), y_2 s \doteq r_2, y_1 s \doteq r_1\}$.

– $\mathfrak{P}^-(7) = \{t \doteq f(y_5 s, g(s, y_6 s)), y_2 s \doteq r_2, y_1 s \doteq r_1\}$.

The two pairs of variables y_3, y_4 and y_5, y_6 occur only once in both sets.

Next, we prove that the supersets of pre-unifiers for e can be restricted by computing pre-unifiers for the problems $\mathfrak{P}^-(i)$ for $0 < i \leq 3m$. This will establish the minimality property which is required for proving termination.

Lemma 22. *For any S and for any $\sigma \in \texttt{PreUnifiers}(S \cup \{e\})$, there is $0 < i \leq 3m$ and $\theta \in \texttt{PreUnifiers}(S \cup \{e\} \cup \mathfrak{P}^-(i))$ such that $\theta|_{FV(S \cup \{e\})} \leq_s \sigma$.*

The following is a corollary of the previous lemma. This result states that termination can be achieved on some problems, even if their sets of solutions is irregular.

Corollary 23. *Given a set S and a cycle e. If, for all $0 < i \leq 3m$, $\texttt{PreUnifiers}(\mathfrak{P}^-(i)) = \emptyset$, then $S \cup \{e\}$ is not unifiable.*

As an example of applying the above corollary, consider the following instance of our running example.

Example 24. Given the cycle $x_0(f(a, g(a, a), a)) \doteq f(a, g(x_0(f(a, b)), a))$, none of the $\mathfrak{P}^-(i)$ for $0 < i \leq 6$ are unifiable. Using the above corollary, we can conclude that this problem is not unifiable.

We will proceed next to the refinement of **PUA** but first, we need to modify unification equations and the predicate $\texttt{cyclic}$. This modification is required in order to apply the refinement at most once per cyclic equation.

Definition 25 (Marked equations). *Given a unification equation $\lambda \overline{z_n}.t \doteq \lambda \overline{z_n}.s$, let $\lambda \overline{z_n}.t \doteq^\bullet \lambda \overline{z_n}.s$ be its marked version. The function $\texttt{cyclic}$ now fails if e is marked.*

The idea of the following procedure is the following. When running on a unifiable problem, the extra equations added by the **(Cycle)** rule will also be unifiable for some $0 < i \leq 3m$ according to Lemma 22. On the contrary, when a problem is not unifiable, the generated sets $\mathfrak{P}^-(i)$ must all be processed before any rule is applied to e. If none is unifiable, we get on all branches of the search **(Symbol-Clash)** failure nodes and therefore will not apply any further rule to e and the procedure will terminate. Corollary 23 also tells us that in that case the problem is indeed not unifiable. In the case the problem is not unifiable but some set $\mathfrak{P}^-(i)$ is, we will proceed with the unifiability of e, which might not terminate.

Definition 26 (RPUA). *The procedure RPUA has the same set of rules as PUA (see Fig. 1) but has, in place of (Imitate) and (Project), the rules in Fig. 3. In addition, all rules apply to marked and unmarked equations in the same way.*

$$\frac{S \cup \{x \doteq u, \lambda\overline{z_k}.x^\alpha(\overline{s_n}) \doteq \lambda\overline{z_k}.f(\overline{t_m})\} \qquad u \in \text{PB}(f, \alpha) \wedge \neg\text{cyclic}(e)}{S \cup \{\lambda\overline{z_k}.x^\alpha(\overline{s_n}) \doteq \lambda\overline{z_k}.f(\overline{t_m})\}} \text{ (Imitate)}^1$$

$$\frac{S \cup \{x \doteq u, \lambda\overline{z_k}.x^\alpha(\overline{s_n}) \doteq \lambda\overline{z_k}.a(\overline{t_m})\} \qquad 0 < i \le k, u = \text{PB}(i, \alpha) \wedge \neg\text{cyclic}(e)}{S \cup \{\lambda\overline{z_k}.x^\alpha(\overline{s_n}) \doteq \lambda\overline{z_k}.a(\overline{t_m})\}} \text{ (Project)}^2$$

$$\frac{S \cup \{\lambda\overline{z_k}.x^\alpha(\overline{s_n}) \doteq^\bullet \lambda\overline{z_k}.a(\overline{t_m})\} \cup \mathfrak{P}^-(i) \qquad 0 < i \le 3m \wedge \text{cyclic}(e)}{S \cup \{\lambda\overline{z_k}.x^\alpha(\overline{s_n}) \doteq \lambda\overline{z_k}.a(\overline{t_m})\}} \text{ (Cycle)}^2$$

1. where $f \in \Sigma$ and $e = \lambda\overline{z_k}.x^\alpha(\overline{s_n}) \doteq \lambda\overline{z_k}.f(\overline{t_m})$.
2. where either $a \in \Sigma$ or $a = z_i$ for some $0 < j \le k$ and $e = \lambda\overline{z_k}.x^\alpha(\overline{s_n}) \doteq \lambda\overline{z_k}.a(\overline{t_m})$

Fig. 3. RPUA- Pre-unification with refined termination

3.3 The Correctness of the Refinement

In this section we prove the soundness and completeness of the procedure. Both are proved relatively to PUA.

Theorem 27 (Soundness of RPUA). *If S' is obtained from a unification system S using RPUA and is in pre-solved form, then $\sigma_{S'}|_{FV(S)} \in \textit{PreUnifiers}(S)$.*

For proving the completeness of RPUA, we need one more definition.

Definition 28 (Imitation blocks). *Let D be a derivation in PUA, and let e be an unmarked cyclic equation. The imitation block for e in D is the following inductive set:*

– *e is in the imitation block.*
– *if there is an application of (Imitate) on an equation in the block, then its principal cycle is also in the block.*

The size of the block is the size of the set plus 1.

The intuition behind this definition is that imitation blocks help us reconstruct, out of some arbitrary derivation, the exact i for constructing $\mathfrak{P}^-(i)$.

Theorem 29 (Completeness of RPUA). *If θ is a pre-unifier of a unification system S, then there exists a pre-solved system S', which is obtainable from S using RPUA such that $\sigma_{S'}|_{FV(S)} \le_s \theta$.*

3.4 Termination and Decidability Results

The most interesting property of RPUA is that it terminates on some cases where PUA does not and, at the same time, has no additional asymptotic complexity. We will investigate these two claims next.

We first prove that RPUA terminates on at least all problems on which PUA terminates.

Theorem 30. *Let S be a unification system, then PUA terminates on it only if RPUA does.*

We now prove that RPUA terminates, in contrast to PUA, on more classes of problems.

As noted above, in order for RPUA to terminate on problems on which PUA does not terminate, one must use the eager strategy of, upon calling (Cycle), attempting to unify all generated sets $\mathfrak{P}^-(i)$ before applying any rule to e. In order for RPUA not to compute unnecessary steps, we will also add a constraint on the calls to (Imitate) and (Project).

Definition 31 (Possible pairs). *Given a problem $S \cup \{e\} \cup \mathfrak{P}^-(i)$, an equation e' derived from e and an equation e'' derived from $\mathfrak{P}^-(i)$ are paired if e' was derived from the set generated by applying (Imitate) i times on e and the generated principle cycles, then applying one (Project) and then following the rule applications used for deriving e'' from $\mathfrak{P}^-(i)$.*

The intuition behind possible pairs, as demonstrated in the following example, is that one can optimize the execution of the procedure by applying the same rules to pairs of equations.

Example 32. Consider the equation from previous examples and consider the application of (Cycle) with $i = 4$, so $\mathfrak{P}^-(4) = \{t \doteq D_3(s), y_2 s \doteq r_2, y_1 s \doteq r_1\}$. After applying 3 times (Imitate) and a (Project) on e and its generated principal cycles (among other rules), we obtain $\{t \doteq D'_3(s), y'_2 s \doteq r_2, y'_1 s \doteq r_1, y'_3 t \doteq y'_1 s\}$ where D'_3 is equal to D_3 except for the renaming of the free variables. Then the following are possible pairs:

- $t \doteq D'_3(s)$ and $t \doteq D_3(s)$.
- $y'_2 s \doteq r_2$ and $y_2 s \doteq r_2$.
- $y'_1 s \doteq r_1$ and $y_1 s \doteq r_1$.

Note that the equation $y'_3 t \doteq y'_1 s$, has no possible pair.

Definition 33 (RPUA strategy). *When running RPUA, we require the following stategies:*

- *Given an unmarked cyclic equation, do the following:*
  - *let $i = 1$.*
  - *apply (Cycle) with i.*
  - *exhaustively apply RPUA on the equations in $\mathfrak{P}^-(i)$.*
  - *if a pre-solved form is found, break. Otherwise increment i by 1 (as long as $i \leq 3m$).*
  - *try to apply (Symbol-Clash) on the current problem.*
- *Always apply the same (Project) or (Imitate) on both equations in a possible pair.*

Theorem 34 (Correctness of the strategy). *RPUA with the strategy is sound and complete.*

We can now define a new class of second-order unification problems and show it to be decidable when using RPUA, in contrast to PUA.

Definition 35 (Projected cycles). *A cycle $x_0 t \doteq C(x_0 s)$ is called a projected cycle if:*

1. t *is ground.*
2. *for all positions p in C which are not on the main path of C:*
   (a) $d(t|_p) < d(s)$.
   (b) $t|_p = C|_p$.

Theorem 36. *PUA does not terminate on problems containing projected cycles.*

In the next theorem, we assert that the unifiability of problems in this class can be decided using RPUA. The idea behind the proof is that, if the problem is unifiable, it is unifiable only by substitutions which map each of the variables y_i^j to terms $\lambda z.s_i^j$ where z does not occur in s_i^j. Such substitutions will always unify the equations $y_i^j t \doteq y_{i-m} s$ and therefore, our computed supersets are actually complete sets of unifiers.

Theorem 37. *RPUA decides the unification problem of projected cycles.*

3.5 Asymptotic Analysis

In the last part of the paper we discuss the complexity of RPUA. We will measure the complexity of both procedures in the number of "don't-know" non-deterministic calls done along the derivation. A naive consideration of RPUA might suggest that it has an asymptotically exponential slow-down, in the number and size of the cyclic equations, on problems on which PUA terminates. We will show next that both procedures have the same complexity.

Theorem 38. *The number of "don't know" non-deterministic choices in runs of RPUA on some problem S when using the strategy is the same as in runs of PUA on S.*

4   Conclusion

Second-order unification problems play an important role within general higher-order unification. Many important theorems, like those in arithmetic which can be finitely axiomatized in second-order logic, require only unification over second-order formulas. Nevertheless, except for few results like the one by Levy [13] for deciding acyclic second-order unification problems, these problems are treated within the general procedures for higher-order problems. In this paper we have attempted to show that these problems are inherently simpler than general higher-order problems and that one can design for them (theoretically) improved unification procedures. We showed that for these problems, one can compute information which is static for arbitrarily long runs and which can be used in order to improve termination.

The fact that the procedure has the same asymptotic complexity does not mean that it is as efficient as Huet's. Indeed, even in the most efficient implementation where different computations of the $\mathfrak{P}$ sets are done using the previous set information, care should be taken to back-track at the right points and those, extra machinery is required. On the other hand, this procedure might also be implemented in a more efficient way that Huet's. This can be achieved by taking advantage of the natural parallelism which is inherent in the procedure in the form of the separate computation of the $3m$ $\mathfrak{P}$ sets. The claims in this paragraph have still to be demonstrated and the implementation of this procedure, both in a sequential form and in a parallel form, is planned using the multi-agent architecture of LEO-III [24].

An extension to higher-order logic is far from being trivial. The main difficulty is that the number of higher-order variables does not decrease when applying projections. One of the consequences is that, in contrary to the second-order case where infinite sequences of cyclic problems can only be generated by applying imitations, such sequences can also be generated using projections. Even if this obstacle can be overcome by detecting these cycles, the fact that the total number of higher-order variables does not decrease at each $\mathfrak{P}$ set, renders our procedure ineffective.

An interesting extension to the work presented in the paper is to consider also the equations in the set $\mathfrak{P} \setminus \mathfrak{P}^-$. This set was considered in this paper only with regard to deciding the unification problem of projected cycles. We have started a very promising work on using this set in order to build finite tree automata which, together with unifiers of the finitely many $\mathfrak{P}^-$ sets, can be used in order to decide the unification problem of far more complex cases than projected cycles.

References

1. Abdulrab, H., Goralcik, P., Makanin, G.S.: Towards parametrizing word equations. ITA **35**(4), 331–350 (2001)
2. Andrews, P., Issar, S., Nesmith, D., Pfenning, F.: The tps theorem proving system. In: Lusk, E.R., Overbeek, R. (eds.) CADE 1988. LNCS, vol. 310, pp. 760–761. Springer, Heidelberg (1988)
3. Barendregt, H.P.: The Lambda Calculus - Its Syntax and Semantics. Studies in Logic and the Foundations of Mathematics, vol. 103. North-Holland, Amsterdam (1984)
4. Benzmüller, C., Paulson, L., Theiss, F., Fietzke, A.: The LEO-II project. In: Proceedings of the Fourteenth Workshop on Automated Reasoning, Bridging the Gap between Theory and Practice. Imperial College, London (2007)
5. Church, A.: A formulation of the simple theory of types. J. Symb. Log. **5**(2), 56–68 (1940)
6. Comon, H.: Completion of rewrite systems with membership constraints. part i: deduction rules. J. Symb. Comput. **25**(4), 397–419 (1998)
7. Farmer, W.M.: A unification algorithm for second-order monadic terms. Ann. Pure Appl. Logic **39**(2), 131–174 (1988)

8. Farmer, W.M.: Simple second-order languages for which unification is undecidable. Theor. Comput. Sci. **87**(1), 25–41 (1991)
9. Huet, G.P.: A unification algorithm for typed lambda-calculus. Theor. Comput. Sci. **1**(1), 27–57 (1975)
10. Jaffar, J.: Minimal and complete word unification. J. ACM (JACM) **37**(1), 47–85 (1990)
11. Le Chenadec, P.: The finite automaton of an elementary cyclic set. Technical Report RR-0824, INRIA, April 1988
12. Levy, J.: Linear second-order unification. In: Ganzinger, H. (ed.) RTA 1996. LNCS, vol. 1103, pp. 332–346. Springer, Heidelberg (1996)
13. Levy, J.: Decidable and undecidable second-order unification problems. In: Nipkow, T. (ed.) RTA 1998. LNCS, vol. 1379, pp. 47–60. Springer, Heidelberg (1998)
14. Makanin, G.S.: On the decidability of the theory of free groups. In: FCT, pp. 279–284 (1985). (in Russian)
15. Miller, D.: Unification of simply typed lambda-terms as logic programming. In: 8th International Logic Programming Conference, pp. 255–269. MIT Press (1991)
16. Paulson, L.: Isabelle: the next seven hundred theorem provers. In: Lusk, E.R., Overbeek, R. (eds.) CADE 1988. LNCS, vol. 310, pp. 772–773. Springer, Heidelberg (1988)
17. Prehofer, C.: Decidable higher-order unification problems. In: Bundy, A. (ed.) CADE 1994. LNCS, vol. 814, pp. 635–649. Springer, Heidelberg (1994)
18. Schmidt-Schauß, M.: A decision algorithm for stratified context unification. J. Log. Comput. **12**(6), 929–953 (2002)
19. Schmidt-Schauß, M., Schulz, K.U.: On the exponent of periodicity of minimal solutions of context equations. In: Nipkow, T. (ed.) RTA 1998. LNCS, vol. 1379, pp. 61–75. Springer, Heidelberg (1998)
20. Schmidt-Schauß, M., Schulz, K.U.: Solvability of context equations with two context variables is decidable. J. Symb. Comput. **33**(1), 77–122 (2002)
21. Schmidt-Schauß, M., Schulz, K.U.: Decidability of bounded higher-order unification. J. Symb. Comput. **40**(2), 905–954 (2005)
22. Snyder, W., Gallier, J.H.: Higher-order unification revisited: complete sets of transformations. J. Symb. Comput. **8**(1/2), 101–140 (1989)
23. Snyder, W.S.: Complete sets of transformations for general unification. Ph.D. thesis, Philadelphia, PA, USA (1988). AAI8824793
24. Wisnieski, M., Steen, A., Benzmüller, C.: The Leo-III project. In: Bolotov, A., Kerber, M. (eds) Joint Automated Reasoning Workshop and Deduktionstreffen, p. 38 (2014)
25. Zaionc, M.: The regular expression descriptions of unifier sets in the typed lambda calculus. Fundamenta Informaticae **X**, 309–322 (1987). North-Holland

Theorem Proving with Bounded Rigid E-Unification

Peter Backeman and Philipp Rümmer[✉]

Uppsala University, Uppsala, Sweden
philipp.ruemmer@it.uu.se

Abstract. Rigid E-unification is the problem of unifying two expressions modulo a set of equations, with the assumption that every variable denotes exactly one term (*rigid* semantics). This form of unification was originally developed as an approach to integrate equational reasoning in tableau-like proof procedures, and studied extensively in the late 80s and 90s. However, the fact that *simultaneous* rigid E-unification is undecidable has limited practical adoption, and to the best of our knowledge there is no tableau-based theorem prover that uses rigid E-unification. We introduce simultaneous bounded rigid E-unification (BREU), a new version of rigid E-unification that is bounded in the sense that variables only represent terms from finite domains. We show that (simultaneous) BREU is NP-complete, outline how BREU problems can be encoded as propositional SAT-problems, and use BREU to introduce a sound and complete sequent calculus for first-order logic with equality.

1 Introduction

The integration of efficient equality reasoning in tableaux and sequent calculi is a long-standing challenge, and has led to a wealth of theoretically intriguing, yet surprisingly few practically satisfying solutions. Among others, a family of approaches related to the (undecidable) problem of computing *simultaneous rigid E-unifiers* have been developed, by utilising incomplete unification procedures in such a way that an overall complete first-order calculus is obtained. To the best of our knowledge, however, none of those procedures has led to competitive theorem provers.

We introduce *simultaneous bounded rigid E-unification* (BREU), a new version of rigid E-unification that is bounded in the sense that variables only represent terms from finite domains. BREU is significantly simpler than ordinary rigid E-unification, in terms of computational complexity as well as algorithmic aspects, and therefore a promising candidate for efficient implementation. BREU still enables the design of complete first-order calculi, but also makes combinations with techniques from the SMT field possible, in particular the use of congruence closure to handle ground equations.

This work was partly supported by the Microsoft PhD Scholarship Programme and the Swedish Research Council.

A.P. Felty and A. Middeldorp (Eds.): CADE-25, LNAI 9195, pp. 572–587, 2015.
DOI: 10.1007/978-3-319-21401-6_39

1.1   Background and Motivating Example

We start by illustrating our approach using the following problem (from [5]):

$$\phi \;=\; \exists x, y, u, v. \; \begin{pmatrix} (a \not\approx b \;\vee\; g(x,u,v) \approx g(y,f(c),f(d))) \;\wedge\; \\ (c \not\approx d \;\vee\; g(u,x,y) \approx g(v,f(a),f(b))) \end{pmatrix}$$

To show validity of ϕ, a Gentzen-style proof (or, equivalently, a tableau) can be constructed, using free variables for x, y, u, v:

$$\frac{\dfrac{\mathcal{A}}{a \approx b \;\vdash\; g(X,U,V) \approx g(Y,f(c),f(d))} \qquad \dfrac{\mathcal{B}}{c \approx d \;\vdash\; g(U,X,Y) \approx g(V,f(a),f(b))}}{\dfrac{\vdash\; (a \not\approx b \vee g(X,U,V) \approx g(Y,f(c),f(d))) \wedge (c \not\approx d \vee g(U,X,Y) \approx g(V,f(a),f(b)))}{\vdash\; \phi}}$$

To finish this proof, both $\mathcal{A}$ and $\mathcal{B}$ need to be closed by applying further rules, and substituting concrete terms for the variables. The substitution $\sigma_l = \{X \mapsto Y, U \mapsto f(c), V \mapsto f(d)\}$ makes it possible to close $\mathcal{A}$ through equational reasoning, and $\sigma_r = \{X \mapsto f(a), U \mapsto V, Y \mapsto f(b)\}$ closes $\mathcal{B}$, but neither closes both. Finding a substitution that closes both branches is known as *simultaneous rigid E-unification* (SREU), and has first been formulated in [9]:

Definition 1 (Rigid E-Unification). *Let E be a set of equations, and s, t be terms. A substitution σ is called a* rigid E-unifier *of s and t if $s\sigma \approx t\sigma$ follows from $E\sigma$ via ground equational reasoning. A simultaneous rigid E-unifier σ is a common rigid E-unifier for a set $(E_i, s_i, t_i)_{i=1}^n$ of rigid E-unification problems.*

In our example, two rigid E-unification problems have to be solved:

$$E_1 = \{a \approx b\}, \qquad s_1 = g(X,U,V), \quad t_1 = g(Y, f(c), f(d)),$$
$$E_2 = \{c \approx d\}, \qquad s_2 = g(U,X,Y), \quad t_2 = g(V, f(a), f(b)).$$

We can observe that $\sigma_s = \{X \mapsto f(a), Y \mapsto f(b), U \mapsto f(c), V \mapsto f(d)\}$ is a simultaneous rigid E-unifier, and suffices to finish the proof of ϕ. In general, of course, the SREU problem famously turned out undecidable [4], which makes the style of reasoning shown here problematic.

Different solutions have been proposed to address this situation, including potentially non-terminating, but complete E-unification procedures [8], and terminating but incomplete algorithms that are nevertheless sufficient to create complete proof procedures [5,11]. The practical impact of such approaches has been limited; to the best of our knowledge, there is no (at least no actively maintained) theorem prover based on such explicit forms of SREU.

This paper introduces a new approach, *bounded rigid E-unification* (BREU), which belongs to the class of "terminating, but incomplete" algorithms for SREU. In contrast to ordinary SREU, our method only considers E-unifiers where substituted terms are taken from some predefined finite set. This directly implies decidability of the unification problem; as we will see later, the problem is in fact NP-complete, even for the simultaneous case, and can be handled efficiently using SAT technology. In our experiments, cases with hundreds of simultaneous unification problems and thousands of terms were well in reach, and future advances in terms of algorithm design and efficient implementation are expected to further improve scalability.

For sake of presentation, BREU operates on formulae that are normalised by means of flattening (observe that ϕ and ϕ' are equivalent):

$$\phi' = \forall z_1, z_2, z_3, z_4. \left(f(a) \not\approx z_1 \vee f(b) \not\approx z_2 \vee f(c) \not\approx z_3 \vee f(d) \not\approx z_4 \vee \right.$$
$$\left. \exists x, y, u, v. \forall z_5, z_6, z_7, z_8. \left(\begin{array}{l} g(x,u,v) \not\approx z_5 \vee g(y, z_3, z_4) \not\approx z_6 \vee \\ g(u,x,y) \not\approx z_7 \vee g(v, z_1, z_2) \not\approx z_8 \vee \\ ((a \not\approx b \vee z_5 \approx z_6) \wedge (c \not\approx d \vee z_7 \approx z_8)) \end{array} \right) \right)$$

A proof constructed for ϕ' has the same structure as the one for ϕ, with the difference that all function terms are now isolated in the antecedent:

$$
\frac{\mathcal{A}'}{\ldots, g(X, U, V) \approx o_5, a \approx b \vdash o_5 \approx o_6} \qquad \frac{\mathcal{B}'}{\ldots, g(U, X, Y) \approx o_7, c \approx d \vdash o_7 \approx o_8}
$$

$$\vdots$$

$$\frac{f(a) \approx o_1 \vee f(b) \approx o_2 \vee f(c) \approx o_3 \vee f(d) \approx o_4 \vdash \exists x, y, u, v. \forall z_5, z_6, z_7, z_8. \ldots}{} \quad (*)$$

$$\vdots$$

$$\vdash \forall z_1, z_2, z_3, z_4. \ldots$$

To obtain a *bounded* rigid E-unification problem, we now restrict the terms considered for instantiation of X, Y, U, V to the symbols that were in scope when the variables were introduced (at $(*)$ in the proof): X ranges over constants $\{o_1, o_2, o_3, o_4\}$, Y over $\{o_1, o_2, o_3, o_4, X\}$, and so on. Since the problem is flat, those sets contain representatives of all existing ground terms at point $(*)$ in the proof. It is therefore possible to find a simultaneous E-unifier, namely the substitution $\sigma_b = \{X \mapsto o_1, Y \mapsto o_2, U \mapsto o_3, V \mapsto o_4\}$.

It has long been observed that this restricted instantiation strategy gives rise to a complete calculus for first-order logic with equality. The strategy was first introduced as *dummy instantiation* in the seminal work of Kanger [13] (in 1963, i.e., even before the introduction of unification), and later studied under the names *subterm instantiation* and *minus-normalisation* [6,7]; the relationship to SREU was observed in [5]. The impact on practical theorem proving was again limited, however, among others because no efficient search procedures for dummy instantiation were available [7]. The present paper addresses this topic and makes the following main contributions:

- we define bounded rigid E-unification, as a restricted version of SREU, and investigate its complexity (Sect. 3);
- we present a sound, complete, and backtracking-free BREU-based sequent calculus for first-order with equality (Sects. 4–6);
- we give a preliminary experimental evaluation, comparing with other tableau-based theorem provers (Sect. 7).

1.2 Further Related Work

For a general overview of research on equality handling in sequent calculi and related systems, as well as on SREU, we refer the reader to the detailed handbook chapter [6]. The following paragraphs survey some of the more recent work.

Our work is partly motivated by a recent line of research on backtracking-free tableau calculi with free variables [10], capturing unification conditions as constraints that are attached to literals or tableau branches. This calculus was extended to handle equality using superposition-style inferences in [11], building on results from [5]. Our work resembles both [5,11] in that we define an incomplete version of SREU, but show it to be sufficient for complete first-order reasoning. Our variant BREU is incomparable in completeness to the SREU solving in [5,11]: BREU is able to derive a solution for the example shown in Sect. 1.1, which [5,11] cannot; on the other hand, the procedures in [5,11] are able to synthesise new terms of unbounded size as unifiers, whereas our procedure only considers terms from predefined bounded domains. The calculus in [11] was further extended to handle linear integer arithmetic in [14], however, excluding functions (but including uninterpreted predicates, to which functions can be reduced via axioms), leading to a further unification problem that is incomparable in expressiveness.

Equality handling was integrated into hyper tableaux in [2], again using superposition-style inferences, and also including redundancy criteria. This work deliberately avoids the use of rigid free variables shared between multiple tableau branches, so that branches can be closed one at a time, and there is no need for simultaneous E-unification. The calculus was implemented in the Hyper prover, against which we compare our implementation in Sect. 7.

2 Preliminaries

We assume familiarity with classical first-order logic and Gentzen-style calculi (see e.g., [8]). Given countably infinite sets C of constants (denoted by $c, d, \dots$), V_b of bound variables (written $x, y, \dots$), and V of free variables (denoted by $X, Y, \dots$), as well as a finite set F of fixed-arity function symbols (written $f, g, \dots$), the syntactic categories of *formulae* ϕ and *terms* t are defined by

$$\phi ::= \phi \wedge \phi \mid \phi \vee \phi \mid \neg\phi \mid \forall x.\phi \mid \exists x.\phi \mid t \approx t, \qquad t ::= c \mid x \mid X \mid f(t, \dots, t).$$

Note that we distinguish between constants and zero-ary functions for reasons that will become apparent later. We generally assume that bound variables x only occur underneath quantifiers $\forall x$ or $\exists x$. Semantics of terms and formulae without free variables is defined as is common using first-order structures (U, I) consisting of a non-empty universe U, and an interpretation function I.

We call constants and (free or bound) variables *atomic terms*, and all other terms *compound terms*. A *flat equation* is an equation between atomic terms, or an equation of the form $f(t_1, \dots, t_n) \approx t_0$, where $t_0, \dots, t_n$ are atomic terms. A *flat formula* is a formula ϕ in which functions only occur in flat equations.

A formula ϕ is *positively flat* (*negatively flat*) if it is flat, and every occurrence of a function symbol is underneath an even (odd) number of negations. Note that every formula can be transformed to an equivalent positively flat (negatively flat) formula; we will usually assume that such preprocessing has been applied to formulae handled by our procedures. This kind of preprocessing is also standard for congruence closure procedures [1], and similarly used in SMT solvers.

If Γ is a finite set of positively flat formulae (the *antecedent*), and Δ a finite set of negatively flat formulae (the *succedent*), then $\Gamma \vdash \Delta$ is called a *sequent*. A sequent $\Gamma \vdash \Delta$ without free variables is called *valid* if the formula $\bigwedge \Gamma \rightarrow \bigvee \Delta$ is valid. A calculus rule is a binary relation between finite sets of sequents (the *premises*) and sequents (the *conclusion*).

A substitution is a mapping of variables to terms, such that all but finitely many variables are mapped to themselves. Symbols $\sigma, \theta, \ldots$ denote substitutions, and we use post-fix notation $\phi\sigma$ or $t\sigma$ to denote application of substitutions. An *atomic substitution* is a substitution that maps variables only to atomic terms. We write $u[r]$ do denote that r is a sub-expression of a term or formula u.

Definition 2 (Replacement relation [16]). *The replacement relation $\rightarrow_E$ induced by a set of equations E is defined by: $u[l] \rightarrow u[r]$ if $l \approx r \in E$. The relation $\leftrightarrow_E^*$ represents the reflexive, symmetric and transitive closure of $\rightarrow_E$.*

3   Bounded Rigid E-Unification

We present *bounded* rigid E-Unification, a restriction of rigid E-unification in the sense that we now require solutions to be atomic substitutions such that variables are only mapped to smaller atomic terms according to a given partial order $\preceq$. This order takes over the role of an occurs-check of regular unification.

Definition 3 (BREU). *A bounded rigid E-unification (BREU) problem is a triple $U = (\preceq, E, e)$, with $\preceq$ being a partial order over atomic terms such that for all variables X the set $\{s \mid s \preceq X\}$ is finite; E is a finite set of flat equations; and $e = s \approx t$ is an equation between atomic terms (the target equation). An atomic substitution σ is called a bounded rigid E-unifier of s and t if $s\sigma \leftrightarrow_{E\sigma}^* t\sigma$ and $X\sigma \preceq X$ for all variables X.*

Note that the partial order $\preceq$ is in principle an infinite object. However, only a finite part of it is relevant for defining and solving a BREU problem, which ensures that BREU problems can effectively be represented.

Definition 4 (Simultaneous BREU). *A simultaneous bounded rigid E-unification problem is a pair $(\preceq, (E_i, e_i)_{i=1}^n)$ such that each triple $(\preceq, E_i, e_i)$ is a bounded rigid E-unification problem. An atomic substitution σ is a simultaneous bounded rigid E-unifier for $(\preceq, (E_i, e_i)_{i=1}^n)$ if σ is a bounded rigid E-unifier for each problem $(\preceq, E_i, e_i)$.*

A solution to a simultaneous BREU problem can be used to close all branches in a proof tree. In Sect. 4 we present the connection in detail.

Example 5. We revisit the example introduced in Sect. 1.1, which leads to the following simultaneous BREU problem $(\preceq, \{(E_1, e_1), (E_2, e_2)\})$:

$$E_1 = E \cup \{a \approx b\}, \quad e_1 = o_5 \approx o_6, \qquad E_2 = E \cup \{c \approx d\}, \quad e_2 = o_7 \approx o_8,$$

$$E = \left\{ \begin{array}{l} f(a) \approx o_1, f(b) \approx o_2, f(c) \approx o_3, f(d) \approx o_4, \\ g(X, U, V) \approx o_5, g(Y, o_3, o_4) \approx o_6, g(U, X, Y) \approx o_7, g(V, o_1, o_2) \approx o_8 \end{array} \right\}$$

with $\{a, b, c, d\} \prec o_1 \prec o_2 \prec o_3 \prec o_4 \prec X \prec Y \prec U \prec V \prec o_5 \prec o_6 \prec o_7 \prec o_8$.

A unifier to this problem is sufficient to close all goals of the tree up to equational reasoning; one solution is $\sigma = \{X \mapsto o_1, Y \mapsto o_2, U \mapsto o_3, V \mapsto o_4\}$.

While SREU is undecidable in the general case, BREU is decidable; the existence of bounded rigid E-unifiers can be decided in non-deterministic polynomial time, since it can be verified in polynomial time that a substition σ is a solution of a (possibly simultaneous) BREU problem (and since an E-unifier only has to consider variables that occur in the problem, it can be represented in space linear in the size of the BREU problem). Hardness follows from the fact that propositional satisfiability can be reduced to BREU, by virtue of the following construction.

3.1   Reduction of SAT to BREU

Consider propositional formulae ϕ_b, which are assumed to be constructed using the following operators:

$$\phi_b ::= p \mid \neg \phi_b \mid \phi_b \vee \phi_b$$

where p is a propositional symbol.

A formula ϕ_b of this kind is converted to a BREU problem by introducing two constants $\mathbf{0}$ and $\mathbf{1}$; two function symbols f_{or} and f_{not}; for each propositional symbol p in ϕ_b, a variable X_p such that $\mathbf{0} \prec X_p$ and $\mathbf{1} \prec X_p$; and for each sub-formula ψ of ϕ_b, a constant c_ψ and an equation:

$$\begin{array}{ll} X_p \approx c_\psi & \text{if } \psi = p, \\ f_{not}(c_{\psi_1}) \approx c_\psi & \text{if } \psi = \neg \psi_1, \\ f_{or}(c_{\psi_1}, c_{\psi_2}) \approx c_\psi & \text{if } \psi = \psi_1 \vee \psi_2. \end{array}$$

The above, together with the set of equations $\{f_{or}(\mathbf{0}, \mathbf{0}) \approx \mathbf{0}, f_{or}(\mathbf{0}, \mathbf{1}) \approx \mathbf{1}, f_{or}(\mathbf{1}, \mathbf{0}) \approx \mathbf{1}, f_{or}(\mathbf{1}, \mathbf{1}) \approx \mathbf{1}, f_{not}(\mathbf{0}) \approx \mathbf{1}, f_{not}(\mathbf{1}) \approx \mathbf{0}\}$ defining the semantics of the Boolean operators, and a target equation $c_{\phi_b} \approx \mathbf{1}$ yields a BREU problem that is naturally equivalent to the problem of checking satisfiability of ϕ_b. Indeed, every E-unifier can be translated to an assignment A of the propositional symbols such that $A \models \phi_b$.

Theorem 6. *Satisfiability of BREU problems is NP-complete.*

3.2 Generalisations

A number of generalisations in the definition of BREU are possible, but can uniformly be reduced to BREU as formulated in Definition 3, without causing a blow-up in the size of the BREU problem.

General Target Constraints. Most importantly, there is no need to restrict BREU to single target equations e, instead arbitrary positive Boolean combinations of equations can be solved; this observation is useful for integration of BREU into calculi. Any such combination of equations can be transformed to a single target equation using a construction resembling that in Sect. 3.1, at the cost of introducing a linear number of new symbols and defining equations.

For the remainder of the paper, we assume that e in Definition 3 can indeed be any positive Boolean combination of atomic equations.

Arbitrary Equations. BREU problems containing arbitrary (i.e., possibly non-flat) equations in E or as target equation can be handled by reduction to equisatisfiable BREU problems with only flat equations, in a manner similar to [1]. Any non-flat equation of the form $t[f(\bar{c})] \approx s$ can be replaced by two new equations $t[d] \approx s$ and $f(\bar{c}) \approx d$, where d is a fresh constant; the symmetric case, and non-flat target equations are handled similarly. Iterating this reduction eventually results in a problem with only flat equations.

Non-atomic E-unifiers. It is further possible to consider partial orders $\preceq$ over arbitrary terms, as long as the set $\{s \mid s \preceq X\}$ is still finite for all variables X. Reduction to problems as in Definition 3 is done by introducing a fresh constant c_t and a (possibly non-flat) equation $t \approx c_t$ for each compound term t occurring in a set $\{s \mid s \preceq X\}$ for some variable X in the BREU problem. A new order $\preceq'$ is defined by replacing compound terms t with constants c_t, in such a way that

$$\{s \mid s \preceq' X\} \;=\; \{s \mid s \preceq X, s \text{ is atomic}\} \cup \{c_t \mid t \preceq X, t \text{ is compound}\}.$$

With this in mind, it is possible to relax Definition 3 by including non-atomic unifiers σ (which might map variables to compound terms) as solutions to a BREU problem, as long as the condition $X\sigma \preceq X$ holds for all variables X.

Example 7. Consider the generalised BREU problem $B = (\preceq, E, e)$ defined by

$$E = \{f(f(a,b),c) \approx g(b), f(X,Y) \approx c, g(b) \approx a\}, \qquad e = a \approx c,$$
$$a \prec b \prec c \prec f(a,a) \prec f(a,b) \prec f(b,a) \prec f(b,b) \prec X \prec Y.$$

Intuitively, the order $\preceq$ encodes the fact that an E-unifier has to be constructed that maps every variable to a term with at most one occurrence of f, and no occurrence of g. A solution is the substitution $\sigma = \{X \mapsto f(a,b), Y \mapsto c\}$.

An equisatisfiable BREU problem according to Definition 3 is $B' = (\preceq', E', e')$:

$$E' = \left\{ \begin{array}{l} f(d_1,c) \approx d_2, f(a,b) \approx d_1, g(b) \approx d_2, f(X,Y) \approx c, g(b) \approx a, \\ f(a,a) \approx d_3, f(a,b) \approx d_4, f(b,a) \approx d_5, f(b,b) \approx d_6 \end{array} \right\},$$
$$e' = e = a \approx c, \qquad a \prec' b \prec' c \prec' d_3 \prec' d_4 \prec' d_5 \prec' d_6 \prec' X \prec' Y,$$

with the E-unifier $\sigma' = \{X \mapsto d_4, Y \mapsto c\}$.

3.3   Encoding of *E*-Unification into SAT

Since satisfiability of BREU problems is NP-complete, a natural approach to compute solutions is an encoding as a propositional SAT problem, so that the performance of modern SAT solvers can be put to use. A procedure for solving a BREU problem will consist of three steps: (i) generating a candidate E-unifier σ; (ii) using congruence closure [1] to calculate the equivalence relation induced by the candidate σ and the equations of the BREU problem; and (iii) checking if the BREU target equation is satisfied by this relation.

Each of these steps can be encoded into SAT. Candidate E-unifiers σ are represented by a set of bit-vector variables storing the index of the term $X\sigma$ that each variable X is mapped to. To guess candidate E-unifiers, it is then just necessary to encode the conditions $X\sigma \preceq X$ as a propositional formula.

A congruence closure procedure can be modelled by representing intermediate results (i.e., equivalence relations) as a sequence of union-find data structures. To represent such a data structure in SAT, it suffices to introduce one bit-vector variable per atomic term t occurring in the BREU problem, storing the index of the parent of t in the union-find forest. Propositional constraints are added to characterise well-formed union-find forests, and to define the derivation of each forest from the previous one.

Lastly, to check the correctness of the candidate σ, it is asserted that the target equation is satisfied in the last union-find structure.

4   A First-Order Logic Calculus with *E*-Unification

We will now introduce our sequent calculus for first-order logic with equality. The calculus operates only on flat formulae, and is kept quite minimalist to illustrate the use of free variables and BREU for delayed instantiation; for practical purposes, many refinements are possible, some of which are outlined in Sect. 6. The BREU procedure is utilised to define a global closure rule that discharges all goals of a proof tree simultaneously. Proof construction is intended to be done in upward direction and backtracking-free manner, following the proof procedures presented in [10,14]; this is possible because all calculus rules are non-destructive and the overall calculus proof-confluent. We will show that fair application of the proof rules is complete.

The propositional, first-order, and equational rules of the calculus are shown in Table 1. Propositional and first-order rules mostly correspond to the classical system LK [8], however, keeping all structural rules implicit (Γ and Δ are sets of formulae). The first-order rules use Skolem symbols $c \in C$ for existential quantifiers in the antecedent, and fresh free variables $X \in V$ for universal quantifiers; and similarly for formulae in the succedent.

The equational rules simplify terms by means of ordered ground rewriting. Given a proof tree, we introduce a strict partial order $\prec \subseteq (C \cup V)^2$ over constants and free variables reflecting the order in which symbols are introduced by the rules $\forall$L, $\forall$R, $\exists$L, $\exists$R: we define $s \prec t$ if the constant or variable t was introduced *above* the symbol s, or if s is a symbol already occurring in the root sequent and

Table 1. Our sequent calculus for first-order logic with equality. In rules $\forall$L and $\exists$R, X is a fresh variable, whereas the rules $\exists$L and $\forall$R introduce a fresh constant c. In $\approx$L and $\approx$R, the equation $(t' \approx s')[t/s]$ is the result of replacing all occurrences of t with s.

$$\frac{\Gamma, \phi, \psi \vdash \Delta}{\Gamma, \phi \wedge \psi \vdash \Delta} \wedge\text{L} \qquad \frac{\Gamma \vdash \phi, \Delta \quad \Gamma \vdash \psi, \Delta}{\Gamma \vdash \phi \wedge \psi, \Delta} \wedge\text{R} \qquad \frac{\Gamma \vdash \phi, \Delta}{\Gamma, \neg\phi \vdash \Delta} \neg\text{L}$$

$$\frac{\Gamma, \phi \vdash \Delta \quad \Gamma, \psi \vdash \Delta}{\Gamma, \phi \vee \psi \vdash \Delta} \vee\text{L} \qquad \frac{\Gamma \vdash \phi, \psi, \Delta}{\Gamma \vdash \phi \vee \psi, \Delta} \vee\text{R} \qquad \frac{\Gamma, \phi \vdash \Delta}{\Gamma \vdash \neg\phi, \Delta} \neg\text{R}$$

$$\frac{\Gamma, \forall x.\phi, \phi[x/X] \vdash \Delta}{\Gamma, \forall x.\phi \vdash \Delta} \forall\text{L} \qquad \frac{\Gamma \vdash \phi[x/c], \Delta}{\Gamma \vdash \forall x.\phi, \Delta} \forall\text{R} \qquad \frac{\Gamma \vdash \Delta}{\Gamma, s \approx s \vdash \Delta} \approx\text{ELIM}$$

$$\frac{\Gamma, \phi[x/c] \vdash \Delta}{\Gamma, \exists x.\phi \vdash \Delta} \exists\text{L} \qquad \frac{\Gamma \vdash \exists x.\phi, \phi[x/X], \Delta}{\Gamma \vdash \exists x.\phi, \Delta} \exists\text{R} \qquad \frac{*}{\Gamma \vdash s \approx s, \Delta} \approx\text{CLOSE}$$

$$\frac{\Gamma, t \approx s \vdash \Delta}{\Gamma, s \approx t \vdash \Delta} \approx\text{ORIENT} \qquad \text{where } t \succ s$$

$$\frac{\Gamma, t \approx s, (t' \approx s')[t/s] \vdash \Delta}{\Gamma, t \approx s, t' \approx s' \vdash \Delta} \approx\text{L} \qquad \begin{array}{l} \text{where } t \succ s \text{ and } t' \succ s', \text{ the term } t \text{ occurs} \\ \text{in } t' \approx s', \text{ and if } t = t' \text{ then } s' \succ s \end{array}$$

$$\frac{\Gamma, t \approx s \vdash (t' \approx s')[t/s], \Delta}{\Gamma, t \approx s \vdash t' \approx s', \Delta} \approx\text{R} \qquad \text{where } t \succ s \text{ and the term } t \text{ occurs in } t' \approx s'$$

$$\frac{\dfrac{*}{\Gamma_1 \vdash \Delta_1} \quad \cdots \quad \dfrac{*}{\Gamma_n \vdash \Delta_n}}{\Gamma \vdash \Delta} \text{BREU}$$

where $\Gamma_1 \vdash \Delta_1, \ldots, \Gamma_n \vdash \Delta_n$ are all open goals of the proof, $E_i = \{t \approx s \in \Gamma_i\}$ are flat antecedent equations, $e_i = \bigvee\{t \approx s \in \Delta_i\}$ are succedent equations, and the simultaneous BREU problem $(\preceq, (E_i, e_i)_{i=1}^n)$ is solvable

t is introduced by some rule in the proof. For instance, for the proof shown in Sect. 1.1, the partial order shown in Example 5 is derived.

By slight abuse of notation, we also write $s \prec f(t_1, \ldots, t_n)$ if s does not start with a function symbol. The rule $\approx$ORIENT moves the bigger term to the left-hand side of an equation. $\approx$L and $\approx$R can be used to replace occurrences of the (bigger) left-hand side term of an equation with the smaller right-hand side term; this rewriting is purely ground and does not unify expressions containing free variables (unification is entirely left to the BREU closure rule discussed in the next paragraph). As a consequence, and since $\prec$ is well-founded, rewriting is terminating and confluent, and in fact implements a congruence closure procedure [1] that eventually replaces every term with a unique representative term of its equivalence class modulo equations in the antecedent.

The BREU rule operates globally and closes all remaining goals of a proof if a global E-unifier σ exists that solves some succedent equation in each goal. The rule makes use of the non-strict partial order $\preceq$ corresponding to $\prec$, with the implication that every variable X can be mapped to symbols that were introduced prior to X in the proof. To encode non-emptiness of the universe, we assume that there is some constant $c_\perp \in C$ below all variables $X \in V$ in a proof

$(c_\perp \prec X$ for all $X \in V)$; if the proof itself does not contain such a constant, it is assumed that $c_\perp$ is some fresh constant with $c_\perp \prec X$ for all variables X.

5   Properties of the Calculus

5.1   Soundness

The soundness of the calculus from Table 1 can be shown by substituting constants for all free variables, and observing the local soundness of each rule.

Lemma 8. *Suppose $\Gamma \vdash \Delta$ is a sequent without free variables. If a closed proof can be constructed for $\Gamma \vdash \Delta$ using the calculus in Table 1, then $\Gamma \vdash \Delta$ is valid.*

Proof. We assume that a proof for $\Gamma \vdash \Delta$ was closed using rule BREU, with a unifier σ that maps every variable X occurring in the proof to a constant $X\sigma \in C$ with $X\sigma \prec X$. In case all goals were closed using $\approx$CLOSE, $X\sigma$ can be some arbitrary constant with $X\sigma \prec X$.

By induction, it can be shown that the instance $(\Gamma' \vdash \Delta')\sigma = \Gamma'\sigma \vdash \Delta'\sigma$ of every sequent $\Gamma' \vdash \Delta'$ occurring in the proof is valid. This is the case for every goal discharged using rule BREU by definition. For all other rules, it is the case that if the σ-instance of the premises is valid, then also the σ-instance of the conclusion is valid. We show two cases, the other rules are verified similarly:

- $\exists$L: assume that the instantiated premise $(\Gamma, \phi[x/c] \vdash \Delta)\sigma$ is valid. Since c is fresh, we know that $X \prec c$ for all free variables X in $\Gamma, \exists x.\phi \vdash \Delta$. Therefore $X\sigma \prec c$, and it follows that $(\Gamma, \exists x.\phi \vdash \Delta)\sigma$ does not contain c. Validity of $(\Gamma, \phi[x/c] \vdash \Delta)\sigma$ then implies validity of $\forall x.(\bigwedge \Gamma \wedge \phi \rightarrow \bigvee \Delta)\sigma$, and equivalently of $(\Gamma, \exists x.\phi \vdash \Delta)\sigma$.
- $\approx$L: assume that $(\Gamma, t \approx s, (t' \approx s')[t/s] \vdash \Delta)\sigma$ is valid. Then the conclusion $(\Gamma, t \approx s, t' \approx s' \vdash \Delta)\sigma$ is valid, too, since the conjunctions $(t \approx s \wedge t' \approx s')\sigma$ and $(t \approx s \wedge (t' \approx s')[t/s])\sigma$ are equivalent.

Since the root sequent $\Gamma \vdash \Delta$ does not contain any free variables, it is implied that $(\Gamma \vdash \Delta)\sigma = \Gamma \vdash \Delta$ is valid.   □

5.2   Completeness

The completeness of the calculus can be shown using a model construction argument (e.g., [8]), which also implies that every attempt to construct a proof of a valid sequent in a "fair" manner will ultimately be successful; this ensures that proofs can always be found without the need for backtracking (although backtracking might sometimes lead to success more quickly, of course).

We call a proof search strategy for the calculus in Table 1 *fair* if the propositional and first-order rules $\wedge$L, $\wedge$R, $\vee$L, $\vee$R, $\neg$L, $\neg$R, $\forall$L, $\forall$R, $\exists$L, $\exists$R are always eventually applied when they are applicable to some formula, and if every proof goal in which one of those rules is applicable is eventually expanded. This implies, in particular, that ∀L and ∃R are applied unboundedly often to every quantified formula. Fairness does not mandate the application of the equational rules, which are subsumed by BREU; eager application of equational rules is in practice cheap and advisable for performance, however.

Lemma 9 (Completeness of fair proof search). *Suppose $\Gamma \vdash \Delta$ is a sequent without free variables, and suppose that a proof is constructed in a fair manner. If $\Gamma \vdash \Delta$ is valid, then eventually a proof tree will be obtained that can be closed using the rule BREU.*

In order to prove this lemma, we first consider a "ground" version GC of our calculus, obtained by removing the rule BREU, and by replacing ∀L and ∃R with the following ground rules:

$$\frac{\Gamma, \forall x.\phi, \phi[x/c] \vdash \Delta}{\Gamma, \forall x.\phi \vdash \Delta} \, \forall L_g, \qquad \frac{\Gamma \vdash \exists x.\phi, \phi[x/c], \Delta}{\Gamma \vdash \exists x.\phi, \Delta} \, \exists R_g$$

where c is an arbitrary constant. GC has the property that systematic application of the rules will either eventually produce a closed proof, or lead to a *saturated* (possibly infinite) branch from which a model can be derived:

Definition 10. *An open proof branch in GC labelled with sequents $\Gamma_0 \vdash \Delta_0$, $\Gamma_1 \vdash \Delta_1, \ldots$ (where $\Gamma_0 \vdash \Delta_0$ is the root of the proof) is called* saturated *if*

(i) the branch is finite and no rule is applicable in the goal sequent $\Gamma_n \vdash \Delta_n$; or
(ii) the branch is infinite, and for the limit sets $\Gamma^\infty, \Delta^\infty$ of formulae occurring on the branch, as well as the sets Γ^p, Δ^p of persistent formulae

$$\Gamma^\infty = \bigcup_{i \geq 0} \Gamma_i, \quad \Delta^\infty = \bigcup_{i \geq 0} \Delta_i, \quad \Gamma_p = \bigcup_{i \geq 0} \bigcap_{j \geq i} \Gamma_j, \quad \Delta_p = \bigcup_{i \geq 0} \bigcap_{j \geq i} \Delta_j$$

it is the case that (a) Γ_p only contains equations and ∀-quantified formulae; (b) Δ_p only contains equations and ∃-quantified formulae; (c) none of the rules ≈ELIM, ≈CLOSE, ≈ORIENT, ≈L, ≈R is applicable in $\Gamma_p \vdash \Delta_p$; (d) at least one constant c occurs on the branch; (e) for every formula $\forall x.\phi \in \Gamma_p$ and every constant c occurring on the branch, there is an instance $\phi[x/c] \in \Gamma^\infty$; and (f) for every formula $\exists x.\phi \in \Delta_p$ and every constant c there is an instance $\phi[x/c] \in \Delta^\infty$.

The ability to construct saturated branches follows directly from the observation that application of the GC-rules other than $\forall L_g$ and $\exists L_g$ terminates (because ≺ is well-founded), and that $\forall L_g$ and $\exists L_g$ can be managed in a fair way using a work queue. The property (ii)–(d) encodes non-emptiness of universes, and is ensured by instantiating every formula $\forall x.\phi \in \Gamma_p$ and $\exists x.\phi \in \Delta_p$ at least once on every branch (e.g., using the ≺-smallest constant $c_\perp$).

Lemma 11. *If a (finite or infinite) GC proof contains a saturated branch, then the root sequent $\Gamma \vdash \Delta$ has a counter-model (is invalid).*

Proof. We use persistent equations to construct a structure $S = (U, I)$. In case of a finite saturated branch, persistent formulae are the ones in the goal; without loss of generality, we assume that also finite branches contain at least one constant. U is chosen as the set of constants that do not occur as left-hand side of some persistent antecedent equation; left-hand side terms are interpreted as the right-hand side constants. In case the value of some function application $f(c_1, \ldots, c_n)$ is not determined by the equations, we set the value to some arbitrary constant $c \in U$:

$$U = \{c \in C \mid c \text{ occurs in } \Gamma^\infty \cup \Delta^\infty\} \setminus \{c \mid c \approx d \in \Gamma_p\}$$

$$I(c) = \begin{cases} d & \text{if there exists an equation } c \approx d \in \Gamma_p \\ c & \text{otherwise} \end{cases}$$

$$I(f)(c_1, \ldots, c_n) = \begin{cases} d & \text{if there exists an equation } f(c_1, \ldots, c_n) \approx d \in \Gamma_p \\ c & \text{otherwise, for some arbitrary } c \in U \end{cases}$$

Since no equational rule is applicable in $\Gamma_p \vdash \Delta_p$, it is clear that $\mathrm{val}_S(t \approx s) = \textit{true}$ for every $t \approx s \in \Gamma_p$, and $\mathrm{val}_S(t \approx s) = \textit{false}$ for every $t \approx s \in \Delta_p$.

By well-founded induction over the equations in Γ^∞, it can then be shown that in fact *all* equations in Γ^∞ evaluate to *true* under S. For this we define a well-founded order $\prec'$ over flat equations (for $c, d \in C$, $\bar{c}, \bar{c}' \in C^*$, $f, g \in F$, and $\prec_{\mathrm{lex}}$ the well-founded lexicographic order induced by $\prec$):

$$(c \approx d) \prec' (c' \approx d') \;\Leftrightarrow\; (d, c) \prec_{\mathrm{lex}} (d', c'), \qquad (c \approx d) \prec' (f(\bar{c}) \approx d'),$$
$$(f(\bar{c}) \approx d) \prec' (g(\bar{c}') \approx d') \;\Leftrightarrow\; f = g \text{ and } (d, \bar{c}) \prec_{\mathrm{lex}} (d', \bar{c}').$$

In particular, note that in any application of rule $\approx$L we have $(t \approx s) \prec' (t' \approx s')$ and $(t' \approx s')[t/s] \prec' (t' \approx s')$; this implies that if all equations $\prec'$-smaller than $t' \approx s'$ hold, then also $t' \approx s'$ holds. In the same way, it can be proven that all equations in Δ^∞ evaluate to *false*.

By induction over the depth of formulae we can conclude that all formulae (not only equations) in Γ^∞ evaluate to *true*, and all formulae in Δ^∞ to *false*. □

Proof. (Lem. 9) Assume that an (unsuccessful) attempt was made to construct a proof P for the valid sequent $\Gamma \vdash \Delta$ by fair application of the rules in Table 1. We define a global mapping $v : V \to C$ of variables occurring in P to constants, and use v to map P to a GC-proof with a saturated branch. The mapping v is defined successively by depth-first traversal of P, visiting sequents closer to the root earlier than sequents further away. Note that for each branch that has not been closed by applying $\approx$CLOSE, fairness implies that $\forall$L ($\exists$R) has been applied infinitely often to every universally quantified formula in the antecedent (existentially quantified formula in the succedent).

When a node is visited where a new variable X is introduced by $\forall$L or $\exists$R for a quantified formula ϕ, set $v(X) = c$ for some constant $c \prec X$ that is $\prec$-minimal among the constants that have *not yet* been assigned for the same formula ϕ on this branch. If no such constant exists, an arbitrary constant $c \prec X$ is chosen. On every infinite branch, this ensures that for every quantified formula ϕ handled via $\forall$L or $\exists$R, and every constant c occurring on the branch, there is some application of $\forall$L or $\exists$R to ϕ such that the introduced variable X is mapped to $c = v(X)$.

The function v can then be used to translate P to a GC-proof P', replacing each variable X with the constant $v(X)$, and inserting exhaustive applications of the equational rules wherever they are applicable. By Lemma 11 and since $\Gamma \vdash \Delta$ is valid, each branch in P' can be closed after finitely many steps through $\approx$CLOSE. This implies that it has to be possible to close the corresponding finite prefix of the original proof P using rule BREU, with the mapping v restricted to the variables occurring in the prefix as E-unifier.   □

6 Refinements of the Calculus

The presented calculus can be refined in many practically relevant ways; in the scope of this paper, we only outline three modifications that we use in our implementation (also see Sect. 7).

General Instantiation. Similar the subterm instantiation method proposed by Kanger [13], our system explicitly generates constants representing all terms possibly required for instantiation of quantified formulae, through application of $\exists$L and $\forall$R. While subterm instantiation is complete, it has been observed (e.g., in [6]) that resulting proofs can sometimes be significantly longer than the shortest proofs that can be obtained when considering arbitrary instances of quantified formulae. Instantiation with new terms can be simulated in our systems by adding a rule TOT representing the totality axiom $\forall \bar{x}.\exists y.\ f(\bar{x}) \approx y$, which iteratively increases the range of terms considered for substitution by the BREU rule. In TOT, f is a function symbol, $X_1, \ldots, X_n$ are fresh variables, and c is a fresh constant (and we set $X_i \prec c$ for all $i \in \{1, \ldots, n\}$):

$$\frac{\Gamma, f(X_1, \ldots, X_n) \approx c \vdash \Delta}{\Gamma \vdash \Delta}\text{TOT}$$

Local Closure. The closure rule BREU can be generalised to operate not only on complete proof trees, but also on arbitrary *sub-trees,* and thus be used to guide proof expansion. For any sub-tree t, it can be checked (i) whether all goals in t contain equations that are simultaneously E-unifiable; as long as this is not the case, proof expansion can focus on t, since rules applied to branches outside of t will not be helpful; and (ii) whether the goals in t are E-unifiable with a unifier σ such that $X\sigma = X$ for all variables X that occur outside of t; in this case, t can be closed permanently and does not have to be considered again. It is also possible to define a notion of *unsatisfiable cores* for E-unification problems, which can further refine the selection of goals to be expanded.

Ground Instantiation. It has also been observed that handling of quantifiers using free variables is very powerful, but is excessively expensive in case of simple quantified formulae that have to be instantiated many times, and provides

Table 2. Comparison of our prototypical implementation on TPTP benchmarks. The numbers indicate how many benchmarks in each group could be solved; the runtime per benchmark was limited to 240 s (wall clock time). All experiments were done on an AMD Opteron 2220 SE machine, running 64-bit Linux, heap space limited to 1.5 GB.

	FOF with eq.	FOF w/o eq.	CNF with eq.	CNF w/o eq.
Princess + BREU	**211**	325	**203**	252
Hyper 1.0_16112014 [2]	119	378	160	**305**
leanCoP 2.2 (CASC-J7)	153	**379**	$-^a$	$-^a$

^{a} leanCoP cannot process benchmarks in the TPTP CNF dialect.

little guidance for proof construction. Possible solutions include the use of connection conditions, universal variables, or simplification rules [3,12]. In our implementation, we use a more straightforward hybrid approach that combines free variables with ground instantiation through *E-matching* [15]; in combination, free variables and e-matching can solve significantly more problems than either technique individually. E-matching can be integrated naturally in our calculus without losing completeness, following [15]; in general this requires the use of the rule TOT shown above.

7   Experimental Results

We are in the process of implementing our BREU algorithm, and the calculus from Sect. 4, as an extension of the Princess theorem prover [14].[1] Our implementation uses the SAT encoding outlined in Sect. 3.3, and the Sat4j solver to solve the resulting constraints; we also include the refinements discussed in Sect. 6. Considered benchmarks were randomly selected TPTP v.6.1.0 problems with status Theorem or Unsatisfiable. To illustrate strengths and weaknesses of the compared tools, the benchmarks were categorised into FOF (first-order) problems with equality, FOF problems without equality, CNF (clause normal form) problems with equality, and CNF problems without equality. 500 benchmarks from all of TPTP were chosen in each group.

We compared our BREU implementation with the tableau provers Hyper and leanCoP from the CASC-J7 competition. Hyper uses the superposition-based equality reasoning from [2], whereas leanCoP relies on explicit equality axioms. The experimental results shown in Table 2 are still preliminary, and expected to change as further optimisations in our BREU procedure are done. However, it can be seen that even our current implementation of BREU shows performance that is comparable with the other tableau systems in all groups of benchmarks, and outperforms the other systems on benchmarks with equality.

---

[1] http://user.it.uu.se/~petba168/breu/.

Conclusion

We have introduced bounded rigid E-unification, a new variant of SREU, and illustrated how it can be used to construct sound and complete theorem provers for first-order logic with equality. We believe that BREU is a promising approach to handling of equality in tableaux and related calculi. Apart from improved algorithms for solving BREU, and an improved implementation, in future work we plan to consider the combination of BREU with other theories, in particular arithmetic, and integration of BREU with DPLL(T)-style clause learning.

Acknowledgements. We would like to thank Christoph M. Wintersteiger for comments on this paper, and the anonymous referees for helpful feedback.

References

1. Bachmair, L., Tiwari, A., Vigneron, L.: Abstract congruence closure. J. Autom. Reasoning **31**(2), 129–168 (2003)
2. Baumgartner, P., Furbach, U., Pelzer, B.: Hyper tableaux with equality. In: Pfenning, F. (ed.) CADE 2007. LNCS (LNAI), vol. 4603, pp. 492–507. Springer, Heidelberg (2007)
3. Beckert, B.: Equality and other theories. In: D'Agostino, M., Gabbay, D., Hähnle, R., Posegga, J. (eds.) Handbook of Tableau Methods. Kluwer, Dordrecht (1999)
4. Degtyarev, A., Voronkov, A.: Simultaneous rigid E-Unification is undecidable. In: Kleine Büning, H. (ed.) CSL 1995. LNCS, vol. 1092, pp. 178–190. Springer, Heidelberg (1996)
5. Degtyarev, A., Voronkov, A.: What you always wanted to know about rigid E-Unification. J. Autom. Reasoning **20**(1), 47–80 (1998)
6. Degtyarev, A., Voronkov, A.: Equality reasoning in sequent-based calculi. In: Handbook of Automated Reasoning (in 2 volumes). Elsevier and MIT Press (2001)
7. Degtyarev, A., Voronkov, A.: Kanger's Choices in Automated Reasoning. Springer, The Netherlands (2001)
8. Fitting, M.C.: First-Order Logic and Automated Theorem Provin. Graduate Texts in Computer Science, 2nd edn. Springer-Verlag, Berlin (1996)
9. Gallier, J.H., Raatz, S., Snyder, W.: Theorem proving using rigid e-unification equational matings. In: LICS. pp. 338–346. IEEE Computer Society (1987)
10. Giese, M.A.: Incremental closure of free variable tableaux. In: Goré, R.P., Leitsch, A., Nipkow, T. (eds.) IJCAR 2001. LNCS (LNAI), vol. 2083, pp. 545–560. Springer, Heidelberg (2001)
11. Giese, M.A.: A model generation style completeness proof for constraint tableaux with superposition. In: Egly, U., Fermüller, C. (eds.) TABLEAUX 2002. LNCS (LNAI), vol. 2381, pp. 130–144. Springer, Heidelberg (2002)
12. Giese, M.: Simplification rules for constrained formula tableaux. In: TABLEAUX, pp. 65–80 (2003)
13. Kanger, S.: A simplified proof method for elementary logic. In: Siekmann, J., Wrightson, G. (eds.) Automation of Reasoning 1: Classical Papers on Computational Logic 1957–1966, pp. 364–371. Springer, Heidelberg (1983). originally appeared in 1963

14. Rümmer, P.: A constraint sequent calculus for first-order logic with linear integer arithmetic. In: Cervesato, I., Veith, H., Voronkov, A. (eds.) LPAR 2008. LNCS (LNAI), vol. 5330, pp. 274–289. Springer, Heidelberg (2008)
15. Rümmer, P.: E-matching with free variables. In: Bjørner, N., Voronkov, A. (eds.) LPAR-18 2012. LNCS, vol. 7180, pp. 359–374. Springer, Heidelberg (2012)
16. Tiwari, A., Bachmair, L., Rueß, H.: Rigid E-Unification revisited. In: CADE. pp. 220–234, CADE-17. Springer-Verlag, London (2000)

SAT/SMT

Expressing Symmetry Breaking in DRAT Proofs

Marijn J.H. Heule$^{(\boxtimes)}$, Warren A. Hunt Jr., and Nathan Wetzler

The University of Texas, Austin, USA
marijn@cs.utexas.edu

Abstract. An effective SAT preprocessing technique is the addition of symmetry-breaking predicates: auxiliary clauses that guide a SAT solver away from needless exploration of isomorphic sub-problems. Symmetry-breaking predicates have been in use for over a decade. However, it was not known how to express the addition of these predicates in proofs of unsatisfiability. Hence, results obtained by symmetry breaking cannot be validated by existing proof checkers. We present a method to express the addition of symmetry-breaking predicates in DRAT, a clausal proof format supported by top-tier solvers. We applied this method to generate SAT problems that have not been previously solved without symmetry-breaking predicates. We validated these proofs with an ACL2-based, mechanically-verified DRAT proof checker and the proof-checking tool of SAT Competition 2014.

1 Introduction

Satisfiability (SAT) solvers can be applied to decide hard combinatorial problems that contain symmetries. Breaking problem symmetries typically boosts solver performance as it prevents a solver from needlessly exploring isomorphic parts of a search space. One common method to eliminate symmetries is to add *symmetry-breaking predicates* [1–3]. However, expressing the addition in existing SAT proof formats is an open problem, which leaves it hard to validate some solver results. We present a method to express the use of symmetry-breaking predicates in the DRAT proof format [4], which is supported by top-tier SAT solvers and was used to validate the results of SAT Competition 2014.

Recent successes of SAT-based technology include solving some long-standing open problems such as the Erdős Discrepancy Conjecture [5], computing Van der Waerden numbers [6], and producing minimal sorting networks [7]. Symmetry-breaking techniques have been applied to each of these problems and allows one to solve them more efficiently. Our new method facilitates the creation of proofs for unsatisfiability results when symmetry-breaking techniques are applied.

The state-of-the-art tool `shatter` [8] performs *static symmetry-breaking* by adding symmetry-breaking predicates to a given problem. Static symmetry-breaking starts by converting a SAT problem into a graph. This graph is used to detect symmetries, which are then transformed into predicates (represented as

---

The authors are supported by DARPA contract number N66001-10-2-4087.

A.P. Felty and A. Middeldorp (Eds.): CADE-25, LNAI 9195, pp. 591–606, 2015.
DOI: 10.1007/978-3-319-21401-6_40

clauses) and added to the SAT problem. *Dynamic symmetry-breaking* [9] adds symmetric versions of learned clauses to the problem. This technique is the most useful when few symmetries exist, which is the case for graph-coloring problems.

We present the expression of adding symmetry-breaking predicates in the DRAT proof format using unavoidable subgraphs. Given an undirected, fully-connected graph G, H is an *unavoidable subgraph* of G if every 2-edge-coloring of G contains H as a monochromatic subgraph. The best-known type of unavoidable subgraphs are cliques (Ramsey numbers), but there are many other types of graphs for which unavoidability has been studied; an online dynamic survey [10] lists over 600 articles on the topic. SAT solvers have severe difficulty solving unavoidable subgraph problems without symmetry-breaking predicates.

Given a satisfiability problem F, we produce a satisfiability-equivalent problem F' that contains symmetry-breaking predicates. The formula F' is similar to the result of applying `shatter` (modulo variable renaming). Additionally, we produce a partial DRAT proof that expresses the conversion from F to F'. We solve F' using an off-the-shelf SAT solver that can emit DRAT proofs. We validate this proof result by merging the partial proof with the SAT solver proof and then we check that the combined result is valid with our `drat-trim` tool [4] or our mechanically-verified checker [11]. We evaluate this method on some hard combinatorial problems.

The remainder of the paper is structured as follows. After some preliminary and background information in Sect. 2, the DRAT proof system is explained in Sect. 3 and the addition of symmetry-breaking predicates is presented in Sect. 4. Breaking a single symmetry may require many clause addition and deletion steps in order to express it in a DRAT proof, as discussed is Sect. 5. In Sect. 6, we detail how to break multiple symmetries. Our tool chain and evaluation are exhibited in Sects. 7 and 8 and we conclude in Sect. 9.

2   Preliminaries

We briefly review necessary background concepts: conjunctive normal form, Boolean constraint propagation, and blocked clauses.

Conjunctive Normal Form (CNF). For a Boolean variable x, there are two *literals*, the positive literal, denoted by x, and the negative literal, denoted by $\bar{x}$. A *clause* is a finite disjunction of literals, and a CNF *formula* is a finite conjunction of clauses. A clause is a *tautology* if it contains both x and $\bar{x}$ for some variable x. Given a clause $C = (l_1 \vee \cdots \vee l_k)$, $\overline{C}$ denotes the conjunction of its negated literals, i.e., $(\bar{l}_1) \wedge \cdots \wedge (\bar{l}_k)$. The set of literals occurring in a CNF formula F is denoted by $\text{LIT}(F)$. A truth assignment for a CNF formula F is a partial function τ that maps literals $l \in \text{LIT}(F)$ to $\{\mathbf{t}, \mathbf{f}\}$. If $\tau(l) = v$, then $\tau(\bar{l}) = \neg v$, where $\neg\mathbf{t} = \mathbf{f}$ and $\neg\mathbf{f} = \mathbf{t}$. Furthermore:

- A clause C is *satisfied* by assignment τ if $\tau(l) = \mathbf{t}$ for some $l \in C$.
- A clause C is *falsified* by assignment τ if $\tau(l) = \mathbf{f}$ for all $l \in C$.
- A CNF formula F is *satisfied* by assignment τ if $\tau(C) = \mathbf{t}$ for all $C \in F$.
- A CNF formula F is *falsified* by assignment τ if $\tau(C) = \mathbf{f}$ for some $C \in F$.

A CNF formula with no satisfying assignments is called *unsatisfiable*. A clause C is *logically implied* by formula F if adding C to F does not change the set of satisfying assignments of F. Two formulas are *logically equivalent* if they have the same set of solutions over the common variables. Two formulas are *satisfiability equivalent* if both have a solution or neither has a solution. Any formula containing the empty clause ϵ is unsatisfiable.

Resolution. Given two clauses $C_1 = (x \vee a_1 \vee \ldots \vee a_n)$ and $C_2 = (\bar{x} \vee b_1 \vee \ldots \vee b_m)$, the resolution rule states that the clause $C = (a_1 \vee \ldots \vee a_n \vee b_1 \vee \ldots \vee b_m)$, can be inferred by resolving on x. We call C the resolvent of C_1 and C_2 and write $C = C_1 \diamond C_2$. Clause C is logically implied by a formula containing C_1 and C_2.

Boolean Constraint Propagation and Asymmetric Tautologies. For a CNF formula F, *Boolean constraint propagation* (BCP) (or *unit propagation*) simplifies F based on unit clauses; that is, it repeats the following until it reaches a fixpoint: if there is a unit clause $(l) \in F$, remove all clauses that contain the literal l from the set $F \setminus \{(l)\}$ and remove the literal $\bar{l}$ from all clauses in F. We write $F \vdash_1 \epsilon$ to denote that BCP applied to F derives the empty clause. A clause C is an *asymmetric tautology* (AT) with respect to formula F if and only if $F \wedge \overline{C} \vdash_1 \epsilon$. Asymmetric tautologies are logically implied by F.

Example 1. Consider the formula $F = (\bar{a} \vee b) \wedge (\bar{b} \vee c) \wedge (\bar{b} \vee \bar{c})$. Clause $C = (\bar{a})$ is an asymmetric tautology with respect to F, because $F \wedge \overline{C} \vdash_1 \epsilon$. BCP removes literal $\bar{a}$, resulting in the new unit clause (b). After removal of the literals $\bar{b}$, two complementary unit clauses (c) and $(\bar{c})$ are created. ∎

Blocked Clauses. Given a CNF formula F, a clause C, and a literal $l \in C$, the literal l *blocks* C with respect to F if (i) for each clause $D \in F$ with $\bar{l} \in D$, $C \diamond D$ is a tautology, or (ii) $\bar{l} \in C$, i.e., C is itself a tautology. Given a CNF formula F, a clause C is *blocked* with respect to F if there is a literal that blocks C with respect to F. Addition and removal of blocked clauses results in satisfiability-equivalent formulas [12], but not logically-equivalent formulas.

Example 2. Consider the formula $(a \vee b) \wedge (a \vee \bar{b} \vee \bar{c}) \wedge (\bar{a} \vee c)$. Clause $(a \vee \bar{b} \vee \bar{c})$ is a blocked clause, because its literal a is blocking it: the only resolution possibility is with $(\bar{a} \vee c)$ which results in tautology $(\bar{b} \vee \bar{c} \vee c)$. ∎

3 Validating DRAT Proofs

A clause C is called *redundant* with respect to a formula F if $F \wedge \{C\}$ is satisfiability equivalent to F. A *proof of unsatisfiability* (also called a *refutation*) is a sequence of redundant clauses, called *lemmas*, containing the (unsatisfiable) empty clause ϵ. There are two prevalent types of unsatisfiability proofs: *resolution proofs* and *clausal proofs*. Several formats have been designed for resolution proofs [13–15], but they all share the same disadvantages. Resolution proofs are often huge, and it is hard to express important techniques, such as conflict clause minimization, with resolution steps. Other techniques, such as bounded variable addition, cannot be polynomially-simulated by resolution. Clausal proof formats [4,16,17] are syntactically similar; they involve a sequence of clauses that are claimed to be redundant with respect to a given formula. It is important that redundant clauses can be checked in polynomial time. Clausal proofs may include deletion information to reduce the validation cost [18,19]. The drat-trim [4] utility can efficiently validate clausal proofs provided in the DRAT format; this format is backwards compatible with earlier clausal proof formats and was used to check the results of SAT Competition 2014.

Resolution asymmetric tautologies (or RAT clauses) [20] are a generalization of both asymmetric tautologies and blocked clauses. A clause C has RAT on $l \in C$ (referred to as the *pivot* literal) with respect to a formula F if for all $D \in F$ with $\bar{l} \in D$, it holds that $F \wedge \overline{C} \wedge (\overline{D} \setminus \{(l)\}) \vdash_1 \epsilon$.

Not only can RAT be computed in polynomial time, but all preprocessing, inprocessing, and solving techniques in state-of-the-art SAT solvers can be expressed in terms of addition and removal of RAT clauses [20]. A DRAT *proof*, short for *Deletion Resolution Asymmetric Tautology*, is a sequence of addition and deletion steps of RAT clauses. Figure 1 shows an example DRAT proof.

CNF formula	DRAT proof
p cnf 4 10	
1 2 -3 0	-1 0
-1 -2 3 0	d -1 2 4 0
2 3 -4 0	2 0
-2 -3 4 0	0
-1 -3 -4 0	
1 3 4 0	
-1 2 4 0	
1 -2 -4 0	

Fig. 1. Left, a formula in DIMACS CNF format, the conventional input format for SAT solvers. Right, a DRAT proof for that formula. Each line in the proof is either an addition step (no prefix) or a deletion step identified by the prefix "**d**". Spacing in both examples is used to improve readability. Each clause in the proof should be an asymmetric tautology or a RAT clause using the first literal as the pivot.

Example 3. Consider CNF formula $F = (a \vee b \vee \bar{c}) \wedge (\bar{a} \vee \bar{b} \vee c) \wedge (b \vee c \vee \bar{d}) \wedge (\bar{b} \vee \bar{c} \vee d) \wedge (a \vee c \vee d) \wedge (\bar{a} \vee \bar{c} \vee \bar{d}) \wedge (\bar{a} \vee b \vee d) \wedge (a \vee \bar{b} \vee \bar{d})$, which is shown in DIMACS format in Fig. 1 (left), where 1 represents a, 2 is b, 3 is c and negative numbers represent negation. The first clause in the proof, $(\bar{a})$, is a RAT clause with respect to F because all possible resolvents are asymmetric tautologies:

$$F \wedge (a) \wedge (\bar{b}) \wedge (c) \vdash_1 \epsilon \quad \text{using} \quad (a \vee b \vee \bar{c})$$
$$F \wedge (a) \wedge (\bar{c}) \wedge (\bar{d}) \vdash_1 \epsilon \quad \text{using} \quad (a \vee c \vee d)$$
$$F \wedge (a) \wedge (b) \wedge (d) \vdash_1 \epsilon \quad \text{using} \quad (a \vee \bar{b} \vee \bar{d})$$

∎

Let F be a CNF formula and P be a DRAT proof for F. The number of lines in a proof P is denoted by $|P|$. For each $i \in \{0, \ldots, |P|\}$, a CNF formula F_P^i is defined below. L_i refers to the lemma (redundant clause) on line i of P.

$$F_P^i := \begin{cases} F & \text{if } i = 0 \\ F_P^{i-1} \setminus \{L_i\} & \text{if the prefix of } L_i \text{ is ``d''} \\ F_P^{i-1} \cup \{L_i\} & \text{otherwise} \end{cases}$$

Each lemma addition step is validated using a RAT check, while lemma deletion steps are ignored as their only purpose is to reduce the validation costs. Let l_i denote the first literal in lemma L_i. The RAT check for lemma L_i in proof P for CNF formula F succeeds if and only if L_i has the property RAT on literal l_i with respect to F_P^{i-1}. Moreover, lemma $L_{|P|}$ must be the empty clause.

4 Symmetries in Propositional Formulas

Two graphs G and H are *isomorphic* if there exists an edge-preserving bijection from the vertices of G to the vertices of H. A *symmetry* (or automorphism) of a graph G is an edge-preserving bijection of G onto itself. Symmetries in the graph representation of SAT problems may cause solvers to explore symmetric parts of the search space again and again. This problem can be avoided by adding symmetry-breaking predicates [1]: the clause-literal graph for a given formula is created, the automorphisms of the graph are computed, and the automorphisms are converted into symmetry-breaking predicates.

A *clause-literal graph* of a CNF formula F is an undirected graph with a vertex for each clause and each literal occurrence in F. A literal vertex and a clause vertex are connected if and only if the designated literal occurs in the corresponding clause. Two clause vertices are never connected. Two literal vertices are connected if and only if the corresponding literals are complements.

A *symmetry* $\sigma = (x_1, \ldots, x_n)(p_1, \ldots, p_n)$ of a CNF formula F is an edge-preserving bijection of the clause-literal graph of F, that maps variable x_i onto p_i with $i \in \{1..n\}$. The sequence $(p_1, \ldots, p_n)$ is a permutation of $(x_1, \ldots, x_n)$ in which each p_i is potentially negated. If x_i is mapped onto p_i, then $\bar{x}_i$ is mapped onto $\bar{p}_i$. Note that the clauses in the clause-literal graph are permuted by a symmetry, but we can ignore this aspect when it comes to symmetry breaking. Breaking σ can be achieved by enforcing that the assignment to literals $x_1, x_2, \ldots, x_n$ is lexicographically less than or equal to ($\leq$) the assignment to literals $p_1, p_2, \ldots, p_n$. This is a choice as $\geq$ could have been used instead.

Example 4. Consider the problem whether a path of two edges is an unavoidable subgraph for graphs of order 3. We name the vertices a, b, and c. The existence of an edge between a and b, a and c, and b and c is represented by the Boolean variables $x_{a,b}$, $x_{a,c}$, and $x_{b,c}$, respectively. The propositional formula that expresses this problem and the labels of the clauses are shown below.

$$\overbrace{(x_{a,b} \vee x_{a,c})}^{C_1} \wedge \overbrace{(x_{a,b} \vee x_{b,c})}^{C_2} \wedge \overbrace{(x_{a,c} \vee x_{b,c})}^{C_3} \wedge \overbrace{(\bar{x}_{a,b} \vee \bar{x}_{a,c})}^{C_4} \wedge \overbrace{(\bar{x}_{a,b} \vee \bar{x}_{b,c})}^{C_5} \wedge \overbrace{(\bar{x}_{a,c} \vee \bar{x}_{b,c})}^{C_6}$$

The clause-literal graph of this formula is shown in Fig. 2 together with three isomorphic graphs that can be obtained by permuting the nodes of the clause-literal graph. The three symmetries are (ignoring the permutation of clauses):

$$(x_{a,b}, x_{a,c}, x_{b,c})(\bar{x}_{a,b}, \bar{x}_{a,c}, \bar{x}_{b,c}); \quad (x_{a,b}, x_{a,c})(x_{a,c}, x_{a,b}); \quad (x_{a,c}, x_{b,c})(x_{b,c}, x_{a,c}).$$

These symmetries can be broken by enforcing $x_{a,b}, x_{a,c}, x_{b,c} \leq \bar{x}_{a,b}, \bar{x}_{a,c}, \bar{x}_{b,c}$ (or equivalently $(x_{a,b} \leq \bar{x}_{a,b}) \equiv \bar{x}_{a,b}$); $x_{a,b}, x_{a,c} \leq x_{a,c}, x_{a,b}$ (or equivalently $x_{a,b} \leq x_{a,c}$); and $x_{a,c}, x_{b,c} \leq x_{b,c}, x_{a,c}$ (or equivalently $x_{a,c} \leq x_{b,c}$), respectively. The clausal representation is: $(\bar{x}_{a,b})$, $(\bar{x}_{a,b} \vee x_{a,c})$, and $(\bar{x}_{a,c} \vee x_{b,c})$. ∎

Expressing the constraint $x_1, x_2, \ldots, x_n \leq p_1, p_2, \ldots, p_n$ in clauses with (only) variables in F can be done as follows:

$$(\bar{x}_1 \vee p_1) \wedge (\bar{x}_1 \vee \bar{x}_2 \vee p_2) \wedge (p_1 \vee \bar{x}_2 \vee p_2) \wedge (\bar{x}_1 \vee \bar{x}_2 \vee \bar{x}_3 \vee p_3) \wedge$$
$$(\bar{x}_1 \vee p_2 \vee \bar{x}_3 \vee p_3) \wedge (p_1 \vee \bar{x}_2 \vee \bar{x}_3 \vee p_3) \wedge (p_1 \vee p_2 \vee \bar{x}_3 \vee p_3) \wedge \ldots$$

The above scheme adds $2^n - 1$ clauses. Using auxiliary variables a_i, it requires only a linear number of clauses, i.e., $3n - 2$, to express this constraint:

$$(\bar{x}_1 \vee p_1) \wedge (a_1 \vee \bar{x}_1) \wedge (a_1 \vee p_1) \wedge (\bar{a}_{n-1} \vee \bar{x}_n \vee p_n) \wedge$$
$$\bigwedge_{i \in \{2..n-1\}} \left((\bar{a}_{i-1} \vee \bar{x}_i \vee p_i) \wedge (a_i \vee \bar{a}_{i-1} \vee \bar{x}_i) \wedge (a_i \vee \bar{a}_{i-1} \vee p_i) \right) \tag{1}$$

So a_i is true if a_{i-1} is true and $p_i \leq x_i$ holds (using a_0 is true). Optionally, the following *blocked clauses* [12] can be added: $(\bar{a}_1 \vee x_1 \vee \bar{p}_1)$, $(\bar{a}_i \vee a_{i-1}) \wedge (\bar{a}_i \vee x_i \vee \bar{p}_i)$ for $i \in \{2..n-1\}$. State-of-the-art SAT solvers, such as Lingeling, remove blocked clauses during preprocessing since they are useless in practice [21].

5   Breaking a Single Symmetry

In this section, we demonstrate how to break a single symmetry within a DRAT proof. Breaking multiple symmetries is more complicated, which will be discussed in Sect. 6. Breaking a single symmetry consists of three steps: adding definitions, redefining involved clauses, and adding symmetry-breaking predicates. Below we discuss these three steps in detail using the following notation. The formula F_0 expresses the initial formula with symmetry σ. Formula F_1 expresses the result

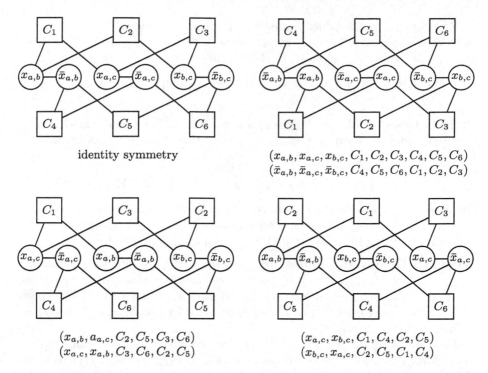

Fig. 2. Four isomorphic clause-literal graphs for the CNF formula in Example 4. Notice that the four graphs are identical modulo the labeling of the nodes. Below each graph, the permutation of the nodes is shown compared to the top-left graph. The permutation of clauses is omitted in symmetries throughout this paper.

of adding definitions (step 1); formula F_2 expresses the result after redefining involved clauses in F_1 (steps 1 and 2); and formula F_3 expresses the result after adding symmetry-breaking predicates to F_2 (all three steps).

Adding Definitions. The first step consists of introducing auxiliary variables s_i. For a given symmetry $\sigma = (x_1, \ldots, x_n)(p_1, \ldots, p_n)$, these variables are defined as follows: $s_i \equiv (x_i, \ldots, x_n > p_i, \ldots, p_n)$ with $i \in \{1..n\}$. Variable s_1 is the only important auxiliary variable and we refer to it as the *primal-swap variable*. The other s_i variables with $i > 1$ are used to efficiently compute s_1. The definitions of s_i with $i \in \{1..n\}$ require $6n - 3$ clauses:

$$\bigwedge_{i \in \{1..n-1\}} ((s_i \vee \bar{x}_i \vee p_i) \wedge (s_i \vee \bar{x}_i \vee \bar{s}_{i+1}) \wedge (s_i \vee p_i \vee \bar{s}_{i+1}) \wedge$$

$$(\bar{s}_i \vee x_i \vee \bar{p}_i) \wedge (\bar{s}_i \vee x_i \vee s_{i+1}) \wedge (\bar{s}_i \vee \bar{p}_i \vee s_{i+1})) \wedge \qquad (2)$$

$$(s_n \vee \bar{x}_n \vee p_n) \wedge (\bar{s}_n \vee x_n) \wedge (\bar{s}_n \vee \bar{p}_n)$$

These clauses can be added to the formula by blocked clause addition in the reverse order as listed in the equation above: first add all clauses containing literals s_n and $\bar{s}_n$, second add clauses containing literals s_{n-1} and $\bar{s}_{n-1}$, etc.

Additionally, we introduce n auxiliary Boolean variables x_i' with $i \in \{1..n\}$ which are defined as follows. If the primal-swap variable s_1 is assigned to false, then $x_i' \leftrightarrow x_i$, otherwise $x_i' \leftrightarrow p_i$. In clauses this definition is expressed as

$$\bigwedge_{i \in \{1..n\}} ((x_i' \vee \bar{x}_i \vee s_1) \wedge (\bar{x}_i' \vee x_i \vee s_1) \wedge (x_i' \vee \bar{p}_i \vee \bar{s}_1) \wedge (\bar{x}_i' \vee p_i \vee \bar{s}_1)). \quad (3)$$

All these clauses are blocked on the x_i and $\bar{x}_i$ literals. The definitions of x_1' and p_1' can be expressed more compactly using only three clauses per definition:

$$\begin{aligned}
(x_1' \vee \bar{x}_1 \vee \bar{p}_1) \wedge (\bar{x}_1' \vee x_1) \wedge (\bar{x}_1' \vee p_1) &\equiv & x_1' := \mathrm{AND}(x_1, p_1) \\
(\bar{p}_1' \vee x_1 \vee p_1) \wedge (p_1' \vee \bar{x}_1) \wedge (p_1' \vee \bar{p}_1) &\equiv & p_1' := \mathrm{OR}(x_1, p_1)
\end{aligned}$$

The more compact definitions are also blocked on the prime literals. All clauses contain only one prime literal and all clauses are blocked on the prime literal. Therefore, they can be added to a DRAT proof in arbitrary order.

Redefining Involved Clauses. In the second step of breaking symmetry $\sigma = (x_1, \ldots, x_n)(p_1, \ldots, p_n)$, we redefine the *involved clauses* C_j, i.e., those clauses in F_0 that contain at least one literal x_i or $\bar{x}_i$ with $i \in \{1..n\}$ by clauses C_j', a copy of C_j with all literals x_i and $\bar{x}_i$ replaced by literals x_i' and $\bar{x}_i'$, respectively.

The clauses C_j' do not have RAT with respect to F_1, the formula resulting after adding definitions. However, the clauses $C_j' \cup \{s_1\}$ and $C_j' \cup \{\bar{s}_1\}$ have AT with respect to F_1, with s_1 referring to the primal-swap variable from the prior step. Using this observation, we express redefining C_j into C_j' with $j \in \{1..m\}$ using $4m$ operations: add $C_j' \cup \{s_1\}$, add C_j', delete $C_j' \cup \{s_1\}$ and delete C_j. Notice that we use $C_j' \cup \{s_1\}$ as an auxiliary clause to add C_j': C_j' has AT with respect to $F_1 \cup \{C_j' \cup \{s_1\}\}$ because $C_j' \cup \{\bar{s}_1\}$ has AT with respect to F_1.

Adding Symmetry-Breaking Predicates. After adding definitions (step 1) and redefining involved clauses (step 2), all assignments for which $x_1', \ldots, x_n' > p_1', \ldots, p_n'$ are eliminated: if $x_1, \ldots, x_n > p_1, \ldots, p_n$, then s_1 is assigned to true which will swap x_i and p_i with $i \in \{1..n\}$. To express this knowledge, i.e., $x_1', \ldots, x_n' \leq p_1', \ldots, p_n'$, in clauses that have RAT with respect to F_2, we first introduce auxiliary variables y_i as follows:

$$(y_1 \vee \bar{x}_1') \wedge (y_1 \vee p_1') \wedge (\bar{y}_1 \vee x_1' \vee \bar{p}_1') \wedge$$

$$\bigwedge_{i \in \{2..n-1\}} ((y_i \vee \bar{y}_{i-1} \vee \bar{x}_i') \wedge (y_i \vee \bar{y}_{i-1} \vee p_i') \wedge (\bar{y}_i \vee y_{i-1}) \wedge (\bar{y}_i \vee x_i' \vee \bar{p}_i')) \quad (4)$$

Notice that all these clauses have the RAT property on their first literal when added in the order as shown in (4). Afterwards, we add the following clauses:

$$(\bar{x}_1' \vee p_1') \wedge \bigwedge_{i \in \{2..n\}} (\bar{y}_{i-1} \vee \bar{x}_i' \vee p_i') \quad (5)$$

The clauses (5) are logically implied by F_2 after the addition of (4). Notice that the clauses (4) and (5) together are the same as (1), but with the blocked clauses. The blocked clauses are required to add (5), but can be removed afterwards.

Partial Symmetry Breaking. To this point, we considered breaking a symmetry fully. However, symmetries can also broken partially. Given a symmetry $\sigma = (x_1, \ldots, x_n)(p_1, \ldots, p_n)$, we refer to *partial* symmetry breaking as using a subset of $x_1, \ldots, x_n$ and the corresponding p_i for the clauses (2), (4), and (5). However, also with partial symmetry breaking, the full set of clauses (3) should be used and also redefining involved clauses should not change.

Example 5. Consider the symmetry $\sigma = (x_1, x_2, x_3, x_4)(x_3, x_4, x_1, x_2)$. We could partially break σ by using only a subset, say $\sigma' = (x_1, x_3)(x_3, x_1)$. This would result in the symmetry-breaking predicate $x_1 \leq x_3$ which is a weakened version of the predicate $x_1, x_2 \leq x_3, x_4$ that would be been created by fully breaking σ.

6 Breaking Multiple Symmetries

Given a problem with k symmetries, tools that add symmetry-breaking predicates add the clauses (1) for each symmetry. However, breaking $k > 1$ symmetries cannot be expressed in DRAT by applying the above procedure (all three steps) only once for each symmetry. Two symmetries are *dependent* if they have at least one overlapping variable. If two symmetries are dependent, it requires more than two swaps to break them both.

Example 6. Consider the formula $F = (x_1 \vee x_2) \wedge (x_1 \vee x_3) \vee (x_2 \vee x_3) \vee (\bar{x}_1 \vee \bar{x}_2 \vee \bar{x}_3)$ and its two symmetries: $\sigma_1 = (x_1, x_2)(x_2, x_1)$ and $\sigma_2 = (x_2, x_3)(x_3, x_2)$. Breaking symmetry σ_1 would result in adding the clauses

$$(x_1' \vee \bar{x}_1 \vee \bar{x}_2), (\bar{x}_1' \vee x_1), (\bar{x}_1' \vee x_2), (\bar{x}_2' \vee x_1 \vee x_2), (x_2' \vee \bar{x}_1), (x_2' \vee \bar{x}_2).$$

Applying the definitions, F can be converted to $F' = (x_1' \vee x_2') \wedge (x_1' \vee x_3) \vee (x_2' \vee x_3) \vee (\bar{x}_1' \vee \bar{x}_2' \vee \bar{x}_3)$. From the definitions, it follows that $x_1' \leq x_2'$, or $(\bar{x}_1' \vee x_2')$. Now, let us break symmetry σ_2 by adding the clauses

$$(x_2'' \vee \bar{x}_2' \vee \bar{x}_3), (\bar{x}_2'' \vee x_2'), (\bar{x}_2'' \vee x_3), (\bar{x}_3' \vee x_2' \vee x_3), (x_3' \vee \bar{x}_2'), (x_3' \vee \bar{x}_3).$$

Again, applying the definitions, F' can be converted to $F'' = (x_1' \vee x_2'') \wedge (x_1' \vee x_3') \vee (x_2'' \vee x_3') \vee (\bar{x}_1' \vee \bar{x}_2'' \vee \bar{x}_3')$. Notice that $(\bar{x}_1' \vee x_2'')$, i.e., $x_1' \leq x_2''$ does not hold. Consider the satisfying assignment $x_1 = 1$, $x_2 = 1$, $x_3 = 0$. Following the definitions, $x_1' = 1$, $x_2' = 1$ and $x_2'' = 0$, $x_3' = 1$. Observe that $0 = x_2'' < x_1' = 1$. In order to break both σ_1 and σ_2, we need to break σ_1 again. ■

The problem in Example 6 is caused by dependent symmetries. Breaking dependent symmetries requires applying the symmetry-breaking procedure, i.e., the three steps to break a single symmetry, again and again. We can limit the number of times the procedure has to be applied if the dependent symmetries have a frequently occurring pattern: a symmetry chain. Consider symmetries of the form $\sigma_i = (x_{i,1}, \ldots, x_{i,n}, x_{i,n+1}, \ldots, x_{i,2n})(x_{i,n+1}, \ldots, x_{i,2n}, x_{i,1}, \ldots, x_{i,n})$. A *symmetry chain* $\langle \sigma_1, \ldots, \sigma_k \rangle$ is a sequence of such symmetries with the additional property that $x_{i+1,j} = x_{i,j+n}$ with $1 \leq i < k$ and $1 \leq j \leq n$. We denote a symmetry chain by $(x_{1,1}, \ldots, x_{1,n})(x_{2,1}, \ldots, x_{2,n}) \ldots (x_{k,1}, \ldots, x_{k,n})$. Breaking such a symmetry chain will add the following constraints to the formula

$$x_{1,1}, \ldots, x_{1,n} \leq x_{2,1}, \ldots, x_{2,n} \leq \cdots \leq x_{k,1}, \ldots, x_{k,n}.$$

We first will explain how to express breaking a symmetry chain wihtin a DRAT proof. Afterwards we will show how to convert dependent symmetries into a chain, which might weaken the symmetry-breaking predicates.

Breaking a Symmetry Chain using Sorting Networks. A *sorting network*, consisting of k wires and c comparators, sorts k values using c comparisons. Values flow across the wires. A comparator connects two wires, compares the incoming values, and sorts them by assigning the smaller value to one wire, and the larger to the other. Figure 3 shows a sorting network of four wires (horizontal lines) and five comparators (vertical lines). The best algorithms in practice for sorting k wires are based on pairwise sorting [22] or Batcher's Merge-Exchange algorithm [23] which produce sorting networks with $\mathcal{O}(k \log^2 k)$ comparators.

Sorting networks can be used to break a symmetry chain. For a symmetry chain $\langle \sigma_1, \ldots, \sigma_k \rangle$, we use a sorting network with k wires. For each comparator in the network we apply the symmetry-breaking procedure once. Since comparators may skip certain wires, such as the first and second comparator in Fig. 3, we need to compute that symmetry.

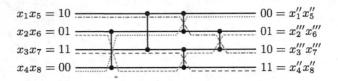

Fig. 3. A sorting network that sorts the assignments of a symmetry chain of length 4.

Example 7. Consider the symmetry chain: $(x_1, x_5)(x_2, x_6)(x_3, x_7)(x_4, x_8)$. The symmetry-breaking predicates will express that $x_1, x_5 \leq x_2, x_6 \leq x_3, x_7 \leq x_4, x_8$. In order to convert any assignment on those variables, the symmetry-breaking procedure is applied five times, i.e., the size of the smallest sorting network on four wires. Figure 3 illustrates such a sorting network. The first comparator in this network connects the wires 2 and 4. This corresponds to applying the symmetry-breaking procedure of the symmetry $(x_2, x_6, x_4, x_8)(x_4, x_8, x_2, x_6)$.

Breaking Multiple Symmetries. Not all dependent symmetries form a chain. If k dependent symmetries cannot be expressed as a chain, it may require more than $\mathcal{O}(k \log k)$ swaps to break them. This is illustrated below.

Example 8. Consider two symmetries: $\sigma_1 = (x_1, x_4, x_2, x_5)(x_2, x_5, x_1, x_4)$ and $\sigma_2 = (x_2, x_4, x_3, x_6)(x_3, x_6, x_2, x_4)$. Notice σ_1 and σ_2 are dependent and cannot be expressed as a symmetry chain. A symmetry chain of length three (a chain of two symmetries) can be broken using three swaps: the size of the smallest sorting network with 3 wires has 3 comparators. However, breaking σ_1 and σ_2 requires four swaps for the assignment $x_1 = x_2 = x_4 = x_6 = 1$ and $x_3 = x_5 = 0$.

$$
\begin{array}{|c|c|c|}
\hline
x_1 & x_2 & x_3 \\
1 & 1 & 0 \\
\hline
1 & 0 & 1 \\
\hline
x_4 & x_5 & x_6 \\
\end{array}
\xrightarrow{\sigma_1}
\begin{array}{|c|c|c|}
\hline
x_1' & x_2' & x_3 \\
1 & 1 & 0 \\
\hline
0 & 1 & 1 \\
\hline
x_4' & x_5' & x_6 \\
\end{array}
\xrightarrow{\sigma_2}
\begin{array}{|c|c|c|}
\hline
x_1' & x_2'' & x_3' \\
1 & 0 & 1 \\
\hline
1 & 1 & 0 \\
\hline
x_4' & x_5' & x_6' \\
\end{array}
\xrightarrow{\sigma_1}
\begin{array}{|c|c|c|}
\hline
x_1'' & x_2''' & x_3' \\
0 & 1 & 1 \\
\hline
1 & 1 & 0 \\
\hline
x_4''' & x_5'' & x_6' \\
\end{array}
\xrightarrow{\sigma_2}
\begin{array}{|c|c|c|}
\hline
x_1'' & x_2'''' & x_3'' \\
0 & 1 & 1 \\
\hline
0 & 1 & 1 \\
\hline
x_4'''' & x_5'' & x_6'' \\
\end{array}
$$

■

Dependent symmetries that do not form a chain can be broken using a sorting network, by breaking them partially, which will be discussed below.

Converting Symmetries into a Symmetry Chain. Some applications, such as computing unavoidable subgraphs, have dependent symmetries that cannot be expressed as a symmetry chain. However, we can still use the above procedure if we partially break such symmetries. We apply the following method: given a set of dependent symmetries, we compute a subset of each symmetry such that they form a symmetry chain. These shorter symmetries are used for partial symmetry breaking as discussed at the end of Sect. 5.

Example 9. Consider the symmetries in Example 8. First, we compute a subset of $\sigma_1' \subseteq \sigma_1$ and $\sigma_2' \subseteq \sigma_2$ such that σ_1' and σ_2' form a chain, say $\sigma_1' = (x_1, x_2)(x_2, x_1)$ and $\sigma_2' = (x_2, x_3)(x_3, x_2)$. Second, we will use σ_1' and σ_2' for the sorting networks and apply the partial symmetry breaking.

7   Tools and Evaluation

Several tools are necessary to produce a DRAT proof for a given formula F that incorporates symmetry breaking. Figure 4 shows an overview of the tool chain. Six tools are used: a formula-to-graph converter, a symmetry extractor, a symmetry-breaking converter, a SAT solver, and a DRAT proof checker. These tools are used in four phases:

  I The symmetries σ of F are computed by transforming F into a clause-literal graph (see Sect. 4). A symmetry-extraction tool, such as saucy [24], can be used to obtain the symmetries.
 II Formula F is converted into a satisfiability-equivalent formula F', a copy of F for which the symmetries σ are broken. F' is equivalent (modulo variable renaming) to adding symmetry-breaking predicates to F using a symmetry-breaking tool. Additionally, the conversion from F to F' is expressed as a partial DRAT proof. Our new tool, sym2drat, implements this second phase, i.e., computing F' and a partial DRAT proof.
III The formula F' is solved by a SAT solver, which produces a DRAT proof. Most state-of-the-art SAT solvers now support emission of DRAT proofs.

IV The last step consists of verifying the result of both the symmetry-breaking
   tool and the SAT solver. The partial DRAT proof and the DRAT proof of
   F' are merged, which is accomplished by concatenating the proofs. A proof
   checker, such as `drat-trim` [4], validates whether the merged proof is a
   refutation for the input formula F.

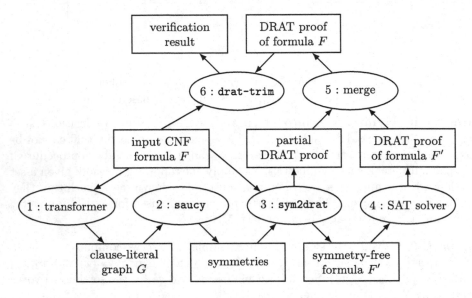

Fig. 4. Tool chain to produce DRAT proofs that incorporate symmetry breaking. The
rectangle boxes are files, while the round boxes are tools. Phase I consists of the tools
transformer and `saucy`, phase II consists of the `sym2drat` tool, phase III uses an off-
the-shelf SAT solver, and phase IV consists of tools to merge and validate the proofs.

Below we will discuss the tools that we developed for phases II and IV.
We used off-the-shelf tools for the phases I and III.

7.1   The Tool `sym2drat`

The main tool that we developed for expressing symmetry-breaking as DRAT
proofs is `sym2drat`. This tool requires two inputs: a CNF formula F and a
set of symmetries of F. Two files are emitted by `sym2drat`: a CNF formula F'
with symmetry-breaking predicates and a partial DRAT proof that expresses the
conversion of F into F'. Our tool `sym2drat` constructs sorting networks based
on the pairwise sorting algorithm [22], which reduces the number of swaps (and
thus the size of the partial DRAT proof) by roughy a factor of two compared to
the bubble sorting network for most problems on which we experimented, i.e.,
problems containing around 20 symmetries. The `sym2drat` tool preprocesses the
input symmetries such that they form a set of symmetry chain, see Sect. 6.

7.2   Improving DRAT Proof-Checking Tools

Apart from implementing `sym2drat`, we improved two tools that validate DRAT proofs. The first tool we improved is `drat-trim`: the fast DRAT proof checker written in C that was used to validate the results of SAT Competition 2014. The current version of `drat-trim` does not support validating *partial proofs*: a sequence of clauses that are all redundant with respect to a given formula, but that does not terminate with the empty clause. Checking partial proofs allows one to validate the output of `sym2drat`. We extended `drat-trim` with the option to validate partial proofs[1]. This feature was very useful for developing our method to express using symmetry-breaking in DRAT proofs. We expect this feature to be helpful to discover how other techniques, such as Gaussian Elimination and cardinality resolution, can be expressed with DRAT proofs.

Our mechanically-verified, RAT validation tool [25], written in ACL2, has undergone significant improvements. This tool was originally designed to demonstrate the soundness of a basic algorithm used to validate RAT proofs. Efficiency of the tool was not a priority. Recent work [11] has been devoted to improving the performance of this tool while maintaining its proof of correctness (soundness). The underlying data structures have been moved from `cons`-based lists to ACL2 STOBJs (Single Thread OBJects) which offer support for LISP arrays, reducing the linear-time cost for accesses and updates to constant-time. This seemingly small change has a large impact on performance but also required a substantial proof effort. A new ACL2 data structure, called `farray`, was developed to facilitate proof development with STOBJs. A mechanical proof of equivalence was established to show that the new tool behaves exactly the same as the original tool, preserving the proof of correctness of the original tool.

8   Evaluation

We evaluated the usefulness of our new method by computing and validating "compact" DRAT proofs[2] on some hard combinatorial problems.

Ramsey Number Four. Ramsey theory addresses unavoidable patterns. The most well-known pattern is unavoidable cliques. The size of the smallest graph that has an unavoidable clique of size k is called Ramsey number k. Ramsey number four is 18. Showing that any graph of size 18 has an unavoidable clique of size 4 can be encoded using a formula consisting of $2 \cdot \binom{18}{4} = 6120$ clauses, each of length 6. The SAT solvers `Lingeling` and `glucose` were unable to determine in 24 hours that the formula is unsatisfiable.

The CNF formula F that encodes Ramsey number four has 18 symmetries: any permutation of vertices and complementing the graph. SAT solvers can determine that formula F', with symmetry-breaking predicates produced by `sym2drat`, is unsatisfiable in less than a second. We merged the proof of F',

---

[1] available at http://www.cs.utexas.edu/~marijn/drat-trim/.
[2] available at http://www.cs.utexas.edu/~marijn/sbp/.

produced by `glucose` 3.0, with the partial DRAT proof, produced by `sym2drat`. This proof can be checked by `drat-trim` in 1.9 seconds. We validated the proof using our ACL2-based, mechanically-verified RAT checker [11] as well. These tools allow one to obtain a mechanically-verified proof in the ACL2 theorem prover that Ramsey number four is 18. We envision that this tool chain is a useful template to obtain trustworthy results of hard combinatorial problems.

Erdős Discrepancy Conjecture. Let $S = \langle s_1, s_2, s_3, \ldots \rangle$ be an infinite sequence of 1's or -1's. Erdős Discrepancy Conjecture states that for any C there exists an d and k such that

$$\left| \sum_{i=1}^{k} s_{i \cdot d} \right| > C$$

Recently, the case $C = 2$ was proved using SAT solvers, resulting in a clausal proof of 13Gb [5]. The problem contains one symmetry (swapping 1's and -1's), but it was not broken in the original approach. We proved the conjecture using our tool chain with `glucose` 3.0 and validated the DRAT proof using `drat-trim`. The size of our proof is slightly more than 2Gb in syntactically the same format as the original proof. Symmetry breaking allowed us to pick a variable which can be added to the formula, similar to the unit $(\bar{x}_{a,b})$ in Example 4. We choose unit $(\bar{s}_{60})$ as it occurs frequently in the original CNF formula. The combination of symmetry-breaking and selecting a good unit variable resulted in a proof a sixth of the size of the original one. The tool `drat-trim` can reduce the new proof to 850 Mb by removing redundant lemmas and discarding the deletion information.

Two Pigeons per Hole. One family of hard problems in the SAT Competitions of 2013 and 2014 are a variation of the *pigeon hole principle.* The Two-Pigeons-per-Hole (TPH) family consists of problems encoding that $2k + 1$ pigeons can be placed into k holes such that each hole has at most two pigeons. Most SAT solvers can refute the problem for $k = 6$, but they cannot solve problems of size $k > 6$ within an hour, unless symmetry breaking or cardinality resolution is applied. It is not yet known how to express cardinality resolution in the existing SAT proof formats. Previously, there was no approach to produce DRAT proofs for the difficult instances of this family ($k > 6$). Using our method, we produced and checked DRAT proofs for this family for $k \leq 12$ within an hour.

A TPH problem of size k contains $2k + 1$ symmetries of length k expressing that the pigeons are interchangeable. After breaking these symmetries, TPH problems become very easy and can be solved instantly. However, the number of involved clauses per symmetry is large and the formulas contain many clauses. As a consequence, expressing a single swap results in many clause addition and deletion steps. For $k = 12$, our method results in a 4Gb proof. The size of the DRAT proof sharply increases with k and so does the time to validate the proof.

9   Conclusions

Validating proofs of unsatisfiability helps one gain confidence in the correctness of SAT solver results, even when some SAT solvers have been shown to contain errors on an implementation [26] and conceptual level [20]. We presented a method to express symmetry breaking in DRAT, the most widely-supported proof format for SAT solvers. Our method allows, for the first time, validation of SAT solver results obtained via symmetry breaking, thereby validating the results of symmetry extraction tools as well.

Symmetry breaking is often crucial when solving hard combinatorial problems. Our method provides a missing link to establish trust that results on these problems are correct. We demonstrated our method on hard combinatorial problems such as Ramsey number four and the Erdős Discrepancy Conjecture. We also constructed DRAT proofs of two-pigeons-per-hole (TPH) problems. Larger TPH problems can only be solved with either symmetry-breaking or cardinality resolution. It was previously unknown how to produce proofs for either technique, but we demonstrate proofs for the former in this paper. Hence, this work brings us closer to validation of all SAT solver results.

References

1. Crawford, J., Ginsberg, M., Luks, E., Roy, A.: Symmetry-breaking predicates for search problems. In: KR 1996, pp. 148–159. Morgan Kaufmann (1996)
2. Aloul, F.A., Ramani, A., Markov, I.L., Sakallah, K.A.: Solving difficult sat instances in the presence of symmetry. In: Proceedings of the 39th Design Automation Conference, pp. 731–736 (2002)
3. Gent, I.P., Smith, B.M.: Symmetry breaking in constraint programming. In: Horn, W. (ed.) ECAI 2000, pp. 599–603. IOS Press (2000)
4. Wetzler, N., Heule, M.J.H., Hunt Jr, W.A.: DRAT-trim: efficient checking and trimming using expressive clausal proofs. In: Sinz, C., Egly, U. (eds.) SAT 2014. LNCS, vol. 8561, pp. 422–429. Springer, Heidelberg (2014)
5. Konev, B., Lisitsa, A.: A SAT attack on the Erdős discrepancy conjecture. In: Sinz, C., Egly, U. (eds.) SAT 2014. LNCS, vol. 8561, pp. 219–226. Springer, Heidelberg (2014)
6. Kouril, M., Paul, J.L.: The van der Waerden number W(2, 6) is 1132. Exp. Math. **17**(1), 53–61 (2008)
7. Codish, M., Cruz-Filipe, L., Frank, M., Schneider-Kamp, P.: Twenty-five comparators is optimal when sorting nine inputs (and twenty-nine for ten). In: ICTAI 2014, pp. 186–193. IEEE Computer Society (2014)
8. Aloul, F.A., Sakallah, K.A., Markov, I.L.: Efficient symmetry breaking for boolean satisfiability. IEEE Trans. Comput. **55**(5), 549–558 (2006)
9. Schaafsma, B., Heule, M.J.H., van Maaren, H.: Dynamic symmetry breaking by simulating zykov contraction. In: Kullmann, O. (ed.) SAT 2009. LNCS, vol. 5584, pp. 223–236. Springer, Heidelberg (2009)
10. Radziszowski, S.P.: Small Ramsey numbers. Electron. J. Comb. #DS1 (2014)
11. Wetzler, N.D.: Efficient, mechanically-verified validation of satisfiability solvers. Ph.D. dissertation, The University of Texas at Austin, May 2015

12. Kullmann, O.: On a generalization of extended resolution. Discrete Appl. Math. **96–97**, 149–176 (1999)
13. Zhang, L., Malik, S.: Validating sat solvers using an independent resolution-based checker: practical implementations and other applications. In: DATE, pp. 10880–10885 (2003)
14. Eén, N., Sörensson, N.: An extensible SAT-solver. In: Giunchiglia, E., Tacchella, A. (eds.) SAT 2003. LNCS, vol. 2919, pp. 502–518. Springer, Heidelberg (2004)
15. Biere, A.: Picosat essentials. JSAT **4**(2–4), 75–97 (2008)
16. Van Gelder, A.: Verifying rup proofs of propositional unsatisfiability. In: ISAIM (2008)
17. Heule, M.J.H., Hunt Jr, W.A., Wetzler, N.: Verifying refutations with extended resolution. In: Bonacina, M.P. (ed.) CADE 2013. LNCS, vol. 7898, pp. 345–359. Springer, Heidelberg (2013)
18. Heule, M.J.H., Hunt Jr., W.A., Wetzler, N.: Trimming while checking clausal proofs. In: Formal Methods in Computer-Aided Design, pp. 181–188. IEEE (2013)
19. Heule, M.J.H., Hunt Jr, W.A., Wetzler, N.: Bridging the gap between easy generation and efficient verification of unsatisfiability proofs. Softw. Test. Verif. Reliab. (STVR) **24**(8), 593–607 (2014)
20. Järvisalo, M., Heule, M.J.H., Biere, A.: Inprocessing rules. In: Gramlich, B., Miller, D., Sattler, U. (eds.) IJCAR 2012. LNCS, vol. 7364, pp. 355–370. Springer, Heidelberg (2012)
21. Järvisalo, M., Biere, A., Heule, M.J.H.: Blocked clause elimination. In: Esparza, J., Majumdar, R. (eds.) TACAS 2010. LNCS, vol. 6015, pp. 129–144. Springer, Heidelberg (2010)
22. Parberry, I.: The pairwise sorting network. Parallel Process. Lett. **2**, 205–211 (1992)
23. Batcher, K.E.: Sorting networks and their applications. In: Proceedings of Spring Joint Computer Conference, AFIPS 1968, pp. 307–314. ACM (1968)
24. Darga, P.T., Liffiton, M.H., Sakallah, K.A., Markov, I.L.: Exploiting structure in symmetry detection for cnf. In: DAC 2004, pp. 530–534. ACM (2004)
25. Wetzler, N., Heule, M.J.H., Hunt Jr, W.A.: Mechanical verification of SAT refutations with extended resolution. In: Blazy, S., Paulin-Mohring, C., Pichardie, D. (eds.) ITP 2013. LNCS, vol. 7998, pp. 229–244. Springer, Heidelberg (2013)
26. Brummayer, R., Lonsing, F., Biere, A.: Automated testing and debugging of SAT and QBF solvers. In: Strichman, O., Szeider, S. (eds.) SAT 2010. LNCS, vol. 6175, pp. 44–57. Springer, Heidelberg (2010)

MathCheck: A Math Assistant via a Combination of Computer Algebra Systems and SAT Solvers

Edward Zulkoski[✉], Vijay Ganesh, and Krzysztof Czarnecki

University of Waterloo, Waterloo, ON, Canada
ezulkosk@gsd.uwaterloo.ca

Abstract. We present a method and an associated system, called MATH-CHECK, that embeds the functionality of a computer algebra system (CAS) within the inner loop of a conflict-driven clause-learning SAT solver. SAT+CAS systems, a la MATHCHECK, can be used as an assistant by mathematicians to either counterexample or finitely verify open universal conjectures on any mathematical topic (e.g., graph and number theory, algebra, geometry, etc.) supported by the underlying CAS system. Such a SAT+CAS system combines the efficient search routines of modern SAT solvers, with the expressive power of CAS, thus complementing both. The key insight behind the power of the SAT+CAS combination is that the CAS system can help cut down the search-space of the SAT solver, by providing learned clauses that encode theory-specific lemmas, as it searches for a counterexample to the input conjecture (just like the T in DPLL(T)). In addition, the combination enables a more efficient encoding of problems than a pure Boolean representation.

In this paper, we leverage the graph-theoretic capabilities of an open-source CAS, called SAGE. As case studies, we look at two long-standing open mathematical conjectures from graph theory regarding properties of hypercubes: the first conjecture states that any matching of any d-dimensional hypercube can be extended to a Hamiltonian cycle; and the second states that given an edge-antipodal coloring of a hypercube, there always exists a monochromatic path between two antipodal vertices. Previous results have shown the conjectures true up to certain low-dimensional hypercubes, and attempts to extend them have failed until now. Using our SAT+CAS system, MATHCHECK, we extend these two conjectures to higher-dimensional hypercubes. We provide detailed performance analysis and show an exponential reduction in search space via the SAT+CAS combination relative to finite brute-force search.

1 Introduction

Boolean conflict-driven clause-learning (CDCL) SAT and SAT-Modulo Theories (SMT) solvers have become some of the leading tools for solving complex problems expressed as logical constraints [3]. This is particularly true in software engineering, broadly construed to include testing, verification, analysis, synthesis, and security. Modern SMT solvers such as Z3 [6], CVC4 [2], STP [12], and

© Springer International Publishing Switzerland 2015
A.P. Felty and A. Middeldorp (Eds.): CADE-25, LNAI 9195, pp. 607–622, 2015.
DOI: 10.1007/978-3-319-21401-6_41

VeriT [4] contain efficient decision procedures for a variety of first-order theories, such as uninterpreted functions, quantified linear integer arithmetic, bitvectors, and arrays. However, even with the expressiveness of SMT, many constraints, particularly ones stemming from mathematical domains such as graph theory, topology, algebra, or number theory are non-trivial to solve using today's state-of-the-art SAT and SMT solvers.

Computer algebra systems (e.g., Maple, Mathematica, and SAGE), on the other hand, are powerful tools that have been used for decades by mathematicians to perform symbolic computation over problems in graph theory, topology, algebra, number theory, etc. However, computer algebra systems (CAS) lack the search capabilities of SAT/SMT solvers.

In this paper, we present a method and a prototype tool, called MATHCHECK, that combines the search capability of SAT solvers with powerful domain knowledge of CAS systems (i.e. a toolbox of algorithms to solve a broad range of mathematical problems). The tool MATHCHECK can solve problems that are too difficult or inefficient to encode as SAT problems. MATHCHECK can be used by mathematicians to finitely check or counterexample open conjectures. It can also be used by engineers who want to readily leverage the joint capabilities of both CAS systems and SAT solvers to model and solve problems that are otherwise too difficult with either class of tools alone.

The key concept behind MATHCHECK is that it embeds the functionality of a computer algebra system (CAS) within the inner loop of a CDCL SAT solver. Computer algebra systems contain state-of-the-art algorithms from a broad range of mathematical areas, many of which can be used as subroutines to easily encode predicates relevant both in mathematics and engineering. The users of MATHCHECK write predicates in the language of the CAS, which then interacts with the SAT solver through a controlled SAT+CAS interface. By imposing restrictions on the CAS predicates, we ensure correctness (i.e. soundness) of this SAT+CAS combination. The user's goal is to finitely check or find counterexamples to a Boolean combination of predicates (somewhat akin to a quantifier-free SMT formula). The SAT solver searches for counterexamples in the domain over which the predicates are defined, and invokes the CAS to learn clauses that help cutdown the search space (akin to the "T" in DPLL(T)).

In this work, we focus on constraints from the domain of graph theory, although our approach is equally applicable to other areas of mathematics. Constraints such as connectivity, Hamiltonicity, acyclicity, etc. are non-trivial to encode with standard solvers [25]. We believe that the method described in this paper is a step in the right direction towards making SAT/SMT solvers useful to a broader class of mathematicians and engineers than before.

While we believe that our method is probably the first such combination of SAT+CAS systems, there has been previous work in attempting to extend SAT solvers with graph reasoning [8,14,22]. These works can loosely be divided into two categories: constraint-specific extensions, and general graph encodings. As an example of the first case, efficient SAT-based solvers have been designed to ensure that synthesized graphs contain no cycles [14]. In [22], Hamiltonicity checks are reduced to *Native* Boolean cardinality constraints and lazy connectivity constraints. While more efficient than standard encodings of acyclicity and Hamiltonicity constraints, these approaches lack generality. On the other hand, approaches such as in CP(Graph) [8], a constraint satisfaction problem (CSP) solver extension, encode a core set of graph operations with which complicated predicates (such as Hamiltonicity) can be expressed. *Global constraints* [8] can be tailored to handle predicate-specific optimizations. Although it can be nontrivial to efficiently encode global constraints, previous work has defined efficient procedures which enforce graph constraints, such as connectivity, incrementally during search [17]. Our approach is more general than the above approaches, because CAS systems are not restricted to graph theory. One might also consider a general SMT theory-plugin for graph theory however given the diverse array of predicates and functions within the domain, a monolithic theory-plugin (other than a CAS system) seems impractical at this time.

Main Contributions:[1]

Contribution I: Analysis of a SAT+CAS Combination Method and the MathCheck tool. In Sect. 3, we present a method and tool that combines a CAS with SAT, denoted as SAT+CAS, facilitating the creation of user-defined CAS predicates. Such tools can be used by mathematicians to finitely search or counterexample universal sentences in the language of the underlying CAS. The current version of our tool, MATHCHECK, allows users to easily specify and solve complex graph-theoretic questions using the simple interface provided. Although our current focus is predicates based in graph theory, the system is easily extended to other domains.

Contribution II: Results on Two Open Graph-Theoretic Conjectures over Hypercubes. In Sect. 4, we use our system to extend results on two longstanding open conjectures related to hypercubes. Conjecture 1 states that any matching of any d-dimensional hypercube can extend to a Hamiltonian cycle. Conjecture 2 states that given an edge-antipodal coloring of a hypercube, there always exists a monochromatic path between two antipodal vertices. Previous results have shown Conjecture 1 (resp. Conjecture 2) true up to $d = 4$ [10](resp. $d = 5$ [9]); we extend these two conjectures to $d = 5$ (resp. $d = 6$).

Contribution III: Performance Analysis of MathCheck. In Sect. 5, we provide detailed performance analysis of MATHCHECK in terms of how much search space reduction is achieved relative to finite brute-force search, as well as how much time is consumed by each component of the system.

2   Background

We assume standard definitions for propositional logic, basic mathematical logic concepts such as satisfiability, and solvers. We denote a graph $G = \langle V, E \rangle$ as a

---

[1] All code+data is available at https://bitbucket.org/ezulkosk/sagesat.

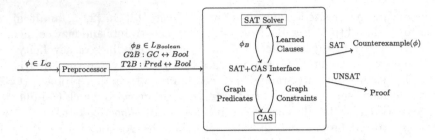

Fig. 1. High-level overview of the MATHCHECK architecture, which is similar to DPLL(T)-style SMT solvers. MATHCHECK takes as input a formula over fragments of mathematics supported by the underlying CAS system, and produces either a counterexample or a proof that none exists.

set of vertices V and edges E, where an edge e_{ij} connects the pair of vertices v_i and v_j. We only consider undirected graphs in this work. The *order* of a graph is the number of vertices it contains. For a given vertex v, we denote its neighbors – vertices that share an edge with v – as $N(v)$. The hypercube of dimension d, denoted Q_d, consists of 2^d vertices and $2^{d-1} \cdot d$ edges, and can be constructed in the following way (see Fig. 3a): label each vertex with a unique binary string of length d, and connect two vertices with an edge if and only if the Hamming distance of their labels is 1. A *matching* of a graph is a subset of its edges that mutually share no vertices. A vertex is *matched* (by a matching) if it is incident to an edge in the matching, else it is *unmatched*. A *maximal matching* M is a matching such that adding any additional edge to M violates the matching property. A *perfect matching* (resp. *imperfect matching*) M is a matching such that all (resp. not all) vertices in the graph are incident with an edge in M. A *forbidden matching* is a matching such that some unmatched vertex v exists and every $v' \in N(v)$ is matched. Intuitively, no superset of the matching can match v. Vertices in Q_d are *antipodal* if their binary strings differ in all positions (i.e. opposite "corners" of the cube). Edges e_{ij} and e_{kl} are antipodal if $\{v_i, v_k\}$ and $\{v_j, v_l\}$ are pairs of antipodal vertices. A *2-edge-coloring* of a graph is a labeling of the edges with either red or blue. A 2-edge-coloring is *edge-antipodal* if the color of every edge differs from the color of the edge antipodal to it.

3   Contribution I: SAT+CAS Combination Architecture

This section describes the combination architecture of a CAS system with a SAT solver, the method underpinning the MATHCHECK tool. Figure 1 provides a schematic of MATHCHECK. The key idea behind such combinations is that the CAS system is integrated in the inner loop of a conflict-driven clause-learning SAT solver, akin to how a theory solver T is integrated into a DPLL(T) system [19]. The grammar of the input language of MATHCHECK is sketched in Fig. 2. MATHCHECK allows the user to define predicates in the language of CAS that express some mathematical conjecture. The input mathematical conjecture is expressed as a set of *assertions* and *queries*, such that a satisfying assignment to the conjunction of the assertions and **negated** queries constitute a counterexample to the conjecture. We refer to this conjunction simply as the input formula in the remainder of the paper. First, the formula is translated into a Boolean constraint that describes the set of structures (e.g., graphs or numbers) referred to in the conjecture. Second, the SAT solver enumerates these structures in an attempt to counterexample the input conjecture. The solver routinely queries the CAS system during its search (given that the CAS system is integrated into its inner loop) to learn clauses (akin to callback plugins in programmatic SAT solvers [13] or theory plugins in DPLL(T) [19]). Clauses thus learned can dramatically cutdown the search space of the SAT solver.

Combining the solver with CAS extends each of the individual tools in the following ways. First, off-the-shelf SAT (or SMT) solvers contain efficient search techniques and decision procedures, but lack the expressiveness to **easily** encode many complex mathematical predicates. Even if a problem can be easily reduced to SAT/SMT, the choice of encoding can be very important in terms of performance, which is typically non-trivial to determine, especially for non-experts on solvers. For example, Velev et al. [25] investigated 416 ways to encode Hamiltonian cycles to SAT as permutation problems to determine which encodings were the most effective. Further, such a system can take advantage of many built-in common structures in a CAS (e.g., graph families such as hypercubes), which can greatly simplify specifying structures and complex predicates. On the other side, CAS's contain many efficient functions for a broad range of mathematical properties, but often lack the robust search routines available in SAT.

Here we provide a very high-level overview, with more details in Sect. 3.2 below. Please refer to Fig. 1, which depicts the SAT+CAS combination. Given a formula over graph variables in the language of MATHCHECK (refer to Sect. 3.1), we conjoin the assertions with the negated queries, and preprocess it as described below. When the SAT solver finds a partial model, additional checks are performed by the CAS using "CAS predicates." The potential solution is either deemed a valid counterexample to the conjecture and returned to the user, or the SAT search is refined with learned clauses. Output is either SAT and a counterexample to the conjecture, or UNSAT along with a proof certificate. Although similar to DPLL(T) approach of SMT solvers in many aspects, we note several important differences in terms extensibility, power, and flexibility: (1) rather than a monolithic theory plugin for graphs, we opt for a more *extensible* approach by incorporating the CAS, allowing new predicates (say, over, numbers, geometry, algebra, etc.) to be easily defined via the CAS functionality; (2) the CAS predicates are essentially defined using Python code interpreted by the CAS. This gives considerable *additional power* to the SAT+CAS combination; (3) the user may *flexibly* decide that certain predicates may be encoded directly to Boolean logic via bit-blasting, and thus take advantage of the efficiency of CDCL solvers in certain cases.

3.1   Input Language of MathCheck

The input to MATHCHECK is a tuple $\langle S, \phi \rangle$, where S is a set of graph variables and ϕ is a formula over S as defined by the abbreviated grammar in Fig. 2. A graph variable $G = \langle G_V, G_E \rangle$ indicates the vertices and edges that can potentially occur in its instantiation, denoted G_I. A graph variable G is essentially a set of $|V|$ Boolean variables (one for each vertex), and $|E|$ Boolean variables for edges. Setting an edge e_{ij} (resp. vertex v_i) to True means that e_{ij} (resp. v_i) is a part of the graph instantiation G_I. Through a slight abuse of notation, we often define a graph variable $G = Q_d$, indicating that the sets of Booleans in G_V and G_E correspond to the vertices and edges in the hypercube Q_d, respectively.

ϕ	$::=$	$(\textbf{assert } \psi \mid \textbf{query } \psi)^+$
ψ	$::=$	$\psi \wedge \psi \mid \psi \vee \psi \mid \neg\psi \mid Atom$
$Atom$	$::=$	$SAT\text{-}Predicate \mid CAS\text{-}Predicate$
$SAT\text{-}Predicate$	$::=$	id '(' $GraphVar^+$ ')'
$CAS\text{-}Predicate$	$::=$	id '(' $GraphVar^+$ ')'
$GraphVar$	$::=$	$\textbf{graph } Id(``\texttt{Set(VertexVariables), Set(EdgeVariables)}")$

Fig. 2. Grammar L_G of MATHCHECK's Input Language.

L_G is essentially defined as propositional logic, extended to allow predicates over graph variables (as in Fig. 2). Predicates can be defined by the user, and are classified as either *SAT predicates* or *CAS predicates*. SAT predicates are blasted to propositional logic, using the mapping from graph components (i.e. vertices and edges) to Boolean variables.[2] As an example, for any graph variable G used in an input formula, we add an `EdgeImpliesVertices(G)` constraint, indicating that an edge cannot exist without its corresponding vertices:

$$\textbf{EdgeImpliesVertices(G)} : \bigwedge \{e_{ij} \Rightarrow (v_i \wedge v_j) \mid e_{ij} \in G_E\}. \tag{1}$$

CAS predicates, which are essentially Python code interpreted by the CAS, check properties of instantiated (non-variable) graphs and are defined as pieces of code in the language of the CAS. In our case, we use the SAGE CAS [23], which for now can be thought of as a collection of Python modules for mathematics.

3.2   Architecture of MathCheck

The architecture of MATHCHECK is given in Fig. 1. The **Preprocessor** prepares ϕ for the inner CAS-DPLL loop using standard techniques. First, we create necessary Boolean variables that correspond to graph components (vertices and edges) as described above. We replace each SAT predicate via bit-blasting with its propositional representation in situ (with respect to ϕ's overall propositional

---

[2] For notational convenience, we often use existential quantifiers when defining constraints; these are unrolled in the implementation. We only deal with finite graphs.

structure), such that any assignment found by the SAT solver can be encoded into graphs adhering to the SAT predicates. Finally, Tseitin-encoding and a Boolean abstraction of ϕ is performed such that CAS predicates are abstracted away by new boolean variables; since these techniques are well-known, we do not discuss them further. This phase produces three main outputs: the CNF Boolean abstraction ϕ_B of the SAT predicates, a mapping from graph components to Booleans $G2B$, and a mapping $T2B$ from CAS predicate definitions to Boolean variables. The CAS predicates themselves are fed into the CAS. The **SAT+CAS** interface acts similar to the DPLL(T) interface between the DPLL loop and theory-plugins, ensuring that partial assignments from the SAT solver satisfy theory-specific CAS predicates. After an assignment is found, literals corresponding to abstracted CAS predicates are checked. The SAT+CAS interface provides an API that allows CAS predicates to interact with the SAT solver, which modifies the API from the programmatic SAT solver Lynx [13].

3.3 Implementation

We have prototyped our system adopting the lazy-SMT solver approach (as in [21]), specifically combining the Glucose SAT solver [1] with the SAGE CAS [23]. Minor modifications to Glucose were made to call out to SAGE whenever an assignment was found (of the Boolean abstraction). The SAT+CAS interface extends the existing SAT interface in SAGE. We further performed extensive checks on our results, including verifying the SAT solver's resolution proofs using DRUP-trim [16] as well as checking the learned clauses produced by CAS predicates, however do not elaborate now due to space constraints.

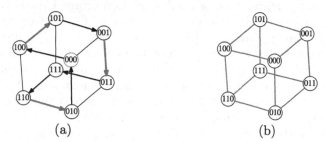

Fig. 3. (a) The red edges denote a generated matching, where the blue vertex 000 is restricted to be unmatched, as discussed in Sect. 4. A Hamiltonian cycle that includes the matching is indicated by the arrows. (b) An edge-antipodal 2-edge-coloring of the cube Q_3. Not a counterexample to Conjecture 2 due to the red (or blue) path from 000 to 111.

4   Contribution II: Two Results Regarding Open Conjectures over Hypercubes

We use our system to prove two long-standing open conjectures up to a certain parameter (dimension) related to hypercubes. Hypercubes have been studied for theoretical interest, due to their nice properties such as regularity and symmetry, but also for practical uses, such as in networks and parallel systems [5].

4.1 Matchings Extend to Hamiltonian Cycles

The first conjecture we look at was posed by Ruskey and Savage on matchings of hypercubes in 1993 [20]; although it has inspired multiple partial results [10,15] and extensions [11], the general statement remains open:

Conjecture 1 (Ruskey and Savage, [20]). For every dimension d, any matching of the hypercube Q_d can be extended to a Hamiltonian cycle.

Consider Fig. 3a. The red edges correspond to a matching and the arrows depict a Hamiltonian cycle extending the matching. Intuitively, the conjecture states that for any d-dimensional hypercube Q_d, no matter which matching M we choose, we can find a Hamiltonian cycle of Q_d that goes through M. Our encoding searches for matchings, and checks a sufficient subset of the full set of matchings of Q_d to ensure that the conjecture hold for a given dimension (by returning UNSAT and a proof). As we will show, constraints such as ensuring that a potential model is a matching are easily encoded with SAT predicates, while constraints such as "extending to a Hamiltonian cycle" are expressed easily as CAS predicates.

Previous results have shown this conjecture true for $d \leq 4$,[3] however the combinatorial explosion of matchings on higher dimensional hypercubes makes analysis increasingly challenging, and a general proof has been evasive. We demonstrate using our approach the first result that Conjecture 1 holds for Q_5 – the 5-dimensional hypercube. We use a conjunction of SAT predicates to generate a sufficient set of matchings of the hypercube, which are further verified by a CAS predicate to check if the matching can **not** be extended to a Hamiltonian cycle (such that a satisfying model would counterexample the conjecture).

Note that the simple approach of generating *all* matching of Q_d does not scale (see Table 1 below), and the approach would take too long, even for $d = 5$. We prove several lemmas to reduce the number of matchings analyzed. In the following, we use the graph variable $G = Q_d$, such that its vertex and edge variables correspond to the vertices and edges in Q_d.

It is straightforward to encode matching constraints as a SAT predicate. For every pair of incident edges e_1, e_2, we ensure that only one can be in the matching (i.e. at most one of the two Booleans may be True), which can be encoded as:

$$\textbf{Matching(G)} : \bigwedge \{(\neg e_1 \vee \neg e_2) \mid e_1, e_2 \in G_E \wedge isIncident?(e_1, e_2)\}. \quad (2)$$

The number of clauses generated by the above translation is $2^d \cdot \binom{d}{2}$, which can be understood as: for each of the 2^d vertices in Q_d, ensure that each of the d incident edges to that vertex are pairwise not both in the matching.

---

[3] We were unable to find the original source of the results for $d \leq 4$, however the result is asserted in [10]. We also verified these results using our system.

A previous result from Fink [10] demonstrated that any perfect matching of the hypercube for $d \geq 2$ can be extended to a Hamiltonian cycle. Our search for a counterexample to Conjecture 1 should therefore only consider imperfect matchings, and even further, only maximal forbidden matchings as shown below. To encode this, we ensure that at least one vertex is not matched by any generated matching. Since all vertices are symmetric in a hypercube, we can, without loss of generality, choose a single vertex v_0 that we ensure is not matched. We encode that all edges incident to v_0 cannot be in the matching:

$$\textbf{Forbidden(G):} \bigwedge \{\neg e \mid e \in G_E \wedge isIncident?(v_0, e)\}. \tag{3}$$

A further key observation to reduce the matchings search space is that, if a matching M extends to a Hamiltonian cycle, then any matching M' such that $M' \subseteq M$ can also be extended to a Hamiltonian cycle.

Observation 1. All matchings can be extended to a Hamiltonian cycle if and only if all maximal forbidden matchings can be extended to a Hamiltonian cycle.

Proof. The forward direction is straightforward. For the reverse, suppose all maximal forbidden matchings can be extended to a Hamiltonian cycle. For any non-maximal matching M, we can always greedily add edges to M to make it maximal. Call the maximized matching M'. If M' is perfect, Fink's result on perfect matchings can be applied. If not, then it is a maximal forbidden matching, and by assumption it can be extended to a Hamiltonian cycle. In either case, the resulting Hamiltonian cycle must pass through the original matching M. $\square$

We encode this by adding the following constraints to MATHCHECK:

$$\textbf{EdgeOn(G):} \bigwedge \{v \Rightarrow \exists_{e \in X} \ e | v \in G_V\}, \tag{4}$$
$$s.t. \ X = \{e | e \in G_E \wedge isIncident?(v, e)\}$$

$$\textbf{Maximal(G):} \bigwedge \{(v_i \vee v_j) \mid e_{ij} \in G_E\}. \tag{5}$$

Equation 4 states that if a vertex is on, then one of its incident edges must be in the matching. Equation 5 ensures that we only generate maximal matchings.

Proposition 1. The conjunction of Constraints 1 – 5 encode exactly the set of maximal forbidden matchings of the hypercube in which a designated vertex v_0 is prevented from being matched.

Proof. It is clear from above that any model generated will be a forbidden matching by Constraints 2 and 3 – we prove that Eqs. 4 and 5 ensure maximality. Suppose M is a non-maximal matching. Then there exists an edge e such that the matching does not match either of its endpoints. By Constraints 1 and 4, no edge is incident with either endpoint. But then edge e could be added without violating the matching constraints, and Constraint 5 is violated. Thus, any matching generated must be maximal. It remains to show that *all* forbidden maximal matchings that exclude v_0 can be generated. Let M be an forbidden maximal matching such that v_0 is unmatched. We construct a satisfying variable assignment over Constraints $1 - 5$ which encodes M as follows:

$$\{e \mid e \in M\} \cup \{\neg e \mid e \in G_E \backslash M\} \cup$$
$$\{v \mid \exists_{e \in M} \; isIncident?(v, e)\} \cup \{\neg v \mid \not\exists_{e \in M} \; isIncident?(v, e)\}. \tag{6}$$

Constraint 2 holds since M is a matching, and therefore no two incident edges can both be in M. Constraint 3 holds since it is assumed that v_0 is not matched, and therefore no edge incident to v_0 can be in M. Constraints 1 and 4 hold simply because they encode the definition of a matched vertex, and the second line of Eq. 6 ensures that only matched vertices are in the satisfying assignment. Constraint 5 holds since M is maximal. □

```
1:  EXTENDSTOHAMILTONIAN()
2:      g ← s.getGraph(G)
3:      q ← CubeGraph(5)
4:      for e in q.edges() do
5:          if e in g
6:              q.setEdgeLabel(e, 1)
7:          else
8:              q.setEdgeLabel(e, 2)
9:      ⟨cycle, weight⟩ ← TSP(q)
10:     if weight == 2 · q.order() − |g|
11:         return True
12:     else
13:         return False
```

```
1:  ANTIPODALMONOCHROMATIC()
2:      g ← s.getGraph(G)
3:      q ← CubeGraph(6)
4:      pairs ← getAntipodalPairs(q)
5:      for ⟨v₁, v₂⟩ in pairs do
6:          if shortestPath(g, v₁, v₂) ≠ ∅
7:              return True ▷ a path exists
8:      return False
```

Fig. 4. CAS-defined predicates from each case study. In ExtendsToHamiltonian, g corresponds to the matching found by the SAT solver. In AntipodalMonochromatic, g refers to the graph induced by a single color in the 2-edge-coloring.

To check if each matching extends to a Hamiltonian cycle, we create the CAS predicate ExtendsToHamiltonian (see Fig. 4), which reduces the formula to an instance of the traveling salesman problem (TSP). Let M be a matching of Q_d. We create a TSP instance $\langle Q_d, W \rangle$, where Q_d is our hypercube, and W are the edge weights, such that edges in the matching (red edges in Fig. 3a) have weight 1, and otherwise weight 2 (black edges).

Proposition 2. A Hamiltonian cycle exists through M in Q_d if and only if $TSP(\langle Q_d, W \rangle) = 2 * |V| - |M|$, where $|V|$ is the number of vertices in Q_d.

Proof. Since Q_d has $|V|$ vertices, any Hamiltonian cycle must contain $|V|$ edges. ($\Leftarrow$) From our encoding, it is clear that $2 * |V| - |M|$ is the minimum weight that could possibly be outputted by TSP, and this can only be achieved by including all edges in the matching and $|V| - |M|$ edges not in the matching.

($\Rightarrow$) The Hamiltonian cycle through M has $|M|$ edges contributing a weight of 1, and $|V| - |M|$ edges contributing a weight of 2. The total weight is therefore $|M| + (2 * (|V| - |M|)) = 2 * |V| - |M|$. From above, this is also the minimum weight cycle that TSP could produce.   □

Finally, after each check of `ExtendsToHamiltonian` that evaluates to True, we add a learned clause, based on computations performed in the predicate, to prune the search space. Since a TSP instance is solved we obtain a Hamiltonian cycle C of the cube. Clearly, any future matchings that are subsets of C can be extended to a Hamiltonian cycle; our learned constraint prevents these subsets (below h refers to the Boolean variable abstracting the CAS predicate):

$$\bigvee \{e \mid e \in Q_{dE} \backslash C\} \cup \{h\}, \text{ where C is the learned Hamiltonian cycle.} \quad (7)$$

Our full formula for Conjecture 1 is therefore:

$$\textbf{assert } EdgeImpliesVertices(G) \wedge Matching(G) \wedge$$
$$Forbidden(G) \wedge EdgeOn(G) \wedge Maximal(G) \quad (8)$$
$$\textbf{query } ExtendsToHamiltonian(G)$$

4.2   Connected Antipodal Vertices in Edge-Antipodal Colorings

The second conjecture deals with edge-antipodal colorings of the hypercube:

Conjecture 2 ([7]). For every dimension d, in every edge-antipodal 2-edge-coloring of Q_d, there exists a monochromatic path between two antipodal vertices.

Consider the 2-edge-coloring of the cube in Fig. 3b. Although the coloring is edge-antipodal, it is not a counterexample, since there is a monochromatic (red) path from 000 to 111, namely $\langle 000, 100, 110, 111 \rangle$. In this case, constraints such as edge-antipodal-ness are expressed with SAT predicates. We ensure that no monochromatic path exists between two antipodal vertices with a CAS predicate. Previous work has shown that the conjecture holds up to dimension 5 [9] – we show that the conjecture holds up to dimension 6.

We begin with a graph variable $G = Q_6$, and constrain it such that its instantiation corresponds to a 2-edge-coloring of the hypercube. More specifically, since there are only two colors, we associate edges in G's instantiation G_I (i.e. edges evaluated to True) with the color red, and the edges in $Q_d \backslash G_I$ with blue. An important known result is that for a given coloring, the graph induced by edges of one color is isomorphic to the other. It is therefore sufficient to check only one of the color-induced graphs for a monochromatic antipodal path.

We first ensure that any coloring generated is edge-antipodal.

$$\textbf{EdgeAntipodal(G): } \bigwedge \{((\neg e_1 \wedge e_2) \vee (e_1 \wedge \neg e_2)$$
$$\mid e_1, e_2 \in G_E \wedge isAntipodal?(e_1, e_2)\}. \quad (9)$$

Table 1. The number of matchings of the hypercube were computed using our tool in conjunction with sharpSAT [24]: a tool for the #SAT problem. Note that the numbers for forbidden matchings are only lower bounds, since we only ensure that the *origin* vertex is unmatched. However, any unfound matchings are isomorphic to found ones.

Dimensions	Matchings	Forbidden Matchings	Maximal Forbidden Matchings
2	7	3	0
3	108	42	2
4	41,025	14,721	240
5	13,803,794,944	4,619,529,024	6,911,604

Note that for every edge there is exactly one unique antipodal edge to it. Since there are $2^{d-1} \cdot d$ edges in Q_d, and therefore $2^{d-2} \cdot d$ pairs of antipodal edges, there are $2^{2^{d-2} \cdot d}$ possible 2-edge-colorings that are antipodal. We can reduce the search space by using a recent result from Feder and Suber [9]:

Theorem 1 [9]. Call a labeling of Q_d *simple* if there is no square $\langle x, y, z, t \rangle$ such that e_{xy} and e_{zt} are one color, and e_{yz} and e_{tx} are the other. Every simple coloring has a pair of antipodal vertices joined by a monochromatic path.

We therefore prevent simple colorings by ensuring that such a square exists:

$$\textbf{Nonsimple(G):} \bigvee \{ (\neg e_{xy} \wedge e_{yz} \wedge \neg e_{zt} \wedge e_{tx}) \vee (e_{xy} \wedge \neg e_{yz} \wedge e_{zt} \wedge \neg e_{tx}) \tag{10}$$
$$| \ e_{xy}, e_{yz}, e_{zt}, e_{tx} \in G_E \wedge isSquare?(e_{xy}, e_{yz}, e_{zt}, e_{tx}) \}.$$

It remains to check whether an antipodal monochromatic path exists, which is checked by the CAS predicate `AntipodalMonochromatic` in Fig. 4. Given a graph g, which contains only the red colored edges, we first compute the pairs of antipodal vertices in Q_d. Using the built-in shortest path algorithm of the CAS, we check whether or not any of the pairs are connected, indicating that an antipodal monochromatic path exists. In the case when predicate returns True, we learn the constraint that all future colorings should not include the found antipodal path P (m abstracts the CAS predicate):

$$\bigvee \{ \neg e \mid e \in P \} \cup \{ m \}, \text{ where P is the learned path.} \tag{11}$$

The full formula for Conjecture 2 is then:

$$\textbf{assert } EdgeImpliesVertices(G) \wedge EdgeAntipodal(G) \wedge NonSimple(G) \tag{12}$$
$$\textbf{query } AntipodalMonochromatic(G)$$

5   Contribution III: Performance Analysis of MathCheck

We ran Formula 8 with $d = 5$ and Formula 12 with $d = 6$ until completion. Since both runs returned UNSAT, we conclude that both conjectures hold for these dimensions, which improves upon known results for both conjectures.

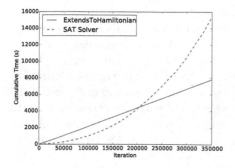 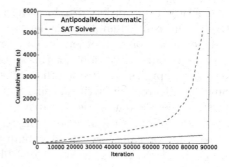

Fig. 5. Cumulative times spent in the SAT solver and CAS predicates during the two case studies. SAT solver performance degrades during solving (as indicated by the increasing slope of the line), due to the extra learned clauses and more constrained search space.

All experiments were performed on a 2.4 GHz 4-core Lenovo Thinkpad laptop with 8 GB of RAM, running 64-bit Linux Mint 17. We used SAGE version 6.3 and Glucose version 3.0. Formula 8 required 348,150 checks of the ExtendsToHamiltonian predicate, thus learning an equal number of Hamiltonian cycles in the process, and took just under 8 h. Formula 12 required 86,612 checks of the AntipodalMonochromatic predicate (learning the same number of monochromatic paths), requiring 1 h 35 min of runtime. We note that for lower dimensional cubes solving time was far less (< 20 seconds for either case study). We find it unlikely that this approach can be used for higher-dimensions, without further lemmas to reduce the search space.

The approach we have described significantly dominates naïve brute-force approaches for both conjectures; learned clauses greatly reduce the search space and cut the number of necessary CAS predicate checks. Given the data in Table 1 and the number of calls to ExtendsToHamiltonian for Q_5, a brute-force check of all matchings (resp. forbidden matchings) of Q_5 would require 39,649 (resp. 20) times more checks of the predicate (i.e. that many more TSP calls) than our approach. Similar comparisons can be made for the second case study.

Figure 5 depicts how much time is consumed by the SAT solver and CAS predicates in both case studies. The lines denote the cumulative time, such that the right most point of each line is the total time consumed by the respective system component. The near-linear lines for the CAS predicate calls indicate that each check consumed roughly the same amount of time. SAT solving ultimately dominates the runtime in both case studies, particularly due to later calls to the solver when many learned clauses have been added by CAS predicates, and the search space is highly constrained. This suggests several optimizations as future work. For example, if SAT solver calls are rapidly requiring more time (e.g., around iteration 75,000 in the second plot of Fig. 5), more sophisticated CAS routines can be used to produce more learned clauses per call (such as by learning constraints corresponding to all cycles *isomorphic* to the found one in case study 1), in order to reduce the number of necessary SAT calls. Alternatively, one can attempt to condense the learned clauses, which are generated independently of each other, into a more compact Boolean representation.

One of our motivations for this work was to allow complicated predicates to be easily expressed, so it is worth commenting on the size of the actual predicates. Since predicates were written using SAGE (which is built on top of Python), the pseudocode written in Fig. 4 matches almost exactly with the actual code, with small exceptions such as computing the antipodal pairs in the second one. All other function calls correspond to built-in functions of the CAS. Learn-functions were also short, requiring less than 10 lines of code each.

6   Related Work

As already noted, our approach of combining a CAS system within the inner-loop of a SAT solver most closely resembles and is inspired by the DPLL(T) [19]. There are also similarities with the idea of programmatic SAT solver Lynx [13], which is an instance-specific version of DPLL(T). Also, our tool MATHCHECK is inspired by the recent SAT-based results on the Erdős discrepency conjecture [18]. Other works [8,14,22] have extended solvers to handle graph constraints, as discussed in Sect. 1, by either creating solvers for specific graph predicates [14,22], or by defining a core set of constraints with which to build complex predicates [8]. Our approach contains positive aspects from both: state-of-the-art algorithms from the CAS can be used to define new predicates easily, and the methodology is general, in that new predicates can be defined using the CAS. Several tools have combined a CAS with SMT solvers for various purposes, mainly focusing on the non-linear arithmetic algorithms provided by many CAS's. For example, the VeriT SMT solver [4] also uses functionality of the REDUCE CAS[4] for non-linear arithmetic support. Our work is more in the spirit of DPLL(T), rather than modifying the decision procedure for a single theory.

7   Conclusions and Future Work

In this paper, we present MATHCHECK, a combination of a CAS in the inner-loop of a conflict-driven clause-learning SAT solver, and we show that this combination allows for highly expressive predicates that are otherwise non-trivial/infeasible to encode as purely Boolean formulas. Our approach combines the well-known domain-specific abilities of CAS with the search capabilities of SAT solvers thus enabling us to verify long-standing open mathematical conjectures over hypercubes (up to particular dimension), not feasible by either kind of tool alone. We further discussed how our system greatly dominates naïve brute-force search techniques for the case studies. We stress that the approach is not limited to this domain, and we intend to extend our work to other branches of mathematics supported by CAS's, such as number theory. Another direction we

---

[4] http://www.reduce-algebra.com/index.htm.

plan to investigate is integration with a proof-producing SMT solver, such as VeriT. In addition to taking advantage of the extra power of an SMT solver, the integration with VeriT will allow us to more easily produce proof certificates.

References

1. Audemard, G., Simon, L.: Predicting learnt clauses quality in modern SAT solvers. IJCAI **9**, 399–404 (2009)
2. Barrett, C., Conway, C.L., Deters, M., Hadarean, L., Jovanović, D., King, T., Reynolds, A., Tinelli, C.: CVC4. In: Gopalakrishnan, G., Qadeer, S. (eds.) CAV 2011. LNCS, vol. 6806, pp. 171–177. Springer, Heidelberg (2011)
3. Biere, A., Heule, M.J.H., van Maaren, H., Walsh, T.(eds.): Handbook of Satisfiability. FAIA, vol. 185. IOS Press (February 2009)
4. Bouton, T., de Oliveira, D.C.B., Déharbe, D., Fontaine, P.: veriT: an open, trustable and efficient SMT-solver. In: CADE (2009)
5. Chen, Y-C., Li, K-L.: Matchings extend to perfect matchings on hypercube networks. In: IMECS, vol. 1. Citeseer (2010)
6. de Moura, L., Bjørner, N.S.: Z3: an efficient SMT solver. In: Ramakrishnan, C.R., Rehof, J. (eds.) TACAS 2008. LNCS, vol. 4963, pp. 337–340. Springer, Heidelberg (2008)
7. Devos, M., Norine, S.: Edge-antipodal Colorings of Cubes. http://garden.irmacs.sfu.ca/?q=op/edge_antipodal_colorings_of_cubes
8. Dooms, G., Deville, Y., Dupont, P.E.: CP(Graph): introducing a graph computation domain in constraint programming. In: van Beek, P. (ed.) CP 2005. LNCS, vol. 3709, pp. 211–225. Springer, Heidelberg (2005)
9. Feder, T., Subi, C.: On hypercube labellings and antipodal monochromatic paths. Discrete Appl. Math. **161**(10), 1421–1426 (2013)
10. Fink, J.: Perfect matchings extend to hamilton cycles in hypercubes. J. Comb. Theor. B **97**(6), 1074–1076 (2007)
11. Fink, J.: Connectivity of matching graph of hypercube. SIDMA **23**(2), 1100–1109 (2009)
12. Ganesh, V., Dill, D.L.: A decision procedure for bit-vectors and arrays. In: Damm, W., Hermanns, H. (eds.) CAV 2007. LNCS, vol. 4590, pp. 519–531. Springer, Heidelberg (2007)
13. Ganesh, V., O'Donnell, C.W., Soos, M., Devadas, S., Rinard, M.C., Solar-Lezama, A.: Lynx: a programmatic sat solver for the rna-folding problem. In: Cimatti, A., Sebastiani, R. (eds.) SAT 2012. LNCS, vol. 7317, pp. 143–156. Springer, Heidelberg (2012)
14. Gebser, M., Janhunen, T., Rintanen, J.: SAT modulo graphs: acyclicity. In: Fermé, E., Leite, J. (eds.) JELIA 2014. LNCS, vol. 8761, pp. 137–151. Springer, Heidelberg (2014)
15. Gregor, P.: Perfect matchings extending on subcubes to hamiltonian cycles of hypercubes. Discrete Math. **309**(6), 1711–1713 (2009)
16. Heule, M.J.H., Hunt, W.A., Wetzler, N.: Trimming while checking clausal proofs. In: FMCAD, pp. 181–188. IEEE (2013)
17. Holm, J., De Lichtenberg, K., Thorup, M.: Poly-logarithmic deterministic fully-dynamic algorithms for connectivity, minimum spanning tree, 2-edge, and biconnectivity. J. ACM (JACM) **48**(4), 723–760 (2001)

18. Konev, B., Lisitsa, A.: A SAT attack on the Erdős discrepancy conjecture. In: SAT (2014)
19. Nieuwenhuis, R., Oliveras, A., Tinelli, C.: Abstract DPLL and abstract DPLL modulo theories. In: Baader, F., Voronkov, A. (eds.) LPAR 2004. LNCS (LNAI), vol. 3452, pp. 36–50. Springer, Heidelberg (2005)
20. Ruskey, F., Savage, C.: Hamilton cycles that extend transposition matchings in Cayley graphs of S_n. SIDMA **6**(1), 152–166 (1993)
21. Sebastiani, R.: Lazy satisfiability modulo theories. J. Satisfiability Boolean Model. Comput. **3**, 141–224 (2007)
22. Soh, T., Le Berre, D., Roussel, S., Banbara, M., Tamura, N.: Incremental SAT-based method with native boolean cardinality handling for the hamiltonian cycle problem. In: Fermé, E., Leite, J. (eds.) JELIA 2014. LNCS, vol. 8761, pp. 684–693. Springer, Heidelberg (2014)
23. Stein, W.A.(et al).: Sage Mathematics Software (Version 6.3) (2010)
24. Thurley, M.: sharpSAT – counting models with advanced component caching and implicit BCP. In: Biere, A., Gomes, C.P. (eds.) SAT 2006. LNCS, vol. 4121, pp. 424–429. Springer, Heidelberg (2006)
25. Velev, M.N., Gao, P.: Efficient SAT techniques for absolute encoding of permutation problems: application to hamiltonian cycles. In: SARA (2009)

Linear Integer Arithmetic Revisited

Martin Bromberger$^{(\boxtimes)}$, Thomas Sturm, and Christoph Weidenbach

Max Planck Institute for Informatics, Saarbrücken, Germany
{mbromber,sturm,weidenb}@mpi-inf.mpg.de

Abstract. We consider feasibility of linear integer programs in the context of verification systems such as SMT solvers or theorem provers. Although satisfiability of linear integer programs is decidable, many state-of-the-art solvers neglect termination in favor of efficiency. It is challenging to design a solver that is both terminating and practically efficient. Recent work by Jovanović and de Moura constitutes an important step into this direction. Their algorithm CUTSAT is sound, but does not terminate, in general. In this paper we extend their CUTSAT algorithm by refined inference rules, a new type of conflicting core, and a dedicated rule application strategy. This leads to our algorithm CUT-SAT++, which guarantees termination.

Keywords: Linear arithmetic · SMT · SAT · DPLL · Linear programming · Integer arithmetic

1 Introduction

Historically, feasibility of linear integer problems is a classical problem, which has been addressed and thoroughly investigated by at least two independent research lines: (i) integer and mixed real integer linear programming for optimization [9], (ii) first-order quantifier elimination and decision procedures for Presburger Arithmetic and corresponding complexity results [3,6,10–13]. We are interested in feasibility of linear integer problems, which we simply call *problems*, in the context of the combination of theories, as they occur, e.g., in the context of SMT solving or theorem proving. From this perspective, both these research lines address problems that are too general for our purposes: with the former, the optimization aspects go considerably beyond pure feasibility. The latter considers arbitrary Boolean combinations of constraints and quantifier alternation or even parametric problems.

Consequently, the SMT community has developed several interesting approaches on their own [1,4,7]. These solvers typically neglect termination and completeness in favor of efficiency. More precisely, these approaches are based on a branch-and-bound strategy, where the rational relaxation of an integer problem is used to cut off and branch on integer solutions. Together with the known a priori integer bounds [11] for a problem this yields a terminating and complete algorithm. However, these bounds are so large that for many practical problems the resulting branch-and-bound search space cannot be explored in reasonable

© Springer International Publishing Switzerland 2015
A.P. Felty and A. Middeldorp (Eds.): CADE-25, LNAI 9195, pp. 623–637, 2015.
DOI: 10.1007/978-3-319-21401-6_42

time. Hence, the a priori bounds are not integrated in the implementations of the approaches.

On these grounds, the recent work by Jovanović and de Moura [8], although itself not terminating, constitutes an important step towards an algorithm that is both efficient and terminating. The termination argument does no longer rely on bounds that are a priori exponentially large in the occurring parameters. Instead, it relies on structural properties of the problem, which are explored by their CUTSAT algorithm. The price for this result is an algorithm that is by far more complicated than the above-mentioned branch-and-bound approach. In particular, it has to consider divisibility constraints in addition to inequalities.

Our interest in an algorithm for integer constraints originates from a possible combination with superposition, e.g., see [5]. In the superposition context integer constraints are part of the first-order clauses. Variables in constraints are typically unguarded so that an efficient decision procedure for this case is a prerequisite for an efficient combined procedure.

Our contribution is an extension and refinement of the CUTSAT algorithm, which we call CUTSAT++. In contrast to CUTSAT, our CUTSAT++ generally terminates. The basic idea of both algorithms is to reduce a problem containing unguarded integer variables to a problem containing only guarded variables. These unguarded variables are not eliminated. Instead, one explores the unguarded variables by adding constraints on smaller variables to the problem, with respect to a strict total ordering where all unguarded variables are larger than all guarded variables. After adding sufficiently many constraints, feasibility of the problem depends only on guarded variables. Then a CDCL style algorithm tests for feasibility by employing exhaustive propagation. The most sophisticated part is to "turn" an unguarded variable into a guarded variable. Quantifier elimination techniques, such as Cooper elimination [3], do so by removing the unguarded variable. In case of Cooper elimination, the price to pay is an exponentially growing Boolean structure and exponentially growing coefficients. Since integer linear programming is NP-complete, all algorithms known today cannot prevent such a kind of behavior, in general. Since Cooper elimination does not care about the concrete structure of a given problem, the exponential behavior is almost guaranteed. The idea of both CUTSAT and CUTSAT++ is, therefore, to simulate a lazy variation of Cooper elimination. This leaves space for model assumptions and simplification rules in order for the algorithm to adapt to the specific structure of a problem and, hence, to systematically avoid certain cases of the worst-case exponential behavior observed with Cooper elimination.

The paper is organized as follows. After fixing some notation in Sect. 2, we present three examples for problems where CUTSAT diverges. The divergence of CUTSAT can be fixed by respective refinements on the original CUTSAT rules. However, in a fourth example the combination of the refinements results in a frozen state. Our conclusion is that CUTSAT lacks, in addition to the refinements, a third type of conflicting core, which we call *diophantine conflicting core*. Theorem 5, in Sect. 3, actually implies that any procedure that is based on what we call *weak Cooper elimination* needs to consider this type of conflicting core for completeness. In Sects. 4–5 we refine the inference rules for the elimination of unguarded variables on the basis of weak Cooper elimination (Sect. 3) and show their soundness, completeness, and termination. Finally, we give conclusions and point at possible directions for future research. For detailed proofs of our Theorems and Lemmas see [2].

2 Motivation

We use *variables* x, y, z, k, possibly with indices. Furthermore, we use *integer constants* a, b, c, d, e, l, v, u, *linear polynomials* p, q, r, s, and *constraints* I, J, possibly with indices. As input *problems*, we consider finite sets of constraints C corresponding to and sometimes used as conjunction over their elements. Each constraint I is either an inequality $a_n x_n + \ldots + a_1 x_1 + c \leq 0$ or a divisibility constraint $d \mid a_n x_n + \ldots + a_1 x_1 + c$. We denote $\operatorname{coeff}(I, x_i) = a_i \in \mathbb{Z}$. $\operatorname{vars}(C)$ denotes the set of variables occurring in C. We sometimes write $C(x)$ in order to emphasise that $x \in \operatorname{vars}(C)$. A problem C is satisfiable if $\exists X.C$ holds, where $X = \operatorname{vars}(C)$. For true we denote $\top$ and for false we denote $\bot$. Since $d \mid cx + s \equiv d \mid -cx + -s$, we may assume that $c > 0$ for all $d \mid cx + s \in C$. A variable x is *guarded* in a problem C if C contains constraints of the form $x - u_x \leq 0$ and $-x + l_x \leq 0$. Otherwise, x is *unguarded* in C. Note that guarded variables are *bounded* as defined in [8] but not vice versa. A constraint is *guarded* if it contains only guarded variables. Otherwise, it is *unguarded*.

Our algorithm CUTSAT++ aims at deciding whether or not a given problem C is satisfiable. It either ends in the state *unsat* or in a state $\langle v, \operatorname{sat} \rangle$, where v is a satisfiable assignment for C. In order to reach one of those two final states, the algorithm produces *lower bounds* $x \geq b$ and *upper bounds* $x \leq b$ for the variables in C. The produced bounds are stored in a sequence $M = [\![\gamma_1, \ldots, \gamma_n]\!]$, which describes a partial model. The empty sequence is denoted by $[\![\,]\!]$. We use $[\![M, \gamma]\!]$ and $[\![M_1, M_2]\!]$ to denote the concatenation of a bound γ at the end of M and M_2 at the end of M_1, respectively.

By $\operatorname{lower}(x, M) = b$ and $\operatorname{upper}(x, M) = b$ we denote the value b of the greatest lower bound $x \geq b$ and least upper bound $x \leq b$ for a variable x in M, respectively. If there is no lower (upper) bound for x in M, then $\operatorname{lower}(x, M) = -\infty$ ($\operatorname{upper}(x, M) = \infty$). The definitions of upper and lower are extended to polynomials as done in [8]. The partial model M is complete if all variables x are *fixed* in the sense that $\operatorname{upper}(x, M) = \operatorname{lower}(x, M)$. In this case we define $v[M]$ as the assignment that assigns to every variable x the value $\operatorname{lower}(x, M)$.

A state in CUTSAT++ is of the form $S = \langle M, C \rangle$ or $S = \langle M, C \rangle \vdash I$, or one of the two *final states* $\langle v, \operatorname{sat} \rangle$, *unsat*. The *initial-state* for a problem C is $\langle [\![\,]\!], C \rangle$. For a state $S = \langle M, C \rangle (\vdash I)$, an inequality $p \leq 0$ is a *conflict* if $\operatorname{lower}(p, M) > 0$. For a state $S = \langle M, C \rangle (\vdash I)$, a divisibility constraint $d \mid ax + p$ is a *conflict* if all variables in p are fixed and $d \nmid ab + \operatorname{lower}(p, M)$ for all b such that $\operatorname{lower}(x, M) \leq b \leq \operatorname{upper}(x, M)$. In a state $S = \langle M, C \rangle \vdash I$, the constraint I is always a conflict. A state is *frozen* if it is not a final state and no rule is applicable.

Via applications of the rule Decide, CUTSAT++ adds *decided bounds* $x \leq b$ or $x \geq b$ to the sequence M in state S [8]. A decided bound generally assigns a variable x to the lower or upper bound of x in M. Via applications of the propagation rules, CUTSAT++ adds *propagated bounds* $x \geq_I b$ or $x \leq_I b$ to the sequence M, where I is a generated constraint, called justification. To this end, the function $\text{bound}(J, x, M)$ computes the strictest bound b and the function $\text{tight}(J, x, M)$ computes the corresponding justification I for constraint J under the partial model M.

We are now going to discuss three examples where CUTSAT diverges. The first one shows that CUTSAT can apply Conflict and Conflict-Div infinitely often to constraints containing unguarded variables.

Example 1. Let

$$C := \{\underbrace{-x \leq 0}_{I_x}, \underbrace{-y \leq 0}_{I_y}, \underbrace{-z \leq 0}_{I_{z1}}, \underbrace{z \leq 0}_{I_{z2}}, \underbrace{z + 1 \leq 0}_{I_{z3}}, \underbrace{1 - x + y \leq 0}_{J_1}, \underbrace{x - y \leq 0}_{J_2}\}$$

be a problem. Let $S_i = \langle M_i, C \rangle$ for $i \in \mathbb{N}$ be a series of states with:

$$M_0 := [\![x \geq_{I_x} 0, y \geq_{I_y} 0, z \geq_{I_{z1}} 0, z \leq_{I_{z2}} 0]\!],$$
$$M_{i+1} := [\![M_i, x \geq_{J_1} i + 1, y \geq_{J_2} i + 1]\!].$$

Let the variable order be given by $z \prec y \prec x$. CUTSAT with a two-layered strategy, after propagating all constraints I_x, I_y, I_{z1}, I_{z2} , applies the rules Decide, Conflict, and Backjump to propagate arbitrarily large lower bounds for the unguarded variables x and y and, therefore, diverges. Notice that the conflicting core $\{I_{z1}, I_{z3}\}$ is guarded, which admits the application of Conflict.

A straightforward fix to Example 1 is to limit the application of the Conflict and Conflict-Div rules to guarded constraints. Our second example shows that CUTSAT can still diverge by infinitely many applications of the Solve-Div rule [8].

Example 2. Let d_i be the sequence with $d_0 := 2$ and $d_{k+1} := d_k^2$ for $k \in \mathbb{N}$, let $C_0 = \{4 \mid 2x + 2y, 2 \mid x + z\}$ be a problem, and let $S_0 = \langle [\![]\!], C_0 \rangle$ be the initial CUTSAT state. Let the variable order be given by $x \prec y \prec z$. Then CUTSAT has divergent runs $S_0 \Rightarrow_{CS} S_1 \Rightarrow_{CS} S_2 \Rightarrow_{CS} \ldots$. For instance, let CUTSAT apply the Solve-Div rule whenever applicable. By an inductive argument, Solve-Div is applicable in every state $S_n = \langle [\![]\!], C_n \rangle$, and the constraint set C_n has the following form:

$$C_n = \begin{cases} \{2d_n \mid d_n x + d_n y, d_n \mid \frac{d_n}{2} y - \frac{d_n}{2} z\} & \text{if } n \text{ is odd}, \\ \{2d_n \mid d_n x + d_n y, d_n \mid \frac{d_n}{2} x + \frac{d_n}{2} z\} & \text{if } n \text{ is even}. \end{cases}$$

Therefore, CUTSAT applies Solve-Div infinitely often and diverges.

A straightforward fix to Example 2 is to limit the application of Solve-Div to maximal variables in the variable order $\prec$. Our third example shows that

CUTSAT can apply Conflict and Conflict-Div [8] infinitely often. The Example 3 differs from Example 1 in that the conflicting core contains also unguarded variables.

Example 3. Let

$$C := \{\underbrace{-x \leq 0}_{I_x}, \underbrace{-y \leq 0}_{I_y}, \underbrace{-z \leq 0}_{I_{z1}}, \underbrace{z \leq 0}_{I_{z2}}, \underbrace{1 - x + y + z \leq 0}_{J_1}, \underbrace{x - y - z \leq 0}_{J_2}\}\}$$

be a problem. Let $S_i = \langle M_i, C \rangle$ for $i \in \mathbb{N}$ be a series of states with:

$$M_0 := [\![x \geq_{I_x} 0, y \geq_{I_y} 0, z \geq_{I_{z1}} 0, z \leq_{I_{z2}} 0]\!],$$
$$M_{i+1} := [\![M_i, x \geq_{J_1} i + 1, y \geq_{J_2} i + 1]\!].$$

Let the variable order be given by $z \prec x \prec y$. CUTSAT with a two-layered strategy, after propagating all constraints I_x, I_y, I_{z1}, I_{z2}, possibly applies the rules Decide, Conflict, and Backjump to propagate arbitrarily large lower bounds for the unguarded variables x and y and, thus, diverges. Notice that the conflicting core $\{J_1, J_2\}$ is bounded in [8] after we fix x and y with Decide to their current respective lower bounds. This in turn admits the application of Conflict.

Applying the fix suggested for Examples 1–3 results in a frozen state. Here, a straightforward fix is to change the definition of conflicting cores to cover only those cores where the conflicting variable is the maximal variable.[1]

The fixes that we suggested for the above examples are restrictions to CUTSAT which have the consequence that Conflict(-Div) cannot be applied to unguarded constraints, Solve-Div is only applicable for the elimination of the maximal variable, and the conflicting variable x is the maximal variable in the associated conflicting core C'. However, our next and final example shows that these restrictions lead to frozen states.

Example 4. Let CUTSAT include restrictions to maximal variables in the definition of conflicting cores and in the Solve-Div rule as described above. Let there be additional restrictions in CUTSAT to the rules Conflict and Conflict-Div such that these rules are only applicable to conflicts that contain no unguarded variable. Let

$$C := \{\underbrace{-x \leq 0}_{I_{x1}}, \underbrace{x - 1 \leq 0}_{I_{x2}}, \underbrace{-y \leq 0}_{I_y}, \underbrace{6 \mid 4y + x}_{J}\}$$

be a problem. Let $M := [\![x \geq_{I_{x1}} 0, x \leq_{I_{x2}} 1, y \geq_{I_y} 0, x \geq 1, y \leq 0]\!]$ be a bound sequence. Let the variable order be given by $x \prec y$. CUTSAT has a run starting in state $S_0' = \langle [\![ ]\!], C \rangle$ that ends in the frozen state $S = \langle M, C \rangle$. Let CUTSAT propagate I_{x1}, I_{x2}, I_y and fix x to 1 and y to 0 with two Decisions. Through these Decisions, the constraint J is a conflict. Since y is unguarded, CUTSAT cannot apply the rule Conflict-Div. Furthermore, [8] has defined conflicting cores

---

[1] The restrictions to maximal variables in the definition of conflicting cores and to the Solve-Div rule were both confirmed as missing but necessary in a private communication with Jovanović.

as either interval or divisibility conflicting cores. The state S contains neither an interval or a divisibility conflicting core. Therefore, CUTSAT cannot apply the rule Resolve-Cooper. The remaining rules are also not applicable because all variables are fixed and there is only one divisibility constraint. Without the before introduced restriction to the rule Conflict(-Div), CUTSAT diverges on the example. For more details see [2].

3   Weak Cooper Elimination

In order to fix the frozen state of Example 4 in the previous section, we are going to introduce in Sect. 4 a new conflicting core, which we call *diophantine conflicting core*. For understanding diophantine conflicting cores, as well as further modifications to be made, it is helpful to understand the connection between CUTSAT and a variant of Cooper's quantifier elimination procedure [3].

The original *Cooper elimination* takes a variable x, a problem $C(x)$, and produces a disjunction of problems equivalent to $\exists x.C(x)$:

$$\exists x.C(x) \equiv \bigvee_{0 \le k < m} C_{-\infty}(k) \vee \bigvee_{-ax+p \le 0 \in C(x)} \bigvee_{0 \le k < a \cdot m} \left[a \mid p + k \wedge C\left(\frac{p+k}{a} \right) \right],$$

where $a > 0$ and $m = \mathrm{lcm}\{d \in \mathbb{Z} : (d \mid ax + p) \in C(x)\}$. If there exists no constraint of the form $-ax + p \le 0 \in C(x)$, then $C_{-\infty}(x) = \{(d \mid ax+p) \in C(x)\}$. Otherwise, $C_{-\infty}(x) = \bot$. One application of Cooper elimination results in a disjunction of quadratically many problems out of a single problem. Iteration causes an exponential increase in the coefficients due to the multiplication with a because division is not part of the language.

Our notion of *weak Cooper elimination* is a variant of Cooper elimination, which is very helpful to understand problems around CUTSAT. The idea is, instead of building a disjunction over all potential solutions for x, to add additional guarded variables and constraints without x that guarantee the existence of a solution for x. We assume here that $C(x)$ contains only one divisibility constraint for x. If not, exhaustive application of div-solve to divisibility constraints for x removes all constraints except one: div-solve$(x, d_1 \mid a_1 x + p_1, d_2 \mid a_2 x + p_2) = (d_1 d_2 \mid dx + c_1 d_2 p_1 + c_2 d_1 p_2, d \mid -a_1 p_2 + a_2 p_1)$, where $d = \gcd(a_1 d_2, a_2 d_1)$, and c_1 and c_2 are integers such that $c_1 a_1 d_2 + c_2 a_2 d_1 = d$ [3,8]. Now weak Cooper elimination takes a variable x, a problem $C(x)$, and produces a new problem by replacing $\exists x.C(x)$ with:

$$\exists K. \left(\{I \in C(x) : \mathrm{coeff}(x, I) = 0\} \cup \{\gcd(c, d) \mid s\} \cup \bigcup_{k \in K} R_k \right)$$

where $d \mid cx + s \in C(x)$, $k \in K$ is a newly introduced variable for every pair of constraints $-ax + p \le 0 \in C(x)$ and $bx - q \le 0 \in C(x)$, and
$$R_k = \{-k \le 0, k - m \le 0, bp - aq + bk \le 0, a \mid k + p, ad \mid cp + as + ck\}$$
is a resolvent for the same inequalities, where $m := \mathrm{lcm}\left(a, \frac{ad}{\gcd(ad,c)} \right) - 1$. Note

the existential quantifier $\exists K$, where all variables $k \in K$ are guarded by their respective R_k.

Let ν be a satisfiable assignment for the formula after one weak Cooper elimination step on $C(x)$. Then we compute a strictest lower bound $x \geq l_x$ and a strictest upper bound $x \leq u_x$ from $C(x)$ for the variable x under the assignment ν. We now argue that there is a value for x such that $x \geq l_x$, $x \leq u_x$, and $d \mid cx + s$ are all satisfied. Whenever $l_x \neq -\infty$ and $u_x \neq \infty$, the bounds $x \geq l_x$, $x \leq u_x$ are given by respective constraints of the form $-ax + p \leq 0 \in C(x)$ and $bx - q \leq 0 \in C(x)$ such that $l_x = \lceil \frac{\nu(p)}{a} \rceil$ and $u_x = \lfloor \frac{\nu(q)}{b} \rfloor$. In this case, the extension of ν with $\nu(x) = \frac{\nu(k+p)}{a}$ satisfies $C(x)$ because the constraint $a \mid k + p \in R_k$ guarantees that $\nu(x) \in \mathbb{Z}$, the constraint $bp - aq + bk \leq 0 \in R_k$ guarantees that $l_x \leq \nu(x) \leq u_x$, and the constraint $ad \mid cp + as + ck \in R_k$ guarantees that ν satisfies $d \mid cx + s \in C(x)$. Whenever $l_x = -\infty$ ($u_x = \infty$) we extend ν by an arbitrary small (large) value for x that satisfies $d \mid cx + s \in C(x)$. There exist arbitrarily small (large) solutions for x and $d \mid cx + \nu(s)$ because $\gcd(c, d) \mid s$ is satisfied by ν.

The advantage of weak Cooper elimination, compared to Cooper elimination, is that the output is still a conjunctive problem in contrast to a disjunction of problems. CUTSAT++ performs weak Cooper elimination not in one step but subsequently adds to the states the constraints from the R_k as well as the divisibility constraint $\gcd(c, d) \mid s$ with respect to a strict ordering on the unguarded variables.

The following Theorem, for which we have just outlined the proof, states the correctness of weak Cooper elimination.

Theorem 5.

$$\exists x.C(x) \equiv \exists K. \left(\{I \in C(x) : \text{coeff}(x, I) = 0\} \cup \{\gcd(c, d) \mid s\} \cup \bigcup_{k \in K} R_k \right)$$

The extra divisibility constraint $\gcd(c, d) \mid s$ in weak Cooper elimination is necessary whenever the problem $C(x)$ has no constraint of the form $-ax + p \leq 0 \in C(x)$ or $bx - q \leq 0 \in C(x)$. For example, let $C(x) = \{y - 1 \leq 0, -y + 1 \leq 0, 6 \mid 2x + y\}$ be a problem and x be the unguarded variable we want to eliminate. As there are no inequalities containing x, weak Cooper elimination without the extra divisibility constraint returns $C' = \{y - 1 \leq 0, -y + 1 \leq 0\}$. While C' has a satisfiable assignment $\nu(y) = 1$, $C(x)$ has not since $2x + 1$ is never divisible by 2 or 6.

For any R_k introduced by weak Cooper elimination we can also show the following Lemma:

Lemma 6. *Let k be a new variable. Let $a, b, c > 0$. Then,*

$$(\exists x.\{-ax + p \leq 0, bx - q \leq 0, d \mid cx + s\})$$
$$\equiv (\exists k.\{-k \leq 0, k - m \leq 0, bp - aq + bk \leq 0, a \mid k + p, ad \mid cp + as + ck\}).$$

That means satisfiability of the respective R_k guarantees a solution for the triple of constraints it is derived from. An analogous Lemma holds for the divisibility constraint $\gcd(c, d) \mid s$ introduced by weak Cooper elimination:

Lemma 7.
$$(\exists x. d \mid cx + s) \equiv \gcd(c, d) \mid s.$$

That means satisfiability of $\gcd(c, d) \mid s$ guarantees a solution for the divisibility constraint $d \mid cx + s$. The rule Resolve-Cooper (Fig. 1) in our CUTSAT++ exploits these properties by generating the R_k and constraint $\gcd(c, d) \mid s$ in the form of strong resolvents in a lazy way. Furthermore, it is not necessary for the divisibility constraints to be a priori reduced to one, as done for weak Cooper elimination. Instead, the rules Solve-Div-Left and Solve-Div-Right (Fig. 1) perform lazy reduction.

4   Strong Conflict Resolution Revisited

Weak Cooper elimination is capable of exploring all unguarded variables to eventually create a problem where feasibility only depends on guarded variables. It is simulated in a lazy manner through an additional set of CUTSAT++ rules (Fig. 1). Instead of eliminating all unguarded variables before the application of CUTSAT++, the rules perform the same intermediate steps as weak Cooper elimination, viz., the combination of divisibility constraints via div-solve and the construction of resolvents, to resolve and block conflicts in unguarded constraints. As a result, CUTSAT++ can avoid some of the intermediate steps of weak Cooper elimination. Furthermore, CUTSAT++ is not required to apply the intermediate steps of weak Cooper elimination one variable at a time. The lazy approach of CUTSAT++ does not eliminate unguarded variables. In the worst case CUTSAT++ has to perform all of weak Cooper elimination's intermediate steps. Then the strictly-two-layered strategy (Definition 13) guarantees that CUTSAT++ recognizes that all unguarded conflicts have been produced.

The eventual result is the complete algorithm CUTSAT++, which is a combination of the rules Resolve-Cooper, Solve-Div-Left, Solve-Div-Right (Fig. 1), a strictly-two-layered strategy (Definition 13), and the CUTSAT rules: Propagate, Propagate-Div, Decide, Conflict, Conflict-Div, Sat, Unsat-Div, Forget, Slack-Intro[2], Resolve, Skip-Decision, Backjump, Unsat, and Learn [2,8].

The lazy approach has the advantage that CUTSAT++ might find a satisfiable assignment or detect unsatisfiability without encountering and resolving a large number of unguarded conflicts. This means the number of divisibility constraint combinations and introduced resolvents might be much smaller in the lazy approach of CUTSAT++ than during the elimination with weak Cooper elimination.

In order to simulate weak Cooper elimination, CUTSAT++ uses a total order $\prec$ over all variables such that $y \prec x$ for all guarded variables y and unguarded variables x. While termination requires that the order is fixed from the beginning for all unguarded variables, the ordering among the guarded variables can be dynamically changed. In relation to weak Cooper elimination, the order $\prec$

---

[2] As recommended in [8], CUTSAT++ uses the same slack variable for all Slack-Intro applications.

describes the elimination order for the unguarded variables, viz., $x_i \prec x_j$ if x_j is eliminated before x_i. A variable x is called *maximal* in a constraint I if x is contained in I and all other variables in I are smaller, i.e., $y \prec x$. The maximal variable in I is also called its *top variable* ($x = \text{top}(I)$).

Definition 8. *Let $S = \langle M, C \rangle$ be a state, $C' \subseteq C$, x the top variable in C', and let all other variables in C' be fixed. The pair (x, C') is a* conflicting core *if it is of one of the following three forms*

(1) $C' = \{-ax + p \leq 0, bx - q \leq 0\}$ and the lower bound from $-ax + p \leq 0$ contradicts the upper bound from $bx - q \leq 0$, i.e., $\text{bound}(-ax + p \leq 0, x, M) > \text{bound}(bx - q \leq 0, x, M)$; in this case (x, C') is called an interval conflicting core *and its strong resolvent is $(\{-k \leq 0, k - a + 1 \leq 0\}, \{bp - aq + bk \leq 0, a \mid k + p\})$.*
(2) $C' = \{-ax + p \leq 0, bx - q \leq 0, d \mid cx + s\}$ and $b_l = \text{bound}(-ax + p \leq 0, x, M)$, $b_u = \text{bound}(bx - q \leq 0, x, M)$, $b_l \leq b_u$, and for all $b_d \in [b_l, b_u]$ we have $d \nmid cb_d + \text{lower}(s, M)$; in this case (x, C') is called a divisibility conflicting core *and its strong resolvent is $(\{-k \leq 0, k - m \leq 0\}, \{bp - aq + bk \leq 0, a \mid k + p, ad \mid cp + as + ck\})$.*
(3) $C' = \{d \mid cx + s\}$ and for all $b_d \in \mathbb{Z}$ we have $d \nmid cb_d + \text{lower}(s, M)$; in this case (x, C') is called a diophantine conflicting core *and its strong resolvent is $(\emptyset, \{\gcd(c, d) \mid s\})$.*
In the first two cases k is a fresh variable and $m = \text{lcm}\left(a, \frac{ad}{\gcd(ad,c)}\right) - 1$.

We refer to the respective strong resolvents for a conflicting core (x, C') by the function $\text{cooper}(x, C')$, which returns a pair (R_k, R_c) as defined above. Note that the newly introduced variable k is guarded by the constraints in R_k. If there is a conflicting core (x, C') in some state S, then x is called a *conflicting variable*. A *potential conflicting core* is a pair (x, C') if there exists a state S where (x, C') is a conflicting core.

Next, we define a generalization of strong resolvents. Since the strong resolvents generated out of conflicting cores will be further processed by CUTSAT++, we must guarantee that any set of constraints implying the feasibility of the conflicting core constraints prevents a second application of Resolve-Cooper to the same conflicting core. All strong resolvents of Definition 8 are also strong resolvents in the sense of the below definition (see also end of Sect. 3).

Definition 9. *A set of constraints R is a* strong resolvent *for the pair (x, C') if it holds that $R \rightarrow \exists x.C'$ and $\forall J \in R. \text{top}(J) \prec x$.*

The rule Resolve-Cooper (Fig. 1) requires that the conflicting variable x of the conflicting core (x, C') is the top variable in the constraints of C'. This simulates a setting where all variables y with $x \prec y$ are already eliminated. We restrict Resolve-Cooper to unguarded constraints because weak Cooper elimination modifies only unguarded constraints.

Lemma 10. *Let $S = \langle M, C \rangle$ be a CUTSAT++ state. Let $C' \subseteq C$ and x be an unguarded variable. Let $R \subseteq C$ be a strong resolvent for (x, C'). Then Resolve-Cooper is not applicable to (x, C').*

For the resolvent R to block Resolve-Cooper from being applied to the conflicting core (x, C'), CUTSAT++ has to detect all conflicts in R. Detecting all conflicts in R is only possible if CUTSAT++ fixes all variables y with $y \prec x$ and if Resolve-Cooper is only applicable if there exists no conflict I with $\text{top}(I) \prec x$. Therefore, the remaining restrictions of Resolve-Cooper justify the above Lemma.

If we add strong resolvents again and again, then CUTSAT++ will reach a state after which every encounter of a conflicting core guarantees a conflict in a guarded constraint. From this point forward, CUTSAT++ will not apply Resolve-Cooper. The remaining guarded conflicts are resolved with the rules Conflict and Conflict-Div [8].

The rules Solve-Div-Left and Solve-Div-Right (Fig. 1) combine divisibility constraints as it is done a priori to weak Cooper elimination. In these rules, we restrict the application of div-solve(x, I_1, I_2) to constraints where x is the top variable and where all variables y in I_1 and I_2 with $y \neq x$ are fixed. The ordering restriction simulates the order of elimination, i.e., we apply div-solve(x, I_1, I_2) in a (simulated) setting where all variables y with $x \prec y$ appear to be eliminated in I_1 and I_2. Otherwise, divergence would be possible (see Example 2). Requiring smaller variables to be fixed prevents the accidental generation of a conflict for an unguarded variable x_i by div-solve(x, I_1, I_2).

Thanks to an eager top-level propagating strategy, as defined below, any unguarded conflict in CUTSAT++ is either resolved with Solve-Div-Right (Fig. 1) or CUTSAT++ constructs a conflicting core that is resolved with Resolve-Cooper. Both cases may require multiple applications of the Solve-Div-Left rule (Fig. 1). We define the following further restrictions on the CUTSAT++ rules, which will eventually generate the above described behavior.

Definition 11. Let $\bowtie \in \{\leq, \geq\}$. We call a strategy for CUTSAT++ eager top-level propagating if we restrict propagations and decisions for every state $\langle M, C \rangle$ in the following way:

1. Let x be an unguarded variable. Then we only allow to propagate bounds $x \bowtie \text{bound}(I, x, M)$ if x is the top variable in I. Furthermore, if I is a divisibility constraint $d \mid ax + p$, then we only propagate $d \mid ax + p$ if:
   (a) either lower$(x, M) \neq -\infty$ and upper$(x, M) \neq \infty$ or
   (b) $\gcd(a, d) \mid \text{lower}(p, M)$ holds and $d \mid ax + p$ is the only divisibility constraint in C with x as top variable.
2. Let x be an unguarded variable. Then we only allow decisions $\gamma = x \bowtie b$ if:
   (a) for every constraint $I \in C$ with $x = \text{top}(I)$ all occurring variables $y \neq x$ are fixed
   (b) there exists no $I \in C$ where $x = \text{top}(I)$ and I is a conflict in $[\![M, \gamma]\!]$
   (c) either lower$(x, M) \neq -\infty$ and upper$(x, M) \neq \infty$ or there exists at most one divisibility constraint in C with x as top variable.

An eager top-level propagating strategy has two advantages. First, the strategy dictates an order of influence over the variables, i.e., a bound for unguarded variable x is influenced only by previously propagated bounds for variables y with

Solve-Div-Left

$\langle M, C \rangle \Rightarrow_{\text{CS}} \langle M, C' \rangle$ if $\begin{cases} \text{divisibility constraints } I_1, I_2 \in C, \\ x \text{ is unguarded and top in } I_1 \text{ and } I_2, \\ \text{all other vars. in } I_1, I_2 \text{ are fixed,} \\ (I'_1, I'_2) = \text{div-solve}(x, I_1, I_2), \\ C' = (C \setminus \{I_1, I_2\}) \cup \{I'_1, I'_2\}, \\ I'_2 \text{ is not a conflict} \end{cases}$

Solve-Div-Right

$\langle M, C \rangle \Rightarrow_{\text{CS}} \langle M', C' \rangle$ if $\begin{cases} \text{divisibility constraints } I_1, I_2 \in C, \\ x \text{ is unguarded and top in } I_1 \text{ and } I_2, \\ \text{all other vars. in } I_1, I_2 \text{ are fixed,} \\ (I'_1, I'_2) = \text{div-solve}(x, I_1, I_2), \\ C' = (C \setminus \{I_1, I_2\}) \cup \{I'_1, I'_2\}, \\ I'_2 \text{ is a conflict,} \\ y = \text{top}(I'_2), \\ M' = \text{prefix}(M, y) \end{cases}$

Resolve-Cooper

$\langle M, C \rangle \Rightarrow_{\text{CS}} \langle M', C \cup R_k \cup R_c \rangle$ if $\begin{cases} (x, C') \text{ is a conflicting core,} \\ x \text{ is unguarded,} \\ \text{all } z \prec x \text{ are fixed and } C' \subseteq C, \\ \text{if } J \in C \text{ is a conflict, then } \text{top}(J) \not\prec x, \\ \text{cooper}(x, C') = (R_k, R_c), \\ y = \min_{I \in R_c} \{\text{top}(I)\}, \\ M' = \text{prefix}(M, y) \end{cases}$

In the above rules, $M' = \text{prefix}(M, y)$ defines the largest prefix of M that contains only decided bounds for variables x with $x \prec y$.

Fig. 1. Our strong conflict resolution rules

$y \prec x$. Furthermore, the strategy makes only decisions for unguarded variable x when all constraints with $x = \text{top}(I)$ are fixed and satisfied by the decision. This means, any conflict $I \in C$ with $x = \text{top}(I)$ is impossible as long as the decision for x remains on the bound sequence. For the same purpose, i.e., avoiding conflicts I where $x = \text{top}(I)$ is fixed by a decision, CUTSAT++ backjumps in the rules Resolve-Cooper and Solve-Div-Right to a state where this is not the case.

Definition 12. *A strategy is* reasonable *if Propagate applied to constraints of the form* $\pm x - b \leq 0$ *has the highest priority over all rules and the Forget Rule is applied only finitely often [8].*

Definition 13. *A strategy is* strictly-two-layered *if:*
(1) it is reasonable, (2) it is eager top-level propagating, (3) the Forget, Conflict, Conflict-Div rules only apply to guarded constraints, (4) Forget cannot be applied to a divisibility constraint or a constraint contained in a strong resolvent, and (5) only guarded constraints are used to propagate guarded variables.

The above *strictly-two-layered* strategy is the final restriction to CUT-SAT++. With the condition 13-(3) it partitions conflict resolution into two layers: While every unguarded conflict is handled with the rules Resolve-Cooper,

Forget

$\langle M, C \cup \{J\}\rangle \quad \Rightarrow_{\text{CS}} \langle M, C\rangle$ if $C \vdash_Z J$, and $J \notin C$

Slack-Intro

$$\langle M, C\rangle \quad \Rightarrow_{\text{CS}} \langle M, C \cup C_s\rangle \quad \text{if} \begin{cases} \langle M, C\rangle \text{ is stuck,} \\ x \text{ is stuck,} \\ x_S \text{ is the slack-variable,} \\ C_s = \{-x_S \leq 0, x - x_S \leq 0, \\ \qquad -x - x_S \leq 0\} \end{cases}$$

Sat

$\langle M, C\rangle \quad \Rightarrow_{\text{CS}} \langle v[M], \text{sat}\rangle$ if $v[M]$ satisfies C

Unsat

$\langle M, C\rangle \vdash b \leq 0 \Rightarrow_{\text{CS}} \text{unsat}$ if $b > 0$

Unsat-Div

$$\langle M, C\rangle \quad \Rightarrow_{\text{CS}} \text{unsat} \quad \text{if} \begin{cases} d \mid a_1 x_1 + \ldots + a_n x_n + c \in C, \\ \gcd(d, a_1, \ldots, a_n) \nmid c \end{cases}$$

A variable x is called *stuck* in state $S = \langle M, C\rangle$ if M contains no bounds for x and there is no inequality $I = ax + p \leq 0 \in C$ that propagates a bound for x [8]. Variables x with a constraint of the form $\pm x - b \leq 0 \in C$ are never stuck as CUTSAT++ is able to propagate at least one bound for x, i.e., either $x \geq -b$ or $x \leq b$. A state S is a *stuck state* if all unfixed variables x are stuck and if the rules Sat, Unsat-Div, Conflict, and Conflict-Div are not applicable [8]. The slack variable x_S is the smallest unguarded variable in $\prec$. As long as Slack-Intro is never applied, we treat x_S as non existent.

Fig. 2. The Forget, Slack-Intro, Sat, Unsat, and Unsat-Div rules

Solve-Div-Left, and Solve-Div-Right (Fig. 1), every guarded conflict is handled with the rules Conflict(-Div) [2]. The conditions 13-(1) and 13-(5) make the guarded variables independent from the unguarded variables. The conditions 13-(2) and 13-(4) give a guarantee that the rules Resolve-Cooper, Solve-Div-Left, and Solve-Div-Right are applied at most finitely often. We assume for the remainder of the paper that all runs of CUTSAT++ follow a strictly-two-layered strategy.

5   Termination and Completeness

The CUTSAT++ rules are Propagate, Propagate-Div, Decide, Conflict, Conflict-Div, Sat, Unsat-Div, Forget, Slack-Intro, Resolve, Skip-Decision, Backjump, Unsat, and Learn [2,8], as well as Resolve-Cooper, Solve-Div-Left, and Solve-Div-Right (Fig. 1). For the termination proof of CUTSAT++, we consider a (possibly infinite) sequence of rule applications $\langle [\![], C_0\rangle = S_0 \Rightarrow_{\text{CS}} S_1 \Rightarrow_{\text{CS}} \ldots$ on a problem C_0, following the strictly-two-layered strategy.

First, this sequence reaches a state S_s ($s \in \mathbb{N}_0^+$) after a finite derivation of rule applications $S_0 \Rightarrow_{\text{CS}} \ldots \Rightarrow_{\text{CS}} S_s$ such that there is no further application of the rules Slack-Intro and Forget (Fig. 2) after state S_s: Such a state S_s exists for two reasons: Firstly, the strictly-two-layered strategy employed by CUTSAT++ is also reasonable. The reasonable strategy explicitly forbids infinite applications of the rule Forget. Secondly, the Slack-Intro rule is applicable only to stuck variables and only once to each stuck variable. Only the initial set of variables can be stuck because all variables x introduced during the considered derivation are introduced with at least one constraint $x - b \leq 0$ that allows at least one propagation for the variable. Therefore, the rules Slack-Intro and Forget are applicable at most finitely often.

Next, the sequence reaches a state S_w ($w \geq s$) after a finite derivation of rule applications $S_s \Rightarrow_{\mathsf{CS}} \ldots \Rightarrow_{\mathsf{CS}} S_w$ such that there is no further application of the rules Resolve-Cooper, Solve-Div-Left, and Solve-Div-Right after state S_w: The rules Resolve-Cooper, Solve-Div-Left, Solve-Div-Right, and Slack-Intro are applicable only to unguarded constraints. Through the strictly-two-layered strategy, they are also the only rules producing unguarded constraints. Therefore, they form a closed loop with respect to unguarded constraints, which we use in our termination proof. We have shown in the previous paragraph that $S_s \Rightarrow_{\mathsf{CS}} \ldots \Rightarrow_{\mathsf{CS}} S_w$ contains no application of the rule Slack-Intro. By Lemma 10, an application of Resolve-Cooper to the conflicting core (x, C') prevents any further applications of Resolve-Cooper to the same core. By Definition 8, the constraints learned through an application of Resolve-Cooper contain only variables y such that $y \prec x$. Therefore, an application of Resolve-Cooper blocks a conflicting core (x, C') and introduces potential conflicting cores only for smaller variables than x. This strict decrease in the conflicting variables guarantees that we encounter only finitely many conflicting cores in unguarded variables. Therefore, Resolve-Cooper is applicable at most finitely often. An analogous argument applies to the rules Solve-Div-Left and Solve-Div-Right. Thus the rules Resolve-Cooper, Solve-Div-Left, and Solve-Div-Right are applicable at most finitely often.

Next, the sequence reaches a state S_b ($b \geq w$) after a finite derivation of rule applications $S_w \Rightarrow_{\mathsf{CS}} \ldots \Rightarrow_{\mathsf{CS}} S_b$ such that for every guarded variable x the bounds remain invariant, i.e., lower$(x, M_b) = $ lower(x, M_j) and upper$(x, M_b) = $ upper(x, M_j) for every state $S_j = \langle M_j, C_j \rangle (\vdash I_j)$ after $S_b = \langle M_b, C_b \rangle (\vdash I_b)$ ($j \geq b$): The strictly-two-layered strategy guarantees that only bounds of guarded variables influence the propagation of further bounds for guarded variables. Any rule application involving unguarded variables does not influence the bounds for guarded variables. A proof for the termination of the solely guarded case was already provided in [8]. We now know that the sequence after S_b contains no further propagations, decisions, or conflict resolutions for the guarded variables.

Next, the sequence reaches a state S_u ($u \geq b$) after a finite derivation of rule applications $S_b \Rightarrow_{\mathsf{CS}} \ldots \Rightarrow_{\mathsf{CS}} S_u$ such that also for every unguarded variable x the bounds remain invariant, i.e., lower$(x, M_b) = $ lower(x, M_j) and upper$(x, M_b) = $ upper(x, M_j) for every state $S_j = \langle M_j, C_j \rangle (\vdash I_j)$ after $S_u = \langle M_u, C_b \rangle (\vdash I_u)$ ($j \geq u$). After S_b, CUTSAT++ propagates and decides only unguarded variables or ends with an application of Sat or Unsat(-Div). CUTSAT++ employs the strictly-two-layered strategy, which is also an eager top-level propagating strategy. Through the top variable restriction for propagating constraints, the eager top-level propagating strategy induces a strict order of propagation over the unguarded variables. Therefore, any bound for an unguarded variable x is influenced only by bounds for variables $y \prec x$. This strict variable order guarantees that unguarded variables are propagated and decided only finitely often.

After state S_u, only the rules Sat, Unsat, and Unsat-Div are applicable, which lead all to a final state. Hence, the sequence $S_0 \Rightarrow_{CS} S_1 \Rightarrow_{CS} \dots$ is finite. We conclude that CUTSAT++ always terminates:

Theorem 14. *If CUTSAT++ starts from an initial state $\langle [], C_0 \rangle$, then there is no infinite derivation sequence.*

All CUTSAT++ rules are sound, i.e., if $\langle M_i, C_i \rangle (\vdash I_i) \Rightarrow_{CS} \langle M_j, C_j \rangle (\vdash I_j)$, then any satisfiable assignment v for C_j is a satisfiable assignment also for C_i. The rule Resolve-Cooper is sound because of the Lemmas 6 and 7. The soundness of Solve-Div-Left and Solve-Div-Right follows from the fact that div-solve is an equivalence preserving transformation. The soundness proofs for all other rules are either trivial or given in [8].

Furthermore, CUTSAT++ never reaches a frozen state. Let x be the smallest unfixed variable with respect to $\prec$. Whenever x is guarded we can propagate a constraint $\pm x - b \le 0 \in C$ and then fix x by introducing a decision. If we cannot propagate any bound for x, then x is unguarded and stuck and, therefore, Slack-Intro is applicable. If we cannot fix x by introducing a decision, then x is unguarded and there is a conflict. Guarded conflicts are resolved via the Conflict(-Div) rules. Unguarded conflicts are resolved via the strong conflict resolution rules. Unless a final state is reached, CUTSAT has always a rule applicable.

Summarizing, CUTSAT++ is terminating, sound, and never reaches a frozen state. In combination with the fact that Sat is applicable only if a satisfiable solution $v[M]$ is found and that Unsat and Unsat-Div detect trivially unsatisfiable constraints, these facts imply completeness:

Theorem 15. *If CUTSAT++ starts from an initial state $\langle [], C_0 \rangle$, then it either terminates in the* unsat *state and C_0 is unsatisfiable, or it terminates with $\langle v, \text{sat} \rangle$ where v is a satisfiable assignment for C_0.*

6   Conclusion and Future Work

The starting point of our work was an implementation of CUTSAT [8] as a theory solver for hierarchic superposition [5]. In that course, we observed divergence for some of our problems. The analysis of those divergences led to the development of the CUTSAT++ algorithm presented in this paper, which is a substantial extension of CUTSAT by means of the weak Cooper elimination described in Sect. 3.

As a next step, we plan to develop a prototypical implementation of CUT-SAT++, to test its efficiency on benchmark problems. Depending on the outcome, we consider integrating CUTSAT++ as a theory solver for hierarchic superposition modulo linear integer arithmetic [5].

Finally, we point at some possible improvements of CUTSAT++. We see great potential in the development of constraint reduction techniques from (weak) Cooper elimination [3]. For practical applicability such reduction techniques might be crucial. The choice of the variable order $\prec$ has considerable impact on the efficiency of CUTSAT++. It might be possible to derive suitable orders via the analysis of the problem structure. We might benefit from results and experiences of research in quantifier elimination with variable elimination orders.

Acknowledgments. This research was supported in part by the German Transregional Collaborative Research Center SFB/TR 14 AVACS and by the ANR/DFG project STU 483/2-1 SMArT.

References

1. Barrett, C.W., Nieuwenhuis, R., Oliveras, A., Tinelli, C.: Splitting on demand in SAT modulo theories. In: Hermann, M., Voronkov, A. (eds.) LPAR 2006. LNCS (LNAI), vol. 4246, pp. 512–526. Springer, Heidelberg (2006)
2. Bromberger, M., Sturm, T., Weidenbach, C.: Linear integer arithmetic revisited. ArXiv e-prints, abs/1503.02948 (2015)
3. Cooper, D.C.: Theorem proving in arithmetic without multiplication. In: Meltzer, B., Michie, D. (eds.) 1971 Proceedings of the Seventh Annual Machine Intelligence Workshop, Edinburgh. Machine Intelligence, vol. 7, pp. 91–99. Edinburgh University Press (1972)
4. Dillig, I., Dillig, T., Aiken, A.: Cuts from proofs: a complete and practical technique for solving linear inequalities over integers. In: Bouajjani, A., Maler, O. (eds.) CAV 2009. LNCS, vol. 5643, pp. 233–247. Springer, Heidelberg (2009)
5. Fietzke, A., Weidenbach, C.: Superposition as a decision procedure for timed automata. Math. Comput. Sci. **6**(4), 409–425 (2012)
6. Fischer, M.J., Rabin, M.: Super-exponential complexity of Presburger arithmetic. SIAM-AMS Proc. **7**, 27–41 (1974)
7. Griggio, A.: A practical approach to satisability modulo linear integer arithmetic. JSAT **8**(1/2), 1–27 (2012)
8. Jovanović, D., de Moura, L.: Cutting to the chase. J. Autom. Reasoning **51**(1), 79–108 (2013)
9. Jünger, M., Liebling, T.M., Naddef, D., Nemhauser, G.L., Pulleyblank, W.R., Reinelt, G., Rinaldi, G., Wolsey, L.A. (eds.): 50 Years of Integer Programming 1958–2008. Springer, Heidelberg (2010)
10. Lasaruk, A., Sturm, T.: Weak quantifier elimination for the full linear theory of the integers. A uniform generalization of Presburger arithmetic. Appl. Algebra Eng. Commun. Comput. **18**(6), 545–574 (2007)
11. Papadimitriou, C.H.: On the complexity of integer programming. J. ACM **28**(4), 765–768 (1981)
12. Presburger, M.: Über die Vollständigkeit eines gewissen Systems der Arithmetik ganzer Zahlen. welchem die Addition als einzige Operation hervortritt. In: Comptes Rendus du premier congres de Mathematiciens des Pays Slaves, pp. 92–101. Warsaw, Poland (1929)
13. Weispfenning, V.: The complexity of almost linear diophantine problems. J. Symb. Comput. **10**(5), 395–403 (1990)

Author Index

Printed in the United States
By Bookmasters